T0344845

Environmental Noise and Management

Wiley Series in Acoustics, Noise and Vibration series list

Environmental Noise and Management

Overview from Past to Present

Selma Kurra

Registered Offices
John Wiley & Sons, Inc., 111 River Street, Hoboken, NJ 07030, USA
John Wiley & Sons Ltd, The Atrium, Southern Gate, Chichester, West Sussex, PO19 8SQ, UK

Editorial Office
The Atrium, Southern Gate, Chichester, West Sussex, PO19 8SQ, UK

For details of our global editorial offices, customer services, and more information about Wiley products visit us at www.wiley.com.

Wiley also publishes its books in a variety of electronic formats and by print-on-demand. Some content that appears in standard print versions of this book may not be available in other formats.

Library of Congress Cataloging-in-Publication Data

Names: Kurra, Selma, author.
Title: Environmental Noise and Management: Overview from Past to Present / Selma Kurra
Description: First edition. | Hoboken, NJ : John Wiley & Sons, Ltd, [2021]
  | Series: Wiley Series in Acoustics, Noise and Vibration | Includes
  bibliographical references and index.
Identifiers: LCCN 2019057705 (print) | LCCN 2019057706 (ebook) | ISBN
  9781118887400 (hardback) | ISBN 9781118887431 (adobe pdf) | ISBN
  9781118887424 (epub)
Subjects: LCSH: Noise control.
Classification: LCC TD892 .K86 2021 (print) | LCC TD892 (ebook) | DDC
620.2/3–dc23
LC record available at https://lccn.loc.gov/2019057705
LC ebook record available at https://lccn.loc.gov/2019057706

Cover Design: Wiley
Cover Images: High Angle View Of Traffic Jam © Thanatham, Piriyakarnjanakul/EyeEm/Getty Images; Frequency Sound Wave © Elerium/iStock.com; Coal Fired Electric Power © landbysea/Getty Images; Six melbourne trains © Kummeleon/Shutterstock

Set in 9.5/12.5pt STIXTwoText by SPi Global, Pondicherry, India
Printed and bound in Singapore by Markono Print Media Pte Ltd

10 9 8 7 6 5 4 3 2 1

Contents

About the Author

Selma Kurra graduated from Istanbul Technical University (ITU) as an architect-engineer with an MS (1968) and a PhD in Building Sciences (1978). As associate professor and full professor in 1988 in the Dept. of Physical Environmental Control at ITU, Professor Kurra developed the acoustical laboratory and established the Noise Control Unit in the Environment and Urbanization Research and Implementation Center. She has conducted various investigations on environmental noise, including field measurements and noise surveys, supported by the government and with international funds.

She has been a visiting professor at various universities, such as ISVR (UK), has worked with Professor C. Rice and conducted research, such as laboratory experiments at Purdue University Ray-Herrick Laboratory with Professor M. Crocker and at the Riverbank Laboratories in the USA. She also conducted the IIT/ITRI and ERIF projects at the Illinois Institute of Technology. She completed a research project at Kobe University, Japan, in collaboration with Professor Z. Maekawa. After 30 years of a career in ITU, Professor Kurra retired and took up a position at Bahcesehir University in Istanbul.

She is the founding member of the Turkish Acoustical Society (TAKDER), and president (1992–93 and 2018–19). Professor Kurra is a member of many professional bodies and institutes of noise control and acoustics internationally and has published a number of congress papers, a book and several book chapters and many articles on noise control and building acoustics. In parallel to the scientific and academic works, she has conducted a number of consultancy projects and reports.

Professor Kurra worked on the development of the first National Regulation on Noise Control (1986) and the recent regulation and guidelines on "Protection of Buildings from Noise." She has organized many acoustic training & certification programs, under the contract with the Ministry of Environment and generated hundreds of Urban Planning.

She has participated in the international collaborations and projects that led to the development of ISO/TS 15666 and ISO/TS 19488. She is a member of several ISO working groups and the National Mirror Committee on Noise of the Turkish Standards Organization.

Professor Kurra recently working as acoustic consultant for dB-KES Engineering Ltd (Turkey) and Hawk Technology (UAE).

List of Figures

List of Tables

Series Preface

This book series will embrace a wide spectrum of acoustics, noise and vibration topics from theoretical foundations to real world applications. Individual volumes included will range from specialist works of science to advanced undergraduate and graduate student texts. Books in the series will review the scientific principles of acoustics, describe special research studies and discuss solutions for noise and vibration problems in communities, industry and transportation.

The first books in the series include those on biomedical ultrasound, effects of sound on people, engineering acoustics, noise and vibration control, environmental noise management, sound intensity and wind farm noise – books on a wide variety of related topics.

The books I have edited for Wiley, *Encyclopedia of Acoustics* (1997), *Handbook of Acoustics* (1998) and *Handbook of Noise and Vibration Control* (2007), included over 400 chapters written by different authors. Each author had to restrict the chapter length on their special topics to no more than about 10 pages. The books in the current series will allow authors to provide much more in-depth coverage of their topic.

The series will be of interest to senior undergraduate and graduate students, consultants, and researchers in acoustics, noise and vibration and, in particular, those involved in engineering and scientific fields, including aerospace, automotive, biomedical, civil/structural, electrical, environmental, industrial, materials, naval architecture, and mechanical systems. In addition, the books will be of interest to practitioners and researchers in fields such as audiology, architecture, the environment, physics, signal processing, and speech.

Malcolm J. Crocker
Series editor

Acknowledgments

I am grateful for the support and encouragement given to me by Professor Malcolm Crocker in writing this book and to Wiley for the accomplishment of its realization. The information in this book has been derived from many sources, such as outstanding reference books, journals, literature, including old classic papers, as well as the recent documents. I wish to express my sincere appreciation to those who kindly permitted me to make use of the results and visual material of their valuable acoustical research and applications and I regret if I omitted to mention some references in the book while attempting a vast overview of each topic.

I thank all those for careful editing of the manuscript and guiding the production stage in Wiley.

My sincere thanks go to Serkan Ayrac who prepared the graphical drawings of all the visual materials in the book and to dB-KES Engineering for its contributions, particularly to Ersin Kivircik at the last stage of proof corrections. Finally, I must add my gratitude to my family for their understanding and support in all those years.

Introduction

Noise is a type of environmental pollution creating a health risk, reducing people's hearing and perception, affecting their physiological and psychological stability, work performance, and reducing the quality of ambience in terms of pleasantness and quietness of the environment. Noise comes from the Latin word "nausea" and was concerned as a problem even in the Middle Ages, for example, the horrifying clattering from the wheels of horse-drawn carriages rolling on the cobblestones or the vast number of workers deafened in mines and in steel factories. One can find sufficiently detailed overviews regarding history of acoustics and noise in literature. In exceptional cases, loud sounds were treated as beneficial in the ancient world, to frighten enemies and to encourage or warn soldiers on the battlefield. As a result of changes brought by the Industrial Revolution, the use of noisy machines and equipment and later in modern times, the great alterations in transportation and human activities have all amplified the severity of the problem. Noise pollution – sometimes called acoustic pollution – has significantly increased in present environments, compared to other kinds of pollution. Noise has been accepted as a danger to human health and a factor in reduced life quality in cities. As Robert Koch declared in 1905, "The day will come when man will have to fight noise as inexorably as cholera and the plague."

The war against noise, first, needs a declaration of tolerance degrees, i.e. the criteria and limits, second, to assess the actual noise conditions through measurements or predictions, and, third, to control the excessive noise by means of proper management systems. Such efforts have been ongoing throughout the world for at least five decades at the scientific and technical levels. Although thousands of reference books, reports, and other documents have been published so far, the issue is still under investigation and is being discussed vigorously at different platforms, e.g. national and international congresses, specific committees, organizations, etc. by scientists, engineers, planners, local, and governmental people, and more importantly by the communities directly exposed to noise. Solutions, especially for environmental noise, are not easy to implement and require the contribution of various stakeholders. Different aspects of environmental noise, such as scientific, technological, educational, social, economic, legislative, planning, etc. matters are considered simultaneously due to the widespread and complex nature of the problem. Noise pollution has been handled in the last three decades under the concept of "Noise Management" and the technical aspects are included in the field of "Noise Control Engineering."

Unfortunately, field studies have proved that noise pollution has not been reduced despite all the efforts, technical successes, and challenges throughout the world. Based on the evaluations made in 2011, it was declared that more than 75 million people are exposed to various types of environmental A-weighted noise levels above 50 dB (at night) in European countries with the numbers still rising at present, i.e. in 2014, 125 million people are affected by road traffic noise L_{den} levels greater than 55 dB and 37 million people are affected by L_{den} levels above 65 dB. The numbers given in 2019 by the European Environment Agency (EEA), are still high among the 33 EEA member countries with some missing data: 113 million people exposed to road traffic noise equal or above 55 dB L_{den} and 78 million people exposed to equal or above 50 dB L_{night}. The demands for efficient noise control in cities have increased greatly, especially for traffic noise, which is generally accepted as the major environmental noise followed by the railway noise.

Evaluation methodologies and management of noise pollution are still being debated globally, however, the EU and some other countries have made great accomplishments with regard to noise mapping and implementations based on action plans and enforcement.

The major aim of this book is to provide basic and up-to-date information about environmental noise aspects, especially for the non-acousticians, such as environmental engineers, urban and transportation planners, architects or those who are involved with noise from the viewpoint of environmental impact and control, along with other pollutions. Covering all the relevant subjects from noise measurements to recent standards and regulations, this book will be useful as a training tool for those who work in such a multidisciplinary field and for those making decisions regarding the management and control of environmental problems.

Based on long-term experience in research, lecturing, and consulting activities, the aspects of environmental noise are described in the book from the past to the present, to remind the younger generation of the pioneering and subsequent researchers who accomplished great achievements in the field of applied acoustics. While giving information about the developments in noise measuring techniques, prediction models, noise mapping samples and the improvements of legislation to control noise pollution, the visual materials displayed in the older documents, have been revitalized and presented in a new format, since some of these works are no longer available in libraries and on the internet. In this way, it may stimulate a better understanding of the recent technology and applications, and the development of new approaches to the problem.

During the writing phase of this book, determining the amount of theoretical background was a difficult issue, since many of the acoustics books had already studied these subjects in a variety of depth. But a textbook or a reference book is required, intended for undergraduate students in general, or those at graduate levels of environmental engineering, urban planning, and architectural engineering, and for those working in government institutions, who are not expert in acoustics. This book is designed to deal mainly with the practical work without requiring a substantial acoustic background, such as writing measurement reports, preparing noise maps, etc., so as to give some basic knowledge but most importantly up-to-date information about basic principles, regulations, etc., which are supported by a rich documentation in each chapter.

One of the aims of this book concerning various aspects of noise pollution as an important environmental issue, is to also create a resource for researchers from various relevant disciplines as well as for those involved with environmental engineering practice. It comprises various topics covering worldwide scientific and technological developments in the field in a comparative manner, providing summaries of the respective standards, and the reference lists annexed to each volume.

This book also aims to enlighten those involved with other environmental problems and to attract potential researchers in this field. It is particularly useful for undergraduate and graduate students pursuing research in this field. As the book meticulously dedicated to both professionals and students, covers both introductory and advanced levels, it will provide detailed information about all the environmental noise aspects and the new developments throughout the world. Each chapter elaborates the topics in general and also in terms of the common noise sources, so that one can find specific information regarding a specific noise source concerned with respect to its characteristics, effects on people, measurement techniques, standards, abatements etc. provided in different chapters.

The Structure of the Book

Chapter 1, after introducing the basic concepts of acoustics and sound physics, contains an overview of sound sources and sound propagation phenomena. The characteristics of noise, spectral and temporal analyses, the physical environmental factors affecting noise propagation, noise units and descriptors are explained in Chapter 2. Environmental noise sources, namely, transportation, industry, construction, and others are examined from the point of view of sound emissions in Chapter 3. Chapter 4 offers an extensive review of the models for predicting immission levels of various types of noise sources that are encountered in the environment.

Chapter 5 gives basic knowledge about acoustical measurements, both in the field and laboratory, by referring to the relative ISO standards, noise measuring and analyzing techniques to be implemented for different purposes, such as occupational noise, noise barriers, and sound insulation measurements as well as for immission levels of environmental noise sources. The systems used in measurements, including equipment and analyzing software and the methodology of measurements to be implemented for different types of noise sources, are discussed. The importance of the preparation of noise maps is currently considered in the world, especially in Europe. The evolution of mapping, applied techniques, mapping software, validity tests, evaluations of results, reporting, and various application examples are explained, in Chapter 6. In addition, legislation and guidance according to EU Directives are annexed.

Chapter 7 covers the issues of community noise effects on people. Providing data on hearing and perception phenomena, the noise dose/response relationships are widely discussed and previous noise impact studies are reviewed. The concept of noise control criteria and environmental noise limits for various types of noise sources and soundscape design which is also associated to creation of a better environment are explained. Noise control techniques are briefly summarized.

Regulations for noise and worldwide noise policies and strategies are outlined and the documents issued by the EU, as well as the international regulations and standards dealing with various aspects of noise, are examined in Chapter 8. The management of environmental noise is discussed principally in Chapter 9, moreover, noise control policies and strategies, implementation problems, and noise control practices are highlighted.

To summarize, this book offers an up-to-date program on environmental noise, detailing its historical background, the present state of affairs, and possible future developments. The data provided will be useful to both undergraduates and those pursuing a career that involves a knowledge of acoustics.

1

Acoustics and Noise Fundamentals

Scope

The first chapter consists of an overview of the basic concepts regarding acoustics and sound physics, with a focus on sound sources and sound propagation phenomenon. It addresses the theoretical background which is important for an understanding of the noise aspects by non-acousticians whose work is relevant to the environment and buildings.

1.1 Acoustics and Brief History: Pioneers and Subsequent Researchers

Acoustics is defined as "The science of sound" and includes physics and mechanics (mechanical waves in gases, liquids, and solids including vibrations). Acoustics is an applied science covering the production, transmission, perception, and effects of noise on humans. Knowledge of the principles of the physics of sound is necessary when considering environmental noise issues as measurements, evaluations, and noise control technologies are involved.

Many authors have reviewed the evolution of acoustics in history and, some pioneers given in chronological order [1–7]. The term "acoustics," which means "hearing," derived from the Greek word "akouein," and was first applied by Sauever to the science of sound in 1701 [4].

Acoustics finds its origins in antiquity, with great achievements made in the field to the present day. A short history of some of the pioneers and their followers, with their discoveries and applications to various fields in acoustics, is outlined in Table 1.1. The first investigations in acoustics began due to an interest in musical instruments and musical theories with experiments on strings, bells, waves in the sea, tinkled glasses, etc. and later continued along with discoveries in optics. The foundation of acoustics was established in the seventeenth century by applying the optical phenomena of refraction, diffraction, and interference to acoustics. The real understanding of acoustics began with wave theory and ray concepts that influenced researchers in later centuries in architectural acoustics and noise control. Increased knowledge in physics, mathematics, and electricity in the eighteenth century produced great advances in theoretical physics and applied mechanics. Mathematicians applied new approaches to create improved theories of sound wave propagation.

The nineteenth century brought numerous research studies leading to further discoveries in the knowledge of acoustics, and these findings and theories allowed expansion of studies on complex

Environmental Noise and Management: Overview from Past to Present, First Edition. Selma Kurra.
© 2021 John Wiley & Sons Ltd. Published 2021 by John Wiley & Sons Ltd.

Table 1.1 Outstanding scientists in acoustics –Pioneers and some of the subsequent researchers [1.1-1.7]

Name	Background	Date	Achievements/discoveries in acoustics
Pythagoras	Mathematician	570–495 BC	• Established mathematics • Conducted studies on producing sound from vibrating strings and musical sounds • Declared "air motion generated by a vibrating body sounding a single musical note is also vibratory the same frequency as the body" • Explained the frequency of sound by dividing a length of string into simple ratios to change "pitch" (octave, the fifth, the fourth) on the proportionality of the length and pitch. • Applied mathematics to music, discovered the musical notes and translated into mathematical equations. • Pitch of notes depends on the rapidity of vibrations.
Marcus Vitruvius	Architect and engineer	80-15 BC	• Demonstrated the resonance of a string • Used spread of circular waves on a water's surface, later explained the sound waves in three dimension (as spherical waves). • Improved acoustic properties of ancient theatres by using large empty vases. • A pioneer of architectural acoustics, understanding and analysis of theatre acoustics • Author of the book *De Arkitectura*
Lucius Seneca	Philosopher and playwriter	4 BC-65 AD	• Wrote physical tragedies • Discovered the requisition of elastic media for sound propagation
Aristotle	Metaphysicist and philosopher	384-322 BC	• Expressed the nature of sound waves with the hypothesis based more on philosophy. • Stated that air motion is generated by a source "thrusting forward in like manner the adjoining air, so that sound travels unaltered in quality as far as the disturbance of the air manages to reach". (Pierce) • Wrote a treatise *On the Soul*.
Farabi	Philosopher, astronomer, musician	872-950	• Worked on mathematics and physics • Explained sound phenomenon and studied on music theories • Worked on music theories in relation to eastern philosophy based on Aristotle • Found the rules in design of musical instruments and invented "kanun". • Worked on musical therapy

Table 1.1 (Continued)

Name	Background	Date	Achievements/discoveries in acoustics
Leonardo Da Vinci	Artist, mathematician, inventor, writer	1452-1519	• Noted that "There cannot be any sound when there is no movement or percussion of the air" • Experimented with the propagation of waves by water waves • Declared that sound wave motion has a velocity by means of experiments using bells
Galileo Galilei	Astronomer physicist	1564-1642	• "Father of modern physics", one of the best understandings of sound frequency and pitch of sound • Theoretical and experimental work in motion of bodies • Discovered the resonance, relationship between pitch and frequency by scraping a chisel at different speeds. • Conducted experiments with pendulums
Marin Mersenne	Theologician, natural philosopher, mathematician, music theorist	1588-1648	• "Father of modern acoustics" • By observing water waves, air motion generated by musical sounds is oscillatory and sound travels with a finite speed. • Conducted experiment and found absolute frequency ratio of (1:2) of two vibrating strings for an audible tone, 84 Hz • Experimented sound bending around corners • Measured velocity of sound by counting heart beats between flash of shot and perception of sound • Explained lowest frequency of a string in relation to length, force and mass per unit length. • Contributed to musical tuning theory.
Pierre Gassendi	Philosopher, astronomer, scientist, mathematician	1582-1655	• Argued for a ray theory whereby sound is attributed to a stream of atoms emitted by the sounding body; the velocity of sound is the speed of atoms in motion, and the frequency is the number of atoms emitted per unit time. • Studied on pitch of sounds • Demonstrated that sound velocity was independent of pitch by comparing results of a rifle with those of a cannon.
Robert Boyle	Philosopher, chemist, physicist, inventor-	1626-1691	• Made experiment by a ticking watch in a glass chamber (jar) and using a vacuum technology, and proved that existence of air was inevitable for sound production. • Declared that sound was a wave rather than particles.

(*Continued*)

Table 1.1 (Continued)

Name	Background	Date	Achievements/discoveries in acoustics
Joseph Sauveur	Mathematician and physicist	1653-1716	• Involved with ancient music and worked on music theory and notation • Named "acoustics" for the science of sound • Introduced acoustical terms like fundamental, harmonics, node, etc. • Improved Mersenne's laws by using beats and metronomes • Used logarithmic division in musical sounds • Discovered new tuning device
Christian Huygens	Mathematician scientist physicist Astronomer	1629-1695	• Developed the wave theory of geometric optics applied in acoustics • Founded the mathematical basis for scalar diffraction theory: successive positions of wavefronts are established by the envelope of secondary wavelets (spherical), known as "Huygens Principle" or "Huygens theory of diffraction"
Isaac Newton	Physicist and mathematician	1642-1726	• "A cornerstone of physical acoustics" • Author of the book*Principia* • Achieved theoretical derivation of the speed of sound and radiation of sound in air • Made numerous experimental measurements • Developed classical mechanics and made mechanical interpretation of sound as pressure pulses • transmitted through neighboring fluid particles • Derived relationship for wave velocity in solids
Daniel Bernoilli, Leonhard Euler, Joseph L.Lagrange, Jean Le Roud d'Alembert	Mathematicians, physicists etc.	1700-1782 1707-1783 1736-1813 1717-1783	• Worked on theoretical mechanics • Among the major contributors to the field theory • Constituted a definite mathematical structure, known as Euler and Lagrange equations, for solving optimization problems in mechanics.
Ernst F.Chladni	Physicist and musician	1756-1827	• Author of the book *Die Akustik* • Established the field of modern experimental acoustics • Conducted experiments on vibrating bars and pipes.
Thomas Young and Augustin Jean Fresnel	Physician and physicist Engineer and physicist	1773–1829 1788–1817	• Developed further explanations on the theory of wave propagation • Elucidated the principle of interference. • Provided better understanding of diffraction by "Huygens- Fresnel Principle"

Table 1.1 (Continued)

Name	Background	Date	Achievements/discoveries in acoustics
B. Joseph Fourier	Mathematician and physicist	1768-1830	• One of the scientists founding the physiological and physical acoustics. • Established the new techniques in mathematical analysis of signals and vibrations by means of Fourier series and Fourier transformations, used in digital signal processing today.
Simon Ohm	Physicist and mathematician	1789-1854	• Worked on pure tones and harmonics and solved new problems. • Made hypothesis about perception of ear and nature of musical sounds
Michael Faraday James Clerk Maxwell, Henrich R. Hertz	Scientists, physicists	1791–1867 1831–1879 1857–1894	• Worked on theory of electromagnetic waves and frequencies, made experiments • Contributed to understanding the physical and physiological aspects of acoustics • The SI unit h*ertz* (Hz) was named in his honor by International Electrotechnical Commission in 1930, also used in sound waves.
Hermann F.L.von Helmholtz	Physician and physicist	1821-1894	• Investigated the relationship between physical stimuli and perception through experiments • Developed foundation of spectral analysis and explained acoustic resonance • Author of the book, *Sensation of Sound* • Contributed greatly to mathematical acoustics • Consolidated the field of physiological acoustics • Explained timbre of complex sounds by using resonators • Produced a device -now called as Helmholtz resonator- to pick out specific frequencies from complex sounds

(*Continued*)

Table 1.1 (Continued)

Name	Background	Date	Achievements/discoveries in acoustics
Lord Rayleigh (J.W. Strutt)	Physicist and mathematician	1845-1919	• Applied the optical phenomena of diffraction, refraction and interference to acoustics. • Real understanding of sound phenomena • Used ray concepts to explain acoustic phenomena by mathematically incorporating it successfully into wave theory • Played important role in transforming acoustics as a science by combining mathematical theories and experiments • Discovered the mathematical theory explaining the radiation of sound in air and gases • Author of the book *Theory of Sound*
Alexander Graham Bell	Scientist, inventor, engineer	1847-1927	• Discovered telephone in 1876 and provided transmission of sound waves to distant areas by means of electrical currents
Thomas Edison	Inventor and businessman	1847–1931	• As known with his great innovations in electricity, • invented phonograph in 1878 and enabled recording of speech- sounds mechanically • Contributed to the field of telecommunication
Wallace Clement Sabin	Physicist, architectural acoustician	1868-1919	• Founder of architectural acoustics as science. • Established the basis of room acoustics • Published articles on sound absorption • Discovered reverberation time and developed "Sabine Reverberation formula" • Established Riverbank Acoustical Laboratories for sound absorption and transmission measurements
Harvey Fletcher	Physicist, psycho-acoustician, acoustical engineer	1884-1981	• "Father of psychoacoustics" • Founded Bell Telephone Laboratories • Described and quantified the concepts of loudness and masking, • Developed the Fletcher-Munson equal-loudness contours for pure tones based on measurements (with Munson) • Invented the accurate audiometers to measure hearing and recorders, electronic hearing aids • Contributed to speech perception, communication and studied on determinants • Winner of Nobel prize in physics

Table 1.1 (Continued)

Name	Background	Date	Achievements/discoveries in acoustics
Vern O. Knudsen	Physicist, acoustical engineer, architectural acoustician	1893–1974	• Carried on Sabine's work by conducting major experiments on sound absorption in various temperature and humidity and sound transmission • Contributed to architectural acoustics and dispersion of sound • Designed the acoustics of many theaters
Harry F.Olson	Electrical and audio engineer and physicist	1902-1982	• A pioneer in acoustical engineering • Worked on sound design and production in the acoustical laboratory at RCA and on design of microphones • Developed modern version of loudspeakers
W.P. Mason	Engineer	1900-1986	• Worked in physical acoustics and established foundations of ultrasonic.
P. Langevin and C. Chilowski	Physicists and scientists	1872-1946	• Found principle of sonar • Investigated echoes from the ocean bottom • Pioneered application of the piezoelectricity in generation and detection of ultrasound waves by employing a vacuum tube amplifier and discovered piezo-electric transducers • Discovered the ultrasonic submarine detector with C.Chilowski • Used acoustic Doppler Shift for measurement of ship velocity
George van Bekesy	Physiologistand biophysicist	1899-1972	• Developed understanding on human hearing mechanism by developing a mechanical model of the cochlea • Winner of Nobel price on Physiology or medicine and received ASA Gold Medal

(Continued)

Table 1.1 (Continued)

Name	Background	Date	Achievements/discoveries in acoustics
Philip M.Morse, K. Uno Ingard Bruce Lindsay	Physicists, scientists, engineers	1903-1985 1921-2014 1900-1985	• Contributed to experimental and theoretical acoustics, non-linear acoustics, fluid dynamics, and noise control engineering. • Authors of major books in physical acoustics such as; *Theoretical Acoustics*, (by Morse &Ingard), *Vibration and Sound* (by Morse), *Noise Reduction Analysis*(by Ingard), *Acoustics (Physics)* (by Ingard) etc. • Received Gold Medal of ASA (Morse)
Leo L.Beranek	Electrical and communication engineer and acoustician, professor	1911-2002 1914-2016	• Former MIT Professor and built first anechoic chamber in Harvard • Founded a major research corporation Bolt, Beranek &Newman (BBN Technologies) and many innovative designs • Authors of many classic books such as *Acoustics*, *Music, Acoustic and Architecture*, *Concert Halls and Opera Houses* and edited books *Noise and Vibration Control* and *Noise and Vibration Control Engineering* (with I. L.Ver) • Conducted design and consultation projects known as best opera houses • Received Gold Medals from the Acoustical Society of America, the Audio Engineering Society and the National Medal of Science, etc.
Carleen Hutchins	Violin innovator, researcher	1911-2009	• Provided great insight into the design and construction of musical string instruments.
Karl D. Kryter	Scientist of Psycho-Acoustics, audition and engineering psychology	1914-2013	• Explained the physiological effects of noise (community noise) on man • Received many awards and honors • Worked on numerous modern-day noise issues including aircraft noise and the sonic boom • Author of two books: *The Effects of Noise on Man* and *Physiological, Psychological, and Social Effect of Noise*.
Per V. Brüel	Physicist and Engineer	1915-2015	• A leader in development of acoustic measurement techniques and analysis and in production of acoustical instrumentation • Founded the Brüel &Kjaer Sound and Vibration measurement A/S (B&K) with V. Kjaer in Denmark.(The company also contributed to education and training in acoustics and acoustical measurements through numbers of publications)

Table 1.1 (Continued)

Name	Background	Date	Achievements/discoveries in acoustics
Sir James Lighthill	Scientist, applied mathematician, acoustician	1924-1998	• Founder of modern aero-acoustics based on earlier research of Lord Rayleigh. • Worked on fluid dynamics and set out theories like Lighthill's eight power Law regarding acoustic power radiated by jet-engine and also known with his Lighthill's report on Artificial Intelligence • Received Gold Medal of the Institute of Mathematics and Its Applications
Isadore Rudnick	Physical acoustician, professor	1912-1997	• Performed major experiments in hydrodynamics involving atmospheric sound propagation and attenuation of sound in sea water • Developed a special type of siren known as most powerful man-made sound source • Received Silver and Gold medals from ASA and number of awards
Cyril M.Harris	Degrees of mathematics and physics, architectural acoustician	1917-2011	• The principal consultant on the acoustics of the Metropolitan Opera House, the Kennedy Center, the Powell Symphony Hall etc. • Co-author of the book *Acoustical Designing in Architecture* with Knudsen, editors of *Handbook of Noise Control* and *Shock and Vibration Handbook*. • Worked on sound absorption and impedance method of rating acoustical materials • Received various awards and held honorary doctorates.
Zyun-Iti Maekawa	Acoustics Engineer	1925-	• Professor Emeritus at the Environmental Acoustics Laboratory in Osaka, Japan, a former Vice-President of INCE/Japan and past member of the International Commission on Acoustics. • Honorary fellow of Acoustical Society of Japan. • Developed calculation method for sound attenuation by screens and the welknown "Maekawa Chart", contributed to knowledge of noise barriers. • Conducted scientific experiments in auditorium acoustics and consultant to design of various concert halls in Japan. • Published numbers of papers and principal author of book *Environmental and Architectural Acoustics* co-authored by J.H. Rindel and P. Lord
Malcolm J. Crocker	Mechanical Engineer, scientist, distinguished professor	1938-	• Research studies in experimental and theoretical acoustics, vibration and noise control. • Many publications and editor-in-chief of several books, i.e. *Handbook of Acoustics*, *Encyclopedia of Acoustics*, *Handbook of Noise and Vibration Control*. • One of the founding directors of International Institute of Noise Control Engineering (I-INCE) and International Institute of Acoustics and Vibration (IIAV). • Received honorary doctorates, various awards and ASME Per Bruel Gold Medal (2017)

sounds. In the early twentieth century, discoveries in electricity and magnetism constituted the foundation for understanding the physical and physiological aspects of acoustics.

The World War I introduced the requirement for measurements in underwater acoustics to detect submarines. In this period, the following developments are as listed [2–4]:

- Phonograph and telephone were invented enabling sound recording and transmission of sound waves to long distance.
- One of the greatest discoveries in acoustics was creating electron tubes and amplifiers with tubes in 1907.
- Sound measurement and analysis by the use of electronics and computers.
- New applications in medicine and industry of ultrasonic frequencies in chemistry.
- During the same period there were great developments in the understanding of hearing mechanisms, psychological acoustics, and the factors affecting sound perception, loudness, intelligibility of speech sounds, and communication.
- New transducers were invented, which played a great role in transforming acoustics as a science by combining mathematical theories and experiments.
- Introduction of the concept of piezoelectric microphones with vacuum tube amplifiers; signal generators enabled the monitoring of sounds.
- Transformation of electrical energy to sound energy enabled the writing and recording of sounds with play-back.
- Acoustics extended its application areas through the discovery of electricity and electronics, leading to the new science called electro-acoustics.
- Electronics applied in underwater sound equipment (hydrophone) were used in underwater echo measurements.
- Electronic amplification was improved, providing better signal processing.
- Research activity and applications in underwater acoustics were heightened during World War II.
- Sabine established the basis of room acoustics, publishing articles on sound absorption, which later became known as architectural acoustics.
- The modern telephone system developed communication acoustics.
- New devices in the field of electronics, such as electronic computers were invented.
- The impedance method for the sound absorption of materials was established.
- Prediction of sound in ducts was developed.
- Later, in the twentieth century, accuracy in measurements and reliability improved.
- The interest in theoretical acoustics became oriented toward applied acoustics.

In the twentieth century, the interest in acoustics shifted more from physics to applied acoustics, and acoustical engineering became a new area in engineering, thanks to the rapid development of computer technology and measurements. For example, real-time signal analyzers facilitated numerical calculations by digital techniques, the development of the finite element method (FEM) and the boundary element method (BEM) provided solutions in acoustic fields and Fast Fourier Transform (FFT) for modal analyses and Statistical Energy Analysis (SEA) in building acoustics, etc.

- At present, advanced electro-acoustics measurement systems, related to ultrasound, are used in many fields, such as aeronautics, underwater systems and medical science technologies which play important roles in research and applications.

- Noise pollution gained importance as a new environmental type of pollution, which led to an interest in noise control technologies, or, in short, noise control.
- Modern acoustical technology provided hearing-aid equipment and ultrasonic devices used widely for medical diagnoses and surgical procedures, testing materials, etc.
- Active noise cancelation was invented and applied to various noise sources for quieter products and operations.

In the late twentieth century up to the present, the outstanding advances have been achieved by scientists in various branches of acoustics (see Table 1.1).

Acoustics is an inter-disciplinary science, which had its beginnings primarily in physics. Later, it became a multi-disciplinary field making contributions to medical, biological, psychological, and chemical sciences and involving practices in various engineering, architecture and planning projects.

Foundation of the Scientific Institutions in Acoustics

Interdisciplinary studies in acoustics have expanded within the past four decades and nowadays acoustics covers a great variety of topics. A new engineering field called "Acoustical Engineering" was discussed by Peutz who divided the field into four disciplines in 1981 [8]:

1) Noise control engineering
2) Architectural or building acoustics
3) Audio engineering
4) Ultrasonic and underwater acoustics.

Acoustical engineering was originally defined in the 1980s as "the application of acoustical scientific and technical knowledge to the realization of installation and manufacturing industry so as to be acceptable considering the noise effects on human beings, by employees and the neighborhood" [9]. In parallel to the activities in Europe mentioned above, the title of Noise Control Engineering was introduced in the USA in 1970s, defining specific noise issues and a new profession. Nowadays matching the theoretical and technical developments, numerous disciplines in acoustics have been initiated and the numbers of researchers involved in applied acoustics have mushroomed. Eventually each discipline has generated its own specific expertise in research, application, technology development, and marketing.

Due to the need for collaboration between acousticians with respect to research, implementation, education, standards development, etc., almost all countries in the world have established their own national societies of acoustics and, in addition – even before some of the national institutions – various international organizations were founded. The topics in the index classification schemes of these institutions differ somewhat over the years according to the interests of researchers, however, these are outlined in Figure 1.1.

The Acoustical Society of America (ASA): http://acousticalsociety.org
The scientific society, established in 1929 by 40 scientists and engineers, joined the American Institute of Physics in 1931. ASA was founded as a national society but also accepts members from other countries in the world.

Objectives: As stated on its webpage: "Premier International scientific society in acoustics, dedicated to increasing and diffusing the knowledge of acoustics and its practical applications ... ASA advocates the development of science, technology, engineering and mathematics (STEM) research and

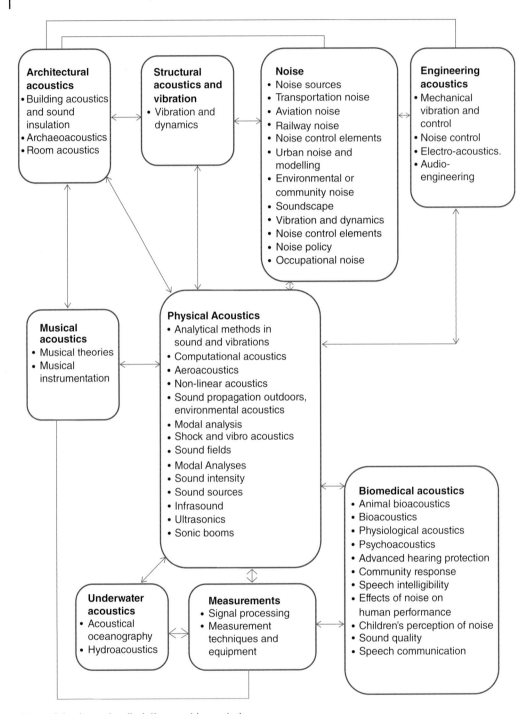

Figure 1.1 Acoustics disciplines and interrelations.

education programs in an effort to strengthen the international STEM workforce" (https://en.wikipedia.org/wiki/Acoustical_Society_of_America).

There are 13 technical committees, whose topics are illustrated in Figure 1.1. The sub-disciplines shown on the webpage have been modified according to the PACS (Physics and Astronomy Classification Scheme).

The International Commission for Acoustics (ICA): http://www.icacommission.org/

The ICA was instituted in 1951 and the new statutes were adopted in Antwerp in 1996.

Objectives: As stated on its webpage: "The purpose is to promote international development and collaboration in all fields of acoustics including research, development, education, and standardization. The ICA convenes the triennial International Congresses on Acoustics, sponsors specialty symposia in acoustics, and coordinates the main international meetings within acoustics."

Among the activities of the Commission are:

- to maintain close contacts with national and regional acoustical societies and associations as well as other relevant professional organizations and seek consensus in matters of mutual interest;
- to provide an information service on societies, congresses, symposia, etc., research and education organizations in the field of acoustics;
- to take a proactive role in coordinating the main international meetings within acoustics.
- to convene the International Congresses on Acoustics.

The topics suggested by the commission and extended yearly according to the needs are included in Figure 1.1.

The European Acoustics Association (EAA): https://www.euracoustics.org/

Established in 1992, the EAA includes in its membership societies predominantly in European countries interested in promoting development and progress of acoustics in its different aspects, its technologies and applications.

Objectives: As stated on its webpage:

- To promote and spread the science of acoustics, its technologies and applications throughout Europe and the entire world.
- To interface with associations whose activities are related to acoustics, in order to promote development and progress of acoustics in its different aspects and applications.
- To establish contacts across member associations from each country, with public and private organizations and enterprises, with associations, science institutions, universities, professional organizations, etc., to assist them in reaching their goals.
- To promote the formation of national acoustical societies in European countries where these do not exist, and to support and strengthen activities of national associations in those countries where they do exist, respecting the principle of subsidiarity.
- To promote acoustics research and application of corresponding technologies.
- To publish a European journal on acoustics, in print as well as in electronic format.
- To organize and promote congresses, to publish books and monographs, and to engage in all those activities that are connected with the diffusion, promotion and development of acoustics.
- To establish agreements for collaboration with European and international entities in order to better serve the objectives of EAA.

- To foster the exchange of knowledge, experience and initiatives present in any one of the member countries, for a better development of and progress in acoustics.
- To stimulate education activities and platforms in acoustics at all educational levels, both academic and professional.
- To promote and divulge the establishment and implementation of norms and recommendations in the various fields of acoustics, especially in the area of environmental acoustics, for a better quality of life.

There are about 7–10 EAA Technical Committees, and their titles are also included in Figure 1.1.

The International Institute of Noise Control Engineering (I-INCE): http://www.i-ince.org

Established first as INCE in the USA in 1971, later I-INCE was founded in 1974 as an international consortium of organizations with interests in acoustics and noise control.

Objectives: As stated on its webpage: "The primary focus of the Institute is on unwanted sounds and on vibrations producing such sounds when transduced." I-INCE is the sponsor of the Inter-Noise Series of International Congresses on Noise Control Engineering held annually in leading cities of the world. Various Technical Study Groups (TSGs) have been organized as principal technical activities of I-INCE. The number of topics included in the Congress technical programs has been increased each year, due to the new noise sources and the growth of interest in noise management, economic and educational issues.

The International Institute of Acoustics and Vibration (IIAV): www.iiav.org

Formed in 1994 and formally incorporated in the USA in June 1995 as a new organization, the IIAV includes the disciplines both of acoustics and vibration.

Objectives: As summarized on its webpage:

"The purpose of the International Institute of Acoustics and Vibration (IIAV) is to advance the science of acoustics and vibration by creating an international scientific society that is responsive to the needs of scientists and engineers in all countries whose primary interests are in the fields of acoustics and vibration. The Institute cooperates with scientific societies in all countries and with other international organizations with the aim of increasing information exchange by sponsoring, cosponsoring or supporting seminars, workshops, congresses and publishing journals, newsletters and other publications. The IIAV organizes an annual premier world event, the International Congress on Sound and Vibration (ICSV), which combines all areas of acoustics, noise and vibration."

In summary, the importance of interdisciplinary work in acoustics studies has been well understood and the field has expanded in the past three decades in terms of theoretical research, implementation, increasing knowledge, standards development, controlling techniques, etc.

1.2 Fundamentals of Sound

Sound can be defined with respect to both physical and physiological aspects. A simple physical definition is the atmospheric pressure variation produced by a physical stimulus. In other words, it is a physical phenomenon resulting from the fluctuations of atmospheric pressure caused by the vibration of a source. Not all simple definitions are sufficient to explain the mechanism of sound

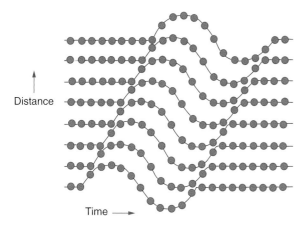

Figure 1.2 Transfer of particle motion in an elastic medium. (*Source:* adapted from Brüel & Kjaer lecture notes, 1970s.)

generation in its entirety, in a more descriptive manner, it can be defined as the variation in the static pressure in an elastic medium by a source oscillating periodically or non-periodically. The propagation of such pressure variations in the medium is at a certain speed and with a delay in time (called the phase-difference) because of the inertia of the medium (Figure 1.2). Even though the pressure changes are very small, the acoustic field is created around sources.

Physiologically, sound is defined as sound waves that excite the hearing mechanism. Pressure variations are transmitted to the ears and the human brain through the complex hearing system that will be described in Chapter 7. The perception of sounds is involved in the science of psycho-physiology. The topics given below are simplified for comprehension at a beginner's level.

The physical context of sound is dependent on the presence of the following factors:

1) A source which makes oscillatory particle motion.
2) A medium having mass and elasticity which transmits this motion (i.e. solid, liquid or gas).

A sound source is defined as an oscillatory motion on a surface or a fluid volume changing its dimensions periodically or randomly. For example, a jet exhaust causing the turbulent mixing of an airflow with the stationary atmosphere or a vibrating diaphragm of a loudspeaker giving aerodynamic sound, are both sound sources. The existence of both factors at the same time might produce complex sounds, such as the noise caused by the interaction between the tires and the road surface.

In acoustics education, some simple mechanical models are useful in demonstrating sound wave generation and types of wave propagation. For instance, sound waves can be simulated by water waves observed when a stone is dropped into a calm water surface and the circles spread outward. These can be used as a simple model for two-dimensional sound waves (Figure 1.3). The waves occur through movements of water particles back and forth about their original positions. Sound phenomena can also be observed by a tuning fork when one taps its arm (Figure 1.4). Some fundamental parameters derived from sound wave propagation can be explained by the back and forth movement of a piston in an infinitely long tube. Figure 1.5 displays a simulation of pressure variations along a tube at different time intervals. The periodic movement of the piston is transferred from one particle to another, causing successive compressions and rarefactions of the particles within the tube. The progress of sound waves can also be visualized in a spring, as shown in

Figure 1.3 Water waves and wavelength. (*Source:* adapted from the teaching software of Mediacoustics, CSTB and Acentech.)

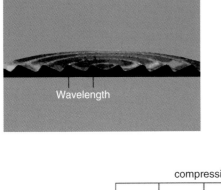

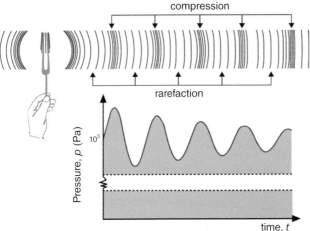

Figure 1.4 Presentation of sound wave and sound pressure by an experiment with diapason. (*Source:* adapted from Brüel & Kjaer lecture notes, 1970s.)

Figure 1.6. When one end of a sufficiently long rope is swung vertically, the shape of the rope resembles a waveform and if the other end is fixed, a returning wave can also be observed as in the early publications in 1940's [10] (Figure 1.7). Longitudinal and lateral propagation of waves can be simulated with a spiral spring in Figures 1.8a and 1.8b. A particle in lateral motion about an equilibrium position and the transfer of the motion to other particles in the same direction, at a time, can also be detected in the oscillations of a series of pendulums placed side by side on the same vertical plane, as shown in Figure 1.9a and b.

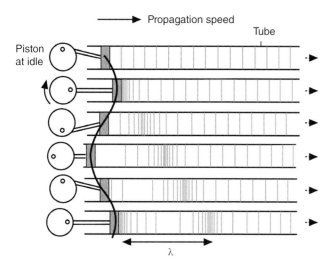

Figure 1.5   Sound propagation model by piston/tube experiment.

Figure 1.6   Spring model and lateral displacement in time. (*Source:* Mediacoustics teaching software, CSTB.)

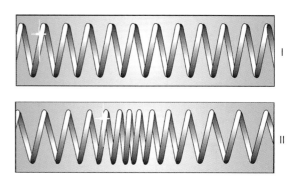

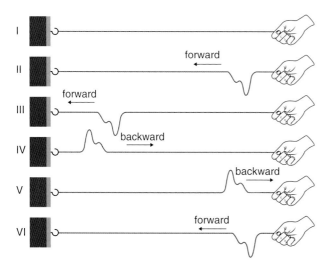

Figure 1.7   Experiment with a rope simulating lateral traveling waves. (*Source:* adapted from [10].)

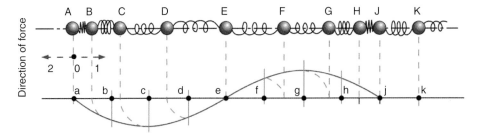

Figure 1.8a Longitudinal wave propagation by spring model. (*Source:* adapted from [10].)

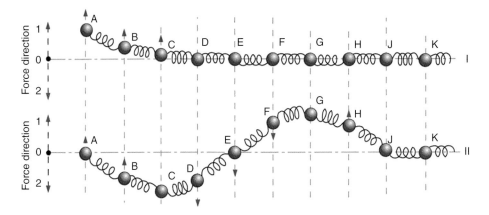

Figure 1.8b Lateral wave propagation by spring model (*Source:* adapted from [10].)

If an object making a circular motion on a plane (like the hand of a clock) is traced on a piece of paper moving horizontally in time at a constant speed, a periodic waveform is obtained (Figure 1.10a). This simplest type of oscillation is called simple harmonic motion whose displacement varies as a sinusoidal function of time.

In summary, the motion has two aspects to examine:

1) A single particle repeats the same vibration (oscillatory motion) in time.
2) Vibration of the particles creates maximum and minimum pressure regions alternately and successively as these regions move forward in time.

The relationship between the motion of a particle implying a small mass of the medium (not a molecule) and wave propagation as a result of compression and rarefaction regions occurring alternately in the medium, can be seen in the piston/tube model in which the piston makes a simple harmonic motion. Figure 1.10b displays the phenomenon supplementary to Figures 1.4–1.6.

Simple harmonic motion can be examined by projection of one-dimensional motion on to a plane (Figure 1.11). The particle is represented by the rotating vector OP (or A) on the x- and y-axis and the motion with time is shown as a sinusoidal wave in Figure 1.12a. In fact, the sinusoidal function of a simple traveling sound wave (i.e. pure tone sound) indicates variation of the atmospheric pressure with time and distance. Sound waves in air are spherical or three-dimensional.

Figure 1.9 Pendulum model for the particle motion.
(a) Series of pendulums oscillating independently;
(b) Transfer of the motion of pendulum (1) to pendulum
(7) and the maximum displacements.

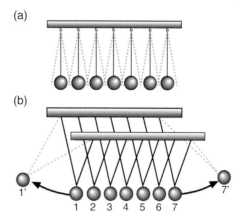

Figure 1.10a Simple harmonic motion and
wave propagation in time. (*Source:* adapted from
Mediacoustics teaching software, CSTB).

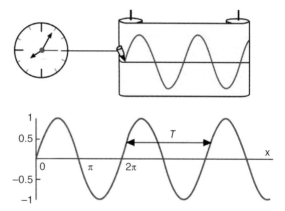

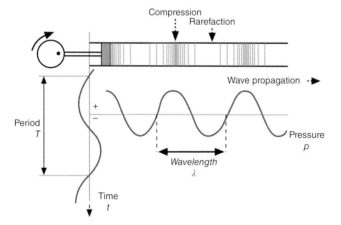

Figure 1.10b Parameters for particle motion and wave propagation in a piston/tube model.

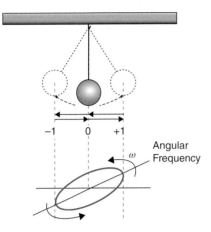

Figure 1.11 Oscillation of a pendulum and conversion of linear motion to circular motion.

The parameters derived from the simple harmonic motion of a particle are described below. The definitions and symbols are given according to the International Standards Organization (ISO) standards [11].

- *Period, T*: Time required for a periodic motion about the origin point or the time for completion of one cycle, if it is projected on a circle to observe the start and end positions (Figures 1.10a and 1.10b). It is the smallest interval of time over which an oscillation repeats itself [12a]. The unit is second.
- *Displacement, ξ*: This is a vector quantity that specifies the change of position of a body (i.e. particle) usually measured from the rest or equilibrium position [12a]. Instantaneous distance between the position of the particle about the origin at any instant projected on to the *x*- or the *y*- axis (Figure 1.12a) (Box 1.1). The unit is the meter.
- *Amplitude, A*: Maximum distance from the origin to the position of the particle, also called maximum displacement. The unit is m, mm, micron (Figures 1.12a and 1.12b).
- *Frequency of a particle motion, f*: Number of cycles per unit time interval (second) expressed in hertz (Hz). Frequency is the inverse of the period: $f = 1/T$, Hz.
- *Particle velocity, u*: Distance moved by a particle in unit time, second, m/s (or rate of change of particle position with time, m/s).
- *Angular frequency, ω*: This is the frequency of a periodic quantity per unit time (rad/s) [12a]. When the particle motion is displayed as a circular motion, the angle which the particle sweeps on the circle in a unit time is known as the angular frequency which is expressed as radian/second (rad/s) (Figures 1.11 and 1.12a). At time *t*, the angle becomes ωt and when the motion repeats itself every time, ωt increases by 2π.
- *Initial phase angle (or phase), ϕ*: If the initial position of particle (*P*) is at P_1 then the angle between the *x* axis and the vector OP_1 is called the initial phase angle [12b] (Figure 1.12b). In Figure 1.12a, $\phi = 0$ and the initial displacement is given in Eq (1.4). When P_1 moves to P_2 after time *t* and the displacement is given in Eq (1.6).
- *Acceleration of motion, a*: Rate of change in particle velocity in a unit time, m/s^2.
- *Force, F*: Force triggered the motion of the particle, newton, N, generally is used for vibrations of machines and structures in industrial noise control problems.

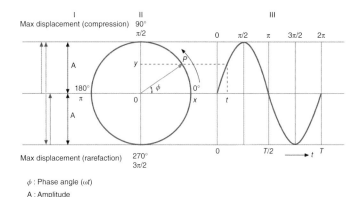

Figure 1.12a Simple harmonic motion and the parameters [12a].

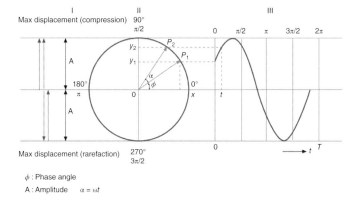

Figure 1.12b Simple harmonic motion with an initial phase angle. (*Source:* adapted from [12a].)

Figure 1.13 shows the relationship between particle velocity, acceleration and displacement. Additional information about particle motion is given in Box 1.1.

The particle velocity u, which is a vector quantity since the motion might occur in different directions, can be determined by the sound pressure measurements and the impedance of the medium, ρc.

$$u = \frac{p}{\rho c} \tag{1.1}$$

ρ: Density of medium, kg/m^3
c: Speed of sound (in air; 344 m/s at 22 °C)
ρc: Impedance of medium (air impedance; 406 mks rayls (N·s/m^3))

Sound pressure and impedance are discussed in Section 1.3.3. The particle velocity is an important parameter used in the definition of sound intensity.

Box 1.1 Equations for Particle Motion

Angular frequency, ω, rad/s:

$$\omega = \frac{2\pi}{T} = 2\pi f \tag{1}$$

Particle velocity, *u*, m/s:

$$Velocity = \frac{distance}{time} = \frac{2\pi A}{T} = 2\pi A f \tag{2}$$

A: Amplitude of the displacement (Figures 1.12a and 1.12b)

Introducing the initial phase angle, ϕ:

$$u = \frac{dy}{dt} = \omega A \cos(\omega t + \phi) \tag{3}$$

Displacement, ξ, *y* or *x*, micron:
(By assuming at *t* = 0, *P* is aligned in the *x* axis)
Projection on the *y* axis:

$$y = A \sin \omega t \tag{4}$$

Projection on the *x* axis:

$$x = A \cos \omega t \tag{5}$$

ωt: The angle displayed as α in Figure 1.12b.

If the initial position of particle *P* (at *t* = 0), is not aligned in the *x* axis (as shown as P_1 in Figure 1.12b), the equation of displacement can be written by introducing the phase angle as:

$$y = A \sin(\omega t + \phi) \tag{6}$$

ϕ : Initial phase angle, radian (When the displacement is on the OY direction, ϕ is positive.)

Acceleration, *a*, m/s²:

$$a = \frac{du}{dt} = -\omega^2 A \sin \omega t = -\frac{4\pi^2}{T^2} r \cos \frac{2\pi}{T} t = -\omega^2 y \tag{7}$$

a: Derivative of velocity with respect to time

Introducing the initial phase angle:

$$a = -\omega^2 A \sin(\omega t + \phi) \tag{8}$$

Force imposed, *F*, newton (N or kg·m/s²): (Newton's Second Law of Motion)

$$F = m\frac{du}{dt} = ma = -m\omega^2 y = -\frac{4\pi^2}{T^2} my \tag{9}$$

m: Mass per unit volume, kg/m³
u: Particle velocity

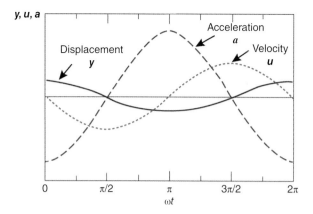

Figure 1.13 Relationship between particle velocity and sound pressure in a simple harmonic motion.

1.2.1 Sound Waves

The theory of sound has been associated with the theories in optics. The transfer of sound waves in a medium is explained in the three disciplines called wave acoustics, geometrical acoustics, and statistical acoustics.

1) *Wave acoustics:* The physical characteristics of sound waves can be expressed using wave equations in which the acoustical parameters are defined as functions of space and time. Parametric solutions can be found for homogeneous and finite size mediums.
2) *Geometrical acoustics (ray acoustics):* Sound waves are displayed by the rays directed from a source to a receiver according to the optical principles (the laws of Snell and Descartes). The complicated wave phenomena in environmental and architectural acoustics can be easily expressed in terms of the ray concept. When the characteristics of the medium vary in space, this approach gives more exact solutions.
3) *Energy acoustics or statistical acoustics:* For enclosed spaces, the energy of the acoustic field is considered and the acoustics problems are solved by means of the statistical approach with the assumption that the acoustic field is uniform.

The choice of which of these approaches to use in order to solve a specific acoustic problem depends on availability (ease), convenience, and accuracy. For enclosed spaces with simple shapes and rigid walls, wave acoustics is useful. However, for large rooms with irregular shapes and absorptive boundaries, the ray acoustic concept, together with the concept of multiple image source, is preferred in room acoustics. The statistical approach is useful in many cases, like duct noise problems where diffuse fields (same energy density in the space) exist to take the time-averaged values.

Types of Waves
Sound waves in geometrical acoustics are three-dimensional vectors, however, for simplicity in the theoretical assessments, two-dimensional and even one-dimensional waves are assumed in the understanding of wave propagation (Figure 1.14). In the simple models given above, the waves are one-dimensional along a string or rope.

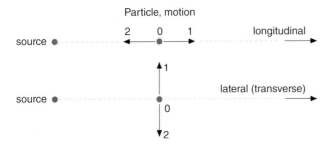

Figure 1.14 Lateral and longitudinal propagation of sound waves.

Longitudinal Waves

Also called compressional waves, longitudinal waves are transferred in the same direction as the particle motion, resulting in alternate compression and rarefaction of the particles as explained above.

Lateral Waves

Also called transverse waves, lateral waves are transferred in the direction which is perpendicular to that of the particle in motion.

Among the mechanical models explained above, such as the motions of a spring, a piston, and a series of pendulums, with the force exerted in the longitudinal direction etc., the particles vibrate back and forth in the same direction of wave propagation, displaying examples of longitudinal waves (Figures 1.4, 1.5, 1.6, 1.8a and 1.9b). On a stretched string, the particles vibrate at right angles to the direction of propagation displaying an example of a lateral wave. The motion of the rope by a vertically exerted force is a simple model of a laterally propagated wave (Figures 1.7 and 1.8b). The water waves indicate both type of waves, however, mostly lateral propagation is found in shallow water [12b] (Figure 1.3).

For liquid and gases, the sound waves are propagated as longitudinal (compressional) waves according to the principles of wave mechanics. The particle motion is more complex in solids, and the longitudinal, bending, and shear waves exist together, depending on the finite size of the solid elastic medium, e.g. a thin sheet can impose all the wave types. The other waveforms are rotational and torsional types which usually occur in solids. This subject is given in Section 1.4.4.

Geometry of Sound Waves

When the pressure disturbances occur uniformly around the source, the waves are called spherical waves. The sound pressure and intensity at any point in the acoustic field can be calculated using the propagation equations. More information will be given in Section 1.3.4. The simple expressions are given in Box 1.2.

When the compressions and rarefactions occur on a plane perpendicular to the direction of the wave propagation, the waves are defined as plane waves in which the pressure changes on the plane surface are uniform. A plane wave is defined by only one dimension in the acoustic field, for example, the wave propagates along a tube in a direction along the duct axis and the plane wave surfaces are perpendicular to this direction [12a, 12b]. Mathematical descriptions (i.e. equations of motion) can be found in the literature, for one-, two- and three-dimensional waves.

The shape of the wavefronts changes with distance. At long distances from a source, a spherical wavefront appears to be a plane surface (see Chapter 2).

Box 1.2   Basic Wave Equations

The sound pressure of a one-dimensional single spherical wave which is moving forward (a forward traveling wave) is given in Figure 1.15):

$$p(r,t) = \frac{A}{r}\cos k(r \pm ct) \tag{1}$$

$p(r,t)$: Instant sound pressure as a function of distance and time
A: Pressure amplitude, N/m
(+): Forward traveling wave (+ X direction), (−): Backward traveling wave
r: Distance from the origin
k : Wave number (See page 31)

Simplified version:

$$p(r,t) = A\cos\omega t \tag{2}$$

Velocity equation for a spherical wave:

$$u(r,t) = \frac{A}{\rho cr}\cos k(r-ct)\left[1 + \frac{1}{kr}\tan k(r-ct)\right] \tag{3}$$

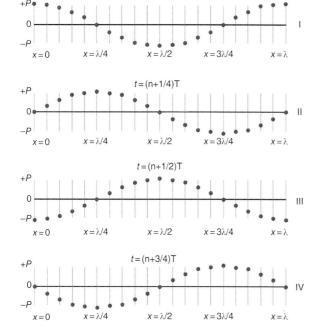

Figure 1.15   Sound pressure variation in a forward traveling (free-progressive) wave at four instants of time, t. (*Source:* adapted from [16].)

Speed of Sound Waves, *c*

The rate at which the motion is transmitted throughout the medium (speed of transmission) is known as the speed of sound. Propagation of sound waves occurs at a speed which depends on the elasticity and density of the medium and, if the medium is a gas, it is only dependent on the absolute temperature. The speed of sound is independent of frequency:

$$c = \sqrt{\frac{D}{\rho}} \ \ \text{m/s} \tag{1.2}$$

c (or c_0 in air): Speed of sound

D: Volume elasticity $= \gamma p_0$ (p_0: Atmospheric pressure, γ: Ratio of specific heat under conditions of constant pressure and constant volume) (also called bulk modulus, K_s)

ρ (or ρ_0 for air): Air density varying with temperature in a constant atmospheric pressure: 1.204 kg/m^3 at 20 °C

The variance of speed with air temperature is given as:

$$c = 331.5\left(1 + \frac{t}{273}\right)^{1/2} \ \ \text{m/s} \tag{1.3}$$

t: Air temperature, °C

The speed of sound in gases varies between 150 and 1500 m/s and the speed of sound in air is 344 m/s (at 22 °C), 343 m/s (at 20 °C) and 331.5 m/s (at 0 °C), however, in environmental noise problems, 344 m/s is generally used.

Speed of sound wave in solids:

$$c_L = \sqrt{\frac{E}{\rho}} \ \ \text{m/s} \tag{1.4}$$

c_L: Speed of sound in solid for a longitudinal wave propagation

E: Elasticity modulus (Young's modulus) of the solid, N/m^2

ρ (or ρ_m): Density of the solid, kg/m^3

The speed of sound waves in solids is related to the direction of the traveling wave. For the longitudinal waves of which the direction of traveling wave is in the same direction of the particle motion, the speed of sound c_L changes between 60 and 6000 m/s. Table 1.2 lists the sound speeds in different media. However, for lateral waves which are called bending waves in solids (e.g. on a panel), the speed of sound c_B is related to the motion of the particle vibrating normal to the direction of wave. The speed of a bending wave in a panel is dependent on frequency and given as:

$$c_B = \sqrt[4]{\frac{\omega^2 B}{\rho_s}} \tag{1.5}$$

c_B: Bending wave speed (phase velocity) on a panel, m/s

ω: Angular frequency

ρ_s: Mass per unit area (surface density of the panel), kg/m^2

B is the bending stiffness of the solid (e.g. homogeneous bar or plate):

$$B = \frac{Et^3}{12(1 - \nu^2)} \text{Nm} \tag{1.6}$$

E: Elasticity modulus of the panel, N/m^2

t: Thickness of the panel in the direction of bending, m

ν: Poisson ratio of the panel

In fact, *B* is a complex number depending on loss factor of the panel.

Table 1.2 Propagation speeds of sound waves in various media.

Type of medium		Longitudinal speed of sound, c_L m/s	Density of the medium, kg/m^3	Characteristic impedance, kg/m^2·s (mks rayl)
Gas	Air (at 20 °C)	343.5	1205	415
Liquid	Water (at 20 °C)	1460	1000	146×10^4
Solid	Aluminum	6420	2580	$16,5 \times 10^6$
	Steel	5790	7850	$45,45 \times 10^6$
	Lead	1200	11 000	$13,2 \times 10^6$
	Concrete	3400 (3500–5000)^a	2300 (2000–6000)^a	$7–13 \times 10^6$
	wood (hard)	3850	850	$3,27 \times 10^6$
	Glass	5200 (4000–5500)^a	2500–5000	$10–25 \times 10^6$
	Gypsum board	1600	650	$1,04 \times 10^6$
	Wood (soft)	3600	450	1.62×10^6
	polyethylene	1950	900	$1,76 \times 10^6$
	Marble	3800	2600	9.9×10^6

^{a} According to different literature. The values differ highly with the physical properties of the materials.

Wavefront

For a progressive wave in space, the continuous surface on which the particles having the same phase are located at a given instant, is called the wavefront. The wavefront of a point source is a spherical surface with radius cT (T: period, c: sound speed) (Figure 1.16).

Wavelength, λ

The distance between the two successive points vibrating with the same pressure amplitude while the sound wave is traveling in a medium, is called the "wavelength." For a sinusoidal wave in a direction of propagation perpendicular to the wavefronts, it is the distance between two successive points where at a given instant of time, the phase differs by 2π [12a] (i.e. the distance that a sound wave travels per period) (Figure 1.10b). The unit is the meter. In terms of geometrical acoustics, the wavelength is the distance between the two successive wavefronts (Figure 1.16). At the two points with $X = cT = \lambda$ apart from each other (where T is period and c is the speed of sound), the pressure changes are equal.

The wavelength of sound is important in geometrical acoustics. As the speed of sound differs from one medium to another, the wavelength varies at the same frequency.

Figure 1.16 Wavefronts and sound rays of a spherical wave.

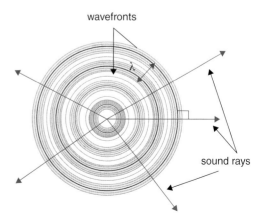

wavefronts

sound rays

Frequency of Sound, *f*

In a periodic motion, the number of cycles that occur per second is called the frequency. The unit is hertz (Hz). The relationship between the wavelength and frequency of a sound in the air is given in Eq (1.7) and illustrated in Figure 1.17.

$$\lambda = c/f \ \text{m} \tag{1.7}$$

λ: Wavelength, m
c: Speed of sound propagation, m/s
f: Frequency, Hz

A simple description of frequency in relation to wave type is seen in Figure 1.18. Frequencies of the perceived sounds cover a large range between 20 Hz and 20 000 Hz, therefore, a logarithmic scale is used in applied acoustics. The audio range of frequencies is divided into various bandwidths whose center frequency is equal to the geometrical average of the upper and lower values. The bandwidth Δf, is the difference between the upper and lower limiting (cut-off) frequencies of a frequency band, containing most of the signal energy [12a]. The bands are assigned with this center frequency according to the International Standards Organization. The most common bands are octave and one-third octave bands and are given in Figures 1.19 and 1.20. The standard mid-frequencies (f_m), are obtained as follows:

$$
\begin{aligned}
&f_m = \sqrt{f_1 \cdot f_2} \\
&f_2 : \text{upper band limiting frequency} \\
&f_1 : \text{lower band limiting frequency} \\
&f_m = 1.414 f_1 \ (\text{for octave bands}) \\
&f_m = 1.12 f_1 \ (\text{for } 1/3 \text{ octave bands})
\end{aligned}
\tag{1.8}
$$

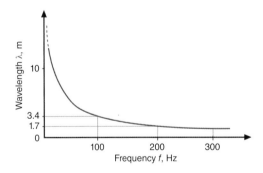

Figure 1.17 Relationship between wavelength and frequency of a sound in air (20 °C).

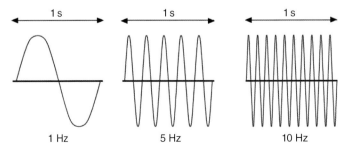

Figure 1.18 Wave types with different frequencies.

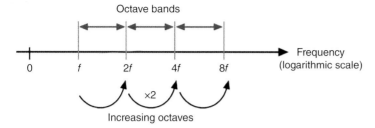

Figure 1.19 Octave band intervals.

Figure 1.20 Standard octave and 1/3 octave (third-octave) bands.

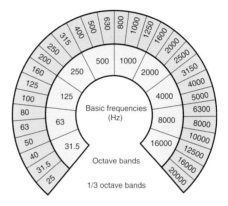

The numbers should be integers.

- *Octave bands*: An octave is a unit of logarithmic frequency interval from f_1 to f_2 defined as $\log_2 (f_2/f_1)$, implying that the upper limiting frequency is twice the lower limiting frequency. The bandwidth of the octave bands (Δf is $\approx 0.70 f_m$). The commonly used mid-frequencies in environmental acoustics, are 63, 125, 250, 500, 1000, 2000, etc. There are eight octave bands between 50 and 5000 Hz.
- *One-third octave bands* (1/3 octave bands): The frequency band whose upper and lower frequency limits are found in the ratio of $2^{1/3}$, so the bandwidth Δf is $\approx 0.23 f_m$, implying that the octave bands are divided into three equal bands. Standard mid-frequencies are 63, 80, 100, 125, 160, 200, 250, etc. There are 24 octave bands between 50 and 5000 Hz.
- *Narrow bands*: Narrow bandwidths can be selected for specific purposes requiring greater resolution of a sound signal, such as analysis can be made at each 5 Hz or even at 1 Hz by using an FFT analyzer (see Section 5.11.3).

For musical sounds, different frequency scales are used. The intervals of the musical sounds within an octave band are divided according to a ratio of the frequencies. In practice, the frequency scale describes the musical sounds subjectively, such as bass sounds (16–160 Hz; $\lambda = 20\text{-}2$ m), moderately bass sounds (160–1600 Hz, $\lambda=2\text{-}02$ m) and high-pitched sounds (1600–16 000 Hz; $\lambda = 0.2\text{-}0.02$ m).

Most noises are complex sounds and include various frequency components and the filtering techniques are used to analyze their frequency characteristics, as explained in Section 2.2.1. The frequency of sound in psychoacoustics implies the tonal characteristics of sound perceived (see Chapter 7) and evaluations of sound with respect to its frequency content can be made by using comparative adjectives, such as simple/complex, harmonic/non-harmonic, intense/weak, high-pitch/low-pitch, etc.

In acoustics, the frequencies are designated as "sonic" in the audio-frequency range between 20 Hz and 20 000 Hz, "ultrasonic" above 20 000 Hz, and "infrasonic" (subsonic) below 20 Hz.

Wave Number, k

The ratio between the angular frequency of particle and the phase speed of sound wave is expressed as:

$$k = \frac{2\pi}{\lambda} = \frac{2\pi f}{c} = \frac{\omega}{c} \tag{1.9}$$

ω: Angular frequency
f: Sound frequency
c: Phase speed of sound (defined as the traveling speed of the phase of a sound wave)

The wave number is a parameter used in wave propagation problems (Box 1.2).

1.2.2 Basic Units and Relations

Sound Power of a Source, P

A sound wave represents a form of energy that emerges when the particles are in motion in the elastic medium. Sound energy is a measure to describe the acoustic field in terms of energy stored in a volume. Box 1.3 gives information about sound energy. Generally defined as the rate at which work is performed or energy is transmitted, total airborne sound energy radiates in all directions by a source, and is known as sound power or acoustic power. Briefly, sound power is the flow of energy emitted by the source and received as sound [12b].

Using the standard definition; sound power through a surface is the product of the sound pressure, p, and the component of the particle velocity, u_n, at a point on the surface in the direction normal to the surface, integrated over that surface [11–15]. It is expressed in watts.

Sound power is a constant value specific to the source. Although the acoustic energy density decreases with distance from the source, the sound power remains constant (Figure 1.21). Sound powers of some sources are given in Table 1.3. Measurements of sound power cannot be made directly but can be measured using sound intensity and calculated from the sound pressures measured in a free-field, i.e. free of any reflective surface or barrier, where the levels are not influenced by the environmental factors (see Chapter 5). Sound power is the basic unit indicating the sound production of a source. However, the human ear is not sensitive to the power or the intensity but can perceive the pressure changes.

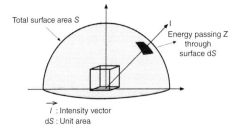

I : Intensity vector
dS : Unit area

Figure 1.21 Radiation of sound in semi-sphere and the energy per unit area: sound intensity.

Sound Intensity, I

The continuous flow of sound energy passing through a unit surface in an acoustic field, at a unit time, is defined as sound intensity. (Figure 1.21) Sound intensity is also defined as the sound energy on a unit area of a surface positioned at a distance c from the source in one second. The unit is watt/m². Sound intensity is a vector quantity, since the surface can be vertical ($I = I_{max}$) or parallel ($I = 0$) to the direction of the sound waves.

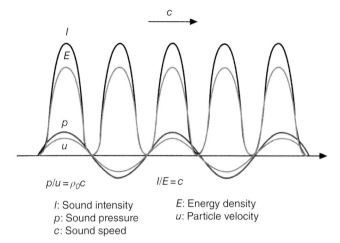

I: Sound intensity   E: Energy density
p: Sound pressure   u: Particle velocity
c: Sound speed

Figure 1.22 Spatial distribution of pressure, particle velocity, energy density, and intensity in a harmonic, progressive plane wave. (*Source:* adapted from [14], permission granted.)

Sound intensity in any other direction varies with the cosine of the angle between the direction of the propagation and the surface.

A spherical sound source radiates sound energy equally at any direction. The acoustic power is distributed homogeneously over the spherical wavefront on which the sound intensity is to be determined. The flow of acoustic energy through a unit surface, i.e. sound intensity, is the ratio of the total energy to the area of spherical surface:

$$I = \frac{W}{4\pi r^2} \quad \text{W/m}^2 \tag{1.10}$$

According to Newton's principle, when a force acts upon an object to cause a displacement, the work done is equal to force times displacement. Thus, the energy passing from a unit surface in a unit time defined as I, is determined in terms of the pressure and the particle displacement. I is a vector quantity specified with direction. Equation (1.11) assumes I is in the same direction as the particle velocity (i.e. $\phi = 0$; $\cos \phi = 1$):

$$\frac{dW}{dt} = Fu = p.\delta s.u = I \quad \text{W/m}^2 \tag{1.11}$$

F: Forces exerted to cause motion
u: Particle velocity (distance of the particle in a unit time)
p: Pressure defined as p_{rms}
δs: Elemental (unit) surface $= 1$

The intensity is written simply as:

$$I = pu \tag{1.12}$$

By taking into account the phase difference between pressure and particle velocity:

$$I = pu \cos \phi \tag{1.13}$$

ϕ: Phase difference between pressure and particle velocity

Box 1.3   Sound Energy

Sound waves cause disturbances in the density and pressure of an elastic medium (liquid) and are transferred in the medium by vibrating particles, i.e. by changing their positions back and forth with a certain velocity. In fact, while transferring the pressure changes, the energy is also passed from the source to the medium. The medium gains kinetic energy in the rest position. During pressure variation, it possesses a potential energy. Thus, sound waves carry both kinetic and potential energies in relation to particle motion and they transmit these energies. Both energies are expressed within a unit volume of the medium.

Kinetic energy, T in relation to particle velocity:

$$T = \frac{1}{2}\rho_0 u^2 \text{ joules} \tag{1}$$

ρ_0: Density of the air, kg/m^3
u: Particle velocity, m/s

Potential energy, U in relation to deviations from static pressure:

$$U = \frac{p^2}{2\rho_0 c^2} \text{ joules} \tag{2}$$

c: Speed of sound, m/s

Kinetic energy is directly related to the square of the particle velocity and the density of the medium whereas the potential energy is directly related to the square of the pressure and indirectly related to the density of the medium.

Sound energy, W or J is the sum of the kinetic and potential energies and defined as energy present in a volume.

$$W = T + U \tag{3}$$

The unit is joules, *J* or watt seconds. It is expressed as the integral of the emitted sound power of a source, *P* over a stated time interval of duration *T*, starting t_1 and ending at t_2 [11, 13]:

$$J = \int_{t1}^{t2} P(t)\mathrm{d}t \tag{4}$$

The quantity is particularly relevant for non-stationary, intermittent sound events.

Sound energy density, w (ε or *E* in earlier documents): Sound energy *W* per unit volume (m^3) is sometimes called sound density. This is defined as the sound energy in a given volume divided by that volume ($w = W/V$) [11]. The unit is J/m^3 or W.s/m^3.

Energy per unit volume for a plane wave with speed c_0 passing through unit area in one second:

$$w = \frac{I}{c_0} = \frac{p^2_{rms}}{\rho c_0^2} \text{ W.s/m}^3 \tag{5}$$

w: The energy stored in a unit volume, J/m^3 or W·s/m^3
c_0: Speed of sound, m/s
I: Intensity of a plane wave traveling a distance, c (i.e. energy in a unit area in unit time = 1 second)

Sound energy density is mainly used for the calculation of acoustic fields in enclosed spaces, e.g. ducts, rooms, etc. In such reverberant rooms where standing waves occur, the relationship between the space-average square of sound pressure and the space-average sound energy density is given in Eq (6) [16, 17]:

$$w = \frac{p^2_{av}}{1.4 p_s} \quad \text{W.s/m}^3 \tag{6}$$

p_s: 1.013×10^5 Pa (atmospheric pressure under normal atmospheric conditions)

Table 1.3 Sound powers of various sound sources.

Sound source	Sound power, watts
Whisper	10^{-9}
Vacuum cleaner	10^{-6}
Human voice at normal level	10^{-5}
Raised voice	10^{-3}
Jigsaw or pneumatic hammer	1
Orchestra	10
Light aircraft	10^2
Jet exhaust	10^4
Large rocket engine	10^8

Figure 1.22 shows the relationship between the pressure, the particle velocity, and intensity for a normal forwarding plane wave [14].

When the particle velocity $u = p/\rho c$ is used, the sound intensity of the sound wave traveling in the free field is written as follows:

$$I = \frac{p^2}{\rho c} \tag{1.14}$$

Sound power for a non-directional source is the sound intensity multiplied by the area of the enclosed surface where the intensities are measured:

$$W = SI = S\left(\frac{p^2_{rms}}{\rho c}\right) \tag{1.15}$$

S: Area of the enclosing surface, m^2

Sound power can be obtained from the space-averaged intensity in a constant acoustic field, by integrating the normal intensity components only, over the surface elements, i.e. to be measured in a direction perpendicular to the enclosed surface:

$$W = \int_S \overline{I_n} \, ds \tag{1.16}$$

W: Total sound energy on a spherical surface

$\overline{I_n}$: Space-averaged sound intensity normal to the surface

ds: Unit area on the enclosed sphere

If the acoustic field is not stationary, i.e. the energy density varies with time, the time-averaging of the instantaneous sound intensities $I(t)$ over an interval during the sound that may be considered stationary, is applied and the time-averaged intensity $\langle I_n \rangle_t$ is used in the above equations.

For a spherical wave, the maximum intensity is determined as:

$$I_{max} = \overline{p.u} \qquad (1.17)$$

Sound Pressure, p

Simply defined as the dynamic variation in atmospheric pressure [12a]. The difference between the instantaneous pressure at a certain point and the atmospheric pressure in a certain time during propagation of sound waves, is known as the "sound pressure" (or the "acoustic pressure" in earlier documents). The unit is pascal, Pa. The sound pressure which varies with time therefore is represented by the root-mean-square of the instantaneous pressure, p_{rms} which is also called the "effective value" (Figure 1.23):

$$p_{rms} = \sqrt{\overline{p^2}}, \quad N/m^2(Pa) \qquad (1.18)$$

p_{rms}: Root-mean-square pressure

p: Instantaneous pressure

Using the sound pressure definition given in Eq (2) in Box 1.2, p_{rms} is calculated by taking into account the variations in time duration T, for a one-dimensional wave according to the initial condition:

$$p_{rms} = \sqrt{\frac{1}{T} \int_0^T (A \cos \omega t)^2 dt} = \frac{A}{\sqrt{2}} = 0.707A \qquad (1.19)$$

p_{rms}: Root-mean-square of the instantaneous pressure (effective pressure)

A: Pressure amplitude (i.e. the maximum pressure in the wave diagram)

T: Time duration, s

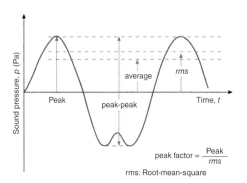

Figure 1.23 Peak, average, and root-mean-square pressures in a sound wave.

The unit of sound pressure in the S.I. system, is $1/1000000 \, \text{atm} = 0.1 \, \text{N/m}^2$ (pascal). The pressure of the minimum audible sound is $2 \times 10^{-5} \, \text{N/m}^2$ by the average young person at 1000 Hz (corresponding to 0.0002 microbar or 0.0002 dyne/cm^2 in the centimeter-gram-second system).

Sound pressure is commonly employed in environmental noise problems since the human hearing reacts to pressure changes (Chapter 7).

Sound Power, Intensity, Pressure and Energy Levels, *L* (the dB Concept)

When the sound power, intensity, and pressure values are expressed as logarithmic values after comparing with a reference quantity (e.g. the minimum audible sound for sound pressure), the results are called "level." The unit is "decibel." For environmental noise, the "sound pressure level" is commonly used whereas the sound power level is preferred to describe source emissions in general. The values in watt, watt/m^2 and Pa are very small quantities, i.e. the range of sound intensity is over a factor of 10^{20} and sound pressure is over a factor of 10^{10}, therefore the logarithmic scale with the compressed numbers by using the ratio 10 : 1 is beneficial to reduce the large exponents in the numbers for practical use (Figures 1.24a and 1.24b). The basic unit in the logarithmic scale is called the "Bell" and 1/10 of Bell (one-tenth) is a "decibel," thus $10 \, \text{dB} = 1$ Bell. For instance, the

Figure 1.24a Relationship between sound power and sound power level (Watt and decibel scales).

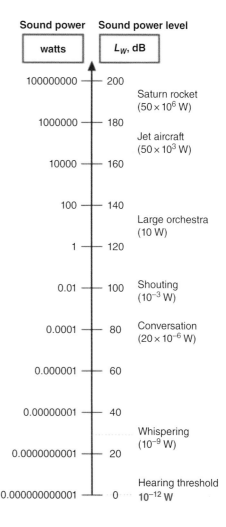

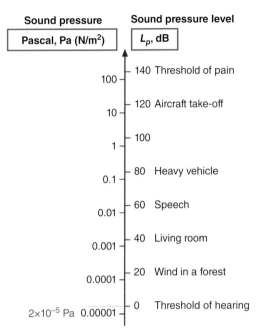

Figure 1.24b Relationship between sound pressure and sound pressure level (Pascal and decibel scales).

sound power level is defined as 10 times the logarithm base 10 (denoted as "lg" according to ISO 80000-8.2: 2019) of the ratio between the measured and referenced powers in watts. The logarithmic functions converting sound power and sound intensity from a linear scale to a logarithmic scale are given in Eqs (1.20) and (1.21):

$$\mathrm{Lw} = 10\lg\left(\frac{W}{W_0}\right), \quad \mathrm{dB}$$

$$W_0 = 10^{-12} \quad \mathrm{watts}$$

(1.20)

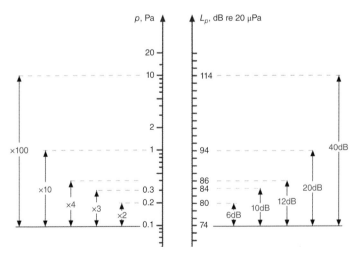

Figure 1.24c Comparison of the increasing rates of sound pressure and sound pressure level (Pascal and decibel scales).

$$L_I = 10 \lg \left(\frac{I}{I_0} \right), \quad \text{dB}$$

$$I_0 = 10^{-12} \quad \text{W/m}^2$$

(1.21)

Sound pressure level in decibel is obtained according to the relationship between the sound intensity and sound pressure given in Eq (1.22):

$$Lp = 10 \lg \left(\frac{p^2}{p_0{}^2} \right), \quad \text{dB} \quad \text{or}$$

(1.22)

$$Lp = 20 \lg \left(\frac{p}{p_0} \right), \quad \text{dB}$$

$$p_0 = 2 \text{X} 10^{-5} \quad \text{N.s/m}^2 (\text{pascal, Pa}).$$

(1.23)

Note that: $p_0 = 0.00002$ Pa $= 20 \ \mu$Pa

L_w, L_I, L_p: Sound power, intensity and pressure levels, dB
W_0, I_0, p_0: Referenced power, intensity and pressure representing the minimum hearing levels

The sound intensity and pressure levels are equal ($L_p \approx L_I$) for most noise measurements, when $\rho_0 c_0 = 400$ mks rayls which is satisfied under various ambient pressure and temperature combinations [17]. If referring to noise rating using linear scales, since the range of hearing is 20 µPa–60 Pa and the dynamic response of the human ear is wide, the intervals of numbers on the linear scale would be so dense that would create difficulties. (e.g. requiring huge numbers and very small intervals). The logarithmic scale on which the threshold of hearing starts from 0 dB is easier to deal with and is well correlated with the subjective responses since it takes into account the pressure squares. As seen in Figure 1.24c, a double increase in pressure causes an increase of 6 dB in the pressure level and a threefold increase corresponds to an increase of 10 dB.

Table 1.4 Summary of acoustic units and their relationships.

Parameter	Linear units	Logarithmic unit (decibel)	Relationship
Sound energy, w	Joule	L_J (dB) Reference energy: J_0: 1pJ $= 10^{-12}$ J	$L_J = 10 \lg (J/J_0)$ dB[a]
Sound power, P	Watt Microwatt	L_w (dB) Reference power; $Wo = 10^{-12}$ watts	$L_w = 10 \lg (W/Wo)$ dB $L_w = 10 \lg W + 120$ dB
Sound pressure, p	Newton/m^2 or Pascal, Pa Micro pascal Bar, millibar, microbar (1/1.000.000 atm) (0.1 N/m^2)	Lp (dB) Reference pressure; $p_o = 2 \times 10^{-5}$ N/m^2 or 20 µPa	$L_p = 20 \lg (p/po)$ dB $L_p = 20 \lg p + 94$ dB
Sound intensity, I	Watt/m^2	L_I (dB) Reference intensity: $I_o = 10^{-12}$ W/m^2	$I = p^2{}_{rms}/\rho c = W/(4\pi r^2)$ $L_I = 10 \lg (I/Io)$ dB $L_I = 10 \lg I + 120$ dB

[a] The logarithmic values are with base 10.

The sound power level is the unique descriptor to define the emission of sound sources, to compare the sound emissions of various sources, for labeling the sources in practice, to confirm the compatibility to the noise limits, to understand the physics of the source and its mechanism, and for noise control at source. The prediction models for environmental noise require sound power levels of noise sources, in addition to the other physical factors that will be explained in Chapter 5.

Table 1.4 provides a summary of the acoustical units and their linear relationships.

Sound energy level in enclosed spaces can be written as in Eq (1.24) [16]:

$$L_J = 10 \lg \left(\frac{J}{J_0} \right) \quad \text{dB} \tag{1.24}$$

J: Sound energy in joule, J
J_0: $1\text{pJ} = 10^{-12}\,\text{J}$

Operations with Decibels

In order to find the total effect of sounds coming from different sources or to find the total spectral level from the octave band levels of a source spectra, the summation process is applied after converting the logarithmical numbers to a linear scale, then the mathematical operation is applied, eventually, the total values are transformed into the logarithmical scale. As explained in Box 1.4, summation and subtraction of the levels can also be made by means of simple diagrams.

Relations between Sound Power, Pressure, and Intensity Levels

Figures 1.24a and 1.24b give comparisons between the sound power and pressure in linear and logarithmic units.

As referred to above, the sound intensity, *I* of a non-directional sound source, which is related to sound pressure, decreases with the distance from the source, corresponding to the radius of the sphere whose surface area is $4\pi r^2$, equal to the ratio of the total power on the spherical surface to the area of the sphere whose diameter is "*r*." *I* is related to acoustic pressure; ($p_{rms}^2 = ZI$) and decreases with the distance from the source (Figure 1.25). The relationship between the sound power and sound pressure levels is given in Eq (1.25):

$$L_p = L_W - 10 \lg \left(4\pi r^2 \right) = L_W - 20 \lg (r) - 11 \quad \text{dB} \tag{1.25}$$

L_p and L_w are sound pressure and sound power levels respectively and the former is dependent on propagation characteristics and the distance, the latter is inherent to the source.

If the sound intensity is uniform on a surface, *S*, from the definition of intensity, the sound power level is calculated as in Eq (1.26) [17]:

$$L_W = L_I + 10 \lg \left(\frac{S}{S_0} \right) \quad \text{dB} \tag{1.26}$$

L_I: Sound intensity level, dB
S: Surface, m^2
S_0: Unit area, $1\,\text{m}^2$

If the source is directional, as will be explained later, the surface on which the intensity is to be determined is divided into sectors, the intensities are measured at each sectors, and then the sub-areas are multiplied by their partial intensities to determine the sound power of the source, *W*.

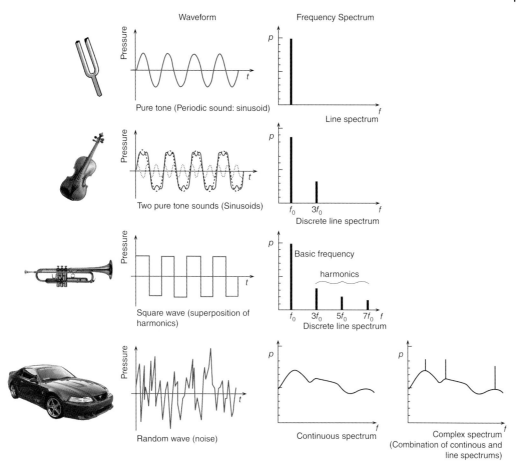

Figure 1.25 Examples of sounds with different spatial and spectral characteristics. (*Source:* adapted from B&K lecture notes, 1970s.)

$$W = \sum_i I_i S_i \tag{1.27}$$

I_i: Intensity on the sector (i)
S_i: The surface area of ith sector (if the sub-areas are equal, the average intensity can be determined)

S_i increases with the size of the sphere and the intensity decreases with the square of the radius. This is called the "inverse square law" (see Chapter 2).

Relations between Frequency and Pressure in Sound Waves

The frequency spectrum: Figure 1.25 shows the relationship between sound pressures, the sound powers or amplitudes versus the frequencies of sound, and is called the "frequency spectrum." The spectrum is used to analyze the sound pressures of the tonal components. The *rms* sound pressure of a frequency component of a complex sound expressed in a certain frequency band in the octave or 1/3 octave bands, is called the "spectral sound pressure level" or the "band pressure level." Sounds containing random components with different frequencies are displayed on a logarithmical scale of frequencies. For the sound spectra without having dominant discrete tones, it is possible to

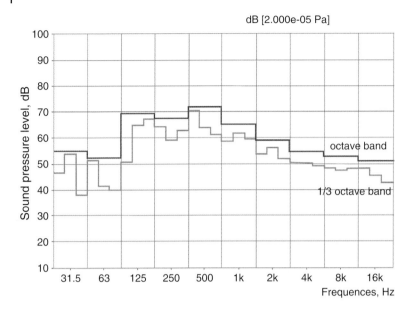

Figure 1.26 Comparison between the octave band and the third-octave band spectrums of sound.

convert the measurement results from 1/3 octave band levels to the octave band levels as shown in Figure 1.26 or vice versa, by using an appropriate filter system.

Sound signals are divided into two groups with respect to the frequency and pressure relationships:

1) Pure tones and harmonics.
2) Complex waves (random sounds).

The sounds in the first group exhibit a simple, linear spectrum shape, whereas in the second group, distribution of sound pressures versus frequency indicates a rather complex shape not easily defined mathematically. The more accustomed noise signals contain various frequencies and pressures randomly mixed in the total sound (see Chapter 2).

Pure tone sound waves: The simplest sound wave is a sinusoidal function with a basic frequency or a simple harmonic motion. It is a sound with a single vibration (Figure 1.27). The equations given above represent the pure tone sound. The pure tone sounds cannot be encountered naturally, but only can be produced by a tuning fork (diapason) or by an electro-acoustic device, i.e. a signal generator.

Complex waves and interference phenomenon

The combination of two sounds generated by two different sources is called "interference." The total effect of two pure tone sounds, arriving at the same point, is obtained by simple arithmetic summation of the time histories of these waves by taking into account the phase differences (Section 1.2). This process is called "superposition." Two sound pressures expressed in waveform can be combined and a new sound pressure function which varies with time and distance is obtained as Eq (1.28) [17]:

$$p(x,t) = p_1(x,t) + p_2(x,t) \tag{1.28}$$

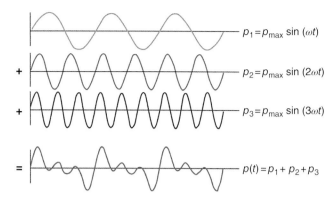

$p_1 = p_{max} \sin (\omega t)$

$p_2 = p_{max} \sin (2\omega t)$

$p_3 = p_{max} \sin (3\omega t)$

$p(t) = p_1 + p_2 + p_3$

Figure 1.27 Superposition of harmonically related pure tones and the resultant wave (*Source:* adapted from Mediacoustics-CSTB, with permission from CSTB).

Summation of the two waves with equal pressures traveling in opposite directions:

$$p(x,t) = 2P_R \cos k (x \pm ct) \qquad (1.29)$$

P_R: Pressure amplitude (max value)

($-$): indicates a right traveling wave.

($+$): indicates a left traveling wave.

Harmonic waves: Sound waves (i.e. periodic complex sounds, such as musical sounds) possess frequency components related to each other. When the components are multiples of a base tone, the lowest component is called the "fundamental sound," the "first harmonic" or the "base tone." The other components at higher frequencies, which are products of the integer (even or odd), are called "harmonics" or "overtones." Sound energies of these components are lower than that of the fundamental sound. The number of harmonics and the ratio of their intensities are the factors characterizing the sound and its source which is produced.

Figure 1.27 shows a complex sound with three components whose sound pressures are p_1, p_2, p_3 and their frequencies are two and three times the fundamental component. The superposition of these periodic waves is also a periodic wave and the resultant is the sound of a trumpet [15]. When the components are odd harmonics (f, $3f$, $5f$, $7f$, ...), the composite wave will be periodic and half-way symmetric. If the components are even harmonics ($0f$, $2f$, $4f$, $6f$, ...), then the composite wave (the resultant) will display an asymmetric waveform, i.e. the top half will not be a mirror image of the bottom.

Musical sounds: Musical acoustics is the branch of acoustics involving the physics of musical sounds. Musical sounds are harmonic sounds produced by various types of instruments (such as strings, wind instruments, or percussion or drums). The form of the harmonic wave and the mathematical relationship which is inherent to a certain instrument are called the "timbre." Timbre is a psychoacoustic characteristic and plays an important role in the perception of complex sounds together with the pitch. Timbre can be determined for an instrument by exciting the resonance regions with different impulses (i.e. intensities) and by the wave speeds. The same musical sounds, even giving the same musical note, can give different timbre perceptions.

Regarding human perception, the term of frequency used for pure tone sounds is called pitch for musical sounds, which is a subjective attribute of auditory sensation related to the repetition rate of the waveform of a sound [18]. A musical composition contains numbers of pitches shown as musical notes corresponding to particular ratios of frequency components ordered on a musical scale. Most music theories use octave bands and with a certain arrangement of pitch classes in sequences separated by certain specific intervals. Scales are described according to the intervals or by the number of pitch classes, such as seven notes per octave for western classical music along with the five half-notes. Combination of these notes and intervals give the melodic characteristics to the sound produced.

Phase relations of sound waves

The phase of a pure tone defines the displacement of the sound in a reference time t. The sinusoidal motion of a pure tone is transmitted between the particles with a certain time delay (phase difference) as explained in Box 1.1. If two particles vibrating with the same frequency pass from the same positions (i.e. the start and finish positions are same), the two waves are called waves "in phase" (Figure 1.12b). If the motions reach the maximum values at opposite directions at the same time, they are in the opposite phase. The phase differences are $T/2$ and π in terms of time and angle respectively. The superposition between the two sound waves with the same frequency in Figure 1.28a, with the same phase (called in phase) and in Figure 1.28b with the opposite phase called the "out of phase," is shown in Figure 1.28. The combination of two sounds with their frequency ratio is 2, is shown in Figure 1.29 for four cases of phase difference. The resultant wave whose components are not correlated with each other respective of frequency and phase (i.e. with random phases) is shown in Figure 1.30.

A combination of two sounds produces a resultant sound wave whose form is dependent on the frequency and the phase of the components. The result can be constructive or destructive, according to the algebraic summation of the pressures at each instant of time.

Summation of sound pressures: When the pure tone sounds have to be combined, the sounds are added by taking into account the phases of each wave. If the frequencies are equal, the pressure of the composite wave becomes as follows:

$$\langle p_t^2 \rangle = \langle p_1^2 \rangle + \langle p_2^2 \rangle + 2\langle p_1 p_2 \rangle \cos (\phi_1 - \phi_2) \tag{1.30}$$

$\phi_1 - \phi_2$: Phase angle difference
$\langle p_i^2 \rangle$: Time-averaged squared pressures

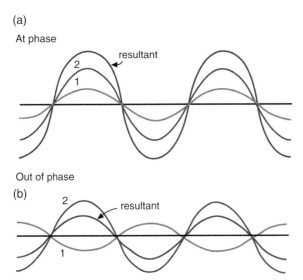

(a)
At phase

Out of phase
(b)

Figure 1.28 Superposition of two waves (1 and 2) which are in phase (a) and out of phase (b) and the resultant waves.

Figure 1.29 Superposition of two harmonically related waves (1 and 2) with phase differences of $0°$, $90°$, $180°$ and $270°$.

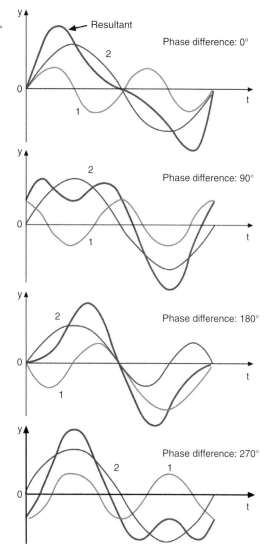

Figure 1.30 Superposition of two pure tones (1 and 2) with random phases.

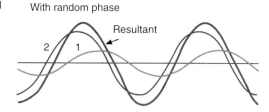

Summation of the waves with random phases:

$$\langle p_t^2 \rangle = \langle p_1^2 \rangle + \langle p_2^2 \rangle \tag{1.31}$$

The pressures above are in pascals, however, when the pressures are given in decibels, the logarithmic summation is applied, which is explained in Box 1.4.

Standing wave: Various types of combinations of two sounds can be of interest, such as given in the following figures. The combination of two waves traveling in opposite directions, with opposite phases and the same amplitude, creates a wave called a "standing wave." The sound pressure of a standing wave does not change with time. Beranek explained this type of wave motion at four instants of time, given in Figure 1.31a [16]. As can be seen, the pressure changes sinusoidally at each point and the total pressure is 0 at points $x = \lambda/4$ and $x = 3\lambda/4$. The maximum values of the pressure of this combined wave are at $x = 0$, $x = \lambda/2$ and $x = \lambda$. The total pressure varying with time and distance is obtained as Eq (1.32):

$$p(x,t) = 2P(\cos kx)(\cos 2\pi ft) \quad \text{N.s/m}^2 \tag{1.32}$$

P: Pressure amplitude (Maximum pressure)

Equation (1.32) represents a wave which does not travel and is called a "standing wave." The pressures of the components are at same phase but they are out of phase at points of x = 0, $x = \lambda/2$, x = λ and $x = 3\lambda/2$. These points are called nodes and antinodes respectively (Figure 1.31b).

When the equation of a normal wave given by Eq (1.29) is compared with Eq (1.32), the standing wave equation has two cosine terms including time and distance variables, implying that there is no repetition of motion and no wave traveling at a time t. An example of a standing wave is given in Figure 1.32 displaying the relationship between resultant sound pressure and particle velocity.

Standing waves in enclosures and room modes: Mode is a characteristic pattern of vibration of a structure in which the motion of every particle is a simple harmonic with the same frequency [12a]. Mostly involved in structure-borne sounds at a natural frequency of the structure, however, the modes also occur in airborne sounds within enclosed spaces since various standing waves can

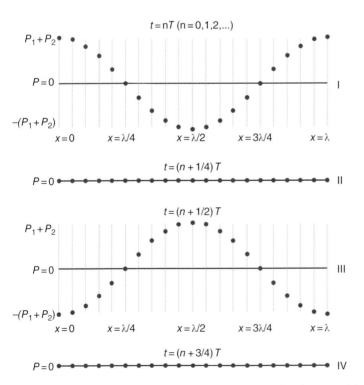

Figure 1.31a Sound pressure of a plane standing wave at four instants of time, *t* occurred by two sources with equal strength. (*Source:* adapted from [11].)

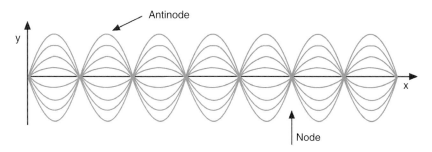

Figure 1.31b Standing wave formation.

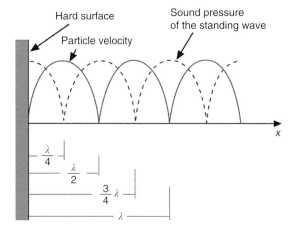

Figure 1.32 Standing wave occurring due to reflection from a hard surface and relationship between sound pressure and particle velocity as a result of superposition of direct and reflected waves.

emerge because of the reflections from the side walls, floor, and ceiling. This phenomenon, which is important in room acoustics, can be analyzed and the frequencies at which the modes as called "room modes" or "normal vibration modes" appear can be calculated. The modes are named by (n_x, n_y, n_z) according to the direction of standing waves on the x, y, z axe (Figure 1.33). The modes are grouped as follows:

- *Axial modes:* Waves existing only in one axis, i.e. between two opposite surfaces, are one-dimensional: $(n_x, 0, 0)$ $(0, n_y, 0)$ $(0, 0, n_z)$.
- *Tangential modes:* Waves parallel to a wall couple and tangent to another wall couple are defined in a two-dimensional surface: $(n_x, n_y, 0)$ $(n_x, 0, n_z)$ $(0, n_y, n_z)$.
- *Oblique modes:* Waves standing at oblique positions to the three surface couples are defined in three dimensions: (n_x, n_y, n_z).

At high frequencies and in large rooms when the multiples of the wavelength are compatible with the room dimensions, numerous standing waves occur, however, some of them cannot be defined.

The lowest frequency at which the resonances occur is calculated as:

$$f = \frac{c}{2d} \text{ Hz} \tag{1.33}$$

d: Distance between the two reflective surfaces, m.

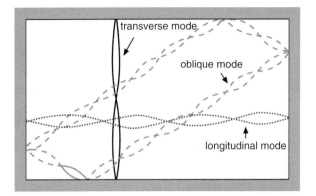

Figure 1.33 Room modes and standing waves occurring in three directions.

Other resonances are determined by multiplying the lowest resonance frequency in the above formula:

$$f = \frac{nc}{2d}$$ (1.34)

n: integers: 1, 2, 3, etc.

The modes are determined in a room, depending on room dimensions as follows [15, 17]:

$$f_n = \frac{c}{2}\sqrt{\left(\frac{n_x}{l_x}\right)^2 + \left(\frac{n_y}{l_y}\right)^2 + \left(\frac{n_z}{l_z}\right)^2}$$ (1.35)

f_n: n^{th} mode (Natural frequency), Hz
n_x, n_y, n_z: Integers ($m = 1$, $n = 1$ imply the lowest mode),
l_x, l_y, l_z: Room dimensions, m,
c: Sound speed, m/s

The number of the room modes excited in a frequency range (from 1 Hz to *f* Hz) is calculated as [19–22]:

$$N = \frac{4\pi f^3 V}{3c^3} + \frac{\pi f^2 S}{4c^2} + \frac{fL}{8c}$$ (1.36)

N: Number of modes
V: Volume of the room m^3
f: Frequency, Hz
S: Total room surface, m^2
L: Total perimeter of the room, m
c: Speed of sound, m/s

The number of modes excited in a narrow frequency band can be obtained from the derivative of Eq (1.37) and is called the modal density [17]:

$$\frac{dN}{df} = \frac{4\pi f^3 V}{c^3} + \frac{\pi f S}{2c^2} + \frac{L}{8c}$$ (1.37)

The parameters are given above.

Generally, in enclosed spaces, the number of modes increases with increasing frequency and almost every frequency band includes many room modes at high frequencies. However, the modes excited at lower frequencies are less crowded with greater frequency intervals and thus cause large fluctuations in sound pressure levels. Even a sound burst can be heard at a modal frequency. This fact is a negative attribute in room acoustics and must be avoided by careful selection of the room dimensions. To simplify the calculations of room modes, to lower limiting frequency (Schroeder's limiting frequency) can be determined based on the statistical methods, in relation to reverberation time and volume of the room [23].

In addition to the resonances in the air, the phenomenon also occurs in solids. For example, building structures and constructional elements have natural frequencies which can be excited by dynamic forces, and thus are capable of causing severe structural vibrations or by sound pressures causing resonant sound transmission. It is possible to analyze standing wave patterns in two-dimensional plates by calculations or measurements. As a benefit of the resonance phenomenon, musical sounds are produced and amplified by musical instruments, such as vibrating strings, open and closed-end air column instruments, and vibrating mechanical instruments, when they are set into vibrational motion at their natural frequencies.

Beats

Beating is a phenomenon caused by the interference between the pure tone sounds (simple harmonic sounds), if the frequency difference between these sounds is very small, such as 2–5 Hz. When the wave equations of two such sounds with equal pressure amplitudes are considered, the displacement in the resultant wave at a time t, can be given as Eq (1.38) (see Box 1.1, Eq (4)).

$$y = A \sin \left[2\pi \left(\frac{f_1 + f_2}{2} \right) t - \frac{\varphi}{2} \right] \tag{1.38}$$

f_1 and f_2: Frequencies of the components, Hz
φ: Phase difference between the waves $(\phi_1 - \phi_2)$, radian
A: Amplitude of the resultant wave:

$$A = 2a \cos \left[\pi (f_1 - f_2) t - \frac{\varphi}{2} \right] \tag{1.39}$$

t: Time
a: Displacement at t

The combined wave is a sound wave with frequency $f = (f_1 + f_2)/2$. The sound is heard as beats (pulses) at instants where the highest pressures are obtained. The number of beats, n is equal to the difference between the frequencies of each sound; $n = f_2 - f_1$ and is called the beat frequency.

If $f_2 - f_1 < 10$ Hz; the beats are heard.

If $f_2 - f_1 > 10$ Hz; two sounds are heard separately.

Sound waves producing beats and the resultant wave are displayed in Figures 1.34a and 1.34b.

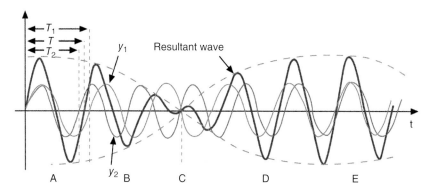

Figure 1.34a Frequency component and occurrence of beats.

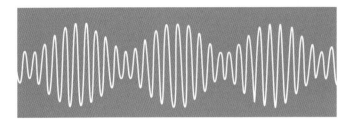

Figure 1.34b   Analysis of beats in the oscilloscope.

Box 1.4   Decibel Operations

Summation of Decibels

In order to obtain the total sound pressure level of two sounds (e.g. generated by two different sources), a summation process is applied by assuming that the sounds have random phases (or they are incoherent). However, for pure tones, this process cannot be applied due to the interference explained above. The logarithm of a summed value is not equal to the summation of the logarithms of each component, therefore, the logarithmic values are converted into pressure units, then an arithmetic addition is applied to the values which are converted into decibels again to find the total pressure level:

$$L_{pt} = 10 \lg \left(\frac{p_t^2}{p_0^2} \right) \quad \text{dB} \tag{1}$$

$$p_t^2 = p_1^2 + p_2^2 + \ldots + p_n^2 \tag{2}$$

$$p_0 = 2 \times 10^{-5} \quad \text{N.s/m}^2 \text{(pascal)}$$

The base of the logarithm is 10.

The above process is simplified by taking the antilogarithms of the sound pressure levels:

$$L_{pt} = 10 \lg \left(10^{L1/10} + 10^{L2/10} + \ldots + 10^{Ln/10} \right) \quad \text{dB} \tag{3}$$

$$L_{pt} = 10 \lg \sum_{i=1}^{n} 10^{L_i/10} \tag{4}$$

L_i: The levels $L_1, L_2, L_3 \ldots L_n$, to be added

For decibel addition, a simple diagram is given in Figure 1.35.

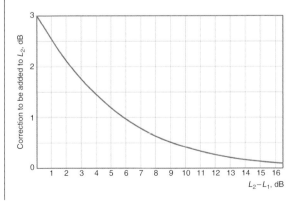

Figure 1.35   Decibel summation chart.

The Total Level from the Band Levels of a Spectrum

When the sound pressure levels at frequency bands (i.e. band pressure levels) are given, the total sound pressure levels are obtained from the band pressure levels:

$$L_{total\ level} = 10 \lg \sum_{i=1}^{n} 10^{L_{i(f)}/10}\ \mathrm{dB} \tag{5}$$

$L_{i(f)}$: Spectral sound pressure level at each octave or 1/3 octave band, dB
n: Number of bands to be added

In practice, the summation process can be made two at a time and the calculation is terminated when the difference between the levels is >10 dB, because a further difference would make no significant change in the total level. Sometimes it is sufficient to find the total value by taking into account only the couple of bands having the highest levels and the rest is compared with the total (Figure 1.36).

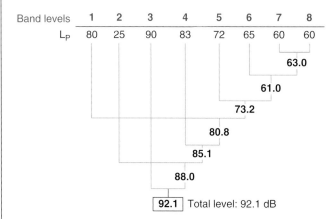

Figure 1.36 Simple method for summation of decibels.

Total Energy Density Level: L_w (or L_E in earlier documents)
In order to obtain the total energy density of the two sound sources whose energy densities are w_1 (= I_1/c) and w_2 (= I_2/c), a linear summation is applied first; $w_1 = w_1 + w_2$. Then the value is transformed into decibels for the total energy density level:

$$L_w = 10 \lg \left(\frac{w_3}{w_0} \right) \quad \mathrm{dB} \tag{6}$$

w_0: 1pJ/m^3 = 10^{-12} J/m^3 (W·s/m^3)

Averaging Decibels
The average energy density level is determined in a room from several measurements at different points:

$$\overline{L_w} = 10 \lg \left(\frac{w_1 + w_2 + w_3 + \dots + w_n}{nw_0} \right) \quad \mathrm{dB} \tag{7}$$

n: Number of measurements

Average Sound Pressure Level
Some noise descriptors require the averaged sound pressure levels, as will be seen later. For this purpose, the arithmetical summation of the squared-sound pressures is divided by the number of sounds:

$$\overline{L} = 10\lg\left(\frac{p_t^2}{np_0^2}\right) \quad \text{dB} \tag{8}$$

$$p_t^2 = p_1^2 + p_2^2 + ... + p_n^2 \tag{9}$$

n: Number of sources or number of sound pressure level 3 to be averaged.

or

$$\overline{L} = 10\lg\left[\frac{1}{n}\left(10^{L1/10} + 10^{L2/10} + ... + 10^{Ln/10}\right)\right] \quad \text{dB} \tag{10}$$

If the difference between the maximum and minimum levels is not greater than 3–5 dB, the arithmetic averaging can be applied within the errors of 0.3–0.7 dB [3]. The time-average-sound pressure level, $L_{AV,T}$ is obtained for the sound levels measured at a certain observation time period:

$$L_{AV,T} = 10\lg\left((1/T)\frac{\int_0^T p^2(t)dt}{p^2{}_0}\right) \quad \text{dB} \tag{11}$$

T: Observation time (about 8–24 hours for a time-varying noise)
$p^2(t)$: Squares of the sample sound pressures measured

Subtraction with Decibels
The logarithmic subtraction is applied to find the difference between the two sound levels for $L_2 > L_1$:

$$L_p = 10\lg\left(10^{L_2/10} - 10^{L_1/10}\right) \quad \text{dB} \tag{12}$$

Subtraction is required, for example, if the level of sound source cannot be discriminated because of the high background noise, then the total sound pressure level is measured and the background sound level is deducted from the measured result, however, this operation is not reasonable when the total noise level is at least 3dB greater than the background noise level (Figure 1.37).

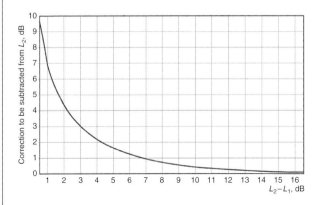

Figure 1.37 Decibel subtraction chart.

Summation of level differences

It may be necessary to obtain the total attenuation level contributed by several factors during sound propagation in an environment. For example, the total barrier performance at a receiver point is obtained by combining the sound level reductions from the three sides, by paying attention to the negative signs:

$$L_p = L_{pr} + NR \tag{13}$$

$$NR = 10 \lg \sum_{i=1}^{n} 10^{-(NR_i/10)} \tag{14}$$

L_p: Sound pressure level at the receiver, dB
L_{pr}: Sound pressure level at the receiver without taking into account any reduction, dB
NR: Total sound level reduction, dB
NR_i: Reduction provided by a factor or through a different sound path, dB

Note: NR is (−), since the signs of attenuations are negative.

1.3   Sound Sources

Sound source is defined as "anything that emits acoustic energy in the adjacent medium" [12a]. The objects or mechanisms generating sound waves into the ambient air, a liquid or a solid directly by periodic or random vibrations in time within auditory range, are called sound sources. Most of the sound sources exhibit a combination of individual sources, some generating simple harmonic sounds. Simple sources are easily modeled and categorized, however, most of the sound or noise sources are complex sources, such as a machine or a vacuum cleaner that need to be examined specifically.

Because of the great variety in structural and operational characteristics, environmental noise sources will be explained in detail in Chapter 3. The basic sound emitters are grouped in terms of their dimensions, geometric properties, and sound radiation mechanisms:

1) Sound sources generating sound into the air or a liquid from their surfaces and changing the volume of the ambient fluid or causing pressure changes through vibrations of their surfaces (example: loudspeaker diaphragm, compressed air out of air sirens).
2) Sound sources radiating sound by continuous rotational movements (example: rotating fans, vibrating guitar strings).
3) Sound sources without a solid surface and creating variable inner forces and irregular transfer of the fluid momentum (example: fluid jets).

These sources transform the mechanical, electrical, or chemical energies into sound energy and their sound powers are related to the acceleration of the excited forces, their velocities, and the frequency of the volume motion, e.g. horn-type loudspeakers. Generally, the acoustic properties of sound sources are analyzed with respect to the following aspects:

- acoustic efficiencies: sound powers and sound pressures at a certain point associated with their operations
- variation of power in time
- variation of power with frequency
- size of the near field and the far field
- strength
- directionality.

Correlations between the mechanical and acoustic powers of specific noise sources in industry and environment have been well established and are implemented in noise control engineering.

1.3.1 Source Types and Sound Powers

Sound sources are divided into different groups according to their geometrical properties (Figures 1.38 and 1.39).The sound powers and the resultant pressures in the near field can be found in the basic literature given in the references [12a, 12b,12c].

Point Source

Sources whose dimensions are smaller than the wavelength of the generated sound and radiating spherical waves are called point sources. They can be either dynamic or static and are categorized as:

1) *Monopole source* (simple source): A solid sphere whose radius changes with time and generating spherical waves is a monopole source (Figure 1.39). It is defined as "An idealized sound source that is concentrated at a single point in space." A monopole source is omnidirectional, i.e. the source radiates the same energy in all directions. Its motion is transferred through the interface between the source surface and the medium. It is the basic sound source explained in Section 1.2. The pressure is dependent on the surface velocity, the radial distance from the origin and the total surface area of sphere. Its dimensions are much smaller than the half-distance between the source and the receiver point, however, an ideal monopole (point) source is assumed to be infinitesimally small in environmental acoustics. The sources that could not satisfy this condition can be considered as comprising several monopole sources, such as a part of mechanical equipment individually radiating noise. The total effect can be determined through summation of individuall sound pressure levels. Noise sources generate sound waves with random phases, i.e. when the phases are compatible to each other, the interference occurs (resonance) as explained above.

2) *Dipole source*: Two monopole sources pulsating simultaneously (with the same strength) but with opposite phases (a phase difference of $180°$) create a dipole source when the distance between them is smaller than the wavelength of the sound that they produce (Figure 1.39). The precise condition of monopole sources is given as: $(kd)^2 \leq 1$, however, an ideal dipole source comprises two monopole sources which are infinitely close to each other (i.e. a compact dipole source). A dipole source is not omnidirectional and the minimum radiation emerges in the direction perpendicular to the line connecting the monopole sources. The sound intensity is maximum on the line passing through their centres (Figure 1.39). A tuning fork or a turbulent flow obstructed by a small barrier can be given as examples of a dipole source. On the other hand, for a non-compact dipole source, complex acoustic fields occur due to the interference

Figure 1.38 Sound sources and sound radiations. (*Source:* adapted from Mediacoustics Teaching Software, CSTB).

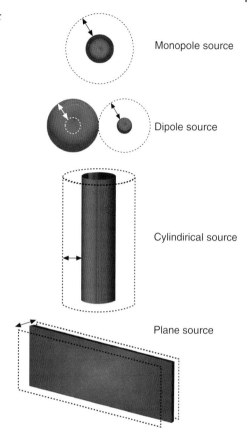

between the point sources depending on source geometries, dimensions, amplitudes, phases, and angular frequencies.

3) *Quadrupole source*: The combination of two dipole sources positioned next to each other, or along a line, forms a quadrupole source. Examples include flow-induced aerodynamic sources, turbulent air from the rotor blades of aircraft or wind turbines. A tuning fork is an example of a longitudinal quadrupole. The sound pressure at a distant point for a quadrupole source which is highly directional can be calculated using its power. Generally quadrupoles are less efficient radiators than dipoles, and much less efficient than monopoles.

4) *Vibrating solid spherical source*: A solid object, whose surface vibrates back and forth in the radial direction, activates the particles in the surrounding medium (Figure 1.39). The radiated sound power is dependent on the diameter of the source, the speed of vibrations (m/s) and the frequency of motion.

Aerodynamic Sound Sources

Aerodynamic sound is caused by the interaction between the high-speed or non-steady gas flow. The sources can be monopole, dipole, or quadrupole. An aerodynamic monopole source occurs when a mass or a heat is given to a non-steady flow or when high-speed air is released periodically from an aperture, like pulsating air jets. For example; a siren generates a steady air flow interrupted

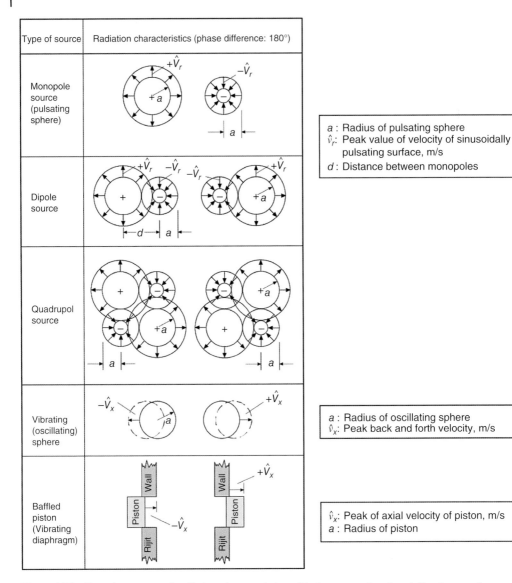

Type of source	Radiation characteristics (phase difference: 180°)
Monopole source (pulsating sphere)	
Dipole source	
Quadrupol source	
Vibrating (oscillating) sphere	
Baffled piston (Vibrating diaphragm)	

a : Radius of pulsating sphere
$\hat{v}_r$: Peak value of velocity of sinusoidally pulsating surface, m/s
d : Distance between monopoles

a : Radius of oscillating sphere
$\hat{v}_x$: Peak back and forth velocity, m/s

$\hat{v}_x$: Peak of axial velocity of piston, m/s
a : Radius of piston

Figure 1.39 Sound sources and radiation characteristics with the parameters in relation to sound powers. (*Source:* adapted from [16 and 17].)

periodically. Similar to a pulsating sphere, the wavefronts of the pulses are in the same phase, causing spherical radiation. An aerodynamic dipole source occurs when an interaction between a solid object and a gas-flow creates irregular forces (in other words, when the flow is obstructed periodically by a solid object). For example, a turbulent flow in a compressor interacting with rotor blades in an electrical motor or the air-flow through the grills or vanes, causes aerodynamic dipole sources. In this type of source, one of the monopoles takes air inside while the other one exhausts air outward reciprocally.

Aerodynamic sounds emerge from the mechanical systems in industry and the acoustic power which is related to frequency, flow speed, gas density, speed of sound in gas, dimensions etc., can be determined through measurement. Calculation of sound pressure levels at a certain point, which is more complicated, is given in various references.

Linear Source (Line Source)

An array of point sources having equal energy and spaced on a line with infinitesimal intervals forms a line source. The point sources vibrate either randomly or at phase and the sound output obeys the rules of interference mentioned above.

Neglecting the phase differences, a line source resembles a pulsating source generating cylindrical waves in which the wavefronts are concentric cylindric surfaces. (Figure 1.38). Traffic on a road is an example of a line source, since the superposition of sounds generated by the individual vehicles on the road forms a cylindrical wave. Trains with their limited lengths are dynamic line sources moving on the rails (i.e. succession of single events). Cylindrical waves are also radiated by ducts and pipes carrying turbulent fluid or by the objects rotating around their axes or by series of machines closely spaced on a line, conveyors etc. These sources are examined with respect to the following aspects (Figure 1.40):

1) Layout of the point sources on the line: (i) Discrete sources with equal distances (intervals) or randomly spaced; (ii) Coherent sources (contiguous).
2) Length of line source respective to a receiver point: (i) Source with infinite length: In practice, the line source extending 20 times the source-receiver's distance on both sides of the projection point of the normal line drawn from a receiver point, is defined as an infinite source. (ii) Source with finite length: The portion of an infinite line which remains within a suspending angle from a receiver point or a sector of a line source, either stationary or moving, is defined as finite-length source (example: a train passage).

The sound power of a line source can be determined simply by adding the acoustic outputs of individual sources. Total energy can be expressed as the power of the unit length of the source.

Plane Source

A vibrating diaphragm or an infinitesimal surface radiating plane waves is a plane (planar) source. A plane source is assumed to be composed of an infinite number of monopole point sources (Figure 1.38). A piston whose radius (a) is much smaller than the wavelength of the sound ($ka \ll 1$, k: wave number) moving back and forth in a tube, can be given as an example of a plane source. (Figure 1.39). Plane waves are radiated by some musical instruments i.e. trumpet, because of the air flow through itself. The sound has a certain limiting frequency as a function of the cross-sectional area. The acoustic power of a plane source (e.g. axially vibrating diaphragm or a baffled piston radiating sound to one side of a rigid wall) can be determined by means of peak vibration velocity, radius of the source [17]. An industrial building radiating sound outside from its walls and roof can be assumed to be a multiple plane source and the total sound energy on the semi-sphere can be calculated.

Sound generation from plane sources and sound propagation in environment are explained in Section 1.3.3.

1.3.2 Directionality Characteristics of Sources

The sound sources mentioned above are assumed to be omnidirectional, i.e. radiating energy equally in all directions and ideally their wavefronts are concentrated spheres. The condition of

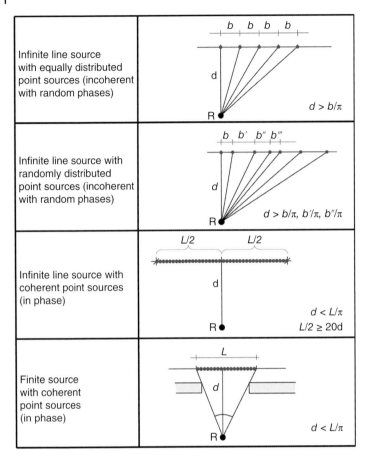

Figure 1.40 Geometrical characteristics of line sources. L_{ps}: Sound pressure level equal for each angle. (*Source:* adapted from [16].)

omnidirectionality is: $d/\lambda \ll 1$, where d is the source dimension and λ is the wavelength. For most sources, the energy varies according to the direction of the sound path and sound sources might radiate more energy in a certain direction (Figures 1.41a and 1.41b). As one of the physical characteristics of sound (or noise) sources, this characteristic is defined as source directivity and depends on the following conditions:

- If the wavelength of the sound is much larger than the geometrical dimensions of the source, the sound is radiated uniformly (equally) in all directions and the source is non-directional (or uniform sources).
- If the wavelength is much smaller than the source dimensions, the sound radiated from the source is limited within a certain range. Generally this channel is narrower at high frequencies.

Directionality characteristics of a source are expressed with a directivity factor and a directivity index. The directivity factor of a source in a free field is the ratio of the mean square pressure (or intensity) at a given frequency in a specified direction from the source, to the average mean square

Figure 1.41a A non-directional source (or omnidirectional source) *Lps : Sound pressure level equal for each angle, θ. (Source:* adapted from [16]).

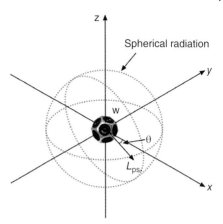

Figure 1.41b A directional source and parameters: $L_{p\theta}$: Sound pressure level at angle θ. (*Source:* adapted from [16]).

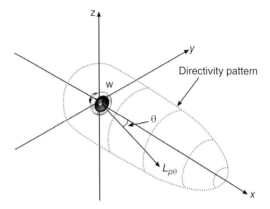

pressure (or intensity) over a sphere of the same radius, centered on the source. The directivity factor by using the symbol of Q_θ (or $D(\theta, \phi)$ in three dimensional field) is calculated as:

$$Q_\theta = \frac{I_\theta}{I_S} = \frac{p_\theta^2}{p_S^2} = \frac{10^{Lp_\theta/10}}{10^{Lp_S/10}} \tag{1.40}$$

$$Q_\theta = 10^{(Lp_\theta - Lp_S)/10} \tag{1.41}$$

Lp_θ: Sound pressure level of a source with power W at a distance r in the free field (without reflections) and at a direction angle, θ

Lp_S: Sound pressure level of a uniform source with power W at the same distance r in the free field

Q_θ: Directivity factor at angle θ (unitless)

Directivity index DI_θ is a measure of directionality factor in dB:

$$DI_\theta = 10 \lg_{10} Q_\theta \quad dB \tag{1.42}$$

$$DI_\theta = Lp_\theta - Lp_S \tag{1.43}$$

For a uniform source (non-directional sound source), the directivity factor at all angles of direction is $Q_\theta = 1$ and $DI_\theta = 0$. The directivity factor is obtained from the directional index:

$$Q_\theta = 10^{DI_\theta/10} \tag{1.44}$$

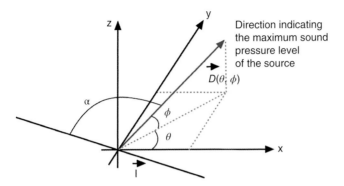

Figure 1.41c Angular parameters defining source directivity $D(\theta, \phi)$.

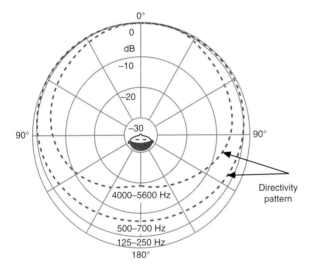

Figure 1.42 Directivity patterns of human voice at different frequencies [15].

The angular parameters are given in Figure 1.41c. The directivity of sound sources can be determined by means of the measurements and the result is presented as diagrams. It is possible to model the directivity of sound sources based on calculations or measurements. Two- or three-dimensional charts, indicating the sound energy distribution at a constant distance from a source, are called "directivity patterns." Since directivity is proportional to the wavelength of sound, the directivity pattern can be presented as frequency-dependent diagrams. Figure 1.42 gives a variation of the directivity of human voice in relation to the frequencies. Generally, noise sources are more directional at high frequencies (with shorter wavelengths) while they are uniform at lower frequencies. Examples of 2D and 3D patterns are given in Figures 1.43a and 1.43b respectively. Figure 1.44 shows the directivity patterns of a loudspeaker at three frequency characteristics.

Directionality characteristics of noise sources are explained in Chapter 3.

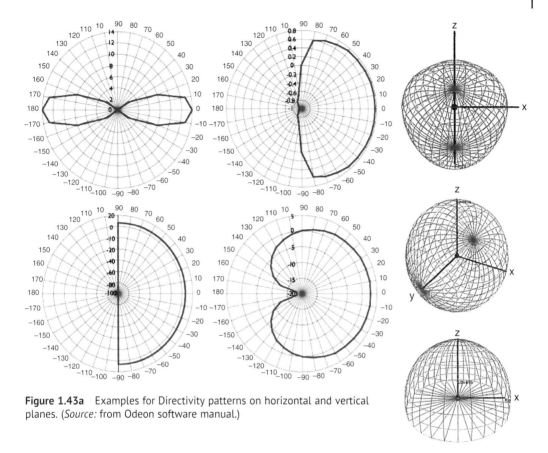

Figure 1.43a Examples for Directivity patterns on horizontal and vertical planes. (*Source:* from Odeon software manual.)

Figure 1.43b Three-dimensional directivity patterns. (*Source:* from Odeon software, permission granted.)

1.3.3 Wave Propagation and Sound Pressures in Relation to Source Type

The waveforms of sound sources display different geometries and the sound pressure at a receiver point in the propagation path, changes with the type of source but more importantly with the distance from the source. The pressure change in relation to distance occurs as a decrease in level, which is called the geometric deviation or the wave divergence. The geometric deviation is normally expressed as a level reduction in dB per unit distance (m) and analyzed in a free field (a medium where no reflections exist) depending on the source geometry (Figure 1.45a).

Point source and propagation of spherical waves: If a receiver point is located at a sufficient distance *d* from a point source (i.e. a distance greater than the wavelength of sound or greater than the source dimensions), the wavefronts are spherical, as explained in Section 1.2.1. If the source is located on a plane surface, the wavefront is semi-spherical. During sound propagation, the surface of the sphere expands four times with doubling of its radius (spread of energy to a greater area) and since the intensity I, is inversely proportional to the square of the distance (which is called the "inverse square law," shown in Figure 1.45b), the reduction in sound intensity level (= sound pressure level if the medium is air) is 6 dB. Figure 1.45c shows the sound pressure level reduction with distance d from a point source. The sound pressure level at the receiver point can be obtained in relation to the sound power:

$$L_p = L_W - 10\lg\left(4\pi d^2\right) \tag{1.45}$$

$$L_p = L_W - 20\lg d - 11 \quad \text{dB} \tag{1.46}$$

For a semi-spherical wavefront (i.e. for the source on the ground), the constant (-11), should be replaced by (-8) in the above equation. Variation of sound pressures between two points located at different distances from the source, can be observed by the following relationship:

$$L_2 = L_1 - 20\lg\frac{d_2}{d_1} = L_1 - 20\lg(n) \quad \text{dB} \tag{1.47}$$

d_1 and d_2: Distances from the receiver points to the source (symbolized by R in Eqs 1.78 and 1.79)
n: Ratio of the distances
L_1 and L_2: Sound pressure levels at two receiver points respectively, dB

For directional sources, the directivity factor, Q in the direction to the receiver point is taken into account:

$$L_p = L_W - 11 - 20\lg d + 10\lg Q \quad \text{dB} \tag{1.48}$$

L_W: Sound power of source, dB
d: Distance from the source, m
Q: Directivity factor given in Eq (1.65)

The sound pressure of the point sources may vary with time according to their operational characteristics. Time-dependent sound pressure levels are determined for the steady or moving (dynamic) sources [17].

Steady Point Source

The sound pressure level of a stationary source giving time-varying energy output can be determined by integrating the instantaneous pressures recorded within the observation period, T:

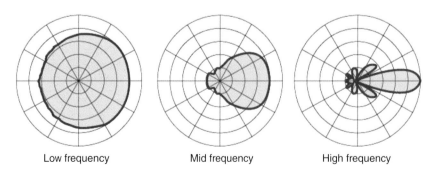

Low frequency   Mid frequency   High frequency

Figure 1.44   Variation of directionality of a loudspeaker in relation to frequency of sound.

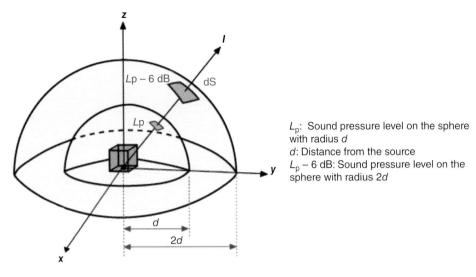

L_p: Sound pressure level on the sphere with radius d
d: Distance from the source
$L_p - 6$ dB: Sound pressure level on the sphere with radius $2d$

Figure 1.45a Variation of sound pressure levels with distance in a semi-spherical radiation.

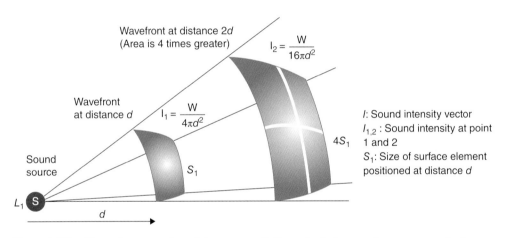

I: Sound intensity vector
$I_{1,2}$: Sound intensity at point 1 and 2
S_1: Size of surface element positioned at distance d

Figure 1.45b Wave-divergence for point sources radiating spherical waves (inverse-square law).

Figure 1.45c Reduction of sound pressure level with distance from a point source.

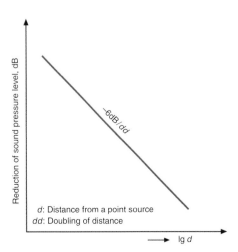

d: Distance from a point source
dd: Doubling of distance

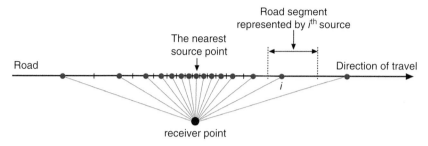

Figure 1.46 Moving point source at instants of time. [17]

$$L_{Tsteady} = 10 \lg \left(\frac{1}{t_2 - t_1} \sum_{1}^{t} 10^{L_{,(t)}/10} \right), \mathrm{dB} \tag{1.49}$$

$L_{(t)}$: Instantaneous sound pressure level at time t, dB
t_1 and t_2: Start and end times of the observation period, T

The above relationship also gives the definition of the L_{eq} descriptor which is commonly used for environmental noises as will be seen later.

Moving Point Source

If a point source is moving on a line, the sound pressure variation can be examined by means of a model. The moving line is divided into numbers of segments each meeting the condition of a point source, then the total sound pressure is obtained by integrating the individual sound pressures at the receiver point (Figure 1.46): [17]

$$L_{Tmoving} = 10 \lg \left(\sum_{1}^{n} 10^{\left[L_{(t),n} + 10 \lg \Delta t_n \right]/10} \right) - 10 \lg (t_2 - t_1) \, \mathrm{dB} \tag{1.50}$$

$L_{(t), n}$: Instantaneous sound pressure level of each segment
Δt_n: Time interval in which the source falls into the nth segment: Length of the segment/speed of source
$10 \lg \Delta t_n$: Time adjustment term, dB
n: Number of segments

When the receiver is far from the source, the length of segments can be larger. If the speed of the source changes, a different adjustment term should be computed for each speed. If there are many other moving sources on the line, the total effect is calculated by taking into account the contributions of each source with their speeds.

Doppler Effect for the Moving Sources

While a source is moving toward or moving away from a receiver point, the sound pressure levels change due to the distance effect, but also the frequency spectrum of the received sound is changed. This phenomenon is called Doppler threshold shift, after Christian Doppler (1803–1853). The effect is seen on the wavefronts associated with the speed of source v_k and the speed of sound c is shown in Figure 1.47.

Sound source approaching receiver: For a steady source, the wavefronts travel by speed c and the waves reach the receiver at a unit time is f. If the source is approaching the receiver with a speed of

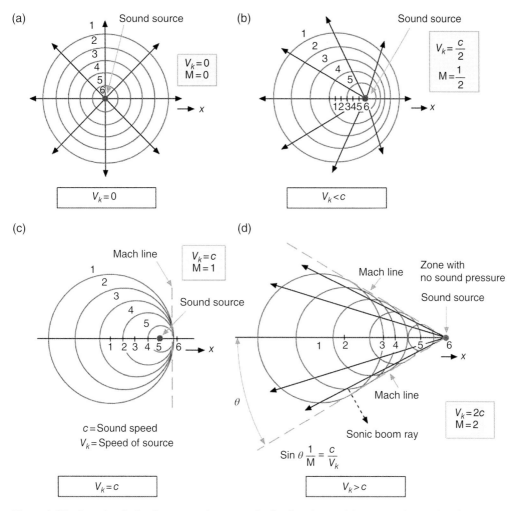

Figure 1.47 Sound radiation from a moving source in the direction *x* with supersonic speed and sonic boom. (a–c) the wavefronts for different conditions of source speed in relation to speed of sound; (d) occurrence of Mach cone. (*Source:* adapted from [19], with permission from Zaveri.)

$v_k < c$, the *f* number of waves are compressed within the distance c-v_k and the receiver takes more wavefronts in a unit of time. The wavelength of the received sound becomes:

$$\lambda' = \frac{c - v_k}{f} \text{ m} \tag{1.51}$$

c-v_k: The distance which the wave takes in a unit time

f: Frequency of the original sound, Hz (from the steady source whose wavefronts travel with constant speed)

v_k: Speed of moving source

The frequency of the received sound, f' becomes $f' > f$ and is calculated in Eq 1.52:

$$f' = f \frac{c}{c - v_k} \tag{1.52}$$

As a summary, when the source is moving towards a receiver, the measured sound frequency becomes higher than the original sound.

Sound source moving away from receiver: Contrary to the above situation, the same number of waves, f, is propagated over a larger distance, $c + v_k$ in a unit time and the imaginary frequency at the receiver becomes lower than the original sound ($f' < f$):

$$f' = f \frac{c}{c + v_k} \tag{1.53}$$

If the receiver is moving away from the source with a speed of v_a, the received frequency is:

$$f_a' = f \frac{c - v_a}{c} \tag{1.54}$$

v_a: Speed of moving receiver

As a result, when either the source or the receiver moves away from each other; the received frequency is lower than the original sound.

Moving source and moving receiver simultaneously to each other:

$$f' = f\{(c \pm v_a)/(c \pm v_k)\} \tag{1.55}$$

v_k: Speed of source, m/s
v_a: Speed of receiver, m/s
c: Speed of sound wave m/s
f: Original frequency, Hz
f': Imaginary frequency at the receiver, Hz

Both the source and the receiver are moving toward each other, so the frequency at the receiver becomes lower than that of the original sound; $f' < f$. In other words, the received frequency becomes $f' > f$ at a fixed point in front of the source, whereas $f' < f$ at a fixed point behind the source.

The Doppler effect is valid for various electromagnetic waves including light. In acoustics this phenomenon is important for the environmental noise sources, such as aircraft, trains, etc. and should be taken into consideration for noise control. For instance, the siren of an approaching train becomes more distinctive due to its high pitch. Sonic booms from high-speed aircraft can be given as another example.

Sonic Boom

The noise from aircraft flying at supersonic speed, i.e. fighter aircraft, causes a shock wave called a sonic boom on the ground [20–22]. For such a source, the wavefronts take different shapes in relation to the ratio between the speed of the moving source v_k and the speed of sound c which is called the "Mach number." Under the condition of $v_k = c$ (i.e. Mach number = 1), the wavefront proceeds together with the moving source, causing an increase in atmospheric pressure. The numbers of wavefronts gradually increase with the increasing speed of the aircraft and accumulate so that they reach the receiver point as a shock wave in a clear *N*-wave form. A shock wave is a wavefront displaying an abrupt pressure change over the Mach cone. Figure 1.47 a,b,c,d display different conditions of supersonic speeds in relation to sound speed. In this case a sound is not heard in front of the source and $f = 0$. In the opposite direction a sound wave with a frequency lower than the original frequency is heard.

For $v_k > c$ (i.e. Mach number >1), the source is faster than the speed of sound. As an example, a supersonic aircraft produces two shock waves from its nose and tail. As shown in Figure 1.47d, the waves lag behind the source and the sound is heard much later after the source is out of sight. The angle θ between the direction of wave propagation and the direction of travel determines the "Mach Cone" according to the ratio of the speeds:

$$Sin\theta = \frac{c}{v_k} \tag{1.56}$$

Various detailed information about the calculation of the maximum sound pressure of blast waves from explosions in addition to the shock waves can be found in the references. Sonic booms from supersonic aircrafts are also explained in Chapter 3, and create a serious noise problem in the environment.

Line Source and Propagation of Cylindrical Waves

The sound intensity I of an infinite line source, radiating cylindrical waves and the resultant sound pressure are explained in Sections 1.2.2 and 1.3.1. The sound pressure level is obtained with respect to the distance between the source and receiver point by assuming that the point sources are incoherent, i.e. in random phases:

$$I = \frac{W}{2\pi d_0} \tag{1.57}$$

W: Sound power per unit length of the line source, watts
d_0: Perpendicular distance from the receiver to the source line, m

Equation (1.57) implies 3 dB reduction with the doubling of distance. The sound intensity level which is equal to the sound pressure level in air under certain conditions at the receiver point, is given in Table 1.5.

The sound pressure at a receiver point, caused by the line source, can be determined by taking into account all the contributions of the point sources with respect to their distances and intrinsic powers. However, this approach assumes that all the contributions are at random phases. Below pressure functions for different types of line sources are given:

Infinite-length line source with point sources at equal distances apart: The total sound pressure caused by the incoherent point sources, equally spaced from each other and generating equal energy, is obtained by the addition of the contributions of each point source:

$$<p^2> = \rho c \frac{W}{4\pi} \sum_{n=-\infty}^{+\infty} \frac{1}{r_n^2} \tag{1.58}$$

$<p^2>$: Average squared total sound pressure, pascal
W: Sound power of the point sources (which is equal), watt

$$r_n^2 = r_0^2 + (nb + d^2) \tag{1.59}$$

n: Number of the point sources which are equal
b: Distance between the point sources
d: Distance between the first point and the projection of the normal
r_0: Perpendicular distance from the receiver to the source line, m

The parameters are given in Figures 1.40 and 1.48. Equation (1.58) can be written according to the subtended angles, as:

$$<p^2> = \rho c \frac{W}{4\pi b r_0} [\alpha_n - \alpha_1] \tag{1.60}$$

The total sound pressure as a function of distance is given in Table 1.5.

Table 1.5 Sound sources and determination of sound pressures in relation to distance from the source.

Source type	Source dimension	Wave type	Source condition	Relationship between pressure, power and distance d: Source receiver distance, d_0: Perpendicular distance from receiver to source	Condition for distance from source	Reduction of sound pressure level per doubling of distance, dd ΔL, dB
Point source	$D < \lambda$	Spherical	Non-directional	$L_p = L_W - 20\lg d - 11$ dB L_W: Sound power of the source	—	6 dB/dd
			Directional	$L_p = L_W - 11 - 20\lg d + 10\lg Q_\theta$ dB Q_θ: Directivity factor	—	6 dB/dd
		Semi-spherical (i.e. on the ground)	Non-directional	$L_p = L_W - 20\lg d - 8$ dB $L_2 = L_1 - 20\lg \dfrac{d_2}{d_1}$ dB d_1 and d_2: Distances from the receiver points to the source L_1 and L_2: Sound pressure levels at two receiver points respectively, dB	—	6 dB/dd
			Discrete point sources with same power and incoherent (with random phase)	$L_p = L_{w_1} + 10\lg\left(\dfrac{\alpha_n - \alpha_1}{d_0 b}\right) + \Delta L - 8$ dB L_{w_1}: Sound power level of each source $\alpha_n - \alpha_1$: Aspect angle ΔL: Geometrical parameter b: Distance between adjacent sources n: Number of sources	For; $\dfrac{d_0}{b\cos\alpha_1} \geq \dfrac{1}{\pi}$ and n≫3; $\Delta L < 1$ dB For; $\dfrac{d_0}{b\cos\alpha_1} < \dfrac{1}{\pi}$ $\Delta L > 1$ dB	3 dB/dd
	Infinite-length with point sources equally distributed	Infinite cylindrical	Infinite number of point sources with same power and incoherent (with random phase) $(\alpha_n - \alpha_1 \to \pi)$	$L_p = L_{W_1} - 10\lg d_0 b - 3$ dB $L_p = L_{W_1} - 10\lg d_0 - 10\lg(4b)$ dB L_{w_1}: Sound power of each source	$d_0 \geq \dfrac{b}{\pi}$	3 dB/dd
				$L_p = L_{W_1} - 20\lg d_0 - 8$ dB $L_p = L_{W_1} - 20\lg d_0 - 11$ dB L_{w_1}: Sound power level of each source point b: Distance between source point	$d_0 < \dfrac{b}{\pi}$	6 dB/dd

Source	Description	Formula and definitions	Condition	dB/dd
Line source	With same power and coherent (at phase)	$L_p = L_{W_1} - 10\lg d_0 - 10\lg b - 8$ dB L_{w1}: Sound power of each source b: Distance between source points	$d_0/\lambda \gg 1$	3 dB/dd
	Infinite number of point sources with same power and incoherent (with random phase)	$L_p = L_{wL} + 10\lg\left(\dfrac{\alpha_2-\alpha_1}{d_0 D}\right) - 8$ dB L_{wL}: Sound power of the finite-line source per unit length (m) D: Finite-length of the source, m	See below	3 or 6 dB/dd
	Finite cylindrical — Finite-length with point sources equally distributed	$L_p = L_{WL} - 10\lg d_0 D - 3$ dB $L_p = L_{W_L} - 10\lg d_0 - 10\lg(4D)$ dB L_w: Total sound power level	Close to the source: $\alpha_2 - \alpha_1 \to \pi$ $d_0 < \dfrac{D}{\pi}$	3 dB/dd
		$L_p = L_{W_L} - 20\lg d - 8$ dB d: Distance from center point of the finite line	Far from the source: $\alpha_2 - \alpha_1 \approx d/d_0$ $d_0 \geq \dfrac{D}{\pi}$	6 dB/dd
	With incoherent point sources — Finite size	$L_p = L_W - 20\lg d_0 - 11$ dB	Very close to source: $\lg d < a/\pi$	0
		$L_p = L_W - 8 - 10\lg\Sigma$ dB $$\Sigma = \int_0^{a/d}\int_0^{b/d} \frac{dX\,dY}{1+X^2+Y^2}$$ $$Y = \frac{b}{d} \qquad X = \frac{a}{d}$$	Far from the source: $d > b/\pi$	6 dB/dd
			For adequate distance: $a/\pi < \log d < b/\pi$	3 dB/dd
Plane source	Plane wave — Single or multiple planes	$L_p = L_{wt} - 10\lg S + 10\lg\left(\dfrac{\rho c}{400}\right)$ S: Surface of the semi-sphere with radius d, L_{wt}: Total sound power from the plane source, watts ρc: Air impedance		6 dB/dd

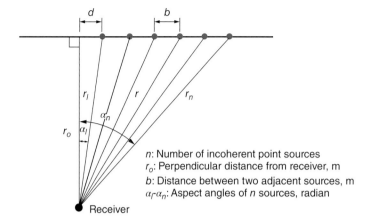

Figure 1.48 Infinite-length line source consisting of discrete point sources. (*Source:* adapted from [16].)

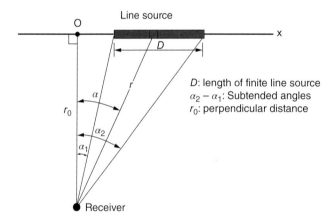

Figure 1.49 Finite-length line source consisting of incoherent point sources.
(*Source:* adapted from [16].)

Finite-length line source with coherent point sources: If there are infinite numbers of point sources with d distance on a line source, having equal energy but at random phase, the sound pressure is calculated as Eq 1.61 (Figures 1.40 and 1.49):

$$<p^2> = \rho c \frac{W_t}{4\pi r_0^2 D}(\alpha_2 - \alpha_1) \tag{1.61}$$

The total sound pressure as a function of distance is given in Table 1.5.

Coherent sources: The point sources are spaced apart as b from each other on the line, and if they are in phase and the condition $r_0/\lambda \gg 1$ is met, the sound waves become cylindrical at distance r_0 (like a pulsating cylinder). For this source, the sound pressure and pressure level are given as Eq 1.62:

$$<p^2> = \rho c \frac{W}{2\pi b r_0} \tag{1.62}$$

Figure 1.50 Attenuation of sound with distance from a finite-length line source.

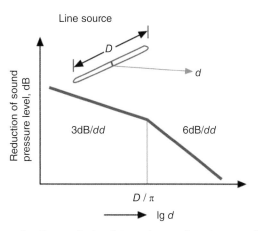

d : Perpendicular distance from reciever to source line
dd : Doubling of distance
D : Length of line source

The total sound pressure as a function of distance is given in Table 1.5. For sources vibrating with equal phase, the sound pressure becomes about 2 dB less than that of the sources vibrating with random phases.

Since the environmental noises are considered as line sources consisting of point sources with random phases, the distance attenuation is taken as 3 dB up to a distance of D/π and as 6 dB for further distances (Figure 1.50).

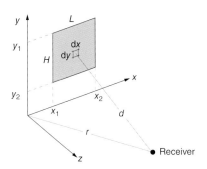

Figure 1.51a Parameters of a plane source with finite size.

Plane Source and Propagation of Plane Waves

The determination of the sound pressure variation of a plane source is composed of incoherent point sources (i.e. with random phases), as explained in Section 1.3.1, and needs to consider the energy density on the finite area of the source. However, by using a simple approach, the pressure function can be expressed by assuming that the wavefronts of plane source are perpendicular to the line of propagation and the amplitudes do not vary during the propagation (Figures 1.51a and 1.51b).

Plane source with finite size: Sound pressure level of such a source at a receiver point is given in Table 1.5 as a function of the perpendicular distance to the source and the source dimensions (Figures 1.51a and 1.51b):

Briefly, the below conditions can be met in practice, regarding attenuation of sound levels from a plane source (see Figure 1.51a and b for the parameters):

1) *Sound pressure at a point very close to the source*: When satisfying the condition of $\lg d < H/\pi$, the wavefront becomes a plane for a receiver point very close to the source and the surface of the wavefront does not change with the distance from source, implying that there is no geometrical divergence (*H*: Smaller dimension, *L*: Longer dimension of the plane source).

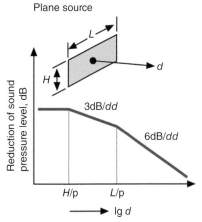

Figure 1.51b Attenuation of sound with distance from a plane source.

H, L: Dimensions of plane source
d: Perpendicular distance from receiver to center of source plane

2) *Sound pressure at a far distance from the source*: Under the condition of $\lg d > L/\pi$; the wavefront resembles a plane wave but it is actually a spherical wave generated from a monopole source. The reduction is 6 dB per doubling of distance.
3) *Sound pressure at a sufficient distance to the source*: Variation of sound pressure level with distance of the plane source depends on the source-receiver geometry, e.g. for $H/\pi < \lg d < L/\pi$, the wavefront becomes a cylindrical wave and the reduction is 3 dB/doubling of distance.

   Multiple plane source: A plane source composed of numbers of point sources located at approximately the same d distance from a receiver (i.e. points on the same surface, such as a wall of an industrial building), radiates sound uniformly over a hemispherical surface of radius d. The sound pressure level is calculated as given in Table 1.5.
   If the plane source is directional, directivity can be taken into account in the pressure function as a correction term. For a multiplane source consisting of several planes, each radiating sound simultaneously, the similar computation is applied for each surface and then the logarithmic summation is performed to obtain the total sound pressure level. For example, this method can be used to find the total noise coming out of the factory building through its side walls and roof.
   Impedance of a plane wave: In mechanics and electricity, the impedance is defined as force/reaction. If a force F is applied to a mechanical vibrating system resulting in a motion with a speed of v, the mechanical impedance is $Z_m = F/v$. Similarly, if the electric voltage E produces a flow I, the electrical impedance becomes $Z_e = E/I$. As explained in Section 1.2.2., for a point source; the sound pressure and the particle velocity are related to each other. In fact, the time-functions of pressure, p and the particle velocity u differ because of the phase difference between the functions (see Figure 1.13). However, this is not the case for a plane wave since the variation of sound pressure with time is similar to the variation of the particle velocity with time at a receiver point, resulting in no phase difference. A pressure p acting on a point, inducing a motion on air particles with velocity u, the ratio between two, is defined as "acoustic impedance":

$$Z_A = \frac{p}{u} = \rho c \tag{1.63}$$

Z_A: Acoustic impedance, N·s/m³ (or mks rayls)
p: Space-averaged sound pressure, N/m²

u: Particle velocity, m/s or speed of mass motion, m^3/s

c: Sound speed, 344 m/s

ρ: Density of air (1.18 kg/m^3)

The impedance of a sound wave generated by a point source is dependent on the density of the medium and speed of sound in the medium (air, liquid or solid). Since the force and the effect (i.e. speed of particle motion) are not in the same phase, the acoustic impedance is a complex number ($Z = R + jX$) (R: real part, X: imaginary part), called a "complex impedance." However, Z is a real number for a plane wave propagating in a free field. Various examples regarding acoustic impedances in different mediums are given in Table 1.6.

There are other types of impedances which are used for the plane waves:

Specific acoustic impedance, Z_s: The complex ratio between the sound pressure and the particle velocity at a constant receiver point.

$$Z_s = \frac{p}{u} \quad \text{N.s/m}^3 \tag{1.64}$$

Mechanic impedance, Z_m: The ratio between the force acting on a surface of a mechanical vibratory system and the linear velocity of the motion (m/s) results in:

$$Z_m = \frac{F}{u} \quad \text{N.s/m (Mechanical ohm)} \tag{1.65}$$

Normal specific acoustic impedance, Z_{sn}: This is the impedance on the boundary between air and a denser medium (i.e. a solid). The motion of the air particles on a solid surface can be in different directions depending on the angle of incidence of the sound wave and the characteristics of the surface material. For this case, the specific acoustic impedance is calculated by taking the normal component u_n of the particle velocity. If the surface material is porous and less dense, u is in the same direction as the wave propagation.

$$Z_{sn} = \frac{p}{u_n} \quad \text{N.s/m}^3 \text{ (mks rayls)} \tag{1.66}$$

Table 1.6 Characteristic acoustic impedances of different mediums.[*]

Medium	Acoustic impedance, N·s/m^3 (mks rayls)
Air (at 20°)	415
Water	146×10^4
Rubber	3.5–28×10^4
Concrete	7–13×10^6
Marble	9.9×10^6
Aluminum	17×10^6
Glass	12.3×10^6
Stainless steel	45.5×10^6
Wood (pine)	157×10^4
Polystyrene	2.47×10^6

[*] Compiled from various literature.

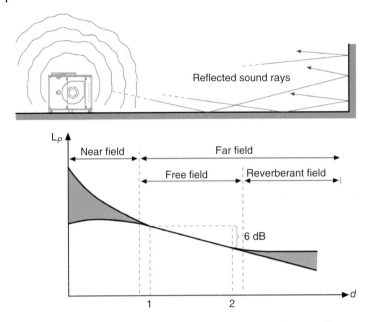

Figure 1.52 Sound fields around a source in an enclosed space. (*Source:* adapted from B&K.)

1.3.4 Acoustic Fields

Practically a sound source is surrounded by radiation fields with different acoustic characteristics. The sound pressure levels vary in these fields according to the source and physical factors affecting the sound propagation (Figure 1.52). Basically, the acoustic fields are divided into two regions: the near field and the far field [20, 21].

Near Field

This is defined as the portion of the radiation field of a sound source that lies between the source and the far field [12a]. The field remaining within several multiples of the wavelength of the sound is the near field whereas the particle velocity may not be in the same direction of wave propagation, because of the phase relations of vibrating particles. The result is interference between the sound waves coming from different points of the plane source, because of that, the sound pressure and particle velocities are nearly out of phase.

In the near field, the sound intensity, I is not simply related to the square root of the average squared pressure (p_{rms}), and the sound pressure may vary a great deal around the source and the pressure/distance relationship is not applied (i.e. the reduction of sound pressure level with distance does not obey the 6 dB rule). Thus, the source cannot be assumed to be a point source in this region and the directionality of the source becomes more complex. As a result, the sound power to be determined from the sound pressure measurements conducted in this region is not reliable and accurate.

The distance of the near field to the source depends on various factors such as the frequency of sound and the size of the source (i.e. the ratio of wavelength to the source dimension), the directivity pattern, and the characteristics of the medium. It is rather difficult to detect the near field around a source, however, it can be determined by means of acoustic measurements.

Far Field

At further distances from the source, where the source field cannot be assumed as radiating plane waves and the acoustic field in this region can be defined by simpler expressions since the sound pressure and particle velocity are approximately in the same phase and the particle velocity is in the direction of sound propagation, this is called the "far field." This is where the relationship between the sound intensity, I and sound pressure, p_{rms} is as given in Eq (1.28) in Section 1.2.2. Although the boundary between the near and far fields is not sharp, the two conditions can be applied to estimate the start of the far field [12b]:

- $kr = \dfrac{2\pi fr}{c} = 10$ (r: distance between source and receiver)
- $r > 10\, d$ (d: source dimension)

The far field is divided into two geometrical sections called the free field and the reverberant field (Figure 1.52).

Free Field

In a steady and homogeneous atmosphere, the acoustic field, which is free from reflective or absorptive surfaces, is called the "free field." In this field only the direct sounds can reach and the waves are called free-progressing waves. The free field cannot be found in built environments but only in certain laboratories called anechoic rooms.

In the free field, the total sound power radiated by the source and the pressure are proportional to the distance and the 6 and 3 dB reductions in sound pressure level are obtained for the point and line sources respectively. Directionality of the source can be determined accurately in this region. Practically, the far field can emerge at distances greater than $\lambda/2$ or at distances four times longer than the maximum dimension of source.

Reverberant Field

The complex acoustic field created by reflections, scattering, diffraction, interference, etc. due to the physical factors of the medium is called a reverberant field (Section 1.4.1). Therefore, the regular reduction of sound pressure levels with distance cannot be obtained after a certain distance in this field (Figure 1.53). Especially for the source located in a space enclosed by walls, floors, and ceiling, the total energy is contributed by the reflected energies, sometimes by the multiple reflections. Direct sound energy at a fixed receiver is always much less than the reflected energies and the sound pressure is rather independent of the distance from the source. Generally, the entire acoustic field is assumed to be reverberant where the energy density at each point is equal and constant. The sound pressure level in a reverberant field can be obtained as Eq (1.67):

$$L_p = L_W + 10\lg\left(\frac{Q}{4\pi d^2} + \frac{4}{R}\right) \qquad (1.67)$$

Q: Directivity factor of the source (Section 1.3.2)
d: Distance from the source to receiver, m
R: Room constant

$$R = \frac{S\overline{\alpha}}{1 - \overline{\alpha}} \qquad (1.68)$$

S: Total room surface, m^2
α: Average surface absorption coefficient in the room, %

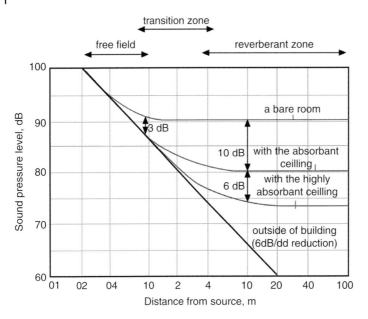

Figure 1.53 Sound fields in a room and effect of absorption on decrease of sound pressure level in relation to distance from the source.

The reverberant field is sometimes called the "perfectly diffuse field" in which the sound pressure level is the same everywhere and the flow of energy is equally probable in all directions [12a]. Diffuse sound is completely random in phase and arrives uniformly from all directions. The energy density is constant throughout the whole room. The degree of diffusion depends on the shape of the room, the amount, and distribution of absorption on the boundary surfaces; the smaller the total absorption, the higher the diffusion. Figure 1.53 shows the relation between the sound pressure level and the distance between the receiver and the source in an enclosed space. As shown, the diffusivity in the field is effective in reducing the sound pressure level and the size of the reverberant zone.

Other deteriorations of sound waves that can be encountered in enclosures, i.e. rooms or halls, are explained in Section 1.4.

1.4 Sound Propagation and Disturbances

When sound waves are radiating in an elastic medium, such as a solid, a liquid, or air, they are influenced by the physical and dynamic properties of the medium, eventually the wave geometry and the sound pressure are changed. The basic phenomena causing wave deformation are important in understanding the noise propagation in an environment to be explained in Chapter 2.

1.4.1 Sound Reflection

A solid object with a hard surface, located in the propagation path of sound waves, acts differently depending on its dimensions and the wavelength of sound. When the dimensions of the hard surface are $\lambda/4$ larger than the wavelength, the sound waves return back into the air, similar to a light

Figure 1.54a   Sound reflection of a plane wave.

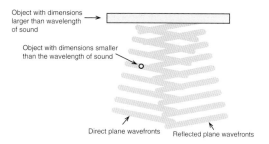

Figure 1.54b   Wavefronts of incident and reflected waves for a point source.

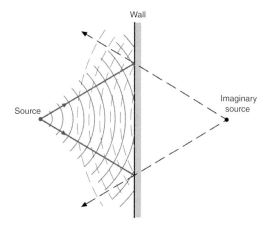

ray, in accordance with the geometrical reflection principle. This phenomenon is called "sound reflection" (Figures 1.54a and 1.54b). When a sound wave in air strikes a surface, the particle velocity component on the normal axis to the surface decreases to zero since it cannot excite motion at the solid media and the pressure is doubled at the interface. The wave returns back into the air as if it comes from an image source behind the surface. In this case, the reflection of a sound wave confirms the Snell optic law and is called specular reflection: the incidence and reflected waves have equal angles with the normal line at the striking point, i.e. $\theta_i = \theta_r$. The angle of incidence is equal to the angle of reflection (Figure 1.55). Pure geometric reflection occurs only when the surface is an infinite plane with a uniform acoustic impedance. For an infinitely large boundary surface and normal sound incidence: $\theta_i = \theta_r = 0$, because of the continuity at the boundary. The relationships between the pressure and particle velocities are given as:

$$p_i + p_r = p_t \tag{1.69}$$

$$u_i + u_r = u_t \tag{1.70}$$

p_i, p_r, p_t: Sound pressures of incident and reflected waves and total pressure respectively
u_i, u_r, u_t: Particle velocity of incident, reflected and total waves

The geometrical parameters affecting the sound pressure of the reflected wave are the sound frequency, the angle of sound incidence, the source-receiver distance, the heights of the source and the

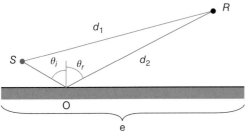

Figure 1.55 Specular reflection and geometrical parameters: (condition of reflection: $e > \lambda$).

d_1 : Direct sound (*SR*)
d_2 : Reflected sound (*SO+OR*)
$\theta_i = \theta_r$: Incidence and reflection angles

receiver. The material properties, such as the complex reflection coefficient of the surface, the flow resistivity of the surface, and the characteristic impedance of the air and the surface are also important. The reflection coefficient r_p of the surface is defined as the ratio of the reflected and incident sound pressures and is related to the impedances of the mediums (Box 1.5):

$$r_p = \frac{\sin\theta - \dfrac{Z_1}{Z_2}}{\sin\theta + \dfrac{Z_1}{Z_2}} \tag{1.71}$$

θ: Reflection angle = incident angle
$Z_1 = \rho_0 c_0$: Characteristic impedance of air: 406 mks rayls [24]
Z_2: Complex characteristic impedance of the surface for spherical sound wave, $Z_2 = R + iX$ (as explained above). For the normal incidence of sound, Z_2 is a real number.

When computing the sound reflection coefficients for a porous medium, the real and imaginary parts of the characteristic impedance of a porous surface are calculated as:

$$R = \rho_0 c_0 \left\{ 1 + 9.08 \, (f/\sigma)^{-0.75} \right\} \tag{1.72}$$

$$X = \rho_0 c_0 \left\{ -11.9 \, (f/\sigma)^{-0.73} \right\} \tag{1.73}$$

σ: Flow resistance of the surface material, Ns/m^4
$\rho_0 c_0$: Air impedance, N·s/m^3
f: Frequency, Hz

In practice, there is no complete reflection ideally and some amount of sound energy is transmitted into the solid medium behind the surface. However, as generally stated, a reflective surface positioned at a distance which is smaller than the wavelength from a sound source, enhances the total sound energy.

Reflection Types in Relation to Surface Geometry

- *Reflection from a single surface:* The wavefront of the sound after reflection from a surface can be obtained by a geometrical drawing of Huygens' principle. As shown in Figure 1.56a, the plane wave (shown as *AB*) is on the surface *MN* with an angle of incidence, the direction of the wavefront is changed since it cannot pass through the surface. Within a time interval, a point on the wavefront of the incident sound, e.g. point *B*, moves to point *D*, however, since point *A* cannot reach point *C*, it arrives at another point on a circle whose center is A and with the radius, *AC*. Similarly

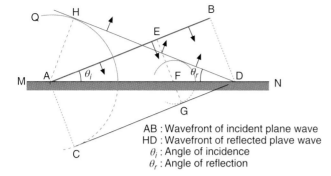

AB : Wavefront of incident plane wave
HD : Wavefront of reflected plave wave
θ_i : Angle of incidence
θ_r : Angle of reflection

Figure 1.56a   Reflection of a plane wave and determination of wavefronts. (*Source:* adapted from [10].)

Figure 1.56b   Reflections from two perpendicular surfaces.

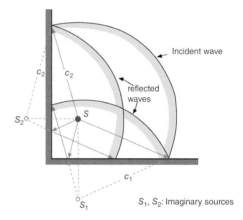

the circle *FC* can be drawn for point *E*. The incident wave is reflected to a space above plane *MN*. It is possible to display the wavefront of the reflected wave by drawing a tangent to the circle.

- *Reflection from two surfaces perpendicular (normal) to each other*: The incident and reflected wavefronts define a sphere with radius *ct* and the result is shown in Figure 1.56b. The imaginary sources of the incident and reflected waves are S_1 and S_2. The energy density at a receiver point is the sum of the energies of all the imaginary sources.
- *Multiple reflections*: When a number of surfaces are present in the acoustic field (as seen in Section 1.3.4), the acoustic field might become more complex because of the interferences between the reflected waves (Figure 1.57a). This subject is involved in room acoustics and ray analysis can be performed by using modeling techniques (Figure 1.57b).

Effect of Reflections on Sound Power

The direct and reflected waves with random phases in a reverberant field have a great influence on the sound power of the source, and the energy densities of the direct and reflected waves are summed to find the total energy density in the space. (In Eq (1.74), the symbol *E* is used in place of *w*.) (See Box 1.3)

Direct sound energy density (explained in Section 1.2.2):

$$E_a = \frac{W}{4\pi r_0^2 c} \tag{1.74}$$

r_0: The path length of the direct ray
W: Sound power

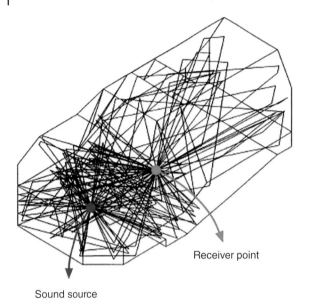

Receiver point

Sound source

Figure 1.57a Multiple reflected rays in enclosures (e.g. a music hall). (*Source:* from Odeon software, permission granted.)

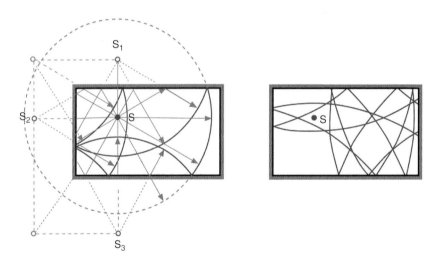

Figure 1.57b Determination of the spherical wavefronts of the reflected sounds in an enclosure. (*Source:* adapted from [10].)

Reflected sound energy density:

$$E_r = \frac{W(1-\alpha)}{4\pi r_1^2 c} \tag{1.75}$$

r_1: The path length of the reflected ray
α: Absorption coefficient of the surface

Total energy density:

$$E_i = E_r + E_a + E_t \tag{1.76}$$

E_t: Transmitted energy density (not considered in this section)

Figure 1.58 Source locations according to reflective surfaces, directivity patterns and directivity factors, Q (source is non-directional).

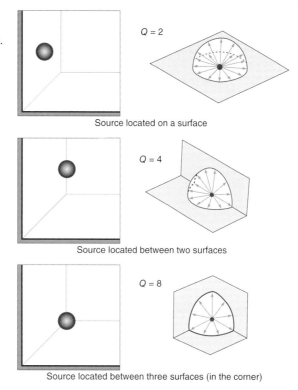

Source located on a surface

$Q = 2$

$Q = 4$

Source located between two surfaces

$Q = 8$

Source located between three surfaces (in the corner)

The effects of the reflective surfaces existing near the source on the sound power can be taken into account in Eq (1.76) while calculating the source power, as below (Figure 1.58):

1) When only one surface exists, radiation of sound energy is on a semi-sphere and the energy density is $E_i = 2\,E_a$. The directivity factor of the source is $Q = 2$, implying a 3 dB increase in sound power level.
2) When the source is closely positioned at two intersecting surfaces, $Q = 4$ corresponding to a 6 dB increase in sound power level.
3) When the source is positioned at the corner of the three surfaces which are perpendicular to each other, then $Q = 8$ corresponding to an 8 dB increase in sound power level.
4) If there is no reflective surface, $Q = 1$, implying no increase in sound power level of the source.

Effect of Reflections on Sound Pressure

For a receiver existing in a reverberant field, the sound pressure is affected by the reflections. If the sound waves are composed of pure tone sounds, the interference inevitably occurs depending on the phases of the contributory sound waves.

The effect of reflections on sound pressure levels according to the position of reflective surface are examined under the following conditions. Parameters are given in Figure 1.59.

The reflective surface is very close to the sound source: In relation to the changes in the radiation characteristics of the source and its directivity, there are two cases:

1) When the normal distance between the source and the reflected surface (ground or a wall), is small compared to the distance between the source and receiver, L (i.e. $h_1 < L$) and for the

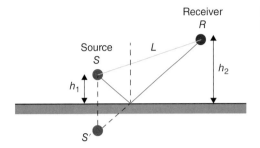

Figure 1.59 Geometrical parameters of source, receiver, and reflective surface.

condition of $h_1 < (\lambda/10)$, the total sound power is radiated into the half space in front of the surface (directivity is 2), implying that the sound pressure in the receiver point is doubled and the increase in sound pressure level, is 6 dB.

2) When the distance between the source and the reflected surface is greater than the distance between the source and receiver, L (i.e. $h_1 > L$) and for the condition of $h_1 > \lambda$, the sound pressure is calculated by adding the separate intensity contributions of the source and the imaginary sources.

The reflective surface is very close to the receiver point: For the condition of $h_2 \leq (\lambda/10)$, (h_2: normal distance from the receiver to the surface); the sound pressure at the receiver is equal to the sum of the direct and reflected wave pressures and the sound pressure level increases 6 dB. For $h_2 > (\lambda/10)$, the increase in sound pressure level is 3 dB. These assumptions are valid for the direct and reflected waves with random phases. If the sound is tonal (i.e. pure tone), the phase differences should be taken into account during summation.

The source and the receiver are on the reflective surface: When the distances of both the source and receiver from the surface are small compared to the distance between them (h_1 and $h_2 < L$), the path difference between the direct and reflected waves is rather small, the sound pressure levels are summed logarithmically, assuming that the direct and reflective waves are at random phases. The result indicates a 6 dB increase at the receiver point in the free field.

Variation of sound power and pressure levels in a reverberant medium, according to the locations of source, receiver, and reflective surfaces, is important for the prediction of indoor and outdoor noise levels.

1.4.2   Sound Scattering (Diffusion)

If the size of the surface on which the sound wave strikes is smaller than the wavelength of sound, and if the surface has an irregular shape or contains elements with equal or slightly same size elements, such as undulated, jagged, ribbed, corrugated, conical, etc., the reflected sound rays are equally distributed around the surface point where the sound strikes. This 3D phenomenon, called "scattering" or "diffusion," implies that the wavefronts are scattered homogeneously and the flow of energy equally probable in all directions near the surface. The diffuse field in an enclosure is an idealized acoustic field in which the sound pressure level is the same everywhere. Figure 1.60 a–e displays the conditions of reflection and scattering in relation to wavelength and surface configurations. Scattering occurs due to the roughness and texture of the surface comparable to the wavelength of sound. The scattering surfaces can be irregular, periodic, or alternating absorbing and reflecting structures. As an example, a vehicle passing a metal balustrade at the edge of a road

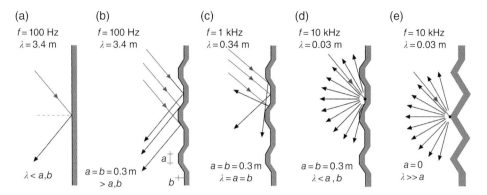

Figure 1.60 (a–e) Conditions at which reflection and scattering occur according to wavelength and dimension of surface indents and projections (*a, b*).

causes a periodic sound like applause, because of the scattering and interference of the frequency components of aerodynamic sound. The result is a sound with different tonal characteristics. Although the sound pressure (amplitude) is low compared to the specular reflection, it contributes to a diffuse field in a space.

In an enclosed space, the maximum distance which both the incident and scattered waves reach, can be determined as:

$$R = \frac{\sin kr}{kr} \tag{1.77}$$

r: Distance between the source and receiver, m
k: Wave number ($2\pi f/c$)

Scattering is an important phenomenon in room acoustics and is beneficial for the good design of halls since it causes an increase in both the sound energy density and the reverberation time. By means of a diffuser, if designed properly, the sound can be perceived as coming from all directions. However, scattering might be undesirable in environmental acoustics and in closed spaces since they create a noise problem. The scattering coefficient (see Box 1.5) of surfaces is defined as the ratio between the acoustic energy reflected in non-specular directions and the totally reflected acoustic energy. It is measured in laboratories either in free-field or in reverberant rooms according to the international standards (Table 1.7).

1.4.3 Sound Absorption

When the sound waves are passing through a medium or at a boundary, conversion of sound energy into heat energy is defined as sound absorption. The ratio of the absorbed sound power to the incident sound power, is called the absorption factor (the absorption coefficient) and represents the absorptive characteristics of the material (Figure 1.61) (Box 1.5 and Table 1.7).

Absorption mechanisms differ according to the medium and boundary conditions. During sound propagation in air, some sound energy is lost due to heat conductivity and viscosity and relaxation processes caused by the molecular structure of gases, of which air consists. The result is sound attenuation in air, as will be seen in Section 2.3.2, or in large halls especially at high frequencies. When the waves meet the objects/walls whose dimensions are larger than

Box 1.5   Reflection, Absorption, Transmission, and Scattering Coefficients

Sound reflection coefficient

$$p_i = A \sin 2\pi f t \tag{1}$$

$$p_r = B \sin 2\pi f \left(t - \frac{2x}{c} \right) \tag{2}$$

$$r_p = \frac{p_r}{p_i} = \frac{Z_2 - Z_1}{Z_2 + Z_1} \tag{3}$$

$\dfrac{B}{A}$: Reflectivity, $\dfrac{B^2}{A^2}$: Reflection coefficient

Sound absorption coefficient

$$\alpha = 1 - |r_p|^2 = 1 - \left| \frac{Z_2 - Z_1}{Z_2 + Z_1} \right|^2 \tag{4}$$

If the first medium is air; $Z_1 = \rho c$. If the second medium is absorptive; $Z_2 = Z$.

$$\alpha = 1 - \left| \frac{Z - 1}{Z + 1} \right|^2 \tag{5}$$

$$Z/\rho c = Z (\text{Acoustic impedance ratio}) \tag{6}$$

Sound scattering coefficient

$$s_\theta = 1 - \frac{E_{rs}}{E_{rt}} \tag{7}$$

E_{rs}: Specularly reflected energy
E_{rt}: Totally reflected energy
θ: Angle of incidence between the incident ray and the normal of the surface

Alternatively, the sound dissipation factor is also defined as the ratio of dissipated sound power to the incident sound power:

$$\delta = \frac{W_d}{W_i} \tag{8}$$

W_d: Sound power dissipated on the source side, Watts
W_i: Incident sound power on the source side, Watts

Sound transmission coefficient

$$\tau(\phi, \omega) = \frac{W_t(\phi, \omega)}{W_i(\phi, \omega)} \tag{9}$$

ϕ: Angle of incidence
ω: Angular velocity = $2\pi f$,
f: Frequency, Hz
W_i: Sound power on the source side, Watts
W_t: Transmitted sound power on the receiver side, Watts

When the medium characteristics are introduced:

$$\frac{p_i}{Z_1} - \frac{p_r}{Z_1} = \frac{p_t}{Z_2} \tag{10}$$

$$\tau = \frac{p_t}{p_i} = \frac{2Z_2}{Z_2 + Z_1} \tag{11}$$

p_i, p_r, p_t: Incident, reflected, and transmitted sound pressures
$Z_{1,2}$: Characteristic impedance of the mediums

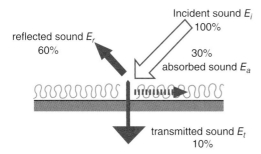

E_i, E_r, E_a, E_t : Incident, reflected, absorbed and transmitted sound energies

Figure 1.61 Reflection, absorption, and transmission phenomena.

Table 1.7 Wave deformation parameters in terms of acoustic energy.

Parameter	Basic definition in energy units	Definition with pressures	Definition with characteristics impedance of mediums
Absorption coefficient	$\alpha = \dfrac{E_i - E_r}{E_i} = \dfrac{E_a + E_t}{E_i}$	$1 - |r_p|^2$	$1 - \left\|\dfrac{Z_2 - Z_1}{Z_2 + Z_1}\right\|^2$
Reflection coefficient	$|r_p|^2 = \dfrac{E_r}{E_i}$	$\dfrac{p_r}{p_i}$	$\dfrac{Z_2 - Z_1}{Z_2 + Z_1}$
Transmission coefficient	$\tau = \dfrac{E_t}{E_i}$	$\dfrac{p_t}{p_i}$	$\dfrac{2Z_2}{Z_2 + Z_1}$
Diffusion (scattering) coefficient	$s_\theta = 1 - \dfrac{E_{rs}}{E_{rt}}$		

$E_i = E_r + E_a + E_t$: E_i: Incident energy; E_r: Reflected energy; E_a: Absorbed energy; E_t: Transmitted energy. E_{rs}: Specularly reflected energy; E_{rt}: Total reflected energy (If no diffusivity it is equal to E_r). p_i: Incident sound pressure; p_r: Sound pressure due to reflected wave (Reflected wave pressure); p_t: Sound pressure after transmission (Transmitted pressure). Z_1: Characteristic impedance of medium 1; Z_2: Characteristic impedance of medium 2.
For absorption coefficient; Z_2 represents the surface material. For transmission coefficient; Z_2 represents the total element neglecting the surface material.

the wavelength, absorption occurs in relation to the frequency, angle of sound incidence, and the wall impedance. The light and thin panels can act as a sound-absorbing boundary whose entire surface is set into motion by the pressure variation in front of it (vibrating panels). Such a panel radiates sound to the outer space, thus absorption is caused only by transmission and depends on the surface mass (mass per unit area) and frequency.

Absorption by porous materials: When the particles of porous material are set into motion by the air molecules within the pores, the sound energy is transformed into heat energy by means of internal friction (viscosity) and by heat conduction between the air and pore walls. Flow resistivity is the important factor in the sound absorption of porous materials and is determined by the pressure differences between the incident sound pressure and the pressure in the pores and the velocity of airflow. The other effective parameters are the surface mass, the dimensions of the material, the angular frequency, the air space between the porous material and the back wall, the layer construction (fiber, void, etc.), and its porosity (volume fraction of voids in the material).

The third type of absorption occurs by the resonators (so- called Helmholtz resonators), special devices enclosing a volume of air which is connected to a short open neck. The air is set into motion when sound waves strike on the neck and absorption occurs due to conversion of acoustic energy into heat as a result of friction on the neck and inside walls of the volume. Absorption characteristics is high but restricted in a narrow frequency band which is called resonance frequency depending on the dimensions of the volume and the neck.

The absorption coefficient is measured in acoustic laboratories (Section 5.9.6) and also by impedance tubes. In the latter method, the absorption coefficient is determined from the standing waves caused by the interference between the incident and the reflected waves in front of the sample material (i.e. the ratio of max. and min. pressures). The laboratory measurements provide more realistic results in the diffuse field.

The sound absorption phenomenon is significant in environmental acoustics and will be discussed in Chapter 2, as air absorption and ground absorption during noise propagation or absorption of barrier surface. This topic is involved with in noise control and in room acoustics to satisfy the reverberation criteria by means of adequate material and proper application.

1.4.4   Sound Transmission into Solid Media

When sound waves strike an element (e.g. a wall), sound energy excites the vibrations of the surface particles and different types of waves are transmitted within the element. Sound waves radiated within solid media are longitudinal (compression and rarefaction), shear, bending, and torsional waves and they cause cross-sectional and longitudinal deformation of the element according to the similar principles in mechanics (Figure 1.62).

Sound transmission through a solid medium (or an element) of finite size is emerged by two wave types (Figure 1.63): Forced waves and free waves (resonant). Forced waves occur due to the incident sound pressure of the airborne sound wave that also excites free bending waves.

Free waves are transmitted readily through the element to the other side at and above the critical frequency (Figure 1.64) and are called the resonant transmission. The element's loss factor (effect of damping), elasticity, and sound radiation efficiency (of free bending waves) in relation to the size of the element are the physical parameters to be taken into account in the resonant transmission. These are of secondary importance in mass-controlled regions below the critical frequency where the force transmission is dominant.

Forced waves are related to the sound frequency and the mass of the element. Below the resonance frequency, there is a region where the sound transmission is governed by the stiffness of the element. Above this frequency, the mass law is applied for simple homogeneous elements up to the critical frequency (i.e. the mass region) and the sound transmission is composed of forced and resonant transmission.

The ratio of the transmitted sound power to the incident sound power, is called the "sound transmission coefficient" (Box 1.5), given as a function of the angle of incidence and frequency, however, its decibel conversion is defined as the "sound transmission loss" or "sound reduction index," and is commonly used in the practice of building acoustics:

$$R(\phi, \omega) = 10 \lg_{10}\left(\frac{1}{\tau(\phi, \omega)}\right) \quad \text{dB} \tag{1.78}$$

$R(\phi, \omega)$: Sound transmission loss in relation to the angle of sound incidence and frequency, dB
$\tau(\phi, \omega)$: Sound transmission coefficient in relation to the angle of incidence and frequency, unitless

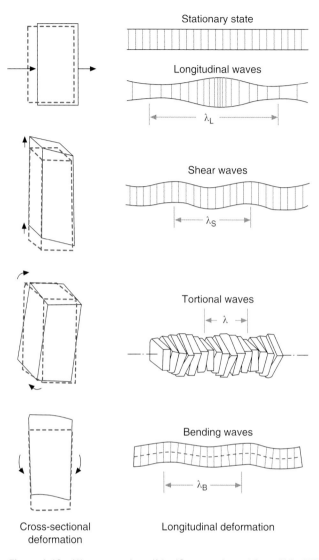

Figure 1.62 Wave types in solids. (*Source:* adapted from [15–17].)

Figure 1.63 Incident and transmitted sounds to and from a solid panel neglecting surface absorption.

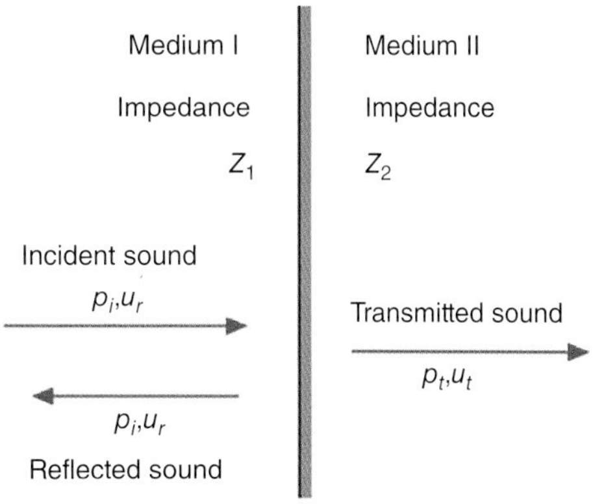

Medium I   Medium II

Impedance   Impedance

Z_1 Z_2

Incident sound

p_i, u_r

Transmitted sound

p_t, u_t

p_i, u_r

Reflected sound

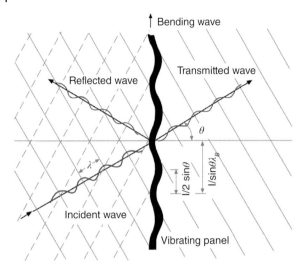

Figure 1.64 Transmission of bending waves at the critical frequency of a panel.

Figure 1.65 displays the variation of a sound transmission loss with a sound frequency for a single homogeneous element.

Sound transmission properties of elements are dependent on various properties, such as homogeneity, bending stiffness, material and constructional properties, and the construction of each layer in multilayered elements with or without an air gap, connections, the size of the element, the additional components on the surface, etc. For thick panels and sandwich panels, the shear waves are taken into account. Finite and infinite plates give different results. Panel size is important in the sound reduction index at low frequencies, e.g. the sound reduction increases with decreasing panel area, whereas at high frequencies above the critical frequency, the larger panels give a higher transmission loss.

In order to determine the sound transmission loss values of elements, the calculation models developed based on the transmission theories and the acoustic measurement methods are applicable (Chapter 5).

Transmission of sound in buildings is an important subject in architectural acoustics and for sound insulation. Sound penetration through walls, ceilings, partitions, floors, etc. is examined in terms of direct and flanking transmissions. Certainly the acoustics phenomena described above are encountered both in the environment and in buildings and should be analyzed in order to provide solutions to acoustical problems and good design (Figure 1.66).

1.4.5 Sound Refraction

The basis of refraction and diffraction was established by the wave motion theories derived by Huygens and Young and the superposition principle. According to Huygens' principle, while a sound wave is propagating in a medium, each point on the wavefront at time t acts as an independent

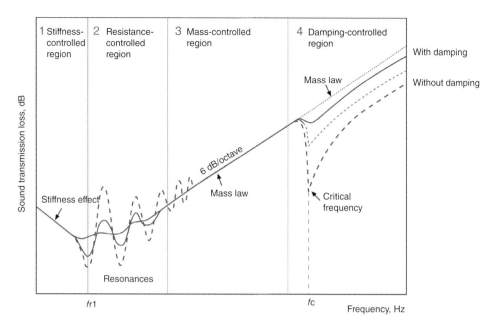

Figure 1.65 Variation of sound transmission loss of a single layer homogeneous panel: (f_{r1}: First resonance frequency, f_c: Critical frequency).

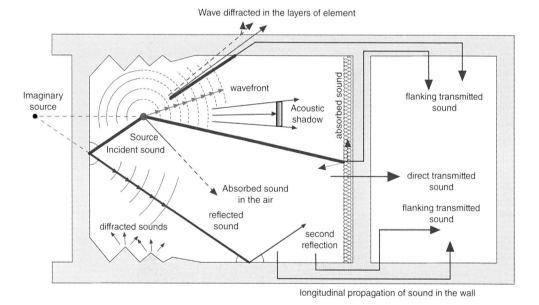

Figure 1.66 Sound propagation in enclosed spaces and sound transmission through building elements.

source radiating sound spherically (Figure 1.67). At a time $t + \Delta t$, the wavefront of the original source becomes the tangent to all the wavefronts of these individual point sources.

The direction of the linear propagation of sound rays is subject to change under different conditions of the medium. If a plane wave strikes a surface between two different mediums with different densities, the direction of the propagation of the sound ray changes due to the change of the speed of

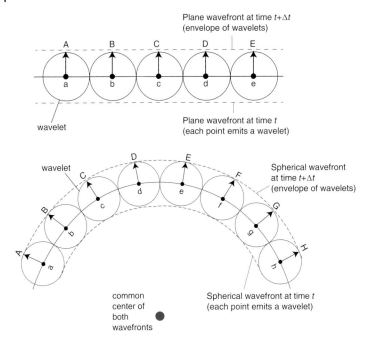

Figure 1.67   Huygens' principle of wave propagation and wavelets.

sound at each medium. This phenomenon is similar to optics and is called refraction. The wave transmitted into the second medium is defined as the refracted sound wave. As shown in Figure 1.68, individual sound sources on the second medium radiate sound with different wavefronts directing toward a different direction in addition to reflection from the surface.

There is a relationship between the angles of incidence and refracted rays and the speed of sounds in each medium. This relationship is given in Eq (1.79) for $c_1 < c_2$:

$$\frac{Sin\theta_1}{Sin\theta_2} = \frac{c_1}{c_2} \tag{1.79}$$

θ_1 and θ_2: Angles of incidence and refraction
c_1 and c_2: Speed of sound waves, in two media

While sound waves are passing from solid to air; $c_1 > c_2$ is valid and the ray bends upwards. The geometry of the wavefronts is given in Figure 1.68.

Refraction of sound waves in the open air occurs because of two reasons:

1) Change in the sound wave speed according to different temperatures of the media: The wave motion is faster in warmer air.
2) Change in the sound wave speed and direction under certain conditions, such as wind. The wave motion is determined by the algebraic summation of the wave and wind speeds.

Variations in temperature and wind speeds with the height from the ground are called temperature and wind gradients and the result of these gradients is in the variation of sound speed,

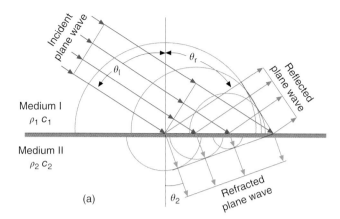

Figure 1.68 Refraction phenomenon for plane waves and wavefronts (Huygens' principle). (*Source:* adapted from [10].)

ultimately bending the sound rays toward the cooler layers in air and toward the wind direction. These refractions constitute the acoustical shadow formation around the sound source. Sound refraction is of importance from the point of environmental noise and will be discussed in detail in Chapter 2.

1.4.6 Sound Diffraction

When a sound wave passes an obstacle and meets an aperture with same size of the wavelength of the sound or smaller, the intensity and propagation direction are changed [12a]. This phenomenon is called sound diffraction. If it is a plane wave striking a surface, all the points of the surface on the direction line radiate sound as point sources. If there is an aperture (hole) on the surface along the direction of wave propagation, the center of the hole becomes the center of a secondary point source according to Huygens diffraction theory (Figure 1.69).

If a hard element (obstacle) is in the wave propagation path, the spherical waves coming from the unobstructed parts of the wavefronts affect each other and the sound rays bend around the edges of obstacle (Figure 1.70).The difference between refraction and diffraction is both involve bending of the sound ray during propagation, but there is no change in the medium characteristics in the diffraction phenomenon. contrary to refraction.

Since Fresnel in 1918 enabled the calculation of the amplitudes of diffracted waves, it has been possible to analyze the effects of obstacles positioned between the source and receiver point. There are two approaches to determine the sound pressure behind an obstacle due to the diffracted waves: the Fresnel diffraction which is applied when the source and the receiver are at a finite distance from the obstruction (barrier), whereas the Fraunhofer diffraction gives the solutions for infinite distances. For barrier problems against noise, as will be explained in detail in Chapter 2, the Fresnel diffraction is usually employed in practice.

If the dimensions of the obstruction (barrier) are equal or smaller than the wavelength, the sound waves continue to propagate behind the barrier. On the contrary, if the sound waves strike an object whose dimensions are greater than the wavelength, the sound waves reach the shadow zone by losing energy due to diffraction, confirming Huygens' principle. The direction of the diffracted

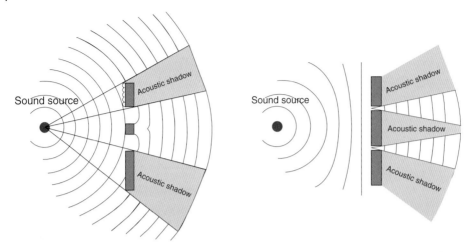

Figure 1.69 Huygens' principle of sound propagation through apertures.

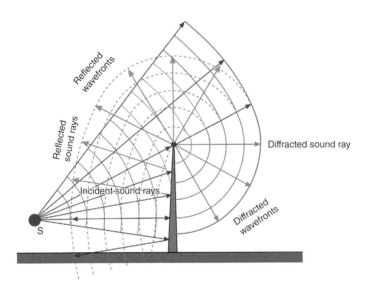

Figure 1.70 Diffraction from a semi-finite barrier and formation of acoustic shadow according to Huygens-Fresnel principles.

ray is determined by geometrical acoustics. If the dimensions are sufficient, a great amount of sound reduction can be obtained at the rear part of barrier. Sound reduction in the shadow zone depends on the dimensions of the barrier and can be obtained by using the simplified wave theory called the Fresnel-Kirschoff approximation, which is also used in optics. Based on this approach, in 1882, Kirschoff enabled computations by incorporating the integral theory into sound physics. In the computations it is necessary to analyze the contributions of the unobstructed sections to be able to find the effect of an obstacle on sound pressures behind the barrier. The Fresnel diffraction theory suggests that only the points near the top edge of a finite barrier contribute to the total acoustic field behind a barrier (Figure 1.70). The analytic expression of both theories enables the calculation

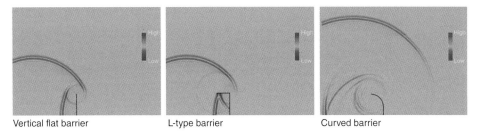

| Vertical flat barrier | L-type barrier | Curved barrier |

Figure 1.71 Wavefronts of diffracted sounds by barriers of different shapes. (*Source:* from software prepared by H. Tachibana, with permission of Tachibana.)

of the acoustic field (i.e. the sound pressures behind barrier) in the shadow zone. In this zone, sound radiates gradually, related to the distance from the barrier and the wavelength of sound. The efficiency and size of shadow zone are determined by the reduction of sound, which is greater at higher frequencies. Also the timbre of sound changes in relation to the distances from the source to the barrier and from the receiver to the barrier.

Mechanism of diffraction: From an observation point remaining in the shadow zone of a barrier, the sound is perceived as if it is coming from an imaginary line source, radiating cylindrical waves along the top edge of the barrier. When the perpendicular distance from the receiver point to the barrier is smaller than the distance between the source and the barrier, the top edge seems to be an imaginary line source radiating cylindrical waves. The acoustic power of this imaginary source is related to the power of the original wave at the edge of barrier reaching from the original source. If the source is a point source, the sound pressure decreases as a linear function of the square root of the source-barrier distance. If the point source is close to the barrier than the receiver point, the effective length of the imaginary source becomes much shorter, this implies a spherical wave radiation from the top edge. In fact, there may be some irregularities near the edge of the barrier (near field) at low frequencies. Also, micro-meteorological disorders can disturb the phase relationships at high frequencies. Since the diffracted wave over the top edge of the barrier is not restricted in the shadow zone, it interferes with the direct wave and a transition region occurs. In optics, the wave lengths of light are much smaller compared to those of sound waves, the shadow zones are quite sharp, however, since the wavelengths of sounds vary between 2 cm–17 m, the size of the transition zone is not much emphasized and changes gradually (less distinctive). Various diffraction effects are shown in Figure 1.71 obtained by different shapes of the barrier edges with two types of head profiles.

Barriers are widely implemented elements against environmental noises. Factors affecting the barrier performance and the principles for barrier design in controlling noise, are explained in Chapter 2.

1.5 Conclusion

It is important to have some knowledge of physical acoustics, the basics of which were founded in antiquity and still apply in the present day. Although the primary concern of this book is environmental noises, the concepts of the sound radiation mechanism, sound sources, and deteriorations of sound waves while transferring in the mediums, and affective parameters have been outlined in this chapter. Physical acoustics covers multidisciplinary fields. Within a five decades it has appeared as an engineering field, namely, acoustical engineering or noise control engineering, which covers an extensive field of work from sound emitters to sound perception and controlling harmful sounds, and has emerged as the individual branches that will be introduced in the following chapters.

References

1 Cuny, D.R. (2000). *The Science and Applications of Acoustics*, 2e. City University of New York/ Springer.

2 Pierce, A.D. (1989). *Acoustics: An Introduction to its Physical Principles and Applications*. American Society of Acoustics.

3 Pierce, A.D. (1998). Mathematical theory of wave propagation. In: *Handbook of Acoustics* (ed. M.J. Crocker), 21–38. Wiley.

4 T.D. Rossing (ed.) (2014), A brief history of acoustics. In: *Handbook of Acoustics* (ed. M.J. Crocker), 11–27. Springer.

5 Lang, W.W. (1996). A quarter century of noise control: a historical perspective. *Acoustics Bulletin* July/Aug: 1–9.

6 Crocker, M.J. (ed.) (1998). Introduction. In: *Handbook of Acoustics*, 3–20. Wiley.

7 Lindsay, B. (1972). *Acoustics*. Dowden Hutchinson and Ross Inc.

8 Peutz, V.M.A. (1981). The all-important practice of noise control engineering. *Inter-Noise 81* 1: 7–15.

9 Mattei, J.P. (1981). Acoustical engineering, Inter-Noise 81, 1: 59–62.

10 Le Moine, M.J. and Blanc, M.A. (1947). Basic and experimental physics, periodic motions: acoustics. In: *Propagation of Vibrations*, (trans. C. Saraç). TC University of Ankara, Faculty of Science.

11 ISO (2012). ISO 80000-8:2007 Acoustics: Quantities and units. Part 8.

12 (a) Crocker, M.J. (2007). Fundamentals of acoustics, noise and vibration. In: *Handbook of Noise and Vibration Control* (ed. M.J. Crocker), 1–16. Wiley.
(b) Crocker, M.J. (2007). Theory of sound-predictions and measurement. In: *Handbook of Noise and Vibration Control* (ed. M.J. Crocker), 19–42. Wiley.
(c) Nelson, P.A. (2007). Sound sources. In: *Handbook of Noise and Vibration Control* (ed. M.J. Crocker), 43–51. Wiley.

13 ISO (2010). ISO 3741:2010 Acoustics: Determination of sound power levels and sound energy levels of noise sources using sound pressure: Precision methods for reverberation test rooms.

14 Fahy, F. (1995). *Sound Intensity*. E & FN Spon.

15 Maekawa, Z., Rindel, J., and Lord, P. (2011). *Environmental and Architectural Acoustics*. Spon Press.

16 Beranek, L.L. (ed.) (1971). *Noise and Vibration Control*. McGraw-Hill.

17 Beranek, L.L. and Ver, I.L. (1992). *Noise and Vibration Control Engineering: Principles and Applications*. Wiley.

18 Moore, B.C.J. (1998). Frequency analysis and pitch perception. In: *Handbook of Aoustics* (ed. M. Crocker), 1167–1180. Wiley.

19 Hassall, J.R. and Zaveri, K. (1979). *Acoustics Noise Measurements*. Brüel & Kjaer.

20 Harris, C.M. (1957). *Handbook of Noise Control*. McGraw-Hill.

21 Harris, C.M. (1979). *Handbook of Noise Control*, 2e. McGraw-Hill.

22 Morse, P.M. and Bolt, R.H. (1944). Sound waves in rooms. *Reviews of Modern Physics* 16: –69. (Erratum, Reviews of Modern Physics 16, 324, 1944).

23 ISO 10534-2:1998 (confirmed in 2015) Acoustics – Determination of sound absorption coefficient and impedance in impedance tubes – Part 2: Transfer-function method.

24 Rindel, J.H. (2018). *Sound Insulation in Buildings*, CRC Press.

Further Reading

Alton Everest, F. (1994). *The Master Handbook of Acoustics*, 3e. McGraw-Hill.

Ando, Y. (1998). *Architectural Acoustics: Blending Sound Sources, Sound Field and Listeners*. AIP Press and Springer-Verlag.

Anon (2001). Odeon Room Acoustics Program Version 5.0, User manual and software.

Benward, B. and Saker, M. (2008). *Music in Theory and Practice*, 8e.

Beranek, L. (1986). *Acoustics*. American Institute of Physics for the Acoustical Society of America.

Beyer, R.T. (1999). *Sounds of Our Times: Two Hundred Years of Acoustics*. Springer-Verlag.

Bies, D.A. and Hansen, C.H. (2002). *Engineering Noise Control Theory and Practice*, 2e. Spon Press.

Brüel, P.V., Pope, J., and Zaver, H.K. (1998). Introduction. In: *Handbook of Acoustics* (ed. M. Crocker), 1311–1316. Wiley.

Conrad, J.H. (1983). *Engineering Acoustics and Noise Control*. Prentice Hall Inc.

Cowan, J.P. (1994). *Handbook of Environmental Acoustics*. Wiley.

Crocker, M.J. (1984). *Noise Control*. Van Nostrand Reinhold Company Inc.

Crocker, M.J. (ed.) (1997). *Encyclopedia of Acoustics*. Wiley.

Crocker, M.J. and Kessler, F.M. (1982). *Noise and Noise Control*, vol. 2. CRC Press.

Fahy, F. (2001). *Foundations of Engineering Acoustics*. Academic Press.

Fahy, F. and Walker, J. (1998). *Fundamentals of Noise and Vibration*. E & FN Spon.

Hunt, F.V. (1978). *Origins in Acoustics: The Science of Sound from Antiquity to the Age of Newton*. Yale University Press.

Ingard, U. (1994). *Notes on Sound Absorption Technology*. Noise Control Foundation USA.

ISO (2014). ISO 17497-2:2014 Acoustics: Sound scattering properties of surfaces. Part 1: Measurement of the random-incidence scattering coefficient in a reverberation room.

Kinsler, L.E. and Frey, A.R. (1982). *Fundamentals of Acoustics*, 2e. Wiley.

Knudsen, V.O. and Harris, C.M. (1978). *Acoustical Design in Architecture*. American Institute of Physics for the ASA.

Lawrence, A. (1989). *Acoustics and Built Environment*. Elsevier Applied Science.

Le Moine, M.J. and Blanc, M.A. (1947). Basic and experimental physics, periodic motions-acoustics. In: *Propagation of Vibrations*, (trans. C. Saraç). TC University of Ankara, Faculty of Science.

Mediacoustics (1980). Software for education in acoustics, CD 01 dB-Stell.

Morse, P.M. and Ingard, K.U. (1968). *Theoretical Acoustics*. McGraw-Hill.

Newton, R.E.I. (1990). *Wave Physics*. Edward Arnold Publishers.

Rathe, E.J. (1969). Note on the two common problems of sound propagation. *Journal of Sound and Vibration* 10: 472–479.

Silverman, J. (1996). Aspects of research on acoustics in the UK. *Acoustics Bulletin* July/August: 11–16.

Sound Research Laboratories (1976). *Practical Building Acoustics*. E & FN Spon Ltd.

2

Noise

Scope

In this chapter, as an introduction to environmental noise issues, the definitions and descriptions regarding noise are briefly presented. Differentiation of noises with respect to spectral, temporal, and spatial characteristics is included and the physical environmental factors affecting the sound pressure levels during noise propagation outdoors are described. Disturbances of sound waves, like divergence, reflection, absorption, refraction, scattering, and diffraction are explained with a special emphasis on the effect of noise barriers. Their design and construction principles are given in the the Appendix. The metrics and descriptors to measure and evaluate environmental noises are outlined.

2.1 Definition and Concepts

Noise as a special type of sound has been of concern in the last century although awareness of noise problems was recognized in the Middle Ages [1, 2]. In the present day, the concept of environmental noise pollution has emerged as an adverse outcome mainly of technological developments, expanded urban transportation, and unplanned urbanization, by creating risks to human health and welfare and by increasing the numbers of people affected. Scientific and technical investigations initially focused on descriptions of mechanical noise sources with particular characteristics, but soon they became interested in traffic noise with its propagation outdoors and the development of new techniques in noise analysis. Parametric studies enabled the prediction of noise by taking into account the intrinsic effects of physical factors, followed by field and laboratory measurements. The results have been used in the development of noise assessment methods to evaluate noise impacts on humans.

Although environmental noise is seen as an urban problem, in reality, it exists anywhere that there are humans, even in rural areas (e.g. where a highway passes virgin land, the animal inhabitants are definitely affected by the continuous high-level noise). If we consider that noise sources also exist in buildings, the concept of the environment can be expanded to cover enclosed spaces. The workspace noise sources create serious health risks, which will be considered separately. Precautions against such noises fall within the field of building acoustics, however, the basic principles of noise propagation are still valid.

The chronological investigation of noise pollution reveals that the most disturbing and most common noise source in urban areas is the transportation systems, specifically motorways, which have been of great concern scientifically and technically.

Environmental Noise and Management: Overview from Past to Present, First Edition. Selma Kurra.
© 2021 John Wiley & Sons Ltd. Published 2021 by John Wiley & Sons Ltd.

After noise and noise control became a huge topic of interest in societies at the scientific and technical levels, noise control engineering emerged as a specific branch of acoustic science in the 1980s. Thus, the standardization process has been accelerated.

Definition of Noise

Noise was defined as "unwanted sound" in the earlier publications by emphasizing its subjective effects, especially if these sounds are not of interest to those who perceive them [3]. However, scientific research (both physiological and psychological), conducted later, revealed the severe adverse effects on human health (Chapter 7), so that noise has been treated as a type of environmental pollution to be controlled. Other definitions for noise can be made from different points of view: for example, it is defined as unexpected and intolerable sound from a psycho-acoustical perspective. It can also be stated as "a residual product of technology," as in the noise caused by industrial sources. The term "environmental noise" is primarily used for those noises emerging from sources located outside buildings, although they do not cause hearing problems immediately, they create health and comfort problems in the short, medium and long term.

Although any pure tone sound can be annoying/unwanted and can be considered as noise, however, generally noise is an assembly of sounds comprising random wave patterns without harmonically related frequency components. It is usually dominated by certain tones, containing high sound energy that might be varying with time, steadily, gradually, or intensely, sometimes characterized by impulsive or impact sounds as repetitive events. Briefly, environmental noise is a complex sound of a random nature causing harm to health and welfare.

The general characteristics of sound sources were explained in Chapter 1, however, it is necessary to examine systematically the parameters that play a role in the evaluation of noise of every kind. Since the environmental noise sources vary a great deal with regard to their constructional and operational aspects, therefore, such parameters are significant in acoustical properties of the produced sounds and while they are propagating in the environment. Descriptions of noises are made generally with respect to the spectral, spatial, and temporal variations explained in a number of publications from the 1950's till today, some are given in the references [1–20].

The acoustic energy within noise and the radiated sound pressure levels can be determined by means of the prediction models and the measurement techniques that will be explained in further chapters, however, the above-mentioned characteristics are explained briefly, regardless of the source type.

2.2 Spectral Characteristics of Noise

The frequency spectrum of complex sounds is a major descriptor of noise (see Section 1.2.2). The frequency content of a noise signal describing its tonality characteristics in musical acoustics is an influential factor on sound perception, which is a physiological process closely related to frequency and the temporal characteristics of noise. This is an impact factor on annoyance, as will be seen in Chapter 7. The further significance of the spectrum depends on the masking phenomenon which is beneficial for noise abatement. The wavelength of sound changes with the use of materials for absorption, transmission, diffraction, etc. Therefore, an efficient noise control needs a detailed spectral analysis of noise.

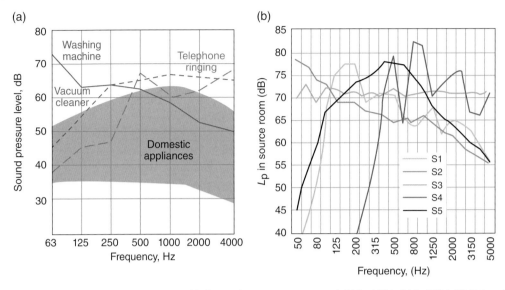

Figure 2.1 Typical noise spectrums of indoor noise sources presented: (a) in 1976; (b) in 2014 (S1:Guitar, S2: Music, S3:Live music, S4:Baby cry, S5:Loud speech, S6:Dog bark). (*Sources:* adapted from [13b, 13c], permission granted by SRL and Hongisto.)

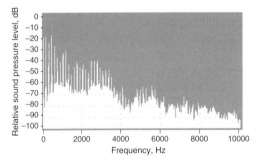

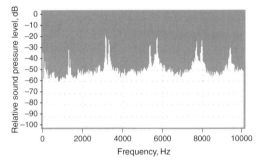

Figure 2.2a Spectrum of a tonal noise with harmonics: trumpet.

Figure 2.2b Spectrum of a tonal noise: ringing.

The range of noise spectra that can be encountered in daily life (indoors) is shown in Figure 2.1a (from 1976) and Figure 2.1b (from data in 2014) [13b, 13c]. The sound energies of the sources, differ considerably in frequency bands, resulting in various shapes of spectra. Some exhibit a linear form only within a narrow bandwidth called the narrow-band noise or display a continuous spectrum in the frequency range concerned, called broadband (or wide-band noise), whereas others can be mixed spectrum with a random shape accompanied by audible discrete tones and harmonics (see Figure 1.25, Figures 2.2a and 2.2b). Some may have fluctuating frequencies (modulations). Sounds indicating variability, within time, require detailed resolution of their spectrums at sufficiently long periods. An earlier diagram from the 1960s emphasizing the importance of pressure/time/frequency analysis is given in Figure 2.3a [21]. The similar analysis made by using the present techniques is shown in Figure 2.3b.

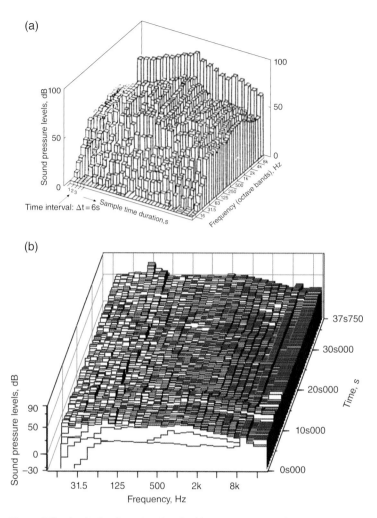

Figure 2.3 Analysis of a noise signal with respect to sound pressure, frequency and time (a) adopted from an earlier document, 1969 [21] (b) waterfall diagram obtained by an analyzing software. (*Source:* Kurra 2017.)

2.2.1 Filters in Frequency Analysis

A filter is an electrical device which allows only a part of sounds within a certain frequency band over the total spectrum range to pass. It is used to measure the sound energy and the spectral content of a given signal and is a simple part of frequency analyzers (see Chapter 5). As mentioned in Section 1.2.2, the filter restricts the signals passing through below and above the limiting frequencies in a selected bandwidth. Figure 2.4a shows the characteristics of a filter for a cut-off frequency (i.e. a center frequency or mid-frequency) of 1000 Hz and the effective range. Among the different types of filters such as low-pass, high-pass, etc., the bandpass filters with upper- and lower-limiting frequencies where the signal is attenuated by 3 dB, are widely used (Figure 2.4b). The bandpass filters can be either constant bandwidth filters or constant-percentage bandwidth filters. Detailed information about filters can be found in the various literature, e.g. [22, 23].

In environmental noise analysis, the bandwidth can be selected as octave, bands, one-third octave bands or narrowbands. Figure 2.5 shows analysis of a road traffic noise signal by using different

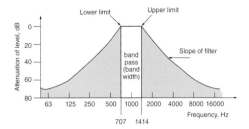

Figure 2.4a General principle of acoustic filters: Frequency-specific attenuation at 1000 Hz (Lower limiting-frequency: 707 Hz and higher limiting-frequency: 1414 Hz).

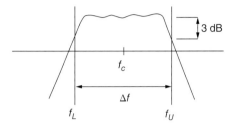

Figure 2.4b Definition of the cut-off frequencies of a bandpass filter (f_L: Lower-limiting frequency, f_U: upper-limiting frequency, Δf: Bandwidth, f_c: Center frequency).

bandwidths. The range of passing signals in one octave band filter is wider than the one-third octave band, therefore, for the same signal, the band level in any octave band is higher than the one-third octave band level.

2.2.2 Fourier Analysis

As seen in Section 1.2, the complex sound waves (i.e. random signals such as environmental noises) are composed of multiple frequency components, either harmonic or non-harmonic, also these components can be either periodic or non-periodic regarding time. To obtain specific information from such signals, i.e. the amplitude characteristics and the distribution of spectral content, the Fourier transformation can be used to process the signal recorded during acoustical measurements.

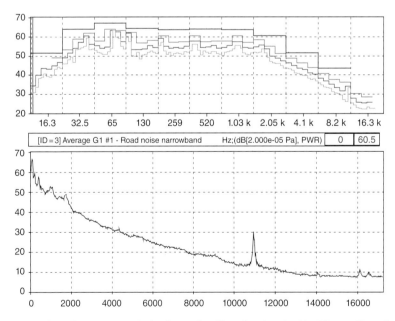

Figure 2.5 Frequency analysis of a road traffic noise signal with different filter characteristics. Upper: Octave, 1/3 octave,1/6 and 1/12 octave bands. Lower: Narrowband analysis in linear frequency scale. (*Source:* from private measurement reports by Kurra in 2016.)

A time-varying signal can be decomposed into an infinite number of single frequency compo-nents of infinite duration (sound pressures of the pure tones) by transforming the signal from the time-domain to the frequency domain using the Fourier integral [5, 8, 24]:

$$f(t) = \frac{1}{2\pi} \int_{-\infty}^{+\infty} F(\omega)e^{j\omega t} d\omega \tag{2.1}$$

$F(\omega)$ is a frequency function called the Fourier transform of a waveform $f(t)$. It consists of several equations that can provide the spectrum of the signal:

$$F(\omega) = \int_{-\infty}^{+\infty} f(t)e^{-j\omega t} dt \tag{2.2}$$

The Fourier theorem in mathematics, which was developed by J.B. Fourier in 1800, states that any periodic function $f(t)$ can be represented by a series of sine and cosine terms (the Fourier series), each of which has specific amplitude and phase coefficients. Thus, a complex sound with multi-frequencies is periodic but not simple harmonic, and can be resolved into a series of instant harmonic sounds (e.g. the individual sinusoidal components with their frequencies, amplitudes, and relative phases) as given in Eq (2.3):

$$f(t) = \sum_{n=1}^{\infty} A_n \cos(n\omega t) + B_n \sin(n\omega t) \tag{2.3}$$

A_1-A_n, B_1-B_n: Pressure amplitudes of individual components
n: Number of samples in the frequency domain
$\omega = 2\pi f$

When the sound pressure is expressed as the *rms* value, the phase parameter (ωt) can be elimi-nated in the summation process given in Eq (2.3), because of the random phases of the components called incoherent sounds.

For non-stationary noises, one needs to analyze the components using different time-domain methods, such as the "wavelet transformation," as shown in Figure 2.6 [25, 26a]. Time histories

Figure 2.6 Fourier transformation of using wavelets of different forms [25]. (*Source:* permission granted by Brüel & Kjaer Sound and Vibration Measurement A/S.)

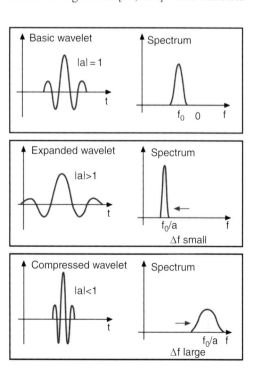

of the signals are divided into short waves of any waveform (wavelet) by applying a window to the signal in a short period of time, according to a criterion and then applying the Fourier transformation to all the wavelets.

The Fourier analysis, which provides information about the components of any time-varying noise and enables time-frequency maps of noise signals to be made, is performed by using the digital analysis technique called Fast Fourier Transformation (FFT) (see Chapter 5). The frequency domain methods are mainly applied to sound emission of single sources, moving or stationary, but are not used widely in environmental noise analyses due to the complexity of the problems.

2.2.3 Test Signals

In applied acoustics, the test signals with specific frequency characteristics are used in acoustical measurements according to the standards. They are generated by the electro-acoustic signal generators, as explained in Chapter 5. The signals are identified with the power spectral density (i.e. sound intensity) per unit of bandwidth differently, in proportion to the frequency. The most common test signals are:

1) *Pure tone sound*: A simple sound wave with a specific center frequency, (Chapter 1).
2) *White noise*: A sound signal with equal sound energy at each frequency, thus it gives a steady and monotonous stream of the entire spectrum of frequencies (flat spectrum). It is defined also as a noise whose power spectrum density is independent of frequency. However, the octave band spectrum displays a 3 dB/octave increase (Figures 2.7a and 2.7b).
3) *Pink noise*: A sound signal (a broadband noise) carrying equal sound energy at each octave band. It is defined also as a noise whose power spectrum is inversely proportional to the frequency, i.e. the decrease is −10 dB/decade (Figures 2.7c and 2.7d). Pink noise is perceived as approximately of equal magnitude on frequencies since the ear is more sensitive to octave bands. Pink noise has less energy than white noise.
4) *Normalized traffic noise*: A signal whose band levels correspond to the road traffic noise spectra normalized during signal processing. Sounds of other noise sources are also modeled by applying a transfer function to the recorded real-time signals and used in the listening tests.

In audio engineering and musical acoustics, other colors of noise have been defined to be used for different purposes, for example: Brown noise contains a spectrum whose power density decreases with increasing frequency, i.e. the energy is higher at the lower end of the spectrum (Figure 2.7e). Decrease is −20 dB/decade corresponding to −6 dB/octave. Blue and violet noises both have energy density increasing with frequency, contrary to the brown noise (Figures 2.7f and 2.7g). Blue noise displays the power density increase of +10 dB/decade throughout the frequency range whereas violet (purple) noise has an increase of +20 dB/decade, corresponding to +3 dB/octave in a limited range of frequency after 1000 Hz. The gray noise spectrum, containing higher energy at both low and high frequencies, provides every frequency equally loud (Figure 2.7h). Black noise represents "silence," which is a barely detectable audio frequency. See https://en.wikipedia.org/wiki/Colors_of_noise and www.wired.co.uk/article/colours-of-noise.

Maximum Length Sequence (MLS)
Different room excitation signals are used to determine the room acoustic parameters, mainly based on the "impulse response" and its post-processing, such as frequency and phase response, reverberation time, energy time-curve (energy/decay plot), cumulative spectral decay (waterfall), etc., which are required in room acoustic measurements, according to ISO 3382 (Section 5.9.6).

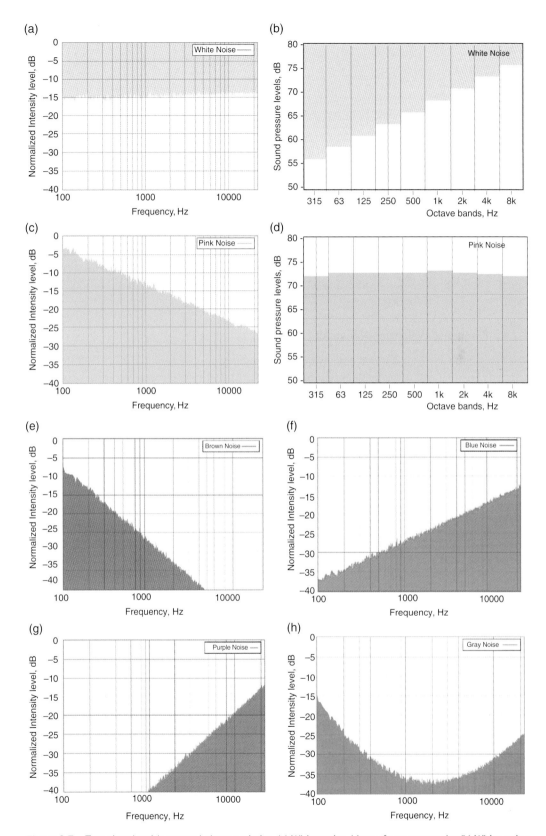

Figure 2.7 Test signals with spectral characteristics, (a) White noise, Linear frequency scale; (b) White noise, Logarithmical frequency scale (octave bands); (c) Pink noise, Linear frequency scale; (d) Pink noise, Logarithmical frequency scale (octave bands); (e) Brown noise, Linear frequency scale; (f) Blue noise; (g) Violet noise, Linear frequency scale; (h) Gray noise, Linear frequency scale.

Along with the "sine-sweep methods" to obtain the impulse response of the room, the MLS technique provides accurate measurements by minimizing the effect of background noise – particularly impulsive noises – and eliminates the distortions in the output signal.

The electrically generated pseudo-random MLS signal is a randomly distributed sequence of the same amplitude, the same positive and negative impulses, symmetrical around 0 [26b, 26c]. These signals, consisting of digitally synthesized binary sequences, are fed into the system input and their cross-correlation with the system output gives exactly the system impulse response. The cross-correlation can be carried out in the time domain. To define and generate the sequence, a filter is needed "feedback shift register".

MLS resembles white noise which is also an artificially generated, non-periodical, and random signal. The advantages of the MLS technique have been discussed in many documents, such as having a complete frequency spectrum, a sequence to be determined, a flat sound spectrum, being periodic with a certain order and an easy-to-calculate frequency response from impulse response signal. The MLS technique, is also implemented in various acoustic tests, such as noise barrier measurements given in Section 5.10.3.

2.2.4 Frequency Weighting Networks

The human ear does not respond equally to each frequency/pressure combinations and this characteristic is explained by the loudness concept (Section 7.1.2). In practice, noises are evaluated not only by using the physical units but also the physiological units by taking into account the human perception based on the loudness concept. The weighting networks that have been internationally standardized, correspond to the reference loudness levels given below:

Weighting network	Reference loudness level (approximately)
A-weighting	40 phons
B-weighting	70 phons
C-weighting	100 phons
Z-weighting	unweighted

Note: The loudness curves are for pure-tone sounds.

To obtain the weighted noise levels, the sound pressure levels (dB) are adjusted according to the selected weighting network shown in Figure 2.8 [27]. The adjustment factors are given in Box 2.1 to

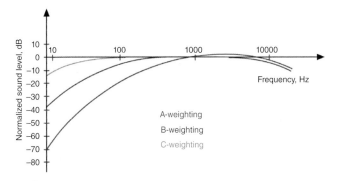

Figure 2.8 Standard A, B, C frequency-weighting networks.

Box 2.1   A-Weighting Conversions (Weighting Procedure)

Table 2.1 shows the adjustment values in the A-weighting network and Table 2.2 presents a worked-out problem.

Table 2.1   Adjustment values to A-weighting network.

Octave bands, Hz	63			125			250		
1/3 octave bands, Hz	50	63	80	100	125	160	200	250	315
A-weighting Adjustment, dB	−30.2	−26.2	−22.5	−19.1	−16.1	−13.4	−10.9	−8.6	−6.6
Octave bands, Hz	500			1000			2000		
1/3 octave bands, Hz	400	500	630	800	1000	1250	1600	2000	2500
A-weighting Adjustment, dB	−4.8	−3.2	−1.9	−0.8	0	0.6	1	1.2	1.3
Octave bands, Hz	4000			8000					
1/3 octave bands, Hz	3150	4000	5000	6300	8000	10 000			
A-weighting adjustment, dB	1.2	1	0.5	−0.1	−1.1	−2.5			

Table 2.2   Example problem to find the A-weighted total noise level from the given band levels.

	Octave bands, Hz								Total level (logarithmic summation of the band levels)
	63	125	250	500	1000	2000	4000	8000	
Linear octave band values, $L(f)$	75.0	76.0	78.0	71.0	68.0	66.0	54.0	50.0	81.9 dB
Adjustment factors, Δf	−26.2	−16.1	−8.6	−3.2	0	+1.2	+1.0	−1.1	–
A-weighted spectral values $L_{A(f)}$, dB	48.8	59.9	69.4	67.8	68	67.2	55	48.9	74.4 dB(A)

apply the sound pressure levels in octave and one-third octave bands to convert them into the A-weighted values. After conversion, the A-weighted spectral sound pressure levels are summed logarithmically to obtain the total weighted sound pressure level (Box 1.4).

The relations between the basic acoustic units and the A-weighted levels are explained in Figures 2.9a and 2.9b for sound power and pressure. A-weighted levels are widely used to evaluate

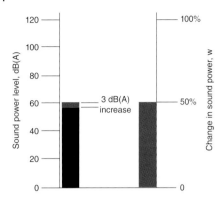

Figure 2.9a Variation of sound power level in dB(A) corresponding to the percentage increase of sound power in Watts.

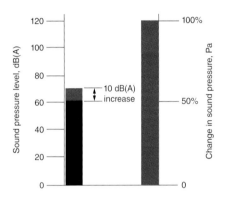

Figure 2.9b Variation of sound pressure level in dB(A) corresponding to the percentage increase of sound pressure, in Pascal (Pa).

general human exposure to noise, however, for the impulsive sounds displaying a sudden change in sound pressure levels, for the sounds having dominant discrete frequencies in their spectra, and for noises with high energies at low frequencies, the A-weighted levels do not accurately describe the negative impact of noise. Different rating units are implemented to evaluate these types of noise.

The common terminology of "A-weighted noise level, dB" recommended in the ISO and IEC standards is mainly used in this book, however, the symbol dB(A), is preferred while discussing the noise and management problems, as in many European documents and reports.

2.3   Spatial Variation of Noise

Sound propagation from noise sources might be from ground to ground, from ground to air or from air to ground with respect to source-receiver geometry. Due to deviations of sound waves in different mediums, namely, wave divergence, absorption, reflection, scattering, refraction, diffraction, interference, and transmission phenomena, as explained in Section 1.4, the sound pressures at a constant point are subject to change in all the paths transferring sound from source to receiver [3, 4, 6, 8, 21–23, 27, 28]. Investigation of spatial variation of sound levels needs an analysis of the characteristics of the overall medium and specifically the propagation path, since the propagation of sound waves in air or a solid medium reveals different patterns associated with the physical characteristics of the medium, which is under the influence of various factors. Special measurements in the field can give the effect of factors individually or in combination with other related factors found in the environment at the same time. The contribution of each factor, which may be significant or negligible, to the noise levels at a certain receiver (observation) point even at a far distance from the source, can be quantified for use in a prediction model to obtain the overall noise levels (Chapter 4). However, most of the parameters are interrelated and some are source-specific, thus it is difficult to make accurate assessments.

Generally sound pressure levels change depending on the position and location of the receiver point in the near field and far field of the noise source. The variations, which need to be determined, can emerge in such a way to increase or decrease in sound pressure levels, or change the frequency or phase characteristics of sounds. The physical environmental factors which influence the sound propagation and result in variation of the received sound levels are given below [14–20, 29–34]:

1) Distance between source and receiver (sound attenuation due to wave divergence).
2) Air absorption (sound attenuation due to molecular absorption).
3) Meteorological factors (both sound attenuation and increase due to refraction and scattering).
4) Ground absorption (sound attenuation due to absorption of porous ground).
5) Forest, woods, and plants (vegetation) (sound attenuation due to diffraction, scattering, and transmission).
6) Natural and built barriers (sound attenuation due to diffraction and scattering).
7) Solid surfaces (sound increase due to reflection).
8) Solid/rigid elements (sound transmission)

2.3.1 The Distance Effect

The wave divergence described in Section 1.3.3 always causes a reduction in the received sound pressure levels depending on the geometrical and acoustical properties of noise sources, as shown in Figures 1.31b, 1.36, and 1.37. Regardless of source type, sound attenuation A_d increases with the distance from the source and can be obtained through Eq (2.4) with respect to a known pressure at distance d_1 from the source by taking into account the sound wave geometry:

$$A_d = L_{p1} - L_{p2} = 20 \lg \left\{ g \left[\frac{p_1}{p_2} \right] \right\} \tag{2.4}$$

A_d: Attenuation due to distance from the source (wave divergence)
L_{p1}: Sound pressure level at distance, d_1 from the source, dB
L_{p2}: Sound pressure level at the receiver located at distance, d_2 from the source, dB
p_1: Sound pressure at distance d_1 from the source, pascal (N/m^2)
p_2: Sound pressure at distance d_2 from the source, pascal (N/m^2)
g: Coefficient (wave type factor):

$g = 0$ (Plane wave)
$g = 1$ (Spherical wave)
$g = \frac{1}{2}$ (Cylindrical wave)

Equation (2.4) gives 6 dB and 3 dB losses per doubling of distance for spherical wave and cylindrical wave propagations respectively.

2.3.2 Effect of Air Absorption

During particle motion and sound wave propagation, some of the sound energy is absorbed due to heat conduction, viscous losses due to the friction of oxygen and nitrogen molecules, and the relaxational process of the molecules in the air. This mechanism is related to sound frequency, the amount of vapor in the air, the temperature and the distance between the source and the receiver. Generally, the sound absorption is rated by the "sound absorption coefficient" which depends on the air temperature, the relative humidity, and the atmospheric pressure. Air absorption is greater at high frequencies and at long distances, and ultimately results in an abrupt decrease in high frequency sound energy in the noise spectrum. This effect is also called atmospheric attenuation, and is investigated as a reduction dB per unit distance (e.g. 1 km):

- reduction in band level of pure tone sounds and noises.
- reduction in A-weighted sound pressure levels.

Sound attenuation for pure tones due to air absorption is given as a function of frequency and distance at 20 °C and under the standard atmospheric pressure in the International Standards (Chapter 5).

Sound attenuation for noises due to air absorption is lower compared to the pure tone sounds. Air absorption also depends on the atmospheric pressure and altitude above sea level and is expressed as attenuation of dB/100 m in the US Standards and as dB/km in Europe and in the ISO Standard [30, 31]. The decrease in sound pressure levels at longer distances is given in Table 2.3 according to temperature and relative humidity. The values shown in the table are the averages of octave band attenuations. The attenuation for other distances is calculated as Eq (2.5):

$$A_{atm} = \alpha d / 1000 \text{ dB} \qquad (2.5)$$

α: Sound attenuation coefficient per 1000 m for pure tone sounds (Table 2.3)
d: Distance between source and receiver, m

Air absorption can be ignored within several hundred meters from the source and at frequencies up to 5000 Hz. At longer distances, attenuation is significant in all frequency bands.

Figure 2.10 displays the variation of atmospheric attenuation with temperature and humidity conditions, based on computations conducted in the 1950s [15]. Further studies resulted in other diagrams to determine air absorption, as given in Figure 2.11, obtained by Beranek, computed using

Table 2.3 Atmospheric attenuation coefficient α for octave bands of noise [32].

		Air attenuation coefficient α, dB/1000 m							
		Octave bands, Hz							
Temperature °C	Relative humidity %	63	125	250	500	1000	2000	4000	8000
10	70	0.1	0.4	1.0	1.9	3.7	9.7	32.8	117
20	70	0.1	0.3	1.1	2.8	5.0	9.0	22.9	76.6
30	70	0.1	0.3	1.0	3.1	7.4	12.7	23.1	59.3
15	20	0.3	0.6	1.2	2.7	8.2	28.2	86.8	202
15	50	0.1	0.5	1.2	2.2	4.2	10.8	36.2	129
15	80	0.1	0.3	1.1	2.4	4.1	8.3	23.7	82.8

the equations presented in ANSI s.1.26 (1978) [30]. Figure 2.11 displays the source-receiver distance for the first 3 dB(A) reduction in noise levels due to atmospheric absorption [13a].

Air absorption can be determined by the field measurements as well as computations using the relationship in Eq (2.6) [25]:

Attenuation due to air absorption at distance *r* [11]:

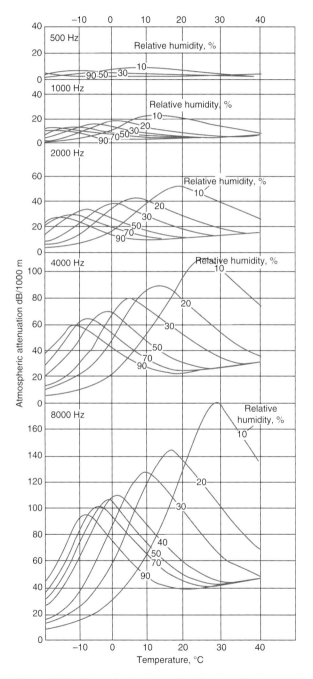

Figure 2.10 Atmospheric attenuation for aircraft -to- ground propagation and effect of humidity and air temperature at different octave band center frequencies. (*Source:* adapted from [15].)

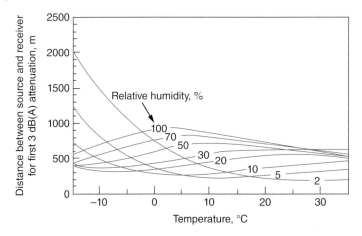

Figure 2.11 Atmospheric attenuation for the typical source spectrums: Determination of distance for the first 3 dB(A) attenuation due to combined effects of temperature and relative humidity. (*Source:* adapted from [13a].)

$$A_e = -20\lg\left[\frac{p_r}{p_o}\right] = -20\lg\left[\exp\left(-\alpha r\right)\right] = ar \qquad (2.6)$$

p_r: Sound pressure at distance r
p_o: Initial sound pressure at $r = 0$
α: Air absorption coefficient in Neper/m
a: Attenuation coefficient dB/m ($=8.686\alpha$)

Cremer gave Eq (2.7) to predict atmospheric attenuation at distances greater than 3 km between 1000 and 2000 Hz [10, 19]:

$$A_e = 7.4\frac{f^2 r}{\phi}10^{-8}\ \text{dB (at 20°C)} \qquad (2.7)$$

r: Distance between receiver and source, m
ϕ: Relative humidity %
f: frequency, Hz

For other temperatures:

$$A_e = \frac{A_e(20°\text{C and }\phi = 50\%)}{1 + \beta\Delta tf}\ \text{dB}\ (\phi = 50\%) \qquad (2.8)$$

β: Constant $= 4 \times 10^{-6}$
Δt: Difference between the actual temperature and 20 °C (Max: ±10 °C)

The equations above are applicable for simple predictions and the isotropic air. However, atmospheric attenuation is a complex phenomenon depending upon air temperature, frequency, or humidity, which are interrelated parameters.

The effect of atmospheric attenuation is emphasized on the higher frequencies of sound. The outdoor propagation models (Chapter 4) take into account the atmospheric attenuation while predicting (simulating) the noise levels in environments by applying certain assumptions.

For example, the aircraft noise prediction models employ the relationships proposed by the International Civil Aviation Organization (ICAO) for two altitudes of aircraft [17]:

Condition 1: $H(1.8t + 32) \geq 4000$

$$\alpha_i = \frac{f_i}{500} \quad \text{dB/1000ft} = \text{dB/305m} \tag{2.9}$$

Condition 2: $H(1.8t + 32) \leq 4000$

$$\alpha_i = \frac{f_i}{750}\left[5.50 - \frac{H[1.8t + 32]}{1000}\right] \quad \text{dB/1000 ft} = \text{dB/305 m} \tag{2.10}$$

H: Relative humidity, %
t: Temperature, °C
α_i: Atmospheric absorption at ith 1/3 octave band
f_i: Geometric center frequency of ith 1/3 octave band

Graphical representations of Eqs (2.9) and (2.10) are given in the ICAO document.

2.3.3 Meteorological Factors

The meteorological factors, such as air temperature, wind speed and direction, turbulence, and humidity affect the propagation of sound from the source to the receiver and the sound pressure levels fluctuate significantly in time at the receiver point. Air temperatures and wind speeds change with height from the ground, represented by temperature and wind gradients (or temperature and wind profiles), causing spatial variation in sound wave speed. Ultimately, the variation of sound pressure levels in time and space occurs due to refraction and scattering of sound waves. The effects of meteorological factors are frequency-dependent and more significant at long-distance propagations, e.g. for aircraft noise.

In fact, these factors exist simultaneously in nature and cause complex phenomena together with the absorption and reflections from the ground. Taking into account the combined effects, environmental sound propagation is investigated under three meteorological conditions:

a) *Homogeneous condition*: Isotropic (stable) air in which the sound rays are linear. There is no influence on sound propagation and sound pressure.
b) *Favorable condition*: Sound propagation is affected due to refraction of sound rays (bending downward) and sound levels are increased.
c) *Unfavorable condition*: Sound levels are decreased due to the refraction of sound rays upward and an acoustical shadow zone is formed at a distance from the source. The reduction in sound pressure levels called the "excess attenuation" can be calculated.

The favorable condition enhancing noise levels is the worst case, which is important in noise control and urban planning. The times when the favorable conditions are met should be determined for use in environmental noise assessments both through the prediction models and the acoustical measurements (Chapters 4 and 5). The situations for favorable conditions are simply described as: during sound propagation over a wide and flat ground with low plant coverage, open field without buildings, propagation at maximum height of 500 m above ground, etc.

Sound propagation is simulated by ray tracing methods under steady meteorological conditions with constant profiles of temperature and wind between the source and the receiver. The effects of meteorological factors on sound propagation have been well explained in the literature based on the experiments and theoretical approaches since 1958, e.g. by Rudnick, and Ingard and continued by Piercy, Embelton, Attenborough, and others [13a, 14–20].

The Temperature Effect

Dry air allows penetration of most of the electromagnetic energy from the sun. During daytime the solar energy heats the ground (as an example, on a hot day, the temperature can be 60 °C on the asphalt surface, 55 °C on sand, 44 °C on a grass surface) and at night time, the lower part of the atmosphere becomes warmer because of re-radiation of ground, whereas the higher layers of the atmosphere are cooler until sun rise (temperature inversion). The difference between the temperatures in relation to altitude is termed the "vertical temperature gradient." Decrease of temperature with height is the negative gradient (night hours) (Figure 2.12b) and increase of temperature with height is the positive gradient (day hours). In a calm and ideal atmospheric condition, there is an air flow from high temperature parts to low temperature parts and the atmospheric temperature tends to keep a steady temperature. (Air pressure and density decrease with height but the temperature remains constant.) Most of the time, the atmosphere is unsettled and remains in a turbulent condition.

Variation of temperature with height is calculated depending on gravitational acceleration, the gas constant (gram) ratio of specific heat and height from ground [3]. Table 2.4 gives the simplified vertical temperature gradients according to the atmospheric conditions.

Since the sound speed increases with the increasing temperature (Eq 1.3), when the temperature gradient exists, sound rays bend upward or downward from warmer to cooler air layers in accordance with the refraction phenomenon mentioned in Section 1.4.5. Figures 2.12a–2.12c display the temperature/sound speed relationships with height above ground and the refraction of the sound rays.

At a positive gradient when the temperature is increasing with height, the sound waves bend to the ground (downward), increasing the sound pressure levels, and no shadow zone occurs (Figure 2.12b). In this situation, sounds can be heard at long distances. The refracted sound paths in long-distance propagation are also influenced by the ground reflections, as shown in Figure 2.13a [5].

At a negative gradient when the temperature is decreasing with height (temperature lapse), the sound rays bend upward, a shadow zone appears, encircling the sound source at a certain distance from the source near the ground. The perimeter of the shadow zone is not sharp and the attenuation of sound is higher near the center (at long distances from the shadow boundary), while it is sharpened at high frequencies. However, the atmospheric turbulence affects the formation of shadow

Table 2.4 Variation of temperature with height (vertical temperature gradient).

Atmospheric condition	Vertical gradient (decrease of temperature °C per 100 m height above ground)
Average	0.65
Most unstable	3.42
Neutral (adiabatic)	0.98–1
Isothermal (homogeneous)	0

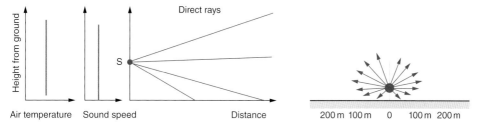

Figure 2.12a   Sound ray propagation in homogeneous conditions: stable atmosphere with no temperature inversion.
Note: Temperature gradient = 0.

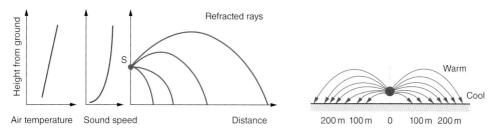

Figure 2.12b   Sound ray propagation under favorable meteorological condition where the lower zones are cooler.
Note: Positive temperature gradient.

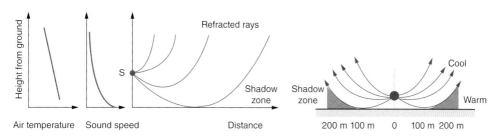

Figure 2.12c   Sound ray propagation under non-favorable meteorological condition where the lower zones are warmer.
Note: Negative temperature gradient.

and should be taken into consideration (Section 2.3.3). There are models to determine the shadow zones, in terms of:

- distance to shadow boundary;
- radius of the first refracted sound path (circular);
- sound pressure level at a receiver point in the shadow zone;
- excess attenuation (difference between the noise levels with and without temperature gradient).

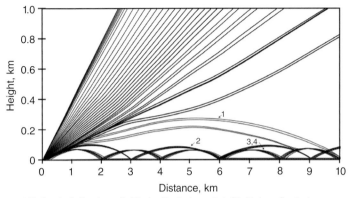

I:Refracted direct ray 2: First reflected ray 3,4: Multiple reflected rays

Figure 2.13a Simulated refraction of sound rays at long distances from the source due to effects of temperature and wind gradients including reflections from ground. (*Source:* adapted from [5], permission granted by Fujiwara.)

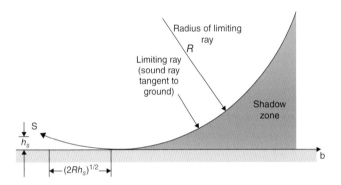

Figure 2.13b Determination of the limiting ray during upward refraction in the case where the sound speed profile is a linear function of height and the acoustic shadow zone. (*Source:* adapted and simplified from [19].)

A graphical method was developed by Rudnick in 1957 to plot sound rays [14]. In the later period, the distance to the boundary of the shadow zone could be calculated as a function of the source and receiver heights and the temperature gradient [20]. The radius of the first refracted wave, which is tangential to the ground (R), is calculated using the sound speed profile, the temperature gradient (temperature profile), and the angle of ray propagation, θ with respect to x direction [14, 19, 20] (Figure 2.13b). As an example, the distance from the source to the perimeter of the shadow was obtained as 1200 m for the positive temperature gradient of 1 °C/100 m and for the curvature with a radius of 56 m, and the shadow zone decreased beyond this distance.

The sound pressure levels within the shadow zones and excess attenuations can be calculated by employing the methods given in the literature [3, 19]. Investigations reveal that significant excess attenuation, i.e. about 20 dB and more, can be obtained at long distances. However, for accurate predictions, it is necessary to provide accurate data based on the meteorological models and after validation by field tests.

The Wind Effect
Generally, the wind speed increases logarithmically with height above ground up to 30–100 m. The increase depends on the topography, physical obstructions, and the height of buildings in urban environments. Such variation is defined as the "profile of wind speed" or simply the "wind

gradient." According to Fermat's principle, "sound rays proceed along the path on which they could move faster rather than on the shortest path" [14]. When the wind blows, the speed of sound is determined by additions and subtractions by considering these as vectors:

$$c = c_0 \pm v_{wind} \tag{2.11}$$

c: Speed of sound in windy air
c_0: Speed of sound in air without wind
v_{wind}: Speed of wind
$\pm$: According to the direction of wind. (Positive sign implies the wind in the same direction as sound propagation and vice versa.)

Two regions are defined according to the wind direction and the direction of the sound ray from source to receiver: The area remaining in front of the wind is called "downwind" and on the opposite side is "upwind" (Figure 2.14a). If the direction of the sound rays and the direction of wind are close to each other within a certain angle, the sound rays bend downward in the downwind region, also called the downward refraction. The direction of the refracted ray can be determined when they are represented by vectors.

In the upwind region, the refraction of the sound waves becomes upwards (upward refraction), creating a shadow zone where the direct sound cannot enter, at a distance from the source depending on the gradient. Since the refraction is associated with the wavelength of sound and less emphasized for the sounds with a longer wavelength, some amount of energy can penetrate into the shadow zone also because of the turbulence scattering the sound energy. Therefore, the shadow perimeter is not as sharp as in light shadow. The shadow zone is an asymmetric form about the wind direction.

The size and location of the shadow zone can be determined by the angle ϕ between the vectors of the direct sound and wind. The upwind and downwind regions are defined by the critical angle, ϕ_c, shown in Figure 2.14b ($2\phi_c$ and $360-2\phi_c$). The limiting angle ϕ_m for the downwind region practically varies within $60-180^0$ and can be determined as given in the literature [15].

The sound level difference between the situations with and without wind at a receiver within the shadow zone, is called the "wind effect" as one of excess attenuation due to the wind gradient. Field investigations have revealed that the excess attenuation could be changed abruptly as high as 10 dB in the shadow zone. Based on investigations in 1958, it was revealed that for the sound pressure level on a sunny day with moderately windy conditions, the excess attenuation can be 20–50 dB higher in the shadow zone at a distance $2x_0$ from the source (x_0 is about 100 m), compared to the upwind region at the same distance. At very low frequencies, the excess attenuation is 0–3 dB in all the zones at a distance of 800 m from the source [15].

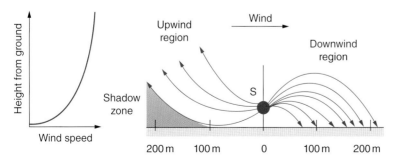

Figure 2.14a Sound refraction due to effect of wind gradient.

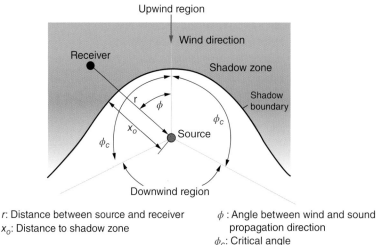

r: Distance between source and receiver

x_o: Distance to shadow zone

ϕ : Angle between wind and sound propagation direction

ϕ_c: Critical angle

$2\phi_c$: Upwind region

$360-2\phi_c$: Downwind region

Figure 2.14b Geometry of shadow zone caused by wind effect. (*Source:* adapted from [15]).

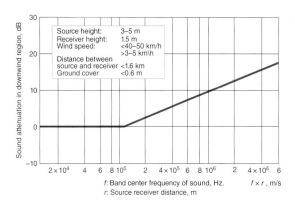

Figure 2.14c Design chart for estimating the excess attenuation for non-shadow (downwind) region on open level terrain, (for frequencies between 300–5000 Hz and distances of 2–4 km) (*Source:* adapted from [15].)

In the prediction models, the excess attenuations in the shadow zone are calculated with respect to the frequency ranges, as: high frequency ($f > 500$ Hz), low frequency ($f < 200$ Hz), and mid- frequency ($200 < f < 500$ Hz) (Chapter 4).

The formation of a shadow zone is important in environmental noise problems. Variation of shadow in relation to distance and the shadow angle can be statistically analyzed to acquire more information in a specific location. Figures 2.14c and 2.14d give the earlier design diagrams in the 1970s to determine the excess attenuations in upwind and downwind regions for level terrains and in the frequency range of 300–5000 Hz. [15]. Field studies near airports have indicated that aircraft noise can propagate into the area with a diameter of 5 miles, without excess attenuation. As wind is intermittent, the shadow zones are also subject to change irregularly in time.

Combined Effect of Temperature and Wind Gradients

Both the wind and temperature gradients can exist at the same time and the combined effect on sound propagation should be considered due to the interaction of these factors. During

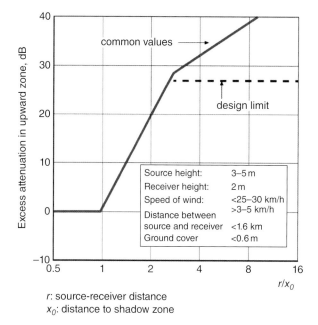

r: source-receiver distance
x_0: distance to shadow zone

Figure 2.14d Design chart for estimating the excess attenuation for upwind region on open level terrain, for frequencies between 300–5000Hz. (*Source:* adapted from [15].)

temperature inversions, the wind speed increases more rapidly with height, and during temperature lapse, the wind speed profile is more gradual under the combined effect, the shadow zone is enhanced in the upwind region and the excess attenuation can be significant at distances up to 50 m from the source on flat ground in certain conditions. In the downwind region, the wind and temperature gradients cancel each other out within the critical angle to be determined and the shadow zone disappears.

By taking into account both the temperature and wind speed gradients, a procedure has been given in the recent ISO standard for the approximation of radius R_{cur} of the curved path based on the sound speed profile [29]:

$$c(z) = c_0 + Az + B\log\left(\frac{z}{z_0}\right) \tag{2.12}$$

$c(z)$: Sound speed profile above ground m/s
c: Speed of sound in air, m/s
z: Height above ground, m
z_0: Roughness length of the ground surface
A: Linear sound speed coefficient, in 1/s (given in the standard for day and night separately)
B: Logarithmic sound speed coefficient, in m/s (given in the standard for day and night)
c_0: Reference sound speed = 331.4 m/s

For a flat terrain, the radius R_{cur} is approximated as [20]:

$$\frac{1}{R_{cur}} = \frac{1}{R_A} + \frac{1}{R_B} \tag{2.13}$$

$$R_A = \frac{A}{|A|}\sqrt{\left(\frac{c_0}{|A|}\right)^2 + \left(\frac{D}{2}\right)^2} \tag{2.14}$$

$$R_B = \frac{B}{|B|}\frac{1}{8}\sqrt{\frac{2\pi c_0}{|B|}}D \tag{2.15}$$

D: Horizontal distance between the source and receiver, m

R_{cur}: Radius of the sound paths caused by atmospheric refraction (combined effect) for nearly horizontal propagation

The distance to the shadow zone for the combined effect, can be calculated as [20]:

$$x = \left[\frac{2c_0}{\frac{du}{dz}cos\beta - \frac{dc}{dz}}\right]^{1/2}\left(\sqrt{h_s} + \sqrt{h_r}\right) \tag{2.16}$$

$\frac{dc}{dz}$: Sound speed gradient

$\frac{du}{dz}$: Wind speed gradient

β: The angle between the direction of the wind and the line connecting source and receiver

Since the meteorological conditions are continuously changing, there are significant differences on the received sound levels in time. Hence, assessment of excess attenuation under different conditions is inevitable in environmental noise predictions (Chapter 4). The combined effects of meteorological factors have been graded into six categories and a series of diagrams for wind profiles at low altitudes, also depending on topography and surface roughness, were prepared by Manning in 1980–1981 [7]. If the meteorological conditions cannot be accurately provided for the calculations, simple approximations are recommended to estimate the effects of meteorological factors on noise levels, as given in Table 2.5.

Table 2.5 Variability of effects of meteorological factors in noise level predictions (both increasing and decreasing effects) [7].

	Distance from source, m			
	100	200	500	1000
Octave bands Hz	Combined effect of wind and temperature gradients, dB			
63	±1	+4, −2	+7, −2	+8, −2
125	±1	+4, −2	+6, −4	+7, −4
250	+3, −1	+5, −3	+6, −5	+7, −6
500	+3, −1	+6, −3	+7, −5	+9, −7
1000	+7, −1	+11, −3	+12, −5	+12, −5
2000	+2, −3	+5, −4	+7, −5	+7, −5
4000	+2, −1	+6, −4	+8, −6	+9, −7
8000	+2, −1	+6, −4	+8, −6	+9, −7

The Turbulence Effect

The atmosphere is an unstable medium because of the variation of the meteorological factors, such as temperature, wind speed, atmospheric pressure, and air density. Variations in wind and temperature gradients during daytimes, which are much enhanced compared to the night time, create turbulent flows of air, i.e. the rotation of air as small vortexes (eddies), in the lower layers over the ground up to 100 m during daytime. While the kinetic energy flows through these air vortexes, the phases of sound waves are influenced and ultimately sound pressures are changed. Turbulence may also occur as a result of local convection when the ground is heated by air and its effect increases with the frequency and strength of vortexes up to a certain limit [19, 20, 28].

Turbulence disturbs the sound paths of direct, reflected, refracted, and diffracted sound waves and causes constructive and destructive interferences resulting in an increase in difference between sound levels for the situations with and without turbulence in air. Fluctuation in overall sound pressure levels in the receiver points can be greater with a standard deviation of about 6 dB from distant sources.

When turbulence is present, the destructive interference between the direct and reflected sound from ground, is eliminated, particularly when the direct and reflected waves are coherent (with no phase difference). When the direct and reflected waves are incoherent (with random phases like noises), the difference in sound levels due to turbulence is taken as 3 dB. Ultimately the atmospheric turbulence decreases the effect of ground absorption.

Turbulence scatters sound energy outwards when it coincides with a group of sound rays (e.g. plane waves) propagating along a direction. The loss of energy due to scattering is small compared to the atmospheric absorption. This effect is negligible for spherical sound waves. The scattered sounds due to turbulence enter the shadow zones, limiting the reduction to 20–25 dB [19]. Estimation of the turbulence effect can be made at short distances based on the wind and temperature parameters and mean propagation height.

Atmospheric turbulence decreases the sound attenuation in the shadow zone of barriers. Also, it causes refractions of the sound rays on the top of a barrier and the rays tend to refract upwards (see Chapter 4).

Other Meteorological Factors

Rain, fog, smoke, and snow have a negligible effect of attenuation, about 0.5 dB per 1000 m. Snow lying on the ground acts as an absorptive layer and minimizes the reflections from ground. The experiments revealed that sound attenuation is about 20 dB greater at low frequencies: 63–125 Hz over snow [20]. In foggy weather, there is no wind and the temperature is homogeneous without the presence of a shadow zone. The dust in the air causes negligible amount of sound attenuation. It was found that the sound levels might decrease about 10 dB due to electromagnetic radiation in the molecules of the humid air [28].

2.3.4 The Ground Effect

The basic principles of the ground effect are explained by the reflection, absorption, and interference phenomena, which were described in Section 1.4.

A number of investigations have been published on the variation of sound levels when sound waves are propagating near the ground or from ground to ground [14–20, 29–34]. Generally, the characteristics of ground are simply categorized as hard (reflective) and soft (absorptive), by specifying the physical parameters, such as ground impedance, reflection coefficient, angle of sound incidence, etc. The intrinsic parameters of soft ground (e.g. a porous surface like grass, etc.) are absorption coefficient, porosity, tortuosity, flow resistivity, and thickness of ground cover.

Ground Reflections

Sound energy is increased over hard ground because of the reflections (Figure 2.15a). When the ground is flat, smooth, and massive, without porous cover (concrete, asphalt, compressed earth, or a water surface), it could be considered a good reflector depending on the difference between direct and reflected paths compared with the wavelength of sound. In the 1970s, some corrections were applied to the direct sound pressure level considering the effect of a reflected path [16]. For instance, the increase in sound pressure levels at almost a great range of frequencies was simply assumed as +6 dB when r_2-$r_1 \ll$ *All* λ and + 3–0 dB when r_2-$r_1 \gg$ *All* λ.

If the source and receiver are not high above the ground, the interaction between the direct and reflected sounds causes a destructive interference because of phase differences (Figure 2.15b) [34]. Ultimately the acoustic pressure decreases abruptly near zero at certain distances from the source. The modes shown in Figure 2.15b imply sudden changes in sound pressure levels for a pure tone sound throughout the frequency range.

The characteristic impedance of the ground is of importance in determining the effect of the ground, whether it is reflective or absorptive. The speed of sound on the ground is much smaller than in the air due to viscous friction. The refraction index of sound from the air to the ground is greater than 1 (locally reacting surface) which means the air-ground interaction is independent of the angle of incidence of the sound waves. Thus, the characteristics of such surfaces can be defined by the relative normal incidence surface impedance, which is zero for perfectly soft ground and infinite for perfectly hard ground [20].

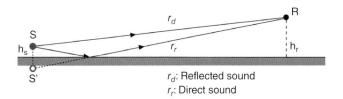

Figure 2.15a   Direct and reflected sounds on the reflective ground.

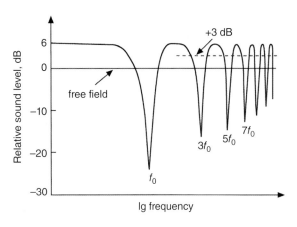

Figure 2.15b   Interference due to reflective ground [34] f_0: Frequency at which difference between the direct and reflected rays is $\lambda/2$. (*Source:* permission granted.)

For normal sound incidence (perpendicular to the ground), the specific impedance of the ground is the same as the characteristic impedance of the ground in a semi-infinite medium, Z_c (which is a complex quantity) given in Eq (2.17) [12, 20]:

$$Z_c = \rho_0 c_0 \left[1 + i \left(\frac{\sigma}{2\pi f \rho_0} \right) \right]^{1/2} \text{N.s/m}^3 \tag{2.17}$$

σ: Flow resistance of the porous layer on the ground, N.s/m^4
$\rho_0 c_0$: Characteristic impedance of air

As shown, the characteristic impedance of the ground is the complex ratio of the acoustic pressure at the surface and the resulting normal component of the particle velocity into the ground [19].

Ground Absorption

When the ground is soft or covered by a porous layer like grass, the sound energy received at a constant point is decreased. There are a number of theoretical models to predict the sound attenuation over a porous surface. Among them, Delany and Bazley in 1970 suggested that Eq (2.17) was also applicable for sound propagations over porous ground and they computed the attenuation on soft ground (grass) based on the impedance model [35–39]. They also developed design charts to predict the characteristic impedance and the absorption coefficient of various materials by using the parameter $f\rho_0/\sigma$. Their results were found to be well correlated with the field measurements [7, 15, 19]. Based on Delany and Bazley's study, the normalized value of the characteristic impedance, which is defined as the surface impedance, is approximated by the empirical expression [35]:

$$\frac{Z_c}{\rho_0 c_0} = \left[1 + 0.0511 \left(\frac{f}{\sigma} \right)^{-0.75} \right] + i \left[0.0768 \left(\frac{f}{\sigma} \right)^{-0.73} \right] \tag{2.18}$$

Flow resistivity of various types of ground have been measured and published. Some examples derived from measurements are given below:

- Concrete surface: 200 000 kN.s/m^4 (or kPa.s/m^2)
- Asphalt: 5000–15 000 kN.s/m^4
- Grass: 125–300 kN.s/m^4
- Sand: 40–906 kN.s/m^4

Calculation of the Ground Effect

Sound attenuation due to the ground effect was widely investigated in the 1970s and the simple estimations were given by tables or charts in the design guides (Table 2.6). The more sophisticated prediction models, as explained in Chapter 4, propose a computation procedure for ground attenuation and obtained results compatible with those obtained by measurements in the field. The factors playing a role in ground attenuation are sound frequency, source-receiver distance, source-receiver heights, acoustic impedance of the ground, characteristic impedance of the air, the complex reflection coefficient of ground, and flow resistivity.

The total sound pressure level in a receiver point is obtained by adding the sound pressures of the direct and reflected waves by taking into account the phase differences [16]. The relationships given in Eqs (2.19) and (2.20), do not cover the surface wave, which also exists as grazing (Figure 2.15a).

Table 2.6 Ground attenuation, A_g in relation to height from ground.

Receiver height H, m	Attenuation A_g, dB
6.0	1
4.5	2
3.0	3
1.5	4
0.7	5

$$p(r_1) = P_d \, e^{-kr_1}/r_1 \qquad (2.19)$$

$$p(r_2) = P_r \, e^{-kr_2}/r_2 \qquad (2.20)$$

$p(r_1)$, $p(r_2)$: Sound pressures of direct and reflected waves

P_d and P_r: Sound pressure amplitudes of direct and ground reflected waves

r_1, r_2: Path lengths of direct and reflected waves respectively

$$k : \text{Wave number} = \frac{2\pi f}{c} \qquad (2.21)$$

Phase difference between the direct and reflected waves $= e^{-i2\pi ft}$

Using the above expressions, the excess attenuation due to, A_g can be obtained as follows:

$$A_g = 10 \lg \left[1 + \frac{r_1^2}{r_2^2}|Q_s|^2 + 2\frac{r_1}{r_2}|Q_s|C_c \right] \text{dB} \qquad (2.22)$$

$$Q_s = \overline{[p\,(r_2)/p(r_1)]} = |Q_s|e^{i\theta} \qquad (2.23)$$

θ: Phase angle

Q_s: Magnitude of reflection factor of the spherical wave (absolute value). $|Q_s|$ denotes time average.

C_c: Time-averaged correlation coefficient between direct and reflected waves given in reference [15].

For certain conditions Q_s is determined as:

$$Q_s = R_p + (1 - R_p)F_w \qquad (2.24)$$

R_p: Plane wave reflection factor of the ground (Chapter 1) which is dependent on surface impedance to be calculated by Eq (2.17) and incidence angle. (The second term is a boundary correction term). [19]

Figure 2.16 shows the frequency-dependent sound attenuation on grass cover versus frequency at various distances from the noise source, for the receiver height of 1.2 m and the source height of 0.30 m (e.g. the height of a car exhaust) [19]. Different diagrams were also published over the years. Sound attenuation is present throughout the frequency range (except at low frequencies) on soft ground and for grazing incidence of sound and is more emphasized at mid-frequencies and increases with the distance from the source.

Reduction in the A-weighted sound pressure level on soft ground is calculated by using the relationships in Eqs. (2.25)–(2.27) for grass-covered ground, based on a Canadian experiment [13a]:

$$A_{grass} = (10G) \lg \frac{r}{15} \geq 0 \qquad \text{dB} \qquad (2.25)$$

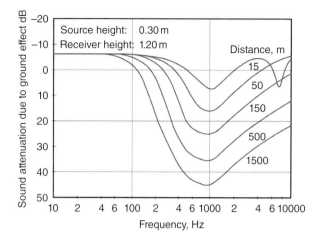

Figure 2.16 Sound attenuation on grass cover as a function of frequency and distance from the source (predicted by using the Delany and Bazley model for the source height=0.3m, 1970) [19, 35].
Note: 0 dB corresponds to propagation in free field.

A_{grass}: Attenuation in grass relative to acoustically hard ground, dB(A)
G: Sound absorption coefficient of the ground (0–1)

$$\text{For} : 0 \le G = 0.75 \left(1 - \frac{h_{eff}}{12.5} \right) \le 0.66 \tag{2.26}$$

$$h_{eff} = \frac{1}{2}(h_s + h_r) \tag{2.27}$$

h_{eff}: Effective height of the path between source and receiver, m
h_s and h_r: Heights of source and receiver respectively, m
r: Distance from source to receiver, m

According to Eq (2.27), the ground attenuation becomes zero within 15 m from the source and with h_{eff} not above 12.5 m.

Generally, attenuation in sound levels due to ground absorption is given as dB per unit distance (30 m or 100 m). Some investigations assume that the attenuation is negligible within 30–70 m from the source and is significant at 70–700 m. For instance, at 250 m from the source and at the receiver height of 2.5 m, the attenuation is 5–10 dB/100 m at 100–6300 Hz. At mid-frequencies (300–600 Hz), it can be as high as 50 dB. Beranek stated that the attenuations on the soft ground should be expected to be about 20–30 dB as a function of the frequency and, on the other hand, the sound level can increase about 6 dB on the hard ground [13a]. It has been found that the difference between the sound levels obtained on the soft and hard surfaces was about 1.5 dB per doubling of distance for transportation noises [43].

Figure 2.17 gives the excess attenuation over thick grass and through shrubbery based on the earlier experiments conducted in 1946 and 1959 [9]. The analytical approximation for the relationship shown in the chart is given in Eq (2.28):

$$A_e = (0.18 \lg f - 0.31)r \ \text{dB}/30 \ \text{m}. \tag{2.28}$$

A_e: Excess attenuation of sound in soft ground and bushland
r: Length of ray path passing along grass or bushy ground, m

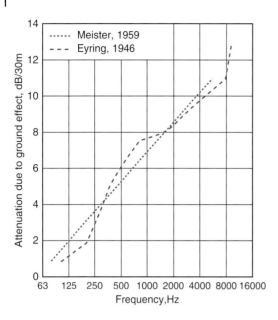

Figure 2.17 Sound attenuation due to ground absorption over shrubbery and thick grass (based on studies in 1946 and 1959). (*Source:* adapted from [15].)

2.3.5 Effects of Forests and Vegetation

Noise levels propagating in a forest area decrease due to the following [15, 19, 36, 40–46]:

1) Sound diffraction over the forest.
2) Scattering by trunks and branches.
3) Sound absorption by the forest layer and foliage.
4) Sound transmission losses while traveling through forest.

Experimental studies revealed that the following factors play a role in the decrease of sound energy:

1) Types of trees (leaf size, evergreen, etc.).
2) Trunk thickness.
3) Density of trees.
4) Width or length of forest area.
5) Distances from forest area to source and to receiver.
6) Frequency of sound.

Various experiments were conducted in the 1970s to provide data about sound attenuation by forests. An empirical relationship has also been developed based on the measurements [15]:

$$A_{forest} = 0.01 \; (f)^{1/3} r \tag{2.29}$$

r: Path length of sound passing through the forest, m

The above relationship does not cover the effect of the forest boundaries, which is defined as the edge effect, i.e. the contribution of the sound ray diffracted at the edge. The field measurements conducted in a forest with four types of trees (cedar, pine, spruce, and other deciduous trees) revealed that sound levels decrease significantly at lower frequencies below 400 Hz including the edge effect. Excess attenuation, A_{forest} decreases at mid-frequencies (500–1500 Hz) and the edge effect is negligible above 1500 Hz. Figure 2.18a (also including the approximation given in Eq (2.28)) and Figure 2.18b display the effect of various trees [9, 15]. However, different measurements conducted in Canada, Russia, and the USA gave different results for the same type of trees. A tree belt with 50 m width can decrease A-weighted sound pressure levels by 10 dB [46]. Attenuation at higher frequencies (>250 Hz) are 4 dB greater than those at lower frequencies (<125 Hz). Excess attenuation provided by the forest floor, which is a thick porous layer, is revealed to be much greater than grass cover at a low frequency [40].

Figure 2.18a Measured excess attenuation provided by forests (based on experiments in the 1960s). (*Source:* adapted from [15].)

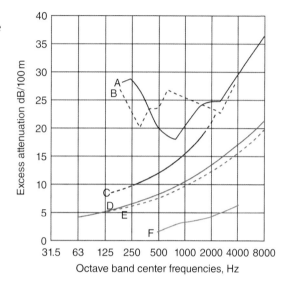

A: Canadian pine, cedar, spruce forests (with edge effect)
B. Same as (A) without edge effect
C: Russian dense pine forest
D: Average forests in the USA
E: Analytical approximation
F: Bare trees

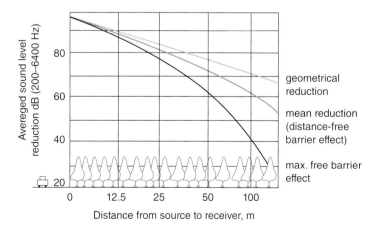

Figure 2.18b Attenuation of exhaust noise in a woodland with trees of 5 m (200–6400 Hz). (*Source:* adapted from [9].)

Besides, experiments on the forest effect, a study on modeling sound propagation in forests was made by adding the attenuations caused by trunks of various sizes and the forest floor [41]. For the purpose of environmental noise predictions, simple approximations regarding the effect of forests on the A-weighted sound pressure levels, were proposed in terms of level reductions per meter, such as given below [38]:

Reduction dB(A)/m

0.1–0.15 for pine trees
0.08–0.1 for trees with wide leaves
0.15–0.17 for bushes

2.3.6 Effect of Barriers

Initial investigations on barriers to be used mainly against traffic noise were based on the diffraction theories explained in Section 1.4.6 and the modeling of the sound field behind obstacles. The first studies involved the thin infinite-length barriers standing on a plane, later the effects of other factors, such as thickness, finite-length, surface absorption, etc., were investigated on the barrier attenuation, leading to the determination of the actual barrier performance. Since the mathematical models of a sound field behind a barrier are rather complicated, the models were simplified for the purpose of environmental noise predictions (as will be seen in Chapter 4) and empirical prediction models and simple diagrams were developed for practical use. After it was realized that barriers can effectively be used in noise control, especially against traffic noise, various experiments were conducted. Later the barrier performance was widely investigated by taking into account the meteorological factors theoretically and experimentally. Various scale model studies were also conducted in the laboratory to discover the effects of various geometrical parameters. Nowadays, detailed studies regarding the excess attenuation of barriers are conducted using computer modeling and complex calculations have been facilitated by means of computer simulations also enabling the optimization of barrier design.

The mechanism of barrier diffraction should be understood as the acoustic shadow forming behind barriers and their characteristics regarding sound attenuation are related to the effective height of the barrier and the wavelength of the incident sound (Section 1.4.6). Unlike the shadow caused by light diffraction, the acoustic shadow zone does not have a distinct boundary due to the transition zone, which can be analytically determined. The models developed for excess attenuation of barriers from the past to date are outlined below. This subject has been included in a great number of research studies published so far, with some given in [47–95].

Infinite Thin Screens
Determination of the diffracted field and the attenuation in sound pressure levels were investigated for thin rigid screens with infinite length and at certain heights. A screen standing on ground is accepted as semi-infinite due to its height and therefore sound is diffracted only over its top edge. The effect of thickness is assumed to be rather small compared to the wavelength of sound. Among the pioneering researchers dealing with thin rigid screens, are Redfearn (1940) [47], Scholes and Sargent (1971) [65], and Pierce (1974) [56]. Keller's (1957) geometric theory of diffraction for barriers and Kirschoff's approximations have been of great benefit in handling barrier problems, i.e. calculating the sound pressure amplitude over a non-reflective ground. As an example, Redfearn's diagram to determine sound attenuation for a point source, by using the geometrical parameters and wavelength of sound, is given in Figure 2.19.

Figure 2.19 Redfearn's model (1940) for point source. (*Source:* adapted from [47], permissions granted by Fujiwara and Maekawa.)

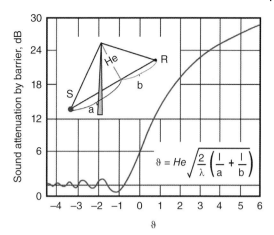

$$\vartheta = He\sqrt{\frac{2}{\lambda}\left(\frac{1}{a}+\frac{1}{b}\right)}$$

Barrier attenuation for point sources: Maekawa (1968) conducted experimental studies and developed a diagram which was used as basic barrier attenuation model for thin-infinite length screens (Figure 2.20a, 2.20b) [47, 48, 53, 62]. The model was found to be compatible with the theoretical results and was validated by the field measurements.

The calculation model based on Kurze and Anderson's research for thin-infinite barriers is still used at present [50]. The parameters involved in the model are: the path difference between the diffracted and direct waves (δ) and the Fresnel number (N) in relation to frequency of sound. The geometrical parameters are shown in Figure 2.20a.

$$\text{Fresnel number}: N = \pm\frac{2}{\lambda}(A + B - d) \text{ (uniteless)} \tag{2.30}$$

A and B: The total shortest path length of the diffracted wave, m
d: The shortest path length of the direct sound traveling from source to receiver, m
$A + B - d =$ Path difference, δ.
$(+)$ sign indicates the receiver in the shadow zone, $(-)$ sign indicates the receiver in the bright zone. Using N, the excess attenuation provided by the barrier can be found in the diagram given in Figure 2.20b.

Kurze and Anderson's basic equation for sound attenuation based on the Fresnel number in the barrier shadow zone and for point sources, which gives a good result with Maekawa's

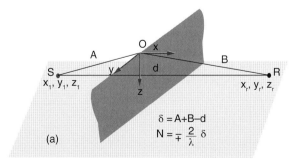

$$\delta = A + B - d$$
$$N = \mp\frac{2}{\lambda}\delta$$

(a)

Figure 2.20a Geometrical parameters for barrier attenuation.
Note: (δ: Path difference, N: Fresnel number).

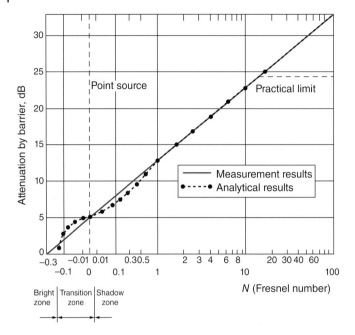

Figure 2.20b Maekawa's model for thin infinite barrier and point source [53, 62]. (*Source:* Permissions granted by Fujiwara and Maekawa.)

findings, has been improved for the three zones behind the barrier. The excess attenuations of pure tone sounds, by a thin barrier of infinite size, $A_{barrier,}$ are computed by the following equations [13a]:

- At transition zone in front of the shadow region ($N \geq -0.2$)

$$A_{barrier} = 20 \lg \frac{\sqrt{2\pi|N|}}{\tan \sqrt{2\pi|N|}} \quad (0 < A_{barrier} < 5\text{dB}) \tag{2.31}$$

- Within the shadow zone:

$$A_{barrier} = 20 C_1 \lg \frac{\sqrt{2\pi N}}{\tanh \left(C_2 \sqrt{2\pi N}\right)} + 5 \quad (5 < A_{barrier} \leq 24 \text{ dB}) \tag{2.32}$$

C_1 and C_2: Constants for point and line sources respectively defined in [13a] (Figure 2.22)
- In the bright zone: ($N < -0.2$)

$$A_{barrier} = 0 \text{ dB}$$

Path-length differences of (0) or (−) imply that the receiver point remains above the line of sight between the source and the receiver. The diffracted wave in the bright zone can be ignored and excess attenuation is assumed to be 0. The practical limit of excess attenuation is 20–24 dB due to other environmental effects. It should be noted that this limit is about 10–12 dB in terms of A- weighted levels.

In the transition zone meeting the condition of $h_r + |x_r/x_s|h_s < 0$, the edge of the barrier approaches the line of sight between the source and the receiver and the excess attenuation is 0–5 dB. Figure 2.22 also provides a comparison between the attenuations of infinite-size barriers for the point and line sources.

In the shadow zone, the excess attenuation starts from 5 dB, which is obtained at low frequencies or at the points very near to the line of sight. The limiting value is influenced by the atmospheric disorderliness and turbulence. Experiments revealed that this value is 24 dB for point sources. If the source and receiver are near the barrier, the shortest path length $(A + B)$ is longer than the direct path length. In this situation, the spherical radiation of the diffracted wave (i.e. the effect of the wave divergence on the diffracted wave) is taken into account by the term of $20\log (A + B)/d$.

If the barrier is positioned not perpendicular to the source-receiver line, the shortest length of the diffracted ray $(A' + B')$ is taken from the geometric projection on the plane perpendicular to the barrier and the diffracted wave is calculated as:

$$A + B = \sqrt{(A' + B')^2 + (y_s - y_r)^2} \tag{2.33}$$

If either or both the source and receiver are not close to the barrier with the condition of $A + B - d \ll d$, it is sufficient to obtain the maximum Fresnel number as:

$$N_{max} = \pm \frac{2}{\lambda}(A' + B' - d') \tag{2.34}$$

Barrier attenuation for line sources: A barrier positioned parallel to a finite-line source gives an excess attenuation that is determined through the angle encompassing both edges of the source. For an infinite- line source, $\Delta\alpha = \pi$. Attenuation of a line source is less than 5 dB than that of a point source due to the cylindrical wavefronts (Figures 2.21 and 2.22). The maximum excess attenuation is obtained on the line perpendicular to the source line from the receiver.

In order to determine the excess attenuation, the line source is divided into segments (finite-length sources) by applying the rule that the difference between the attenuations provided by each segment and by the nearest source point to the receiver, is 1 dB [15] (See Section 1.3.3). Then the excess attenuation of the line source, $A_{e,line}$ is calculated by adding the attenuations of each point (at the center of the segments) by considering the ($-$) sign of the attenuation values:

$$A_{e,line} = -10\lg \sum_i \frac{1}{\Delta\alpha_i} 10^{-\left[A_{i,point}/10\right]} \quad dB \tag{2.35}$$

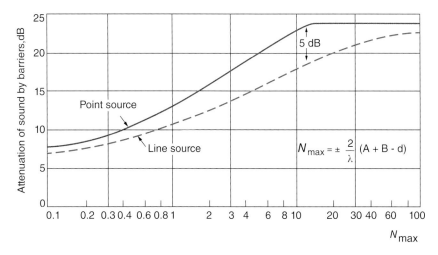

Figure 2.21 Comparison of barrier attenuations for point and line sources based on report VDI 2720, 1987 [13a].

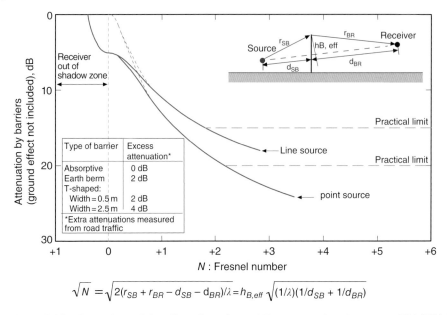

$$\sqrt{N} = \sqrt{2(r_{SB} + r_{BR} - d_{SB} - d_{BR})/\lambda} = h_{B,eff}\sqrt{(1/\lambda)(1/d_{SB} + 1/d_{BR})}$$

Figure 2.22 Comparison of the effects for point and line sources based on report VDI 2720,1987. (*Source:* adapted from [13a].)

$\Delta\alpha_i$: Angle of view of ith segment encompassing the line source from receiver

$\Delta\alpha_1 = \alpha_2 - \alpha_1$: Aspect angle of the nearest point of the line source

$\Delta\alpha_2 = \alpha_3 - \alpha_2$: Aspect angle of the next sector

$A_{i,point}$: Attenuation of barrier for a point source in the direction of α_l

The above angle parameters are given in Table 2.7 according to the number of the sector *i*. The similar methodology can be applied for different source and barrier configurations. An infinite- line source (i.e. roadway) obstructed by a barrier of finite-length, the angle of obstruction and when there is a gap on barrier, the angle of view at the receiver points are shown in Figures 2.23a and 2.23b.

The fundamental studies for barriers have been improved and are used in the prediction models in Chapter 4.

Table 2.7 Aspect angles of a line source in radians [15].

i	α_i (radian)	$\Delta\alpha_i$ (radian)
1	0	1.05
2	1.05	0.20
3	1.25	0.12
4	1.37	0.067
5	1.237	0.067
6	1.504	0.067

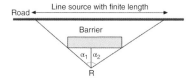

Figure 2.23a Angle of barrier obstruction from receiver to infinite line source.

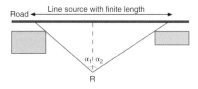

Figure 2.23b Angle of view from receiver to finite line source.

Effect of Thickness

When the thickness of the barrier is not negligible compared to the wavelength of sound, naturally the path of the diffracted sound will be longer, implying an increase in the sound attenuation due to distance effect [53–56]. The thickness of the barrier also has an additional influence on the barrier performance depending on the frequency of the sound, the angles of diffraction and incidence according to the barrier geometry. The parameters are shown in Figure 2.24a. Fujiwara developed a diagram in 1974, to determine the effect of thickness b with respect to the profile angle [55] (Figure 2.24b). The constant K obtained from the chart is applied in Eq (2.36) to find the effect of thickness:

$$\text{Additional reduction} = K \lg(kb) \tag{2.36}$$

K: Constant to be determined from the chart (Figure 2.24b)
k: Wave number
b: Thickness of barrier

The above additional reduction due to thickness of barrier is added arithmetically to the attenuation provided by the infinite-thin barrier.

Effect of Top Edge

Various edge profiles can be designed along the top of barriers by changing the diffraction angle and the respective path difference between the diffracted and direct sound paths [53–57]. The basic shape experimented, was the wedge or triangular shape similar to the earth-berms (Figure 2.25a). The investigations by Fujiwara (1973) revealed a 5 dB additional reduction provided by a wedge or a top edge device with a wedge shape, depending on the angle of sound incidence θ, the diffraction angle ϕ, and the wedge angle, Ω. The attenuation increases by the smaller incidence angle and the larger wedge angle. When the incidence angle is greater than 30^0, the effect is negligible (about 2 dB) independent of the wedge angle. The additional attenuation by the wedge can be obtained through the chart given in Figure 2.25b [55]. A number of innovative geometrical configurations

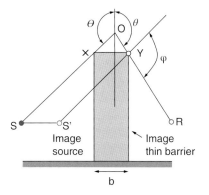

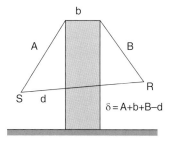

Figure 2.24a Geometric parameters affecting diffraction by thick barrier.

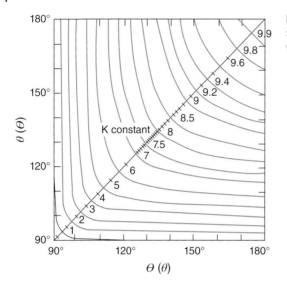

Figure 2.24b Additional effect of thickness on sound attenuation [54, 55]. (*Source:* Permissions granted by Fujiwara and Maekawa.)

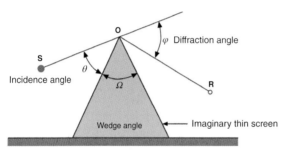

Figure 2.25a Geometrical parameters for diffraction by wedge-shaped barriers.

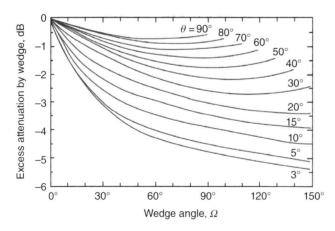

Figure 2.25b Additional effect on sound attenuation by wedge-shaped barrier [53, 55]. (*Source:* Permissions granted by Fujiwara and Maekawa.)

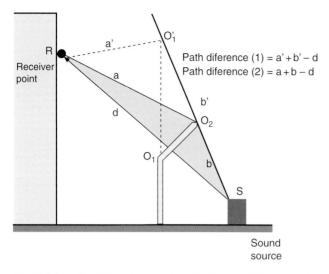

Figure 2.26 Geometrical parameters for barrier with a slanted top.

have been widely tested both by scale model studies and computer modeling by using Boundary Element Method, BEM as given in the references (Appendix 2.A).

Experiments on the effect of various shapes of barrier headings are referred to in Appendix 2.A. Special profiles like zig-zag or cylindrical modules designed along the edge of barrier have been found to have an increase of a few decibels by diffusing the diffracted sound above the barrier [58, 60]. For example, a 2 dB increase in barrier performance was obtained at 500–4 kHz for the mushroom profile. Different wave formations caused by the top edges can be observed in Figure 1.71 in Chapter 1. Generally, the effects of edge profiles can be as high as 4–8 dB, which is important in some noise control problems. By means of a slanted heading, it is possible to reduce the required height of the barrier to obtain shadow zone as can be seen in Figure 2.26.

Effect of Sound Absorption on the Barrier Surface

The absorption or reflection characteristics of the barrier surface, especially on the source side, have an influence on the sound attenuation performance of the barrier. Reflections from the surface are added to the diffracted sounds above the barrier, resulting in an increase of 6 dB in the received sound levels. On the other hand, an absorptive treatment on the barrier surface increases the barrier attenuation particularly for the greater angles of diffraction. The sound attenuation of a barrier which is completely covered by an absorptive treatment is related to the angle of diffraction and the sound absorption coefficient of the material (Section 1.4.3). The additional sound attenuation that can be obtained by a completely absorptive barrier surface was given by Fujiwara and Maekawa in 1970's [55]:

$$\text{Attenuation} = 20 \lg\left(\left[\sin 0.5(\theta + \varphi) - \cos 0.5(\theta - \varphi)\right]/\left[\sin 0.5(\theta + \varphi) + \cos 0.5(\theta - \varphi)\right]\right) \text{ dB} \tag{2.37}$$

θ: Angle of incidence
ϕ: Angle of diffraction

Figure 2.27 can be used to determine the effect of absorption [54, 55]. Obviously, the greater the angle of diffraction and the absorptivity of the treatment imply a greater additional effect as high as 15 dB. However, success in achieving this performance depends on the continuity in material condition with time, i.e. maintenance and protection from humidity and dust when the barrier is constructed in an outdoor environment.

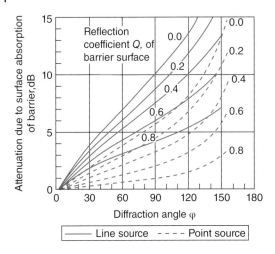

Figure 2.27 Effect of surface absorption of barrier [54, 55]. (*Source:* Permissions granted by Fujiwara and Maekawa.)

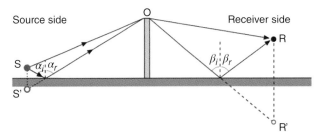

Figure 2.28 Addition of reflected and diffracted sound waves (α_i and α_r: incident and reflection angles on the source side, β_i and β_r: incident and reflection angles on the receiver side).

Effect of Ground Reflection on Barrier Attenuation

As explained in Section 2.3.4, when the barrier does not exist, the total sound level at a receiver point is affected by the constructive interference between the direct sound and the reflected sound from the ground. If a barrier is inserted between the source and receiver, the sound pressure level at the same point is obtained by the superposition of the reflected and diffracted waves on the source and receiver sides of the barrier (Figure 2.28). If the source and the receiver are both on the hard ground, great variations can be expected on the sound pressure levels behind the barrier in relation to the frequency of sound. The effect is greatly emphasized for the pure tone components.

Effect of Ground Absorption on Barrier Attenuation

Sound propagation on absorptive ground reveals a different phenomenon depending on the ground impedance. The destructive interference emerging due to phase differences between the direct, reflected and refracted waves, particularly in the 300–600 Hz range, cannot occur when a barrier is inserted between the source and the receiver. In that case, the sound levels at the receiver can be higher compared to the case without a barrier, in other words, the barrier attenuation on a sound absorptive ground can be smaller than expected (i.e. the insertion loss is a negative value). Figure 2.29 displays an experimental result regarding the interference effect behind a barrier [18, 34, 38]. Particularly great attenuations or increases at certain tones (i.e. pure tone sounds) can be obtained because of the modes.

Figure 2.29 Variation of spectral levels at a receiver point on a reflective ground (from an experimental study) [34, 38]. (*Source:* Permission granted.)

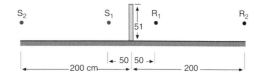

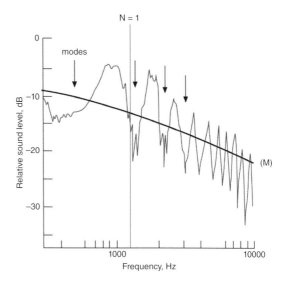

Effect of Wind on Barrier Attenuation

Barrier performance is influenced by the substantial wind because of the following reasons [51, 56]:

1) Sound rays are refracted due to the wind profile and the angle of sound incidence to the barrier change. This effect is independent of the barrier shape but depends on the wind direction (Section 1.4.5).
2) Sound rays are scattered into the shadow zone due to turbulence caused by winds. This effect is independent of the wind direction but dependent on the barrier shape. An experiment revealed that the barrier performance decreased about 2 dB in comparison with the case without wind, when the direction of wind is from the source (e.g. a road) to the barrier and the wind speed is 2–3 m/s. The wind effect on the barrier performance is determined by the prediction models as given in Chapter 4.

Effect of Finite-Length on Barrier Attenuation

The barrier performances explained so far consider the diffraction of sound only over the top of the barrier. However, in practice, since the lengths of barriers are limited, the diffractions from the vertical sides should also be determined according to the source-receiver-barrier configuration given in Figure 2.30 [62]. The total barrier attenuation by the finite-length barriers is computed by adding the attenuations from each sound path, considering their negative signs. Implementation of this approach to the acoustically thick barriers compared to the wavelength of incident sound (e.g. a building on a road with traffic), is shown in Figure 2.31. The computerized model in 1977 enabled the prediction of traffic noise levels behind buildings acting as finite-length barriers and investigation of the effect of various dimensional ratios of buildings. The calculations were based on Kurze's

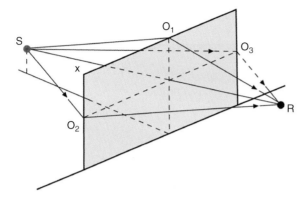

Figure 2.30 Thin and finite-length barriers [62]. (*Source:* Permissions granted by Fujiwara and Maekawa.)

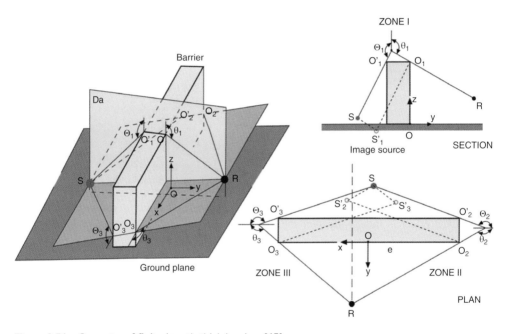

Figure 2.31 Geometry of finite-length thick barriers [63].

approach and the Monte Carlo snapshot technique for the vehicles assumed as moving point sources and distributed randomly on the road with different sound powers [63]. Validation of the model with the laboratory experiment is shown in Figure 2.32 displaying the effect of height of a sample finite-length barrier on sound attenuation related to octave bands. Variation of sound pressure levels behind a barrier while a vehicle passing by, is shown in Figure 2.33 based on another scale-model experiment in the context of the same study in 1982 [64]. At present, such investigations are readily carried out by noise mapping techniques as explained in Chapter 6.

If the finite-size barrier is not parallel to the road, or not perpendicular to the source-receiver line, the geometrical parameters are shown in Figure 2.34. The computations should take different geometries into account by means of a 3D coordinate system with the origin on the ground or on top of the barrier.

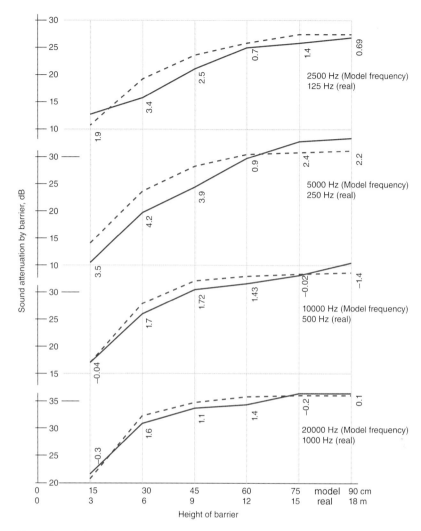

Figure 2.32 Comparison of the calculated and experimented attenuations of a finite-size barrier and variation of attenuations with the height of barrier. (Point source: 14m and microphone position: 12m from the barrier respectively. Thickness of barrier: 3m) [63].

Expression of Barrier Performance

Considering the aspects explained above, the barrier performance can be presented in terms of the following:

1) *Predicted sound reductions*: The difference between the sound pressure levels "with barrier" and "without barrier" calculated by the prediction models or by Maekawa's chart (for pure tone sounds) regardless of the effects of other interrelated physical factors. (To be given as frequency band levels.)

2) *Attenuation (level difference)* $A_{barrier}$: The difference between the sound pressure levels at a receiver point before and after the barrier only due to the diffraction effect. The "after" sound pressure levels are those measured at the site and the "before" sound levels are calculated by conducting measurements at a reference point on the top edge of the barrier and then by applying the distance effect (inverse square law).

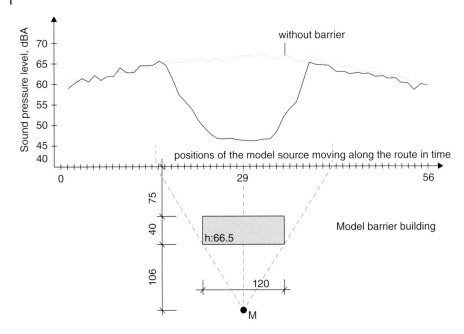

Figure 2.33 Variation of instantaneous sound level from a vehicle passing by behind a building barrier (from a scale model experiment) [64] (Scale: 20:1. Dimensions in cm).

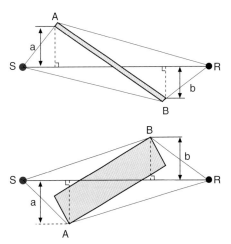

Figure 2.34 Geometric parameters for a barrier slantly positioned between source and receiver.

$$A_{barrier} = L_{before\ barrier}(\text{calculated}) - L_{after\ barrier}(\text{measured}) \tag{2.38}$$

3) *Insertion loss*: The difference between the sound pressure levels at the receiver before and after the barrier constructed at the site. These values are based on the field measurements, thus giving the actual performance values since they are influenced by the wind and ground:

$$IL = L_{without\ barrier} - L_{with\ barrier} \tag{2.39}$$

$$IL = A_{barrier} - \left[A_{ground\ without\ barrier} - A_{ground\ with\ barrier}\right] \tag{2.40}$$

Attenuation and insertion losses are given as frequency band levels and as A-weighted levels, particularly for marketing purposes.

Nowadays the declaration of barrier performance (DoP) is made based on a European Norm 1793 which includes the measurement methods and the design principles [67a,b,c] (Chapter 5).

Appendix 2.A provides some information about design principles of barriers and historical perspective.

2.3.7 Vertical Surfaces

The principles of reflection, diffraction, and absorption, that were referred for the ground, are also valid for the other surfaces which can be vertical or inclined and whose dimensions are greater than the wavelength of sound (e.g. building façades). When they are positioned near the source, there is a certain increase in sound pressure levels that should be expected in comparison with the case of no surface (Figure 2.35). The magnitude of the effect varies with the position of the receiver point with respect to the surface as well as the absorption coefficient of the surface material. As simply stated, if the receiver point is on a vertical reflective surface, the increase in sound pressure levels is 6 dB and if the receiver is 0.5–1 m high above the ground, the increase is +3 dB [65]. If there are many surfaces near the source (Figure 2.36), the complex sound field created by the numbers of reflections with the imaginary sources, can be modeled to determine the sound pressure levels (Section 1.4.1). The parallel surfaces with a narrow distance in between, positioned on opposite sides of a point sound or along a line source, generate multiple reflections, thus increasing the noise levels [69]. This effect is defined as the "canyon effect" and is common in urban environments in which the buildings are closely located (Figure 2.37).

2.3.8 Total Effect of the Physical Environmental Factors

If the basic aim is to obtain sound pressure levels in a physical environment, which is under the influence of various physical factors, a reliable model should be employed to take into account all the individual effects of each factor referred above and the interrelations between them. In this process, the total effect which is called the "total excess attenuation" A_E is simply determined by the summation of the individual effects contributing to the total noise levels:

$$A_E = A_a + A_b + A_c + A_d + ... + A_n \quad \text{dB} \tag{2.41}$$

A_a-A_n: The attenuations provided by the physical factors with a positive or negative sign, such as distance, air absorption, meteorological factors, ground absorption, vegetation, forest and woodland, noise screens (barriers), vertical reflective surfaces, etc. These effects might be

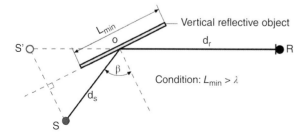

Figure 2.35 Reflection from a vertical surface.

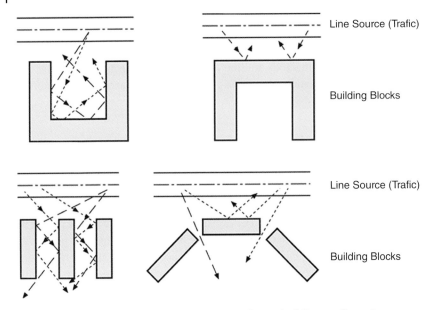

Figure 2.36 Repetitive random reflections according to building configuration.

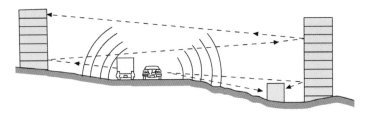

Figure 2.37 Multiple reflections (the canyon effect).

predicted based on theories or derived from the experiments and presented as empirical relationships.

Since the effects are mostly frequency dependent, A_E should be obtained in each octave or one-third octave bands, as well as the A-weighted attenuations. In fact, most of the factors are interrelated to each other and it is difficult to easily discriminate their individual contributions to be used in Eq (2.41). Modeling outdoor sound propagation has been studied by many researchers since the 1970s and the computer models were initiated in the 1980s employing various algorithms, also enabling comparisons between different meteorological conditions under different temperatures and wind gradients. Nowadays, simulation of sound propagation has almost reached a satisfactory quality and accuracy. The prediction models and noise mapping based on these models will be explained in Chapters 4 and 6.

Due to the difficulty in predicting the outdoor noise propagation, which is rather complex issue, some factors can be treated as "design parameters" for practical implementation purposes. An outline of the principles regarding noise propagation in the environment, which are important for noise control in urban areas, is given in Box 2.2.

Box 2.2   Principles of Outdoor Noise Propagation

- Sound pressure levels decrease with the increasing source-receiver distance.
- Sound waves move faster in warmer temperatures.
- Sound pressure levels increase in the early morning rather than in the afternoon.
- Sound pressure levels decrease more with increasing temperature and decrease with the lower humidity at high frequencies.
- Sound rays bend in the direction of the wind.
- Sound pressure levels increase in the downwind region and decrease in the upwind region.
- High frequency sound pressure levels decrease more due to atmospheric (air) absorption.
- Bass sounds are heard louder than other sounds at far distances.
- Barriers yield about 10–15 dB A-weighted sound reduction, depending on their configuration.
- Barrier reduction is smaller when the source and receiver are at the same height and greater when the receiver is lower.
- Sound waves are absorbed more while propagating at lower levels above the absorptive ground (grass, vegetation, etc.) and the sound levels decrease.
- Sound waves are reflected from the hard ground (concrete, asphalt, etc.) and the sound levels increase.
- Sound pressure levels increase near building façades due to reflections.
- Parallel walls increase the sound levels due to repetitive reflections (the canyon effect).

2.4   Temporal Variation of Noise

Sound pressure levels due to the environmental sources generally fluctuate temporarily or continuously in time. Temporal variation is an important characteristic of noise, displaying various patterns, as seen in Figure 2.38 with descriptions of the signal characteristics, some of which are important for machinery noise sources. Such variations with short or long duration are presented by "time-history diagrams" that can be obtained by the techniques explained in Chapter 5. Usually the spectral and temporal analyses are made simultaneously to check the instant data and to obtain more detailed information about the noise (Figures 2.3a and 2.3b).

Noises are divided into two groups according to the temporal variations:

- *Steady (stationary) noises*: The average noise levels and the spectral characteristics are constant for long periods. They can be broadband or narrowband sounds, some have audible tones, such as a chainsaw, a transformer, or a turbojet engine (Figure 2.38a).
- *Non-steady (non-stationary) noises*: Average noise levels and spectral characteristics change with time. They are categorized as follows:

    – Fluctuating noises: Variation of sound pressure levels of about 5–6 dB and equal to background noise at least once within an observation period, such as heavy traffic (Figure 2.38b).
    – Intermittent noises: Sound pressure levels within observation period are equal to background noise twice or more and with two to three seconds, such as aircraft, a train or a single truck passing by (Figure 2.38c).
    – Impact or impulsive noises: Sound pressure levels change abruptly and rise above the background noise levels at each time and define individual (single) impacts which lasts less than a second, such as a hammer, a pistol, a door banging, an electrical shutter (Figure 2.38d). Such sounds can be defined as;

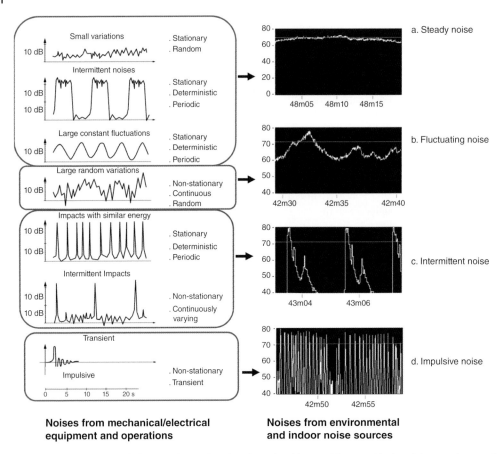

Figure 2.38 Examples of time-varying noise signals emitted by machines and industrial operations and the corresponding environmental noises.

- Isolated bursts which have certain and constant (less varying) wave shape and with less than 1 impulse/second.
- A series of bursts called quasi-steady noises composed of successive bursts with similar amplitudes with intervals of 0.2 second or defined as the noises when the numbers of bursts are more than 10/second. Figure 2.38d shows successive impulsive noise events.

Temporal characteristics of environmental noises have been also described in the ISO standards which are outlined in Chapter 5.

Since sound radiations change with the structural and operational characteristics of noise sources, it is necessary to analyze the long-term temporal variation of noise in accordance with the type of source. Some noise descriptors characterizing specific noise sources have been derived from the statistical analysis averaging the varying levels in the time-history data, the maximum and minimum levels in a standard reference time, the level distribution diagrams, histograms, and cumulative diagrams shown in Figures 2.39a and 2.39b. In addition to the instantaneous levels, sometimes it is necessary to determine hourly, weekly, daily or seasonal variations in noise levels.

The techniques used in temporal analyses and the equipment are explained in Chapter 5. Generally, the noise from a specific source in the environment is usually accompanied by

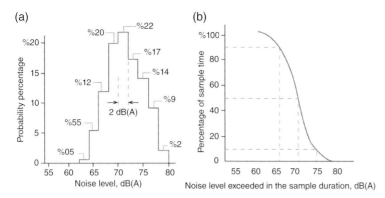

Figure 2.39 Statistical analysis of noise: (a) histogram; (b) cumulative distribution curve.

a lower background noise (or ambient sound) which is composed of the nearby or distant sources. Sometimes background noises are detected clearly due to their high noise levels and intruding characteristics. Similar temporal and spectral analyses have to be made for background noises as the specific noise concerned and the elimination of background noise is made by applying the procedure given in Box.1.4.

2.5 Noise Metrics and Descriptors

Description of noise to be made either for source emissions or environmental assessments, necessitates some quantities, in addition to the physical measures of sound. The quantities should measure the magnitude of noise from sources as single or in various combination, and describe the special characteristics, such as impulsiveness, tonality, or low-frequency content.

Some of the noise metrics used in environmental noise measurements are the basic units in relation to the hearing and perception of sound and have been defined in the ISO. Others are the special noise descriptors that have been developed by taking account of the constructional, operational, and exposure characteristics of specific noise sources and are well-correlated with the results of the noise impact analyses. Table 2.8 gives the most common metrics and descriptors widely used for evaluation of environmental noises.

Table 2.8 Common terminology used for noise indicators.

Title	Definition	Symbol
Unit	A-weighted noise level C-weighted noise level	dB(A) dB(C)
Scale	Instantaneous weighted or unweighted sound pressure level	SPL, L_p, or L_{pA}
Metric	Equivalent A-weighted noise level Sound exposure level	L_{Aeq} L_{AX} or SEL
Index/Indicator/ Descriptor	Day, evening, and night level Night level	L_{den} L_n

2.5.1 Basic Noise Units

Expression of the noise levels (sound pressure level) using the physical unit of a decibel is made in terms of the following scales:

- Frequency-weighted sound pressure level, dB:
  A- and C-weighted levels (IEC 61672–1) [27]

- Time-weighted sound pressure levels (Chapter 5)
  F-weighting: Fast sampling
  S-weighting: Slow sampling

- Maximum time-weighted and frequency-weighted level (L_{max}): Highest time-weighted and frequency-weighted sound pressure level within a stated time interval, dB.
- N% exceedance level: Time-weighted and frequency-weighted sound pressure level that is exceeded for N% of the time interval considered (e.g. $L_{AF95,1h}$ or simply denoted as L_{10}, L_{50} or L_{90})
- Peak sound pressure level (L_{peak}): Maximum instantaneous level in a reference time period by measuring a frequency-weighted or in a bandwidth as calculated in Eq (2.42):

$$L_{peak} = 10 \lg \left(\frac{p_{peak}^2}{p_0^2} \right) \text{ dB} \tag{2.42}$$

p_{peak}: Peak sound pressure which is the maximum absolute value of the instantaneous sound pressure during a stated time interval with a standard frequency weighting or measurement bandwidth.

p_0: Reference sound pressure (= 20 μPa)

The definitions of the basic units recommended in ISO 1996-1 are outlined below [85].

Frequency-Weighted Sound Pressure Level, dB
The sound pressure level is measured by taking (p_{ref}) = 20 micropascal as the reference value, according to IEC 61672–1, and by using the A- or C-weighting network (Section 2.2.4 and Section 7.1.2), The A-weighted level is commonly used to measure the environmental noises of any kind except impact noises requiring other metrics such as C-weighting or using peak sound pressure level. C-weighting can also be preferred for assessing low frequency noises (rumble).

Equivalent Continuous Sound Pressure Level, L_{eq} dB or L_{Aeq} dB
For noise sources whose sound levels are fluctuating in time, the equivalent continuous sound pressure level is also defined as the steady sound pressure level, which, over a given period of time, has the same total energy as the actual fluctuating noise. It is sometimes called the mean energy level (Figure 2.40). In order to determine the equivalent continuous sound pressure level, the instantaneous levels of time-varying noise are measured within short periods like 0.5 seconds, then the logarithmic average of the instantaneous sound pressure levels over the total time period is taken. The average sound pressure is then converted into decibels as given in Eq (2.43):

$$L_{eq,T} = 10 \lg \left\{ (1/T) \int_0^T 10^{Lp(t)/10} dt \right\} \text{ dB} \tag{2.43}$$

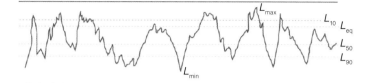

Figure 2.40 Statistical metrics in a fluctuating noise.

$L_p(t)$: Instantaneous sound pressure levels, dB
T: Sampling or measurement period: t_2-t_1 according to ISO 1996-1 [85]
t: Sampling time, second, minute or hour

$$L_{p(t)} = 20 \lg \left[p\,(t)/p_0 \right] \quad \text{dB}$$
$$p_0 : 2 \times 10^{-5} N/m^2 = 20\mu Pa \tag{2.44}$$

$p(t)$: Sound pressure in sampling time, t, Pa
p_0: Reference sound pressure (minimum hearing level), Pa

Equation (2.43) can also be given as below:

$$L_{eq} = 10 \lg \left(\frac{1}{T} \sum_{i=1}^{n} 10^{\frac{Lpi}{10}} \right) \quad \text{dB} \tag{2.45}$$

i: Number of instantaneous sound samples
L_{pi}: Sound pressure levels of each instantaneous noise sample, dB

When the equivalent sound pressure level is expressed as an A-weighted level;

$$L_{\text{Aeq } T} = 10 \lg \left[\frac{1}{T} \int_T p_A^2(t)/p_0^2 \ dt \right] \quad \text{dB} \tag{2.46}$$

T: Reference period of time (observation period), e.g. part of day, full day, or a week (For long-term evaluations, e.g. three months, six months, or a year)
$p_A(t)$: A-weighted instantaneous sound pressure which is continuous in a short time t, Pa
p_0: Reference sound pressure ($=20 \ \mu$Pa)

Sound Exposure Level, L_E dB or L_{AE} dB

Discrete noise events, such as an individual impact of a mechanical noise source or an aircraft or a train passing by, are described by different indexes (Figure 2.41). The most common descriptor is "sound exposure level" L_E, which is defined as the total sound pressure level during a time interval, or during a noise event, T:

$$L_E = 10 \lg \left(\frac{E}{E_0} \right) \quad \text{dB} \tag{2.47}$$

E: Sound exposure which is time integral of the time-varying square of the frequency-weighted instantaneous sound pressure over a stated time interval, T or an event, as given by:

$$E = \int_T p^2(t) dt \quad \text{Pa}^2\text{s} \tag{2.48}$$

E_0: Reference sound exposure which is the square of the reference sound pressure of 20μPa multiplied by the time interval of 1 second taken as; 400 $(\mu Pa)^2$s.

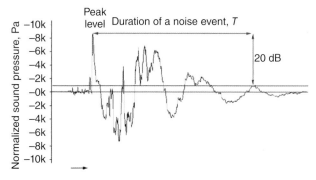

Figure 2.41 Example of a noise event.

As seen, the sound exposure level is similar to L_{eq}, however, the total sound energy is integrated over the measurement period, then averaging it by using a reference duration of 1 second. L_E of a noise event can be expressed as an unweighted level in a certain band or A-weighted level. The symbols that were used in the past for sound exposure level varied as SEL, L_{AE}, L_{Ax} or L_A that can still be found in the literature.

Figure 2.42 displays the determination of noise events within a time-history of a noise. The noise events are described by their lowest and highest values (range), the number of noise events are determined and the average L_E during the total observation period is calculated as shown in Figure 2.43. Sound exposure level is a metric considering both time and level. The contribution of the interrupted or impulsive noises called "noise events" is taken into account according to their effective values throughout their existence.

The relationship between L_E and L_{eq} is given in Figure 2.43 and in Eq (2.49):

$$L_{eq} = 10 \lg \left(\sum_{i=1}^{N} 10^{LE_i/10} \right) - 10 \lg T, \mathrm{dB} \tag{2.49}$$

T: Observation time, s
L_{Ei}: Sound exposure levels of ith noise event, dB

$$LE_i = 10 \lg \int_{t1}^{t2} 10^{LA(t)/10} \, dt, \mathrm{dB} \tag{2.50}$$

$LA(t)$: Instantaneous A-weighted sound pressure levels

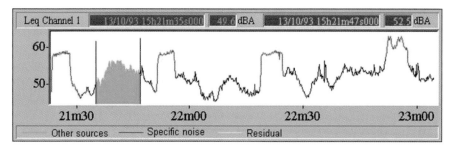

Figure 2.42 Identification of noise events in time-varying noise and determination of L_{AE} (SEL).

Figure 2.43 Comparison of A-weighted L_{eq} and L_{AE} (*SEL*).

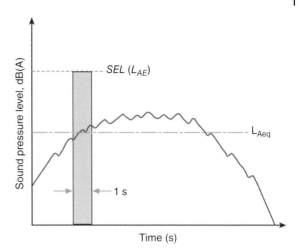

t_1–t_2: Time needed for an impulsive sound to drop 10 dB from its peak level

N: Number of noise events (of short duration and impulsive)

T: Observation time duration, second (e.g. eight hours, one hour, or several minutes)

If the equivalent continuous sound pressure level of a single event is known, *SEL* (L_{AE}) is determined as in Eq (2.51):

$$SEL = L_{eq}(t) + 10\lg(t/t_0), dB \qquad (2.51)$$

$L_{eq}(t)$: Equivalent sound level in t

t: Duration of a single noise event (impulse)

t_0: Reference duration of one second

Note: For the noises whose exposure duration is less than 5 minutes (e.g. industrial operations, train and aircraft passages)

2.5.2 Noise Descriptors

A noise descriptor is used for environmental noise in relation to the harmful effects of noise on humans and to determine the extent of noise pollution and to specify noise limits in regulations. The specific indexes have been developed by taking account of the type of noise source and the physical characteristics of the sound, particularly the time-varying characteristics that can be obtained through statistical analyses. There are several indexes in the world that have been used for various types of noise sources since the 1950s. Some of those widely used in the past have been abandoned at present and the new descriptors have been standardized internationally to provide coordination between different countries.

Source-specific descriptors were widely employed by taking into account the different manufacturing and operational characteristics of noise sources, e.g. the number of noise events and the standard deviation of noise levels, etc. Some of the environmental noise descriptors which are specific to the noise sources are given in Table 2.9. Those described in ISO 1996-1 and proposed in the noise prediction models recommended by the EU Directive, particularly for traffic, railway, aircraft, and industrial noises, are outlined below [85, 86].

Descriptors of Road Traffic Noise

L_{den} (Day-evening-night equivalent level) was proposed in Directive 2002/49/EC and first defined in ISO 1996-2:1987 to be obtained from long-term A-weighted average sound pressure levels. L_{den} to be measured over 24 hours with penalties applied for three periods of day, is determined for the use of noise maps, the levels should be computed at 1.5 m above ground and about +3 dB should be considered due to the reflections from nearby buildings. However, in the USA, L_{dn} (DNL), recommended by the Environmental Protection Agency (EPA), is still used as a metric for environmental noises taking the two periods of day-time (day and night-time) (Table 2.9).

Table 2.9 Noise indexes and descriptors used from past to present for environmental noise sources.

Noise source	Descriptor	Symbol	Relationship (all levels are A-weighted noise level, dB)
Traffic	Percentage levels (exceeding 10, 50, and 90% of time)	L_{10}, L_{50}, L_{90}	Determined from time-history and statistical evaluations
	Traffic noise index	TNI	$4(L_{10} - L_{90}) + L_{90} - 30$
	Noise pollution level	L_{NP}	$L_{eq} + 2.56\,\sigma$ σ: Standard deviation
	Long-term noise level	L_{LT}	$L_{LT} = 10\lg\left(p \times 10^{Lpf/10} + (1-p) \times 10^{Lph/10}\right)$ p: Percentage of heavy vehicles L_{pf} and L_{ph}: Noise levels of heavy and light vehicles
	Day and night level	L_{dn}	$L_{dn} = 10\lg\dfrac{1}{24}\left\{15*10^{Ld/10} + 9*10^{(10+Ln)/10}\right\}$ L_d = Equivalent A-weighted noise level in daytime, i.e. L_{eq} (07:00–22:00), dB L_n = Equivalent A-weighted noise level in night-time, i.e. L_{eq} (22:00–07:00), dB
	Day, evening, night level	L_{den}	$L_{den} = 10\lg\dfrac{1}{24}\left\{12*10^{Ld/10} + 4*10^{(5+Le)/10} + 8*10^{(10+Ln)/10}\right\}$ L_d: A-weighted equivalent noise level, L_{Aeq} at 07:00–19:00 hours L_e: A-weighted equivalent noise level, L_{Aeq} at 19:00–23:00 hours L_n: A-weighted equivalent noise level, L_{Aeq} at 23:00–07:00 hours
		L_{AX}	$L_{Amax} + 10\lg(\tau/\tau_{ref})$ $\tau = \dfrac{\tau_2 - \tau_1}{2}, \tau_2 - \tau_1 = \tau_{10}$ τ_{10}: Initial and termination time of the level 10 dB lower from the peak level $\tau_{ref} = 1$ second
Aircraft	Perceived noise level	PNL (L_{pn})	$40 + 33.3\lg(N_t)$ $N_t = N_m + 0.3\left[\sum N - N_m\right]$
	Effective perceived noise level	$EPNL$	$PNL + D + F$ $D = 10\lg(t_{10}/15)$ t_{10}: Given above.

Table 2.9 (Continued)

Noise source	Descriptor	Symbol	Relationship (all levels are A-weighted noise level, dB)
			F: 0.34 x (level difference on the tone frequency) D: Time correction
	Noise exposure forecast	*NEF*	$EPNL_{ij} + 10 \lg (N_{dij} + 17N_{nij}) - 88$ i: Type of aircraft j: Number of flight path N_{dij} and N_{nij}: Number of flights for a specific type of aircraft
	Community noise exposure level	*CNEL*	N: $NL + \lg (N_d + 3N_e + 10N_n) - 38.4$ N_d, N_e, N_n: Number of flights during day, evening and night times
	Weighted effective community perceived noise level	*WECPNL*	$10 \lg ((5/8)10^{0.1ECPNL_D} + (3/8)10^{0.1ECPNL_N + 10}) + S$ $ECPNL_D$ and $ECPNL_N$: Effective community-perceived noise levels during day and night $ECPNL = TNEL - 10 \lg (T/t_0)$ T: Specific observation time, second $t_0 = 1$ second $TNEL = 10 \lg \left(\sum_{i=1}^{n} 10^{0.1EPNL_i} \right) + 10 \lg (T_0/t_0)$ T_0: 10 seconds, t_0: 1 second $EPNL_i$: Effective perceived noise level for ith noise event (after correction for tone and duration)
	Noise and number index	*NNI*	$PNL + 15 \lg N - 80$ N: Total number of aircraft during daytime
Railway noise	Noise exposure level for trains	*NEL$_T$*	$10 \lg \left(10^{NEL_C/10} + 10^{NEL_L/10} \right)$ $NEL_C = L_{AC\,max} + 10 \lg T_{EC}$ $NEL_L = L_{AL\,max} + 10 \lg T_{EL}$ $L_{AC max}$ and $L_{AL max}$: Peak levels during passing of locomotive and cars T_{EC} and T_{EL}: Effective passing durations of locomotive and cars
	Community noise exposure level for trains	*CNEL*	$NEL_T + 10 \lg (N) - 49$ N: $N_d + 3N_e + 10N_n$ (given above)

L_{LT} (Long-term noise level) is a composite noise indicator proposed in the prediction models for road traffic noise (Chapter 4) and takes into account the types of different vehicles on the road with their numbers and individual sound pressure levels.

Descriptors of Aircraft and Airport Noise

The noise descriptors defined by International Civil Aviation Organization (ICAO) in noise simulation model are given below:

L_A(AL): A-weighted sound pressure level, dB
L_{AMAX} (MAL): Maximum value of L_A, dB
L_{AE} (SEL): Sound exposure level, dB
L_p: Sound pressure level in 1/3 octave bands, dB

EPNL: Effective perceived noise level, dB

NNI: Noise and number index, dB

Descriptors of Railway Noise

The descriptors in the railway noise model in the European Community for railway noise in dB(A) are given below:

L_d, L_e, L_n: Day-time, evening and night-time levels respectively

L_{den}: Day-evening-night levels

L_E: Train exposure level or third octave band level

L_{Aeq}: Equivalent train noise level

Descriptors of Industrial Noise

Impulsiveness and tonality are the important factors in evaluating industrial noises. The most common descriptors are:

L_C: C-weighted level (dB(C))

L_{CE}: C-weighted event (or operation) level, dB(C)

L_{Amax}: A-weighted maximum level

L_{AE}: A-weighted exposure noise level

L_{eqT}: Equivalent noise level in time T (in 1/3 octaves)

The EU Directive 2002/49/EC has clarified the conditions in which the supplementary descriptors can be used for evaluation of environmental noises:

- If the noise source is activated only for a short duration of time (i.e. less than 20% of day, evening and night times).
- If the average number of events is very low (i.e. less than 1 in an hour).
- If the low frequency components of noise are strong enough.
- If noise protection at night is important (L_{Amax} and *SEL*).
- If additional abatement is necessary during day and evening.
- If additional abatement is necessary during weekends and other times.
- When different noise sources exist.
- When quiet spaces exist in rural areas to be protected.
- If the noise consists of strong tonal components.
- If the noise has an impulsive character.

2.6 Conclusion

Within the wide context of acoustics, noise and noise-related issues cover a great range since they are directly related to individual and community health problems. Definition of noise simply as "unwanted sound" does not minimize its importance as a type of environmental pollution and noise has been investigated for two centuries, in terms of physical properties, techniques for measurements and analysis, assessment and evaluations, and management for noise control strategies, as will be discussed in later chapters.

Noise is a complex phenomenon when it is generated by environmental noise sources of great variety, hence it requires physical characterization with respect to spectral and temporal variations during the stages of radiation from sources and transmission into the environment. While noise is propagating outdoors, significant deviations occur in the sound paths, changing the characteristics of the sound waves due to the various physical factors in the environment. Of these, some factors causing the spatial variation of noise (e.g. shadow zones) are beneficial from the point of view of noise control in environmental planning. The methods for analysis and evaluations of noise have developed enormously over the years since the 1950s, by means of the experimental studies yielding empirical results and theory-based empirical approaches to understanding the principles of sound radiation and propagation.

To facilitate the measurements, evaluations, and comparisons of environmental noises, various descriptors (metrics) have been developed over the years in addition to basic acoustic units, some are used for specific sources and for specific noise types, whereas others are used for general evaluations. Many countries declared different descriptors to specify their noise limits in their national regulations and guidelines in the 1970s and the 1980s, however, at present, the ISO standards have resolved the conflicts emerging from different evaluations in the world, and the most appropriate source-specific descriptors are recommended for assessing, measuring, and evaluating the environmental noises.

Appendix 2.A

2.A.1 Barrier Design, Constructions, and Samples

Noise barriers, as one of the noise reduction devices (NRD), are the physical elements with a surface mass of not less than 20 kg/m², which are rigid, of low porosity and of the appropriate height. When they are placed between the noise source and the receiver, they cause a significant amount of reduction in sound pressure levels contributing to environmental noise control. Noise barriers have been widely implemented since the 1970s against road traffic noise, railway noise, aircraft ground-based operations, and industrial noises, and the knowledge based on this experience has increased a great deal, in line with the technological advances in manufacturing. A vast number of publications can be found in the literature regarding an overview of barriers and focusing on their design principles [87–95]. Any element causing sound diffraction around its edges can behave as an obstacle, however, if it is intended to be used for noise abatement, it should be designed according to the principles of diffraction and calculations and, at present, by using computer modeling and ray-tracing techniques (Chapters 4 and 6). Barrier performance, expressed in quantitative terms, depends on a variety of acoustical and non-acoustical parameters, which are significant in practice. For example, the dimensions of barriers need to be greater than the wavelength of the sound to be attenuated (note that the wavelength of the environmental sounds can vary between 0.02 and 17 m). Thus, the noise of a truck, whose spectrum contains the dominant lower components, is less attenuated by a barrier compared to the noise emitted by a car.

Designing a barrier should primarily take into account the formation of a shadow zone, however, a complete acoustic shadow cannot readily be determined, since it does not have a distinctive boundary, contrary to the light shadow. The size and pattern of the shadow zone are associated with the effective height and length of the barrier in addition to the wavelength of sound. Therefore, roadside barriers cannot protect the receivers in the upper floors of tall buildings near the roads since they are not efficient above the shadow zone. The efficient shadow is obtained when the barrier is positioned appropriately between the source and the receiver (source-barrier-receiver geometry).

2.A.2 Overview of the Design Principles of Noise Barriers

Barrier designs for specific noise situations and environments should satisfy some criteria:

- Performance and regulatory compliance (acceptable noise levels and noise limits).
- Applicability.
- Economic concerns and cost-effectiveness.
- Visual effect and aesthetic considerations.
- Compatibility with environment (climate, landscape, culture, social life, etc.).

Selection of Barrier Location

At the preliminary stage, the line of sight and distance between the source and the receivers (users), the topographical conditions, the number of buildings and their heights to be protected and the soil properties for barrier foundations, should be analyzed. The decisions regarding the placement of the barrier should also be made by consulting the public and the responsible authorities. Some practical rules to be remembered prior to design are:

a) A barrier designed to provide at least 5 dB reduction must be tall enough to block the line of sight between a source and receiver. If there are buildings with different heights (floors), the receiver points are selected on the levels of the highest floor levels.
b) The barrier should be placed close to the road to provide more benefit from the shadow formation.
c) The receivers on the top of a hill, or on the hillside, cannot get much benefit from the barrier.
d) Lower height barriers are not effective for the most distant traffic lanes.
e) Trees and bushes themselves are not considered barriers, however, they can provide ground absorption.
f) Computer models can display the potential effect of the barrier on the building façades.
g) The computer models enable the optimization of the barrier heights in relation to the receiver positions (Chapter 5).

Type of Barriers

Various types of barriers with different characteristics can be designed to protect communities from transportation noise sources:

Natural barriers and earth berms:

- Topographical configurations may restrict the sound propagation between the source and the receiver depending on their length along the road and provide an obstruction against noise propagation (Figure 2.A.1).
- Depressed roads or embankments along traffic roads or railways can form natural barriers and can give about 3–5 dB (A) reduction in environmental noise levels (Figure 2.A.2). They can be combined with a relatively lower height of noise barrier if an additional shield is required.
- Earth berms are man-made mounds that look natural in appearance by using filling material (rocks, stones, waste but non-toxic material, recycled materials, etc.), a filter layer of gravel, and natural soil over the entire surface to grow grass or vegetation (Figure 2.A.3). They need a large area at the base to attain the required height. Earth embankments can be used to install the solar panels on one side without disrupting their acoustic performance.

Tunnels and semi-tunnels: If excessive height of a barrier is required, different shapes including semi-tunnels are applicable. Complete cover of roadways and railways by tunnel structures, although not specifically constructed for noise abatement, are the most efficient solution to isolate

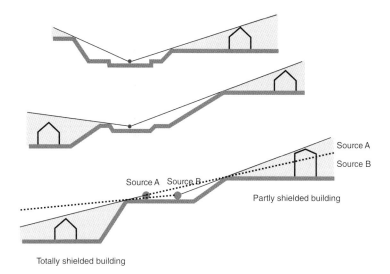

Source A
Source B

Partly shielded building

Source A   Source B

Totally shielded building

Figure 2.A.1   Topographical barriers.

(a)

(b)

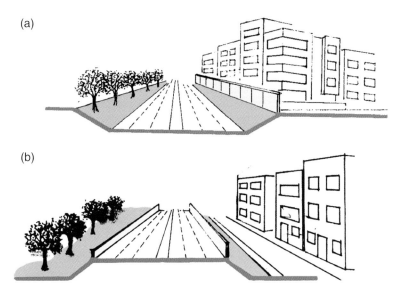

Figure 2.A.2   (a) Screening with depressed roads and embankment (road in cuttings), (b) road on the embankment with side barriers.

the noise source from the environment. However, considering the increased noise levels inside the tunnel due to multiple reflections and sound focusing, absorptive treatment is applied on the long tunnel walls (Figure 2.A.4).

Wall-type barriers: Single or multiple barriers also called "sound walls" are built along motorways and railway tracks at-grade or at the sides of the elevated structures (Figures 2.A.5a and 2.A.5b). Sound absorptive treatment on the roadside surfaces provides extra attenuation in addition to the diffraction effect.

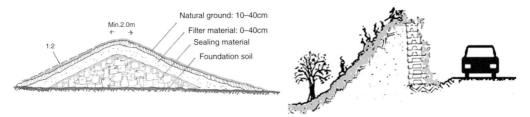

a) Cross-section of a typical earth berm

Min.2.0m

1:2

Natural ground: 10–40cm

Filter material: 0–40cm

Sealing material

Foundation soil

b) Earth-berm and planted masonry block retaining wall

Figure 2.A.3   Earth berms (cross-sections).

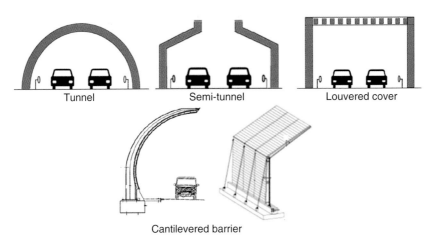

Tunnel

Semi-tunnel

Louvered cover

Cantilevered barrier

Figure 2.A.4   Tunnels, semi-tunnels, louvered and cantilevered barriers with examples of constructional details.

Figure 2.A.5a   Noise screens along the road.

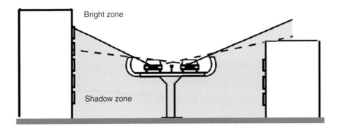

Bright zone

Shadow zone

Figure 2.A.5b   Barrier on elevated road structure.

Effective Parameters on Wall-Type Barrier Design

Referring to the information on the scientific basis of diffraction given in Section 2.3.6, performance of the noise barrier is closely related to the following parameters that should be taken into consideration in their design:

Geometrical parameters:

- Length of barrier
- Height of barrier
- Thickness of barrier
- If double or multiple barriers, the distances in between
- Form and shape of barrier and top edge profile
- Distance from barrier to source and barrier to receiver
- Height of noise source
- Maximum height of receivers to be protected.

Acoustic parameters:

- Type of noise source: point source, line source (if finite: angle of view) or plane source
- Spectral and temporal characteristics of noise-sound power of source
- Directivity pattern of noise source (road vehicles: monopole, railway noise strongly directive at the horizontal plane)
- Wavelength of sound (i.e. the dominant frequency)
- Sound transmission loss of the barrier structure
- Sound absorption coefficient of surfaces

Physical environmental parameters:

- Ground characteristics:
  - Acoustically reflective or absorptive: absorption coefficient, characteristic impedance, flow resistivity of surface
  - Configuration of different ground types between the source and the receiver
  - Soil properties which are important for foundations

- Meteorological factors:

  - Ambient temperature, winds (directions and speeds), turbulence, relative humidity of air
  - Air absorption

Construction properties:

- Structural systems and materials
- Surface cover: reflective or absorptive
- Surface texture, holes, cracks, and gaps on the surface and access openings
- Plant characteristics and arrangement if incorporated with barriers

Construction Materials for Noise Barriers

Various construction materials are used in the manufacture and applications of noise barriers; the basic commercial materials are: concrete, brick, wood, metal, glass, plastics and composite materials. The choice of materials is dependent on the type of barrier structure system that must resist vertical (self-weight) and horizontal loads (dynamic loads such as wind, vehicle impact, snow-load, stone impacts, etc.).

Load-bearing structures (free-standing walls seated on concrete pads or supported by concrete foot-ings): Massive and rigid elements, such as concrete walls built-in-place, pre-cast concrete panels, masonry walls made of bricks, lightweight concrete blocks or gas-concrete units, box-type hollow units with plants, etc.

Frame structures: Post and panel systems either built in-place as a "separate post-and-panel sys-tem" or a prefabricated "integral post-and-panel system." The vertical posts and horizontal beams transmit loads to the ground, are made of concrete, metal (steel or extruded aluminum) or wood. The filling elements between the posts and fixed by a different connection assembly within the frame, can vary such as: prefabricated panels composed of single layer or multilayered systems. Examples from common practice are:

- Metal sheeting of different types (perforated or non-perforated, corrugated, ribbed, etc.) produced as modular panels, such as aluminum sheets, which are preferred due to their static strengths, recyclability, durability against corrosion and salt, possibility of color selection
- Metal panels in combination with plants and vegetation (green barriers)
- Metal panels in combination with transparent materials
- Lightweight concrete-cemented fibrous panels
- Aerated concrete units as filling material
- Transparent materials, such as glass, acrylic or polycarbonate resins or other plastic materials such as plexiglass
- Composite panels including a layer resistant against sound transmission through an absorbing treatment outside
- Other panel materials: such as ceramics, sintered metals, cement bonded wood-wool or chip, panels from recycled materials

Materials for surface absorption: Absorption on the barrier surface facing the road reduces the noise in front of the barrier and eliminates multiple reflections from vehicle bodies and other sur-faces across the road so as to reduce the noise to be diffracted over the barrier. Surface absorption can be provided by using the following measures:

- Systems with cavities incorporating absorbing materials: e.g. mineral fibers or rock wool behind a perforated or slit facing made of metal or plastics, or behind horizontal louvers, etc.
- Panels of open textured porous materials
- Hollow concrete units, hollow bricks, or other lightweight units
- Use of foliage (integrated walls with plants)

Holes and gaps: Barriers usually accommodate drainage holes or gaps on the surfaces of small size (i.e. less than 3% of the surface), however, they do not considerably affect the barrier performance.

Acoustic tightness at the joints between the panels or posts during fixing must be checked to prevent sound leakage. Gaps at the bottom edge of the barriers must be filled with solid material (i.e. concrete) and the gaps and cracks, due to errors in construction, should be sealed.

Large openings for pedestrian passages, emergency exits, safety doors and access alleys must be planned so as not to destroy the continuity of the barrier and reduce its acoustic efficiency. The positions of gantry signs, lamp posts, etc. must be in front of recessed barriers also providing con-tinuity [87, 91].

Surface texture: The roughness of the surface increases the barrier performance when the dimensions of irregularities are in the order of the wavelength of the dominant sound of noise (i.e. for 100 Hz, the wavelength is 3.4 m). Textured patterns on the concrete surface are made while casting the concrete and should be considered along with the visual concerns. A variety of surface finishes, textures, geometric patterns, and reliefs scattering sound are used as a part of barrier design [88].

Barrier Configurations

Different type of barriers can be designed as single and double barriers, multiple-edge barriers causing multi-diffractions over the top, etc. The shape of the barrier and the top edge profiles (caps) have been widely investigated, since they play a role in increasing the total sound attenuation [58–84, 96–106].

Massive barriers can be either simple vertical walls or designed in eye-attracting forms such as in serpentine, castellated (undulated), fan-shaped and staggered forms that give extra strength to barriers against horizontal loads. Sloped-faced or tilted barriers reflect noise upwards and reduce noise, also decreasing the collision impact (Figure 2.A.6) [74]. The tilt angle can be changed according to the width of the road and must be selected so as not to reduce the barrier insertion loss.

A number of innovative designs regarding the edge-modified barriers have been experimented and the additional attenuations through absorption, interference, and resonances on the cap devices, were published in the literature. For example, a cylindrical unit with about 0.30 m diameter, installed on the top edge, could give an additional reduction by as much as 2–4 dB(A). Example barrier configurations are shown in Figures 2.A.7a and 2.A.7b [75–84, 96] and the complex designs applying new technologies (e.g. solar or photovoltaic barriers) are given in Figures 2.A.8–2.A.12 [72, 73].

When designing barriers, some negative impacts should also be considered, for example, tall barriers reduce sunshine and daylight and this is not desirable by the buildings behind the barriers. If the barrier contains plants (vegetation), they need trimming and maintenance. The feasibility study is a must (i.e. applicability, availability of materials, total and partial costs, etc.) before the final decision about barrier design is made.

Buildings: As three-sided thick barriers, building blocks provide quieter zones behind them selves, depending on their dimensions and spacings. If the longer and higher blocks are present in the neighborhood along a traffic line, the acoustic shadow can protect the larger areas against the excessive noise. Even the lower dwellings and garden walls can provide sufficient acoustic privacy in their courtyards and atriums. Computer models enable tracing the multiple diffractions and reflections in complex built-up areas and produce noise reduction contours (Chapter 6).

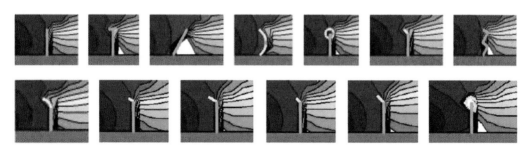

Figure 2.A.6 Experimental study for various barrier configurations and sound attenuation [74]. (*Source:* Permission granted by Bite.)

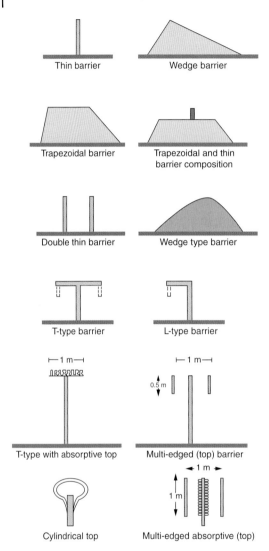

Figure 2.A.7a Example structures tested as noise barriers.

Other Characteristics of Barriers

In addition to the requirements for sound reduction performance, noise barriers must have strength and stability as well as being durable under different environmental conditions, such as winds, rain, snow, exhaust gases, and chemicals. The European Norms giving the specifications regarding the acoustic and non-acoustic characteristics of roadside barriers, are outlined below [66a,b, 67a,b,c]:

- *Mechanical performance and stability requirements (EN 1794-1: 2018)*: Design loads to be taken into account in structural calculations are self-weight, impact of stones, collision, load from snow clearance and the dynamic loads (wind load) [66a].
- *General safety and environmental requirements (EN 1794-2: 2011):* Resistance to brushwood fire, risk of falling debris, environmental protection (materials and products not imposing negative effects on the environment and use of recycled materials), means of escape in emergency, light reflection, effect of transparency, are explained [66b].

Figure 2.A.7b Various types of edge-modified barriers.

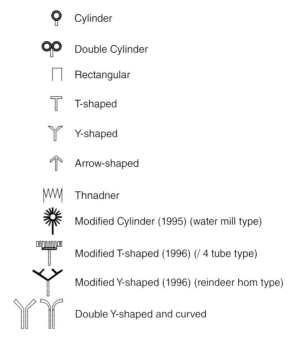

Cylinder

Double Cylinder

Rectangular

T-shaped

Y-shaped

Arrow-shaped

Thnadner

Modified Cylinder (1995) (water mill type)

Modified T-shaped (1996) (/ 4 tube type)

Modified Y-shaped (1996) (reindeer hom type)

Double Y-shaped and curved

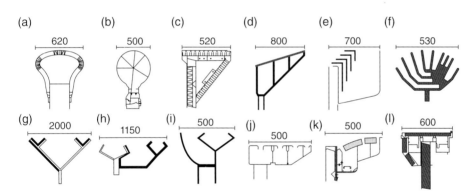

(a) 620 (b) 500 (c) 520 (d) 800 (e) 700 (f) 530

(g) 2000 (h) 1150 (i) 500 (j) 500 (k) 500 (l) 600

Figure 2.A.8 Examples of edge-modified barriers marketed in Japan [72, 73] (a–d) Absorption type; (e–i) interference or multi-edge types; (j–k) resonance types, l: ANC type. (*Source:* Permission granted by Yamamoto.)

Figure 2.A.9 Mushroom type with planting box [73]. (*Source:* Permission granted by Yamamoto.)

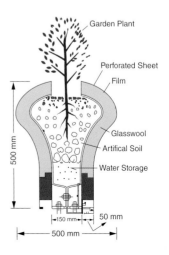

Garden Plant

Perforated Sheet

Film

Glasswool

Artifical Soil

Water Storage

500 mm

150 mm

50 mm

500 mm

Figure 2.A.10 Sonic crystal barrier [97, 98]. (*Source:* Permission granted by Perez.)

Figure 2.A.11 Diffracting pavement scatters [100, 101]. (*Source:* Permission granted by Wijnant.)

Figure 2.A.12 Solar noise barriers [103]. (*Source:* Permission granted by Hammer.)

- *Durability:* Materials used in barrier constructions and absorptive treatments are subject to aging, thus the effectiveness of the barriers reduces in time. Procedures for assessing the long-term performance of barriers are described in EN 14389: 2015 in terms of the durability of the acoustical characteristics in Part 1 and the durability of the non-acoustic characteristics in Part 2, to be quantified as "working lifetime in years" [107a, 107b]. Sustainability is one of the global criteria at present implemented for all types of buildings including noise barrier constructions. The standard's Part 1 also describes the testing of acoustic performance.

EN 14388:2015 is a "Specifications standard" and declares the criteria and requirements associated with the EU certification for marketing purposes [68].

Visual effect and aesthetic considerations in barrier design: Since noise barriers are in the proximity of the receivers (drivers, passengers, and residents in a neighborhood), their design needs to evaluate the visual effect and aesthetic concerns, in addition to the other requirements. This subject has been discussed in various documents giving guidelines for barrier design [60, 91]:

- Design should be visually interesting and attractive, however, it should not disturb visual conformity in the natural and built environments.
- The community should be informed about the design alternatives and their approval should be obtained.
- The design must not cause a reduction in barrier efficiency. A simple design comprising large and easily detectable shapes and forms should be preferred.
- The aesthetic art works should be cost-effective and it is worthwhile when they are viewed by a large number of people and give a specific characteristic to the environment.
- Designs which are eyesores and those causing difficulty in implementation should be avoided.
- Sophisticated designs at the lower part of barrier surface cannot be perceived by drivers.
- The design should consider the effect of the size and the length of the barrier on users. Long and tall parallel concrete walls along the roads evoke a prison sensation for drivers.
- Long continuous barriers evoke a monotonous visual impression and can be modified by colors, textures, or by supplementary elements creating a strong contrast.
- Visual orientation: The design should enhance a dominant visual direction, for instance, a horizontal pattern on the surface tends to direct the eye forward. The barriers and embankments parallel to the road provide a guiding effect for the road users and enhance the continuity of nature.
- Transparent barriers are beneficial for viewing the natural and cultural environment.
- Light reflection causing glare for drivers creates the possible risk of accidents and can be avoided by use of color, surface treatment, or a sloped barrier.
- Barrier design should be compatible with the environment, thus blending in with nature or the landscape is more desirable. An artistic form can be created by using line, shape, color, texture, and artistic expression within the context of its surroundings.
- The visual effect of tall noise barriers can be diminished by means of texture (horizontal lines) or painting, e.g. vertical color strips with light colors at the top and dark colors at the bottom.
- Planting low vegetation in front of the barrier and tall vegetation at the back is more convenient.

Traffic Safety Considerations
One of the important aspects in barrier design and planning is the requirements for ensuring traffic safety [87, 88]:

- Barriers must not obstruct the vision of users (drivers, pedestrians, etc.) especially at road crossings, curvatures, and intersections.
- It should be remembered that the shadow of barriers on the road may be hazardous, since dark zones create a risk of ice and slippery conditions, in addition to the difficulty in eye-adjustment from light to shadow.
- In case of collisions, the work to be done – removing the crash barrier and replacing it – must be considered at the design stage by taking precautions, such as the use of modular elements.
- The existence and replacement of signposts, lamp-posts, electric pylons, etc. should be considered during planning and design of barriers.
- Emergency access doors (equipped with signs, lights and emergency telephones) or exits on the barriers, must be arranged at certain intervals.

Modeling of Noise Barrier Design

The environmental noise prediction models also include barriers. The basic International Standard model is ISO 9613-2: 1996 (confirmed in 2017) which describes a calculation process for barrier attenuation [31]. The relatively more detailed calculation procedure is given in the Cnossos model which is mandatory for use in EU countries at present (Chapter 4). The procedures consist of various calculation steps taking into account all the configurations of barrier, source and receivers in addition to other physical parameters influencing the barrier attenuation. Since the 1980s, such computations of barrier performance and the improvement studies have been facilitated by means of computerized simulations, as long as the theoretical basis is accurately established. The models also enable optimizations to determine the suitable barrier heights for a fixed receiver point: based on a noise control criterion (i.e. the outdoor noise limit in regulations), the height of the barrier is increased stepwise according to a scenario until the required sound attenuation is obtained. The process can be extended over the entire area concerned. Barrier design has become a part of noise mapping software, as an efficient tool for road traffic noise abatement (Chapter 5).

The validation of the calculated performances of barriers through field measurements is necessary after the barrier construction is in place. The outdoor noise tests, to compare the computed and measured results, are conducted according to the ISO 1996-2:2017 at the selected receiver points (Section 5.5.1). Besides, the EN 1793 series describe the specific test methods for barrier performance, in terms of sound absorption in the laboratory and in-situ (Parts 1 and 5), the airborne sound insulation index, both in the laboratory and in situ (Parts 2 and 6), and sound diffraction in situ (Part 4) [67]. These standards for measuring acoustic performances of barriers are explained in Section 5.10.3.

Design of Roadside Noise Barriers in Historical Perspective

A state-of-the-art review of the design of noise barriers based on the experimental and theoretical studies is presented in detail in various publications [60, 70, 73, 102]. A summary regarding the advances in barrier design is given below:

- 1965: The first barrier was constructed in Japan. They were concrete panels only applicable for flat terrain and roads, some combined with earth berms, the vertical surfaces on the roadside were later covered by absorptive material.
- 1975: Barriers with light-weight panels, so-called metallic absorptive panels with a melix box with slits, started to be constructed and have been used up to the present.
- 1976: Fujiwara applied an absorptive material at the top edge and found 4 dB(A) increase in barrier attenuation, based on his predictions.

- 1980: Tall barriers up to 8 m were built especially in Japan and in Europe, however, it was realized that they cause disturbance due to the problems, such as visual obstruction of landscape and blocking sunshine.
- 1980: May et al., in Canada, carried out scale model experiments and found additional attenuation of maximum 4 dB for T-shaped, Y-shaped, and arrow-shaped barriers and experimented on them in the following years [75, 76].
- 1984–1985: Mizuni et al. first developed a novel barrier profile called the "interference type" [77].
- 1989: Fujiwara et al. invented the cylindrical tube mounted on the top edge of a barrier [78].
- 1990: Planting boxes and empty tubes (of 0.4 m diameter) filled with plants considering aesthetically appropriate in urban areas, were tested and an extra attenuation of 2 dB was obtained at 500–4 kHz for $h = 5$ m barrier [73] (Figure 2.A.9).
- 1993: Resonance types consisting of acoustic tubes with a length of $\lambda/4$, were investigated and called "acoustically soft."
- 1994–1995: Hothershall conducted numerical simulations using BEM for three types of barriers [82, 83].
- 1994–1996: Watts experimented with the interference type barrier, estimating the attenuation less than 2 dB(A) [80, 81].
- 1995–1998: Various shapes called "edge-modified barriers" were tested.
- After 2000: Further studies were performed and modern designs were realized in practice, such as:

  – A variety of sophisticated top edge devices were tested (absorptive, multi-edge, and resonance types) [72, 73] (Figure 2.A.8).
  – Green walls made of recycled materials filled with fibrous material, the so-called "ecologic barriers."
  – More complex devices, such as quadratic residue diffusers (QRD) at the top edge [96] (Figure 2.A.8).
  – A periodic array of scatters (SCAS) called "sonic crystals" are still under investigation, however, they are presented as potential noise barriers because of their sound absorption characteristics and aesthetically pleasing structures in the landscape [97, 98, 108] (Figure 2.A.10).
  – Active noise control barriers (ANC) with microphones and sound emitters placed inside the barrier elements to control the sound pressure at the edge position [72].
  – Diffracting road surface elements, also called "silent pavements", are developed. Consisting of concrete units with cavities, they provide about 4 dB because of the sound waves diffracted upwards resulting from the resonance inside the cavities [100, 101]. (Figure 2.A.11).
  – Solar noise barriers are integrated with the photovoltaic (PV) energy system (or solar cell panels) acting as a power plant [103–106] (Figure 2.A.12).

Figure 2.A.13 presents examples of the common roadside noise barriers from the 1960s until the present.

2.A.3 Noise Barriers (Screens) for Other Sources

Noise barriers (or sound screens) can be designed and constructed not only in the open air, to protect against road and railways, but also in enclosed spaces against noise from machinery and equipment in workplaces and against various types of noises in open-plan offices, exhibition areas, etc. The acoustic performance of the enclosures can be calculated by applying the same

Figure 2.A.13 Examples of implemented barrier constructions.

principles if the area of the open part is greater than 10% of the total enclosure surface. However, their construction and materials differ from the roadside barriers, as will be briefly summarized below:

1) *Sound screens in open-space offices*: They are used to provide a visual obstruction as well as for noise abatement. ISO 17624:2004 (confirmed in 2017) gives guidelines to design and assess the performance of office screens, which are free-standing, demountable, movable if required, pre-manufactured lightweight panels, sometimes integrated with furniture (Figure 2.A.14). Their positions in space, dimensions (heights and lengths), and the surface materials are the parameters in performance calculations, in addition, the absorption properties of ceilings above them. Similar design principles based on geometrical diffractions apply, e.g. the screen should be positioned close to the sources to get the best efficiency. Performance is determined through measurements either in situ (free-field or direct field) or in the laboratory (diffuse field). Performance rating descriptors are: sound pressure level differences without and with screen situations at the same receiver position, expressed as D_p dB (in-lab), D_z dB (free-field) respectively and the insertion loss: D_i, dB defined as "the difference between the sound power levels radiated in the room by the sound sources to be shielded without and with the screen installed" [109]. All the values are given in octave or one-third octave bands, however, the A-weighted level differences D_{pA} can also be used for simplicity in marketing.

   Free-field screen sound attenuation, $D_{z:}$ dB can be determined by using Eq (A.1), which is valid for screens with a hard core and covered by textile and mineral wool of 50 mm and in a room with a low and partially absorptive ceiling [109]:

   $$D_z = 10 \lg \left(3 + 40\frac{z}{\lambda}\right) \text{dB} \tag{A.1}$$

   z: Path difference, m between the longer sound propagation path around the least effective diffractive edge of the screen and the direct path
   λ: Wavelength, m
   D_z is reduced about 3–5 dB due to the reflections from the nearby walls around the noise source
   The required sound insulation, which the screens should satisfy, is determined by the sound reduction index R, defined as the ratio of transmitted energy through the screen and the incident energy onto its surface and measured according to ISO 10140 series both in the laboratory and in situ (Section 5.9.2).

Figure 2.A.14 Sound screens in open-plan offices (combined from the market).

Figure 2.A.15 Enclosures and cabins for outdoor mechanical equipment (combined from the market).

2) *Enclosures and cabins*: The protection elements are used for operators in workspaces (indoors) and for residents of nearby buildings exposed to machinery and equipment located outdoors, such as cooling towers, rooftop equipment, generators, air-conditioning units, etc. The enclosures and cabins are generally lightweight metal constructions with absorptive treatment inside and can be either portable or fixed elements [110]. Noise radiation from air-supply and exhaust openings and ducts should be carefully eliminated by silencers as well as the doors to be heavily isolated. (Figure 2.A.15).

References

1 Pierce, A.D. (1989). *Acoustics: An Introduction to its Physical Principles and Applications*. McGraw-Hill.

2 Lang, W.W. (1996). A quarter century of noise control: a historical perspective. *Acoustics Bulletin* July/Aug: 1–9.

3 Harris, C. (ed.) (1957). *Handbook of Noise Control*. McGraw-Hill.

4 Harris, C. (1979). *Handbook of Noise Control*, 2e. McGraw-Hill.

5 Maekawa, Z., Rindel, J.H., and Lord, P. (1997). *Environmental and Architectural Acoustics*, 2e. Spon Press.

6 Piersol, A.G. (1992). Data analysis. In: *Noise and Vibration Control Engineering: Principles and Applications* (eds. L.L. Beranek and I.L. Ver), 45–74. Wiley.

7 Bies, D.A. and Hansen, C.H. (2002). *Engineering Noise Control Theory and Practice*, 2e. Spon Press.

8 Fahy, F. (2001). *Foundations of Engineering Acoustics*. Academic Press.

9 Schaudinisky, L.H. (1976). *Sound, Man and Building*. Applied Science Publishers.

10 Beranek, L.L. (1971). *Noise and Vibration Control*. McGraw-Hill.

11 Beranek, L.L. (1986). *Acoustics*. American Institute of Physics for the Acoustical Society of America.

12 Morse, P.M. and Ingard, K.U. (1968). *Theoretical Acoustics*. McGraw-Hill.

13 (a) Andersen, G.S. and Kurze, U.J. (1992). Outdoor sound propagation. In: *Noise and Vibration Control Engineering: Principles and Applications* (eds. L.L. Beranek and I.L. Ver), 113–144. Wiley.
(b) Sound Research Laboratories Ltd (1976). *Practical Building Acoustics*. Sudbury: Sound Research Laboratories Ltd.
(c) Hongisto, V., Oliva, D., and Keranen, J. (2014). Subjective and objective rating of airborne sound insulation: Living sounds. *Acta Acustica, United with Acustica* 100: 848–863.

14 Rudnick, I. (1957). Propagation of sound in the open air. In: *Handbook of Noise Control* (ed. C.M. Harris). McGraw-Hill.

15 Kurze, U. and Beranek, L.L. (1971). Sound propagation outdoors. In: *Noise and Vibration Control* (ed. L.L. Beranek). McGraw-Hill.

16 Piercy, J.E. and Embelton, T. (1979). Sound propagation in the open air. In: *Handbook of Noise Control* (ed. C.M. Harris). McGraw-Hill.

17 ICAO (1993). Annex 16, vol. 1, International Civil Aviation Organization, 3, July.

18 Embelton, T.F.W. (1982). Sound propagation outdoors: improved prediction schemes for the 80s. *Noise Control Engineering* Jan.–Feb: 30–39.

19 Sutherland, L.C. and Daigle, G.A. (1998). Atmospheric sound propagation. In: *Handbook of Acoustics* (ed. M.J. Crocker), 305–336. Wiley.

20 Attenborough, K. (2007). Sound propagation in the atmosphere. In: *Handbook of Noise and Vibration Control* (ed. M.J. Crocker), 67–78. Wiley.

21 De Lange, P.A. (1969). Sound insulation of glazing with respect to traffic noise. *Applied Acoustics* 2: 215–236.

22 Crocker, M.J. (ed.) (2007). Fundamentals of acoustics, noise and vibration. In: *Handbook of Noise and Vibration Control*, 1–16. Wiley.

23 Herlufsen, H., Gade, S., and Zaveri, H.K. (2007). Analyzers and signal generators. In: *Handbook of Noise and Vibration Control* (ed. M.J. Crocker), 470–485. Wiley.

24 Burgess, J.C. (1998). Practical considerations in signal processing. In: *Handbook of Acoustics* (ed. M. J. Crocker), 1063–1082. Wiley.

25 Gade, S. and Hansen, K.G. (1996). Non-stationary signal analysis using wavelet transform, short time Fourier transform and Wigner-Ville distribution, Brüel & Kjaer Technical Review, No. 2

26 (a) Newland, D.E. (2007). Wavelet analysis of vibration signals. In: *Handbook of Noise and Vibration Control* (ed. M.J. Crocker), 585–597. Wiley.
(b) Policardi, F. (2011). MLS and sine-sweep technique comparison in room-acoustic measurements. *Elektrotehniski Vestnik* 78 (3): 91–95. (English edition).
(c) Guidorzi, P., Barbaresi, L., D'Orazio, D. et al. (2015). Impulse responses measured with MLS or swept-sine signals applied to architectural acoustics: an in-depth analysis of the two methods and some case studies of measurements inside theatres, 6th International Building Physics Conference, IBPC 2015, Energy Procedia 78

27 IEC (2013). IEC-061672-1:2013 Edition 2. 2013–09, Electroacoustics: Sound level meters. Part 1: Specifications.

28 Piercy, J., Embleton, T., and Sutherland, L. (1977). Review of noise propagation in the atmosphere. *Journal of Acoustical Society of America* 61: 1403–1418.

29 ISO (2017). ISO 1996-2:2017 Acoustics: Description measurement and assessment of environmental noise. Part 2: Determination of sound pressure levels.

30 American National Institute (1978). ANSI s.1.26 Method for the calculation of the absorption of sound by the atmosphere.

31 ISO (1993). ISO 9613-1:1993 (confirmed in 2017) Acoustics: Attenuation of sound during propagation outdoors. Part 1: Calculation of the absorption of sound by atmosphere.

32 ISO (1996). ISO 9613-2:1996 (confirmed in 2012) Acoustics: Attenuation of sound during propagation outdoors. Part 2: General method of calculation.

33 Daigle, G.A. (1979). Effects of atmospheric turbulence on the interference of sound waves above a finite impedance boundary. *Journal of Acoustical Society of America* 65 (1): 45–49.

34 Embelton, T.F.W., Piercy, J.E., and Olson, N. (1976). Outdoor propagation over ground of finite impedance. *Journal of Acoustical Society of America* 59: 267–277.

35 Delany, M.E. and Bazley, E.N. (1970). Acoustical properties of fibrous absorbent materials. *Applied Acoustics* 3: 105–116.

36 Aylor, D. (1972). Noise reduction by vegetation and ground. *Journal of Acoustical Society of America* 51: 201–209.

37 Rasmussen, K.B. (1972). Sound propagation over grass covered ground. *Journal of Sound and Vibration* 51: 201–209.

38 Embelton, T.F.W., Piercy, J.E., and Daigle, G.A. (1983). Effective flow resistivity of ground surfaces determined by acoustical measurements. *Journal of Acoustical Society of America* 74: 1239–1244.

39 Mobile, M.A. and Hayek, S.I. (1985). Acoustical propagation over an impedance plane. *Journal of Acoustical Society of America* 78: 1325–1336.

40 Parry, G.A., Pyke, J.R., and Robinson, C. (1993). The excess attenuation of environmental noise sources through densely planted trees. *Proceedings of the Institution of Acoustics* 15: 1057–1065.

41 Price, M.A., Attenborough, K., and Heap, N.W. (1988). Sound attenuation through trees: measurements and models. *Journal of Acosutical Society of America* 84: 1836–1844.

42 CETUR (1983). Acoustique et végétation: effets de la végétation sur la propagation de bruit routier ou ferroviaire, Les Dossiers du CETUR, February.

43 Attenborough, K. (1982). Predicted ground effect for highway noise. *Journal of Acoustical Society of America* 81 (3): 413–424.

44 Embelton, T.F.W. (1963). Sound propagation in homogenous deciduous and evergreen woods. *Journal of Acoustical Society of America* 35: 1119–1125.

45 Watts, G., Chinn, L., and Godfrey, N. (1999). The effects of vegetation on the perception of traffic noise. *Applied Acoustics* 56: 39–56.

46 Pal, A.K., Kumar, V., and Saxena, N.C. (2000). Noise attenuation by green belts. *Journal of Sound and Vibration* 234 (1): 149–165.

47 Maekawa, Z. (1968). Noise reduction by screens. *Applied Acoustics* 1–3: 157–173.

48 Maekawa, Z. (1974). Environmental sound propagation, 8th ICA Conference.

49 Attenborough, K., Li, K.M., and Horoshenkov, K. (2007). *Predicting Outdoor Sound*. Taylor and Francis.

50 Kurze, U.J. and Andersen, G.S. (1971). Sound attenuation by barriers. *Applied Acoustics* 4 (1): 35–53.

51 Jonasson, H.G. (1972–). Sound reduction by barriers on the ground. *Journal of Sound and Vibration* 22 (1): 113–126.

52 Kurze, U.J. (1974). Noise reduction by barriers. *Journal of Acoustical Society of America* 55 (3): 504–518.

53 Maekawa, Z. (1993). Recent problems with noise barriers, *Proceedings of Noise 93 Congress*, St Petersburg, Russia.

54 Fujiwara, K., Ando, Y., and Maekawa, Z. (1973). Attenuation of a spherical sound wave diffracted by a thick plate. *Acoustica* 28 (6): 341–347.

55 Fujiwara, K., Ando, Y., and Maekawa, Z. (1974). Effect of thickness of barriers on noise reduction, *Proceedings of 8th ICA Congress*.

56 Pierce, A.D. (1974). Diffraction of sound around corners and over wide barriers. *Journal of Acoustical Society of America* 55 (5): 941–955.

57 Jonasson, H.E. (1972). Diffraction by wedges of finite acoustic impedance with applications to depressed roads. *Journal of Sound and Vibration* 25 (4): 577–585.

58 Okubo, T. and Fujiwara, K. (1998). Efficiency of a noise barrier on the ground with an acoustically soft cylindrical edge. *Journal of Sound and Vibration* 216 (5): 771–790.

59 Jean, P. (1998). A variational approach for the study of outdoor sound propagation and application to railway noise. *Journal of Sound and Vibration* 212 (2): 275–294.

60 Kotzen, B. and English, C. (1999). *Environmental Noise Barriers: A Guide to Their Acoustic and Visual Design*. E and EF Spon.

61 FHWA (2000). *Highway Noise Barrier Design Handbook*, FHWA-EP-00-005 DOT -VNTSC-FHWA-00-0. U.S. Department of Transportation, Federal Highway Administration.

62 Maekawa, Z. (1966). Noise reduction by screens of finite size. Kobe University Report, 12: 1–12.

63 Kurra, S. (1980). A computer model for predicting sound attenuation by barrier buildings. *Journal of Applied Acoustics* 13: 331–355.

64 Kurra, S. and Crocker, M. (1982). A scale model experiment of noise propagation at building sites. Research report, Department of Mechanical Engineering, Ray Herrick Laboratory, Purdue University, USA.

65 Scholes, W.E. and Sargent, J.W. (1971). Designing against noise from road traffic, BRS current paper 20.

66 (a) EN 1794-1:2018 Road traffic noise reducing devices – Non-acoustic performance. Part 1: Mechanical performance and stability requirements;
(b) EN 1794-2:2011 Part 2: General safety and environmental requirements.

67 (a) EN 1793-1:2017 Road traffic noise reducing devices – Test method for determining the acoustic performance. Part 1: Intrinsic characteristics of sound absorption under diffuse sound field conditions;
(b) EN 1793-2:2018 Road traffic noise reducing devices – Test method for determining the acoustic performance. Part 2: Intrinsic characteristics of airborne sound insulation under diffuse sound field conditions;
(c) EN 1793-3:1998 Road traffic noise reducing devices – Test method for determining the acoustic performance. Part 3: Normalized traffic noise spectrum.

68 EN 14388:2015 Road traffic noise reducing devices.

69 Horoshenkov, K.V., Hothersall, D.C., and Mercy, S.E. (1999). Scale modelling of sound propagation in a city street canyon. *Journal of Sound and Vibration* 223 (5): 795–819.

70 I-INCE (1999). Final report: Technical assessment of the effectiveness of noise walls. I-INCE publication 99–1, Sept.

71 Yamamoto, K., Taya, K., Yamashita, M. et al. (1989). Reduction of road traffic noise by absorptive cylinder adapted at the top of a barrier, Inter-Noise 1989, Newport Beach, USA.

72 Okubo, T. and Yamamoto, K. Procedures for determining the acoustic efficiency of edge-modified noise barriers. *Applied Acoustics* 68 (7): 797–819.

73 Yamamoto, K. (2015). Japanese experience to reduce road traffic noise by barriers with noise reduction devices, Euronoise 2015, Maastricht, The Netherlands.

74 Bite, M., Nagy, S.D., Bite, P. et al. (2015). Computer modelling and site investigations of noise barriers with complementary noise reduction element, Euronoise 2015, Maastricht, The Netherlands.

75 May, D.N. and Osman, M.M. (1980). Highway noise barriers: new shapes. *Journal of Sound and Vibration* 71: 73–101.

76 May, D.N. and Osman, M.M. (1980). The performance of sound absorptive, reflective and T-profile noise barriers in Toronto. *Journal of Sound and Vibration* 71 (1): 65–71.

77 Mizuno, K., Sekiguchi, H., and Iida, K. (1985). Research on a noise control device: second report, fundamental design of the device. *Japan Society of Mechanical Engineers International Journal Series A: Solid Mechanics and Material Engineering* 28 (245): 273743.

78 Fujiwara, K. and Fruta, N. (1989). Sound shielding efficiency of a barrier with a cylinder at the edge, International Conference Noise and Vibration '89, Singapore (August).

79 Fujiwara, K., Hothersall, D.H., and Kim, C. (1998). Noise barrier with reactive surfaces. *Applied Acoustics* 53: 255–272.

80 Watts, G.R., Crombie, D.H., and Hothersall, D.C. (1994). Acoustic performance of new designs of traffic noise barriers: full-scale tests. *Journal of Sound and Vibration* 177: 289–305.

81 Watts, G.R. and Morgan, P. (1996). Acoustic performance of interference type noise barrier-profile. *Applied Acoustics* 49 (1): 1–16.

82 Hothersall, D.C. (1994). The mathematical modelling of the performance of noise barriers. *Building Acoustics* 1 (2): 91–104.

83 Crombie, D.H., Hothersall, D.C., and Chandler-Wilder, S.N. (1995). Multiple edge noise barriers. *Applied Acoustics* 44 (4): 357–367.

84 Oldham, D.J. and Egan, C.A. (2011). A parametric investigation of the performance of T-profile highway noise barriers and the identification of a potential predictive approach. *Applied Acoustics* 72: 803–813.

85 ISO (2016). ISO 1996-1:2016 Acoustics: Description, measurement and assessment of environmental noise. Part 1: Basic quantities and assessment procedures.

86 EC (2002). Directive 2002/49/EC of 25 June 2002 of the European Parliament and Council relating to the assessment and management of environmental noise.

87 Parker, G. (2006). Effective noise barrier design and specification. *Proceedings of Acoustics 2006*, Christchurch, New Zealand (20–22 November).

88 Farnham, J. and Beimborn, E. (1990). Noise barrier design guidelines. Part 2: Noise barrier design principles, Final report. Center for Urban Transportation Studies, University of Wisconsin-Milwaukee, USA. https://www4.uwm.edu/cuts/noise/noiseb.htm.

89 Bendtsen, H. (2010). Noise barrier design: Danish and some European examples. Reprint report: UCPRC-RP-2010-04, The Danish Road Institute/Road Directorate and University of California Pavement Research Center (May).

90 Klingner, R.E., McNerney, M.T., and Busch-Vishniac, I. (2003). Design guide for highway noise barriers. Research report 0–1471, Effective Noise Barrier Solutions for TxDOT (conducted for the Texas Department of Transportation in cooperation with the U.S. Department of Transportation Federal Highway Administration by the Center for Transportation Research Bureau of Engineering Research), the University of Texas at Austin.

91 Flemming, G.G., Kauner, H.S., Lee, C.S.Y. et al. (2017). *Noise Barrier Design Handbook*. FHWA https://www.fhwa.dot.gov/ENVIRonment/noise/noise_barriers/design_construction/design/design04.cfm.

92 Clairbois, J.P. and Garai, M. (2015). Noise barriers and standards for mitigating noise, CEDR Conference on Road Traffic Noise, Hamburg (8–9 September).

93 I-INCE (1999). Technical assessment and effectiveness of noise walls. I-INCE Publication 99–1.

94 Rongping, F., Zhongqing, S., and Li, C. (2013). Modeling, analysis, and validation of an active T-shaped noise barrier. *Journal of the Acoustical Society of America* 134 (3): 1990–2003.

95 Ekici, I. and Bougdah, H. (2003). A review of research on environmental noise barriers. *Building Acoustics* 10 (4): 289–323.

96 Monazzam, M.N. and Lam, Y.W. (2005). Performance of profiled single noise barriers covered with quadratic residue diffusors. *Applied Acoustics* 66: 709–730.

97 Koussa, F., Defrance, J., Jean, P., and Blanc-Benon, P. (2013). Acoustical efficiency of a sonic crystal assisted noise barrier. *Acta Acustica United with Acustica* 99 (3): 399–409.

98 Sanchez-Perez, J.V., Michavilla, C.R., Garcia-Raffi, L.M. et al. (2015). Noise certification of a sonic crystal acoustic screen design using a triangular lattice according to the standard EN 1793(-1,-2,-3):1997, Euronoise 2015, Maastricht, The Netherlands, paper no 000193

99 Kleinhenrich, C., Weigler, T., and Krahe, D. (2015). The real-time performance of a two-dimensional ANC barrier using a DSP and common audio equipment. Euronoise 2015, Maastricht, The Netherlands.

100 Wijnant, Y.H. and Hooghwerff, J. (2015). A model for diffracting elements to reduce traffic noise, Euronoise 2015, Maastricht, The Netherlands.

101 Hooghwerff, J., Reinink, H.F., and Wijnant, Y.H. (2015). Whisstone, a sound diffractor: does it really affect traffic noise?, Euronoise 2015, Maastricht, The Netherlands.

102 Clairbois, J.P. and Garai, M. (2015). The European standards for roads and railways noise barriers: state of the art, 2015, Euronoise 2015, Maastricht, The Netherlands, paper no. 000607.

103 Heijmans (2017). Heijmans to undertake Solar Highways Project. https://www.heijmans.nl/media/filer_public_thumbnails/filer_public/95/a0/95a08972-ab7f-481b-8d5f-6973ddf2a7d2/heijmans_sonob_solar_noise_barriers_2.jpg__1600x0_q69_crop-scale%20crop_subsampling-2.jpg.

104 Treacy, M. (2015). Solar power generating noise barriers go up in the Netherlands, Technology/Solar Technology (1 July).

105 Treacy, M. (2015). Solar power generating noise barriers go up in the Netherlands. https://www.treehugger.com/solar-technology/solar-power-generating-noise-barriers-go-netherlands.html.

106 Kanellis, M., de Jong, M.M., Slooff, L., and Debije, M.G. (2017). The solar noise barrier project: 1. Effect of incident light orientation on the performance of a large-scale luminescent solar concentrator noise barrier. *Renewable Energy* 103: 647–652.

107 (a) EN 14389-1:2015 Road traffic noise reducing devices. Procedures for assessing long-term performance. Acoustical characteristics.
(b) EN 14389-2:2015 Road traffic noise reducing devices. Procedures for assessing long-term performance. Non-acoustical characteristics.

108 Krynkin, A., Umnova, O., Chong, A.Y.B. et al. (2010). Sonic crystal noise barriers made of resonant elements. *Proceedings of 20th International Congress on Acoustics*, ICA 2010, Sydney, Australia (23–27 August).

109 ISO (2004). ISO 17624:2004 (confirmed in 2017) Acoustics: Guidelines for noise control in offices and workrooms by means of acoustical screens.

110 ISO (2000). ISO 15667:2000 (confirmed in 2015) Acoustics: Guidelines for noise control by enclosures and cabins.

3

Environmental Noise Sources

Scope

This chapter provides a review of the noise sources in categories, that contribute to the environmental noise pollution with respect to their importance in noise impact and management. The manufactoral and operational characteristics of the sources and the factors which are effective in their acoustic energy radiations are summarized, based on the research and implementations widely documented from the past to the present. Detailed information regarding acoustical modeling and measurement of each type of source is discussed further in the later chapters.

3.1   Introduction to Environmental Noise Sources

Sound sources that were explained in Chapter 1, generate sound waves which are perceived as undesirable, unpleasant or disturbing due to the inherent acoustical parameters, are defined as noise sources. However, this definition, emphasizing the unwanted characteristics of noises since the 1950s, is inadequate to describe the negative impacts, since noise creates serious health risks in addition to disrupting hearing, as explained in Chapter 7. Noise pollution – a concept developed in the second half of the last century – is caused by environmental noise sources, affecting many people throughout the world.

  Noise sources, either singly or in combination, generally exhibit complex structures and usually radiate high-level sounds. They are widely spread not only in urban settlements but also in rural areas. For example, a highway passing through a woodland, an aircraft taking off above a forest or a lake or pile-driving in an offshore station radiating underwater noise, all these sources adversely affect the peaceful natural and marine life. In general, the term "environmental noise" implies outdoor noises, however, apart from those transferred to indoors, a variety of noise sources are found in buildings, sometimes having a greater impact on the occupants. Hence one must also consider indoor noise sources within the context of environmental noise management.

  Environmental noise sources are generally composite sources radiating sounds from different parts of their structures through subsequent or irregular processes. The noise emissions are associated not only with the intrinsic features of sources, but also with the physical elements in close proximity to the source. In addition, urban areas contain different sources radiating different types of sounds, -coherent or incoherent- and operating simultaneously or randomly, hence this creates a "mixed noise environment." Basic information about the common noise sources and their acoustic characteristics, depending on their manufactoral and operational parameters, which are mostly

Environmental Noise and Management: Overview from Past to Present, First Edition. Selma Kurra.
© 2021 John Wiley & Sons Ltd. Published 2021 by John Wiley & Sons Ltd.

interrelated, are summarized below to facilitate a better understanding of the emission and immission models and source-specific measurements.

Noise sources have evolved from the past to the present in terms of varieties and numbers, which have increased in line with the extended impact size. Specific investigations according to the source type are carried out, due to the diversity of their structural configurations and the operating techniques influencing the acoustic features, the nature of their impacts and the remedial measures to be taken against them.

3.1.1   Outdoor Noise Sources

Urban noise sources, affecting great numbers of people, can be categorized as follows:

- *Transportation noise sources:* Motor vehicle traffic, composed of light, moderate, and heavy vehicles; railway traffic with locomotives, wagons, etc. at grade, elevated or underground structures; air traffic with aircraft, helicopters, etc., airport operations; and waterway transportation (or vessel traffic) with passenger or cargo ships, boats, submarines, etc.
- *Industrial noise sources:* Mechanical systems, equipment and operations, installation, manufacturing and processes (power stations, plants, workshops, open-air manufacturing and equipment, maintenance and repair activities); all types of machinery, engines, outdoor equipment and installation of building services, such as fans, HVAC (heating, ventilation, and air conditioning) units, cooling towers, chillers, energy generators, etc. and wind turbines, etc.
- *Construction noise sources:* Road and building construction sites employing equipment, i.e. machinery and heavy vehicles for demolition, erecting, assembling, transporting, lifting, etc.
- *Noise sources in relation to outdoor human activities:* Electric and electronic equipment/toys generating noise, public announcements, and emergency systems, fire alarms, recreational activity spaces, such as sports fields, children's playgrounds, shooting ranges, amusement parks, open-air entertainment with amplified music, motor-race tracks, open-air cinemas, and all types of open-air equipment in the neighborhood, such as lawnmowers and loud human voices, shouting, etc.

The noise sources in the first three groups, which are generally composed of various individual sources in a system, have been widely investigated scientifically and technologically. The sources have been categorized and prediction models for noise emission and immission levels have been developed according to the inherent characteristics of each type of noise source. Those in the last group display great diversity and distinctiveness which make simple modeling rather difficult and are described through acoustic measurements to be performed in specific cases.

Table 3.1 outlines the environmental noise sources and their components which are important in the radiation of noise.

3.1.2   Indoor Noise Sources

Noise sources located in buildings generate continuous, intermittent, or impulsive noises to be treated as noise events from individual sources:

- *Neighborhood noises:* Loud conversations and shouting, amplified music, TV/radio, sounds of steps, doors banging, impacts, noises from household equipment and devices, dragging furniture, dropping objects, running children, electronic toys, etc.
- *Mechanical devices and installation noises:* Airborne and structure-borne sounds and mechanical vibrations from HVAC systems, such as boilers, air conditioning (AC) units, cooling towers, terminal boxes, motors, fans, ducts, dampers, diffusers, grills, sanitary system equipment and

Table 3.1 Outdoor noise sources and their components contributing to noise emission of total system.

Generic noise	Specific group	Level/scale	Noise-producing component
Transportation noise	Road traffic	Individual vehicles	Motor vehicles: heavy vehicles (trucks), moderate vehicles (vans), light vehicles (cars) and motorcycles
		System	Traffic and roads (urban, intercity, and motorways)
	Railway traffic	Individual vehicles	Locomotive and wagons passing-by or stop-start at stations
		System	Railway traffic: at grade, elevated or underground and tracks (rails, joints, ballast, sleepers, bridges, etc.), train yards
	Aircraft traffic	Individual vehicles	Aircraft and helicopters, flying over, taking-off or landing
		System	Airport traffic, routes (taxiways, runways, flight paths), ground test zones
	Waterway traffic	Individual vehicles	Waterway vehicles (boats, ships, cargo vessels, etc.) transiting or mooring at quay
		System	Waterway traffic density (number of vessels), ports, harbors, wharfs, and shipyards
Construction noise	Road and building construction	Individual vehicles	Construction equipment, stationary or moving: bulldozer, backhoe, cranes, etc. and operations
		System	Construction site: Number of equipment, operations according to time schedule
Industrial noise	All types of manufacturing processes	Individual sources	Mechanical and electrical equipment and operations
		System	Number and type of equipment in open air or closed premises
Entertainment/ recreational noise	All types	Individual sources	Amplified sound emitters (LS), electronic devices, machines like roller coaster, etc.
		System	Outdoor premises and sports areas
Wind farm noise	All types	Individual sources	Wind turbines and components
		System	Wind farms

installation, such as water flowing through pipes, water pumps, tanks filling and emptying, toilet reservoirs, etc., and waste grinders, garbage collectors.

- *Electrical system noises:* Noises from power generators, transformers, alternators, lighting appliances, etc.
- *Vertical circulation/transportation systems noises:* Noises caused by elevators (lifts), escalators, staircases, moving platforms, elevator machine rooms, auxiliary systems, installation devices, cables, elevator cabin doors, etc.
- *Specific noises:* Other noises in relation to building type and usage.

Acoustic emissions of the mechanical and electrical equipment in buildings are determined through acoustic measurements, as explained in Chapter 5, or for some specific equipment, such as fans, air handling units (AHUs), electric generators, duct noise, etc., the noise outputs (sound powers) can be calculated by using their technical and operational parameters. For the other types such as neighborhood noises, which are usually temporary, it is rather difficult to predict their sound emissions and indoor acoustic measurements are needed. Transfer of indoor noises from one space to another or to outside of building is involved with in the field of "building acoustics" and once the emission levels are obtained, the immission levels can be calculated to obtain noise levels (see Chapters 5 and 6).

3.2 Transportation Noise Sources

Noises generated by different transportation systems are the most common types and these have primarily been investigated in the field of environmental acoustics since the 1950s [1]. They are confirmed as the noises that disturb the most, based on field studies in many countries (see Chapter 7). Emission of transportation noise sources is calculated at two scales (Figure 3.1):

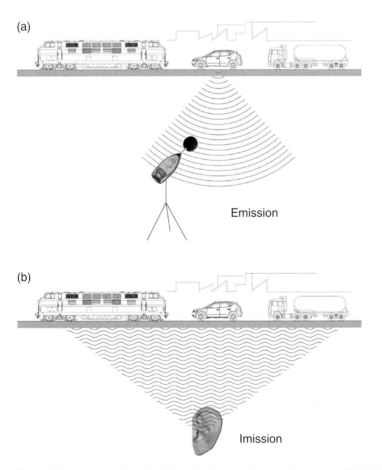

Figure 3.1 Investigation of traffic noise. (a) Individual vehicle noise; (b) Traffic noise. (*Source:* adapted from [2]. Permission granted by VDA.)

1) individual noises of vehicles;
2) the transportation system noise as a whole.

The emission levels of motor vehicles depend on numerous factors associated with the manufacturing and operational characteristics of the vehicle. Vast numbers of investigations have been conducted in the past on noise radiation from individual vehicles, e.g. motor vehicles, trains, aircraft, etc. and numerous scientific and technical documents have been published, with ongoing interest in the topic. Eventually great success in reducing the emission levels was accomplished in parallel to the developments in manufacturing and material technology, thus motor vehicles and aircraft have been much quieter in the twenty-first century compared to the last century. However, the same success cannot be achieved for the complete transportation noise systems due to the increasing speeds, variety of sources, and the increasing spread of transportation routes as a result of urban development. Determination of the acoustic energy of transportation systems needs to examine the contributions of various interrelated factors separately. The results of the experimental studies conducted in the past and the present are summarized in the following sections.

3.2.1 Road Traffic Noise

Of all the transportation systems, road traffic noise is the most widespread in urban and suburban environments, hence, the adverse effects have mostly been conducted through field studies (Chapter 7). A wide range of literature, since the 1950s up to the present, is given in the references, with some providing an extensive overview of specific noise sources [1–52]. As explained in Chapter 1, road traffic is assumed to consist of point sources with different acoustic powers, randomly distributed on the road and moving at different speeds. In the earlier investigations, they could also be treated as a line source with infinite and finite length with respect to receiver position.

A) Noise Emission of Single Vehicles
Acoustically defined as moving point sources, the vehicles radiate acoustic power from various parts of their body, hence it can be considered as point source only at sufficiently long distances. The total noise from a running motor vehicle comprises the following noise components (Figure 3.2) [1–10]:

- engine and transmission noise (air intake, cylinder block, fan, gearbox, and transmission);
- exhaust noise;
- aerodynamic noise;

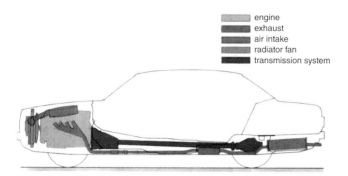

engine
exhaust
air intake
radiator fan
transmission system

Figure 3.2 Noise sources in automobiles [2]. (*Source:* Permission granted by VDA.)

- rolling noise (tire-road interaction noise);
- brake noise;
- warning signals, sirens.

Recent documents divide the vehicle noise components into two groups [30]:

- propulsion noise (power unit noise) generated by the engine and its accessories (the air inlet and the exhaust), the cooling fan and transmission (the gearbox and rear axle), and controlled by engine speed;
- rolling noise generated by tire-road interaction, aerodynamics, body rattle, and controlled by the vehicle speed on the road.

The quantity and quality of the noise from moving vehicles vary with the properties presented in Box 3.1. The motor vehicles have been categorized using different classification systems according to the acoustic output. The earliest grouping of vehicle types was simply categorized as light vehicles, heavy vehicles, and, in some models, moderately heavy vehicles. Acoustically,

---

Box 3.1 Factors Affecting Noise from Road Traffic

Single vehicles:

- Type (weight, no. of axles) and model: heavy, moderately heavy, and light vehicles
- Motor power and structure: gas or diesel
- Speed and acceleration (rpm)
- Radiator, fan, gearbox, transmission system, brakes
- Tire types
- Exhaust and silencers
- Age and maintenance
- Warning signals

Traffic:

- Type of traffic flow (continuous: free-flowing, accelerating/decelerating, and interrupted)
- Volume of traffic (vehicle/unit time: hour or day)
- Traffic composition: heavy vehicle percentage
- Average speed

Roads:

- Width of road
- Gradient (at grade, climbing, or descending)
- Crossings and traffic lights
- Curvatures: radius of curve
- Height of road relative to ground (depressed or elevated structures)
- Embankments and trenches
- Road pavement (asphalt, concrete, stabilized, etc.)
- Maintenance and quality
- Tunnels
- Bridges

trucks of various size are considered heavy vehicles, while cars, vans, small buses, and motorcycles were defined as light vehicles. The acoustic equivalence between light and heavy vehicles was also defined in various documents in the 1980s. For instance, a heavy vehicle was accepted as being equivalent to the sound power of ten light vehicles in France (the Cetur model) [11], while it is equivalent to seven cars in the USA (the Federal Highways Administration (FHWA) model) [12]. In order to be used in the immission models, further descriptions were made with respect to engine type (diesel and gasoline), number of axles, weight, usage, etc. Recently, a standard categorization for motor vehicles has been applied in noise prediction models, in emission measurement, and also in the declaration of noise emission limits, as explained in later chapters.

The past experimental studies in the 1970s and the 1980s resulted in various empirical relationships and diagrams displaying vehicle noise levels according to operational parameters. Speed is an important factor in the noise emission of single vehicles. The effect of speed on the A-weighted noise levels (both in tire-road noise and total noise) varies with the type of vehicle, however, increasing speed after 50 km/h causes an increase in noise levels. Figures 3.3a–3.3c present the relationship of motor vehicle noise to speed, as presented by the Transport Road Research Laboratory (TRRL) in the UK (1977), the Centre Scientifique et Technique du Bâtiment (CSTB) in France (1980), and the US FHWA (1978) [6, 11, 12]. The interrelations between noise level and the speed of vehicles, in different gears, were investigated for different types of vehicles, as shown in Figures 3.4a–3.4c [7, 13].

The dominant noises from single vehicles are the power unit noise at lower speeds, the rolling noise (tire-road interaction noise) at higher speeds, and the aerodynamic noise at moderate speeds. The type of engine has a strong effect on noise levels, which are higher for diesel vehicles. Figure 3.5

Figure 3.3a Reference energy mean emission levels for trucks and cars as a function of speed. (*Source:* adapted from [12].)

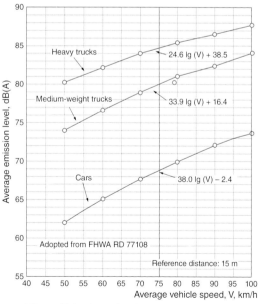

1. Cars: with 2 axles and 4 wheels
2. Medium-weight trucks with 2 axles and 6 wheels
3. Heavy trucks with 3 axles and more

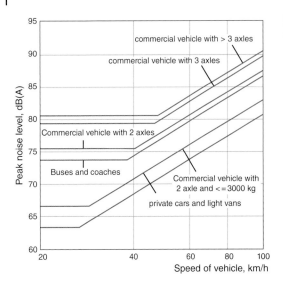

Figure 3.3b Peak noise levels for different vehicle categories in relation to speed in 1977. [6].

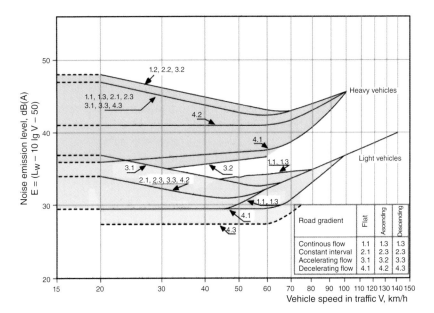

Figure 3.3c Variation of emission levels of light and heavy vehicles with speed under different traffic conditions. *Note:* The numbers display types of traffic flow on three road gradients (on-grade, acceleration and deceleration) shown in the table (Cetur, 1980) [11].

displays the variation of A-weighted sound pressure levels with a motor load and engine rotation frequency per minute (revolutions per minute, *rpm*) presented in [13]. The current measurement standards recommend calculating the A-weighted noise levels as "cruise-by noise" and "coast-by noise" (only tire-road noise) separately. The noise levels increase due to other factors, such as the weight, the age, and the maintenance level of vehicles.

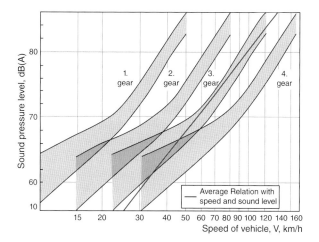

Figure 3.4a Noise levels at 7 m from the road edge as a function of speed in different gear positions [13]. (*Source:* Permission granted by Umweltbundesamt.)

Figure 3.4b Relation between noise level and vehicle speed at different accelerations [7]. (*Source:* Permission granted by Buna.)

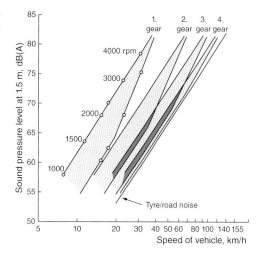

The generation of tire-road interaction noise, which has been investigated since the 1970s, is described as occurring due to a vibration-related mechanism (structure-borne noise) and an aerodynamic mechanism (airborne noise) [37, 38]. The former is due to the impact of tire treads and adhesion properties and the latter is due to air pumping (air displacement and the amplification mechanism causing the horn effect and the resonances that occur in the tire cavity, the tire tread, and the profile canals). Contributions of these two mechanisms to total noise depend on the type of tire, the type of road, and the operating conditions. Tire-road noise levels logarithmically increase with vehicle speed. The effect is stated as 25 dB(A) when the speed increases from 30 to 130 km/h.

Spectral and temporal characteristics: When the frequency-dependent noise levels are analyzed, the low frequency content of heavy and light vehicles is dominant and the spectrums of light and

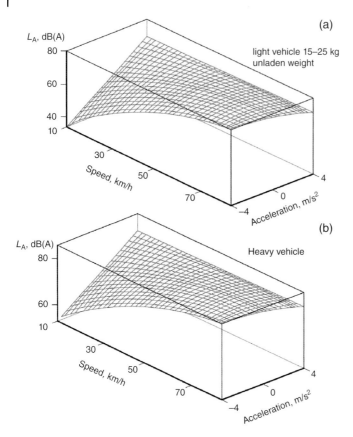

Figure 3.4c Three dimensional representation of A-weighted sound pressure level functions for (a) light vehicles (b) Heavy vehicles [7]. (*Source:* Permission granted by Buna.)

heavy vehicles have different patterns, as shown in Figure 3.6 [8]. The reference noise level of heavy vehicles at 7.5 m from the roadside is 85–90 dB at low frequencies (63 Hz) and drops to 45 dB at high frequencies, i.e. 8 kHz. The spectral noise levels from both type of vehicles are also affected by the type of road surface, such as dry smooth asphalt and porous asphalt called drainage asphalt (Figure 3.7a) [28]. Earlier experiments displayed the frequency-dependent levels versus engine capacity (Figure 3.7b) [9].

Motorcycle noise has intrinsic spectral and temporal features which create a higher annoyance. Due to the frequency modulation in 100–200 Hz, the noise emission needs to be evaluated by using a kind of sputtering index.

Time-dependent noise levels for a single vehicle passing by in front of a constant microphone position and passage of three different type of cars are displayed in Figures 3.8a and 3.8b. The spectrum of a moving vehicle with respect to a fixed receiver point also is subject to change, conforming to the Doppler effect, as explained in Chapter 1.

Warning sirens, particularly on emergency vehicles, contain intensive sound pressure about 100 dB(A) at 7 m with the high frequency components (e.g. at 2000 Hz) and have been regulated due to severe noise disturbance.

Figure 3.5 A-weighted noise levels at 7.5 m in relation to engine speed for loaded and unloaded (idle) conditions [13]. (*Source:* Permission granted by Umweltbundesamt.)

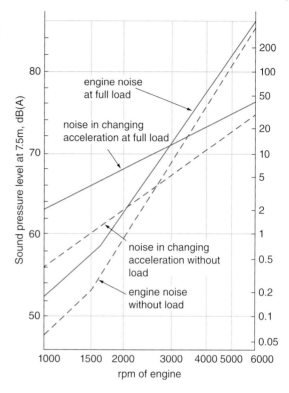

Figure 3.6 Comparison of A-weighted noise level, for light and heavy vehicles moving under freely flowing conditions (Adapted from Lewis diagram, 1973) [8].

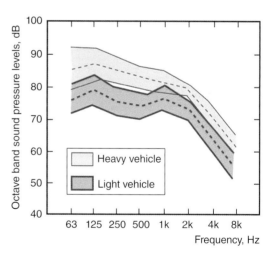

Vehicle indoor noise: Manufacturers are concerned about interior noise in vehicles as this is a quality indicator in marketing. External noise is transmitted through the vehicle floor, doors, and windows, depending on the thickness, density, and absorption coefficient of the material of which the vehicle body is made. Interior noise increases with the speed of the vehicle and the rotation speed (rpm) of the engine. A recent experiment in Europe showed that about 80% of the cars

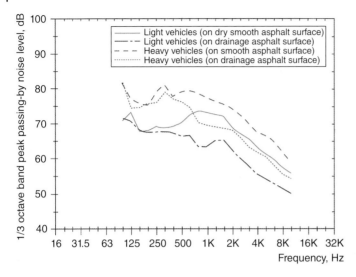

Figure 3.7a Frequency spectra (peak pass-by) for light and heavy vehicles traveling at 80–90 km/h and 70–80 km/h respectively on dry smooth asphalt and porous asphalt called drainage asphalt surfaces, (Adapted from Sandberg, 1987) [28]. (*Source:* Permission granted by Sandberg.)

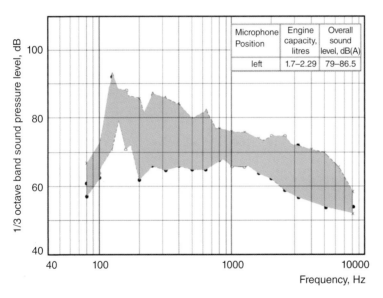

Figure 3.7b Range of noise spectrums of light commercial and passenger vehicles measured at 7.5 m. (*Source:* adapted and simplified from [9]).

had noise levels of 65 dB(A) inside [15]. The statistical distribution of interior levels measured in various cars at 100 km/h is given in Figure 3.9.

B) Noise Emission of Road Traffic

Experimental studies on traffic noise intensified between 1960 and the 1980s and focused on the determination of the basic A-weighted noise levels for free-flowing traffic at a reference distance (e.g. 7.5 or 15 m) in relation to some traffic parameters. Empirical relationships were derived

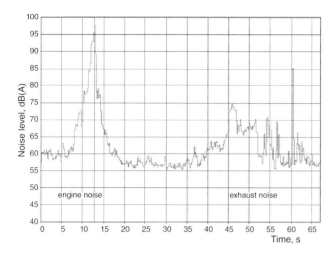

Figure 3.8a Noise of a racing car (Mustang 1994) passing by at 15 m. (*Source:* Kurra, 2008.)

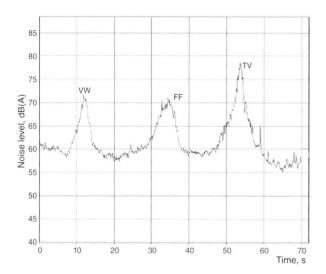

Figure 3.8b Noises of different cars passing by (Volkswagen Polo, Ford Focus, and Toyota van) at the same speed at 15 m from road. (*Source:* Kurra, 2008.)

and recommended in the design guides prepared for use of transportation planners and urban designers (see Chapter 4). The acoustic energy radiated from a road traffic (L_w) is expressed as sound power per unit length of road and used in the prediction models and in noise mapping.

Traffic Parameters Affecting Rolling Noise

Factors affecting the traffic noise emission are summarized in Box 3.1.

- *Road categories:* Types of roads are defined in relation to road geometry (width of the road and the number of lanes, etc.), traffic volume and permitted speed, such as urban or rural roads, motorways (highways), local access roads, etc. Different countries implement different programs to

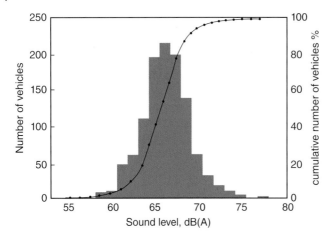

Figure 3.9 Distribution of measured interior noise levels for passenger cars at the same speed of 100 km/h [15].

classify the road network, based on a defined function associated with urban planning, engineering, and safety management purposes (functional classification). Geometrical design standards for each road category have been developed at national and international levels.

- *Type of flow:* Interrupted or uninterrupted (free-flowing) traffic increases and decreases the total noise. The stop-start condition at traffic lights and acceleration or deceleration due to gear changes result in a considerable increase in total noise levels, especially when heavy vehicle traffic is dominant. Figures 3.10a and 3.10b show traffic noise levels recorded in free-flowing and interrupted city conditions, displaying both temporal and spectral variations.

- *Traffic volume:* The traffic density has a huge impact on the total noise, described as vehicle per unit time, e.g. hour, vehicle per 18, 24 hours [18–24]. Some guidelines define the "effective value" with certain weights for daytime and night-time traffic [20, 21]. Figures 3.11a–3.11e display the relationships between noise and traffic volume derived from various sources. In terms of the statistical noise levels; L_{eq} and L_{10} increase up to a certain level with traffic volume, while L_{90} decreases at the same time. Doubling of traffic volume causes a 2.5–3 dB(A) increase in noise levels outside urban areas. For highways, the computations have resulted in a linear increase shown in Figure 3.11c using traffic volume on different road configurations [25]. The rate of increase is reduced due to the slower and even concentrated traffic in cities and the noise becomes consistent, as displayed in Figures 3.11d and 3.11e [18, 26].

- *Heavy vehicle percentage:* A great number of heavy vehicles in traffic implies higher noise levels, as shown in Figures 3.11e and 3.12 [18, 22]. The doubling of the percentage of heavy vehicles increases the noise levels about 7 dB(A) at lower speeds [27]. In fact, traffic volume, average speed of traffic, and heavy vehicle percentage are the parameters significantly correlated with each other.

- *Average speed of traffic:* Under free-flowing conditions, the noise levels display a linear increase with the average speed of traffic. Correlation between the average speed and noise levels has been established in various models, however, the results can reveal a great discrepancy as shown in Figure 3.13 [25]. The recent models given in Chapter 4 provide better relationships based on the long-term experiments.

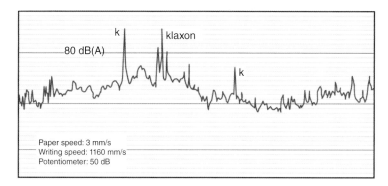

Figure 3.10a Earlier sample recording of traffic noise in the free-flowing condition. (*Source:* Kurra, 1978.)

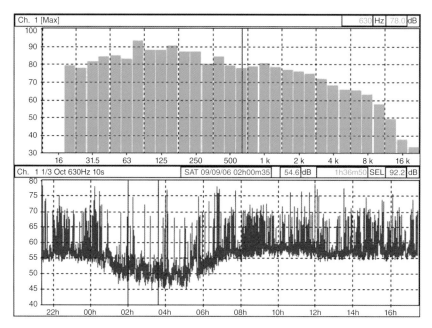

Figure 3.10b Analyses of city traffic noise in 24 hours. Upper: Max. spectrum levels, lower: Time-history diagram Kurra, 2012 [53].

Road Parameters Affecting Noise

- *Road surface:* The effect of the road surface material, e.g. macadam, asphalt-concrete, cobble-stone, etc., on noise generation has been widely experimented on in the past, as an additional factor increasing traffic noise levels, also applicable in the modeling of traffic noise [1, 9, 28–42]. Different classification systems defining road surface types are developed in Europe and in other countries by specifying their acoustic properties derived from the acoustic measurements, as explained in Chapter 6 [17]. Psychoacoustic surveys have revealed the relationship between community disturbance from traffic noise in built-up areas and the type of road surface

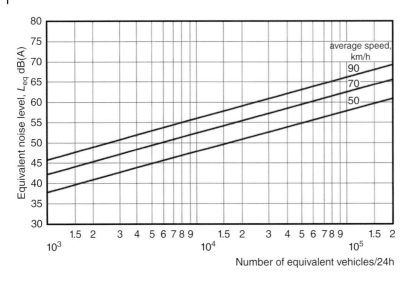

Figure 3.11a Relationship between speed of traffic, traffic volume (as equivalent vehicle/24 h) and noise level in L_{eq}, dB(A) (Adapted from a design guide, 1970's) [20, 21]. (*Source:* Permission granted by Ljuggren.) *Note:* The diagram is valid at ground level at a distance of 100m from the straight road and for open terrain.

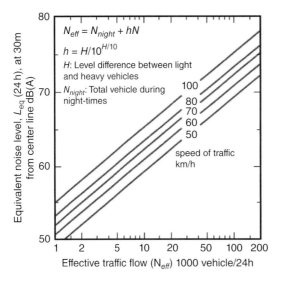

Figure 3.11b Relationship between traffic speed, effective traffic flow, and noise level in $L_{eq,24h}$ at 30 m from the centerline [21]. (*Source:* Permission granted by Ljuggren.)

which, eventually, has been emphasized as a parameter changing the sound quality, such as loudness, sharpness, or roughness of the sound [35].

- The experiments in the 1980s revealed that the surface type (such as dry, smooth, dense asphalt, or concrete) had a positive or negative influence (+6 to -5 dB(A)) on the reference A-weighted noise levels in free-flowing traffic. In a study conducted in the UK in 1979, the variation of noise levels on different traditional road surface types with the same skidding resistance was found to be 4 dB(A) for heavy vehicles and 5 dB(A) for light vehicles [29].

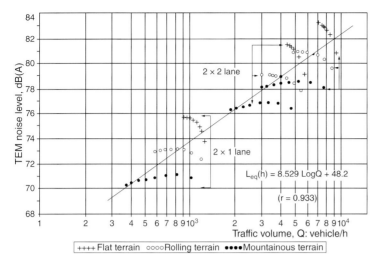

Figure 3.11c Predicted noise levels of Trans-European Motorways with respect to the traffic volumes calculated for different lanes [25].

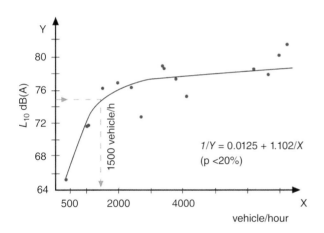

Figure 3.11d Relationship between the traffic volume and L_{10} based on measurements in an urban area (1980) [26].

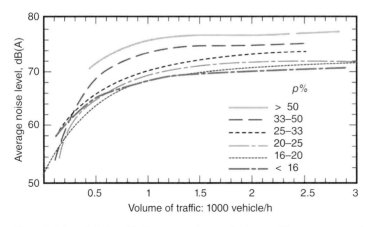

Figure 3.11e Relationship between volume of urban traffic, average speed, and the average noise level [18].

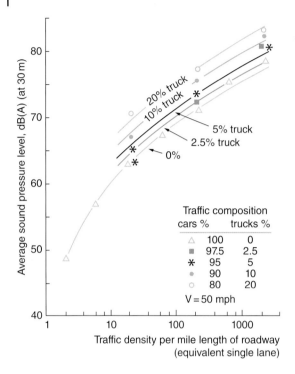

Figure 3.12 Effect of traffic composition on average noise level, dB(A) at 30 m [27].

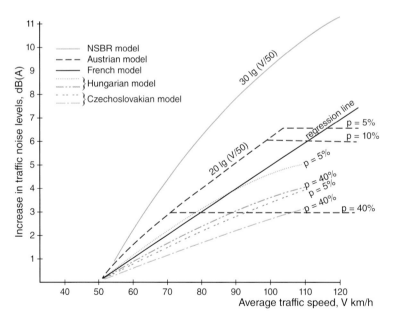

Figure 3.13 Effect of average traffic speed on noise levels according to various models [25].

– Surface properties influencing the tire-road interaction noise are mainly material characteristics, including size and distribution of aggregates, the thickness of the covering layer, and the surface texture (roughness). The noise is caused due to air-pumping called the horn effect, which is strongly related to the surface texture (defined in terms of microtexture, macrotexture, and megatexture, according to the horizontal and vertical dimensions of irregularity and its depth). Roughness is also an important factor in driving safety (increasing skidding resistance) and wet grip (to remove water) in road design, however, roughness can be a factor increasing noise levels. Macrotexture is good since it prevents air pumping and increases safety, and megatexture, corresponding to greater irregularity on the surface, such as worn asphalt pavement, increases noise emission (induced tire vibrations). It was found that the corrugated (grooved concrete) surface, used in some northern countries in Europe, increased noise levels as high as 6 dB(A), as shown in Figure 3.14a [30]. In further experiments in the 2000s, the change in tire-road noise levels, according to the depth of texture was shown in Figure 3.14b for light vehicles [38]. Texture profile can be measured by using standard methods.

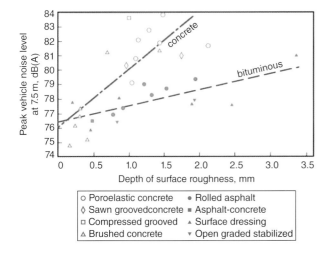

Figure 3.14a Effect of surface roughness on traffic noise with average speed of 70 km/h [30].

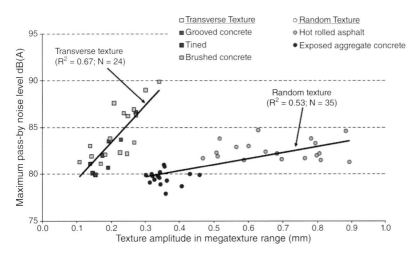

Figure 3.14b Variation of passing-by level (at 7.5 m) of light vehicles at a speed of 90 km/h, with surface texture [32b].

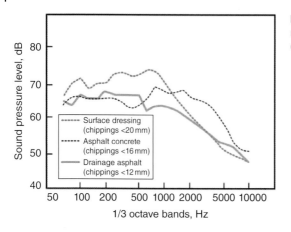

Figure 3.14c Effect of road surface on spectral noise levels for a car at 90 km/h [14]. (*Source:* Permission granted by Sandberg.)

- In the earlier studies in 1987, the difference in noise levels on dry and wet surfaces was found to be significant, i.e. about 1–10 dB(A) [28]. Based on relatively recent studies, wet conditions increase tire-road noise about 5 dB(A) and it is even higher in heavy rain [37, 38].
- In the past, attempts were made to build noise-reducing road surfaces, such as open-graded asphalt or drainage asphalt with chippings, as shown in Figure 3.14c [14]. At present, the search for new materials and applications, called "silent roads" or "low-noise pavements," is continuing in the USA, Sweden, and Japan to reduce tire-road noise [40, 41]. These low noise pavements are grouped as [30]: (i) Porous surface with air void content greater than 20%, such as porous asphalt, which is widely used in Europe and Japan. A porous surface to absorb sound reduces the aerodynamic noise at the source about 5–8 dB, particularly at low and high frequencies. (ii) Asphalt rubber pavements: small rubber granules mixed into a binder or in concrete. (iii) Poroelastic pavements: The porous material including a soft aggregate and rubber particles can reduce the tire-road noise level about 10–12 dB(A) for normal tires [28, 29].
- Comparisons of the sound-reducing performance of different types of surfaces, like mastic asphalt, with surface dressings, cement-concrete with exposed aggregates, pavements with various surface treatments such as chip seals, etc. are shown in Figure 3.14c. The recent results indicate a difference of 17 dB between tire-road noise emissions varying with the surface type, e.g. between the quietest (best porous asphalt) and the noisiest (worst stone) surfaces. The durability of the material and the decrease in acoustic performance over time due to aging, are of importance for long-term urban noise control [42].

- *Tire type:* The effect of the road surface on traffic noise levels is strongly related to the vehicle tires [43–46]. The type of tires with respect to tread pattern is important in noise radiation – also called rolling noise – which is due to the forces interacting at the contact point. The combined effects of tire characteristics, road surface, and speed of heavy vehicles on noise levels are displayed in Figures 3.15a and 3.15b, based on the earlier experiments in the 1970s [2, 23, 36]. The radial tires (e.g. with longitudinal grooves) are about 1–2 dB(A) quieter than cross-ply (or cross-bar) tires. The recent measurements indicate about 10 dB(A) difference between noise levels of "all new" tires. The tire properties that play a role in tire-road noise levels are the tire tread pattern and the road texture, implying smooth tread patterns and smooth roads are the best solutions for a "silent road" criterion and a compromise between quietness and safety in traffic should be achieved. Various investigations have been presented in the literature regarding the "silent tires"

Figure 3.15a Combined effect of tire type and road surface on heavy vehicle noise level at 30 m [36].

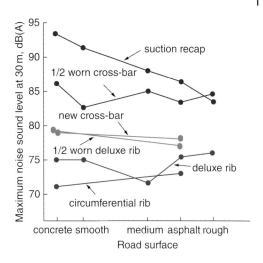

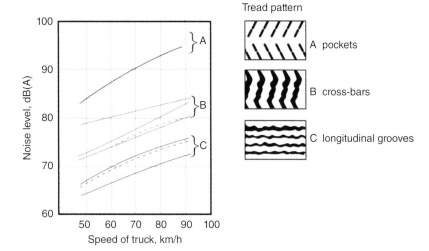

Figure 3.15b Effect of tire type on heavy vehicle rolling noise level at 15 m from the truck moving on concrete road surface as a function of speed. (*Source:* adapted from 2 and 23).

some with innovative designs and discussions about their cost-effectiveness in Europe, like the field study conducted by the road safety operations in The Netherlands [47, 48]. Tire labels displaying the noise level are mandatory in European countries. Noise limits for new tires are given in the EU Directive 2000/43/EC to be based on the measurements according to international standards (see Chapter 5).

- *Road gradient and curvature:* Heavy vehicle noise is significantly increased with the slope of the roadway. Various measurements have evidenced an increase in noise levels of about 5–6 dB(A) on climbing roads for heavy vehicles, due to the lower gears (Figures 3.16a and 3.16b) [19, 21].

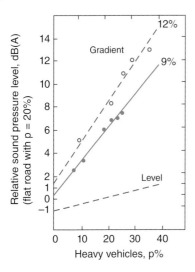

Figure 3.16a Effect of road gradient on noise level [19].

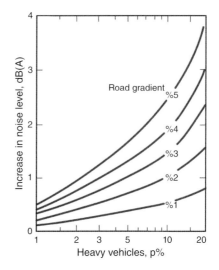

Figure 3.16b Increase in noise levels with road gradient [21]. (*Source:* Permission granted by Ljuggren.)

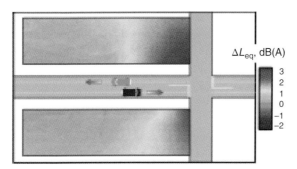

Figure 3.17 Variation of noise levels with distance from a road crossing [49].

Similarly, road curvatures, level crossings, and roundabouts increase the noise levels because of the interrupted flow of accelerations (Figure 3.17) [49]. Elevated road structures and depressed roads affect sound generation in the environment so that the height of the source from the ground associated with the topography and the infrastructure should be accurately determined for reliable prediction of the noise levels (Figure 3.18) (see Chapter 2).

- *Bridges*: Road traffic bridges can increase traffic noise levels at a low frequency range similar to railway bridges. The type of bridge, its material, and the structural connections are the factors affecting the noise generation from bridges [50, 51]. The search for new designs called "silent bridge," e.g. by using an absorptive layer in the cavity underneath the bridge floor, are continuing in Europe, such as in The Netherlands [52]. The prediction models take into account the construction properties by defining a "bridge correction factor" (see Chapter 4).

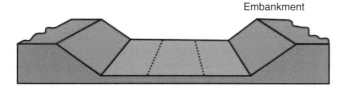

Embankment

Figure 3.18 Depressed road configuration.

To summarize, the following elements imply higher noise levels, a wider spread of noise, and more people disturbed:

- greater number of heavy vehicles
- higher speeds of traffic
- rough road surfaces
- curvatures and crossings
- traffic lights
- climbing roads
- elevated road structures (bridges).

By means of the earlier efforts on quantifying the effects of the above-mentioned factors, modeling traffic noise and accurate predictions have been accomplished at present, as will be explained in Chapters 4 and 6.

3.2.2 Noise from Railways

Although not as extensive as motorways, the rail transportation systems, including various types of trains such as passenger and freight trains, trams, light-rails passing through or along densely popu-lated areas, constitute a kind of environmental noise source sometimes causing serious noise impact. Railway vehicles comprise wagons (railway carriages), pulled by a locomotive, and generate a noise signature which is composed of a discrete series of noise events with similar characteristics, unlike continuous traffic noise [54–64]. As the single noise events are repeated during each train passing by, the railways have been assumed acoustically to be a finite-length line source (see Chapter 1).

Railways generate both airborne and structure-borne noise and also mechanical vibrations occurring due to the rolling wheels on rails (Figure 3.19a).

A) Noise Emission of Railway Vehicles (Stationary and Passing By)
In most of the literature, the three major components of railway vehicle noise are described as: (i) power unit noise; (ii) aerodynamic noise; and (iii) rolling noise. The noise sources during a train passing by are investigated for locomotive and wagons separately. The noise generated from locomotives is composed of the following components [54–59]:

- traction noise (radiated by the engine enclosure and contributed to by the traction engine, the auxiliary systems and the exhaust noise)
- rolling noise (wheel and rail interaction)
- brake noise
- auxiliary equipment noise

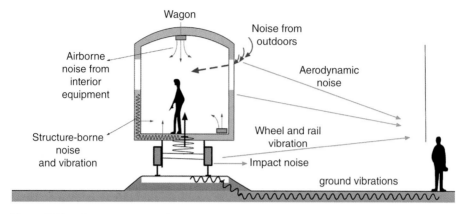

Figure 3.19a Components contributing to railway noise.

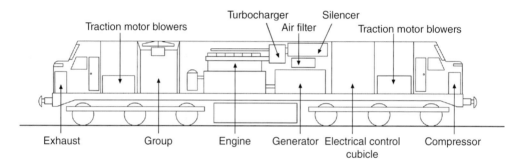

Figure 3.19b Noise sources of a diesel electric locomotive [54].

- aerodynamic noise
- noise from the body during operation
- noise from untightened doors and windows
- siren or whistling.

Noise sources of diesel electric locomotive and in fast trains are shown in Figures 3.19b and 3.19c. respectively. The dominant noise differentiates in relation to the source heights which are standardized for different train categories. For trains of lower source height, rolling and traction noises are emphasized [65]. The secondary noise sources contributing to total emission levels during a locomotive operation and for high-speed trains are shown in Figure 3.19c.

The factors affecting railway noise emission are outlined in Box 3.2.

- *Traction engine noise:* One of the major factors in relation to noise emission of trains is the energy supply system for locomotives, i.e. the electrical power to be provided by diesel generators or supplied by electricity energy lines. The electricity is transmitted to the traction motor that controls the wheels [54, 65]. Experiments reveal that the traction noise levels of diesel locomotives are related to the capacity of the diesel engine and the engine speed and they change according to different operational phases (stationary, acceleration, and

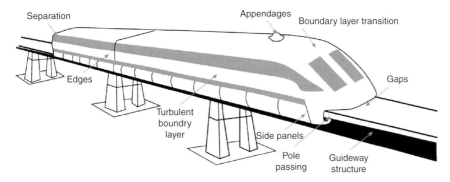

Figure 3.19c Noise sources in fast trains [59]. (*Source:* Permission granted by Hanson.)

Figure 3.19d Components contributing to rolling noise.

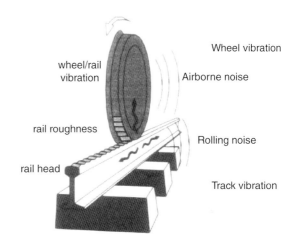

braking). Cooling fans removing the excessive heat also generate noise. Exhaust noise of diesel locomotives is independent of speed and can reach up to 98 dB(A) depending on the type of silencers. Air-intake noise is also significant in diesel locomotives and is heard as a whining noise. Electric locomotives receive their energy from a third rail or supply system on the top wiring that makes them quieter. Noise levels are dependent on the electrical motor, cooling fans, sometimes on gears and speed [54, 60, 65]. Electric locomotives have cooling systems incorporated, with hydraulic pumps. Also, the compressors and exhaust systems removing the pollutant air from the interior and the units controlling brakes contribute to the total noise levels.

Based on the measurements, the maximum sound levels were obtained as 87–96 dB(A) for diesel trains at 30 m from the rails, whereas it was 6–7 dB(A) lower noise for electric trains.

Wagons (railcars), which are less noisy than locomotives, radiate a dominant rolling noise which is a broadband noise caused by wheel-rail interaction. The air-conditioning system and all the auxiliary equipment connected to wagons, according to the mounting properties, contribute to the overall noise. Investigations conducted in the USA revealed that the auxiliary equipment noise in wagons was dominant for the trains at lower speeds (slower than 15 km/h) [54, 55, 65].

Box 3.2 Factors Affecting Railway Emission Noise Levels

Trains:

- Type of trains: train categories (i.e. passenger and freight transportation)
- Locomotives:
  - Traction motor: diesel engine and electric trains
  - Air-compressor
  - Number of axles
  - Electrical equipment
  - Gearbox
  - Brakes
  - Pantograph of electric trains
  - Wheels of locomotive
  - Sleeper support systems
  - Ventilation shaft fans
  - Sirens
- Wagons:
  - Weights
  - Type of wheels
  - Air conditioning and refrigeration equipment
  - Electrical equipment
  - Type of brakes
  - Pantograph
  - Gearbox

Railway line (track):

- Type of rails and rail pad, design of rail head
- Rail surface roughness and maintenance
- Railway joints (bolted, welded) and switches
- Ballast (material, thickness), lateral connections
- Material of sleepers for railway track (concrete or wooden) and under sleeper pads
- Horizontal curvature radius
- Embankments
- Type of railway structure (at grade, elevated, tunnel, and cuttings)
- Elevated structures and constructions
- At-grade crossing
- Tunnels and geometry
- Viaducts, bridges, and structure types
- Stations and stop and waiting points

Railway operations:

- Single or double track or more
- Number of locomotives and wagons (length of train) at each track
- Total train volume in unit time
- Train composition (density of each type trains)
- Average speed of each train category
- Sirens and restrictions if exist
- Stop-start conditions (acceleration/deceleration)
- Train yards and service operations (maneuvering and maintenance)

The type of brakes also has an effect on noise output and the trains. Noise from brakes while stopping is a serious problem for freight wagons, especially when the cast-iron brake-blocks are used, however, nowadays they are made of composite materials throughout Europe, reducing the noise levels significantly. Trains with disc-brakes are about 10 dB(A) less noisy than those with cast-iron blocks.

- *Rolling noise:* Wheel-rail interaction noise, called rolling noise, is associated with train speed, type of wheels (e.g. damped or not), and rail parameters, e.g. rail joints (bolted or welded), rail roughness, and the curvature of the railway. The total noise level of trains passing by is contributed by:

  Rumble due to uneven contact when wheels and rails are interacting during operation.
  Clashing due to rail joints.
  Curve squeal while the vehicle travels on a curvature.

Wheel and rail roughness (irregularity of vertical profile of the rail) is an important factor in the airborne and impact noises from both rail and wheel. Squeal noise is caused due to vibrations of the wheel and track by the train traveling on the curves [56].

Rolling noise is a broadband, random noise in the range of 100–5000 Hz, mainly over 250 Hz with peaks at 500–1250 Hz. However, at speeds less than 80 km/h, the noise spectrum is dominated by the lower frequencies below 250 Hz with the peaks at 40–125 Hz. Figure 3.19d shows the components contributing to rolling noise. Increase in rolling noise with train speed is shown in Figure 3.20. Prediction models for rolling noise have been developed and incorporated into the railway noise models, which are explained in Chapter 6. Squeal noise is intermittent noise mainly above 5 kHz, however, the stick–slip noise generated on a large amount of rail roughness, is a lower frequency noise.

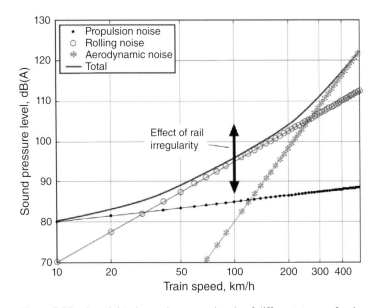

Figure 3.20 A-weighted sound pressure levels of different types of noises as a function of train speed [60].

- *Sirens and warning sounds:* A secondary but important noise source is the train signals, consisting of intense high frequency tones. The total sound pressure levels can be 105 dB(A) at 30 m in front of the train, although the levels are 5–10 dB(A) lower at the side of the railway at the same distance. The acoustic characteristics of permissible train signals are described in the standards.

B) Noise Emission from Railway Operations (Traffic)

Train categories: Trains are grouped in order to investigate the effects of the inherent factors on noise generation which are important in noise simulation models. The main categories are passenger trains and freight trains. In comparison, the available categories are, in the USA, the freight trains, which are longer and heavier than those in Europe, forming the majority, whereas in Europe, passenger trains dominate with a great diversity. Other categorization of in-service trains, such as suburban trains (commuter/regional), intercity trains, and high-speed trains (rapid transit trains), light rails (trolleys), express trains, etc., might change in different countries. Regarding noise emissions, basically two categories were accepted in the 1980s: The light trains with weights ≤1.5 tons and the heavy trains with weights of >1.5 tons [54, 55, 65]. However, at present, ten categories with respect to speed, length in addition to weight, are standardized for European trains as applied in the prediction models (see Chapter 4) [58]. Modern trains are getting faster and the interiors are quieter, however, efforts are continuing to control noise generated outside, particularly by the rapid transit systems. The relative spectral levels of different speeds are compared in Figure 3.20.

Speed effect: Train speed is an important factor determining railway noise. All the secondary noise components, such as propulsion noise, wheel-rail interaction noise, and aerodynamic noise are speed-dependent. As seen in Figure 3.20, the steepest rise in the A-weighted noise level is observed for aerodynamic noise with increasing speed, and the rolling noise reaches the same level at high speeds. Aerodynamic noise is not important for slower trains with less than 240 km/h. [56, 59, 60, 65]. The effect of train speed on the A-weighted noise levels and on the spectral levels is also shown in Figures 3.21a and 3.21b respectively, based on experiments conducted on Chinese trains in 2015 [61]. Relations between train speed and A-weighted peak noise levels for different passenger trains, with different types of brakes, are shown in Figure 3.22, which indicates that the cast-iron brakes

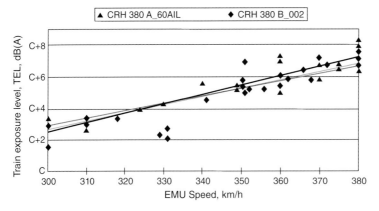

Figure 3.21a Diagram of high speed trains' electrical multiple unit (EMU) radiation noise with change of train speed [64].

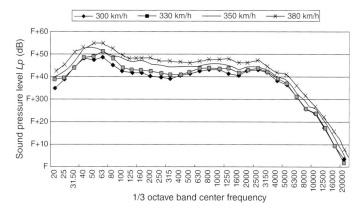

Figure 3.21b Comparison chart of high speed trains' (EMU) noise frequency spectrum at different running speed [64].

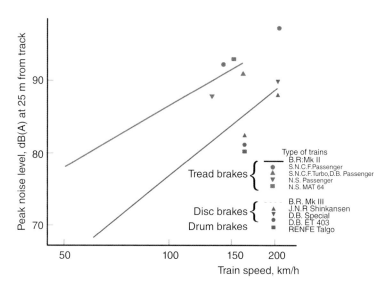

Figure 3.22 A-weighted peak noise levels of various passenger trains with respect to brake types (at 25 m from the rail) [55a, 55b].

are noisier than other types [55a, 55b]. A number of experimental studies have been conducted on the noise sources of high-speed trains especially in the USA, the UK, China, and Japan where railway noise is a serious problem (Figure 3.23) [59, 62–64].

For moving trains, the rolling noise from wheel-rail interaction is correlated with the term of "30$\log v$" while aerodynamic noise is correlated with "60$\log v$" (v: train speed in mph). When the speed is greater than 150 mph (241 km/h), the aerodynamic noise becomes dominant [54, 55a, 55b, 60a, 60b, 65] (Figure 3.24). L_{Amax} of rolling noise increases 9 dB(A) by the doubling of speed which is important up to 30 m on both sides of the railway. The total noise increases equally at higher speeds while the contribution of the rolling noise remains stable. Rolling noise for lower speeds is less important while the train is decelerating.

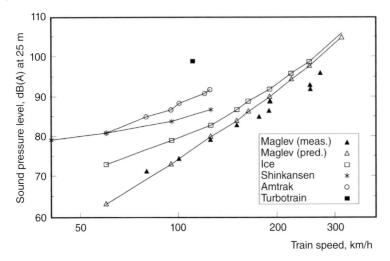

Figure 3.23 A-weighted noise levels of various freight and passenger trains according to train speed [62]. (*Source:* Permission granted by Hanson.)

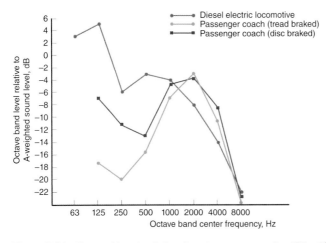

Figure 3.24 Spectral levels of diesel and passenger trains [55a, 60a].

The experiments revealed that for train speeds less than 50 km/h, the rolling noise is dominant, but for speeds above 50 km/h, the propulsion noise is dominant, if the propulsion engine is air-conditioned. When the total train noises of the low-speed freight trains and the passenger trains at about 100 km/h are analyzed, the locomotive noises are found to be dominant [65]. However, for the high-speed trains traveling at speeds more than 200 km/h, such as the Train Grande Vitesse (TGV) in France and Shinkansen in Japan, the rolling noise is prominent. For the transit trains in the USA, the variation of noise levels in relation to the speed of trains is given in Figure 3.25, at different railway structures (e.g. at grade and on concrete bridges) [65].

Figure 3.25 Effects of speed and railway structure on noise level of transit trains (at 15 m from the center of the track). (*Source:* Adapted from [65].)

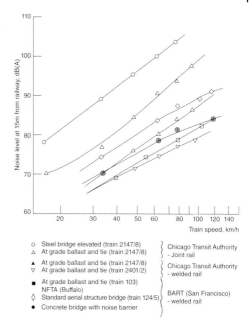

Steel bridge elevated (train 2147/8) ○
At grade ballast and tie (train 2147/8) △ } Chicago Transit Authority - Joint rail

At grade ballast and tie (train 2147/8) ▲
At grade ballast and tie (train 2401/2) ▽ } Chicago Transit Authority - welded rail

At grade ballast and tie (train 103) ■
NFTA (Buffalo)
Standard aerial structure bridge (train 124/5) ◊ } BART (San Francisco) - welded rail

Concrete bridge with noise barrier ●

Railway configuration: The effect of railway (track) infrastructure on railway noise generation is examined with the factors increasing noise levels, however, these are also important in noise abatement technology:

- *Sleepers and ballast bed*: The major railway tracks are composed of ballast on the same plane and rails standing on traverses which are lateral (transverse to the rail) connections, called sleepers. Sleepers transferring the load to the ballast can be made of various materials, e.g. wood (timber), metal, cast iron, or concrete. Wooden and concrete sleepers have identical noise conditions. Ballast is a layer of stones or gravels placed below and around the sleepers and provides lateral stability and keeps the verticality of the rail track, in addition, it absorbs noise. However, the material can reradiate noise by its vibration during trains passing. The thickness of the ballast bed influences the impact of the structure-borne noise and the vibrations propagating into nearby buildings. The effect of the ballast thickness on the spectral sound power levels, generated by the ballast vibrations, is shown in Figure 3.26 as prepared based on calculations [66]. Greater thickness of ballast increases sound power significantly at frequencies below 300 Hz. In some cases, concrete slabs are used as a base to support the rails, however, they increase the rolling noise by about 2–10 dB(A).
- *Rail roughness:* The roughness of the rails increases the noise about 10–15 dB(A) [65]. The rough surface of the railway tracks causes higher noise emission and ground-borne vibrations. Rail roughness can be monitored and the corrugations are removed by grinding the rail head [67]. The recent "Quiet track" project, supported by the EU, developed a model to assess the "rail unevenness profile" for the prediction models [68–70]. Smoothness of the wheel tread is another factor reducing noise. Rail joints that can be bolted or welded change the noise levels, i.e. continuously welded rail joints are about 5 dB(A) less noisy than bolted ones.
- *Curvature:* Wagons moving on curvatures, whose diameter is less than 100 m, cause a noise up to 89 dB(A), consisting of one or more tonal components. For smaller radius of track curvature, the stick–slip motion of wheels generates intense high-frequency sounds called "curve-squeal"

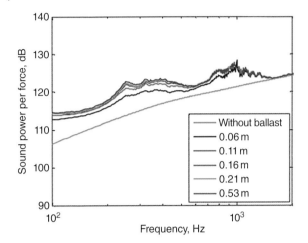

Figure 3.26 Effect of ballast vibration on radiated sound power of the sleeper embedded in ballast on rigid ground for different ballast widths on one side of the sleepers, m (based on the calculated results) [66].

which increase the disturbance [67]. The prediction models take into account this effect which increases the total noise levels up to 8 dB(A) as a correction value.

- *Bridges:* The elevated structures for trams and regular or transit rails, running on viaduct structures or bridges, influence the noise generation and spread in the environment. When the rails are laid on the elevated steel structures without using ballast and if the rails are directly fastened to the viaduct construction or steel bridges, the noise levels increase significantly. Two types of bridges made of different materials can be examined: Light-weight steel bridges or heavy-weight concrete constructions. Light structures increase the noise and vibration problems, both in the environment and inside wagons, due to the rail joints and the ballast support systems. Experiments revealed that the noise levels are higher about 16 dB(A) in steel bridges compared to concrete bridges [57]. However, the problem is controllable by various techniques. Combinations of concrete and steel structures are beneficial because of the damping effect of the rail bases. Using heavy ballasts, concrete surfaces on the bridge, and resilient rail connections minimize the total sound [60a,b].

Passing-by noise: The acoustic characteristics of railway vehicles, while moving freely or during stop-start operations near stations, vary considerably [53–68, 71]. Noise impacts during trains passing by are taken into account by the length of the train and the length of time it takes the train to pass. The determination of time-varying patterns of passing-by noise is displayed in Figures 3.27a and 3.27b for locomotive and wagons, respectively [58]. The length of the locomotive and the wagons and the passing-by time play a role in modeling the railway noise, as explained in Chapter 4. Attenuation of the rolling noise in relation to distance from the railway is shown in Figure 3.28 as a function of the train length [55a].

Examples of field measurements are given in Figure 3.29a (1980) and in Figure 3.29b (2012), displaying temporal variations in the noise levels of suburban and freight trains when they pass by [26, 53]. As seen, the noise levels increase and decrease while the locomotive is running. The noise levels are lower and more consistent during longer periods when the wagons are passing.

A number of investigations have been conducted in Japan where railway noise is given prominence in the entire transportation system [43]. In these studies, the relationships were obtained between the related parameters referred to above, and the noise levels expressed in L_{Amax}, L_{AE}, and L_{50}, also by considering the background noises.

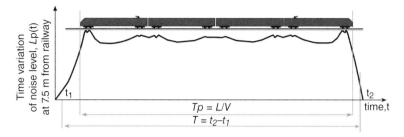

Figure 3.27a Variation of noise level at a receiver point when a locomotive is passing by [58].

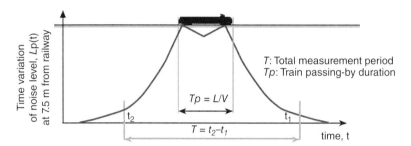

Figure 3.27b Variation of noise level at a receiver point when wagons are passing by [58].

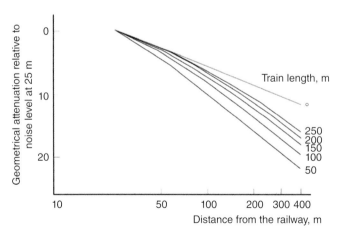

Figure 3.28 Geometric attenuation of rolling noise level, dB(A) relative to train length and distance [55a].

- *Acceleration and deceleration and stop-start at stations:* The noise levels increase because of the diesel locomotive engine or propulsion engine, cooling fans, indoor noise in cars, ancillary equipment (which is not relevant to train movement), air conditioning fan, chutes (shafts), cooling system, etc. Type of fans, size, and fan silencers are important in the total noise [54, 60a, 65].

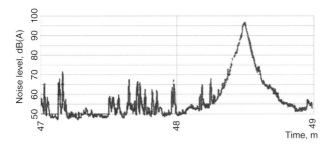

Figure 3.29a Train noise recorded during the passage of an electric train at 25m. (In 1980's) [26].

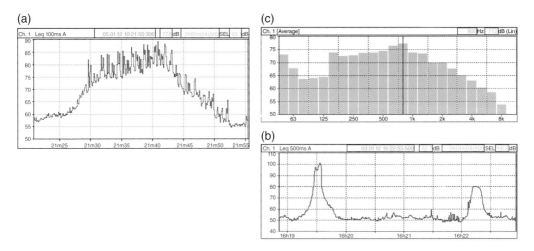

Figure 3.29b (a) Train noise recorded during a passage of a suburban train at 30m from a railway; (b) the time-history of noise levels and (c) average noise spectra [53].

- *Effect of tunnels:* When trains are passing through the tunnels, noise levels are increased because of the multiple reflections from concave surfaces also causing a focusing problem inside the tunnel (Figure 3.30). Structure-borne sounds and vibrations are transmitted to the soil around the tunnel and to the building walls. When spectral levels are analyzed, the noise levels are higher about 20 dB at 125 and 250 Hz, compared to the lower frequencies [72]. For the high-speed trains entering a tunnel, the problem called "tunnel boom" arises and is important to consider in environmental planning [73]. Noise at the tunnel entrances can be modeled for use in noise mapping and can be controlled by using appropriate shielding (Chapter 6). Figures 3.31a and 3.31b show the time-history and average spectral levels recorded inside the metro tunnel near a station The noise levels are high due to the lack of absorption inside the tunnel and station [53].
- *Ground-borne noise and mechanical vibrations:* In addition to airborne sounds, the impact noises and mechanical vibrations are generated from trains passing by. Therefore, specific measures are needed to protect the buildings located at 10–15 m from the railways. The noise spectrums contain very low frequency components between 10–200 Hz and the noise levels reach 50–70 dB(A) at 15–30 m from the rail. The peak frequency is 16–32 Hz [74].

Railway noise abatement has been an issue of concern for railway designers since the 1970s and various solutions have been sought and are being implemented. As an example of an innovative

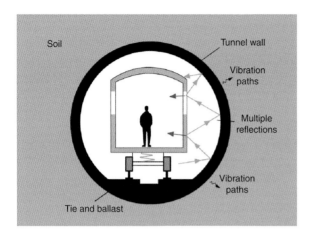

Figure 3.30 Noise and vibration components in tunnels.

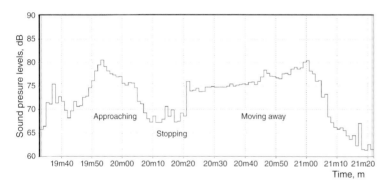

Figure 3.31a Noise measurements in a metro station: Variation of A-weighted noise levels during metro train passing by [53].

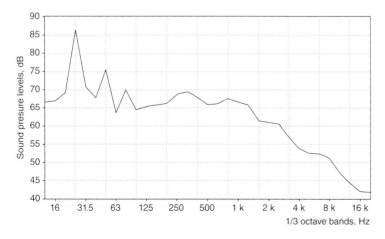

Figure 3.31b Noise measurement in a metro station: Average one-third octave band spectrum [53].

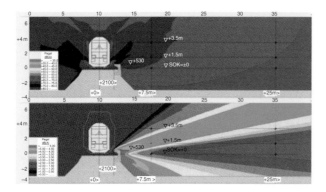

Figure 3.32 Screening effect of a special noise-reducing device. Upper: Overall sound pressure level, Lower: Sound reductions compared to free-field levels [75].

design, a low height shielding element shown in Figure 3.32 was tested to reduce significant rolling noise and structure-borne sound [75].

C) Noise Emissions from Service Operations

Train operations during stopping at stations, assembling and uncoupling trains, maneuvering and maintenance in train yards, generate high-level complex sounds which are difficult to estimate, since they are related to various factors, such as number of trains, operational state, technical aggregates installed on trains, power supply, the safety standards, etc. Usually the noises that can spread over a large area are at high frequencies, e.g. intense sounds about 120 dB at 30 m, thus cause a great disturbance in nearby urban buildings [76].

As a summary, this section has discussed the following:

- trains with diesel locomotives
- trains with greater number of locomotives and wagons
- high-speed trains
- bolted rail connections
- rough/worn rails and wheels
- railways on metal construction cause higher railway noise pollution and greater disturbance in the environment.

3.2.3 Aircraft and Airport Noise

Aircraft noise is of an intermittent nature with high noise levels and causes serious pollution in urban environments near airports and within their services. Increases in global air traffic in the last two decades because of lower airfares, shorter travel times, etc. compared to the past, have resulted in an expansion in the size and capacity of airports, however, this increases the adverse effects and arouses community reactions, particularly against aircraft noise.

Developments in civil aviation and jet technology have resulted in larger, faster, and noisier aircraft in the last century, nevertheless, attempts to reduce noise emissions of civil aircraft have been successful over the last three decades. Technological improvements enabled 75% quieter aircrafts between 1966 and 2000s (https://www.icao.int/SAM/Documents/2014-ENV/3.2.Noise%20 TechnologyV3_notes.pdf).

Unlike motor and rail vehicles, the movements of aircraft are not restricted in a line or a plane, therefore the noise can spread over a larger area under the flight paths. The projection of an aircraft on a ground plane is called the operation path, defined by the coordinates of aircraft positions with the angles of height and azimuth at each instant of time [77–84]. Modeling aircraft operation paths for prediction noise is given in Chapter 4.

Airway transportation noise can be investigated at two stages: (i) noise emission of single flights; and (ii) noise emission from airfields.

A) Noise Emission of Aircraft

The International Civil Aviation Organization (ICAO) has issued noise standards for subsonic jet aeroplanes, propeller-driven aeroplanes, and aeroplanes driven both by subsonic jet and heavy propellers. Categorization of aircraft is made with respect to various aspects that are detailed by the US Federal Aircraft Administration (FAA) and the ICAO and other documents:

a) Size of aircraft.
b) Take-off gross weight (small and light aircraft with a single engine for short-range operations and large and heavy aircraft with turboprops for long-range operations).
c) Wake vortex classification (wake turbulence categories).
d) Passenger capacity (number of seats).
e) Aircraft performance categories (A–D) with respect to speed given in knots.
f) Aircraft approach category with respect to maneuvering and approach speed.

Regarding noise emissions, the earlier aircraft noise models simply grouped air vehicles into civil aircraft (passenger and freight) and military aircraft. With respect to manufacturing characteristics influencing noise radiation from various components, detailed classifications of aircraft are made according to engine size and speed: turbo-prop, turbo-jet, turbo-fan, supersonic aircraft, and helicopters. Although the noise emissions of air vehicles differ significantly, the overall noise contains distinct components that contribute to the overall noise during departure, landing, and flyover:

- air intake noise (ventilator, compressor, etc.)
- engine noise (engine jet creating turbine noise, engine fan, combustion chamber, afterburner, undercarriage, etc.)
- exhaust noise (aerodynamic jet noise)
- airframe noise (structural noise and vibrations and aerodynamic noise).

Figures 3.33a and 3.33b display these noise sources from the earlier documents in 1980's [82].

An aircraft's power source is a single or multi-engine of different types, which use a turbine. The energy is produced when the fuel is ignited in the compressed air within the turbine which is heated in the combustion chamber. The thrust is produced from the expansion of gas through a jet muzzle (exhaust). Turbo-jets are examples of turbine engine and are used for long-range aircraft and supersonic flights. The noise components are intake, vibration of the engine body, and exhaust noise (aerodynamic jet noise), which is the dominant noise. Exhaust noise is generated by the interaction (turbulence) of the high velocity jet stream and the still air in the atmosphere and dominates at low frequencies. The most efficient engine is a turbo-fan which drives a fan at the front of the engine and provides some air passing around the engine which aids thrust. The fan and compressor noise are dominant noises with broadband and tonal characteristics. As a third type of engine, turbo-props drive a propeller at the front of the engine, adjusting the air entering by gears and a shaft drives the compressor; some turboprops can reverse the flow direction in time. They are efficient

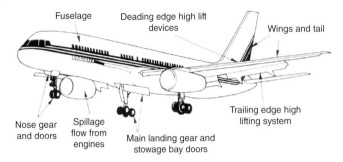

Figure 3.33a Sources of airframe noise. (*Source:* Adapted from [82].)

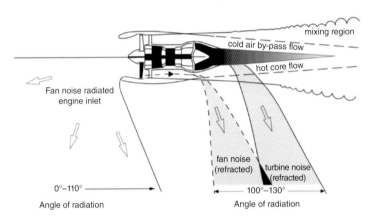

Figure 3.33b Noise radiation from internal engine sources in aircraft (*Source:* Adapted from) [82].

at lower speeds and are also used in military aircraft since they provide a compromise between efficiency and speed. The dominant noise source in turbo-prop aircraft is the propeller noise which consists of a fundamental frequency with harmonics. Figure 3.33c displays the noise sources in three types of aircraft engines, and the spectral characteristics of radiated noises based on the earlier and recent documents [77–84].

Airframe noise from moving aircraft causes friction and turbulence and the wind is an important factor in the radiation of noise into the air. The structural noise, depending on the type of aircraft, is generated by various parts, such as the power systems, the flap condition, and the air speed management systems. The primary factors affecting the noise output of a single aircraft are shown in Box 3.3.

According to the noise standards issued by the ICAO, since 1971, all aircraft need to obtain ICAO certification based on the noise emission measurements, explained in Chapter 5. Strategies for the mitigation of aircraft emission noise are still of great concern to manufacturing companies.

Aircrafts are noisier than other transportation vehicles. For example, the acoustic power radiated by a jet aircraft is 30 kW higher, compared to a human voice which has acoustic power less than 1 mW. The total noise level near a jet engine is above 130 dB(A) [78a]. Analysis of the spectral characteristics reveals that the dominant engine noise is a constant broadband noise ranging from the lower frequencies up to 2.3 kHz and it is higher than the aerodynamic noise [77] (Figure 3.34).

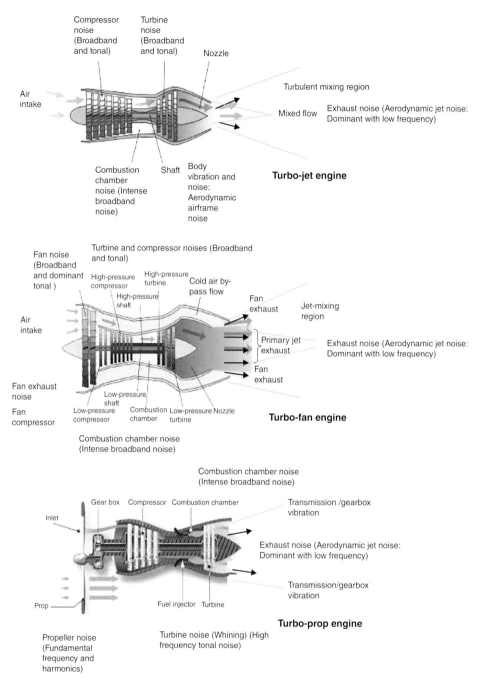

Figure 3.33c Noise radiation from different aircraft engines (compiled from the literature).

Box 3.3   Factors Affecting Noise from Aircraft and Airports

Aircraft:

- Type: Aircraft categories with respect to manufacture and operational characteristics
- Engine type and jet exhaust
- Operation procedures (speed and variation, power usage, e.g. half or full power)
- Take-off weight and power
- Flight profiles (take-off and landing angles)

Flight procedures:

- Layout of take-off and landing positions on runway
- Number of runways, layout, and distance to built-up areas
- Coordinates of flight routes and footprints

Airport operations:

- Volume of flights: Daily total, number of landings and take-offs for each aircraft category
- Variation of volume with season, day, and night
- Percentages of different aircraft categories in total aircraft volume
- Location of ground operations and maintenance areas
- Type of ground tests and screening

Note: The environmental factors are not included. Operational conditions are strongly related to wind profiles.

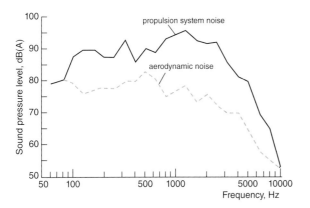

Figure 3.34   Frequency spectra of a turbofan aircraft during approach (measured at an altitude of 150 m) [77].

Aircraft noise has a strong directivity and the max energy is reached at 135° and 225° on both sides of the aircraft axis depending on the number of jet engines (Figure 3.35) [78a]. The lateral and vertical directivity patterns of aircraft noise can be examined by using 3D models [79, 80]. Figure 3.36 displays the directivity of noise during a flight test [84].

Figure 3.35 Directivity pattern for overall sound pressure levels of a 6-engine jet aircraft on the ground and during flyovers with different speeds [78a].

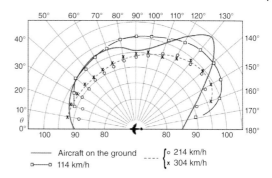

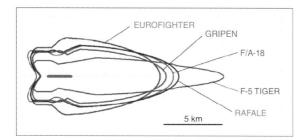

Figure 3.36 EMPA flight tests: noise immission and directivity evaluations for military aircrafts [84].

B) Noise Emission from Aircraft Operations

Aircraft operate according to flight procedures approved by the ICAO. Flight noise can be examined in three stages of flight operation regardless of the aircraft type [84, 85]:

a) Take-off (departure) (Figure 3.37a).
b) Landing (approach) (Figure 3.37b).
c) Overhead flying (flyover) and maneuvering.

The sound pressure levels and spectral characteristics of noise during these operations are significantly different, hence requiring special concern of each [77–84, 86].

- *Take-off:* Jet noise is at a maximum level because of the high engine power required during departure and decreases with the height of the aircraft. The low frequencies are dominant in the spectrum [26, 85, 87] (Figures 3.38 and 3.39a).
- *Landing:* Engine noise decreases while landing, however, the compressors and turbines generate strong narrowband noise, consisting of a high frequency component at 2000 Hz, perceived as a whining sound [85]. Figure 3.38 compares the spectrums for landing and take off [87].
- *Overhead flying:* Noise levels received at an observation point on the ground change with the position of the aircraft (i.e. the distance from the aircraft to the receiver and the angle of view). The spectral shape of aircraft noise also varies during flyover; while approaching the receiver or microphone, the high frequency air intake noise is high, while moving away, the lower frequency jet exhaust noise is substantial [78a]. During straight flying above the receiver, the maximum noise is obtained and the aerodynamic noise becomes dominant. Regarding the temporal variation of

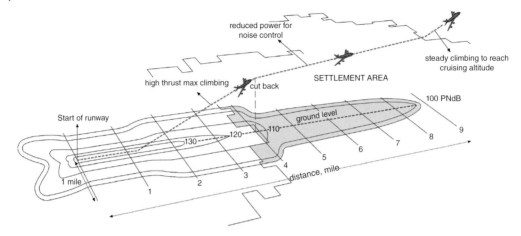

Figure 3.37a Flight profile during aircraft take-off [86].

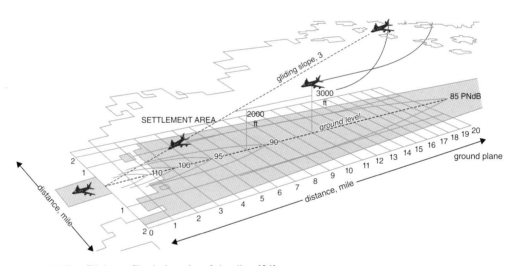

Figure 3.37b Flight profile during aircraft landing [86].

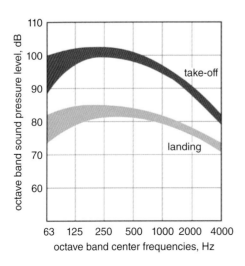

Figure 3.38 Typical spectrum range for aircraft during landing and take-off [87].

Figure 3.39a Noise spectra recorded for various aircraft measured in 1981 [26].

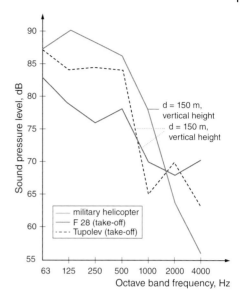

Figure 3.39b Noise recorded during aircraft flyover in the 1980's [26].

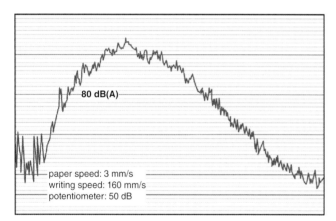

noise levels during the flyovers, two examples are given in Figures 3.39b and 3.39c: the first was recorded in 1981 for a single aircraft and the second was obtained during the long-term measurements in 2012 at a busy airport. The aircraft altitudes were about 150 m and 800 m in two different measurements [26, 53]. Advanced measurement techniques enable detailed analyses of temporal and spectral characteristics of aircraft movements, as shown in Figure 3.40 [78b].

- *Supersonic flights:* Military aircraft and passenger jets like Concorde, which has stopped operations at present, sometimes reach supersonic speeds. As explained in Section 1.3.3, an aircraft flying at supersonic speed creates a shock wave [77, 78a, 83]. The shock wave is a pressure variation defined as "*N*-type" and the pressure rises rapidly above atmospheric pressure and then drops rapidly to a negative value. In this condition the consecutive booms (about 2–3 or more blasts) are heard within 0.1–0.2 second intervals. The two types of shock waves are: (i) During the straight flight, a constant shock wave is received on the ground (Figure 3.41a); (ii) During the flight with increasing and decreasing speeds or diving, the sonic wave is radiated from aircraft

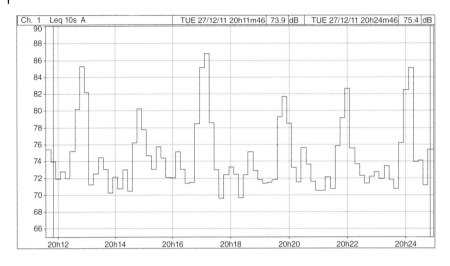

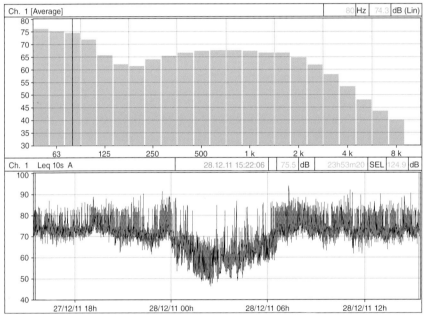

Figure 3.39c Time-history diagrams during different aircraft flyovers (of short and long durations) and the average spectra [53].

toward the ground with greater amplitude (Figure 3.41b). In each situation, the noise blasts might have equal energies, however, the largest area on the ground, called the sonic carpet, is exposed in the first case. Due to the shock waves, the mechanical vibrations in addition to the blast noise can emerge at certain locations on the ground and can have negative impacts on humans and buildings.

- *Helicopter noise:* Due to the unsteady flight conditions (e.g. maneuvering, changing the angle of descent, acceleration, deceleration, etc.) and the variation of speed, flight path, slope angle, and radius of turn, helicopters (rotary aircraft) are very noisy air vehicles even while flying steadily.

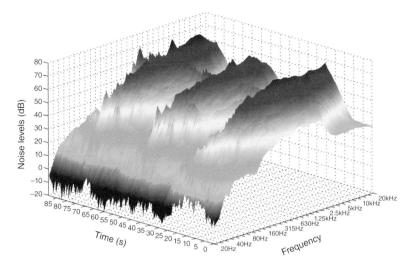

Figure 3.40 Time-frequency spectrum of three aircraft approaches [78b]. (*Source:* Permission granted by Khardi.)

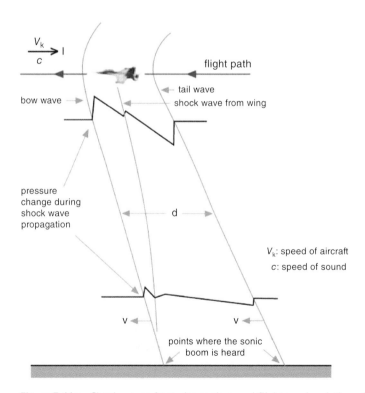

Figure 3.41a Shock waves from ultrasonic speed flights and variation of sound pressure as *N*-wave. (*Source:* adapted from [78a].)

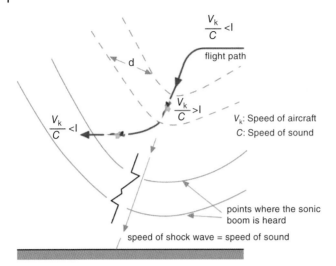

Figure 3.41b Shock wave radiation from a military aircraft at supersonic speed. (*Source:* adapted from [78a].)

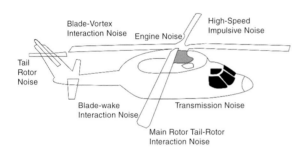

Figure 3.42 Noise sources in helicopters.

The dominant noise is radiated from the rotor blades which is called the blade/vortex interaction noise or simply the vortex noise. The rotor (body and tail) produces aerodynamic forces on the fan vanes similar to aircraft with propellers and radiates noise as monopole, dipole, and quadrupole sources. When a helicopter is as low as 10 meters above a receiver point, a strong whistle is heard. Total radiated sound varies in terms of magnitude and the directivity pattern. Figure 3.42 shows the noise sources in helicopters [88, 89]. Modeling helicopter noise is rather difficult.

C) Noise Emission of Airports

Airports, sometimes called airfields or aerodromes in the past, that might accommodate one or more runways, are categorized according to their size and volume of aircraft traffic (i.e. the numbers of take-offs and landings). Noise emission from a total airfield consists of three dimensional point sources, some moving on the ground plane, some in the three dimensional space, generating a great variety of sound power, and can be determined through modeling (see Chapter 4) or monitoring (Chapter 5). Currently governments are very involved in airport noise mitigation strategies through airport management systems and aviation noise policies (see Chapter 9). The airport noise sources are investigated at two scales which are also concerned in airport noise management:

Overall Flight Noise Emission at Airports

Airport noise is a complex phenomenon contributed by a number of independent sources, very time- and space-dependent, and is relative to the airport capacity, flight procedures, flight management, airport operation rules for civil airports, and the design of the airport, including runways, taxiways, apron areas, terminals, protection areas (safety), etc. Time-varying noise contours based on the real-time noise measurements can provide emission levels, which are important not only for the surrounding settlements but also for passengers, crew, and terminal personnel, etc.

Noise from Service Operations at Airports

Ground tests conducted before aircraft departures generate considerably high-level noises depending on the test procedure and type of aircraft. The noises are directive according to the aircraft position and the type of equipment used in the process, as shown in Figure 3.36 [84]. Generally, service noise (from tests and maintenance), which is composed of different components created by various movable equipment and vehicles, is very high, with the intense tonal components that cause hearing risk for the operators and ground service staff. They are required to use ear protectors and silencers for test devices to reduce noise [86]. Figure 3.43 shows the ground service noise sources documented in the 1980's. It is difficult to assess the noise emission of such complex noise sources operating simultaneously or individually, depending on the airport facilities and the service pattern, therefore, specific noises can be determined by acoustic measurements in service areas during preparation, maintenance, and repairs. The overall noise at airports is a combination of all the operational noises in addition to the departures and landing noises explained above.

Examples of runway arrangements with the facilities are shown for two airports in Figure 3.44.

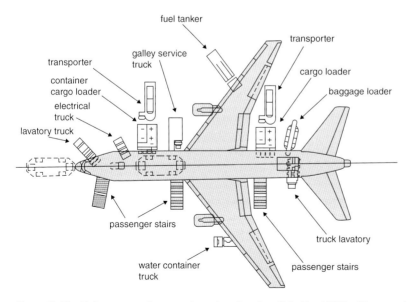

Figure 3.43 Noise sources in ground services for aircraft in the 1980's. (*Source:* adapted from [86].)

Ataturk airport, Istanbul   O'Hare airport in Chicago

Figure 3.44 Runway layout of two airports, in Istanbul and Chicago.

3.2.4 Waterway Traffic Noise

Waterway traffic is not as common in urban areas as the other transportation systems, however, the local communities close to transit lanes on seas, rivers, lakes or canals, particularly near the piers, wharfs (docks), seaports, harbors, marinas, or shipyards, can be seriously affected by the noise pollution caused by transportation and naval operations. Noise from vessels, including military or cruise ships (simply called "ship noise"), previously was not a concern, except in marine engineering and in underwater acoustics, as an example, the underwater noise radiated by submarines had significance in World Wars I and II. At present, the numbers of studies on waterway noise as an environmental problem have increased to protect people and marine life, because of the increase in cruise traffic and the emergence of greater tonnage of freight vessels all over the world. From the standpoint of noise impact assessment, waterway transportation noises are investigated at different scales [90–99]:

1) Waterway vehicle noise.
2) Waterway traffic noise (transiting or free-sailing noise).
3) Port noise (delivery, loading/unloading, embarking, equipment operating, service processes, etc.).
4) Noise from shipyards (manufacturing, production, maintenance, tests, etc.).
5) Underwater radiated noise.
6) Off-shore construction noise (pile-driving, etc.).

Waterway vehicle noises cause serious disturbances for crew and passengers on board, in addition to those living and working in nearby environments. Noises inside passenger cabins interfere with sleep and rest and relaxation and with other activities (like communication) in dining and living areas, corridors, etc., thus, negatively influencing the quality of ambient atmosphere. The excessive mechanical engine noises can cause noise-induced hearing loss due to high noise levels, prevent speech intelligibility, reduce work performance, furthermore, noise in staff cabins causes sleep disturbance and creates health risks. Problems are increased because of high levels of noise containing low frequencies and weak insulation of the interior partitions in all kinds of vessels.

A) Noise Emission of Waterway Vehicles

Waterway vehicles can be classified according to various aspects, which are important in marine engineering, such as ocean-going vessels, river-going vessels, and sea vessels, however, with respect of noise emissions, they are categorized as:

- passenger carriers (short trips or pleasure cruises): Cruise ships, ocean liners, barges on canals, boats, ships, sailboats, sea buses, leisure craft, etc.
- freight carriers (for trade and military purposes): Cargo and container ships, ferries, heavy tonnage vessels, all-use diesel engine refrigerated roll-on, roll-off, some carrying wheeled cargo.

Identification of noise sources in vessels is rather difficult because of their large size, and the distribution of discrete noise sources on board radiating different type of noise. The primary and secondary noise sources in a vessel, contributing to the total acoustic energy, have been explained in various publications [90–94]. The parameters which are important in noise emission of individual vessels are summarized in Box 3.4.

Box 3.4   Factors Affecting Noise from Waterway Transportation

Single vehicles:

- Type of vessel (category)
- Weight in tonnage (gross tonnage)
- Hull dimensions, ship structure, interior design, partition materials
- Number and type of engines, mounting details, and bases
- Cooling, ventilation (AC), and exhaust system and installation
- Other ancillary systems
- Age and maintenance
- Speed of transiting
- Cruising direction
- Ship horn (siren)

Transportation:

- Total volume of waterway traffic in unit time
- Composition of vessel categories
- Seasonal differences in vessel flow
- Average transiting speed of each category
- Routes of different vessels
- Width of waterway
- Size of port/harbor and capacity (number of ships moored)
- Density of loading and unloading processes in ports and harbors
- Location and size of shipyards
- Service buildings and configurations
- Topography and shape of coastline (land profile, elevation of hills, etc.)
- Urban buildings and open-air relaxation areas in the proximity

---

Noise Sources in Vessels

Figure 3.45 displays some of the noise sources on board in a moored cruise ship [93].

1) Machinery noise sources:

a) Propulsion machinery including diesel engines, steam turbines, gas turbines, main motors, gearbox, etc. [91]: The airborne noise radiated by the engine is proportional to the engine speed and the combustion pressure and the noise from scavenger and exhaust housing, the noise from ventilating ducts, are of importance. The acoustic power of the sources and emission levels are determined based on their weight. The steam turbines in the largest ships (with gross tonnage greater than 60 000) are very noisy, although they might be less noisy compared to the internal combustion engine and the steam valves radiating an intense sound at high frequency. Noise from ventilating fans, which is generated from the casing or cabin, can be as high as 120 dB(A). Engine exhaust noise is composed of low frequency pure tones, broadband noises, and high frequency sounds.

b) Auxiliary equipment, including pumps, compressors, generators, air-conditioning equipment, hydraulic control systems.

The engine room noise may be contributed by more than one engine when in full sail and the sound is amplified by the reflections from the room surfaces, increasing the noise levels up to 100 dB(A).

The relationship between engine speeds and the spectral levels in 1/3 octave bands, is given in Figure 3.46a for a cargo ship [90].

In addition to the acoustic waves, the mechanical vibrations which are caused by the propulsion machinery, the ship's services, and the auxiliary installations, including steam, water, and hydraulic piping systems need to dealt with separately. Vibratory forces due to the mechanical imbalance of equipment depend on their weight, power, and the bases as well as the connections with ducts and pipes.

2) Hydro-acoustical noise sources: Hydro-acoustical noise generated underwater is caused by propellers, cavitation, vortex shedding, and turbulent boundary layer flow-induced noise and transmitted to the ship hull by flow interaction with attachments, cavities, and other discontinuities [94]. Propeller noise is a kind of turbulence noise created by

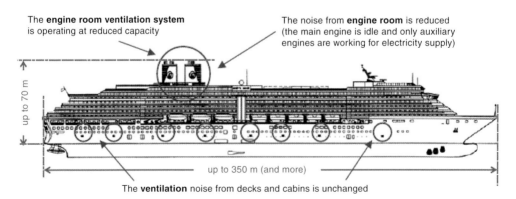

Figure 3.45 Noise sources on board of a moored cruise ship [93]. (*Source:* Permission granted by Di Bella.)

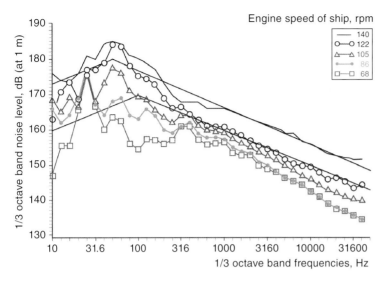

Figure 3.46a Spectral levels of noise (in dB re 1 micropascal), generated by a cargo ship at various diesel engine speeds [90].

cavitation on the propeller blades and depends on the blade characteristics (number, type, surface).

3) Other vessel-specific noise sources: The additional noise sources in relation to the type of vessel, ship design, and operation are summarized below:

a) Noise sources in ferries, cruise vessels, and yachts:
- HVAC for cabins
- ventilation plants for garages; ventilation noise is a broadband noise
- electrical power supply units (diesel generator sets)
- ancillary machinery (e.g. winches, hydraulic motors).

In addition to airborne sound generated by the above sources, the impact noise and vibrations from engine departments and from mounts of engines and auxiliary motors, are propagated on board, depending on the ship structure.

b) Noise sources in passenger ships:
- Noises on board and in passenger cabins are composed of noises coming from various parts of vessels, such as the machine room, the restaurant, the kitchen, etc. The measurements indicate a great variation in noise levels in the passenger sitting and relaxing areas. The main noise sources, which are important for the passengers' and crew's health and comfort, are the propulsion mechanism, propelled by diesel internal combustion engines, the exhaust systems of motors, ventilators, and appliances, such as hydraulic generators, steam valves, etc. Figure 3.46b displays a histogram of the measurements in workspaces on a ferry, based on a study conducted by Borelli, Gaggero, et al. [92]. The noise levels in engine rooms were measured as being as high as 107 dB(A) and 123 dB(C) at peak level. The spectral levels of noise on board recorded at the stern for different operating conditions are given in Figure 3.46c.

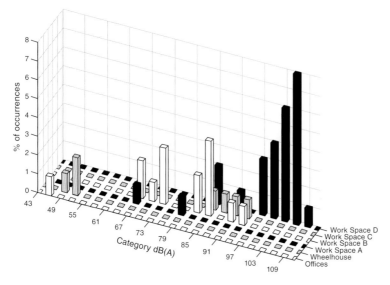

Figure 3.46b Histogram of the measurements in the work spaces in a ferryboat [92].

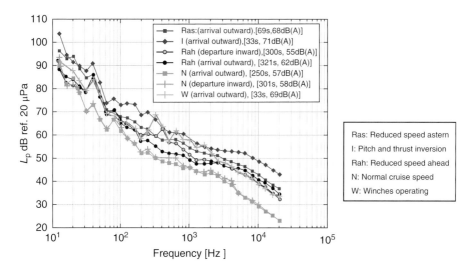

Figure 3.46c Mean spectra recorded at the stern for different operating conditions [92].

- Time histories of the engine room noise levels of a passenger ship while sailing and docking, are given in Figures 3.47a and 3.47b.
- In addition to the airborne sounds, the structure-borne sounds and mechanical vibrations are generated from the light-weight structures on which the engines and installations are mounted, and transmitted to the cabins and other rooms. A ship's structure has usually very low insulation, especially the dividing elements on board, i.e. between the engine room and the corridors, between cabins, service areas, and other functional spaces. The interior partitions, such as walls, floors, and ceiling, made of light-weight constructions and the air-conditioning ducts transmit the airborne and structure-borne sounds between the interior spaces even to the distant rooms.

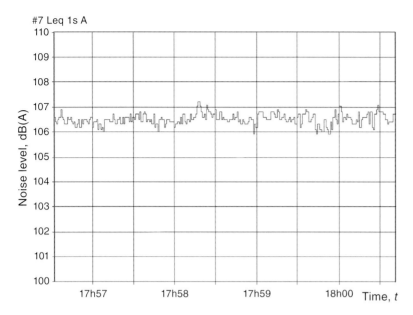

Figure 3.47a Noise measurements in the engine room of a passenger boat in free sailing. (*Source:* Kurra, 2008.)

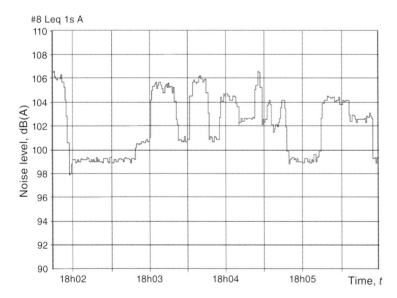

Figure 3.47b Noise measurements in the engine room of a passenger boat while docking at the pier. (*Source:* Kurra, 2008.)

c) Noise sources in cargo ships:
- Cargo-handling equipment (from the ship or from ashore): Grabbers, conveyors, gantry cranes, ramps for vehicles
- Plants on board.

Generally, noises in vessels are a combination of various types of noises [91]:

Steady noises from machinery, gears, etc. containing both broadband noise (or modulation of broadband noise) having a continuous spectrum, and tonal noise containing discrete frequency or line components generally dominated at low frequencies of 10–1000 Hz.

Transient and intermittent noises caused by impacts, loose equipment, or rudders and unsteady flow.

The spectral characteristics of noise from specific parts of cargo ships vary in a broad range, e.g. diesel electric propulsion systems: 10–1000 Hz, turbines: 50–200 Hz, propeller cavitation noise: 50 Hz to ultrasonic frequencies, propeller blade noise is 5–50 Hz and auxiliary units: 15–5000 Hz [91].

Sound power levels of various specific equipment contributing to the overall noise level, such as the diesel generator exhaust, the sound power of fans, etc., can be determined through calculations and acoustic measurements, then the total sound emission of ships is obtained [90, 94]. Sound radiation from the ship's structure and the propagation of noise outdoors, i.e. seashore or riverside, can be investigated by modeling the "source-transmission path-receiver" system, as explained in Chapter 4. In addition to airborne sound transmission, the structure-borne sounds throughout the ship's structure which are re-radiated as airborne sounds, must be be taken into account. Surface ships in transit are treated as dipole sources because of reflections from the water surface.

B) Noise Emission of Waterway Transportation

The important factors affecting noise generated by water traffic and services as a whole, are given in Box 3.4. The flow of transit vessels can be changed daily, weekly, particularly for the passenger cruise ships in the tourist seasons. Freight traffic for deliveries and good supply may be increased on weekdays in river ports.

Total emission of vessel noise can be obtained simply by categorizing them based on their tonnage and cruising speeds, then introducing these into the emission models to assess the noise impact in surrounding areas (Chapter 4). The total waterway traffic noise, which is composed of contributions from the transit flow of all kinds of vessels, is a broadband noise containing dominant low frequency components of high levels. They are heard as "roaring" at far distances because of the reflections from the sea surface. Since ships at sea travel by leaving sufficient distance between them, according to the rules of safety and do not follow the same lines, they are examined separately as moving point sources in simulation models. Calculation of the noise levels at a point outside a vessel can be made by the methods given in Chapter 4.

Figures 3.48a and 3.48b give two examples of noise levels measured at a distance of 15 m from a fishing boat while docking and leaving. The measurements regarding seaway traffic in a dense cruising lane, including various types of vessels and boats with different speeds, are shown in Figure 3.48c as average spectral levels and time-varying sound pressure levels [53]. One of the problems generated by tourist (leisure) boats or cruising night clubs is the amplified music which

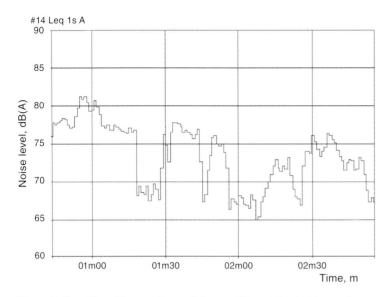

Figure 3.48a Time-history of a small boat at 15 m while docking and maneuvering. [53]

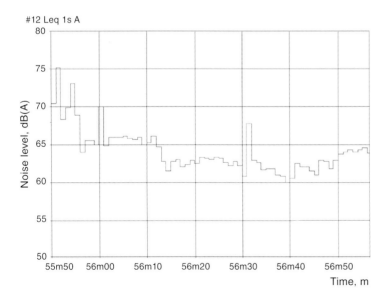

Figure 3.48b Time-history of a small boat at 15 m while leaving the dock. [53]

causes great disturbance to the seashore settlements during day and night while they are passing slowly and at anchorage in the quiet bays. The time-history displaying the overall noise recorded at the façade of a residential building just on the seashore, contains such loud music generated by one or two leisure boats (Figure 3.48c).

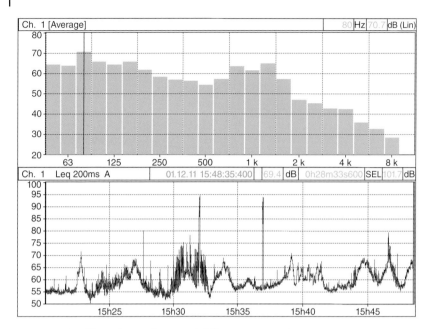

Figure 3.48c Time-history diagram of different sea vessels cruising and the average spectra at 20 m [53].

- *Port noise*: Overall noise from harbor areas is contributed by different noise sources depending on the operations of the vessels, such as loading/unloading, etc. The noise emerging due to acceleration from zero knots to full speed in a very short time, from rundown of engine to zero speed with zero pull, noise from engines running idle (with 600 rpm), and the noise of the propellers with the clutch out, are independent components of the total noise. When the ships are moored at port or at the quay, a number of machines are activated, such as ventilation, heating, AC for cabins, etc. Port noise which is a disturbance for visitors, ship, and port staff, can be treated as a plane source if the noise sources are randomly distributed on the area, or as mixture of point sources whose emission should be determined individually. For ships in port, the sound power level of the diesel generator exhaust can be 135–142 dB(A) and the ventilation fans can be 81–110 dB(A), and a ship horn is intensely high at 1–2 kHz [99]. A shipyard, which is another source of pollution, needs to be investigated separately, similar to industrial premises or construction site. Assessment of noise levels from a port and shipyard are made by proper identification of noise sources and by taking into account the physical environmental factors [95].

- *Underwater radiated noise (URN)*: The overall noise radiated by surface marine vehicles (i.e. commercial ships, military craft, and fishing research ships) and commercial vessels, mainly due to the propulsion power, turbulence in the water flow around the ship's hull (i.e. propeller cavitation), and the machinery on board, is transmitted from the engine into the water, mostly as periodical sounds of various tones. The sonar systems generating underwater signals during naval operations are also significant noise sources. The underwater noise propagates at long distances because of the greater speed of sound (1500 m/s), the greater acoustic impedance (1.5×10^6 Pa·s/m), and the lower absorption coefficient (about 0.06 at 1–500 Hz) of sea water [96]. URN is a special concern in environmental acoustics. The adverse effects of vessels on marine life are of great concern globally, since it has been proved that the noise could mask the acoustic signals of marine mammals and can harm them by obstructing their communication, breeding, etc. At present, underwater acoustic maps are becoming more common to investigate the expansion of shipping noise for the protection of the environment (see Chapter 6).

- One type of *URN* is the offshore construction work noise. Marine construction operations, such as pile driving, wind farm construction, hammering during oil or gas exploration, drilling, etc. produce low frequency impulsive sounds with high amplitudes under the sea. Underwater noise levels are associated with a number of factors, such as pile size, the energy of the hammer strike, the nature of the sea bed, etc. The off-board measurements have revealed that the underwater noise levels were as high as 210–250 dB (ref. $1\ \mu Pa^2 m^2$) at 1 m and containing the dominant frequencies below 1 kHz [100].

3.3 Industrial and Mechanical Noise Sources

Industrial premises involve manufacturing processes at different phases. A vast variety of devices and machines are used in manufacturing processes, creating significant noise pollution if they are close to residential areas and even in rural areas. Industrial facilities may be in large buildings (e.g. power plants or petro-chemical plants associated with storage and delivery and transportation of materials or products) or in open-air fields (e.g. rock-crushing plants) or small workshops that might be located near urban buildings.

Although in modern urban planning systems, industrial zones are organized at a good distance from noise-sensitive areas, however, due to unplanned urbanization in some countries, these buildings may be distributed randomly within urban areas. Noise characteristics differ widely according to type of industrial activity, operation type, machinery, devices, and installations. In their own way as different as those of transportation vehicles, industrial noise sources are operated according to a specific work schedule at certain intervals, repeating the same job in consecutive phases. The location of industrial premises, the distance from urban buildings, the size and shape of the area, the outdoor equipment and installations mounted on external elements of buildings (walls, roofs, etc.), are important in noise assessments. Noise emissions can be investigated at two scales:

1) Single vehicle, machine, or installation.
2) Industrial area, array of machines, and operations.

3.3.1 Noise Emission from Single Machines and Installations

Types of operations performed for different processes generate different noises. Some of these processes are shown below:

a) Impact (pressing and compression)
b) Mechanical processes (power transmission, cutting, etc.)
c) Fluid flowing (fans, air compressors, pumps, etc.)
d) Combustion (burners, flame shooting, blasts, etc.)
e) Electromagnetic processes (electric motors, generators, transformers, etc.).

Three types of vibrations are radiated from individual devices [101–105]:

- Airborne sounds radiated from the machine bodies and exhaust systems.
- Airborne sounds, structure-borne sounds, and impact noises radiated from machines and transmitted by the connections, mountings and bases, into the building structures and constructional elements.
- Mechanical vibrations generated by their body and parts and transferred through bases and connections.

The factors affecting the acoustic power during the operation of machines or devices are given in Box 3.5. As referred in Chapter 1, rotating and reciprocating machines create tonal sounds

---

Box 3.5 Factors Affecting Industrial Noise

Machines and equipment:

- Structure of machine (compatibility of gears and bases, type of material, rotation speed of fans, etc.)
- Engine power and rotation frequency, fan, exhaust and transmission systems, installation connected to machines
- Surface size, parts, and other segments vibrating independently
- Material of machine enclosure, stiffness, and thickness
- Age and maintenance
- Job performed and technique implemented
- Operation time and intervals (operation mode), number of repetitions, and duration of each
- Layout of machines and equipment in the production area
- Mounting characteristics (shape, bases, connections)

Manufacturing area (production stations):

- Number of machines working at the same time, types of each
- Operation modes of machines
- Positions of machines in relation to each other
- Architectural and structural properties of the building
- Layout (location) of building in relation to environment (nearby buildings)
- Reflective and absorptive surfaces around, reverberation time of space
- Sound barriers around machines
- Outdoor openings on the walls and roof (for AC and other purposes, such as ducts, pipes, for lighting, or doors)
- Layout of machine within the space
- Mounting surface and properties

---

comprising harmonic components. Any device causing an air flow radiates broadband sound dominated at low frequencies. The highest sound pressures are caused by the high-speed gas flow, for instance, fans, vapor pressure vanes, etc. Processes covering mechanical impacts, such as hammering, riveting, stamping, breaking, directly generate structure-borne sounds that can also be radiated into airborne sounds. Nowadays, thanks to technological improvements, many of those processes generating high noise levels in the past have been abandoned.

Beyond the complex equipment and installation in large plants, there are various devices and machines used in buildings and outdoor environments, such as domestic appliances, road sweeping machines, rubbish containers, lawnmowers, etc. as other sources of noise, although they are in use temporarily, at certain hours of the day. Causing permanent noise sources in neighborhoods are the building services placed outdoors, such as heat pumps, exhausts of AC systems, cooling towers, and other installations as examples. The ducts for heating and ventilating, pipes for sanitary and plumbing systems transferring fluids like air, gas, vapor, or water in buildings also transmit noise from engine rooms into the noise-sensitive spaces in buildings. The new technologies have produced quieter devices and machines which are of great benefit from the standpoint of environmental noise control.

The problem of high levels of noise in workplaces from machinery or engine sets, which differ greatly with respect to their structural and operational features, emerges due to the following deficiencies in the sources:

- high speeds of moving parts
- high speeds of moment changes
- weak silencers in air intakes and exhausts
- type of cooling system and size
- irregular power of engine
- loose parts and faults in some part of the machine
- unstable movement of parts
- the attitude of the machine operator and the driver
- lack of lubrication
- weak damping on components, etc.

Depending on these factors, the characteristics of noise from a machine or a process display a great temporal variability, for instance, the noise levels of some industrial processes are steady, whereas some operations cause consecutive noise events with a sudden rise to a peak level and drop, repeating the same pattern periodically (e.g. cyclic). An example is given in Figure 3.49 that was recorded in a car factory. Also, the spectral contents of different type of noises, such as intermittent, impulsive, and impact noises, may vary, for instance, in addition to broadband, some discrete pure tones or harmonics can accompany the spectrum of a machine, like fan noise, of which the tonality is apparent in the middle and high frequencies. Generally, noise levels are rather high at low frequencies. The directivity patterns of sources indicate a great variability to be determined along with the noise emissions.

In the past, a categorization in respect of acoustic emmission could have been made for the typical equipment or machinery according to the usage and operation, as given in Figure 3.50 [1]. Since at present, the industrial equipment and installations reveal great

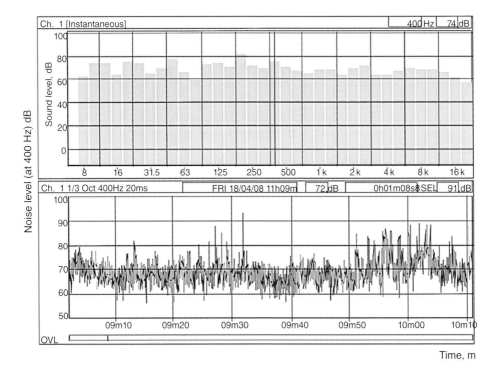

Figure 3.49 Noise measurements in a car factory. Upper: Average spectrum of indoor noise. Lower: Time-history at 400 Hz. (*Source:* C. Kurra, 2008.)

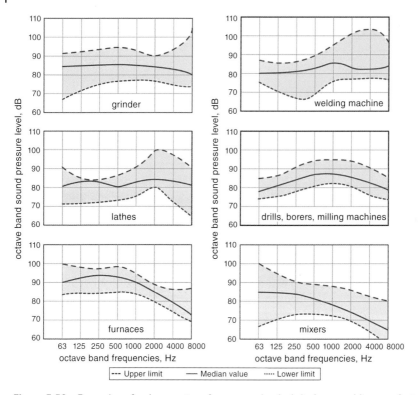

Figure 3.50 Examples of noise spectra of some mechanical devices used in manufacturing industry in the 1950s [1].

diversity regarding physical and acoustic characteristics in relation to their structural properties, locations, and mounting conditions, it is inevitable to deal with them separately to determine their noise emissions for each specific case through measurements or applying calculation models.

Figures 3.51, 3.52a and 3.52b give examples of temporal and spectral analyses for noise generated from boiler chimneys placed on a roof, a noise output from an AC unit, and a water pump, respectively.

3.3.2 Noise Emission from Industrial Premises

Industrial areas can be found as collective industrial estates of small businesses or large single plants including different types of equipment. Individual machines and installations are located in the open air, in semi-enclosures, in workshops, and factory indoors. Generally, where machine groups are gathered in a space for an industrial activity, the total noise is considered to be generated from a plane source, consisting of different processes as industrial activities. As will be seen in Chapter 6, all the point sources in relation to the premise, radiating different noises, their positions, noise emissions, and directivities have to be defined separately. Total sound power is obtained by summation of the point and plane sources and described as $dB(A)/m^2$ (see Chapter 6). While examining the noise propagation according to physical factors, it is necessary to take into account the spectral and temporal factors, which make it rather difficult to evaluate the industrial noises, however, modeling industrial noise

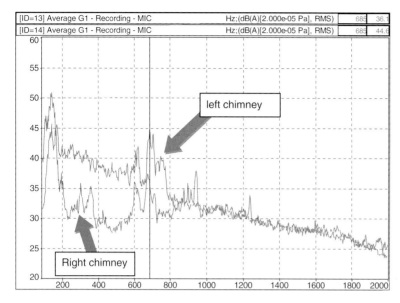

Figure 3.51 Spectral analysis of noise from boiler chimneys at 2 m. (*Source:* C. Kurra and F. Bonfil in 2008.)

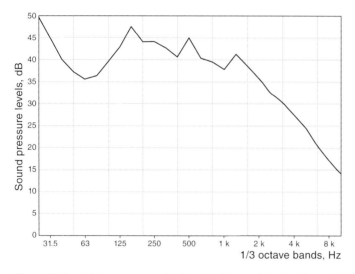

Figure 3.52a Noise spectrum of a household type AC unit. (*Source:* C. Kurra, 2007.)

even from large plants has been accomplished (Chapter 6). Also, the measurement standards for industrial noises have been greatly improved to facilitate the monitoring and assessment of noise in actual cases (see Chapter 5).

Many examples of the field studies can be found in the literature, acquiring emission levels of specific industrial operations and buildings. For example, a metal process factory gave 73–75 dB(A)

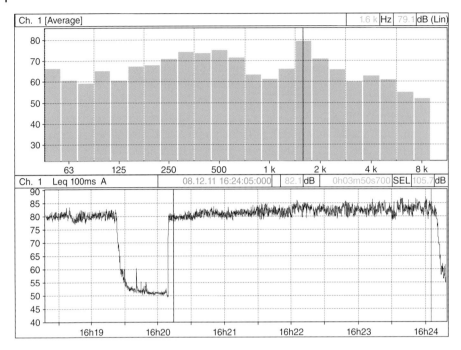

Figure 3.52b Water-pumping noise in a service room of a high-rise building. (*Source:* Kurra, 2015.) Upper: Average spectrum of noise; Lower: Time-history of noise levels.

on the façades of a nearby building while the indoor noise levels were measured as 95–98 dB(A) near the arc-furnace where workers are all around. Other studies revealed the levels exceeding $L_{eq} = 40$ dB(A) at night and 45–60 dB(A) for day-time use for outdoor noise from a cellulose and paper factory and 55–60 dB(A) from a steelworks [105]. The noises from such places are frequently masked in the day-time by the background noises, such as the other mechanical noises and nearby traffic noise, however, the ongoing activities at night are rather disturbing in the environment and can be perceived at long distances.

3.4 Wind Farm Noise

As one of the global sustainable energy resources, wind turbines are widely scattered in the rural areas of many countries. Based on the principles of windmills, the wind turbine is a device that converts the kinetic energy of the wind into electric power. Capacities and types have varied over the years; from smaller turbines which are used for battery charging to provide auxiliary power, back in the 1990s and the medium-sized turbines for domestic power supply, to the wind farms consisting of arrays of large turbines to supply energy for industries and urban settlements nowadays.

Currently growing numbers of onshore or offshore wind farms have been constructed, normally in vast empty areas or at sea, however, some are located rather close to urban areas, increasing the adverse effects and complaints, particularly at night when the background noises, e.g. traffic noise, are lower. It has been shown that the wind turbine noise perceived by people was more annoying than road traffic noise, not because of the magnitude of noise, but more importantly because of temporal and spectral variations [106].

3.4.1 Emission of Noise from Wind Turbines

Wind turbine noises are composed of mechanical noise from mechanical and electrical components and aerodynamic noises due to the air motion around the blades and speed regulation. The most common types are the horizontal-axis wind turbines, including the rotor shaft, blades, gearbox, generator, sensor-motor, all assembled on top of a tower whose height ranges from 20–40 m in the earlier ones and up to 80 m or more in the modern turbines. The turbine is directed toward the wind direction by a simple vane in smaller ones and by a wind sensor connected to a motor in larger ones. Gearboxes manage the rotation of blades (usually three blades) to drive an electric generator. Blade speeds are up to 320 km/h in relation to the wind speeds and the sound generated by wind turbines depends on the blade rotation [106–110]. The major noise is created due to the turbulence and releasing of vortexes from the surfaces and borders of the blades. The interaction between the blades and the tower has an important role in the occurrence of beats.

The noise characteristics of wind turbines are broadband, low frequency, tonal, infrasound, and impulsive and are described as a thumping noise, a stalling noise, a rumbling noise, bursts, beats, a swishing noise, etc. The total noise has a time-varying nature, because of the factors, such as meteorological conditions, the blade rotation variations, directivity, or interaction between the other turbines. More explicitly, random variations are due to meteorological factors and regular (periodic) variations are dependent on the modulation factors at the source, i.e. the rate at which the blades pass the tower [106].

The characteristics of the total noise which is perceived at a given location are more important than their magnitude:

- The highest acoustic emission occurs at low frequencies. Emission levels are affected by the meteorological factors changing the inflow turbulence incident on the blades, affecting blade loading and hence aerodynamic noise production [106].
- Frequency variation is equal to the rotational speed of the blades. Harmonics are at about 28 Hz, even at infrasound range, below the threshold of hearing [107, 108].
- Regular variation in the pressure amplitude of noise signal is referred to as "amplitude modulation" and perceived as "swishing."
- Transient bursts of coherent turbulent energy in turbine inflow are caused by blade pulses.
- Thumping noise at low frequency occurs due to the thumping of the local stall on the turbine blades (caused by the high wind shear at the top of the blade trajectory) and can easily propagate over long distances, i.e. several km from the wind turbine.
- Rumbling noise is radiated from the gearbox and also heard at far distances.
- Because of the above characteristics of wind turbine noise, the appropriate metrics should be used for the quantification of the amplitude variation, in corporation with the detection of infrasound frequencies.

3.4.2 Noise Emission of Wind Farms in Relation to the Environment

The total noise from wind farms is contributed by groups of wind turbines located in the area with a certain distance between them. However, the individual noises of these turbines cause random variations in the total noise emission of the wind farm, because of the different blade rotations and changes in the weather conditions. Generally, the total noise level increases at high wind speeds. The interaction between the noise signals from two or more turbines due to phase differences of pure tones results in constructive and destructive reinforcement of sound waves, creating a complex noise environment. Although the A-weighted noise levels near a wind farm are dominated by the mid-high frequencies, such as 400–1 kHz, they are greatly attenuated by the ground and air absorption, whereas the low frequency components propagate readily at far distances [106–110]. The resultant sound heard in and near the wind farm zones is also time-dependent, low frequency noise, as shown in Figure 3.53 [107]. Therefore, rating wind farm noises by using only A-weighted levels is not appropriate, since even though the environmental limits given in dB(A) are not exceeded, the complaints of residents living at several kilometers distant from the turbines continue.

Background noises usually mask the wind farm noises reaching urban areas and it is difficult to detect them in the field recordings, since it is not possible to shut down the plant during the experiments. Although there are discussions on the issue, the recent technical standards are applicable for the measurement and evaluation of wind farm noise (Chapter 5). The prediction models enable computations of noise in a certain location by taking into account the environmental data and the engineering characteristics of the plant (Chapter 4). Modeling wind turbine noise and noise mapping around wind farms have become widely implemented at present (Chapter 6). Box 3.6 shows the factors affecting wind farm noise.

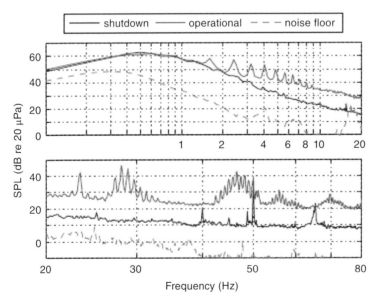

Figure 3.53 Narrowband spectra of wind farm noise measured outdoors [107]. Top: Frequency range of 0–20Hz. Bottom: Frequency range of 20–80Hz. (*Source:* Permission granted by Hansen.)

Box 3.6 Factors Affecting Wind Farm Noise

Wind turbines:

- Types: motor shaft, blades, rotation speed of blades, gearbox, electrical generator
- Hub-height
- Age and maintenance
- Mounting characteristics (shape, bases, connections))

Wind farms and environmental conditions:

- Number of wind turbines in the same area
- Layout of wind turbines
- Configuration of residential area or building with respect to dominant wind direction
- Altitude and topography
- Wind speed and other meteorological factors
- Reflective and absorptive surfaces around, reverberation time of space
- Ground cover

3.5 Construction Noise

One of the main environmental problems in cities has emerged due to the construction activities employing heavy machines and equipment, also manpower. The road and building construction activities in built-up areas, including public works, like pipe-laying, street water mains, and sewers or large constructions like bridges, are performed in a certain period of time following a scheduled pattern of operations. Although these works are temporary, they cause risk of hearing damage for workers and great disturbance because of the excessive noise levels at a distance up to several hundred meters from the construction site. The other aspects of construction noise are given below [111–122]:

- Noisy activities are conducted in the open air.
- Noise events do not occur at predetermined times throughout a day.
- Construction work is intensified in spring and summer in which the outdoor life is active.
- Noise is generated by different kinds of operations.
- Placement of noise source cannot be detected and predetermined.
- Noises have transient characteristics, lasting different durations and intervals.
- Noises are intermittent (as noise events occurring with intervals or phases) periodically or randomly and with high levels.
- Noise generated from a construction site is also related to heavy vehicle transportation.

3.5.1 Noise Emissions from Construction Equipment and Operations

Construction equipment (stationary or moving) noises are identified with their specific acoustic properties associated with the factors given in Box 3.7. Most road construction equipment generates its own power through internal combustion, which is the predominant noise source. Diesel machines produce high-level low frequency exhaust noise and the secondary noise sources are

---

Box 3.7 Factors Affecting Construction Noise

Construction equipment:

- Type of machine and engine (internal combustion, impact, etc.)
- Power, exhaust, brakes, transmission systems, fans
- Type of work and material they process
- Operation period
- Operation technique (directionality and process mode)
- Maintenance and age

Construction site:

- Size and shape of site
- Number of pieces of equipment and types existing at site
- Operational mode of each piece of equipment and operation program
- Distance from equipment to the perimeter of site
- Location and position of equipment at construction site
- Earth and material stacks within site
- Type of earth
- Large reflective surfaces at site
- Barriers and obstructions at site
- Physical conditions of adjacent environment (topography, woods, vegetation, etc.)

---

engine enclosure, air intake, transmission, and hydraulic systems. For loaders, the total noise is combined by fan noise 50%, exhaust 35% and other parts 15% [113].

The acoustic characteristics of the main noise sources of construction equipment (off-road vehicles) and movable compressors powered by diesel engines and some by electric motors, are given as [114]:

- engine body: 100–110 dB(A) dominant at high frequencies
- engine exhaust with muffler: 91–98 dB(A) dominant at low frequencies
- engine intake: 85–90 dB(A) dominant at low frequencies
- compressor: 97–100 dB(A) dominant at high frequencies
- vibrating roller: 100–105 dB(A) dominant at high frequencies
- cooling system fan: 98–102 dB(A) with moderate frequency range components

Construction equipment, which is rather noisy due to manufacturing characteristics and operational systems, is generally operated at maximum power and for long periods. The measured noise levels can be 7–85 dB(A) at 15 m during its operational cycle [114]. Their capacity, size, and the materials they process vary a great deal, hence it is difficult to obtain average data regarding their noise emissions for the purpose of environmental impact assessment. The experiments conducted in the past to collect acoustic data regarding various construction equipment have provided sufficient information [1, 111–117]. Figure 3.54a gives the range of A-weighted sound pressure levels of typical construction equipment published by the Environment Protection Agency (EPA) in the USA in the 1970s, followed by the noise limits declared in the standards. Further documents cover a more detailed database for all sizes of equipment and all operational modes for each equipment group. Table 3.2 displays examples of

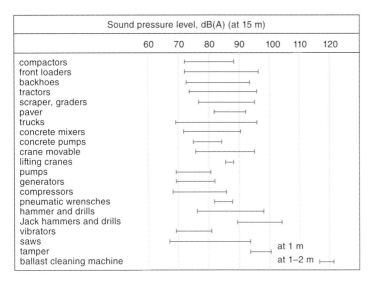

Figure 3.54a Range of noise levels from various construction equipment used in the 1970's [111].

Table 3.2 Noise levels of some construction equipment from the documents published in the 2000's.

Equipment	Noise level dB(A)	Reference
Pile driver	112	Construction Safety Association, Ontario [115]
Impact wrench	108	
Bulldozer	107	
Crane	102	
Circular saw and hammering	96	
Jack hammer	96	
Compactor	94	
Loader	87	
Grinder	86	
Welding machine	85–90	
Pneumatic chip hammer	102–113	Occupational Safety and Health Administration OSHA (USA) [121]
Jackhammer	102–11	
Crane	90–96	
Front end loader	86–94	
Backhoe	84–93	
Bulldozer	93–96	
Backhoe	78	Federal Highway Administration FHWA Road Construction Noise [122]
Chain saw	84	
Compactor (ground)	83	
Crane	81	
Dozer	82	
Excavator	81	
Generator	73	
Impact pile driver	101	

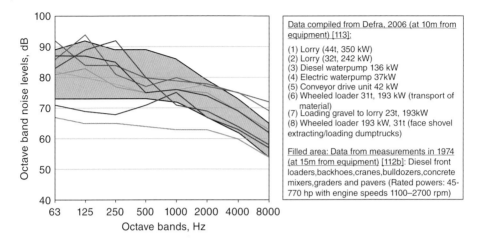

Figure 3.54b Spectral characteristics of various construction equipment compiled from different measurements conducted in 1974 and 2006 [112b, 113].

the A-weighted noise levels for building and road construction equipment, measured at a reference distance of 15 m, in the the 2000's. The spectral contents of various road construction noises which are compiled from the earlier and recent measurements conducted in 1974 (at 15 m from the source) and 2006 (at 10 m) respectively, are shown in Figure 3.54b [112b, 113]. As seen, due to advances in the equipment manufacturing technology over the years, the range of spectral sound pressure levels (at 10 m) corresponds roughly to those obtained at 15 m from the source in the 1974 experiment.

Field measurements are the common method to determine the temporal variation of construction equipment and to display the noise patterns of each type of equipment according to different operational phases (Figure 3.55) [116]. The standard measurement techniques for the noise emission of each bit of equipment is similar to single industrial noise sources, are explained in Chapter 5. Since each operation phase has its own characteristics (duration, level, spectrum), the units used in the measurements are L_{max} and L_{eq} of the operation at full power at a reference distance at both dB(A) and spectral levels.

Construction noise sources are acoustically moving point sources and some devices like conveyors, are considered as a line source with semi-cylindrical sound radiation. Their noise emission can be predicted according to their position if the movements are periodic within certain intervals. This is valid particularly for road construction equipment. Some equipment moves along a straight line (path) or a narrow or wide restricted space, like excavators, lifts, and earth movers. Some are located at a constant point (like concrete mixers) or are dynamic in a smaller area. Some equipment is coherently located so that they can be considered as plane sources. These can be divided into several number of parallel strips to ease calculation. If the distance to those sources is five times larger than their width, the plane source is assumed to be composed of a number of line sources. In order to assume construction equipment as a point source, the distance which the equipment moves forward and backward between two points should be equal or smaller than d/π. (d: Distance from receiver to the operation line, m) [113].

Figure 3.55 Time-history of noise from road construction equipment during the operation phases (from measurements in Hungary, 1996) [116]. (*Source:* Permission granted by Buna.)

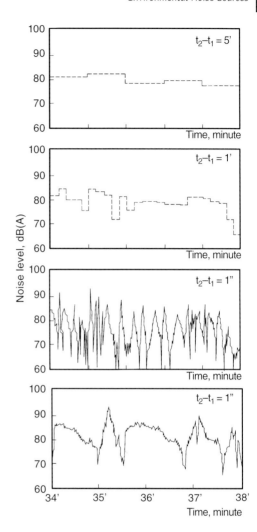

3.5.2 Noise Emission of Construction Sites

Construction site noise as a type of environmental noise is composed of noises from each piece of equipment and transmitted from the construction site. Generally various pieces of equipment are distributed randomly or in a certain order on a construction site, and are employed independently or in combination, thus, the noise generated outside of the area is increased with the number of pieces of equipment and operations [111–122]. Dominant noise sources and patterns of operations have been categorized for use in the prediction models. Emission levels of the total construction site are determined by the prediction models (Chapter 4).

Some of the operations and noise sources at road and building construction sites are:

- Demolition and clearing land
- Site preparation (construction of offices, storage, delivering materials and equipment, loading and emptying by trucks, lifting by means of bulldozer, grader, forklifts, etc.)
- Excavation (digging earth)
- Earth-moving (loading, transporting, and filling)
- Pile driving

- Drainage works (digging, delivery of pipes, placing, earth-filling, leveling)
- Stabilization of earth (delivery, expanding material, compressing)
- Covering and isolation work
- Layering installation
- Signaling
- Removal of materials
- Erection and modification
- Maintaining and repairing
- Delivery of material, transporting, lifting
- Expanding drainage
- Cleaning up, landscaping
- Renovation and replacement.

Building and road construction activity noises are higher also because they are older and worn in some situations. As shown in Table 3.2, the noise levels are very high, therefore, they create a great disturbance, especially at night.

The speed of operation, density, frequency of movement, the number of deliveries of equipment at the same time vary according to construction site management (work program). Thus, the noise levels vary from very high to very low levels. Several earlier investigations have been published since the 1970s in the UK and the USA. Figure 3.56 gives an example recording in 1976 by the Building Research Establishment (BRE, in the UK) [119]. Figures 3.57a–3.57c indicate example

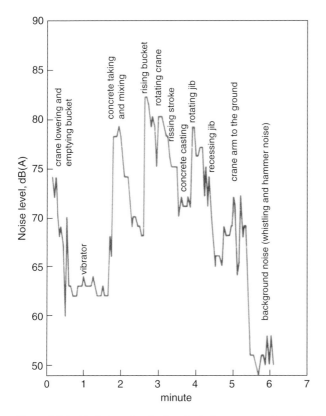

Figure 3.56 Variation of noise levels during construction of main drainage system (from measurements of BRE in 1975) [119].

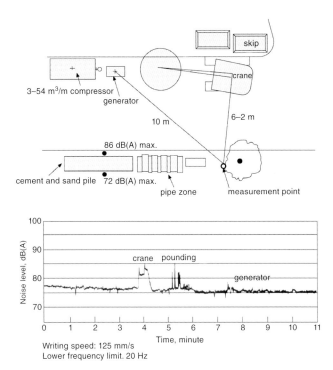

Writing speed: 125 mm/s
Lower frequency limit. 20 Hz

Figure 3.57a Variation of noise levels during concrete preparation and casting (from measurements of BRE in 1975) [112a].

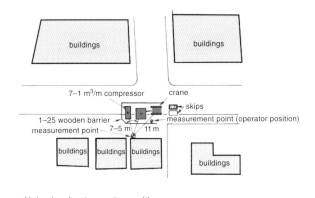

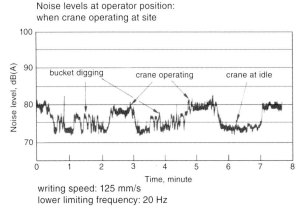

writing speed: 125 mm/s
lower limiting frequency: 20 Hz

Figure 3.57b Variation of noise levels in the preparation of a manhole chimney (from measurements of BRE in 1975) [112a].

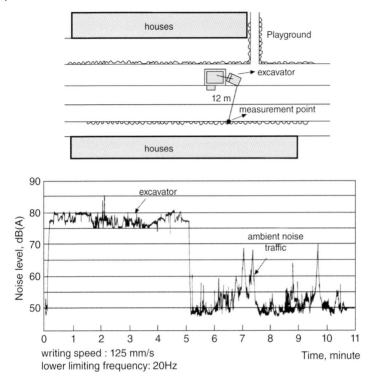

Figure 3.57c Variation of noise levels during road construction (from measurements of BRE in 1975) [112a].

measurements also adopted from the BRE design guides prepared for planning construction sites. The measurement results indicated that road construction noise in day-time was about 17 dB(A) higher above the background noise (25 dB(A) on average) in rural areas and about 6 dB(A) in urban sites [112a]. Some of these studies conducted in the UK in 1978–1980 by TRRL yielded the empirical prediction models as given in Chapter 4.

Noise conditions near construction sites are associated with the factors outlined in Box 3.7. When the contributions of these factors to the total acoustic energy generated by the entire construction site are quantified, the emission levels outside the area can be determined. However, since the building industry has been growing fast in some countries, construction activities have become more complex due to the size of buildings, particularly for high-rise building sites and due to the necessity of completing construction in a short time by employing increasing quantities of multifunctional equipment and machines with less manpower, the situation is much more complicated for noise predictions, compared to the past. Hence, monitoring the noise both day and night is necessary to protect the neighborhoods which are negatively affected by the construction site.

Some examples regarding spectral and temporal diagrams obtained by measurements at about 100 m from the site are given in Figure 3.58.

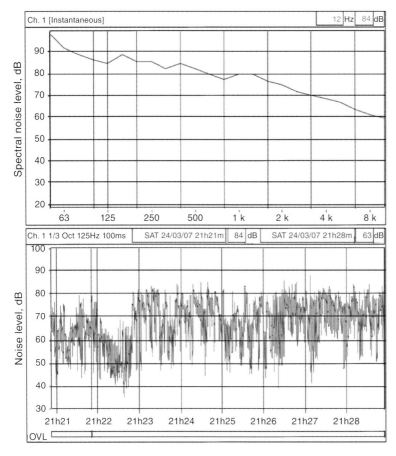

Figure 3.58 Noise measurements at 100 m from a construction site of a high-rise building. Upper: Spectral analysis; Lower: Temporal analysis for 125 Hz. (*Source:* Kurra, 2012.)

3.6 Entertainment Noise

As a basic human need to socialize in a community, gatherings such as meeting, crowds, assemblies, talks, are usually accompanied by high amplified music in outdoor cafés, restaurants, etc. particularly at night in summer, mostly to attract young people. Also, live performances, such as concerts, shows, and events, are frequently organized as leisure or recreational activities in the city centers or in the suburbs, in line with the growth of the entertainment sector, however, this has other side effects like traffic congestion. In order to create a perfectly diffused acoustic environment in the premises, the sound emitters, i.e. different types of loudspeakers, are positioned all around the premises above head level and on the stage, whose sound pressure levels are amplified according to the desire of the musicians or the DJs. Such places which are called Movida (the Spanish word for movement), can be either temporary or permanent, however, in both cases, they can constitute a source of noise in the vicinity, causing disturbance with high levels of sounds, particularly rhythmic bass sounds, sometimes creating an unbearable situation for those living in the neighborhood, since the noise interferes with relaxation, sleeping, and other activities [123–125]. At present, a number of surveys have been performed to search for the effects of Movida on public health and comfort, to draw the attention of the local authorities to find solutions, however, without destroying people's right of enjoying themselves.

3.6.1 Noise Emission from Sound Amplification Systems

Electronically amplified sound is produced by the prerecorded (CD, DVD, radio, or TV) music or live concerts. The equipment set to generate sound and for amplification, which is used for musical events, is composed of an equalizer console, microphones, synthesizers, power amplifiers, and sound emitters (i.e. loudspeakers) of various types. Particularly the bass speakers (woofers) are more powerful and sometimes uncontrollable devices.

Loudspeakers can be defined as point sources with different acoustic characteristics and their acoustic outputs are certified based on the laboratory tests in anechoic rooms, in terms of maximum acoustic power, dynamic range, frequency response, and directionality characteristics. Generally, the sound power levels display a great variation on both sides of their symmetrical axis and the energy is concentrated within the narrower angles for the high-frequency sounds and in the much broader angles (almost omni-directional) for the low frequencies.

The acoustic characteristics of music vary greatly in terms of spectral and temporal aspects; for example, pop music is rhythmic and tonal (mainly with dominant bass tones) with high level, intermittent, and of an impulsive nature. Noise emitted from some open-air places (like restaurants) even without music, contains mainly crowd noise (shouts, singing, cries) or babble noise from loud voices at relatively constant level and this sometimes can be amplified due to the surrounding reflective walls of the premises. Figure 3.59 shows a recorded passage of speech sounds in an open-air tea garden next to residential buildings.

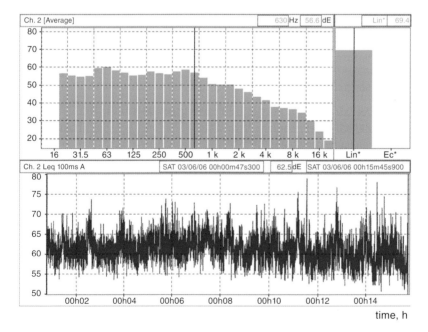

Figure 3.59 Speech conversation noise recorded in a cafeteria. Upper: Average spectrum of noise. Lower: Time-history of noise levels in 15 minutes. (*Source:* Kurra, 2007.)

3.6.2 Noise Emission from Entertainment Premises

Both for noise mapping and impact assessment around Movida, it is necessary to determine the total noise emission of the premises by assuming the entire space or only the stage platform as a plane source with reasonable height. The emission is calculated by logarithmical summation of the acoustic powers of all the sound emitters distributed in the premise, by taking into account the associated factors given in Box 3.8 [123–125].

Since sound emission levels and spectral characteristics are subject to change at each instant of time depending on the music, it is difficult to determine the typical emission levels, therefore, they can be determined by field measurements based on the measured sound pressure levels around the source, as explained in Chapter 5.

Sound examples, recorded in an open-air night club and in a restaurant with music, can be seen in Figures 3.60a and 3.60b.

To determine the environmental noise levels due to the amplification system or live performance, systems, the effects of environmental factors should be taken into account. As modern music contains a high amount of low frequency components, they cannot be attenuated by air absorption and are heard as bass impulses at far distances. Normally, the surface materials in premises are less absorptive and the sound insulation of building elements is weaker at low frequencies. The screens

---

Box 3.8   Factors Affecting Entertainment Noise

Sound sources:

- Acoustic properties of sound emitters (LS): Dynamic response, amplified level, frequency and directivity pattern
- Location and position of sound emitter
- Cabins and type of mounting
- Reflective surfaces around LS (vertical and horizontal)
- Type of music, temporal variation, and spectral content
- Playing duration and intervals
- Amplification power set by DJ or sound technician

Entertainment premises:

- Layout plan of the premise (stage design, dance floor, and audience areas)
- Total number of sound emitters (LS) and types
- Distribution of LSs in the premises, positions of each
- All reflective surfaces in the place and obstructions (i.e. furniture, bars, etc.)
- Capacity of the premise (number of people)
- If open-air space, the outdoor physical elements causing focusing, multiple reflections, and echo
- If open-air space, atmospheric conditions, temperature gradient, humidity, and wind characteristics
- If enclosed space (indoors), acoustic properties of the indoor environment, size, dimensions, height, surface materials, reverberation times
- If enclosed space (indoors), sound insulation performances of building elements transmitting sound throughout the building
- Topography and ground cover around the premises

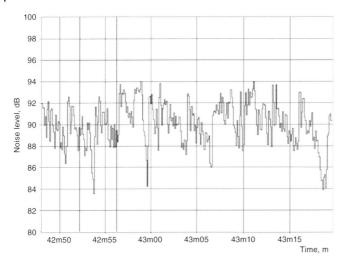

Figure 3.60a Noise measurement in an open-air night club. (*Source:* Kurra, 2006.)

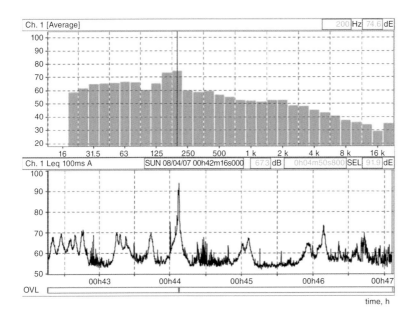

Figure 3.60b Characteristics of noise in a restaurant with live music. (*Source:* Kurra, 2007.)

which may be constructed on the perimeter of premises, unless specifically designed, are not effective at such low frequencies. Field studies and the complaints to local authorities have proved that the disturbance from impulsive bass sounds is more intense rather than steady sounds with flat spectrum. It is difficult to control these sounds especially in summertime when the windows in buildings without air-conditioning are open and when balconies and gardens are frequently used.

When such a business is operated in a building, e.g. a public or private club, bar, pub, disco, restaurant, café, cinema, casino, ballroom in a hotel, etc., the sound transmitted through the walls and windows and radiated outdoors can be a problem, similar to the case from an industrial building. Assessment of noise levels can be made by physical modeling, as given in Chapter 6.

3.7 Shooting Noise

One of the most severe noise pollution sources caused by recreational activities is shooting practice fields (shooting ranges), where blast sounds are perceived at far distances. These areas can be used for civil or military training purposes. Various investigations regarding the propagation of blast noises from firearms have been conducted in the past and at present [126, 127]. Box 3.9 outlines the factors affecting the shooting range noise levels.

3.7.1 Noise Emission of Firing Weapons

Hand-held military weapons used for firing practices generate high-level impulsive sounds intermittent or consistent, depending on the type of firing and the weapon, such as rifle, pistols, revolvers, shotguns, etc. with different calibers [126]. The acoustic energy of the bursts is dependent on the ballistic parameters, such as the muzzle pressure and speed and bullet weight, and are measured according to specific technical standards (see Chapter 5). Some firearms are equipped with muzzle brakes that increase the noise emission. The emission levels can be measured by applying technical standards (see Chapter 5) or predicted by means of the propagation models (see Chapter 4). Figure 3.61 gives the results of the temporal and spectral analyses of the tests using different types of weapons [127].

Box 3.9   Factors Affecting Shooting Noise

Weapons:

- Type of firearm (rifle, pistol, sniper, revolver, etc.)
- Calibers
- Muzzles (muzzle pressure, muzzle time, muzzle speed)

Shooting range:

- Layout of the premises, range of shooting
- Purpose: military training or civil practice
- Number of firing positions and number of shooters
- Sheds and their design
- Height of gun level
- Distribution of shooting positions in the range
- Direction of shooting
- All reflective surfaces in the place and obstructions around the shooting range
- Ground cover (absorption coefficient)
- Meteorological conditions (i.e. winds)

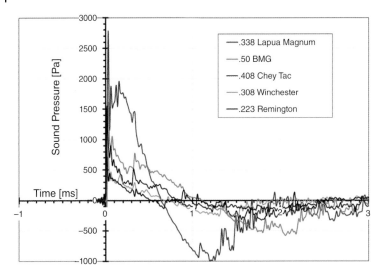

Figure 3.61 Time-history of peak sound pressure for rifles [127]. (*Source:* Permission granted by Zurek.)

3.7.2 Noise Emission of Shooting Ranges

Shooting ranges in the open air or semi-covered spaces are not normally in the vicinity of urban areas since the effect of shooting noise extends at a long distance if the weapons are not equipped with mufflers. The acoustic energy of the bursts can be measured individually or an average emission of such fields can be determined through modeling, as given in Chapter 4. The entire shooting range can be considered as a planar source by taking into account the factors, such as types of weapons and number of rounds using each weapon, intermittence, the maximum energy of each burst, time of shooting, and nearby structures (e.g. reflectors and obstructions), and the form and construction of noise shields if they exist around shooting areas. Among the meteorological factors, particularly wind direction is of importance in variation of shooting noise levels.

3.8 Other Public Noise Sources

Apart from the sources mentioned above, various community noises listed in 3.1.1 create severe problems, although not widely spread, because of their high levels, time-varying or tonal characteristics. Some noises are emitted by recreational activities, such as sports, commercial, religious or other human activities performed in the open air or semi-open spaces. Such activities can adversely affect not only other people but also the individuals themselves. Misbehaviours like shouting or yelling, producing noise disregarding other people, can not be tolerated in modern societies.

The noise sources in this group that require specific involvement as case studies, are given below:

- Sounds of crowds, festivals, carnivals, parades, etc.
- Sounds for religious purposes (chimes, gongs or electronically amplified praying sounds)
- Public announcement systems, sirens.
- Children's playgrounds and school gardens (Figure 3.62)
- Electronic and mechanical toys
- Dogs barking from boarding kennels (Figure 3.63) [128]

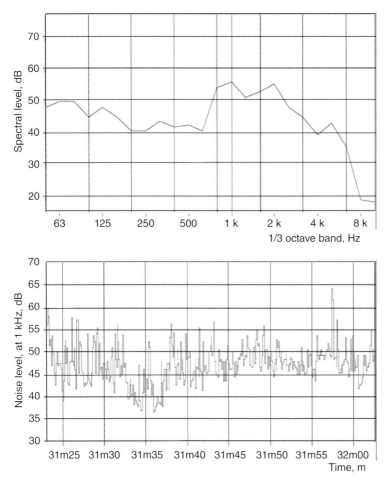

Figure 3.62 Noise measurements at a distance of 100 m from a children's playground. (*Source:* Kurra, 2007.)

- Some natural habitats (crickets, birds, etc.)
- Parking lots
- Delivery of furniture, collection, storage facilities, lifting, etc. in neighborhoods
- Rubbish garbage containers, garden appliances (e.g. lawnmowers)
- All machines and coolers located in the open air (e.g. rooftop, backyards, etc.) and shutes
- AC units on external walls, air inlet and exhaust openings on the façades of commercial and trade facilities
- Public sport fields (stadiums, swimming pools, tennis courts, etc.) and cheers of fans
- TV broadcasting, sports channels from TV or radio from open windows
- Model airplane or drone flying areas
- Amusement parks and mechanical equipment like roller coasters
- Car and motorcycle racing fields

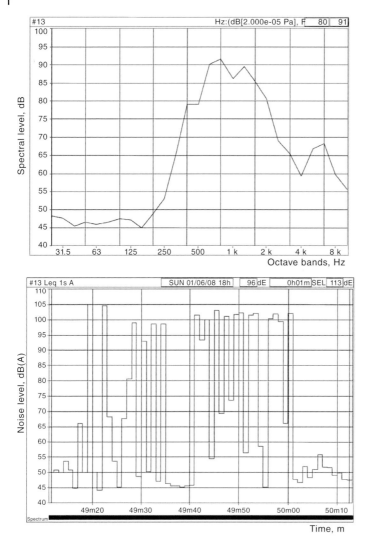

Figure 3.63 Noise recording of a dog barking. (*Source:* Kurra, 2008). Spectrum diagram (upper) and time-history sound pressure levels (lower).

In addition to the above sources, at present, it is guaranteed one will encounter unanticipated new noise sources within dynamic societies continuously coping with new technological products, some satisfying and exciting people, others facilitating daily life. They may generate short-term or intense noises, either temporary or permanent and their effects are dependent on factors, such as detectability, predictability, empathy, level of pleasing, and degree of tolerance. Involvement with these specific noises is a matter of noise policy and management, e.g. licensing rules, inspections, and legislative actions (see Chapters 8 and 9). On the other hand, some of the noise sources explained so far may represent the intrinsic and cultural nature of the local community, so that they can be treated as the components of the soundscape, unless the noise level exceeds the noise criteria (Chapter 7).

3.9 Noise Sources in Buildings

Various types of noise sources are located inside buildings where different activities are performed in daily life. Generally, the indoor spaces in buildings of various kinds, such as houses, apartments, offices, hospitals, schools, administration buildings, etc. are grouped according to the purpose of use in architectural design, however, regarding building acoustics, they can be classified according to their sensitivity to noise and their noisiness, i.e. the possibility of being a noise source due to the activity inside, due to the equipment or installations, as declared in some building regulations [129]. Some of the indoor noise sources generate so-called neighborhood noises, others are mechanical sources, all creating considerable disturbances for occupants, as listed below:

- Loud speech conversations, shouts, cries
- Footsteps noise, children running, furniture dragging, repair noise, and other impact noises
- Radio, TV sounds, and live or electronically amplified music
- Domestic appliances (mechanic or electrical)
- Domestic animals
- Garbage chutes and rubbish grinders
- Garages and car parks, delivery of goods
- Elevators and escalators
- Common spaces (corridors, staircases)
- Doors banging
- Room-type Air-conditioning units
- Workshops or small manufacturing spaces adjacent to a building or a unit
- Kitchen noises, delivery, storing, and preparation of foods
- Bathrooms and sanitary appliances, installations, water fillings, plumbing
- Bowling halls and other indoor sports spaces (Figure 3.64)
- Electronic toys
- Music studios, practice rooms

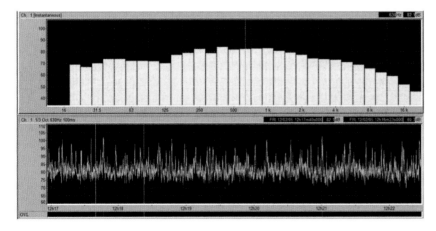

Figure 3.64   Noise measurement in a bowling alley in a shopping center. (*Source:* Kurra, 2003.)

In addition to the above sources, the mechanical and electrical service rooms in buildings (especially in public buildings or high-rise residential buildings) contain a vast variety of equipment, machines, water installations, etc., each constitutes a different type of noise source to be considered for noise control (e.g. furnaces, burners, air handling units [AHU], air or gas compressors, coolers, heat exchangers, condensers, pumps, diesel generators, transformers, etc.) [103, 104, 130]. They are treated as industrial noise sources, as explained in Section 3.3.

3.9.1   Noise Emission of Individual Sources in Buildings

The variety of noise sources requires source-specific investigations analyzing the sound propagation paths in buildings. Factors affecting noise emissions of indoor noise sources are numerous, depending on the types of sources, as given in Box 3.10. The sound power levels of equipment or machinery could be documented by the manufacturer, however, the certain increase in emissions should be considered due to aging, maintenance, and lack of repair. The acoustic characteristics of noises from temporary human activities can be evaluated as time-varying indoor noise levels to be measured in the problematic spaces, since it is difficult to realize modeling to predict the acoustic energy radiated.

3.9.2   Noise Emission of Specific Buildings (e.g. an Industrial Building)

Many features regarding building constructions influence the transmission of indoor noises to the outside. This issue is important especially when dealing with excessive noise in plants or in other noisy buildings for the protection of nearby environments. Radiation of noise from a building itself, as an environmental noise source and its overall sound energy can be determined through field measurements applying technical standards (see Chapter 5) or the emission levels can be obtained by modeling the buildings (as point, line, area, or three-dimensional noise sources) and the sound transmission paths, according to architectural configurations (see Chapters 4 and 6).

Box 3.10   Factors Affecting Indoor Noises

Noise sources:

- Manufacturing and operational properties
- Acoustic characteristics (peak and average noise levels, temporal, and spectral)
- Types of noise radiated: airborne or structure-borne and mechanical vibrations
- Operational period and activation time

Building and indoor spaces:

- Proximity of source rooms to the sensitive rooms
- Orientation of receiver rooms according to outdoor noise sources
- Number of noise sources, types of sources
- Layout of sources in certain spaces and distribution in buildings
- Sound transmission paths (direct or flanking)
- Sound insulation properties of building elements (against airborne and impact noises)
- Connections with building elements and other transmission lines
- Indoor sound absorption (surface materials) and reverberation time

---

3.10 Conclusion

The major input to the assessment and management of environmental noise is the quantification of source emissions in terms of sound power levels. Noise sources radiating frequency-dependent acoustic energy exhibit a great variety with respect to their structural and operational characteristics, resulting in significant differences between the acoustic properties, so that categorization of individual sources is rather difficult.

The main characteristics of the noise sources discussed in this chapter emphasize the transportation noise sources. The physical and source-related factors influencing the emission levels are approached in two stages, which is applicable to all environmental noise problems: Individual noise sources and a group of sources existing in the same place and operating simultaneously or randomly. Thanks to the earlier investigations, some experimental studies conducted as early as the 1950s, the correlations between the emission levels and the structural and operational parameters of noise sources could be elucidated. As explained in Chapter 4, prediction of noise levels (or emission levels) in the environment and the decision on noise abatement depend on the determination of the emission levels of the individual sources.

The sound power levels of specific sources can also be determined by standard measurement techniques conducted either in the field or in the laboratory (see Chapter 5). In situations where categorizations of noise sources could be achieved in terms of certain parameters – as realized for road traffic in relation to the source and the road features – the total emitted power can be computed by analytical models and the results are implemented in rating environmental noise pollution. However, the multi-source and dynamic source situations, such as the human-originated sources, create complex problems and the energy outputs are unpredictable.

Nowadays noise control for some types of noise sources has been rather successful in the production or manufacturing phase, with the aid of technological development, material technology, improved operational techniques, and so on, and hence the noise emissions of various products have been significantly reduced. The interest in silent products on the market is growing. Despite that, the environmental noise problem is expanding and cannot be completely solved, thus, it continues to be a major concern globally.

References

1 Harris, C.M. (ed.) (1957). *Handbook of Noise Control*. McGraw-Hill.

2 VDA (1978). *Urban Traffic and Noise*. Verband der Automobilindustrie E.V. (VDA).

3 Sharp, B.H. and Donavan, P.R. (1979). Motor vehicle noise. In: *Handbook of Noise Control*, 2e (ed. C.M. Harris), 32-1–32-21. McGraw-Hill.

4 Nelson, P. (ed.) (1987). *Transportation Noise Reference Book*. Butterworth.

5 Hickling, R. (1998). Surface transportation noise. In: *Handbook of Acoustics* (ed. M.J. Crocker), 897–907. Wiley.

6 Nelson, P. and Piner, R. (1977). Classifying road vehicles for the prediction of road traffic noise, Transport and Road Research Laboratory, Department of the Environment, LR 752.

7 Buna, B. (1987). Some characteristics of noise from single vehicles. In: *Transportation Noise Reference Book* (ed. P. Nelson), 6/3–6/13. Butterworth.

8 Lewis, P.T. (1973). The noise generated by single vehicles in freely flowing traffic. *Journal of Sound and Vibration* 30 (2): 191–206.

9 Tyler, J.W. (1987). Sources of vehicle noise. In: *Transportation Noise Reference Book* (ed. P. Nelson), 7/3–7/39. Butterworth.

10 Sandberg, U. (2001). Noise emissions of road vehicles, effect of regulations, final report 01–1. I-INCE Working Party on Noise Emissions of Road Vehicles International Institute of Noise Control Engineering.

11 Anon (1980). *Guide du Bruit des Transports Terrestres, Prévision des Niveaux Sonores*. Ministre des Transports, CETUR.

12 FHWA (1978). FHWA highway traffic noise prediction model, FHA, RD-77-108. U.S. Dept. of Transportation.

13 Anon (1981). *Lärmbekampfung 81, Entwicklung-Stand-Tendenzen*. Erich Schmidt Verlag.

14 Sandberg, U. (1984). Reduction of tyre/road noise by drainage asphalt, *Proceedings of the International Seminar on Tyre Noise and Road Construction*, ETH, Zurich (February).

15 Young, J.C. and Jordan, P.G. (1981). Road surfacings and noise inside saloon cars. *TRRL Digest* Supplement Report 655.

16 Sliggers, J. (2015). A noise label for motor vehicles: towards quieter traffic. Informal document GRB-61-01, agenda item 9, 61st GRB (27–29 January). https://www.unece.org/trans/main/wp29/wp29wgs/wp29grb/grbinf61.html.

17 Zambon, G., Benocci, R., and Brambilla, G. (2016). Cluster categorization of urban roads to optimize their noise monitoring. *Environmental Monitoring and Assessment* 188 (26) online).

18 Stephens, R.J. and Vulkan, G.H. (1968). Traffic noise. *Journal of Sound and Vibration* 7 (2): 247–262.

19 Johnson, D.R. and Saunders, E.G. (1968). The evaluation of noise from freely flowing road traffic. *Journal of Sound and Vibration* 7 (2): 287–309.

20 Ljundgren, S. (1973). *A Design Guide for Road Traffic Noise*. National Swedish Building Research, D10.

21 Ingemasson, S. and Ljundgren, S. (1970). Buller problem vid trafikleder, Byggforskningen Report 20.

22 Leong, R.K. (1975). Noise from motor vehicles. In: *Noise in the Human Environment*, vol. 2 (ed. H.W. Jones), 123–152. Environmental Council of Alberta, Canada.

23 Leasure, W.A. and Bender, E.K. (1975). The tire-road interaction noise. *Journal of Acoustical Society of America* 58 (1): 39–50.

24 Halliwell, R.E. and Quirt, J.D. (1980). Traffic noise prediction. Building Research Note 146. Ottawa: National Research Council of Canada.

25 Kurra, S., Gedizlioğlu, E., and Yayla, N. (1988). Final recommendations for Trans-European Motorway traffic noise analysis, project report, UNDP/ECE, TEM /CO /TEC/16, Turkey, February.

26 Kurra, S., Aksugur, N., and Arık, A. (1981). Analysis of environmental noise and determination of highest acceptable noise levels with regard to noise control in Istanbul. Project no: 524/A, supported by Turkish Scientific and Technical Research Establishment (TUBITAK), February.

27 Galloway, W.J., Clark, W.E., and Kerrick, J.S. (1969). Highway noise: Measurement simulation and mixed reactions. NCHRP Report 78, FHWA.

28 Sandberg, U. (1987). Road traffic noise: the influence of the road surface and its characterization. *Applied Acoustics* 21: 97–118.

29 Franklin, R.E., Harland, D.G., and Nelson, P.M. (1979). Road surfaces and traffic noise. Department of the Environment and Department of Transport, TRRL Report LR 896.

30 Nelson, P. and Ross, N.F. (1981). Noise from vehicles running on the open textured road surfaces. TRRL Supplement Report 696.

31 Hamet, J.F., Pallas, M.A., Doisy, S. et al. (2004). Modeling noise emission of heavy trucks: do a power unit and rolling noise suffice? *Proceedings of Inter-Noise 2004*, Prague, the Czech Republic.

32 (a) Sandberg, U. and Decornet, G. (1980). Road surface influence on tyre road noise: Parts I and II, Proceedings of Inter-Noise 80, Miami, Florida, USA;

(b) Morgan, P. (ed.) (2006). Sustainable road surfaces for traffic noise control: guidance manual for the implementation of low-noise road surfaces. FEHRL Report 2006/02.

33 Nelson, P.M. and Abbot, P.G. (1987). Low noise road surfaces. *Applied Acoustics* 21: 119–137.

34 Hamet, J.F. and Klein, P. (2000). Road texture and tire road. Inter-Noise 2000, Nice, France.

35 de Freitos, E.F., da Cunha, C.A.C., Lamas, J. et al. (2015). A psychoacoustic based approach to pavement classification, Euronoise 2015, Maastricht, The Netherlands.

36 Leasure, W.A. (1972). Truck noise-1, Peak A weighted sound levels due to truck tires. Report OST/TST-72-1. U.S. Department of Transport.

37 Sandberg, U. and Ejsmont, J.A. (2007). Tire/road noise-generation, measurement and abatement. In: *Handbook of Noise and Vibration Control* (ed. M.J. Crocker), 1054–1071. Wiley.

38 Sandberg, U. and Ejsmont, J.A. (2002). *Tire/road noise*. Informex.

39 Van Keulen, W. and Li, M. (2015). Influence of changes in surface properties on rolling noise; measurements and model, the 22nd International Congress of Sound and Vibration, ICSV 22, Florence, Italy (12–16 July).

40 Tsujiuchi, N., Koizumi, T., Maeda, Y. et al. (2002). Application of operational analysis to a rolling tire noise prediction (11.7.1,75.5), Inter-Noise 2002, Michigan, USA.

41 Ledee, F.A., and Toussaint, L. (2015). Three approaches to study the reduction of pavement noise performances over time, Euronoise 2015, Maastricht, The Netherlands.

42 Van Bochove, G.G. (2015). Self-healing asphalt: the match between noise reduction and durability, Euronoise 2015, Maastricht, The Netherlands.

43 Beckenbauer, T. (2013). Road traffic noise. In: *Handbook of Engineering Acoustics* (eds. G. Müller and M. Möser). Springer.

44 Donavan, P. (2002). Examination of tire/road noise at frequencies above 630 hertz (11.7.1, 30), Inter-Noise 2002, Michigan, USA.

45 Stenschke, R. and Vietzke, P. (2001). Tyre/road noise emissions, rolling resistance and wet braking behaviour of modern tyres for heavy-duty vehicles (state of the art). (I 423, cl. 13) 165, Inter-Noise 2001, The Hague, The Netherlands.

46 Bekke, D., Wijnant, Y., Schipper, D. et al. (2015). Silent and safe traffic project: an optimization of the tyre-road interaction on noise and wet grip, Euronoise 2015, Maastricht, The Netherlands.

47 Dietrich, M., Roo, F.D., Van Zyl, S. et al. (2015). Triple-A tyres for cost-effective noise reduction in Europe, Euronoise 2015, Maastricht, The Netherlands.

48 Wolfert, H. (2015). Deceleration on better tyres, Euronoise 2015, Maastricht, The Netherlands.

49 Watts, G. (2005). Harmonoise prediction model for road traffic noise. PPRO 034 project report.

50 Chiles, S. (2014). Selection of state highway bridge expansion joints in noise sensitive areas, Inter-Noise 2014, Melbourne, Australia (16–19 November).

51 Glaeser, A.K.P., Schwalbe, G.A., and Zöller, M. (2012). Mitigation of noise emissions from vehicles passing bridge expansion joints. *Noise Control Engineering Journal* 60 (2): 125–131.

52 Van den Dool, P., Gardien, W., and van Vliet, W.J. (2015). Making road traffic bridge silent, Euronoise 2015, Maastricht, The Netherlands.

53 Kurra, S. (2012). Derivation of reference spectrums for transportation noise sources to be used in rating sound insulation, Euronoise 2012, Prague, the Czech Republic (10–13 June).

54 Stanworth, C. (1987). Sources of railway noise. Chapter 14. In: *Transportation Noise Reference Book* (ed. P. Nelson), 14/3–14/13. Butterworth.

55 (a) Hemsworth, B. (1987). Prediction of train noise, Chapter 15. In: *Transportation Noise Reference Book* (ed. P. Nelson), 15/3–15/15. Butterworth.

(b) Hemsworth, B. (1979). Recent developments in wheel rail noise research. *Journal of Sound and Vibration* 66: 297–310.

56 Thompson, D.J. (2007). Wheel-rail interaction noise, prediction and its control. In: *Handbook of Noise and Vibration Control* (ed. M.J. Crocker), 1138–1147. Wiley.

57 Hemsworth, B. (2007). Rail system environmental noise prediction, assessment and control. In: *Handbook of Noise and Vibration Control* (ed. M.J. Crocker), 1438–1445. Wiley.

58 Anon (2001). Reken–Meervoorscrift Railverkeer Slawaai, 96, AR Interim – CM (B4–3049/2001/ 329750.

59 Hanson, C.E. (1993). Aeroacoustic sources of high speed Maglev trains, Noise-Con1993, Virginia.

60 (a) Lotz, R. and Kurzweil, L.G. (1979). Rail transportation noise, Chapter 33. In: *Handbook of Noise Control* (ed. C. Harris), 33-1–33-22. McGraw-Hill.
(b) Lotz, R. (1977). Railroad and rail transit noise sources. *Journal of Sound and Vibration* 51: 319–336.

61 Liu, L., Chen, Y., He, C., and Xing, X. (2015). Characteristics of noise sources in high-speed railways, Euronoise 2015, Maastricht, The Netherlands.

62 Hanson, C. (1990). High speed rail system noise assessment. TRB Annual Meeting Committee A2 M05, paper no. 89 0359.

63 Walker, J.G. (1990). A European high speed railway network: the noise implications, Inter-Noise 1990.

64 Liu, L., Chen, Y., Caissong, H.E. et al. (2015). Experimental study on the characteristics of noise sources in high speed railway, Euronoise 2015, Maastricht, The Netherlands.

65 Wolfe, S.L. (1987). Introduction to train noise, Chapter 13. In: *Transportation Noise Reference Book* (ed. P. Nelson). Butterworth.

66 Zhang, X., Thompson, D., and Squicciarini, G. (2015). Effects of railway ballast on the sound radiation from sleepers, Euronoise 2015, Maastricht, The Netherlands.

67 Wettschureck, G.R., Hauck, G., Diehel, R.J. et al. (2013). Noise and vibration from railroad traffic. In: *Handbook of Engineering Acoustics* (eds. G. Müller and M. Möser), 393–487. Springer.

68 Kalivoda, M., Danneskiold-Samsoe, U., Kruger, F. et al. (2003). EU rail noise: a study of European priorities and strategies for railway noise abatement. *Journal of Sound and Vibration* 267 (3): 387–396.

69 Schwanen, W., Kuijpers, A., and Torbijn, J. (2015). Ten years of rail roughness control in the Netherlands: lessons learned, Euronoise 2015, Maastricht, The Netherlands.

70 Höjer, M. and Almgren, M. (2015). *Monitoring system for track roughness, Euronoise 2015*. The Netherlands: Maastricht.

71 Broadbent, R.A., Thomson, D.J., and Jones, C.J.C. (2009). The acoustic properties of railway ballast, Euronoise 2009, Edinburgh.

72 Carman, R. (2004). Prediction of train noise in tunnels and stations, InterNoise 2004, Prague, the Czech Republic.

73 Krylov, V.V. and Bedder, W. (2015). Calculations of sound radiation associated with tunnel boom from high-speed trains, Euronoise 2015, Maastricht, The Netherlands.

74 Okumura, Y. and Kuno, K. (1992). Statistical analysis of field data of railway noise and vibration collected in urban areas. *Applied Acoustics* 33: 263–280.

75 Lanz, G. and Jaksch, M. (2015). Phonoblock rail track in-situ tested low noise barriers in platform design made of concrete, Euronoise 2015, Maastricht, The Netherlands.

76 Iser, N., Lutzenberger, S., Craven, N. et al. (2015). Managing noise from parked trains, Euronoise 2015, Maastricht, The Netherlands.

77 Raney, J.P. and Cawthorn, J.M. (1979). Aircraft noise, Chapter 34. In: *Handbook of Noise Control* (ed. C.M. Harris), 34-1–34-18. McGraw-Hill.

78 (a) Von Gierke, H.E. (1957). Aircraft noise sources, Chapter 33. In: *Handbook of Noise Control* (ed. C.M. Harris), 33-1–33-23. McGraw-Hill.
(b) Khardi, S. (1992). An experimental analysis of frequency emission and noise diagnosis of commercial aircraft on approach. *Journal of Acoustic Emission* 26: 290–310.

79 Zaporozhets, O., Tokarev, V., and Attenborough, K. (2011). *Aircraft Noise: Assessment, Prediction and Control*. CRC Press.

80 Krebs, W. (2008). Lateral directivity of aircraft noise. *Journal of Acoustical Society of America (JASA)* 123 (3126), online).

81 Collin, D. (2002). Aircraft noise engineering: issues and challenges, Inter-Noise 2002 (13.1, 52.2), Michigan, USA.

82 Smith, M.J.T. and Williams, J. (1987). Aircraft noise, Chapter 18. In: *Transportation Noise Reference Book* (ed. P. Nelson), 18/3–18/36. Butterworth.

83 Warren, C.H. (1987). Subsonic travel and sonic boom, Chapter 21. In: *Transportation Noise Reference Book* (ed. P. Nelson), 21/3–21/10. Butterworth.

84 Hodel, W. (2010). "Viel Lärm um nichts? in Cockpit-das Schweizer Luftfahrt-Magazine, Nr.1, Jan 2010.

85 Bragdon, C.R. (1987). Control of airport impact, Chapter 20. In: *Transportation Noise Reference Book* (ed. P. Nelson), 20/3–20/21. Butterworth.

86 House, M.E. (1987). Measurement and prediction of aircraft noise, Chapter 19. In: *Transportation Noise Reference Book* (ed. P. Nelson), 19/3–19/37. Butterworth.

87 Sound Research Laboratories, Ltd. (1976). Practical building acoustics, Sound Research Laboratories Ltd.

88 Papa, A. (2015). Capri island helicopter noise control, Euronoise 2015, Maastricht, The Netherlands.

89 Bernardini, G., Anobile, A., Serafini, J. et al. (2015). Methodologies for helicopter noise footprint prediction in maneuvering flights, the 22nd Congress on Sound and Vibration, ICSV 22, Florence, Italy (12–16 July).

90 Arveson, P.T. and Vendittis, D.J. (2000). Radiated noise characteristics of a modern cargo ship. *Journal of Acoustical Society of America* 107 (19): 118–129.

91 Fischer, R. and Collier, R.D. (2007). Noise prediction and prevention on ships. In: *Handbook of Noise and Vibration Control* (ed. M.J. Crocker), 1216–1232. Wiley.

92 Borelli, D., Gaggero, T., Rizzuto, E. et al. (2015). Analysis of noise on board a ship during navigation and manoeuvres. *Ocean Engineering* 105: 256–269.

93 Di Bella, A. (2014). Evaluation methods of external airborne noise emissions of moored cruise ships: an overview, ICSV 21, Beijing, China (13–17 July).

94 Collier, R.D. (1998). Ship and platform noise, propeller noise. In: *Handbook of Acoustics* (ed. M.J. Crocker), 407–424. Wiley.

95 Bakogiannis, K., Argyropoulos, D., Dagres, P. et al. (2015). Residential exposure to port noise, mapping and source identification: a case study of Piraeus, Greece. 22nd Congress on Sound and Vibration, ICSV 22, Florence, Italy (12–16 July).

96 Dambra, R. and Firenze, E. (2015). Underwater radiated noise of a small vessel, 22nd Congress on Sound and Vibration, ICSV 22, paper no. 524, Florence, Italy (12–16 July).

97 Tsouvalas, A. and Metrikine, A.V. (2015). A three-dimensional semi-analytical model for the prediction of underwater noise from off-shore pile driving, Euronoise 2015, Maastricht, The Netherlands.

98 Merchant, N.D., Pirotta, E., Barton, T.R. et al. (2014). Monitoring ship noise to assess the impact of coastal developments on marine mammals. *Marine Pollution Bulletin* 78: 85–95.

99 Lloyd's Register ODS (2010). Noise from ships in ports: Possibilities for noise reduction. Environmental project No. 1330. Danish Ministry of Environment, Environmental Protection Agency.

100 Jegaden, D. (2013). Textbook of Maritime Medicine. http://textbook.ncmm.no/index.php/textbook-of-maritime-medicine/51-textbook-of-maritime-medicine/18-noise/917-main-noise-sources-on-board-ships-1.

101 Gibbs, B.M., Petersson, B.A.T., and Qui, S. (1991). The characterization of structure-borne emission of building services machinery using source description concept. *Noise Control Engineering Journal* 37 (2): 53–61.

102 Camp, S. (2002). Transformer noise. *Acoustics Bulletin* Sept.–Oct: 18–21.

103 Zindeluk, M., Medeiros, J.B., Souza, A.P. et al. (2002). Exhaust fan noise control by speed in power and industrial plants: technical and economical assessment, InterNoise 2002 (11.4, 38.4), Michigan, USA.

104 (a) Miller, L.N. (1979). Machinery. In: *Handbook of Noise Control* (ed. C. Harris), 26-1–26-16. McGraw-Hill.
(b) Moreland, J.B. (1979). Electrical equipment. In: *Handbook of Noise Control* (ed. C.M. Harris), 25-1–25-10. McGraw-Hill.
(c) Kingsbury, H.F. (1979). Heating, ventilating and air conditioning systems. In: *Handbook of Noise Control* (ed. C. Harris), 28-1–28-13. McGraw-Hill.

105 Van Diepen, T.J.M., Granneman, J.H., and Van Wijk, A.M. (2015). Benchmark indicators for industrial noise emission, Euronoise, 2015, Maastricht, The Netherlands.

106 Hansen, K.L., Zajanssek, B., and Hansen, C.H. (2015). Quantifying the character of wind farm noise. The 22nd Congress on Sound and Vibration, ICSV 22, Florence, Italy (12–16 July).

107 Hansen, K.L., Zajamsek, B., and Hansen, C.H. (2014). Comparison of the noise levels measured in the vicinity of a wind farm for shutdown and operational conditions, Inter-Noise 2014, Melbourne, Australia (16–19 November).

108 Lee, S., Kim, K., and Choi, W. (2011). Annoyance caused by amplitude modulation of wind turbine noise. *Noise Control Engineering Journal* 59 (1): 38–46.

109 Van den Berg, G. (2004). Effects of the wind profile at night on the wind turbine sound. *Journal of Sound and Vibration* 277 (4–5): 955–970.

110 Bowdler, D. (2008). Amplitude modulation of wind turbine noise. *Acoustics Bulletin* July/August: 31–35.

111 Leasure, W.A. (1979). Construction equipment, Chapter 31. In: *Handbook of Noise Control* (ed. C.H. Harris), 31-1–31-14. McGraw-Hill.

112 (a) Akama, E.A.A. and Lawson, P. (1975). Construction site noise. BRE current paper, 57/1975;
(b) Hersh, A.S. (1974). Construction noise: its origins and effects. *Journal of the Construction Division* 100: 433–448.

113 Defra (2006). Update of noise database for prediction of noise on construction and open sites. Phase 3: Noise measurement data for construction plant used in quarries.

114 Drozdova, L., Ivanov, N., and Kurtsev, G.H. (2007). Off-road vehicle and construction equipment exterior noise prediction and control. In: *Handbook of Noise and Vibration Control* (ed. M.J. Crocker), 1490–1500. Wiley.

115 Gilchrist, A., Allouche, E.N., and Cowan, D. (2003). Prediction and mitigation of construction noise in an urban environment. *Canadian Journal of Civil Engineering* 30 (4): 659–672.

116 Buna, B. (1996). Construction noise. Institute for Transport Sciences Report, Budapest.

117 Buna, B. (1988). Draft Standard for construction noise control for Trans-European Motorways. UNDP Project Report.

118 Martin, D.J. (1978). A comparison between road construction noise in rural and urban areas. TRRL Digest Report.

119 Lawson, P. (1976). Prediction and measurement of the L_{eq} during the construction of a main drainage system. BRE current paper no. 9/76.

120 Greene, R., Pirie, R., and Greene, M. (2002). Comparison of pile-driver noise and vibration from various pile-driving methods and pile types, Inter-Noise 2002 (12.2.3,12), Michigan, USA.

121 Nipko, K. and Shield, C. (2003). *OSHA's approach to noise exposure in construction*. OSHA.

122 FHWA (2006). FHWA Roadway Construction Noise Model. User's Guideline. U.S. Department of Transportation.

123 Kurra, S. (2003). Environmental noise problems in cities with a special emphasis on entertainment noise, *Proceedings of International Symposium on Quality of Life*, organized by ITU.

124 Merge, C.W. (1999). Noise from amusement park attractions: sound level data and abatement strategies. *Noise Control Engineering Journal* 47 (5): 166–172.

125 (a) Fimiani, F. and Luzzi, S. (2015). Monitoring and reducing noise related to Movida: Real cases and smart solutions, the 22nd Congress on Sound and Vibration, ICSV 22, Florence, Italy (12–16 July); (b) Gallo, E., Ciarlo, E., Santa, M. et al. (2018). Analysis of leisure noise levels and assessment of policies impact in San Salvario district, Torino (Italy), by low-cost IoT noise monitoring network, Euronoise 2018, Crete, Greece.

126 Schomer, P., Brandy, M., Lamb, J. et al. (2000). Using fuzzy logic to validate blast noise monitor data. *Noise Control Engineering Journal* 48 (6): 193–205.

127 Swieczko-Zurek, B., Ejsmont, J., and Ronowski, G. (2015). Impulse noise associated with firing handheld military weapons, 22nd Congress on Sound and Vibration, ICSV 22, Florence, Italy (12–16 July).

128 Brosnan, D. and Pritchard, J. (2016). The assessment of dog barking noise from boarding kennels. *Acoustics Bulletin* 41: 43–48.

129 Bayazit, N.T., Kurra, S., Şentop, A. et al. (2016). Proposed methodology for new regulation and guidelines on noise protection for buildings and sound insulation in Turkey, Inter-Noise 2016, Hamburg, Germany.

130 Weber, L. and Leistner, P. (2004). Water installations in buildings: a system of vibro-acoustics sources, Inter-Noise 2004, Prague, the Czech Republic (22–25 August).

4

Prediction Models for Environmental Noises

Scope

Dealing with acoustic problems requires quantification by using physical metrics. The magnitude of the noise is determined for the purpose of noise impact assessment and noise management either by calculations using appropriate models or by measurements or by employing both techniques. In this chapter the development of models, from the 1970s to the present, is overviewed with emphasis on the international standardized models employed in noise mapping (Chapter 6). The source-specific models which are based on the interrelationships between the noise levels and the associated parameters, some of which were developed as national models in the past, are introduced for motorway traffic noise, railway noise, aircraft noise, industrial noise, as well as for other types of community noises, such as waterway transportation noise, construction noise, shooting range noise, and wind turbine noise. The models comprising different structures (analytical, empirical, and guideline models) are referred to, along with the figures adapted from earlier documents, some of which are out of date, but worth mentioning to enable an understanding of the background of the advanced techniques.

4.1   Development of Acoustic Modeling

A description of environmental noises as quality and quantity is the first phase of noise management involving noise assessment and noise control. In a general sense, the magnitude of noise is determined according to what the information is intended to be used for, at two scales:

1) Noise source levels: (sound emission): The total sound energy radiated by a specific noise source in the free field where environmental factors do not exist.
2) Environmental levels (immission): Sound pressure levels at a specified receiver point in the environment where a number of physical factors exist and are effective on noise.

Emission and immission levels can be ascertained through acoustical measurements using the methodology and equipment recommended in the technical standards (see Chapter 5). When noise problems become a serious issue in urban areas, the development of the prediction methods becomes a priority in order to acquire data for the solutions against the negative effects of noise and for future planning. As is known, a prediction model covers the subset of calculations, applying

Environmental Noise and Management: Overview from Past to Present, First Edition. Selma Kurra.
© 2021 John Wiley & Sons Ltd. Published 2021 by John Wiley & Sons Ltd.

certain algorithms to obtain the future sound pressure levels at a receiver location by using the measured and predicted sound power levels and sound attenuation data.

A wide range of models have been developed since the 1960s, from complex mathematical algorithms to simple design guides. The state of the art in environmental noise modeling has been evaluated in the various literature under different categorizations. Considering the historical background, the models can be divided into three groups:

A) Theoretical models and semi-analytical models
B) Empirical models with use of graphical techniques (Design Guides)
C) Computer-aided modeling for noise mapping.

4.1.1 Analytical Models

The mathematical models that were developed in the earlier stages of acoustic science provide computations of sound pressures in sound fields of different characteristics [1–9]. They were based on the theories regarding the propagation of sound waves, such as the Helmholtz wave equation, Kirschoff's laws, Huygens' principles, Kurze's approaches, etc. Verification of the mathematical models could be made through field measurements under specified conditions with limited parameters. Theoretical models were too complicated and difficult for manual calculations, however, in the later period, simplified analytical solutions have been derived from the complex acoustic wave equations and "approximate semi-analytical methods" have been prepared. Defining certain variables, they aimed to derive the meaningful relations between sound pressure levels and the physical parameters which differentiate the specific source and environments. By means of advanced computer techniques, direct application of algorithms from theories has now been achieved at the present time (Chapter 6).

4.1.2 Empirical Models

Prediction models in this group, based on experimental results, are applicable for practical use, able to deliver noise levels in environments under certain conditions. Since many countries issued noise regulations in the 1980s, including outdoor noise limits to check the compliance with existing or future environments, it has been necessary to perform predictions using simplified methods also by non-acousticians. Parallel to the advances in measurement technologies, it was possible to obtain sufficient data to be able to construct statistical relationships and to investigate the effects of physical factors on noise levels. However, the number of parameters included in the models was limited, due to the difficulty in handling all the data, especially for complex source and environments.

The empirical methods require data on the characteristics of noise sources, generally displaying great variability regarding structural and operational characteristics. Therefore, a reference sound pressure level (called the basic noise level) was accepted at a certain reference distance from the source and the sound pressure levels were computed at the other receiver positions by using the regression equations derived from the statistical analyses of the measurement results.

The standardized guideline models covering design diagrams, which were easy for the non-acousticians to use, had a number of benefits at that time, e.g., they were quite appropriate for noise assessments with a certain approximation, compared to the field measurements requiring a laboratory facility with expensive equipment and trained personnel. In fact, the noise measurements describing the snapshots of the sound field at a certain time and under the prevailing physical factors, could not be generalized to represent all the other conditions, taking into account the varying parameters. However, the simple prediction methods were reasonable enough to obtain future noise levels for urban and transportation planning and for decisions over different scenarios, at that time.

In the 1980s, in many countries, in parallel to the noise regulations, national design guides were published by government institutions, such as the Ministry of the Environment, the Department of Transportation, the Ministry of Building, the National Building Research Center, etc. Figure 4.1 displays some examples of design guidelines that were widely used at that time by road designers, transportation planners, source owners, operators, builders, to make decisions and determine whether the noise limits were exceeded, for noise impact surveys, for transportation and urban planning, and to decide whether sound insulation was required for existing buildings. Initially they were prepared for road traffic noise, then extended to the other transportation noise sources. Some guidelines presented design charts giving the minimum distances to roads, facilitating the decisions in urban planning, such as the width of buffer areas between the roads and buildings.

The prediction models specified the range of physical environmental conditions in the estimation of noise levels. Although these conditions are subject to vary in time, the aim was to discover the average noise levels under the average conditions in the environment.

In the later phase, after the 1990s, the semi-analytical models called the "practical engineering methods" were developed and standardized. The engineering methods are also based on the empirical results obtained by measurements and take into account the average meteorological conditions. The technique adopted by these models involves the calculation of noise levels by adding the separate contribution of each parameter which has an effect on noise propagation. The operation is faster using computers, however, the accuracy is poor or moderate, as all the meteorological conditions cannot be included. The models have improved recently, to enable planners to deal with all the varying conditions with increasing number of sub-parameters affecting the results. Two examples are the Harmonoise model and lately the Cnossos-EU model, which has been accepted as the standard model throughout Europe. The engineering methods are currently used for noise mapping (Chapter 6) and the results are taken as the basis for noise management and action plans.

4.1.3 Numerical Methods with Advanced Computer Techniques

With the advances in computer technology, numerical methods, although they required longer time for calculations at the beginning, have found their way into noise mapping. As is the case with the similar characteristics of the analytical models explained above, this group of methods is also based on numerical solutions of wave equations. The best-known methods currently used are the Fast Field Program (FFP), the Parabolic Equation (PE), and the Boundary Element Method (BEM). They provide a great opportunity to analyze a wide range of noise problems in complex environments, however, each of them has its advantages and shortcomings. They are useful in analyses for certain meteorological conditions, however, they cannot be used for all possible environmental conditions and frequencies. Their accuracy is dependent on the specific situation in practical applications [1].

The FFP (wave number integration method) gives the exact solution for horizontal sound fields, but is restricted to homogeneous and flat ground surfaces and a layered atmosphere, without taking into account the different surface impedances and longer propagation distances. It is accurate but needs a longer computation time [10].

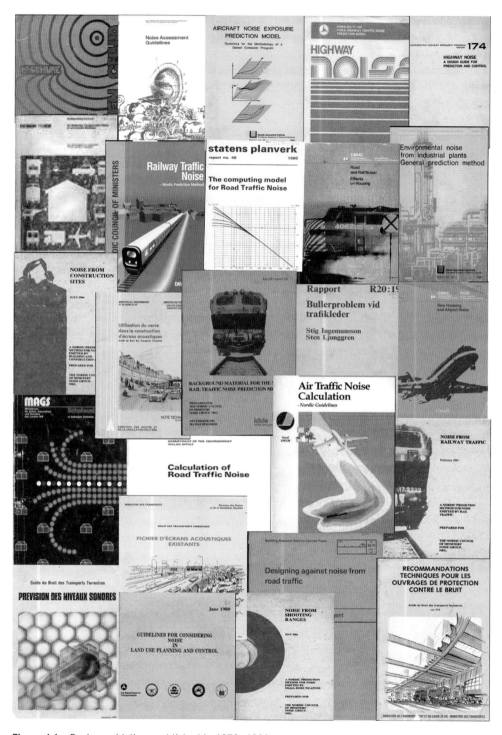

Figure 4.1 Design guidelines published in 1970–1980.

The BEM is capable of analyzing sound propagation over more complex level terrains under specific meteorological conditions and includes the effects of sound diffraction from large obstacles [10, 11].

The PE algorithm is restricted by an elevation angle because of the parabolic equation it employs, i.e. it takes only the sound paths from source to receiver [12]. Although it is slow in computation, its accuracy is high.

For longer-range propagations, the hybrid models combining different type of solutions are currently used. In the last few decades, models using the ray tracing technique have provided a visual tool and are implemented in noise mapping, as explained in Chapter 6. Although they are fast, however, there are some restrictions, e.g. they cannot include all the meteorological conditions as others and the prediction of the acoustic shadow requires some corrections to be applied. The numerical models are useful to derive some approximations.

4.2 Prediction Models for Road Traffic Noise

As the major source of noise, the first prediction models were developed for motorway noise, the characteristics of which are given in Section 3.2.1.

4.2.1 Development of the Models

Attempts to make assessment for traffic levels noise have been continuing since the 1950s. The analytical models first assumed free-flowing traffic, consisting of vehicles as point sources, with equal acoustic energy positioned on the road with equal spacings, later assuming the vehicles randomly distributed on the road, with at least two different acoustic powers (for light and heavy vehicles). The total acoustic energy is related to the number of point sources traveling at different speeds. Other models dealt with the prediction of the probability distribution of traffic noise and the statistical parameters of the noise levels. The earlier models enabled analyses of simply road and building layouts, later, more complicated configurations became applicable. These simulation models accepted that the noise distribution was in Gaussian form, the distribution of the number of vehicles along a certain road length was Poisson (random), and the spacing between the vehicles was displayed as a negative exponential distribution. The time-consuming processes like the use of special sampling techniques and the iteration of calculations for each vehicle condition could be facilitated by using advanced computer techniques. Simulation studies were good for complex road configurations and where acoustic measurements were not feasible. Thus, investigation of the interrelated effects of different measures and for comparison of performance became possible. In the last few decades, the engineering models have facilitated all the computations and are found to be more accurate than the previous models.

The recent standard ISO 1996-2:2017 (Annex L) gives a list of national and European source-specific calculation models, however, some of those which were developed in the earlier periods are still being used at present, and are outlined below in chronological order.

Theoretical Models

The first models for road traffic noise were developed in the 1950s and the number of studies escalated in the 1970s [2–12]. Some of them are the following:

- *K.N. Stevens and S.J. Barruch (1957)*: The cumulative noise levels in a time period, were modeled for the traffic with non-directional vehicles equally distributed on the road moving at a constant

Figure 4.2 Computed cumulative distributions of sound pressure levels measured at distance *d* from a road line (*s*: Spacing between the moving sources along the line, *d*: Perpendicular distance from the receiver to the line) [13].

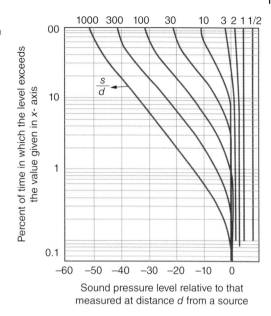

speed [13]. The model was given in Figure 4.2 prepared using the parameter, *s/d* (*s*: distance between vehicles which is determined from the speed and number of vehicles per minute, *d*: the perpendicular distance from the receiver point to the road, m). The percentage of time the level exceeds the level given in the *x*-axis is shown on the ordinate axis.

- *Kurze (1971):* A statistical traffic model was developed to calculate the sound pressure levels generated in homogeneous free-flowing traffic conditions, including randomly distributed vehicles (Pearson type III distribution). In the first part of the study, the model gave the probability distribution of normalized sound intensity. The second part involved complex road traffic, and the distribution of noise in terms of peak intensity, peak sound pressure levels, so A-weighted mean sound pressure level and mean energy level were given for various mixtures of traffic and the approximation was derived for the standard deviation of the sound pressure levels. The parameters were the types of vehicle, in relation to the traffic composition, the reference levels of vehicles and the distances [3, 4].
- *Show and Olson (1972):* Considering the urban area as a plane source emitting a constant noise consisting of randomly distributed noise sources, an approach was developed to determine the average energy density by taking account of different acoustic powers, the distances between the receiver and the road, the atmospheric absorption, and the barrier screening effect [14].
- *Kessler (1976):* The cumulative distribution of noise levels was investigated for traffic noise, taking into account the background noise levels [7].
- *Jonasson (1973):* A theory for predicting the excess attenuation of traffic noise from individual vehicles was developed. Applying the basic sound attenuation approaches to the traffic noise spectrum, the excess attenuations in dB(A) were calculated as corrections for different road geometries, the ground effect and the barriers by calculating the half-plane sound pressures [8].
- *Galloway, Clark, Kerrick, and others (1969):* Emphasizing the moving sources on the road, a simulation model was described to compute the sound pressure levels at points near a highway. The model was based on a simulation procedure within a time period, employing the Monte Carlo snapshot technique [15]. The vehicles were assumed to be randomly distributed on the road,

moving at different speeds, and the categories of vehicles were given by spectrums, as cars and heavy diesel trucks. The calculations were made in octave bands, eventually enabling the A-weighted mean noise levels.

Empirical Models

Simple approximations that could be derived from the theoretical models or the field measurements were presented in the national guidelines in 1970–1980. A number of empirical models displaying the diversity in terms of assumptions regarding traffic and environmental conditions (e.g. traffic volume, percentages of heavy vehicles, road surface type, road gradient) established the relationship between the traffic parameters and the noise levels based on the statistical analyses [15–20]. Some of them are outlined below:

- *Johnson and Saunders (1968):* The relationship in Eq (4.1) was given to determine the L_{50}, dB(A) for a free-flowing road and at the receiver points at a distance of four times the spacing between vehicles [16]:

$$L_{50} = 51.5 + 10\lg\frac{Q}{d} + 30\lg\frac{V}{40} \quad \text{dB(A)} \tag{4.1}$$

  d: Distance from the road side, ft.
  Q: Vehicle/hour
  V: Speed of traffic, mph

  The heavy vehicle percentage, p was assumed as 20%.

- *Schreiber (1969):* Calculation of the equivalent continuous sound level, L_{eq}, for constant traffic with the vehicles as point sources was made as a function of various parameters [17]. The simple approximation equation for a stream of sound sources is given as Eq (4.2):

$$L_{eq} \approx 10\lg\frac{N\overline{P}\rho c}{2avp_0^2} \quad \text{dB(A)} \tag{4.2}$$

  N: Vehicles per unit time
  $\overline{P}$: Average sound power level of vehicles, dB (A)
  a: Distance from road, m
  v: Average speed of traffic, km/h
  ρc: Characteristic impedance of air, $p_0 = 2.10^{-5}$ Pa

- *Ingemasson and Benjegard (1969):* Equation (4.3) was developed for L_{eq} (24h) as a function of certain traffic parameters [18]:

$$L_{eq} = 63 + 30\lg\frac{V}{70} + 10\lg\frac{FE}{25000} - 20\lg\frac{X}{50} \tag{4.3}$$

  V: Speed limit, km/h
  FE: Number of equivalent vehicles in 24 hours (number of light vehicles $+k$. heavy vehicles)
  k: Weighting factor given according to speed and road gradient.
  X: Distance from road, m

- *Lewis (1973 and 1977):* Based on the measurements both in rural and urban areas, the noise spectrums were presented for light and heavy vehicles, as given in Figure 3.6. His empirical models to determine the noise levels as a function of speed of vehicles were widely implemented for years in the past, as the basic relationship describing traffic noise [19, 20]. The average peak sound pressure levels of heavy and light vehicles were calculated in octave bands and in A-weighted levels at a reference distance of 7.5 m from the road:

$$L_{car} = 32.8 \lg V + 14.9 \qquad (4.4)$$

$$L_{truck} = 26.9 \lg V + 34.2 \qquad (4.5)$$

L_{car} and L_{truck}: Peak sound pressure levels at 7.5 m for light and heavy vehicles respectively, dB(A)
V: Vehicle speed, km/h

- *Delany and NPL (1972):* The National Physical Laboratory in the UK (NPL) proposed the regression equation for traffic noise level, L_{10} at a reference distance of 10 m from the center of traffic flow over hard ground [21, 22]:

$$L_{10} = 18.1 + 16.2 \lg V + 8.9 \lg Q + 0.117p \quad \text{dB(A)} \qquad (4.6)$$

V: Mean traffic speed, km/h
p: Percentage of heavy vehicle
Q: Traffic volume per hour

The traffic was assumed to be free-flowing in a straight line. Predictions for grassland, shielding by barriers, propagation along side-roads, at exposed façades, etc. were explained in the document with design diagrams prepared for different road configurations.

- *Nelson (Transport and Road Research Laboratory) (1970–1978):* Statistical distribution of maximum and minimum noise levels from moving vehicles in road traffic was investigated based on the field experiments and cumulative levels were obtained [23–25]. The relationships between the speed and noise levels were presented according to the vehicle type. Some results were derived, such as that the traffic noise increased 9 dBA with doubling of the speed above 50 km/h [26] (Figure 4.3). Further studies were conducted regarding the effects of road surface on the noise level in relation to vehicle categories [27]. The noise levels were investigated by dividing the vehicle types into three groups as [28]:
  - Light vehicles: Those with unloaded weight equal to 3000 kg or less (cars, vans, etc.).
  - Medium-weight vehicles: Commercial vehicles with two axles.
  - Heavy vehicles: Vehicles with three or more axles.

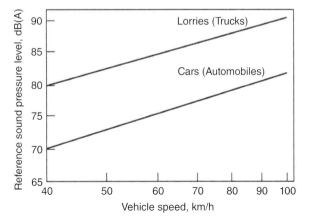

Figure 4.3 Prediction of traffic noise in relation to speed and vehicle type [26].

Harland, for the TRRL, investigating the variation of noise levels in rural areas (open ground, farmland, woodland), presented different equations to obtain the sound attenuation characteristics of ground in relation to the distance to the edge of the road [29].

- *Buna (1984):* Concerning the effects of acceleration of vehicles on noise levels, a model was developed to compute noise levels under stop-start conditions and in the deceleration of vehicles [30]. In the model, the noise level distribution with time was calculated at a certain reference point, in relation to increasing and decreasing speeds of the vehicles.

Besides those given above, a great number of research studies conducted in the 1980s focused on the secondary noise sources (i.e. engine noise, aerodynamic noise, surface noise, etc.) and the effects of various vehicle and traffic parameters on noise levels could be individually determined [31]. The consequences have been widely used for advanced traffic noise modeling in the later period.

Design Guide Models

Building Research Station (BRS) Methods in the UK (1969, 1970): Based on the investigations by Scholes and Sargent, the prediction of traffic noise was presented as set of diagrams for design purposes. Various metrics for traffic noise such as *TNI*, L_{NP}, and L_{10} (18h) between 06:00 and 24:00 hours, were proposed [32, 33]. It was declared that the traffic noise problem was arising where the traffic volume was above 10 000 vehicles per 18 hours.

The BRS model was developed to calculate the road traffic noise levels by assuming free-flowing traffic at an average speed of above 50 km/h, the roads were without junctions and curvatures, the ground was flat, and the maximum wind velocity was 12.3 m/s. The algorithm of the model was designed first to give the basic traffic noise level of L_{10} (18h) for an average velocity and with no heavy vehicles. Later, the parameters such as average speed, heavy vehicle percentage, road surface, gradient and crossings were introduced into the model.

The basic formula was given for the L_{10} at 30 m from the road only as a function of the traffic volume over 18 hours and at the average traffic speed of 75 km/h. The relationship was displayed as a diagram, given in Figure 4.4a, i.e. the noise level increase of 2.5 dB(A) per doubling of volume. Corrections recommended in the design guide to apply the basic level for traffic speeds other than the reference speed and for the distance from the road, are given in Figures 4.4b and 4.4c respectively. When the road gradient was greater than 8%, the traffic was assumed to be non-free-flowing and the corrections were + 1 dB(A) for the range of 2–4% and +2 dB(A) for

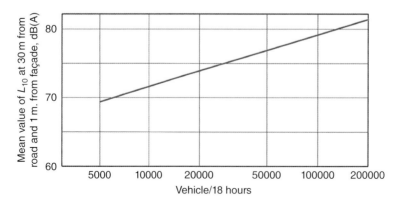

Figure 4.4a Basic chart for prediction of traffic noise levels in L_{10} at 30 m in relation to total flow of vehicles per 18 hour at average speed of 75 km/h (BRS 1971) [32].

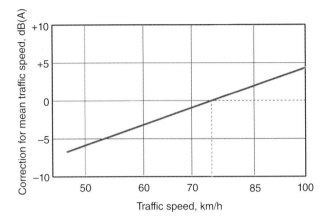

Figure 4.4b Correction values to basic L_{10}(18h) according to the average speed (BRS 1971) [32].

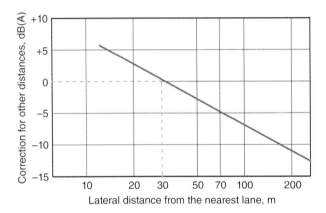

Figure 4.4c Correction to basic L_{10}(18h), dB(A) at 30 m according to other distances from road (BRS 1971) [32].

4–8% respectively, to increase the basic noise levels. The design guide presented other diagrams to determine the barrier attenuations, some as shown in Figures 4.4d and 4.4e [32].

- *Department of the Environment and Department of Transport Models (1975 and 1980):* BRS (later changed to Building Research Establishment, BRE) investigations were improved and published by the Department of Environment, DOE in the UK, as the national design guide model [34]. In this model, the basic noise levels in L_{10}(h) and L_{10}(18h) were given at 10 m from the nearest lane as a function of traffic volume:

$$L_{10}(18h) = 28.1 + 10 \lg Q \quad dB(A) \tag{4.7}$$

Q: Vehicles/18 hours (between 06 : 00 and 24 : 00)

$$L_{10}(h) = 41.2 + 10 \lg q \quad dB(A) \tag{4.8}$$

q: Vehicles/hour

In the above formulas, the assumptions were: average speed 75 km/h, road gradient 0, heavy vehicle percentage (HV)% 0, and soft/absorptive ground. The basic level was then corrected according to traffic speed, percentage of heavy vehicles, road gradient, and surface type with the aid of

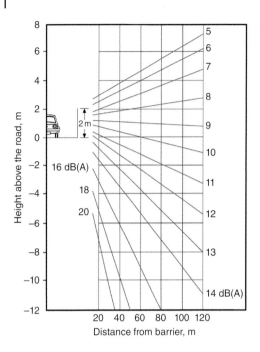

Figure 4.4d Reduction of L_{10} (h) by very long barrier of 2 m height (BRS 1971) [32].

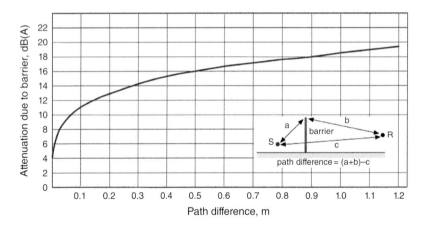

Figure 4.4e Reduction of L_{10} (h) from a long barrier (BRS 1971) [32].

graphic sets and equations. As an example, the correction for $HV\%$ was calculated according to the average speed of traffic:

$$Correction = 33 \lg \left\{ V + 40 + \left(\frac{500}{V} \right) \right\} + 10 \lg \left\{ 1 + \frac{5p}{V} \right\} - 66.8 \quad dB(A) \tag{4.9}$$

p: Heavy vehicle percentage
V: Average traffic speed, km/h

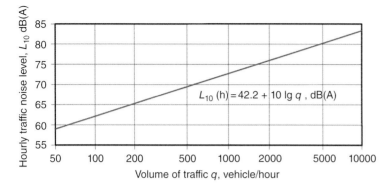

Figure 4.5a Prediction of basic noise level, L_{10}(h) in relation to total flow; hourly traffic volume, q [35].

The relationship between noise levels and volume of traffic is given in Figure 4.5a. The correction according to heavy vehicle percentage in relation to average trafic speed, is shown in Figure 4.5b. The additional parameters in the model are: distance between the receiver and the road, hard ground, barriers, angle of view, and reflections from nearby buildings. The model (DOT) was further improved based on the results of the experiments and published in 1980 by the Department of Transport, preserving approach in terms of L_{10} (h), however, more information was provided, e.g. the distance effect as a function of the height of the reception point, a correction according to the

Figure 4.5b Correction chart for average traffic speed and heavy vehicle percentage [34].

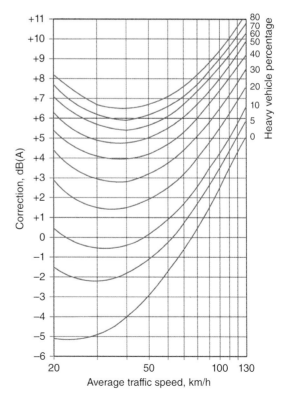

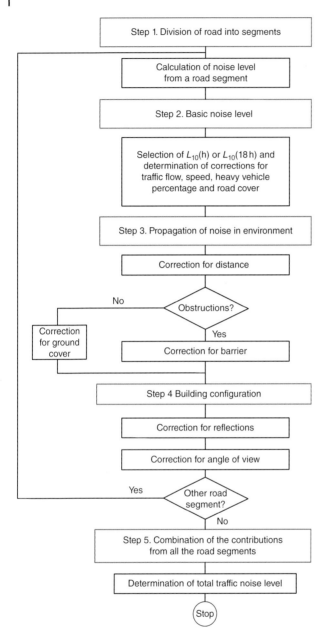

Figure 4.5c Flow chart of DOT Model [35].

type of traffic flow and barrier configurations (e.g. multiple screening) [35]. The flow chart of the prediction model is given in Figure 4.5c. This version covered a wide range of applications of the guidelines which recommended the field measurements to compare the results approximated by the design model.

- *Byggforskningen Model (1968, 1973, 1985)*: The first studies were conducted by the National Swedish Building Research Institute, in Sweden [36], and a guideline was published for traffic

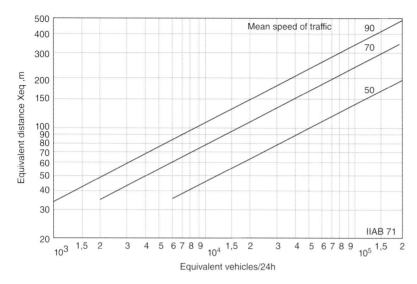

Figure 4.6a Distance between the road and a receiver point at ground level at which the equivalent steady level (L_{eq}) is 55 dB(A) [37]. (*Source:* Permission granted by Ljunggren.)

noise in 1973 [37, 38]. In the model, the concepts of "equivalent distance, X_{eq}" and "equivalent vehicle" as in Eq (4.3), which were related to the average speed of traffic, were introduced. Determination of an equivalent vehicle was based on the assumptions that the noise output of a heavy vehicle, with the speed equal or less than 50 km/h, generated 10 times greater energy than a car, whereas it generated three times more at the speed above 70 km/h. Figure 4.6a gives the equivalent distance as a function of an equivalent vehicle and average speed [36]. The empirical results were used to construct the design diagrams, for instance, to determine the allowable distance between the road and residential buildings in a design chart. The correction factors for other parameters, namely, the combination of distance and height from the ground, ground attenuation, road gradient, unscreened and screened land, winding roads, vertical reflecting surfaces, were presented in the consecutive diagrams (Figures 4.6b–4.6d).

- *Scandinavian Model (Nordtest Model) (1980)*: Later the above model was improved by the Nordic Council of Ministers and published as a design guide [39]. The L_{eq} (24h) was accepted as the basic descriptor and determined on a flat ground, then corrected according to gradient, $HV\%$, and barriers. The percentage of heavy vehicles was obtained from a database prepared for the Nordic countries [40].
- *Nord2000 Road Traffic Noise Model (2002)*: Based on the earlier prediction models, the Nordic environmental noise prediction method for road traffic has been improved by giving the sound power and sound pressure levels in one-third octave bands (25 Hz–10 kHz) [41, 42]. The propagation model takes into account complicated terrains, nine weather classes, and applies ray theory for propagation. Refraction is modeled by using curved sound rays in relation to the sound speed profile determined by an analytical approach. Compatibility has been verified with the Harmonoise model although it takes atmospheric refraction in a different way.
- *CETUR Model (1974)*: This comprehensive design guide was prepared by the Centre Scientifique et Technique du Bâtiment (CSTB) and later published by the Ministère de l'Environnement, in France. The noise metric was average hourly mean energy level, L_{eq}(h) between 08:00 and 20:00 hours at 2 m

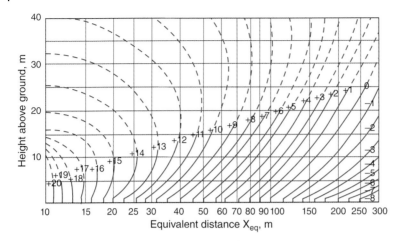

Figure 4.6b Design diagram for corrections with respect to distance and height from the ground, to be applied to L_{eq} relative to the sound level on the ground at a distance of 100 m [37]. (*Source:* Permission granted by Ljunggren.)
Note: Dashed lines show uncertain curves.

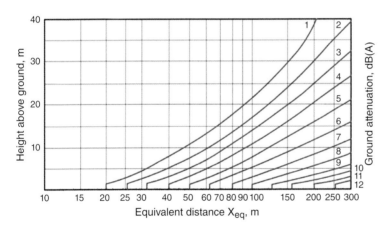

Figure 4.6c Effect of ground attenuation on L_{eq} as a function of receiver position given by height and equivalent distance to source, X_{eq} [37]. (*Source:* Permission granted by Ljunggren.)

from the building façade. The basic level was calculated according to the traffic and environmental parameters [43, 44]:

$$L_{eq}(h) = 20 + 10 \lg (Q_{VL} + EQ_{PL}) + 20 \lg V - 12 \lg \left(d + \frac{l_c}{3} \right) + + 10 \lg \frac{\theta}{180^0} \ \ \text{dB(A)} \qquad (4.10)$$

V: Traffic speed, km/h
d: Distance from road curb, m
l_c: Width of the road, m

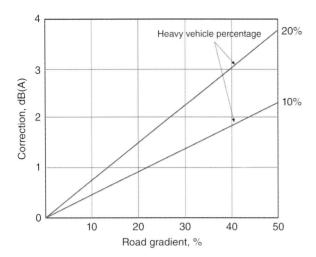

Figure 4.6d Effect of road gradient on noise level, L_{eq}, dB(A) [37]. (*Source:* Permission granted by Ljunggren.)

θ: Angle subtended, °
Q_{VL}: Volume of light vehicles/hours
Q_{PL}: Volume of heavy vehicles/hours
E: Equivalency factor for heavy and light vehicles (1 *heavy vehicle* = 10 *light vehicles*) to be determined according to gradient and type of road (e.g. highway, city traffic, with fast and normal speeds respectively) (e.g. for gradient $r \le 2\%$; $E = 4$ for highways and $E = 10$ for urban roads)

The model also gave Eq (4.11) for city streets along with buildings called U-type roads, to predict noise level on the façade of buildings:

$$L_{eq}(h) = 55 + 10\lg (Q_{VL} + EQ_{PL}) - 10\lg l + k_h + k_v + k_r + k_c \qquad (4.11)$$

l: Distance between the buildings in the U-type of streets, m
k_h, k_v, k_r, k_c: Factors taking into account the effect of receiver and source heights, traffic speed, road gradient, and crossings respectively

The representative traffic volume per hour was taken as 1/17 of the average total daily volume of traffic. Average traffic speed, distance, and road width were the other parameters in the basic logarithmic equation. Buildings, finite-length of road, and variation of noise levels with other factors, were taken into account in a simple algorithm. The nomogram given in Figure 4.7a displays the noise levels on the façades of buildings. Sound level variation in U-type roads is given in Figure 4.7b.

- *CMCH Model (1977–1978):* This guideline, published by National Research Center of Canada (NRCC), proposed a series of tables to determine the $L_{eq}(24h)$ for traffic noise [45, 46]. Some assumptions were made, such as the traffic speed limit was 112 km/h and the road was a single lane unless the distance from the road axis to receiver was less than the total width of the road. The basic noise level in $L_{eq}(24h)$ was calculated at 30 m from the road, according to the volume, composition, and average speed, then the corrections for further distances, road gradients, interrupted flow and barriers, were provided to apply to the basic noise level. Later the modified guideline was published following the investigations by Halliwell and Quirt [47].

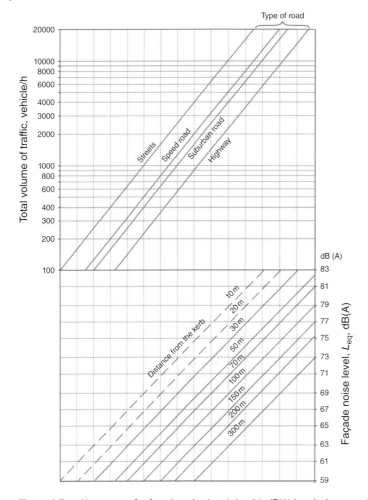

Figure 4.7a Nomogram for façade noise level, L_{eq} (h), dB(A) in relation to traffic parameters and road type [43].

- *National Cooperative Highway Research Program Report 117 (1971):* The prediction model was prepared in the USA as a hand tool for road engineering design [48]. By using the spectral values of emission data of the vehicles which were accepted with equal power and distributed uniformly on the road, the total power level of traffic line was computed using the Galloway, Clark, et al. model. The simple formula for traffic noise was proposed in terms of the statistical metric L_{50}:

$$L_{50} = 20 + 10 \lg V - 10 \lg D + 20 \lg S \tag{4.12}$$

V: Hourly volume of traffic (vehicle per hour)
D: Distance from center line to receiver, ft.
S: Average speed of vehicles, mph

The model was described in short and detailed versions. In addition to L_{50}, the reference levels in L_{10} were obtained by taking into account the traffic parameters at the receiver positions like the other models explained above. However, discrete equations were given for automobiles and trucks at 100 ft., as a function of volume of traffic flow and average speed, then the adjustments

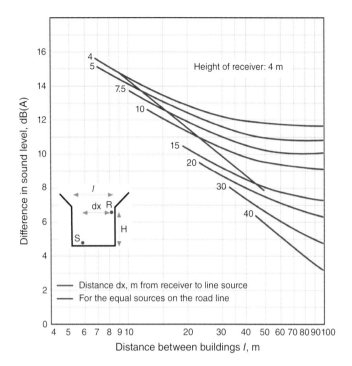

Figure 4.7b Variation of traffic noise level L_{eq}, dB(A) in *U*-type road and crossings [44].

were made for road width, observation distance, road gradients for trucks, shielding, interrupted flow, road surface. Ultimately, the evaluations according to the noise control criteria could be made in the guideline, including numbers of graphics and tables.

- *US Housing and Urban Development (HUD) Guide (1983):* The prediction was made for roadway noise based on various diagrams for day-night noise level (*DNL or L_{dn}*) given to evaluate noises of automobiles and truck traffic separately by taking into account their volumes in 24 hours at various distances from highways [49]. Adjustment factors were presented for speed, distance to stop-sign (in ft.), road gradient, night-time fraction, stop and go traffic, barriers, in the work charts and tables. A diagram displaying the relationships between the effective parameters and noise levels for truck noise is shown in Figure 4.8.
- *National Bureau of Standards (NBS) Guideline (1978):* In another guideline prepared in the USA, the highway noise prediction nomogram was presented, based on the work of Kruger, Commins, and Galloway [50]. The noise levels could be determined by using the traffic parameters (traffic volume for heavy and light vehicles) and the distance from the road, as given in Figure 4.9 [51].
- *Highway Research Group Guideline, USA (1976):* In the design guide prepared for highway designers and planners for the prediction of noise levels near highways, a nomogram, as in the NBS guideline, was given for L_{10}. The procedure introduced to divide roads into segments is shown in Figure 4.10 [52].

During this period, various laboratory studies were conducted in the USA by Donavan in 1974 and the results derived were included in the prediction process of traffic noise [53]. Figure 4.11a shows the noise level variation at the crossroads, according to the distance from the road [54]. Also,

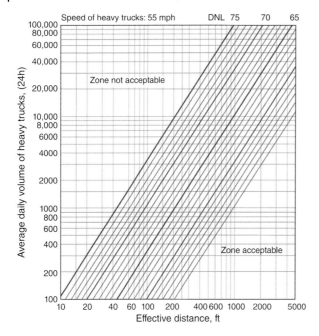

Figure 4.8 Workchart to predict truck traffic noise (L_{dn}) at the speed of 55 mph [49]. (*Source:* Originally published by the U.S. Department of Housing and Urban Development, Office of Policy Development and Research, and is reproduced here with the Department's permission.)

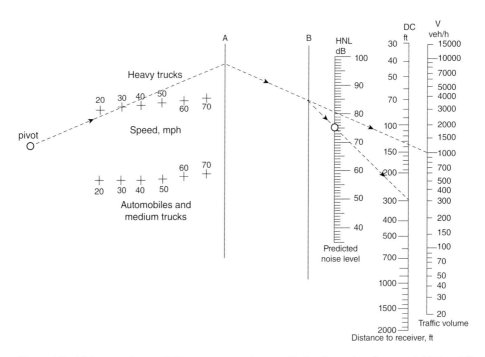

Figure 4.9 Highway noise prediction nomogram (arrows display the order of process) (National Bureau of Standards, NBS) [51].

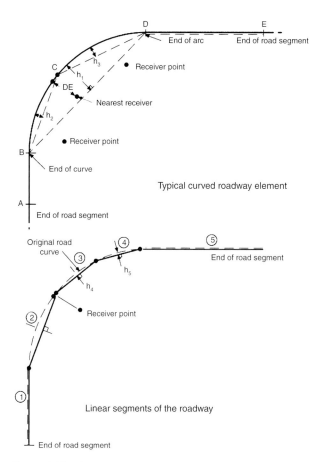

Figure 4.10   Segmentation of the curved roadway for calculation [52].

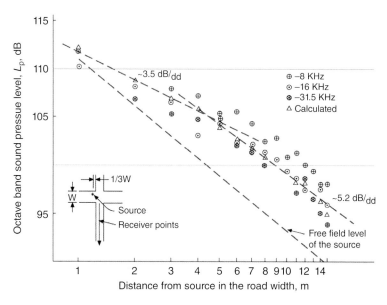

Figure 4.11a   Variation of noise level with distance to the road crossing [54].

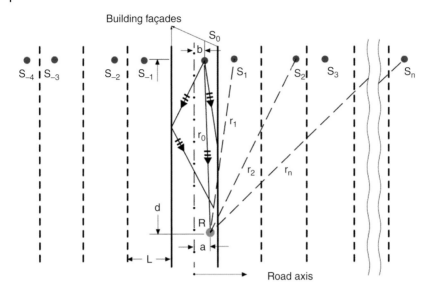

Figure 4.11b Imaginary sources and path lengths due to multiple reflections from the buildings along the road [55]. (*Source:* Permission granted by MIT libraries.)
Notes: R: Receiver point in front of the façade, S_0: Sound source position on the road. Original: Images and path lengths for contribution to sound received by observer. Walls are assumed smooth with energy reflection coefficient *R*, *L*: width of the street, *d*: distance along street from source to observer, r_n: distance from n^{th} image source to observer.

an algorithm was developed by Lyon,1974 to predict the effects of multiple reflections from the facades of buildings along the road, as given in Figure 4.11b [55].

- *Federal Highway Administration Highway Traffic Noise Prediction Model (USA) (1978):* The standard prediction model comprising an algorithm was developed by the Federal Highway Administration and vehicle noise emission levels were collected in a database [56, 57]. Three types of vehicles were defined as: automobiles, medium trucks, and heavy trucks. The A-weighted noise level in L_{eq} (h), at 15 m from the centerline of the traffic lane, was accepted as the basic level as a function of the vehicle speed. Traffic noise levels could be determined for different types of road surface, such as heavy asphalt, concrete with dense gravel, concrete with Portland cement, open-graded asphalt-concrete, and mixed surface. The secondary factors on highways, such as stop-signs, toll boxes, ramps, etc. were considered as noise increasing factors, due to interrupted flow and their effects were quantified in the model. The corrections were given for actual traffic flows using the "traffic flow adjustment factor," distance, finite roadway, gradient, and shielding. The effects of the physical factors (i.e. atmospheric absorption, distance, ground absorption according to the flow resistivity, topography, barriers (parallel screens, walls, earth berms, or combined barriers), buildings and vegetations were calculated in one-third octave bands. For the propagation model, the reflective and absorptive ground and effects of different types of barriers were taken into account as corrections to be applied to the reference level. Figure 4.12 shows the nomogram for barrier attenuation presented in the first document published by the FHWA [57]. In this comprehensive model, the road was divided into segments according to ground type between the road and the receiver, the finite length of barriers (angle of view), and the contributions of each segment were combined. The guideline made it possible to employ the other descriptors, i.e. day and night

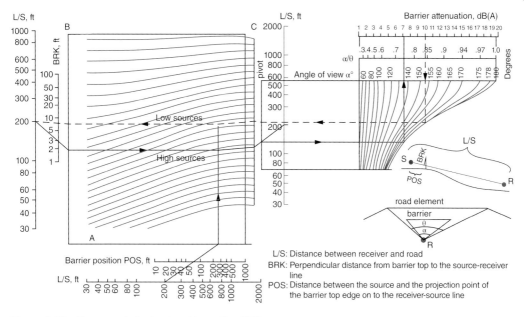

Figure 4.12 Nomograph for barrier attenuation [57].

level, L_{dn} (*DNL*) and the community noise equivalent level (CNEL). The FHWA Model was revised in 1998 by categorizing the vehicles as [58]:

– Cars (Automobiles): Two axles, four wheels, and with weights less than 4500 kg.
– Moderately heavy vehicles (trucks): Two axles with six wheels of all loading conditions.
– Heavy vehicles (trucks): All cargo vehicles with 3×6 axles and weights of more than 12 000 kg, trucks and buses with more than nine passengers, motorcycles with two to three wheels and all passenger vehicles. In the version published in 2004, three more standard vehicle types were added: buses, motorcycles (as independent type), and user-defined vehicles.

- *Stamina 2.0/OPTIMA (1982):* The FHWA model was later prepared as computer software enabling the noise contours to be obtained. The physical map of the area could be imported from CAD drawings and the measured results could be inserted for calibration purposes. Analyses of multiple diffractions from parallel barriers, detailed barrier design, and barrier optimization, were provided through the software. The simulation of real-time traffic flow was possible. Different types of noise contours, not only those representing the noise levels, but also noise reduction and barrier performance contours were available. The outputs were similar to those of other noise mapping programs (Chapter 6) [59a]. In the later version published in 1998, entitled the FHWA TNM, and updated as different versions over the years, the noise emission database and the acoustical algorithm were extended. Different pavement types, effects of graded roadways, calculations in one-third octave bands, graphically interacting noise barrier design and optimization, multiple diffraction analysis, parallel barrier analysis, etc. were included. TNM Version 3.0 was released in 2017 with the new features in data entry and analysis with the new acoustical algorithms increasing flexibility and accuracy [59b].

- *German guidelines (1971–1983):* The first guideline was prepared by the Bundesminister für Verkehr in 1971, based on the investigations of Reinhold. The model comprised diagrams to predict road traffic noise like other models [60]. In the second document, published in 1975, the average and maximum levels could be computed by taking into consideration the buildings along the road [61]. As explained in Chapter 6, the numerous examples regarding building and

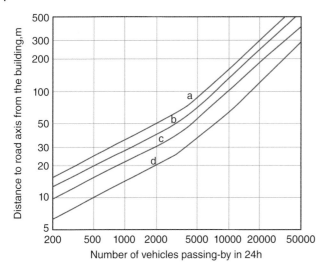

Figure 4.13a Minimum distance between roads and residential buildings in relation to traffic volume and average speed. Average speeds (from upper curve): (a) 100km/h; (b) 80km/h; (c) 60km/h; and (d) 40km/h [63].

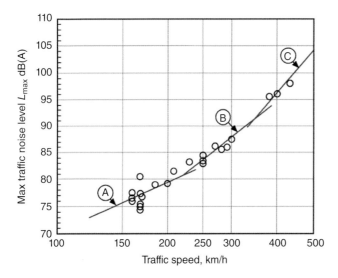

Figure 4.13b Maximum traffic noise levels in relation to traffic volume and distance (A, B, C are the partial regression lines) [63].

road configuration were included in the guideline to predict the screening effect based on the measurements. The Ministry issued improved guidelines over the years, enabling predictions of the basic noise level at 25 m from the road [62]. The noise levels as a function of the vehicle volume in 24 hours and the maximum noise levels as a function of vehicle speed, could be determined by means of the diagrams given in Figures 4.13a and 4.13b. Another document, published in Germany in 1983, also presented $L_{eq}(h)$ for 06:00–22:00 hours by taking the reference level of 32 dB(A) from the road, then applying correction factors for other traffic conditions [63].

- *Guidelines in The Netherlands (1981):* In the guideline prepared by the Volksgezondheid en Milienhygiene, the simple calculations for traffic noise levels could be made by using a hand-held

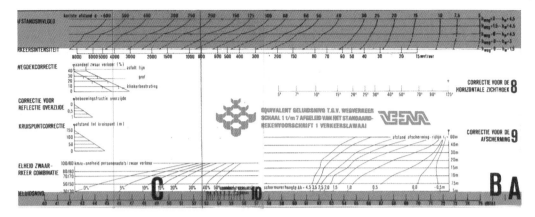

Figure 4.14a Guideline tool for prediction of traffic noise, L_{Aeq} developed in the Netherlands [65].

calculator (ruler), as shown in Figure 4.14a [64, 65]. The basic noise level L_{Aeq}(h) could be determined as a function of the hourly number of vehicles categorized as light, moderate, and heavy vehicles and the distance from the road. The corrections were applicable for the road surface, the meteorological effect, the angle of view, etc. The model could provide predictions of noise levels in urban areas considering the reflections from building façades, ground effect, attenuation by barriers, and enabled comparisons with the noise limits.

The Netherland prediction model, that was developed by Moerken, was issued in the later period (1986), covering the procedures for national roads, railways, and industrial noises [65]. In the model, the line sources were divided into moving point sources with equal powers, equal heights, and similar propagation conditions. For a plane source corresponding to an industrial area, a single point source was assumed at the center of plane. The calculations were made at the receiver positions satisfying the condition of $d > 1.5D$ (d: Source-receiver distance and D: the maximum length or diameter of the source plane). The basic equation for the sound pressure level for the average downwind condition was given based on the free-field sound power level of the source to be obtained first. Excess attenuations were determined along the propagation paths. In order to calculate the long-term L_{Aeq}, the table sets were available to determine the effects of meteorological factors.

Nordic Council of Minister Computing Model (1980): Calculation procedures for L_{Aeq} and L_{Amax} by means of the nomograms were described in the document [39]. The basic value L_1 was obtained at 10 m from the center of an infinite length of straight road, as a function of speed, number of vehicles both per hour and in 24 hours and proportion of heavy vehicles (Figure 4.14b). The geometrical attenuation and the effects of terrain type, embankment, barriers, excavated roads, soft ground, and road segmentation that could be obtained, under favorable meteorological conditions, based on the meteorological reports, were explained in the document.

4.2.2 The NMPB Method

The French national computation method for specifically road traffic noise (National Méthode de Prediction du Bruit) (NMPB), was developed in 1996 based on the earlier CETUR (1980) model [66]. The NMPB method was recommended by EU Directive 49 (END) in 2012 for strategic noise mapping and is still in use by many countries [67]. The emission data for vehicles were taken from the earlier document. Sound pressure level L_p is calculated both for the favorable conditions (L_{pF}) and

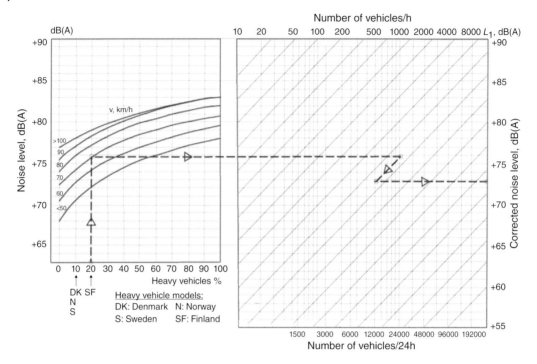

Figure 4.14b Nomogram for prediction of traffic noise level L_{eq}(h) and L_{eq}(24h), dB(A) in Scandinavian countries, 1980 [39]. (*Source:* Permission granted by Nordic Council of Ministers.)

for the homogeneous condition (L_{pH}), and the long-term noise levels are calculated by the energy summation of the contributions of each according to their probability of occurrences.

The Basic Characteristics of the Model

The model aims to calculate the motorway noise by taking into account all the physical factors and particularly the meteorological conditions during sound propagation. The basic unit is long-term, A-weighted sound energy level per meter length, determined in octave bands. This is obtained for two periods of time: L_{night} (22 : 00–06 : 00) and L_{day} (06 : 00–22 : 00), if requested, $L_{evening}$ and L_{den} are defined in END (see Chapter 6). Long-term levels take into account the yearly traffic flow and meteorological conditions. It is stated that the effect of the meteorological factors is important at source-receiver distances greater than 100 m. The percentages of long-term occurrences of favorable meteorological conditions for day and night are given in tables and maps for different regions of France based on meteorological reports. The calculations are performed for the frequency range of 125–4000 Hz in octave bands. The algorithm of the model is given in Figure 4.15.

Source and Receiver Descriptions

The length of the road as a line source (each lane or assumed to be a single lane) is divided into the elementary sources according to certain principles and the midpoints are taken as point sources with the height at 0.5 m from the road surface. The receiver point is at least 2 m above from the ground (4 m as required for noise mapping). The model is applied for the receiver points at a maximum distance of 800 m perpendicular to the road. The validity of the model increases with increasing the height of the receiver and increasing the distance to the road.

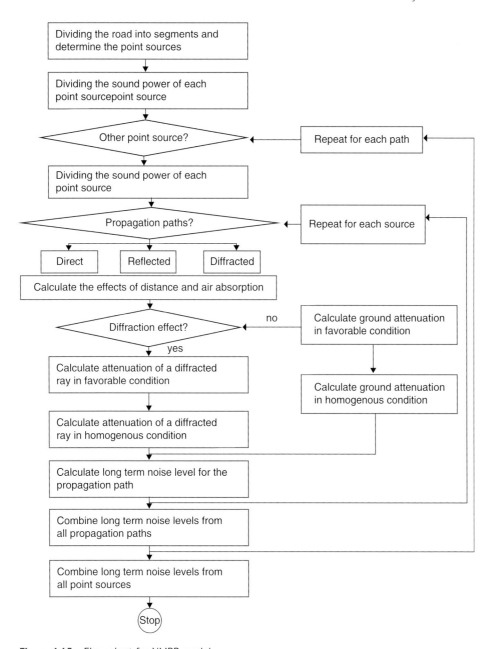

Figure 4.15 Flow chart for NMPB model.

The sound emission of each lane of the road to be defined with the geometric properties is calculated as A-weighted sound power level per meter length of the road:

$$L_W = L_{WL} + 10 \lg \left(\frac{Q + Q.\ \%PL.\ (EQ-1)/100)}{V_{50}} \right) - 30 \quad \text{dB} \tag{4.13}$$

L_{WL}: Sound power level of light vehicles $= 46 + 30 \lg V50 + C$
V_{50}: Average speed of traffic, km/h (for $V_{50} < 50$; $V_{50} = 50$ km/h)
C: Correction factor for traffic flow;
$C = 0$ (free flowing traffic), $C = 2$ (interrupted traffic), $C = 3$ (accelerating traffic)

Q: Traffic volume (vehicle/hour/lane)

PL%: Percentage of heavy vehicles

EQ: Equivalent light or heavy vehicles (in relation to road gradient and speed of traffic)

When the traffic parameters are not available, the emission levels L_w, dBA/m for the roads categorized as highways, main roads, primary and secondary arteries, are determined using the database provided for *Q*, *PL*, *V* in the document.

When the power of sound source S_i is known, octave band sound pressure levels are calculated as in Eq (4.14):

$$L_{i \text{ (F or H)}} = L_{Awi} - A_{i \text{ (F or H)}} \tag{4.14}$$

L_{Aw}: Sound power of the sources at each octave band as A-weighted level, dB

i: Source indices

F and *H* imply favorable and homogeneous conditions respectively.

$A_{i,F}$ and $A_{i,H}$ are the total of sound attenuations to be calculated in octave bands as:

$$A_{i,(\text{F or H})} = A_{\text{div}} + A_{\text{atm}} + A_{\text{grd}(\text{F or H})} + A_{\text{dif (F or H)}} \tag{4.15}$$

A_{div}: Attenuation due to wave divergence or distance effect (the concept of equivalent distance is introduced)

A_{atm}: Attenuation due to air absorption in relation to average temperature and humidity (for France; 15° and 70%).

$A_{\text{grd,F}}$ and $A_{\text{grd,H}}$: Attenuations due to ground under favorable and homogeneous conditions respectively (a calculation procedure is given using the absorption coefficient of the ground, e.g. if G = 0; attenuation is – 3 dB.)

$A_{\text{dif,F}}$ and $A_{\text{dif,H}}$: Attenuations due to diffraction under favorable and homogeneous conditions (when diffraction exists, ground effect is not considered).

Long-term Octave Band Level

The total continuous equivalent energy level in each octave band, and at a certain receiver point *R*, is calculated by summation of all the contributions from the real and imaginary sources due to ground reflections, and by applying a certain weight according to the occurrence of the favorable conditions. The procedure is as follows:

$$L_{i,LT} = 10 \lg \left(p_i \, 10^{L_{i,F}/10} + (1 - p_i) \, 10^{L_{i,H}/10} \right) \tag{4.16}$$

$L_{i,LT}$: Long-term sound pressure level for i^{th} source

p_i: Time percentage in which the favorable conditions are observed

i: Indices for real and imaginary sources

L_F: Band pressure levels in favorable conditions during the ratio *p* of time

L_H: Band pressure levels in unfavorable conditions during the ratio (1–*p*) of time

Total A-weighted long-term noise level, dB is obtained by summation of all octave band sound pressure levels at receiver point *R*:

$$L_{Aeq,LT} = 10 \lg \left[\sum_{j=1}^{6} 10^{0.1 L_{eq,LT}(j)} \right] \tag{4.17}$$

j: Octave bands between 125 and 4000 Hz

The NMPB model has been revised several times by minor changes and was issued in 2008, entitled NMPB-Routes 2008, with some modifications [68], such as the source height is reduced and two different spectra are introduced. The ground attenuation, which was given for downward conditions, was replaced by the homogeneous condition considering the turbulence effect. Road configurations were given by adding cuttings/embankments. The validation of the model by experiments is presented, revealing better results more compatible to those measured.

4.2.3 The Harmonoise Model

A project supported by the European Union in 2005, Harmonoise, has delivered prediction methods for both motorway traffic and railway noises [69]. The source emission model for road traffic noise is constructed for the sources generating rolling and propulsion noises, at two heights from the road surface: 0.3 m for light vehicles and 0.75 m for trucks. The basic relationships for calculation of rolling and propulsion noise levels are given in terms of the A-weighted sound exposure levels (*SEL*) by using some coefficients, according to the vehicle categories and speeds. The relationship between the *SEL* (measured at a microphone position) and the sound power level of the vehicle L_w is given in Eq (4.18):

$$SEL = L_W - 10 \lg v + 10 \lg d + 10 \lg \alpha - 10 \lg \left[4\pi \left(d^2 + (h_r - h_s)^2 \right) \right] - \Delta L \tag{4.18}$$

v: Vehicle speed, m/s
d: Distance from receiver point (or microphone) to the road, m
α: Angle subtended, radian
h_r and h_s: Heights of receiver and source above ground respectively
ΔL: Correction term for reflections

Using the *SEL* calculated above, the weighted equivalent continuous level L_{Aeq} is obtained based on the number of vehicle flow per hour for each vehicle category:

$$L_{eq} = 10 \lg \left[\frac{\sum_c n_c 10^{SEL_c/10}}{3600} \right] \quad dB(A) \tag{4.19}$$

c: Category of vehicles
n_c: Number of vehicles per hour in each category

Variation of speed is defined statistically in the model and a coefficient is defined for the accelerated vehicles. Corrections for type of tires and type of the road surface according to the size of gravel are proposed. A similar approach of road segmentation is applied as in the previous models. Sound propagation in the physical environment is modeled by using two approaches; the BEM and the two-dimensional PE method for mixed conditions (Chapter 6). Calculations are made for each of the 1/3-octave bands in the range of 20–10 kHz. The specific procedure is applied for barrier attenuations, and for more complicated barrier configuration, the PE is employed with an emphasis on top profiles.

Validity tests of the model were conducted in Germany and France. The comparisons were made in terms of the "ambition level" defined as the square root of the averaged square levels of differences between the calculated and the measured results. It was reported that the difference was not more than 1–5 dBA. Parallel to the Harmonoise model development, the IMAGINE project, supported by the EU in 2005, has also been completed with the aim of collecting and organizing data for traffic and railway noise sources throughout Europe [70]. Both methodologies have constructed the basis for the latest EU model which will be described below.

4.2.4 The Cnossos-EU Traffic Noise Model

The attempts to develop a harmonized method enabling noise assessment complying with the EU Environmental Noise Directive requirements, were accomplished with a document called Cnossos-EU (Common Noise Assessments Methods in Europe) in 2012 [71]. It gives the assessment procedures for all transportation and industrial noise sources that will be employed in noise mapping and action planning in the European countries, in terms of the noise indicators of L_{den} and L_{night} to be determined in the range 63–8 kHz (see Chapter 6).

For road traffic noise, the model proposes source emission with the effective parameters to establish a database. As in the other models, the road is assumed to be a line source, applying the segmentation technique, the power parameter per frequency band is determined by using the equivalent point source at each segment.

The parameters to be taken into account in the emission computation procedure are determined according to the criteria: in which the range of values cause a variation in L_{den} or L_{night} more than ± 2 dB 95% C.I. The four categories of vehicles, symbolized by m, are accepted as (the letters in parentheses correspond to the vehicle types defined in the EU Directives): Light vehicles (M1, N1), medium vehicles (M2, M3, N2, N3), heavy vehicles (M3, N3), powered two-wheelers, and an open category. The emission levels of each vehicle category are to be determined through measurements in the semi-free-field (see Chapter 5).

Noise emission from the traffic flow, which is expressed as the sound power level per meter in octave bands between 125 Hz–4 kHz, is obtained through summation of the sound powers of all individual vehicles by taking into account their directional characteristics:

$$L_{W,eq,line,i,m} = L_{W,i,m} + 10 \lg \left(\frac{Q_m}{1000 v_m} \right) \tag{4.20}$$

$L_{W,i,m}$: Instantaneous directional sound power in the semi-field of a single vehicle in category, m

$L_{W,eq,line}$: Directional sound power of line source per meter per frequency band in dB (re. 10^{-12} W/m)

Q_m: Vehicles of category m per hour for steady traffic flow (yearly average per day, evening and night periods)

v_m: Average speed of flow of vehicles per category m (maximum legal speed)

The emission model for individual vehicles is established by accepting that the noise emission of individual vehicles is composed of two types of noises generated by road traffic:

A) Rolling noise and aerodynamic noise due to tire/road interaction
B) Propulsion noises from engine, exhaust, etc.

Eventually the general mathematical expression is given as:

$$L_{W,i,m}(v_m) = A_{i,m} + B_{i,m} \cdot f(v_m) \tag{4.21}$$

$L_{W,i,m}(v_m)$: Instantaneous directional sound power of a single vehicle from category m, as a function of speed

$A_{i,m}$ and $B_{i,m}$: Coefficients for rolling and propulsion noises respectively

$f(v_m)$: Speed function

The speed function is represented by two expressions, for each type of noise, at each frequency band for each vehicle category and for the reference speed of 70 km/h: a logarithmic function for rolling and aerodynamic noises, and a linear function for propulsion noise. The partial sound power levels calculated for two types of noise are logarithmically added to find the total emission level of a vehicle. The reference conditions are described in the document, including road surface, tire type, and meteorological conditions. Then the corrections are applied according to the actual conditions, e.g. for temperature, road surface, tire type, including the studded tires of light vehicles. The effects of the road gradient on vehicle speed, both on rolling and propulsion noise emission, engine load, and the engine speed are formulated for each vehicle category.

Since the noise emission of the accelerated and decelerated vehicles reveals higher uncertainty and hence makes accurate estimations rather difficult, certain corrections are proposed for inter-sections (road crossings). In order to take into account the effect of the road surface when the rolling noise is predominant, the spectral emission coefficients in 125–4 kHz are applied in relation to the speed of the vehicles, if the road surface is different from the reference surface of averagely dense asphalt concrete. The age of the road surface and its maintenance condition are also counted as correction factors. The document provides sets of tables to simplify the calculations of all the coefficients for each category of vehicles at each octave band.

The sound propagation model is applicable in octave bands between 63 and 4000 Hz and for the receiver points up to 800 m distant from the road and at least 2 m high from the ground, in rural and urban areas, including canyon (*U*-type) streets. Calculation of the sound pressure levels by using the directional sound power level of the road traffic noise is performed under homogeneous and favorable conditions for each sound path and by combining the energies of all the other paths. For the long-term sound level, the mean occurrence, p of the favorable conditions is considered in the direction of each path i. Determination of the geometrical divergence and atmospheric absorption is similar to the model given in ISO 9613-2 (Section 4.5.2). The ground effect is determined, associated with the type of ground identified with the absorption properties, G. The complex grounds are represented by G_{path}, defined as the fraction of the absorbent ground between the source and the receiver. Ground attenuation is calculated for homogeneous and favorable conditions.

The effect of diffraction from barriers, A_{diff}, is calculated according to the path difference and the wavelength. When the path difference is $\delta < -(\lambda/20)$, no calculation is needed, otherwise the attenuation due to diffraction is calculated for the direct and reflected paths on the source and receiver sides of the barrier.

Cnossos-EU was mandated by the Commission Directive issued in May 2015 and its implementation in EU countries started in 2019 [72]. Recently studies are continuing for the conversion of the existing road emission database to adapt to the new standard model, emphasizing the differences on road pavement categorizations and the national emission models [73a, 73b].

4.3 Prediction Models for Railway Noise

Noise radiation from various trains, including metros, light trains, etc. has been extensively investigated in different countries, especially where the railways are the major transportation system. The models developed to estimate the emission levels of train noise with specific characteristics referred to in Section 3.2.2 and the emission levels in the environment are outlined below.

4.3.1 Development of the Models

Since the earliest times to the present, the studies on railway noise modeling have been based on the field measurements and data analyses. Hanson, in the 1990s, conducted a series of investigations for high speed trains in the USA (Maglev) and revealed that the rolling and aerodynamic noise levels were well correlated with the train speed, *V* by the factors of 30lg*V* and 60lg*V* respectively [74, 75]. In the 1990s, various guidelines for the prediction of railway noise were developed, however, requiring the train emission data to be acquired from field measurements at 10–30 m to the railways. Such measurements are important, especially for the new trains, to observe the influence of changes in rail conditions and speeds on the generated noise.

Okumuro and Kuno, in Japan, derived the relationships between the train noise levels (in $L_{A\text{max}}$ and L_{AE}), the background levels (L_{50}), and a number of railway parameters, such as speed, length of trains, distance from tracks, type of railway structures (steel and concrete bridges), at grade or elevated structures, cuttings, retaining walls, rail type (standard or welded rail), type of traverse (wood and precast stressed concrete) [76]. The outputs from their model were compared with the measurement results conducted for Shinkansen trains.

Two different approaches are employed in immission models for railway noise:

1) Trains are assumed to be coherent point sources. The contributions from the individual sources, determined at a receiver point, are combined to find the total railway noise.
2) The entire railway line is assumed as a finite-length moving line source and the levels are computed at a receiver point.

In both approaches the effects of the physical environmental factors on the source-receiver path are taken into account in the simulation model.

4.3.2 Analytical Models

Some of the computation models using both approaches are explained below by taking into account the train parameters which have an effect on the emission and the immission levels as given in Section 3.2.2.

The Earlier CSTB and Mithra Model (1994)
To determine the sound power level of a railway, Eq (4.22) was given mainly for railway noise simulation in the environment [77]:

$$L_W = 18 + 10 \lg \left(a \sum \frac{n_i L_i}{b_i} v_i^2 \right) \tag{4.22}$$

L_W: Equivalent sound power level for each meter of track, dB(A)
n_i: Number of trains' pass-by per hour for i^{th} train type
L_i: Length of train for i^{th} type, m
v_i^2: Speed for i^{th} train, km/h
a: Noise factor of rail:
  a = 1 (Continuously welded trains)
  a = 2 (Rails with cross-connections)
  a = 3 (Short rails or 20 m-long rail around a maneuvering station)

b_i: Quietness factor of train
  b = 0.5 Noisy wagon
  b = 1 Normal wagon
  b = 3 Quiet wagon
  b = 4 Very quiet wagon

The Ringheim Model (1984)

The guideline model for the prediction of railway noise was developed by Kilde in Norway, derived from the research conducted for the Ministry of the Environment and Communication [78, 79a, 79b]. The railway line was considered as a line source at 0.5 m above rails and the reference noise level in $L_{eq}(24h)$ was calculated as:

$$L_{eq(24\,h)} = 50 + 10\lg\left(l_{24}/1000\right) - 10\lg\left(a/100\right) \quad \text{dB(A)} \tag{4.23}$$

$L_{eq}(24h)$: Reference noise level for a reference train length (1000 m/24h) at a distance of 100 m from the railway
a: Perpendicular distance to the railway projected on the horizontal plane, m
l_{24}: Total length of trains within 24 hours, m

The equations for calculating the maximum noise levels L_{max} were also presented for the reference train speed of 80 km/h and for continuous welded railways. In this earlier model, train types were not considered and the equations could be used for all types of trains (ignoring the difference between diesel or electric engines). The emission levels had to be measured for locomotives and wagons, then the total emission of the train could be calculated by applying a logarithmic summation. The model provided a correction according to the angle of view (subtending angle) from the receiver to apply to the reference level (Figures 4.16a and 4.16b). For the railway curvature, the method of dividing the curvature into finite straight lines (segments) was applied.

Environmental factors, such as topographical conditions, tunnels, speed, and acceleration, ground cover, screen effect, reflections from nearby buildings, rail conditions, and bridges, could

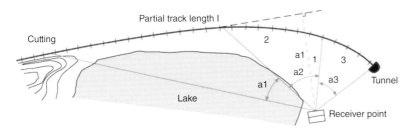

Figure 4.16a Modeling a railway by dividing into straight segments [79a]. (*Source:* Permission granted by Ringheim.)

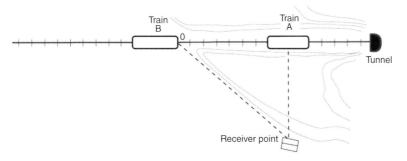

Figure 4.16b Modeling a railway vehicle entering a tunnel [79b].

be taken into account in the immission model and various equations were provided to obtain the correction factors. The corrected values for locomotive and wagons were combined to obtain the total railway noise levels at a receiver point.

Nordic Countries Railway Prediction Model

The guideline published by the Nordic Council of Ministers in 1996, which was developed in a collaborative study, intended for use in planning, noise impact, and noise control studies [80]. It was based on the Kilde model in principle, however, it improved it by taking into account the properties of the land around railways. The noise levels were calculated in octave bands and in statistical units ($L_{eq,T}$ and L_{max}) for the passing trains according to an algorithm. The predictions were restricted to the airborne sounds of trains with speeds more than 30 km/h, ground conditions of summertime, downwind zones, and the receiver points at distances greater than 50 m to the rails.

In order to provide noise data for trains of different types, a wide range of measurements were conducted, and the sound power levels were obtained as A-weighted *SEL* (or L_{AX}). The noise level variation as a function of train length and distance from the railway is shown in Figure 4.17a and the correction values in relation to train speed are given in Figure 4.17b.

The sound power level for a given train type was expressed as:

$$L_{Wo} = a \lg (v/100) + 10 \lg (l_{24}) + b \quad dB(A) \tag{4.24}$$

L_{Wo}: Sound power level per meter track length (in octave bands)
v: Speed of a specific train, km/h
l_{24}: Total train length of a specific train type within 24 hours, m a, b: Coefficients at octave bands according to train type (obtained from measurements)

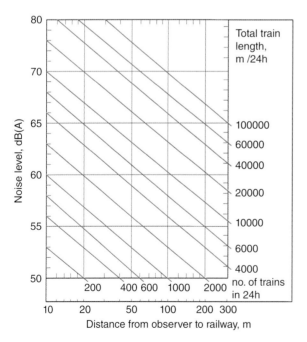

Figure 4.17a Determination of noise levels at various distances in relation to number of trains and total train length in 24 hours [78, 79a, 79b, 80]. (*Source:* Permission granted by Ringheim.)

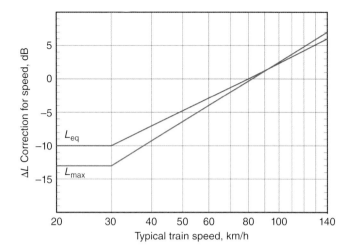

Figure 4.17b Correction to be applied on L_{eq} and L_{max} according to pass-by speed of trains [79a, 80]. (*Source:* Permission granted by Nordic Council of Ministers.)

Assuming each unit of train length generates the same amount of energy, the sound power radiated along the train length was calculated as:

$$L_{wt} = a \lg (v/100) + 43.8 + b \tag{4.25}$$

L_{wt}: Sound power level per unit length of train, dB(A)
v: Speed of the given train, km/h

The emission levels calculated above are inserted into the immission model to obtain the sound pressure levels in L_{eq} (24h), dB(A) at a receiver point in octave bands between 63 and 4000 Hz, by taking the environmental factors into account, e.g. the effects of distance, air absorption, ground, and screens. Correction factors between 0 and 6 dB were applied, according to the condition of the rails. Corrections were: train type, finite train length, track conditions, façade correction, ground. A correction was also applied for the angle of view of the railway (Figures 4.16a and 4.16b). The guidelines included various design charts prepared for electrical trains under different environment and source conditions (Figures 4.18a and 4.18b). The algorithm of the procedure was later developed as a computerized model.

BRE Railway Noise Model
Cato developed a model to determine the spectral levels of high-speed electric trains in the UK in 1976 [81]. The model was applicable to railways with concrete and wood supports and for continuously welded rails.

The Canadian CMCH model
The Canadian guideline, published in 1977, described a simple method for measuring railway noise levels in L_{24h}, dB(A) by providing table sets [82]. The model covers the separate determination of three types of noise: locomotive engine noise and rolling noise (wheel–rail noise) of locomotive and cars, and whistles. The steps of the procedure were:

- Determination of locomotive engine noise level for a train at 50 mph (80 km/h) speed and at 100 ft. (30 m) from the railway center line. Corrections for the actual speed.

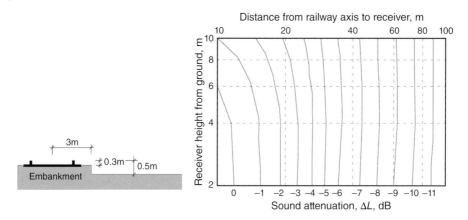

Figure 4.18a A chart to predict sound attenuation at a receiver point in relation to distance from railway [80]. *Notes:* Electric trains. Normal ballasted track. Acoustically soft ground. Track surface 0.5m above ground surface. Edge of embankment 3m from track center line, 0.3m below ballast top surface. Distance from track centerline = 4m. Height above ground level = ΔL. Sound level reduction, dB. (*Source:* Permission granted by Nordic Council of Ministers.)

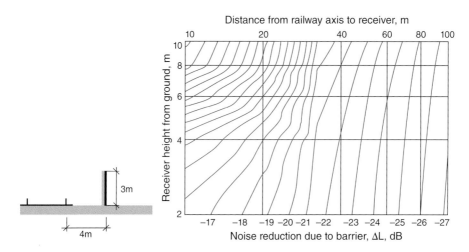

Figure 4.18b A chart to predict noise levels at a receiver point in relation to distance from railway when a noise barrier exists [80]. *Notes:* Soft screen-track surface level with ground surface. Absorbing surface on the side facing track. Screen distance from track centerline = 4m. Screen height =3m above the track surface. (*Source:* Permission granted by Nordic Council of Ministers.)

- Determination of rolling noise at 100 ft.
- Determination of whistle points within 1/2 mile of the site for a train at 50 mph and for a single train on hard ground. Corrections of siren noise for soft ground, according to train speed and number of train pass-bys.
- Correction for actual distance between receiver and railway and height of receiver.

- Determination of effective barrier height and barrier attenuation and correction of barrier attenuation according to "effective barrier length" ratio.
- Combining the three components for each train.
- Obtaining 24 hours noise level in dB(A) by taking into account the number of trains.

The UK Railway Prediction Model 1995

A process was developed to predict the train noise immission levels in *SEL* and L_{Aeq} for daytime, evening, and night-time, by using the emission database prepared for the UK trains [83]. The input data required in the guideline was train speed and operational characteristics. The model was revised after the EU Directive 49 was issued and the increasing effect of rail configuration, defined as Acoustic Rail Quality Value (ATQ), was introduced into the model with a correction function [84a,b].

The German Railway Noise Prediction Model (SCHALL 03)

The calculation model, issued in 1990 and updated in 2014, aimed to be used for planning new rail lines, or for modified railways [85–87]. In the first document, the noise emission level of trains was calculated using Eq (4.26):

$$L_{m,E} = 51 + D_{Fz} + D_D + D_l + D_v + D_{Fb} + D_{Br} + D_{Bü} + D_{Ra} \tag{4.26}$$

$L_{m,E}$: L_{Aeq} in one hour of a train pass-by at a distance of 25 m, dB(A)

The term 51 represents the reference emission level of a disc-brake train with a length of 100 m at a train speed of 100 km/h under average conditions of rail surface. The other terms in Eq (4.26) calculate the effects of train parameters:

D_{Fz}: Type of vehicle
D_D: Brakes (disc or cast-iron)
D_l: Length of train
D_v: Speed of train
D_{Fb}: Track type
D_{Br}: Bridges
$D_{Bü}$: Railroad crossings
D_{Ra}: Squeal noise

The revised standard model gives the calculation of the noise emission by using the sound power levels in octave bands:

$$L_{WA,f,h,m,Fz} = \alpha_{A,h,m,Fz} + \Delta a_{A,h,m,Fz} + 10 \lg \frac{n_Q}{n_{Qo}} + b_{f,h,m} \lg \left(\frac{v_{F,z}}{v_o}\right) + \sum c_{f,h,m} + \sum K \tag{4.27}$$

$a_{A, h, m, Fz}$: A-weighted overall level of the sound power emitted from one vehicle ($v_0 = 100$ km/h) under average conditions of rail and on a ballasted straight track
h: Source height, m
m: Type of vehicle

Fz: The vehicle unit

n_Q: Number of sound sources of a vehicle unit

$b_{f,h,m}$: Velocity factor

v_{Fz}: Train velocity, km/h

$\Delta a_{A,h,m,Fz}$: Level difference between overall level and the octave band level in the octave band, *f*, dB

$\sum c_{f,h,m}$: Level corrections for type of track and rail surface condition, dB

$\sum K$: Level corrections for bridges and particular nuisance of noise, dB

Trains with different heights, different vehicles, and different track types, wheel and rail roughness, noise sources and partial noise sources (rolling, aerodynamic, propulsion noise, etc.) are all included in the emission model.

4.3.3   The RMRS Model

The Netherland National Computation model (the RMRS) was prepared in 1996, for railway noise prediction, and is still in use in many countries as a standard model recommended by EU Directive 49 [88]. The model aims to determine the railway noise levels for all types of railway systems, to be used for strategic noise mapping. Two procedures are described in the model:

A) The simple model giving noise levels in dB (A) (SRM I).
B) The advanced model for the calculation of noise levels in octave bands and then combining the spectral values to give the A-weighted sound pressure levels (SRM II).

The Simple Model

The A-weighted sound pressure levels are obtained directly in terms of various metrics such as L_{eq}, L_{den}, L_d, L_n, and L_e from the database prepared for all trains in EU countries. The following steps are implemented in the model:

1) Definition of input data
   Inputs related to noise source:
   - Train categories: Ten categories of trains are defined according to propulsion systems and wheel brake systems.
   - Railway types: Nine categories of railway tracks and rails are specified according to ballast beds, sleepers, rail fixation, rail joints, switch types, etc.
   - Railway structures: Various types of bridge constructions are defined.
   - Source heights: Four different train heights are introduced in relation to train categories, as 0.5, 2, and 5 m above the rails.
   - Brake types: Two conditions are accepted; using or not using the brake during the passing-by.
   - Maximum speeds: Ten different speeds are specified according to the train categories.
   - Train volume: Hourly number of passing trains are to be determined separately for the trains braking and not braking.
     The database covers information about the noise emissions of each type of train and the corrections regarding the effective factors. Emissions for each train category are given in octave bands and identified as emission index codes (the noisiest trains are in groups of 2,

9, and 8). The corrections to modify the reference emission level are given by particular equations for train and railway parameters:

- Train speeds
- Rail type
- Brake type for each train category
- Train speed
- Structural connections, switches, and rail types
- Bridge structures (concrete and steel).

Inputs related to the physical environment:

- Geometrical configuration of source, receiver, and physical elements in the environment: Horizontal and vertical distances, heights (position of rail upper edge at horizontal and vertical planes), angle of view, position of railway axis (as a line source)
- Sector definition (division of railway line into segments)
- Definition of reflective surfaces
- Definition of ground types (ballasts, grass, agricultural ground, sandy surface or uncultivated ground without vegetation)
- Determination of meteorological parameters

2) Calculation of emission (sound power):

Contribution of the input parameters related to the source and the environment, the partial noise levels are obtained by using a series of tables given in the guideline. The total emission level is calculated by integrating the partial noise levels:

$$E = 10 \lg \left(\sum_{c=1}^{y} 10^{E_{nr,c}/10} + \sum_{c=1}^{y} 10^{E_{r,c}/10} \right) \tag{4.28}$$

$E_{nr, c}$: Emission of the non-braking trains (sound power level), dBA
$E_{r, c}$: Emission of the braking trains (sound power level), dBA
c: Train category
y: Total number of train categories

Calculation of the standard emission levels, for each category of rail vehicles, is given in the model by using the average quantity of non-braking and braking trains, average speeds, and brake type:

$$L_{E,i}^{bs} = 10 \lg \left(\sum_{c=1}^{8} 10^{E_{bs,nr,i,c}/10} + 10^{E_{bs,r,i,c}/10} \right), \text{dBA} \tag{4.29}$$

$L_{E,i}^{bs}$: Emission level for the ith train type
b: Index code for track
bs: Indicates the emission value on the railhead (for 0.5 m above railhead; *as*)
n: Number of points or junctions
c: Index code for non-braking trains

For the trains in the ninth category, the contributions from the brake, the motor, and the diesel engine are separately determined and then combined to obtain the total noise level. As shown in Eq (4.29), the emission levels are calculated in relation to the type of track, the speed, the average number of passages, and the brake operations.

3) Calculation of physical environmental levels (immission levels)

The model is composed of a number of equations to calculate the effects of various physical factors, including air absorption, which are called "global correction values." These values are introduced into the basic formula given below to obtain the immission levels as A-weighted equivalent continuous level, L_{Aeq}, dB, at a certain point. For example, the distance correction as a function of ground type and meteorological factors, the correction for air absorption as a function of source-receiver shortest distance, the correction for the reflection effect according to the distances between the source line and the reflective surface and between the source line and the receiver.

$$L_{Aeq} = E_s + C_{reflection} - D_{distance} - D_{air} - D_{ground} - D_{meteo} \quad \text{dB} \tag{4.30}$$

$C_{reflection}$: Correction value for reflections from buildings and other vertical surfaces
$D_{distance}$: Attenuation due to distance between receiver and rail
D_{air}: Attenuation due to air absorption
D_{ground}: Attenuation due to ground absorption
D_{meteo}: Correction according to meteorological conditions
E_s: Calculated emission level

The Advanced Model

This is the procedure to reiterate the above process at each octave band and then obtain the total sound pressure level by combining the partial contributions. The steps of the procedure are outlined below:

1) Calculation of emission levels: the railway line is divided into segments depending on the total extension angle and the segments are defined so that the max angle of view is 5°. Then the emission values at an observation point are calculated for each segment at each of the eight octave bands between 63 and 8000 Hz. Ultimately, the following summation is made:

$$L_{Aeq} = 10 \lg \sum_{i=1}^{8} \sum_{j=1}^{J} \sum_{n=1}^{N} 10^{\Delta L_{eq,i,j,n}/10} \tag{4.31}$$

$\Delta L_{eq,\,i,\,j,\,n}$: Specifies the contribution of each source, depending on the source position within a sector, in octave bands, dB
i: Index code of octave band
j: Index code of a sector
N: Index code of a source point

$$\Delta L_{eq,i,j,n} = L_E + \Delta L_{GU} - \Delta L_{OD} - \Delta L_{SW} - \Delta L_R - 58.6 \quad \text{dB} \tag{4.32}$$

L_E: Emission value in an octave band, dB/octave
ΔL_{GU}: Distance effect, dB
ΔL_{OD}: Total attenuation (atmospheric absorption, ground effect, and effects of meteorological factors), dB/octave
ΔL_{SW}: Attenuation due to barriers, dB/octave
ΔL_R: Effect of reflection, dB/octave

2) Calculation of the effects of the physical environment: The parameters to be defined in the model with the geometrical properties, are given below:

- Specifying the railhead as vertical and horizontal plane
- Defining the emission sectors with the distances of each, not less than 100 m
- Determination of the correction factors for ground types (hard and soft)
- Determination of the average heights from ground
- Calculation of the effects of earth berms, embankments, and walls along the railway
- Calculation of the effects of barriers and screens in relation to effective height, angle of view, masses, U-type, horizontal floor, barrier absorptive treatment, etc.
- Determination of the corrections for types of bridge constructions, geometries, sector and ground factors (reflective or absorptive), and barriers on the bridges
- Determination of the platform effect (height of the platform is accepted as 0.8 m from the rail bed)
- Calculation of the effect of reflection of vertical elements
- Determination of the receiver points on the building façades

The RMRS model is applied for downwind conditions. Three types of ground are defined between the source and the receiver and different equations are provided for the ground effect, depending on the ground absorption properties. For long-distance propagation, a correction factor is given in the model in relation to the meteorological parameters. Barrier effects are calculated by means of an algorithm similar to the NMPB model. A correction is given for the rail roughness based on the specific measurement.

4.3.4 The Cnossos-EU Railway Noise Model

As for traffic noise, the prediction of railway traffic is a part of the Cnossos-EU methods which have been implemented by the EU countries since 2018 [71]. The model recommends the emission and immission calculations with the basic approach similar to the other models, however, with more detail in the organization of the source parameters. The sound power emission of railway vehicles are determined by considering:

Emission of individual railway vehicles: L_{w0} per each vehicle passing by at each frequency band. Emission of traffic flow: $L'_{w,line}$ per each meter of the railway track.

It aimed to establish an emission database associated with effective parameters described in the method. Defining the "vehicle" as a single unit of a train (locomotive, coach wagon) that can be detached, the acoustical directional sound power per meter length of the equivalent source line is calculated in the model.

Emission of single vehicle: Noise emission from various noise sources of trains is calculated independently for rolling, impact, squeal, traction, aerodynamic noises and, in addition, the contribution of railway structures, i.e. bridge noise. Different source heights are taken into account, according to the relevant noise sources:

0.5 m for rolling noise,
0.5–4.0 m for tracking noise,
0.5 m for noise from louver, cooling outlet and fans, for aerodynamic and impact noises,
0.5 m for sequel noise,
0.5 m for bridge noise.

The sound power level of a single vehicle is determined with the directional characteristics defined by the angles ψ and φ with respect to the direction of the train operation. These angles are given for the sound power levels at 1/3-octave bands between 100 Hz and 5 kHz. The power level of a single vehicle is denoted by $L_{W,0,dir\ (\psi,\ \varphi)}$. The model provides the independent functions to estimate the noise contribution from each of the secondary noise sources by using the inherent parameters:

- for rolling noise: as a function of rail roughness, wheel roughness, etc., expressed as vehicle-track transfer function
- for aerodynamic noise of high-speed trains: as a function of source height and speed
- for traction noise: depending on the operation condition
- for squeal noise: as a function of curvature and friction condition for each type of rolling stock
- for impact noise: as a function of crossings, switches and junctions, number of impacts per unit length or joint density and roughness level.

The tracks and support structures are classified according to the track base (ballast, slab track, ballasted bridge, embedded track, other, etc.), rail type, railhead roughness (smooth, not well maintained, normally maintained, etc.), rail pad type (with soft, medium, high stiffness), additional measures (none, rail damper, absorber plate, etc.), rail joints (none, single joint or switch, two or more than two joints), and curvature (straight track, low, medium, high).

Briefly, the sound power level of a train, $L_{W,0,dir}$ (ψ, φ) is determined by the energy summation of all factors contributing, for all vehicle types at each frequency band, each truck section, each source type with its height and directivity. It is recommended that a database is set up regarding combinations of "vehicle-track-speed-running conditions" for European trains. The studies are continuing for conversion of the existing railway source emission data for use in the new model [89].

Emission of train traffic flow: The sound power level of a vehicle flow on a railway track section is defined as the directional sound power level per meter length, denoted by $L_{W',eq,line}$ (ψ, φ) and expressed as:

$$L_{W,eq,line}(\psi,\varphi) = L_{W,0,dir}(\psi,\varphi) + 10\lg\left(\frac{Q}{1000v}\right) \qquad (4.33)$$

Q: Average number of vehicles per hour
v: Average speed of vehicle, km/h
Q and v are to be determined for each track section for each vehicle type and average speed.

The Cnossos-EU method employs the propagation model for railway noise which is similar to other types of noises as summarized in Section 4.2.4 for road traffic noise.

4.4 Prediction Models for Aircraft Noise

Various models have been developed for aircraft noise which are more complicated compared to those for other noise sources and require different analyses due to the characteristics of aircraft noise given in Section 3.2.3.

4.4.1 Development of the Models

Studies on developing the single aircraft models around airports increased in the 1980s. The conventional methods comprised a procedure to determine the instantaneous noise levels during aircraft movement and to obtain the total noise level by integration in a certain period. The noise metrics used in the models were single event or cumulative noise metrics as defined in

Chapter 2, such as L_{AE} (*SEL*), L_{EPN} (*EPNL*), *NNI* or L_{max}, L_{eq}(24h), L_{dn}, L_{den} (DNEL), L_{WECPNL}, L_{NEF}. One of the first prediction models was developed in the UK by using *NNI* [90]. Simple relationships were presented to calculate the noise levels taking into account the flight parameters, e.g. power setting, engine thrust, and slant distance between the aircraft and the receiver and those required for the noise emission data (or the reference levels) based on the measurements. In the earlier models, some limitations were accepted, such as linear or straight flight path, constant speed, constant thrust setting, however, by taking the directivity of the aircraft into account. The ground effect was not included in the initial phase of modeling. Based on the reference noise levels, the calculations for each aircraft type at a certain receiver point were conducted and the total levels were obtained by integrating the contributions of each aircraft. Since computations were made using a certain performance data, the validity tests on site were rather difficult. Upon the increased quality of the basic data, the prediction models have been improved, dealing with various factors relative to the flight operations and the physical environment, for example, the effects of flight procedure, aircraft turnrounds, etc. on the ground noise levels. At this stage, the simulation models have emerged which are more feasible for the noise contours.

4.4.2 Guidelines and Simulation Models

An overview of the models developed in various countries is given below.

The Danish Aircraft Noise Exposure Prediction Model (DANSIM) (1983–1990)
What was different in this model, compared to the previous traditional models, is the implementation of a simulation technique whose validity was confirmed by measurements [91–93]. The basic noise metric was L_{AE} for individual flights considered as noise events. The measured reference levels were inserted into the model and the results were presented with the multiple event indexes of L_{Aeq} and L_{den}, that could be adjusted to L_{EPN}, L_{WECPNL} which were used in some countries.

Initial calculations were made by the traditional method, introducing the following parameters for a single aircraft, into the procedure:

Geometric parameters: the aircraft performance data (position and direction as a function of time). A procedure is applied to subdivide the flight path into segments.
Acoustical parameters: the noise level as a function of distance, direction, velocity, and thrust setting of the aircraft.

Steps of the model:

1) Calculation of SEL (L_{AE}) from a single aircraft following a straight track, at a certain receiver: The equations were given to correct the reference noise levels to calculate the following:
   - geometrical parameters varying with time (i.e. position of aircraft with coordinates; altitude profile) (Figure 4.19a);
   - directional parameters according to the position of the aircraft and the time by means of a simulation model;
   - the effect of the thrust setting (correction for forward flight effect as a function of relative jet velocity), flight speed (from take-off roll to lift off) and directivity during ground roll and applied as a correction for the actual trust setting;
   - effect of jet noise emission on the forward flight;
   - effect of thrust reversal in landing position as a function of directivity, distance and ground attenuation by using the maximum noise levels;
   - effect of lateral dispersion of aircraft;
   - total effect for a specific flight position and the noise level.

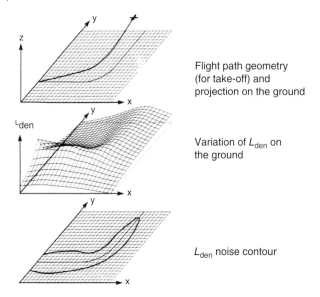

Flight path geometry
(for take-off) and
projection on the ground

Variation of L_{den} on
the ground

L_{den} noise contour

Figure 4.19a Simulation model (DANSIM) for calculation of noise contour on the ground [93].

2) Calculation of sound pressure levels (immission level) at a receiver point as a function of distance, flight direction, thrust, and speed.
   • coordinates of receiver and source positions;
   • directional characteristics;
   • attenuation due to distance;
   • attenuation due to ground effect as a function of distance and the angles of flight profile;
   • obtaining contributions from all the segments by applying the integration procedure for each position of the aircraft.

3) Computation of total noise exposure, time-of-day weighting process and integration for total aircrafts and flight paths.

4) Obtaining noise contours on the ground, as explained in Chapter 6.

The Nordic Guideline Calculation Model for Aircraft Traffic (1993)
The model was developed in a joint study by Sweden, Denmark, Norway, and Finland intended for use by environmentalists, aircraft management authorities, planners, aircraft companies, flight clubs, consultants, etc. [94]. It was stated that the model could be taken as a basis for complaints near airports as well as to calculate noise levels at certain points to develop noise contours around airports. Thus, a single point calculation method is presented (by the manual method), in addition to an algorithm for a computer program.

The validity tests implemented by comparing the results of the model with the other models, such as DANSIM and INM3, revealed good compatibility with a difference of a maximum 5 dB. The parameters employed in the model are given below:

• Air traffic: Traffic categories, traffic volume, yearly, daily, hourly, and weekly distribution of flights, the number of operations should be given as a statistical distribution according to aircraft categories, distribution according to time of year, day and weekend.

- Aircraft operation procedure: Configuration of runways (direction, length, surface types, extension forecasts), usage of runway under different wind and environmental conditions, segmentation of flight path (flight track), horizontal and vertical projections of the flight path, air traffic distribution at each flight path or on specific tracks or sectors in airports.
- Performance and noise data: All propeller-type aircraft were divided into four categories based on the noise certification data. The flight profiles according to the flight speed, and the power setting during arrival and landing, should be determined for the typical take-off weights and distances to take-off point. A noise data set has to be prepared for each flight in terms of L_{AE} and L_{max} as a function of the shortest distance to the flight path and as a function of the engine power setting. For take-off; the "altitude/speed/engine power" data are determined as a function of distance to take-off ground roll. The model calculates the effect of vertical and horizontal dispersion (deviation from the nominal track during actual departures and arrivals) and the effect of the aircraft turns.

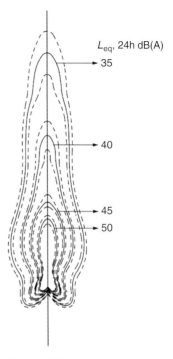

Figure 4.19b Example noise contours in L_{eq}, 24 hours, dB(A) for B767 take-off path (calculated by Nordic Guideline) [94]. (*Source: Permission granted by Nordic Council of Ministers.*)

The computations consisted of determination of the effects of meteorological factors on sound propagation, called lateral attenuation. Ultimately, the noise levels were obtained in terms of the noise descriptors to construct the noise contours. Examples of aircraft noise contours are given in Figures 4.19b and 4.19c.

The U.S. National Bureau of Standards Model
In this preliminary guideline developed by the U.S. National Bureau of Standards, noise contours were predicted as *NEF* values around the airports in 1978 [51]. The guideline covered a layout scheme to approximate *NEF* 30 and *NEF* 40 contour lines in relation to the length of runways for airports with different flight densities. Figure 4.20 describes the geometric parameters and a

Figure 4.19c Example noise contours in L_{eq}, 24 hours, dB(A) for B737 making a turn after take-off (calculated by Nordic Guideline) [94]. (*Source: Permission granted by Nordic Council of Ministers.*)

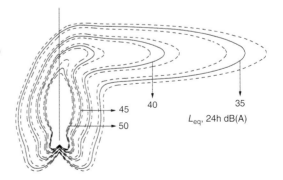

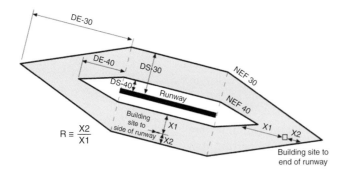

Figure 4.20 Approximation of NEF contours and geometrical parameters in NBS Model [51].

procedure for categorization of airports based on the effective number of jet operations and for assessment of the layout of the noise contours with respect to runways.

The UK Aircraft Noise Contour (ANCON) Model (1999)

The UK model determined the aircraft noise levels in L_{eq} (16h) by using aircraft noise data collected over a long period of time [95]. The contours were obtained from the grid levels calculated from the *SEL* of each aircraft for each flight path segment, as in the other models. The model used the data on flight speed, engine power, and thrust setting. Data for flight position and speed were provided through the radar system, and the power settings were determined based on the aircraft weight and aircraft performance data. Such information was employed to obtain the engine power versus the distance diagrams. The outputs of the model were verified against the measurements and comparisons of other models.

The model was improved as Version 2 in 1999 and the Noise-Power-Distance tables were presented as a function of engine thrust to calculate the noise contributions to L_{AE} from each segment of the flight path [96].

The U.S. FAA (Federal Aviation Association) Integrated Noise Model (INM, 1978)

The model, developed by the U.S. Federal Aviation Association Environment and Energy Bureau, calculates the effects of aircraft noise in and around airports [97, 98a]. As stated in the document, the aims for the model, were: (i) to clarify the differences in noise levels due to alteration or renovation of airports, e.g. extension or changing runways and configurations; and (ii) to determine the effects of new transportation requirements, change in aircraft types, change in routes and airport structures, decisions on alternative flight profiles and new procedures.

The FAA model has become the standard model in the USA since 1978 and the commercial software INM was developed in 1996 revised in 2002. The model is still in use with some modifications.

The inputs of the model are: the flight path information covering the flight profiles as standard or user-defined, the directivity pattern of each aircraft, the land configuration, buildings, barriers, etc. The land use maps are now available through the Geographic Information System (GIS). Eventually the model gives the noise contours around airports that can be evaluated for various purposes, e.g. enabling comparisons, describing noise conditions in noise-sensitive areas and buildings such as hospitals, schools, etc., displaying locations where the noise levels are at the highest and the noisiest flights are observed. The results can be obtained as A-weighted and C-weighted levels as well as in

16 different metrics such as *EPNL* and *NEF,* etc. Based on the statistical analysis, the cumulative noise levels, maximum levels, and time-histories of instant sound pressure levels can be determined.

Since the INM model gives the average aircraft noise levels, it is essential to introduce the information about the long-term meteorological factors (e.g. yearly averages of temperature and wind gradients, average humidity, average ground absorptions) and furthermore, their yearly variations of these parameters. The validity of the model is strongly related to the accuracy of meteorological data.

German Aircraft noise model (AzB, 2008):
This model covers not only the aircraft noise in the air, but also the ground noise from taxiing operations and noise from the use of auxiliary power units [98b]. The data consists of 36 aircraft classes derived from the measurements in terms of sound power levels in octave bands along with the triple directivity factors. The propagation model is based on segmentation approach and air traffic scenarios in Germany.

4.4.3 The ECAC Model

For strategic noise mapping, the EU Directive 49 in 2002 recommended employing the model entitled "Standard Computation Model of Noise Contours around Civil Airports" developed by the International Civil Aviation Organization (ICAO), first published in 1986. The document called ECAC CEAC Doc. 29, which was revised in 1997, has been defined as the "best methodology" of that time. The original model is based on the model described by the Society of Automotive Engineers (SAE) in the USA and on the earlier ICAO documents that were initially published for the purpose of International Aircraft Certification [99–102]. The flow chart regarding the inputs and the procedure of the 1997 version of the model is given in Figure 4.21.

The Database Relative to the Noise Source
The model requires detailed input data for the computation process:

A) *Data relative to aircraft*: Aircraft should be categorized with respect to noise and performance characteristics (e.g. types of aircraft, flight procedures, numbers, etc.).
　　A.1. Types of aircraft: Aiming to reduce the amount of calculations and to employ a limited number of noise/performance data, the aircraft are grouped according to the following parameters influencing their noise emission and flight performance:
　　a) Aircraft propulsion
　　b) Number of engines (one to four engines)
　　c) By-pass ratio for fan motors
　　d) Maximum take-off weight

　　A.2. Measured noise emissions: The noise data are provided through the measurements conducted at each of the 24 third octave bands between 50 and 10 000 Hz and at 0.5 second intervals, according to ICAO certification requirements, as explained in Section 5.4.2. The measured emission data are adjusted to the actual situation by using the relationships given in the model for:
　　● correction according to special airport conditions;
　　● correction according to distance to be normalized for 100 m;
　　● correction according to atmospheric conditions;
　　● acquisition of noise/power/distance functions.
　　　　Noise emission data are determined primarily in terms of L_A and L_{Amax} dB at the distances of 80, 100, ... 800 m, then normalized by taking into account the attenuations in 1/3-octave bands which are expressed as dB/100 m.

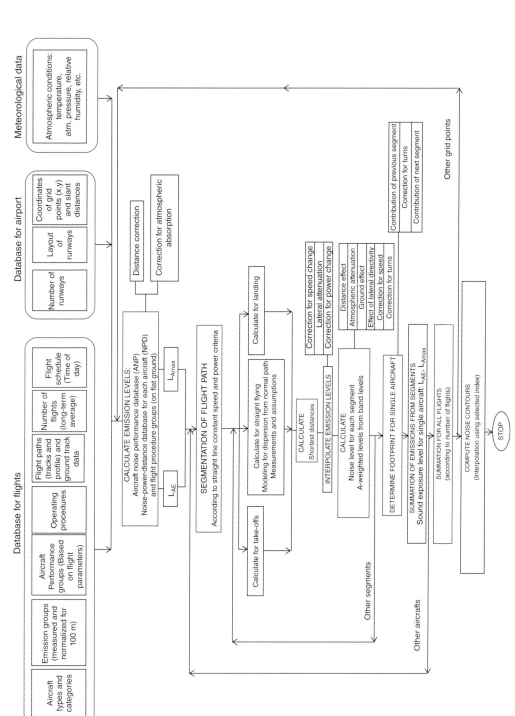

Figure 4.21 Flow chart for prediction of aircraft noise according to ECAC model (1997 version).

The database for variation of sound power/distance: Based on the measurements, the variation of the emission values, with aircraft type and engine power, is determined and adjusted according to the actual distance. The sound/power versus distance database is prepared analytically and graphically by taking into account the atmospheric attenuation (air absorption). The distance parameter implies the distance perpendicular to the flight path projection of the actual slant distance. The geometric parameters of the flight path, i.e. direction, aircraft profile, noise footprint, and height (altitudes), can be seen in Figures 4.22a–4.22c.

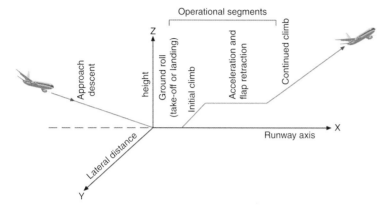

Figure 4.22a Geometrical parameters on vertical plane related to the aircraft operation [102]. (*Source:* Permission granted by ICAO.)

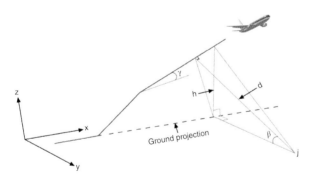

Figure 4.22b Parameters related to flight path geometry [102]. (*Source:* Permission granted by ICAO.)

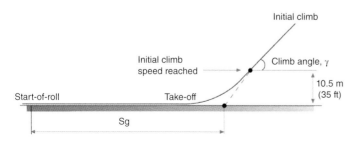

Figure 4.22c Parameters related to flight path geometry on vertical plane [102]. (*Source:* Permission granted by ICAO.)

A.3. Grouping flight performance (basic performance values).

A.4. Grouping flight procedures: Operational data and flight procedures for each road (mass of aircraft, power, speed, flap angle during flight, angle of slat, gear position during landing, weight, wind speed in knots).

A.5. Flight paths: Routes to be defined according to the origin point.

A.6. Ground projection of flight path (footprint).

A.7. Flight profile: Flight path indicating variation of altitude of aircraft along the ground projection.

A.8. Number of flights: Number of aircraft movements per aircraft type at each route to be given as long-term averages (daily, monthly) and within selected time periods.

A.9. Temporal distribution of flights: Time in which the aircraft movement occurs depending on the selected noise descriptor.

B) *Data relative to airport and environment*

B1. Number of runways and configurations.

B2. Geometry of the land, grid system and distance between the receiver points (about 300 m).

B3. Slant distance from the receiver point to the footprint of flight path.

C) *Meteorological data*

Atmospheric conditions: Temperatures, relative humidity, winds.

ECAC Doc 29 presents an aircraft performance model in which the source location is calculated as a function of aircraft type, weight, and operation procedure. The steps in the calculation model are given below:

1) Preparation of the database.
2) Determination of the coordinates of grid points on the ground.
3) Dividing the flight paths into segments and calculation of slant distances for each segment.
4) Calculation of immission levels for each segment at all the receiver points by using the noise emission database. Calculations are made also by using ISO 9613 that will be explained in Section 4.5.2, and applying the below corrections in two groups:

Corrections for aircraft operations according to:

- flight speed
- difference between the flight procedure and power management
- engine power
- turnarounds
- lateral attenuation (due to the directivity pattern).

Corrections for environmental factors:

- distance attenuation
- attenuation due to atmospheric absorption
- attenuation due to ground effect
- topographic factors
- the screening effect.

Different calculation procedures have been proposed for take-off, landing and normal straight flights in the model also by taking into account the horizontal and vertical divergences from the flight route.

5) Obtaining the noise footprint: Line charts of a constant noise level around a runway during landing and take-off of an aircraft under the defined conditions of air and atmosphere, flight profiles, and operational conditions, etc.

6) Combination of contributions of all segments for a single aircraft at a certain receiver point, $L(x, y)$ in terms of spectral values and the frequency-weighted levels (L_{AE} or L_{Amax}).

7) Reiteration of the procedure for each aircraft and determination of the total noise levels $L_{Aeq,w}$ by taking into account the contributions of all flights: the *SEL* (L_{AE}) levels for each group of flights including different types and operations are combined according to the selected index (such as hourly, daily, day-time, and night-time noise levels or L_{den}), as given in Eq (4.34): (The equation is also applicable for the L_{Amax} values.)

$$ L_{Aeq,W} = 10 \lg \left[\frac{1}{T} \sum_{j=1}^{N} W . 10^{L_{AEj}/10} \right] \quad \text{dB(A)} \tag{4.34} $$

$L_{AE, j}$: Sound exposure level due to j^{th} aircraft movement
T: Reference time for L_{Aeq},s (for 24 hours: 86400 seconds)
W: Weighting factor according to day and week

8) Computation of noise contours in terms of the selected index by means of interpolation.

Aircraft noise contours or, in other words, noise maps around airports are explained in Chapter 6.

ECAC Doc29, 4th Edition (2016): The latest edition has been issued following the progress in scientific achievements in the aircraft noise reduction technology, however, the basic computation procedure explained above has not been changed [102a,b, 103]. The report comprises three volumes. Volume 1 for noise model users, policy-makers or planners, to give information about non-technical matters, applications, limitations, modeling options, etc. The detailed description of the model, described as the best practice methodology, is given in Volume 2, which explains the algorithms for computer programming to obtain noise contours around airports. Volume 3 concerns the verification process to be used while developing noise software in compliance with the best practice method, and for validation of the existing noise models employing their specific emission database.

4.4.4 The Cnososs-EU Aircraft Noise Model

The recent European standard method, which is compulsory after 2018 for the purpose of noise impact assessment in Europe, accepts the basics of the ECAC model (3rd edition) for aircraft noise predictions and the document (ICAO ANP) comprising noise and performance database [71]. The model is applicable for general aviation aircraft, some types of military aircrafts and helicopters, and describes the fixed-wing aircraft ground noise calculation methodology.

As explained in the ECAC model, the location of the aircraft at each instant of time, which is important to determine the distance to the receiver point, is dependent on the aircraft type, weight, and operation procedure. This information is provided through Noise-Power- Distance (NPD) data with the corrections to be applied for the flight path parameters varying in time (i.e. speed, height, power, distance) and the meteorological conditions. The document suggests some changes in the ANP database, e.g. heights of receiver points fixed as 4 m, corrections for ground reflections, screening effects, and reflections from the vertical obstacles.

The document supports the field measurements for validation of the predicted results in residential areas to check the accuracy of the model. The Cnossos-EU method employs the sound propagation model for aircraft noise similar to other types of noises, as summarized in Section 4.2.5 for road traffic noise.

4.5   Prediction Models for Industrial Noise

Noise generated from industrial premises generally covers complex mechanical systems and can be determined by means of the calculation models.

4.5.1   Development of the Models

Assessment of noise from mechanical and aerodynamic sources located in industrial premises and within associated buildings, such as machinery, equipment, and installations as given in Section 3.3, is rather difficult because of the variety of noise sources and acoustical characteristics and the transmission characteristics of the environment. The models that have been developed to calculate the acoustic fields in general can be applied for mechanical equipment and installations, such as ventilation ducts, fluid flows in sanitary installation systems (pipes, appliances). However, due to the complexity of such sources, the theoretical models remain inefficient. The mechanical noise sources located both outdoors and indoors contribute to the indoor noise levels of plants or service rooms in buildings. The outdoor noise generated by the entire industrial building, or open-air industrial activity, can be computed by the simple or advanced simulation models from the standpoint of environmental noise impact assessment.

Modeling indoor noises: Determination of indoor noise levels from manufacturing activities, etc. is important because of the protection of workers from excessive noise. Also, these noise levels are transmitted to outdoors through enclosures, walls, roofs and windows, or open doors of the premises, creating noise pollution in the nearby environment. The models employ the equipment emission levels, generally obtained through measurements (Chapter 5) and the number of noise sources in groups. The immission models compute the total sound power levels and the corrections according to the indoor acoustical parameters, such as the size of room, the surface properties, screens and other objects in the proximity of sources. Determination of interior noise, which is a subject for building acoustics, is made by using different techniques, such as the imaginary source model, ray tracing techniques or both. Chapter 6 gives an overview of the modeling techniques and developments of sound maps for workspaces. The acoustic modeling software has been available since the 1970s [104, 105].

Modeling outdoor noises: Various models have been developed to determine sound radiation from industrial buildings and activities and open-air sources propagated in environments. The models assumed the industrial area or buildings as plane or line sources and enabled the calculations of noise levels at a receiver point in terms of a single value L, dB(A). An example of the simple guideline model was published in Austria in 1973 and shown in Figures 4.23a and 4.23b [106].

In the advanced models, it is necessary to analyze all the outdoor equipment, installations attached to buildings, noisy activities, even heavy vehicle transportation in addition to the indoor noises transmitted outside. Sound distribution between the buildings within the premises and enhancement of noise, due to reflections and scattering, are taken into consideration. Some of the models for assessments of industrial noise levels are outlined below.

The Danish Industrial Noise Guideline (1982)
This is one of the earliest guidelines, published by Denmark, to determine the sound power level $L_w(\varphi)$ in the receiver-source direction, which is described as spectral sound powers at 63–8000 Hz after correction according to the source directivity [107]. The normalized values at 45° and 110°

Figure 4.23a Noise level reductions with distance from rectangular-shaped and square-shaped industrial buildings [106]. *Note:* Boundaries providing attenuations of 10, 15, and 20 dB(A).

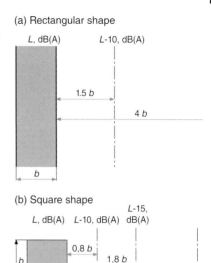

(a) Rectangular shape

(b) Square shape

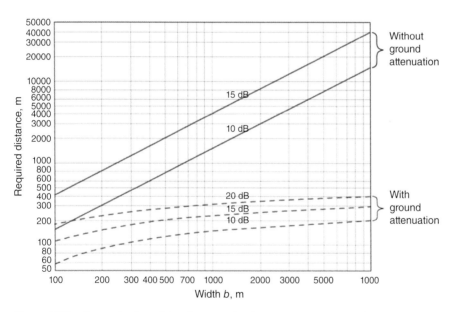

Figure 4.23b Determination of the distances yielding 10, 15, and 20 dB noise reductions for the square-shaped industrial premises [106].

were taken into account for the directivity correction (Figures 4.24a and 4.24b). The model required the layout plans from the architectural drawings of the premises and an accurate description of noise sources and locations, i.e. the coordinates of the source positions, the propagation paths, the angles between the perpendicular line to the center of the source plane, and the receiver direction. The equipment and installation were assumed to be monopole point sources and variation of the sound levels from the source to the receiver was obtained via a transfer function. A supplementary table given in addition to the model enabled the calculation of all the attenuations, due to physical factors, such as distance attenuation, the air absorption effect, reflective surfaces, barriers, vegetation, the ground effect, the scattering effect, etc.

The air absorption was calculated at each octave band with a function of relative humidity, static pressure, and air temperature, as described in ISO 9613-1 [108]. A decrease of 12 dB per octave band could be taken for a random noise spectra. The document recommended for calculating the attenuation due to air absorption at each octave band for a noise spectrum containing narrow band sounds or pure tone components and for the situations where the higher accuracy is required, for instance, at greater distances or for high frequency sources. Effects of noise barriers were calculated also for multi-screen situations. Diffusion of the surfaces within the site was considered in the model.

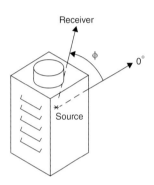

Figure 4.24a Definition of the directivity characteristics of mechanical noise sources [107].

The calculated spectral values of industrial noise under consideration were converted into the total A-weighted levels. Time-averaged levels, expressed as L_{Aeq}, could be obtained after checking for the existence of dominant frequency components, particularly while averaging noise levels in a certain period. The guideline referred to the uncertainties due to the diversities from the long-term averaged values while obtaining the L_{Aeq} values. It was required to make assessments for the uncertainties resulting from the actual source heights and meteorological factors. The expected standard deviations in the computed noise levels were listed in the document. The implementations of the procedure for various geometrical configurations were displayed with diagrams.

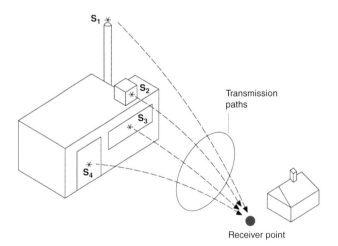

Figure 4.24b Individual noise sources (S_1, S_2, ...) positioned in industrial building and sound transmission paths for calculation of industrial noise level in Danish Design Guide [107].

4.5.2 The ISO 9613 Model

The computation procedure, which was first published in 1993 as the international standard can be used for traffic noise and for the other point sources. The standard later was accepted as the mandatory standard in EU Directive 49, and is comprised of two parts:

1) *ISO 9613-1:1993 (confirmed in 2015):* The first part involves mainly the atmospheric absorption influencing the propagation of sound in the environment [108]. The excess attenuations in one-third octave bands are displayed as dB/km in the tables for pure sounds in relation to the parameters of distance between the source and the receiver, the air temperature, the atmospheric pressure, and relative humidity. The atmospheric attenuation is also given for broadband noises and for the A-weighted levels and for sounds with complex spectrums (comprising both broadband noises and pure tones).
2) *ISO 9613-2:1996 (confirmed in 2017):* The engineering method described in the standard can be used for all kinds of environmental noises. However, most are employed for industrial noise predictions. The model enabling calculations for point sources and all sound transmission paths was presented in this part of the standard [109]. The environmental noise sources, whether they are line or plane source characteristics (like traffic noise as a line source, industrial noise, railways and construction activities as line or plane sources), are transformed into the sets (arrays) of acoustic point sources, stable or dynamic. Calculations are made in octave bands in the range of 63 Hz–8 kHz. The necessary inputs of the model are: locations of sources and receivers with distances, ground type, barriers in the propagation path, and reflective surfaces. Sound pressure levels are calculated according to the basic procedure given in Eq (4.35):

$$L_p = L_w - A_{div} - A_{atm} - A_{gro} - A_{scr} - A_{abs} \tag{4.35}$$

L_p: Sound pressure level, dB
L_w: Sound power level of the source, dB (emission level)
A_{div}: Attenuation due to geometrical divergence, dB
A_{atm}: Attenuation due to atmospheric absorption, dB/km
A_{gr}: Attenuation due to the ground effect, dB (including meteorological factors)
A_{bar}: Attenuation due to a barrier, dB (including meteorological factors)
A_{misc}: Attenuation due to miscellaneous other effects, dB

The reference emission levels of the individual sources to be introduced in the model are determined through the acoustic measurements explained in Chapter 5. However, emission of some mechanical noise sources can be obtained through calculations using the technical properties of the sources. For the noise sources with a directionality of sound radiation, certain corrections are applied to the emission level.

The environmental factors influencing the sound propagation from industrial areas or activities are introduced in Eq (4.35) as excess attenuations. Besides the atmospheric absorption (A_{atm}), which is given in ISO 9613-1, effects of other factors are obtained from the relationships given in this part of the standard. Ground attenuation (A_{gr}) is determined by dividing the ground between the source and receiver into three zones in relation to the geometrical configuration of both, then the values are calculated using the acoustical properties of ground (i.e. reflective, absorptive, and mixed grounds).

The model is applicable in downwind conditions, which is described by the angle between the wind direction and the receiver-source line within ±45° and by the wind speed of 1–5 m/s at the height of 3–11 m from the ground.

Barrier attenuation (A_{scr}) is calculated in each octave band for the objects with a mass greater than 10 kg/m^2 and with dimensions greater than the wavelength of sound. The combined effect in relation to meteorological factors and the ground characteristics are taken into account for barrier diffractions. Calculation procedures are available for single and wide barriers.

The increasing effect of reflections due to vertical surfaces is calculated for the surfaces greater than the wavelength, and the calculation process, by introducing the surface reflection coefficient, is reiterated for imaginary and real sources. When the reflection coefficients are unknown, some approximations given in the standard can be used. For the structures or equipment in cylindrical form, like chimneys, towers, storage tanks, etc., partially reflecting and scattering sounds, the model presents a procedure to acquire the sound reflection coefficients of such surfaces.

In addition to the above factors, the extra attenuations are given in the annexes of the standard, such as for the sound waves propagating over an industrial zone or over dense building sites. The equivalent continuous A-weighted downwind sound pressure level is obtained by taking into account all the factors causing extra attenuations:

$$L_{AT}(DW) = 10 \lg \left\{ \sum_{i=1}^{n} \left[\sum_{j=1}^{8} 10^{0.1[L_{ft}(ij) + A_f(j)]} \right] \right\} \text{dB} \qquad (4.36)$$

n: Total number of real and imaginary sources
i: Index for each point noise source
j: Index for each of eight octave bands between 63 and 8000 Hz
$L_{ft(i,j)}$: Calculated sound pressure level for source i at j^{th} octave band
$A_{f(j)}$: Standard A-weighting at j^{th} octave band

If only the A-weighted levels are needed, it is recommended using the attenuations at 500 Hz. The long-term A-weighted levels (LT) are calculated with the parameter C_{meteo} corresponding to the yearly averaged meteorological data.

For a receiver point, with the contribution of all the individual sources, the process is reiterated and then the noise levels calculated for each source and propagation paths are combined in the energy basis. At present, application of the model for complex elements and for multi-source plants has been accomplished through computer models (Chapter 6), by reducing the uncertainties associated with computation errors.

This standard model was mandated as the reference method to be used by the EU countries for noise mapping around industrial establishments in 2002 [67].

4.5.3 The ISO 13474 Blast Noise Model

This international standard, first issued in 2003, includes an engineering method to determine the impulsive sounds of moderate size blasts emerging from mining, shooting, and bomb explosions, at distances up to 30 km and under different weather conditions [110]. Source emission levels, to be used in the calculation of the immission levels, can either be measured or calculated applying the procedure given in the standard. The basic principles of the model are similar to the analytical method described in ISO 9613-2 [109]. The environmental levels are calculated for the downwind conditions, however, the long-term average levels can be determined by using a correction factor C_{met} derived from the statistical analysis of meteorological conditions as a weighted average. The standard recommends a database to be prepared by the countries for meteorological parameters and ground properties. The procedure to apply the standard prediction method is outlined below:

1) Estimation of emission level of explosion: Two approaches are proposed to obtain the source emission levels:
   the calculation procedure;
   determination based on the measured reference levels.
   In the first approach; the unweighted band exposure level for each direction angle, $L_{E,s(j)}$ is calculated at a distance of 1 km according to the equivalent TNT mass (denoted by Q). The average blast size is taken as equivalent to a TNT of 0.05–1000 kg. The calculation steps are given below and the flow chart is presented in Figure 4.25. The parameters used in the calculation are distance, positive peak over pressure, positive phase duration, and frequency. If the blast mass Q for confined (underground) and unconfined explosions, is not known, it can be computed from the measured C-weighted *SEL*, L_{CE} or from the measured peak level L_{pk} and the results are inserted, as inputs to the model.

   In the second approach to obtain emission data for blasts; the acoustic measurements are performed as explained in Section 5.5.8, by defining the below parameters prior to the measurements:

   - height of the center of the blast source above the ground, m
   - height of the microphone point above the ground
   - distance between the source and microphone, km
   - direction angle of the microphone
   - equivalent mass of TNT as kg in an unrestricted explosion in open air (assuming that 1 kg TNT gives a blast energy of 4.26 MJ)
   - atmospheric pressure, Pa
   - sound speed of the directed path as a function of height, m/s
   - temperature, °C
   - relative humidity, %

   The measured levels expressed as unweighted sound exposure level, L_{Emeas}, unweighted peak sound pressure level, $L_{pk,m}$ or C-weighted exposure level, L_{CEmeas} are inserted in the procedure described in the standard as seen in Figure 4.25.

2) Calculation of immission level: The emission levels determined above are introduced into the immission model to obtain the sound pressure levels or sound exposure levels at any receiver position after calculation of the effects of environmental factors (sound attenuations):

   a) Band exposure levels, $L_E(j)$ at a certain receiver point:

   $$L_{E(j)} = S_\phi(j) - A_{total}(j) \tag{4.37}$$

   $S_\phi(j)$: Reference band exposure level in the jth octave band at 1 km from the blast point relative to the direction angle, ϕ
   j: Index for frequency band
   ϕ: Direction of receiver point in relation to the source location
   $A_{total}(j)$: Total sound attenuation on sound propagation path of the impulse at j^{th} band, dB

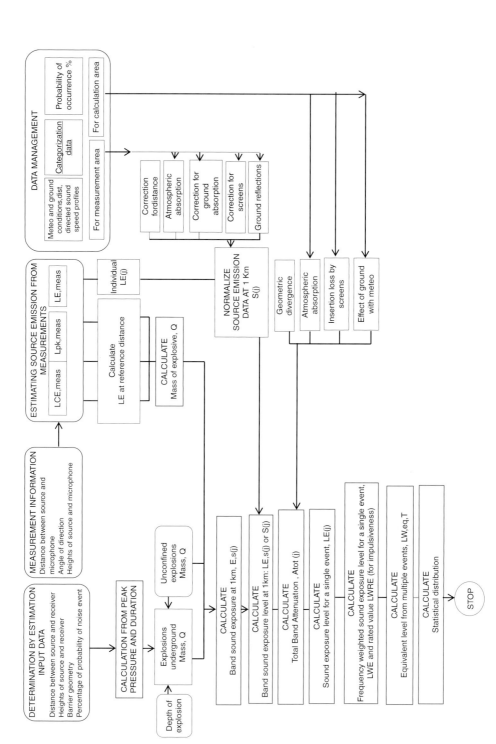

Figure 4.25 Flow chart for blast noise prediction according to ISO 13474 [111].

The attenuations due to the environmental factors are calculated by using ISO 9613-2. Since the atmospheric absorption and the resultant excess attenuation depend on the meteorological conditions, the standard recommends determining the probability of the occurrence of each atmospheric attenuation class by applying the procedure given in the model. Excess attenuation due to the ground effect is calculated by using the transfer functions in octave bands relative to the directed sound speed profile, the ground type, the height of the source and the receiver, the distance from the blast point and the probability functions. To apply these functions, it is necessary to establish a database containing the computed attenuation values according to the directed sound speed profiles and ground conditions. Directed sound speed profiles are computed in relation to height from the ground, temperature, wind speed and direction, and angle between the source and the receiver. Based on the probability of the occurrence, the directed sound speed profile is assigned by computing the maximum height above ground. The reflection from the ground during the measurement is considered in the model and the increase of sound energy is calculated according to the blast height. Determination of the other attenuation values is explained in the standard.

b) Calculation of frequency-weighted SEL of a single noise event, L_{WE}:

$$L_{WE} = 10 \lg \left[\sum_{j=1}^{N_{band}} 10^{0.1[L_E(j) + w(j)]} \right] \text{dB} \qquad (4.38)$$

$L_E(j)$: Unweighted band exposure level for the jth band, dB
$w(j)$: Frequency weighting for the jth band, dB (1/3 octave bands) (A- or C-weighting)
N_{band}: Number of frequency bands

Since the impact noises contain low frequency components below 200 Hz with higher sound pressure levels, they should be taken into account, particularly when the difference between the levels of low frequency component and the maximum level is greater than 20 dB.

ISO 13474:2009 (Last revised and confirmed in 2020)

The standard has been republished under a different name describing the "method to obtain sound exposure levels from high energy impulsive sound sources." The main characteristics of high energy impulsive sources are: they contain high levels of noise, dominant low frequencies, short duration, and ability to propagate over long distances [111]. The method is more descriptive compared to the premier standard and requires the statistical analyses of different parameters, placing emphasis on the atmospheric factors and the ground properties.

The sources of demolition and muzzle blasts and the projectile sound (high-velocity supersonic projectiles with flat trajectories) are defined and their calculations and measurements are explained. The effects of ground reflections, meteorological effects, or topography on immission levels are taken into account based on statistical analyses. In addition to the parameters in the earlier standard, some new parameters have been introduced, such as "roughness height" in relation to wind velocity. The detailed descriptions are made for atmospheric conditions defined as "replica atmosphere" in relation to ground conditions and the probability occurrence. The band sound exposure level used in the earlier standard, is given as a function of the atmospheric absorption class and the excess attenuation classes and in addition to the relevant azimuth and the elevation angles characterizing the source direction. The equations given in the premier standard are somewhat the same with differences in notations and with the parameters more descriptive. For a long-term average single-event sound exposure level, the rating adjustments are proposed and the calculation of the statistical distribution of the frequency-weighted levels is described based on the probability density function. The excess attenuation due to ground is determined for different surface types by applying the impedance models.

The standard also gives the calculation of equivalent sound power level $L_{W,eq,T}$, for the situations where multiple noise events exist. Some examples regarding the estimation of the statistical distribution of single-event *SEL* and the calculation of uncertainty are included.

4.5.4 The Cnossos-EU Industrial Noise Model

The Cnossos-EU method covers the assessment of industrial noise similar to the other environmental noise sources explained above [71]. For the prediction of the emission levels of industrial noise sources, the document recommends using the pre-defined database or the measurements to be conducted in semi-anechoic rooms, as explained in Chapter 5.

The industrial sources of varying size, e.g. from large plants to small industrial tools or operating machines in factories, are modeled as point, line, or plane sources depending on their dimensions. When the single sources are distributed in an area, the concept of the "equivalent sound source" is implemented with the aid of the wavelength and distance relationships.

The input data for noise sources comprise the sound power levels of the individual sources by taking into account the major source and its operation period during working hours (or day, evening, night as yearly averages). The other parameters are location of sources with three dimensional coordinates, type of source, dimensions, orientation, and operation conditions. Directivity of sources should be given in one-third octave bands and be relative to the position of the source (or equivalent source), since the reflections from nearby surfaces influence the source directivity. It is recommended to check the effect of reflections at the site according to the presence of any surface less than 0.01 m from the source. The directivity is expressed by the factor of $\Delta_{LW,dir,xyz}$ to correct the sound power measured in the laboratory.

For the point source, the sound power level is given as (L_W) with directivity values in a three-dimensional coordinate system. For a line source, the sound power level is expressed as per meter $(L_W/m$ or $L_W')$ and the directivity is required as a function of two orthogonal coordinates to the axis of the source line. For a source moving on a line, the speed and the number of sources traveling on the line during day, evening, and night periods have to be specified. For area sources, the sound power levels are needed per square meter $(L_W/m^2$ or $L_W'')$ without regarding directivity or by giving horizontal or vertical directivities if available.

The sound propagation model recommended by Cnossos-EU is described as a basic procedure applicable to all the environmental noises including industrial noise, as given in Section 4.2.4 for road traffic noise,

4.6 Prediction Models for Construction Noise

Noise emissions from construction activities vary significantly due to the differences in operational and acoustical characteristics of industrial noise sources (Section 3.5). Although the activities are temporary, however, the annoyance from construction noise might be serious in nearby communities, thus assessment of noise levels is of importance for the management of environmental noise.

4.6.1 Development of the Models

The first models were developed for road construction noise during activities such as clearing, demolition and removal, earthwork, paving, signing, etc. The immission levels are predicted when the emission of the noise sources, that can be assumed as point, line, and plane source, are known. Some procedures are based on moving source approximation. The immission levels are presented as continuous mean energy level L_{eq}, dB(A). The reference noise levels at the site can be measured as peak sound pressure level or noise exposure level (sound energy level) generated by each operation type.

4.6.2 Analytical Models

The methods that were applied for the assessment of construction noise in 1970–1980 are explained below [112–115]:

1) *Prediction using the maximum sound pressure levels:*
   The general theory for the prediction of the noise from a non-haul line source (stationary) is: [112, 113]

$$L_{eq} = L_{max}(d_0) + 10\lg \frac{d_0{}^2 \pi \beta}{d.180.l} \tag{4.39}$$

L_{max}: Maximum sound pressure level at a reference distance from the source, dB or dB(A)
d_0: Reference distance, m
β: The angle subtended by the process at the receiver position
l: Length of the line source, m
d: Normal distance from the receiver to the source line, m

2) *Prediction using sound power level:* [114]

$$L_{eq} = L_{wa}(d) - 38 + 10\lg \frac{Q}{vd} \tag{4.40}$$

$$L_{wa} = L_{max}(d_0) + 10\lg 2\pi d_0{}^2 \tag{4.41}$$

Q: Numbers of equipment passing during one hour

$$Q = \frac{VT}{l}$$

$T = 3600$ seconds
d: Normal distance from the receiver to the source line, m
d_0: Reference distance, m
v: Average speed of the moving equipment, km/h
l: Length of line source

3) *Prediction using sound energy level:* [115]

$$L_{eq} = L_{ax} + 10\lg \left[\frac{d_0 N \pi \beta 180}{180 \pi \phi T d} \right] \text{dB} \tag{4.42}$$

L_{ax} (or L_{AE}): Average energy level in one second or in a cycle of operation at a reference distance
N: Number of repetition of work in time, T
β: The angle of view subtended by the process at the receiver position
ϕ: Angle of view of the operation process from the reference point at reference distance, d_0

T: Period which the L_{eq} is calculated, s

Buna proposed a computation model for A-weighted sound pressure level from a point source (1986) [112]:

$$L_{Aeq} = L_{max}(d_0) + 10\lg \frac{t}{T} + 20\lg \frac{d_0}{d} + C_1 + C_2 \qquad (4.43)$$

d_0: Reference distance = 15 m

d: Distance from the receiver to the source, m

C_1: Propagation correction term taking into account the ground effect and the reflections

C_2: Adjustment factor for the screening effect

Q: Numbers of equipment at the site

T: Period which the L_{eq} is calculated, s

t: Time of an operation cycle, s

For a non-haul line sound source (stationary):

$$L_{Aeq} = L_{max}(d_0) + 10\lg \frac{d_0^2 \pi \beta}{d\ 180\ l} + C_1 + C_2 \qquad (4.44)$$

d: Distance from the receiver to the source, m

l: Length of the operation line, m

β: The angle of view subtended by the process at the receiver position

From the above equations, the equivalent continuous sound pressure level L_{eq}(h) of each piece of equipment at a reference distance, can be calculated by using the average emission level to be measured within the same period or in a typical operation cycle. The analytical models have been compared in the past and the compatibility between the results was found to be rather high. However, it was stressed that the accuracy of the models was dependent on the reference data and its reliability. The predictions have been verified by the measurements at construction sites [113].

4.6.3 Guideline Models

For the prediction of construction site noise, some empirical models were provided in the guidelines published in different countries, employing the measured data.

The BRE Model (1975–1977)

As one of the earlier investigations on construction site noise, the BRE, in the UK, published some useful documents, including an approach aimed at predicting the noise levels in $L_{eq(24h)}$ by using the reference noise levels obtained from the noise measurements [116, 117]. The guideline covered an extensive database regarding the short-term highest noise levels relative to various construction equipment and activities. It was accepted that the machines were consistent and operating in steady conditions.

The ultimate noise levels in the environment were determined for multi-source situations, by considering the duration of the operations, the effects of distance, the barriers at the construction site, or reflections from objects, by means of the table sets given in the document. The model was verified by comparing the results with measurements. The following simple relationship between sound power levels and sound pressure levels was presented in the document: [117]

$$\text{Sound power level, dB (A)} = \text{Sound pressure level, dB (A)} + 8 + 20\lg R \qquad (4.45)$$

R: Radius at which the sound level is measured from the center of the machine, m

The UK TRRL Model (1977)

Holland and Martin conducted various investigations on construction noise in TRRL, and presented a basic equation to calculate the sound energy from an array of point sources moving along a straight line on hard ground [118]:

$$L_{eq} = L_{ax} - 10\lg(\pi.\phi/180) + 10\lg R + 10\lg n - 10\lg T + 10\lg(\pi.\theta/180) - 10\lg d \qquad (4.46)$$

$$L_{ax} = L_{eq}(R) + 10\lg t(R) \qquad (4.47)$$

L_{eq}: Reference sound pressure level for a process line, dB(A)
L_{ax}: Reference sound pressure level at distance R from a work cycle, dB(A)
n: Number of times the process is repeated during the time period, T
T: Time period over which the L_{eq} is calculated, s
d: The shortest distance from the receiver position to the line of the process trajectory, m
ϕ: Angle subtended by the process at the receiver position
θ: Angle of view of process line from the reference point
R: Radius at which the sound level is measured from the center of the machine, m

For noise propagation on soft ground such as grass, gravel, and cultivated land, the excess attenuation was determined in the model as:

$$\Delta_{Leq}(\text{for soft ground}) = 5.2\lg[6(h-0.25)/e] \qquad \text{for } 075 \le h \le 0.25 + e/6 \qquad (4.48)$$

$$\Delta_{Leq} = 0 \quad \text{for } h > 0.25 + (e/6)$$

e: Distance, m from the reception point to the nearest point on the process trajectory
h: Average height above the ground of sound traveling from the nearest point to the reception point

If several construction machines were simultaneously operating at the time of prediction, the energy-based summation of the L_{eq} levels was obtained. The model results were compared with the measurements conducted for various operation processes on site and the "*rms*" standard deviation was found acceptable with the maximum divergence of 3.3 dB(A).

The Nordic Prediction Model (1984)

The joint guidelines, published by the Scandinavian countries, proposed two different approaches: a basic model and a detailed (advanced) model [119]. Sound propagation was handled similar to traffic and industrial noise models and the sound pressure levels for the construction noise were predicted in L_{eq} and L_{max}. It was noted that the uncertainty could be expected at longer distances with higher standard deviations.

A list displaying noise emission levels of various types of machines and construction activity, at the reference distance of 10 m, was included in the document. For the construction work operating on hard ground, corrections for the effect of reflections and screening were applied to the predicted noise levels with the aid of a simple algorithm.

The guideline presented some examples for manual applications of the method by describing the geometrical parameters relative to the site and workstation during the noisiest period of the entire work. It was stressed that the layout plan displaying the construction site, operation line (paths), and the heights from a reference plane, should be carefully examined prior to the study. An example of a construction site is given in Figure 4.26 [119].

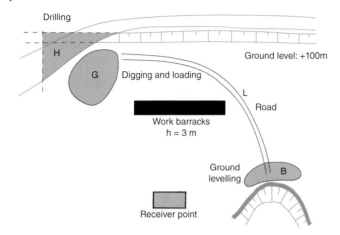

Figure 4.26 Definition of construction site and noise sources in Nordic Model [119]. (*Source:* Permission granted by Nordic Council of Ministers.)

The US FHWA Construction Noise Models

The earlier version of the US national method was developed for highway construction noise in 1977 to introduce noise abatement measures [120]. The model required the following information:

- Construction schedule: Discrete phases of construction and type of activities at each phase
- Equipment schedule: A list indicating type and numbers of equipment at each phase
- Equipment noise levels: Maximum A-weighted noise levels, L_{Amax} and measurement distances for each equipment
- Usage factors (UF): The time percentage for the equipment operating at the loudest mode which is determined as:

$$UF = F_1 \times F_2 \tag{4.49}$$

F_1: Operating factor defined as the portion of the typical work cycle when the equipment is emitting its maximum noise (i.e. at full power)
F_2: Utilization factor defined as the portion of the work period (e.g. 8-hour work day) when the equipment is on site and is being used (number of work cycles)

The average noise level was estimated after normalizing all the equipment noise levels to a common distance for smaller construction sites as given below:

$$L_{eq} = 10 \lg \sum_i^n UF_i.N_i.10^{L_{pi}/10} \tag{4.50}$$

L_{pi}: Maximum sound level of equipment type i
UF_i: Usage factor of equipment type i
N_i: Number of equipment units type i

The above relationship is applicable if the dimensions of site are smaller than the distance to the sensitive area (1 : 5 ratio). The large construction areas could not be considered point sources, therefore, the average noise levels for each piece of equipment was extrapolated and the summation was applied to determine the total value.

Moving equipment on site during the construction process was grouped as:

Category 1: Equipment staying stationary at certain locations, the calculations were made for each location independently.

Category 2: Equipment moving in a simple pattern, the calculation was made for a point described as the acoustic center of the paths determined according to a criterion based on the longest straight path.

Category 3: Equipment moving randomly or along a complex path, partly stationary, the calculations were performed at the geometrical center of the construction site.

The 2006 version of the national method called the Roadway Construction Noise Model (RCNM) has been prepared as a computer program using a spreadsheet tool, providing a wide range of construction noise data and acoustical usage factor for each piece of equipment [121]. The measured data were given as L_{Amax} at a distance of 50 ft (15m). Calculation of the noise level at a certain point in terms of L_{Amax} comprised the shielding effect that was given as simple reduction values in dB(A) in the document. The relationship between the maximum calculated level and the equivalent continuous level of equipment was given as:

$$L_{eq} = L_{maxCalc} + 10 \lg (UF \text{x} 100) \tag{4.51}$$

UF: Time averaging equipment usage factor in percent

The noise limits could be inserted into the model and the results could give the limit exceeded in terms of the required noise index for each construction machine, (e.g. L_{Amax}, L_{Aeq}, and L_{10}.)

The UK BS 5228:2009 (Amended in 2014)

The UK standard, proposing noise control measures against construction noise, was first published in 1997 and revised in 2014 with an amendment [122]. The document covers the manual methods for stationary and moving equipment in plants by using database tables and design graphics. The noise levels at receiver points can be obtained in terms of L_{Aeq} (in 1 hour, 12 hours, etc.) or L_{Aeq} (in a time as short as 5 minutes) and for isolated events, the maximum level, $L_{PA(max)}$ and the percentile level, L_{A01}. The model introduces the concepts of "percentage of-time," implying the percentage of time in which the equipment works at full power and the "percentage of assessment period," implying the time to be selected for calculation of noise levels during the total work period of construction.

For stationary plants, two methods are described in the standard based on:

• the sound power levels of equipment given in the tables according to the percentage on-time at maximum level,
• the equipment L_{Aeq} levels directly given in the databases obtained from the measurements at 10 m from the equipment.

Employing the equipment sound power levels, the actual noise level is calculated at the actual distance from the source with the corrections for the distance effect on soft and hard ground and for the screening effect. The actual levels from all the stationary items of plant with the corresponding percentage of on-time, are combined to obtain the total L_{Aeq}. The next step for both methods is to estimate the percentage of the assessment period over the time in which the activity is performed and to calculate L_{Aeq} during this assessment period for that specific activity (operation) of the equipment. The calculated noise levels are combined to obtain the overall noise level at site.

The document also gives the prediction of noise for the mobile equipment on site in a limited area by using a parameter called the distance ratio (i.e. the traverse length to the minimum distance from the source line to receiver) with corresponding on-time correction factor, to estimate the equivalent on-times. Then, an approach similar to that used with the stationary equipment or plant is applied to calculate the combined construction site noise L_{Aeq} for the assessment period at a receiver point.

The uncertainty and diversity in construction noise assessments are still a point of discussion among researchers [123].

4.7 Prediction Models for Other Sources

There have been various modeling studies for noise sources other than the conventional types explained so far, however, they frequently create severe noise problems in local communities, tourist areas, and in restricted parts of the environment.

4.7.1 Waterway Noise Prediction Models

Noise generated from waterway vehicles, activities from ports, etc. and the propagation of sound over rivers, lakes, or seas, has been investigated in some studies, particularly for assessment of noise impact in the settlements along rivers, seashores, and near ports. The prediction of noise can be made by employing the ISO 9613 model for immission levels by assuming that the waterway vessels are point sources moving randomly or on a line, and radiating typically broadband low frequency noises. The prediction of the environmental levels of seaway traffic requires the emission data for different types of water vehicles, like fishing boats, cargo ships, passenger boats, ferries, speedboats, military vessels, etc. Some are available based on the measurements, although others are still not published widely for various reasons, such as being military or commercial craft.

Propagation models should take into account the reflections from the sea surface and the multiple reflections from nearby buildings, harbor constructions, screening objects, and the topography. In addition, the predictions should include the transmission of underwater noise radiated from engines and other parts of vessels and the diffractions on the water surface while sounds are transferred from the sea to air.

Seaway noises are involved with in the field of underwater acoustics. mainly with two different objectives: (i) to provide good underwater telecommunication (sonars, radar, etc.), which should not be prevented by excessive noise, and (ii) to protect marine and freshwater habitats. Predictions and measurements for the second aim cannot be made in terms of A-weighted sound pressure levels, instead, the decibel levels with specific reference pressure of 1 micropascal (dB re 1 µPa), the different filter characteristics and the signal/noise ratios, are implemented. Recently there have been more investigations in which the new methodologies have been presented, such as deriving underwater noise emissions from vessel activity with technical data and developing noise modeling for the purpose of the impact assessment of underwater noise [124a, 124b, 124c].

4.7.2 Shooting Noise Prediction Models

Shooting practice, generating high energy impulsive noises, is performed in certain areas outside settlements, however, due to the expansion of urbanization over the years, it is inevitable that assessment of shooting noise for planning purposes to protect communities around these areas will have to be made.

Construction Engineering Research Laboratory (CERL), US Model (1978)

Various studies and technical documents were produced in the USA in the 1970s on shooting noise. Some mathematical models were developed to predict the peak noise levels for military weapons in the far field and the lists were given for the peak levels of different firearms. Based on the measurements, variations of noise levels with distance to muzzle point were investigated and the correction factors were derived for different weapons and accessories [125–127].

This first guideline, published in 1984, presented a manual method by using the impulse time-weighting denoted by dBA(I) for sound pressure levels together with correction for weapon type, shooting direction, distance, screens, topography, ground surface, enclosures, and reflections. The reference levels were given in dB(A) by diagrams and various practices were included in the document.

Included with numbers of studies regarding army blast noise impacts, the CERL developed statistical models to assess the blast noise levels from weapons used in military or civil practice fields, based on the long-term field measurements [127]. The broadband noise levels, maximum levels and the temporal variations of blast noises, were investigated and a computer model was developed providing *NEF* contours near the shooting ranges by taking into account parameters, such as shooting positions, number of shots, type of weapons and the physical environmental conditions, relative to the meteorology and the land configuration. CERL published a blast noise prediction manual and computer program to develop a noise management strategy in 1979 (see Chapter 6).

Scandinavia: the Nordtest Method for Shooting (1984)

As one of the few earlier models published by the Scandinavian countries, the design guide was presented as a manual prediction method for shooting noise in the environment with the aid of design charts [128]. The physical conditions of shooting fields (a shooting range) that could be in the open air or within enclosures of different constructions (with or without walls or with a roof only or completely enclosed with openings on the wall or with inner walls covered by absorptive material, etc.) were considered in the predictions. The geometrical parameters in the model are shown in Figures 4.27a and 4.27b. For open-air shooting ranges, the terrain profile had to be specified with the topography, ground surface, vegetation, and screens. The Nordtest prediction model describes the general principles of the industrial noise prediction by using the measured emission levels of different weapons. A list displaying the reference levels of various types of weapons in L_0, dBA(I) was given with the shapes of spectra measured in the reference directions.

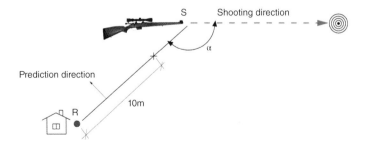

Figure 4.27a Geometrical parameters in the model to assess the shooting noise [128].

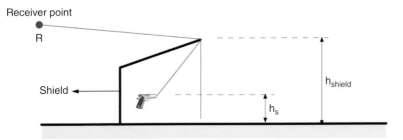

Figure 4.27b Geometrical parameters in semi-open shooting range [129].

The improved model in 1997 included supersonic bullets, multilayered barriers for noise control, the design of different shooting shed forms, and the correction functions for reflections in octave bands [129]. The noise emissions were determined by the acoustic measurements for different types of weapons specifying the shooting direction in the guideline, for example; the muzzle noise to be measured at 10 m away from the rifle and in at least five different directions, i.e. 10°, 45°, 90°, 135°, and 180°.

The calculations of noise levels were given for the weapons with caliber smaller than 20 mm. The contributions of muzzle noise and the noise from supersonic bullets could be calculated independently. The results were obtained in octave bands from 31.5 Hz to 8 kHz and as well as in A-weighted sound pressure levels dB(A). A correction was proposed for the reference sound pressure levels according to the type of shooting spectrum. For supersonic shooting, the muzzle noise level (bullet noise) was found in relation to distance and Mach cone. The total noise in a shooting range was assessed by summation of the highest levels of noise from all shots. The position of the shooting direction with respect to the receiver points is shown in Figures 4.28a and 4.28b.

In order to obtain the sound pressure level at a receiver point, various factors, namely, distance to the receiver, air absorption, indoor absorption of the enclosure or shooting range, geometric properties of open-air sheds, barriers and screens, ground cover, and tree groups higher than 1 m above the propagation line of sound, wind speed and direction, and reflections were taken into account in the model. Eventually the A-weighted total levels could be obtained.

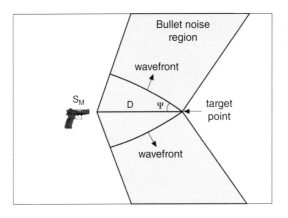

Figure 4.28a Parameters in supersonic shootings and noise zones [129]. (*Source:* Permission granted by Nordtest.)

Figure 4.28b Calculation parameters in supersonic shots and bullet noise region [129]. (*Source:* Permission granted by Nordtest.)
Note: SB: Bullet noise source point.

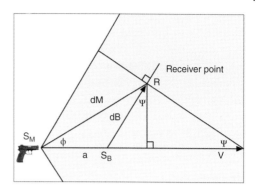

Following the earlier guidelines, various recent studies on modeling shooting noise have been conducted and can be found in the literature, especially for larger caliber weapons and machine guns.

ISO 17201 (2006–2010): The standard, comprising five parts, gives the methodology for determination of the noise from small civilian weapons with directivities depending on the shooting direction [130, 131]. The measurement method described in Part 1 of the standard was improved according to ISO 13474 given below, to measure the angle-dependent sound exposure in one-third octave bands, by applying the corrections for reflections from the ground. Part 4 of 17201 includes the prediction of the projectile sound following flat trajectories, by referring to some constraints [132].

ISO 13474 : 2009 (last confirmed in 2020): The standard which was explained in detail in Section 4.5.3 among the industrial noise models, it is the only standard for shooting noise assessments. It describes an engineering method to calculate the sound exposure levels (*SEL*) of impulsive sounds propagating in a wide range in an environment up to 30 km [111]. The muzzle blasts and projectile sounds are described and the distribution of *SEL* in relation to the direction angles is given in terms of the directivity coefficients. Calculations of the sound attenuations caused by the meteorological conditions during sound propagation and the terrain properties are explained by emphasizing the high elevated trajectories.

4.7.3 Wind Turbine Noise Prediction Models

The aero-acoustic source emissions (sound pressures) of wind turbines can be calculated by using semi-empirical and analytical models or other simulation models [133a]. However, prediction of noise from wind farms is difficult due to various environmental factors, long distances, extreme height of source (hub height), changing wind speeds and directions (i.e. effect of wind shear), and temperature gradients. The point source models and the extended source models for wind turbine noise propagation, with some approximations, were developed at least two decades ago and validated by the field measurements mostly under homogeneous conditions. They are based on the measured sound power levels of wind turbines and are published as engineering methods or guideline methods, some are declared in the national standards of different countries [133b, 134]. The accuracy of the models depends on the atmospheric conditions, hub heights, the precision of the terrain model (landscape), and the ground cover with different acoustical properties.

The NPL Model

This simple model enables the calculation of the sound power level for point sources with a spherical radiation of sound in a free field or hemispherical radiation over a reflecting surface, by taking into account the distance on the flat terrain and air absorption [135]. Atmospheric attenuation is calculated by assuming the noise is broadband. The total noise level is obtained by the logarithmical addition of the noise levels from each turbine. There are some restrictions in the model; the ground is assumed to be flat and the large objects, refraction of sound, wind speed and direction, or the other factors changing the frequency spectra cannot be taken into account.

Implementation of ISO 9613-2:1996 (Confirmed in 2017) for Wind Turbine Noise

The standard model, explained in Section 4.5.2, is commonly used for wind turbine noise assessment as well as the other noise sources, however, it is applicable for point sources and at the distances up to 1000 m from the source [109]. The accuracy of the method has been declared as ± 3 dB for sources with heights more than 30 m. Since the wind turbine noise reaches long distances and has a low frequency noise, which is dominant below 200 Hz, the researchers still debate developing certain modifications in the ISO 9613 model for the infrasounds and the low frequency noise generated by wind turbines.

The NZ Concawe Method

Developed for high-level noise sources in New Zealand, the rather simplified prediction model is applicable to wind turbines whose hub height is more than 25 m and at distances up to 2000 m from the receiver [136, 137]. The model uses six different weather categories and the effect of air absorption is calculated by using ISO 9613-1. [108]

The Nord2000 Method

The advanced method, developed as a joint research study in the Nordic countries, involves a procedure for prediction of wind turbine noise, with a special concern for the effects of ground and meteorological factors on propagation [138]. The topography of the terrain is divided into segments according to the surface type characterized by the parameter called the "acoustic roughness of ground." The "roughness classification" for the terrain is made with respect to the flow resistivity of various surfaces, such as forest ground, crop area, grass, frozen ground, summer, and winter conditions separately.

The other important feature of the model is that the vertical sound speed profiles can be selected according to the weather classes. The document recommends preparing a database for the vertical wind speed profile and the temperature profile for the area to be selected for the calculation of long-term noise levels. Since the wind speeds are generally measured at a fixed height (at 10 m from the ground), the model converts them into the speeds at actual hub-height (which is 80–100 m for modern wind turbines) by extrapolation. A database for a "noise value matrix" displaying the "noise level/wind speed/hub height" is presented in the model. As a result, the sound emission of a turbine (L_W) is calculated at the actual hub-height at each octave band.

The calculations of the immission levels are made at the octave or one-third octave bands at the receiver position and then, for each turbine, they are combined to determine the total windfarm noise as A-weighted sound pressure levels. The calculations in frequency bands comprise also the evaluations at low frequencies which are strongly related to the annoyance from the noise. The model has been validated by the field measurements.

4.8   Validation of Models

Prediction models enabling noise assessments in the environment, with specific characteristics and with dominant noise sources and operations, have to ensure the reliability and accuracy of the results as explained in Section 5.6. Three approaches have been used in the investigations for verification of the models.

4.8.1   Comparison of Analytical Models

Prediction models can be compared with the results obtained from the other models, by using similar parametric values under identical source and environment conditions. Discrepancies between the results due to the differences in methodologies, restrictions, assumptions, approximations, etc. are evaluated based on the statistical analyses.

Some of these models were compared in a study for different highway configurations in the past and it was revealed that the deviations could be as much as 10–15 dB(A) for the same road and traffic conditions, as shown in Figure 4.29 [139]. This result underlines the possible errors in the selection of a model arbitrarily without seeking its validity under actual traffic and vehicle conditions. Various scientific documents can be found in the literature regarding the comparisons of the European prediction models for other environmental noise sources in addition to road traffic noise, such as railway noise, aircraft noise, wind turbine noise, etc. The comparisons have been made both for emission and immission levels as well as for the noise attenuations provided by the physical elements [133b, 140–142].

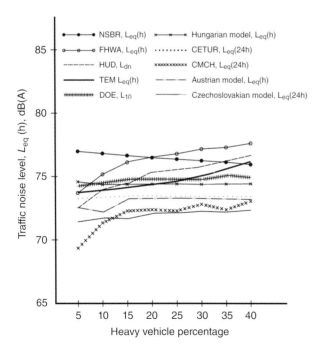

Figure 4.29   Comparison between the calculated traffic noise levels by using different national standard models for Trans-European Motorways (TEM) [139].

4.8.2 Comparison with the Field Measurements

Calculated noise levels are compared with the measured results under conditions similar to those assumed in the models, however, certain discrepancies are expected between the results due to:

- variation in meteorological parameters during measurements under the reference specified in the prediction model. (Thus, a calibration process might be necessary for the measurement results.)
- presence of background noise (continuous or intermittent, sometimes intensive) during measurements, causing difficulty in analysis of the noise concerned.

As summarized in Chapter 5, the measurement uncertainties given in the measurement standards differ with the type of noise source, and should be determined and declared while using the measurement results in comparisons with the predicted values.

4.8.3 Validation with the Laboratory Experiments

The predicted results based on the theoretical or empirical methods can be verified by the experimental studies conducted in laboratories. As explained in Section 5.8, the measurements can be conducted under controlled conditions and by using scale models. However, there are some restrictions to be considered, such as size limitation of the physical model, difficulty in simulation of some sources and multi-noise situations, difficulty in modeling the different meteorological conditions in the laboratory, for comparison of the predicted results.

4.9 Conclusion

Determination of the existing and future noise conditions in a specific environment is vital to make an assessment of the noise impact, prior to management programs e.g. decisions about the measures, quantifying the performance and the feasibility of measures, to evidence the economic impacts of noise pollution, etc. The prediction models have been developed, especially between 1970 and 1990, when the concept of noise mapping was not widely common, unlike today, and the measurements could not be performed extensively due to various reasons. The earlier prediction models, some based on scientific studies or derived from field measurements and statistical regressions, others manually applicable from design guide procedures, made use of sound emission levels (the sound power level) of sources that can be either single or multiple, stationary or dynamic, or used a basic noise level (a reference sound pressure level), which was valid for some restricted conditions, then corrections were applied for other situations. Immission models need to introduce the attenuation values due to various environmental factors, which are calculated by means of empirical formulae. The result is the sound pressure levels at a specific receiver point, acquired in terms of various noise units (descriptors), according to the source type.

In all the prediction models, it is necessary to provide the construction and operational data relative to the noise sources, with the aid of a noise emission database, as well as a database for the meteorological factors. Great efforts were made in the past and extensive field measurements were conducted to derive the expressions between noise levels and the effective parameters associated to the source and the environment, which are also interrelated with each other.

Today's concept of noise management requires more accurate and precise predictions, thus the models have significantly evolved, however, maintaining the basic structure. It is vital to obtain the results not only at a single receiver point but throughout the area concerned and, most importantly, as fast as possible. The advanced noise prediction models, which are currently used for noise mapping, were established, based on the simple or complex prediction models in the past, which are worth knowing by those working in this field, since they have provided enlightenment for the present modeling techniques that have been improved through experience over the years. Nowadays, the models have become much more complicated, involving complex interrelations of a great number of variables, some of those that could not be accounted before. The source-specific models, with greatly increased precision, can be readily comprehended by everyone and implemented at present with fast computer technologies (see Chapter 6). Recently the engineering methods have also yielded the uncertainties of the calculated results, some of which have been standardized to harmonize various methods for comparability in the international arena.

It can be noted that the empirical models are beneficial in the preliminary assessments of potential noise impact at a certain extent and for the decisions about future planning in local communities without requiring the assistance of acoustic experts.

References

1 NPL (1976). Guide to predictive modelling for environmental noise assessment, NPL. www.npl.co.uk.
2 Lamure, C. (1965). Niveaux de bruit au voisinage des autoroutes, *Proceedings of the Fifth International Congress on Acoustics*.
3 Kurze, U.J. (1971). Statistics of road traffic noise. *Journal of Sound and Vibration* 18 (2): 171–195.
4 Kurze, U.J. (1971). Noise from complex road traffic. *Journal of Sound and Vibration* 19 (2): 167–177.
5 Jonasson, H.G. (1973). A theory of traffic noise propagation with application to Leq. *Journal of Sound and Vibration* 30 (3): 289–304.
6 Maekawa, Z. (1974). Environmental Sound Propagation, 8th International Congress on Acoustics, London.
7 Kessler, F.M. and Gottlieb, P. (1976). Method to convolve sound level distributions for prediction+ business media, BV, 2001g of community sound levels. *Journal of the Acoustical Society of America* 60 (5): 1108–1110.
8 Jonasson, H.G. (1973). A theory of traffic noise propagation with applications to L_{eq}. *Journal of Sound and Vibration* 30 (3): 289–304.
9 Clayden, A.D., Culley, R.W.D., and March, P.S. (1975). Modelling traffic noise mathematically. *Applied Acoustics* 8 (1): 1–12.
10 Salomons, E.M. (2001). *Computational Atmospheric Acoustics*. Springer Science.
11 Chandler-Wilde, S.N. (1997). The boundary element method in outdoor noise propagation. *Proceedings of the Institute of Acoustics* 19 (8): 27–50.
12 Gilbert, K.E. and White, M.J. (1989). Application of the parabolic equation to sound propagation in a refracting atmosphere. *Journal of the Acoustical Society of America* 85: 630637.
13 Stevens, K.N. and Baruch, J.J. (1957). Community noise and city planning. In: *Handbook of Noise Control* (ed. M.C. Harris). McGraw-Hill.
14 Shaw, E.A.G. and Olson, N. (1972). Theory of steady state urban noise for an ideal homogenous city. *The Journal of the Acoustical Society of America* 51 (6): 1782–1792.
15 Galloway, J.W., Clark, W.E., and Kerrick, J.S. (1969). Highway noise measurement simulation and mixed reactions, National Cooperative Highway Research Program Report 78, Appendix.
16 Johnson, D.R. and Saunders, E.G. (1968). The evaluation of noise from freely flowing road traffic. *Journal of Sound and Vibration* 7 (2): 287–309.

17 Schreiber, L. (1969). On the calculation of the energy equivalent of continuous sound level traffic noise from a street. *Acta Acustica United with Acustica* 21 (2): 121–123.

18 Ingemasson, S. and Benjegard, S.O. (1969). Physical scales of traffic noise, minimum distance between road and dwellings in level unshielded terrain, National Swedish Building Research Center Report, 52.

19 Lewis, P.T. (1973). Noise generated by single vehicles in freely flowing traffic. *Journal of Sound and Vibration* 30 (2): 191–206.

20 Lewis, P.T. (1977). Noise generated by single vehicles in freely flowing traffic: some further comments. *Journal of Sound and Vibration* 55 (3): 472–473.

21 Delany, M.E. (1972). Prediction of traffic noise levels, NPL Acoustics Report AC 56.

22 Delany, M.E. (1972). A practical scheme for predicting noise level (L10) arising from road traffic. NPL Acoustics Report AC 57, NPL Dept of Trade and Industry.

23 Anon (1970). A review of road traffic noise. TRRL Report LR 527.

24 Nelson, P.M. (1972). The combination of noise from separate time varying sources. TRRL Report LR 526.

25 Nelson, P.M. (1973). A computer model for determining the temporal distribution of noise from road traffic. TRRL Lab Report 611.

26 Duncan, N.C. (1973). A method of estimating the distribution of speeds of cars on motorways. TRRL Report LR 598.

27 Anon (1975). Road surface texture and noise. TRRL Report.

28 Nelson, P.M. and Piner, R. (1977). TRRL classifying road vehicles for the prediction of road traffic noise.TRRL Report.

29 Harland, D.G. (1978). Rural traffic noise prediction: an approximation. TRRL Supplementary Report No. 425.

30 Buna, B. (1984). Predicting the noise of accelerating road traffic. *Proceedings of the Federation of Acoustical Societies of Europe* 84: 265–268.

31 Sandberg, U. (1987). The influence of the road surface and its characterization. *Applied Acoustics* 21: 97–118.

32 Scholes, W.E. and Sargent, J.W. (1971). Designing against noise from road traffic, BRS current paper no. 20/71.

33 Scholes, W.E. (1971). Traffic noise criteria. *Applied Acoustics* 3 (1): 1–21.

34 Anon (1975). Calculation of Road Traffic Noise, Department of the Environment, Welsh Office, HMSO.

35 Anon (1988). Calculation of Raod Traffic Noise, Department of Transport, HMSO.

36 Anon (1968). Traffic noise in residential areas: study by the National Swedish Institute for Building Research and the National Swedish Institute of Public Health, 36E/68, Rapport fran Byggforskningen.

37 Ljunggren, S. (1973). A design guide for road traffic noise, D10:1973. National Swedish Building Research, Stockholm.

38 Bennerhult, O., Lundqvist, B., Nilsson, N.A. et al. (1977). A new method for rating fluctuating noise, D 12. Swedish Council for Building Research, Stockholm.

39 Nordic Council of Ministers (1980). The computing model for road traffic noise, Statens Planverk Report No. 48. Nordic Council of Ministers.

40 Tema Nord (1996). Road traffic noise, Nordic Prediction Model, Nordic Council of Ministers, Tema Nord Environment.

41 Kragh, J. (2000). Nord 2000. State-of-the-art overview of the new Nordic prediction methods for environmental noise, Inter-Noise 2000, Nice, France (August).

42 Kragh, J., Plovsing, B., Storeheier, S.A. et al. (2002). Nordic environmental noise prediction methods, Nord2000 – Summary Report, Denmark (May).

43 Favre, B. (1974). Méthode de calcul automatique des niveaux de bruit de circulation routière 657–41, Centre d'Evaluation et Recherche des Nuisances. Institute de Recherche des Transports, IRT, Lyon.

44 CETUR (1980). Guide du bruit des transports terrestres prévision des niveaux sonores. CETUR, Ministère de l'Environnement et du Cadre de Vie.

45 CMCH (1977). Canada road and rail noise effects on housing. CMCH, Central Mortgage and Housing Corporation.

46 CMCH (1978). Design guide for reducing transportation noise in and around buildings. CMCH. Central Mortgage and Housing Corporation.

47 Halliwell, R.E. and Quirt, J.D. (1980). Traffic noise prediction. Building Research Note No. 146, Division of Building Research, National Research Council of Canada, Ottawa.

48 Gordon, C., Galloway, W., Kugler, B.K. et al. (1971). Highway noise: A design guide for highway engineers. National Cooperative Highway Research Program Report 117. US Federal Highway Administration Highway Research Board.

49 HUD (1983). *Noise Assessment Guidelines*. US Dept of Housing and Urban Development.

50 Kugler, B.A., Cummins, D.E., and Galloway, W.J. (1974). Establishment of standards for highway noise levels. Vol. 1, Design guide for highway noise prediction and control. BBN Report no. 2739.

51 Pallet, D.S., Wherli, R., Kilmer, R.D. et al. (1978). *Design Guide for Reducing Transportation Noise in and Around Buildings*. National Bureau of Standards.

52 Kugler, A., Cummins, D., Galloway, W. et al. (1976). Highway design guide for prediction and control, No. 174. National Cooperative Highway Research Report, Transportation Research Board.

53 Donavan, P.R. and Lyon, R.H. (1974). Sound propagation near ground level in the vicinity of street intersections II. Interagency Symposium in Transportation Noise, University of North Carolina, USA (5–7 June).

54 Lyon, R.H., Holmes, D.G., Donavan, P.R. et al. (1974). *Sound Propagation in City Streets*. MIT Acoustics and Vibration Laboratory.

55 Lyon, R. (1974). Role of multiple reflections and reverberation in urban noise propagation. *Journal of Acoustical Society of America* 55 (3): 493–503.

56 Anderson, G., Lee, C.S.Y., Fleming, G.G. et al. (1978). FHWA traffic noise model.

57 Barry, T.M. and Reagan, J.A. (1978). FHWA Highway traffic noise prediction model, FHWA-RD-77-108, USA.

58 FHWA (1998). *Traffic Noise Model User's Guide*. Federal Highway Administration, US Dept. of Transportation.

59 (a) FHWA (1982). Stamina/Optima User's Manual Report, FHWA-DP 58–1, April;
(b) FHWA (2017). https://www.fhwa.dot.gov/Environment/noise/traffic_noise_model/tnm_v30/.

60 Reinhold, G. (1971). Bau- und Verkehrstechnische Massnahmen zum Schutz gegen Strassenverkehrslärm, 119, Strassenbau und Strassenverkehrstechnik, Köln.

61 Anon (1975). Schallausbreitung in bebauten Gebieten. Bericht über das Ergebnis einer modellmässigen Untersuchung zur Schallausbreitung in Städten, Minister für Arbeit, Gesundheit und Soziales des Landes Nordrhein Westfahlen.

62 Der Bundesminister für Verkehr (1981). Richtlinien für den Lärmschutz an Strassen, RLS81, Revised in 1990.

63 Anon (1983). Dienstanweisung betreffend Lärmschutz an Bundesstrassen, Bundesministerium für Bauten und Technik.

64 Haskoning, B.V. (1981). Verkeerslawaai, Ministerie van Volksgezondheid en Milieuhygiene. Den Haag (and Stedebouw en Geluid, Broshure).

65 Moerkerken, I.A. (1986). The Netherland Prediction Model for Outdoor Noise Propagation. Directoraat-general Voor de Milieuhygiene, Ministere van Volkshuisvesting Ruim telijke Ordening en Milieubeheer.

66 NMPB Routes (1996). 96 Road traffic noise. New French calculation method including meteorological effects (Bruit des Infrastructures Routiers, Méthode de calcul incluant les effets météorologiques).

67 EC (2002). Directive 2002/49/EC of 25 June 2002 of the European Parliament and Council relating to the assessment and management of environmental noise.

68 Dutilleux, G., Defrance, J., Gauvreau, B. et al. (2008). The revision of the French method for road traffic noise prediction. *The Journal of the Acoustical Society of America* 123: 3150.

69 Watts, G.R. (2005). Prediction model for road traffic noise. PPR 034, Harmonoise.

70 Anon (2015). Improved methods for the assessment generic impact of noise in the environment. The Noise Emission Model for European Traffic, (IMAGINE) Deliverable 11, Deliverable 4: Specifications for GIS-Noise Databases.

71 Kephalopoulos, S., Paviotti, M., and Anfosso-Lédée, F. (2012). Common Noise Assessment Methods in Europe (CNOSSOS-EU), to be used by the EU Member States for strategic noise mapping following adoption as specified in the Environmental Noise Directive 2002/49/EC. Luxembourg: Publication Office of the European Union.

72 EU (2015). Commission Directive (EU) 2015/996 of 19 May 2015, establishing common noise assessment methods according to Directive 2002/49/EC of the European Parliament and of the Council.

73 (a) Shilton, S.J., Anfonso, F., and van Leuwen, H. (2015). Conversion of existing road source data to use Cnossos-EU, Maastricht;
(b) Duilleux, G. and Soldano, B. (2018). Matching Directive 2015/996/EC (CNOSSOS/EU) and the French emission model for road pavements, Euronoise 2018, Crete, Greece.

74 Hanson, C. (1990). High speed rail system noise assessment. TRB1990, Annual Meeting Committee A2M05 Paper no. 890359.

75 Hanson, C. (1993). Aeroacoustic sources of high speed Maglev Trains. Noise Con 93.

76 Okumura, Y. and Kuno, K. (1992). Statistical analysis of field data of railway noise and vibration collected in urban areas. *Applied Acoustics* 33: 263–280.

77 Anon (1994). Logiciel Mithra-Fer, version 2.1 Manuel technique. Centre Scientifique et Technique du Bâtiment, Grenoble, France, (in French).

78 Ringheim, M. (1984). Noise from railway traffic. KILDE Report 67, prepared for the Nordic Council of Ministers Noise Group.

79 (a) Ringheim, M. (1984). Background material for the Nordic rail traffic noise prediction method. Kilde Report 130;
(b) Ringheim, M. (1988). Sound propagation and the prediction of railway noise. *Journal of Sound and Vibration* 120 (2): 363–370.

80 Nielsen, H.L. and the Railway Noise Group (1996). Railway traffic noise, the Nordic prediction method. The Nordic Council of Ministers, Tema Nord 524.

81 Cato, D.H. (1976). Prediction of environmental noise from fast electric trains. *Journal of Sound and Vibration* 46 (4): 483–500.

82 CMCH (1977). Road and rail noise: effects on housing. CMCH: Central Mortgage and Housing Corporation, NHA 5156, Canada.

83 Anon (1995). Calculation of railway noise. The Department of Transport, London, UK.

84 (a) Craven, N.J., Bewes, O.G., Fenech, B.A. et al. (1997). Investigating the effects of a network-wide rail grinding strategy on wayside noise levels. In: *Noise and Vibration Mitigation for Rail Transportation Systems* (eds. E. Nielsen, D. Anderson, P.E. Gautier, et al.), 369–376. Springer. (b) Anon (1995). Calculation of Railway Noise, The Department of Transportation, Her Majesties stationary office, London, United Kingdom.

85 Schall 03 (1990). Richtlinie sur Berechnung der Schallimmissionen von Schienenwegen, Information Akustik 03 der DB. [Guidelines for the calculation of sound emission near railway lines].

86 Moehler, U., Kurze, U.J., Liepert, M. et al. (2008). The new German prediction model for railway noise "Schall 03 2006": an alternative method for the harmonised calculation method proposed in the EU directive on environmental noise. *Acta Acustica United with Acustica* 94 (4): 548–552.

87 Anon (2015). Erläuterungen zur Anlage 2 der Sechzehnten Verordnung zur Durchführung des Bundesimmissionsschutzgesetzes (Verkehrslärmschutzverordnung – 16. BImSchV), Berechnung des Beurteilungspegels für Schienenwege (Schall 03) Teil 1: Erläuterungsbericht. Bundesministerium für Verkehr und digitale InfrastrukturStand 23.

88 Nederland Ministerie van Volkshuisvesting (1996). Reken-Meervoorscrift Railverkeerslawaai 96, AR Interim–CM B4–3049/2001/329750.

89 Paviotti, M., Shilton, S.J., Jones, R. et al. (2015). Conversion of existing railway source data to use Cnossos-EU, Euronoise 2015, Maastricht, The Netherlands.

90 L.I.C(1981). A guide to the calculation of NNI. DORA Communication, 7908 Civil Aviation Authority (2nd edition).

91 Plovsing, B. and Svane, C. (1983). Aircraft noise exposure model: guidelines for the methodology of a Danish computer program. Danish Acoustical Institute, Technical Report No. 101.

92 Danish Acoustical Institute (1984). Methodology for calculation of noise exposure around general aviation airfields, Report no. 120. Danish Acoustical Institute.

93 Plovsing, B. and Svane, C. (1990). DANSIM, Danish airport noise simulation model: basic principles, experiencs and improvement, Inter-Noise 90.

94 Nordic Council of Ministers (1993). Air traffic noise calculation: Nordic guidelines, Nord 1993:38. Nordic Council of Ministers, Copenhagen.

95 Ollerhead, J.B.(1992). DORA report 9120. The CAA Aircraft Noise Contour Model: ANCON Version 1.

96 Anon (1999). R&D report 9842. The UK Civil Aircraft Noise Contour Model, ANCON: Improvements in Version 2.

97 Zaporozhets, O., Tokarev, V., and Attenborough, K. (2012). *Aircraft Noise: Assessment, Prediction and Control*. Spon Press.

98 (a) FHWA (2002). Integrated Noise Model INM version 6.0, FAA-AEE-99-03. Federal Highway Administration. (b) Anon (2008). Bekanntmachung der Anleitung zur Datenerfassung über denFlugbetrieb (AzD) und der Anleitung zur Berechnung vonLärmschutzbereichen (AzB) vom 19.11.2008, Instructions on the Acquisition of Data on Flight Operations and the Calculation of Noise Protection Areas, Bundesanzeiger Nr. 195a.

99 SAE International (1995). SAE-AIR1845: procedure for the calculation of airplane noise in the vicinity of airports.

100 ICAO (1993). ICAO Annex 16, vol. 1, 3rd edition. International Civil Aviation Organization.

101 ECAC (1997). ECAC -CEAC Doc 29 Standard method of computing noise contour around civil airports.

102 (a) ECAC (2005). Report on standard method of computing noise contours around civil airports, Doc 29, 3rd edition, European Civil Aviation Conference.
(b) ECAC. CEAC Doc 29 4th Edition Report on Standard Method of Computing Noise Contours around Civil Airports Volume 1: Applications Guide As endorsed by DGCA/147 on 7 December 2016.

103 Zaporozhets, O., Tokarev, V., and Attenborough, K. (2011). *Aircraft Noise: Assessment, Prediction and Control*. CRC Press.

104 Shield, B.M. and Corlett, E.U. (1975). A simulation of factory noise. *The Production Engineer* September: 489–492.

105 Orlowski, R. (1990). Scale modelling for predicting noise propagation in factories. *Applied Acoustics* 31: 147–171.

106 Bruckmayer, F. and Lang, J. (1973). *Lärmschutz und Stadtplanung*. Institut für Stadtforschung.

107 Danish Academy of Technical Sciences (1982). Environmental noise from industrial plants general prediction method, Report No. 32, Danish Acoustical Laboratory, The Danish Academy of Technical Sciences.

108 ISO (1993). ISO 9613-1:1993 (confirmed in 2015) Acoustics: Attenuation of sound during propagation outdoors. Part 1: Calculation of the absorption of sound by atmosphere.

109 ISO (1996). ISO 9613-2:1996 (confirmed in 2012) Acoustics: Attenuation of sound during propagation outdoors. Part 2: General method of calculation.

110 ISO (2003). ISO 13474:2002 Acoustics: Impulse sound propagation for environmental noise assessment.

111 ISO (2009). ISO 13474:2009 (Last confirmed in 2020) Acoustics: Framework for calculating a distribution of sound exposure levels of impulsive sound events for the purposes of environmental noise assessment.

112 Buna, B. (1986). Prediction of road construction noise, *Proceedings of Institute of Acoustics*.

113 Buna, B. (1986). Construction noise, Institute for Transport Sciences, Budapest.

114 Anon (1980). Contrôle des bruits aériens et des vibrations mechaniques emis dans l'environnement par les chantiers de genie civil et de bâtiment. Env.28/80, Ministère de l'environnement et du Cadre de Vie, Paris.

115 Martin, D.J. and Solaini, A.V. (1976). Noise of earthmoving at the road construction sites, TRRL report SR 190 UC.

116 Akam, E.A.A. and Lawson, P. (1975). Construction site noise. BRE current paper 57/75.

117 Lawson, P. (1976). Prediction and measurement of the equivalent continuous sound level (L_{eq}) during the construction of a main drainage system, BRE current paper no. 9/76.

118 Harland, D.G. and Martin, D.J. (1977). The prediction of noise from road construction sites, TRRL Digest Report LR 756.

119 Nordic Council of Ministers (1984). Noise from construction sites: a Nordic prediction method for noise emitted by building and construction activities. The Nordic Council of Ministers, Noise Group.

120 FHWA (1977). Highway construction noise. Report of 1977 Symposium on Highway Construction Noise, FHWA-TS-77-211, USA.

121 FHWA (2006). FHWA roadway construction noise model user's guide. Federal Highway Administration, US Department of Transportation.

122 BS (2014). BS 5228-1:2009+A1: 2014, Noise and vibration control on construction and open sites: Part 1 Code of practice for basic information and procedures for noise and vibration control.

123 Tompsett, R. (2014). Uncertainty and diversity in construction noise assessment. *Acoustics Bulletin* Jan./Feb.: 45–47.

124 (a) Wittekind, D.K. (2014). A simple model for the underwater noise source level of ships. *Journal of Ship Production and Design* 30 (1): 1–8.
(b) Jalkanen, J.-P., Johansson, L., Liefvendahl, M. et al. (2018). Modelling of ships as a source of underwater noise. *Ocean Science* 14: 1373–1383.
(c) Farcas, A., Thompson, P.M., and Merchant, N.D. (2016). Underwater noise modelling for environmental impact assessment. *Environmental Impact Assessment Review* 57: 114–122.

125 CERL (1978). Predicting noise impact in the vicinity of small arms ranges, CERL, USA.

126 Anon (1983). Data base for assessing the annoyance of the noise of small arms, Technical Guide No. 135, United States Army Environmental Hygiene Agency.

127 Constructor Engineering Research Laboratory (1981). Technical blast noise prediction, vol. I: Data bases and computational procedures, Technical Report N-98.

128 Nordic Council of Ministers (1984). Noise from shooting ranges: a Nordic prediction method for noise emitted by small-bore weapons, The Nordic Council of Ministers, Noise Group.

129 Anon (1997). Shooting ranges: Prediction of noise, Nordtest method, 1997–05 NT ACOU 099. Finland.

130 ISO (2006). ISO 17201-2:2006 (confirmed in 2013) Acoustics: Noise from shooting ranges. Part 2: Estimation of muzzle blast and projectile sound by calculation.

131 ISO (2010). ISO 17201-3:2010 (revised by ISO 17201-3:2019) Acoustics: Noise from shooting ranges. Part 3: Sound propagation calculations.

132 ISO (2006). ISO 17201-4:2006 (confirmed in 2014) Acoustics: Noise from shooting ranges. Part 4: Prediction of projectile sound.

133 (a) Cotte, B. (2018). Coupling of an aeroacoustic model and a parabolic equation code for long range wind turbine noise propagation. *Journal of Sound and Vibration* 422: 343–357.
(b) Evans, T. and Cooper, J. (2012). Comparison of predicted and measured wind farm noise levels and implications for assessments of new wind farms. *Acoustics Australia* 1: 28–40.

134 Hansen, K., Hessler, G., Hansen, C. et al. (2015). Prediction of infrasound and low frequency noise propagation for modern wind turbines: a proposed supplement to ISO 9613-2, Sixth International Meeting on Wind Turbine Noise, Glasgow (20–23 April). https://www.researchgate.net/publication/275964097.

135 NPL (1994). Wind turbine noise model. National Physical Laboratory, UK. http://resource.npl.co.uk/acoustics/techguides/wtnm.

136 Manning, C.J. (1981). The propagation of noise from petroleum and petrochemical complexes to neighbouring communities, CONCAWE, ATL Report No. 4/81.

137 Malcolm Hunt Associates and Marshall Day Acoustics (2007). Stakeholder Review & Technical Comments: NZS 6808:1998 Acoustics – Assessment and measurement of sound from wind turbine generators. Report prepared for New Zealand Wind Energy Association and EECA, Wellington, New Zealand.

138 DELTA (2009). Acoustics and vibration noise and energy optimization of wind farms: Validation of the Nord2000 propagation model for use on wind turbine noise. Report PSO-07 F&U, project no. 7389, Hørsholm, Denmark.

139 Kurra, S. and Gedizoglu, E. (1987). Traffic noise generated by Trans-European motorways. *Acoustics Letter (England)* 11 (3): 40–46.

140 Van Leeuwen, H.J.A. (2000). Railway noise prediction models: a comparison. *Journal of Sound and Vibration* 231 (3): 975–987.

141 Steele, C. (2001). A critical review of some traffic noise prediction models. *Applied Acoustics* 62 (3): 271–287.

142 Isermann, U. and Vogelsang, B. (2010). "AzB-2008 and ECAC Doc.29 – two modern European aircraft noise calculation models". *Noise Control Engineering Journal* (58): 455–461.

5

Noise Measurements

Scope

In this chapter, the subject of acoustic measurements is reviewed, primarily with respect to the quantification of environmental noises both at source and in the propagation path, which are important in noise impact assessment and noise management. Information about the measurement and analysis systems is given, with a historical background up to recent times, however, but not too much emphasis is put on the technical details required in acoustical engineering. The major principles of the emission and immission measurements, the basic and advanced equipment are described, and an overview of the technical standards is given. Examples of source-specific measurements, which are of practical use to environmental engineers, are explained with the earlier and recent implementations published in the literature. Since the measurement systems and the methodologies, particularly those presented in the international standards, have constantly been under development, thus the present situation. which is described in this chapter, might need some revision in the future.

5.1 Brief History of Advances in Noise Measurements

Acoustic measurements have a long history. Lord Kelvin stated, "We can start learning when we discover how to measure" [1]. The World Health Organization (WHO) described the purpose of acoustic measurements simply as "Measuring is knowing" [2]. Brüel defined acoustic measurements as "Physical measurements characterizing the acoustical phenomenon" [3]. Research in the sound field needs an appropriate physical unit which is measurable by the appropriate instruments. However, using only a single unit or scale is not sufficient for the environmental noises and complex measurement quantities and more descriptive indicators are employed to describe both sound field and impacts of noise. Acoustic measurements should take into account a number of parameters; acoustical and non-acoustical, according to the selected methodology and the measurement conditions, hence this can be accomplished by employing more advanced and precision instruments, to acquire reliable data which represents actual acoustic conditions.

The history of sound and audio measurements has been well documented in the literature [4–6]. In antiquity, the theories of Aristotle could not be proved by measurements until the seventeenth century, at that time the experiments focused on the speed of sound and frequency (See table 1.1). The earliest measurements in applied acoustics were realized in musical and architectural acoustics. Observation of the sound field through some physical models could be made, however, without quantification, particularly to demonstrate the sound distribution in amphitheaters and auditoriums. Figure 5.1a,b show an experiment using water waves as a simulated wave propagation in two-dimensional media [6, 7] and the set-up of the Ripple Tank used in such experiments, respectively.

Measuring loudness was of concern even in the nineteenth century, later the scale of the loudness sensation was established by scientists, such as Fletcher in the 1900's, later Stevens in the 1950s, based on laboratory experiments [8, 9]. In the nineteenth century, quantitative measurements of sound pressure and particle velocity were achieved and at the beginning of twentieth century, in addition to sound power measurements, the parameters of environmental noise were widely investigated through experimental studies outdoors. W.C. Sabine examined the basic principles of architectural acoustics in experimental methods in his pioneering acoustic laboratory, founded in 1918 [10] (Figure 5.2). He also invented some instruments to measure sounds for different purposes. In 1920, the Bell Laboratories emerged as the leading institute for researchers and scientists by the construction of the first anechoic room. In general, the acoustical measurements were conducted to validate the hypotheses relating the physical principles and theory and also to provide data to derive empirical relationships between sound pressures and relative parameters.

In field measurements, specially equipped mobile laboratories were used to transport the large measurement devices and the equipment sets, in which different types of instruments were combined into one measurement system. The instruments used the electrical networks, for the amplification of sound signals [5]. Figures 5.3a–5.3c, display the traffic noise measurements in the 1930s and the equipment used by the National Physical Laboratory (NPL) in the UK [11]. After the 1960s, the electro-acoustics and measurement techniques made great advances, thus increasing interest in the field surveys comprising the noise measurements along with the subjective tests. An example experiment is shown in Figure 5.4 for traffic noise conducted by the Transport and Road Research Laboratory (TRRL) in the UK in 1975. Figure 5.5 shows an earlier example of aircraft noise

(a)

(b)

Figure 5.1 Physical models for visualization of sound wave propagation (a) Modeling sound waves simulated as water waves to explain the interference field produced by two identical point sources [7], (b) Set- up of Ripple Tank first used by Michel (1921) and Davis Kaye (1927), for auditorium acoustics modeling (Scale 1:50) [*Sources:* Berry,2006 and Rindel, 2019: https://www.youtube.com/embed/ES-lWISLg3A].

Figure 5.2 Riverbank Acoustical Laboratory, constructed in 1919, and the Sabine Museum in the USA. (*Source:* photo taken by S. Kurra in 1998.) (See also Figure 5.55.)

Figure 5.3a National Physical Laboratory (NPL) in the UK and measurement of vehicle noise in 1930 [1, 11]. (*Source:* permission granted by B. Berry.)

Figure 5.3b Equipment set for measurement of vehicle noise used at NPL in the UK, 1930 [11]. (*Source:* with permission of B. Berry.)

Figure 5.3c Noise measurement and analysis systems used in the 1930s at NPL in the UK [11]. (*Source:* with permission of B. Berry.)

measurement in 1978 by using the heavy equipment carefully located on the site [12]. Since the equipment did not have internal memory and sufficient storage capacity, all the analyses had to be done in situ and the sound signals were traced by graphics obtained in the field using paper rolls, marking by pencil, etc. As described later in this chapter, the evolution from analogue to digital

Figure 5.4 Noise emission measurements and the annoyance survey conducted at TRRL in the UK, in 1975. (*Source:* photo: S. Kurra.)

Figure 5.5 Measurement of aircraft noise at Yesilkoy airport in Turkey in 1978 [12]. (*Source:* photo: S. Kurra.)

techniques was a great step in acoustical measurements facilitating signal receiving, recording and analyses.

Over the years, the technical specifications of the measurement equipment were standardized, the methodologies of measurements to be conducted in the field and in the laboratory were clarified according to the type of source, describing the general principles of measuring the basic quantities. Advances in measurement techniques up to the present time can be highlighted as follows [13]:

- Increased measurement accuracy and consistency.
- Reduced number of people required to take measurements and for data processing.
- Facilitated measurement task in a better way, e.g. with touch screens.
- Improved efficiency of the portable instruments, providing more possibilities for data receiving and analyzing.
- Unattended and remote-controlled measurements possible.
- Increased flexibility of measurement applications.
- Increased number of measurement parameters measured simultaneously and in terms of different metrics.
- Increased measurement periods of the portable (hand-held) instruments with increased battery life for outdoor measurements.
- Reduced weight of the instruments after using transistors.
- Portable instruments enabled post-processing with a laptop.
- Delivery of the temporal and spectral information and results of statistical analyses as colored graphical representations instead of typewritten reports, even using portable devices.
- Improvement in the quality of the display.
- Flexibility in use by means of the modular devices with their own software according to the user's need.
- Unlimited storage capacity for all recordings and the processed data, compared to the sound level meters that were capable of storing data for only a few hours.
- In monitoring noise, much faster transfer without errors in data transmission, compared to earlier remote connections with dial-up modems and telephone lines.

Virtual techniques like acoustic holography were developed at the end of the twentieth century and are widely employed for source identifications in industry at present. Eventually, it is noted that the problem of the intangibility of sounds has been overcome by means of the modern visual and audial techniques. However, it should be stressed that the microphones and other instrumentation can not fully represent the hearing mechanism, hence, the rating of environmental noise based on the objective measurements also needs careful subjective evaluations; as stated, "After all, it is the biological analysis system that will be the final judge of our success or failure."

Besides the great innovations in noise measurements, the scientists, technicians, manufacturers, etc. have been pursuing new ideas, such as low-cost acoustic signal receivers, new area-based monitoring with a wireless sensor network, web-based applications, smart devices, software applicable on multiple platforms, converting smart phones into a sound level meter with greater accuracy, etc.

5.2 Objectives of Noise Measurements

Acoustics measurements were categorized in the earlier documents, with respect to the following aspects [3, 14–17]:

1) Physical quantities to be measured (sound pressure and its variation, particle velocity, energy density, etc.).
2) Measurement parameters: amplitude, frequency, phase, modulation, etc.
3) Practical use of results:

- physical evaluations (source identification, sound path analysis, material properties, evaluation of acoustic field)
- biological evaluations (determination of negative impacts in terms of physical, physiological, psychological, and performance effects).

Regarding environmental noise, acoustic measurements are conducted at the noise source and the propagation path to obtain quantified data for various purposes. For example, studies on the impact of noise on individuals and the community need physiological and psychological evaluations based on data derived from the measurements of actual noise conditions. In general, acoustic measurements are important for rating noise pollution, for protecting human hearing, improving living quality, or implementation of noise control, e.g. physical planning. Figure 5.6 outlines the major objectives of noise measurements [18].

Measuring methodologies, instrument properties, recording, storing and analyzing techniques, evaluating outputs, all have been described in the international standards as well as in some national standards pioneering in specific topics, like those of the American Society for Testing and Materials (ASTM) (founded in 1898) in the USA. This chapter will emphasize the general principles recommended by the International Standards Organization (ISO) (1947), since most countries have accepted so-called ISO standards, which are revised every five years.

The context of noise problems has a great variability. Different noise conditions, noise sources, physical environments necessitate different approaches for managing data according to the objectives. Therefore, it is difficult to find a simple method applicable for all situations. Due to a number of factors influencing the field measurements, accuracy and repeatability of the results are important. The measurement errors defined as uncertainty can be reduced by selecting the appropriate measurement system, methods, units, and descriptors. The noise measurements can be grouped with respect to the objectives given below [2, 6, 14–17]:

1) *Hearing and perception measurements:* Audio tests for hearing risks loudness measurements and other psychoacoustic tests for annoyance and speech communication to be conducted in a laboratory.

2) *Noise measurements and analyses:* Targeting noise control for both sources and community, acoustic measurements can be grouped as;
   - sound radiation of noise source (i.e. emission measurements for acoustical definition of noise sources);
   - sound propagation in the environment (i.e. immission measurements aiming to investigate environmental noise conditions and the effects of noise on people, the determination of noise abatement measures and rating their performance, etc.).

3) *Measurements of building performance against noise, or sound insulation measurements:* Mainly involved in architectural acoustics, however, the following parameters are of importance also in noise control:
   - reverberation times of indoor spaces;
   - absorption coefficients of surface materials;
   - airborne and structure-borne sound reductions of building elements and the rating of insulation.

4) *Performance measurements of noise abatement devices* (noise barriers, silencers, etc.) both in the field and in the laboratory.

Investigations on environmental noise may require almost all the types of measurements mentioned above, according to the scale of the problem and the intended use of the results. The

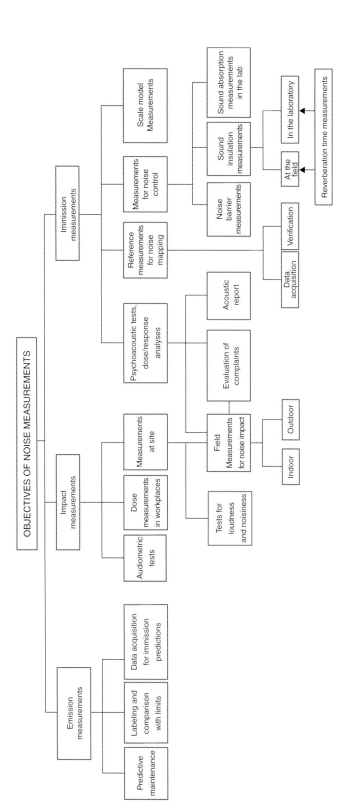

Figure 5.6 Objectives of noise measurements [18].

measurements given in items 3 and 4 are conducted both in laboratories and in-situ (actual conditions). As more complicated type of sounds compared to pure tone sounds and musical sounds, the noises are measured and rated by using specific metrics, indexes, and descriptors that were discussed in Chapter 2. In addition to noise measurements, the mechanical vibrations, occurring due to noise sources, are measured in the field by using different units and specific techniques, however, they are not included in this chapter.

The concepts of emission and immission were introduced into the vocabulary of environmental acoustics a long time ago, by the international consensus with the contribution of the Institute of International Noise Control Engineering (INCE).

5.3 Basic Acoustic Measurements

Acoustical measurements are conducted for the quantification of the basic acoustic parameters explained in Chapter 1:

- sound pressure
- sound power and sound intensity
- particle velocity
- sound speed
- sound frequency
- loudness.

In practice, all the basic quantities above are used in noise management except sound speed and particle velocity, which are included in theoretical acoustics, building acoustics, and mechanical vibrations. The sound pressure level and the sound power level are the most common units which are directly related to environmental noise assessment and noise management.

5.3.1 Measurement of Sound Pressure

Since the human ear and the man-made sensors like the microphones receiving sound signals are sensitive to sound pressure caused by motion of particles of the medium (see Chapter 1), the measurement of sound pressure facilitates the evaluation of noise problems. As will be seen below, sound power and the sound intensity of a noise source are obtained through sound pressure measurements. In environmental acoustics, there are numbers of objectives for measuring sound pressure levels, such as investigating the variation of sound pressures according to the operational and physical properties of the sound sources, analyzing the temporal and spatial variations of sound pressures (i.e. for immission measurements), investigating the effects of noise on people and to develop noise control technologies. The measurement of sound pressure level was standardized by the American National Standards Institute (ANSI) in the 1970s as the earliest standard (revised in 2005) aiming to determine the sound power levels of products [19a, 19b]. It was followed by the ISO standards to measure sound pressure levels of various noise sources. Different methodologies that are applied to environmental noise sources the American National Standards Institution (ANSI) are given in Section 5.5.

5.3.2 Measurements of Sound Power and Sound Energy

Since sound pressure varies with the source directivity, in addition to the physical factors in the environment, the primary acoustic characteristics of a source are represented by the acoustic power (i.e. sound energy radiated by the source), which is independent of the effects of the test environment. However, sound power cannot be obtained directly but is determined by means of:

a) sound pressure level measurements
b) intensity measurements
c) vibration measurements.

   Different standards describing the procedures of measurements under specified conditions are given in Table 5.1. For most noise sources, i.e. machinery and equipment -if sound power levels are not available-, the sound pressure levels measured at reference points are used as representative value of noise emission particularly in environmental noise problems.

Table 5.1 International standards regarding noise emission measurements in terms of sound power level, sound energy level, and sound pressure level as emission descriptor.

		ISO standard	
Noise source	Content of the standard	Laboratory	Field
Machinery and installation	Precision method in reverberation test room	ISO 3741 [25a, 25b]	—
	Engineering method in reverberation rooms (comparison method)	ISO 3743-1 [26a]	—
	Engineering method in special reverberation test rooms	ISO 3743-2 [26b]	—
	Engineering method for emission measurement of machinery and equipment in free field over a reflective plane	ISO 3744 (free field over a reflective plane) [27b]	ISO 3744 (free field over a reflective plane) [27b]
	Precision method in anechoic rooms or semi-anechoic rooms	ISO 3745 [28]	—
	Observation (survey) method in the field (enclosed space) and machinery used in open air over reflecting plane	—	ISO 3746 [29]
	Engineering/survey methods in situ (in reverberant field)	–	ISO 3747 [30]
	Intensity method in the lab and field for sound power measurements	ISO 9614-1 ISO 9614-3 [38a, 38c]	ISO 9614-2[38b]
	Determination of sound powers from vibration measurements	—	ISO 7849 [49a, 49b]
	Presentation of the data and accuracy	ISO 12001 [31]	
	Statistical methods for noise emission value of machinery	ISO 7574 [145a–d]	
Appliance noise	Laboratory measurements for noise emission of appliances	ISO 3822 [90a–d]	—

(Continued)

Table 5.1 (Continued)

Noise source	Content of the standard	ISO standard	
		Laboratory	Field
Industrial building and open-air activity	Determination of total emission level of an industrial area including building	—	ISO 8297 [53]
	Measurement of emission values at work station	—	ISO 11204 [50a, 50b]
Blasts	Measurement of sound energy from muzzle-blast of weapons and mines, with directivity	—	ISO 17201 [128a, 128b]
Motor vehicles	Emission measurements for road traffic vehicles (engineering method)	—	ISO 362-1 and 362-2 [57a, 57b]
	Emission measurements for stationary road traffic vehicles (observation method)	ISO 362-3 [57c]	ISO 5130 [64a, 64b]
	Noise emitted by passenger cars under urban driving condition	—	ISO 7188 [60]
	Measurements for M and N categories at standstill or low speed (for certification)	—	ISO 16254 [61]
	Measurement of sirens	—	ISO 6969 [71]
Railway vehicles	Measurement of emission levels of railway vehicles	—	ISO 3095 [75] (on a particular track)
	Warning devices measurements	—	EN 15153-2 [77]
Aircrafts	Immission measurements near airports	—	ISO 3891 [79] withdrawn
	Emission certification measurements	—	ICAO Annex 16 [80]
	Measurement of sonic boom	—	ISO 2249 w. BS 5331 w. [82]
Waterway vehicles	Measurements for vessels	—	ISO 2922 [83b]
	Measurement of noise on board	—	ISO 2923 [88]
	Measurement for recreational small craft	—	ISO 14509 [87]
Construction vehicles	Measurements at operator places, limits and accuracy	—	ISO 4872 (revised by 3744:2010) ISO 6396
		—	ISO 6395 ISO 6393
	Measurements for earth-moving machines		ISO 6394

The sound power determination through sound pressure level measurements can be conducted either:

1) In the laboratories where the environmental factors are eliminated and sound propagation conditions are controllable, such as anechoic/semi-anechoic rooms that are free from reflections (the free-field method) or in reverberant/semi-reverberant rooms where the diffuse reflections exist (reverberant-field method). The required properties of the acoustical test environments are given in the recent standard ISO 3740:2019 [20].

2) In the actual site (open or closed) where the source is operating, however, it is necessary to correct the results according to the physical conditions in the field.

Different methodologies to implement the above techniques have been reviewed and discussed in a number of publications since the 1950s [6, 17, 21, 22]. The technical standards describe the sound power determination, based on measurements involved with different types and sizes of noise sources, emitting all types of noise (steady, non-steady, fluctuating, isolated bursts of sound energy, etc.) [23–30]. The International Standard Organization published series of standards (ISO 3740 series) on sound power measurements and categorized the measurement methods based on the accuracy grades, as defined: (i) Precision method; (ii) Engineering method (applicable both in the laboratory and in the field with the restricted conditions); and (iii) Survey method in the field under actual conditions (in situ).

ISO 3740:2019 gives a guideline for the use of basic standards for the determination of sound power levels through measurements and facilitates selection of the appropriate standard for implementation according to the objective [20]. Below the laboratory methods are summarized and the methods to be applied in the field are discussed in Section 5.4.

Sound Power Determinations in Laboratories

Reverberation room methods: The diffuse sound field conditions and the other properties of reverberant rooms to be used for sound power measurements, were explained in the earlier documents published in the 1950s and the 1960s. The American National Standards Institute (ANSI) issued a standard on sound power measurement in reverberation rooms in 1962, preceding the international standards published by the International Standards Organization (ISO) in the 1990s [23]. Later the American Society for Testing and Materials (ASTM) recommended a different methodology which was applicable also in the field in 1997, as given below.

ASTM E 1124-97 (1997 republished in 2016): As one of the primary technical standards regarding acoustic power measurements both in a reverberant room and in the field, the ASTM standard proposes the double-surface method for sound powers of machine and equipment in workspaces [24]. Two measurement surfaces are assumed to be parallel to the reference surface enveloping the source. The surfaces can be a rectangular prism or a cylindrical semi-sphere whose surface areas should be easily computed. The microphones are placed on the outer and inner surfaces as shown in Figures 5.7a–5.7c. Simultaneous measurements on each surface at octave bands are made also with two microphones mounted on a bar or with a microphone boom rotating in a circular motion.

Figure 5.7a Double skin measurement according to the earlier standard (ASTM E 1124-97) [24].

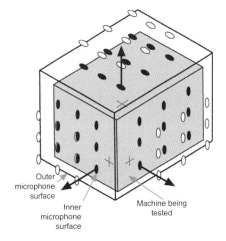

Outer microphone surface

Inner microphone surface

Machine being tested

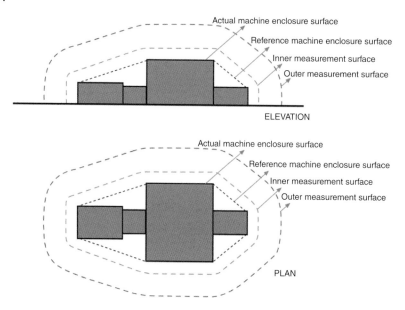

Figure 5.7b Measurement surfaces according to the earlier standard (ASTM E 1124-97) [24].

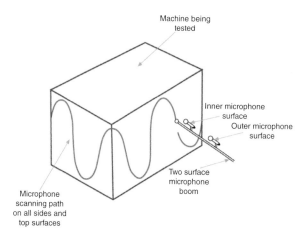

Figure 5.7c Microphone path in two surface measurement method according to the earlier standard (ASTM E 1124-97) [24].

The sound power level is calculated from the difference between the recorded sound pressures at each point and the average values, by also taking into account the effects of the reverberant field and the nearby equipment surfaces. However, when large reflectors are present in the enclosed spaces, this model cannot be implemented.

ISO 3741 first issued in 1999, revised in 2010 and confirmed in 2015, describes a precision method with accuracy grade 1 according to ISO 12001:1996 (confirmed in 2017) [25a, 25b, 31] and with the approximation that sound power and sound energy are directly proportional to the mean-square sound pressure averaged in space and time in the test room. The sound power level and the energy level for the sources generating time-varying sounds are determined in 1/3 octave bands, from which the octave band and A-weighted levels can be computed.

The criterion for the background noise in the reverberant room is specified in the standard. The properties of the reverberation test room in terms of volume and shape, in relation to the frequency range of interest, the required sound absorption (defined as $T_{60} = (V/3)$ where V is the total volume of the test room and S is the total surface area), the required diffuse field conditions, the criteria for background noise in relation to the frequency bands and the requirements for the meteorological conditions (temperature, humidity, and atmospheric pressure) are given in the standard [25a, 25b]. The principles for locating the source in the room, the installation and mounting conditions, the auxiliary equipment, the operating conditions during the tests, are also described in the standard. Different source positions and the minimum number of microphone positions are suggested to obtain the time- and space-averaged sound pressure levels in the test room. The microphones should be located in the room with a minimum distance from the source which is determined according to the room volume and reverberation time. The arrangement of the fixed and moving microphones is based on the pre-measurements and in relation to source type and size, the wavelength of the sound, and the directivity pattern of the source radiation. This is important especially for the sources emitting discrete frequency components.

The frequency range of the measurements is 100–10 000 Hz in one-third octave bands. The standard explains two methods:

1) The direct method using the equivalent sound-absorbing area of the test room.
2) The comparison method using a reference sound source.

The common procedure for measuring sound pressure levels in both methods is given below.

Direct method: In the former standard ISO 3741:1999, which could be conducted in free or semi-free rooms whose acoustic conditions were known, the measurement and calculation procedures were rather simple. After the volume and reverberation time of the test room are determined, the sound pressure levels were measured at a hypothetical surface in each frequency band and the levels were simply averaged and called "space-averaged sound pressure levels." The sound power level of the test source could be obtained by applying the following formula, at each frequency [14, 25a]:

$$L_w = \overline{L_p} + 10 \lg V - 10 \lg T_{60} + 10 \lg [1 + (S\lambda/8V)] - 3.5 \quad \text{dB} \quad \text{re}10^{-12} \, \text{Watt} \qquad (5.1)$$

L_w: Sound power level of the source at each frequency, dB
$\overline{L_p}$: Average sound pressure level measured in the reverberant room with the source, dB
λ: Wavelength of the sound for the concerned frequency band, m
V: Volume of the room, m^3 (excluding the volume of the source)
S: Total absorption area of the room, m^2
T_{60}: Reverberation time of the room, s

In this early standard, if the condition of background noise is not met during the measurements, a simple correction was proposed to the measured level. However, the revised standard proposes a more detailed procedure as given below [25b]:

1) Measured time-averaged sound pressure levels, $L'_{pi}(ST)$ at each microphone position at each one-third octave band which are averaged over the number of source positions.

$$L'_{pi \, (ST)} = 10 \lg \left\{ \frac{1}{N_S} \sum_{j=1}^{N_S} 10^{0.1 \left[L'_{pi \, (ST)} \right]_j} \right\} \text{dB} \qquad (5.2)$$

$\left[L'_{pi\,(ST)}\right]_j$: Time-averaged one-third octave band sound pressure levels at ith microphone and at jth source position (while the source is operating), dB

N_S: Number of source position

2) Correction for background noise: After the test, the background noises are measured at the same microphone position, $L_{pi\,(B)}$ and the measured levels L_{pi} which are obtained for i^{th} microphone position or i^{th} microphone traverse, are corrected at each third octave band:

$$K_{1i} = -10\lg\left(1 - 10^{-0.1\Delta L_{pi}}\right)\ \text{dB} \tag{5.3}$$

K_{1i}: Background noise correction, dB

$$\Delta L_{pi} = L'_{pi\,(ST)} - L_{pi(B)} \tag{5.4}$$

Practically if $\Delta L_{pi} \geq 15$ dB, K_{1i} is assumed to be zero.

3) The corrected measured levels at each microphone are:

$$L_{pi\,(ST)} = L'_{pi\,(ST)} - K_{1i} \tag{5.5}$$

4) Calculation of mean time-averaged sound pressure level in the test room by taking average of the $L_{pi\,(ST)}$ over the number of microphone positions (N_M):

$$\overline{L_{p(ST)}} = 10\lg\left\{\frac{1}{N_M}\sum_{i=1}^{N_M} 10^{0.1\,L_{pi\,(ST)}}\right\}\text{dB} \tag{5.6}$$

N_M: Total number of microphone positions

$L_{pi\,(ST)}$: The corrected one-third octave band sound pressure level at the ith microphone position or microphone traverse

$\overline{L_{p(ST)}}$: The mean corrected one-third octave band time-averaged sound pressure level in the test room while the source is activated

5) Calculation of sound power levels based on the principle that the average sound pressure levels are constant and related to sound power in a reverberant field:

$$L_W = \overline{L_{p(ST)}} + \left\{10\lg\frac{A}{A_0} + 4.34\frac{A}{S} + 10\lg\left(1 + \frac{Sc}{8Vf}\right) + C_1 + C_2 - 6\right\}\text{dB} \tag{5.7}$$

A: Equivalent absorption area, m^2 of the reverberation room

S: Total surface area of the reverberation test room

$A_0 = 1\ \text{m}^2$

c: Speed of sound in the reverberation test room conditions

V: Volume of the reverberation test room

f: One-third octave band frequency, Hz

C_1: Reference quantity correction, dB as a function of characteristic impedance of the air [14]

C_2: Radiation impedance correction, dB to change the actual sound power into the power under the reference meteorological conditions [25b]

Comparison method: Sound power measurements are conducted in the diffuse field (i.e. the reverberant room) by using a reference source whose power is known at each frequency band. The acoustical characteristics of the reference source, e.g. an omnidirectional source, are specified in the standard. The sound pressure levels are measured at the same microphone positions as for the test source, then the sound power levels at each one-third octave bands are calculated according to Eq (5.8) [15a, 15b, 26a]:

$$L_W = L_{W(RSS)} + \left(\overline{L_{p(ST)}} - \overline{L_{p(RSS)}} \right) + C_2 \tag{5.8}$$

$L_{W(RSS)}$: One-third octave band power level of the reference source, dB

$\overline{L_{p(ST)}}$: The mean corrected one-third octave band time-averaged sound pressure level in the test room from the noise source under test, dB

$\overline{L_{p(RSS)}}$: The mean corrected one-third octave band time-averaged sound pressure level in the test room from the reference noise source under test, dB

C_2: Radiation impedance correction, dB to change the actual sound power into power under the reference meteorology conditions same as in Eq (5.7) [25b].

Sound events: For the sources emitting noise in a certain operation period (in a cycle) or for intermittent or impulsive noises, the sound energy level, L_E is determined and integrated in an operation cycle time or event period (from $t = 1$ to $t = T$) to obtain single event, time-integrated, sound pressure level, $L'_{Ei(ST)}$), at each microphone position (i) for each source position (j) [26b]:

$$\left[L'_{Ei(ST)} \right]_j = \left[L'_{Ei,N_e(ST)} \right]_j - 10 \lg N_e \mathrm{dB} \tag{5.9}$$

$\left[L'_{Ei(ST)} \right]_j$: the measured uncorrected one-third octave band single event time-integrated sound pressure level at the ith microphone position and at the jth source position (while the source is operating), dB

N_e: The number of the sound events

After correction for background noise applying the same procedure given above, ultimately the mean corrected one-third octave band single event, time-integrated sound pressure level in the test room is calculated for the noise source under test in operation.

The standard describes the calculation of the measurement uncertainty (Section 5.6).

Anechoic room methods: The free-field model for sound power level measurements requires open air or an echo-free room, where the acoustic and geometric properties comply with certain acoustical principles [6, 14, 22, 32]. However, the standards explained below are applicable both in the free-sound field and the free-sound field over a reflective plane, as produced in anechoic and semi-anechoic rooms. Some example of different type and size of anechoic rooms are shown in Figures 5.8a–5.8c. The history of the anechoic chambers whose inner surfaces are covered with large fiberglass wedges, goes back to the early 1940s. The earliest example was constructed by Beranek in the USA for sound power measurements of noise sources located on the rigid floor in the anechoic room. Since then, various standards have been published on the measurement methodologies in anechoic rooms. Determination of the sound power level in this method is also based on measuring sound pressure levels. An earlier measurement scheme for sound power measurement in an anechoic chamber (1987) is shown in Figure 5.9 [18].

The earlier documents, published in 1970s, assumed a hypothetic measurement surface enclosing the source, like a hemisphere, rectangular prism or a multi-façade prism (i.e. an icosahedron) in the far field, as shown in Figures 5.10a–5.10c [14, 15a, 24]. The number of microphone positions to be

Figure 5.8a Noise emission measurement in an anechoic room. (*Source:* photo: S. Kurra in 2015.)

Figure 5.8b Aircraft noise emission measurements in a large anechoic room. (*Source:* EMC Testing solutions; photo by T.Powel, US Air Force.)

located on the measurement surface was determined according to the required precision and the directivity of the source. Determination of the total radiated sound power was made by space integration of the sound intensity over the measurement surface. The measured sound pressure levels (L_{pi}) on the microphone points are converted to p_{rms} values and multiplied by partial surface areas S_i (normally equal size), then the average sound pressure is obtained as below [14]:

$$S_i \left(p_i^2 / p_{ref}^2 \right) = S_i 10^{Lpi/10} \tag{5.10}$$

Figure 5.8c Small-anechoic room. (*Source:* IAC acoustics A/S.)

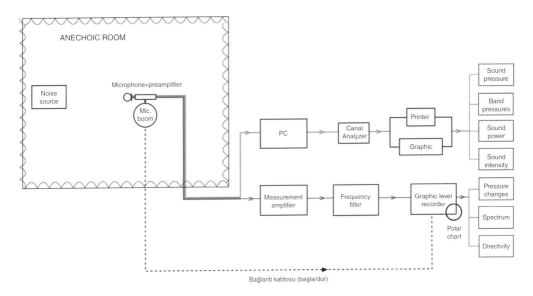

Figure 5.9 Procedure for noise emission measurement in an anechoic room [18]. (*Source:* from an earlier study by S. Kurra, 1987).

$$\left(\overline{p_H}/p_{ref} \right)^2 = \left\{ \sum S_i \left(p_i/p_{ref} \right)^2 \right\} / 2\pi r^2 \quad \text{dB} \tag{5.11}$$

$$\overline{L_{pH}} = 10 \lg \left(\overline{p_h}/p_{ref} \right)^2 \quad \text{dB} \tag{5.12}$$

(a)

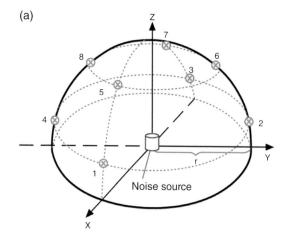

Mic. position	X	Y	Z
1	0.97r	0	0.25r
2	0	0.97r	0.25r
3	−0.97r	0	0.25r
4	0	−0.97r	0.25r
5	0.63r	0	0.78r
6	0	0.63r	0.78r
7	−0.63r	0	0.78r
8	0	−0.63r	0.78r

(b)

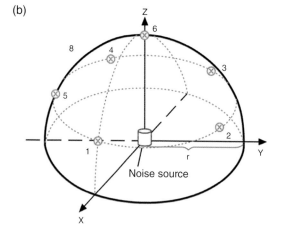

r: Radius of hemisphere
⊗ Microphone position

Mic. position	X	Y	Z
(1)	0.89r	0	0.45r
(2)	0.28r	0.85r	0.45r
(3)	−0.72r	0.53r	0.45r
(4)	−0.72r	−0.53r	0.45r
(5)	0.28	−0.85r	0.45r
(6)	0	0	r

Figure 5.10a Sound power measurement with semi-sphere measurement surface (ASTM E 1124-97) (the sphere method) (a) with eight measurement points; (b) with six measurement points [15a, 24]. (*Source:* with permission of B&K.)

p_i: Measured sound pressures, N/m²
p_{ref}: Reference sound pressure, $2 \times 10{-}5$ N/m²
$\overline{p_h}$: Spatial average of sound pressure levels, p_i measured at microphone positions
S_i: Area of the partial test surfaces, m²
$\overline{L_{pH}}$: Space-averaged sound pressure level, dB
r: Diameter of the test semi-sphere, m

The sound power level in each frequency band L_w is calculated from the space-averaged sound pressure level, for a semi-spherical space in an anechoic room:

$$L_w = \overline{L_{pH}} + 20 \lg r + 10 \lg 2\pi \quad \text{dB} \tag{5.13}$$

Figure 5.10b Sound power measurement with rectangular prism measurement surface as described in ASTM E 1124-97) (the box-method) [15a, 24]. (*Source:* with permission of B&K.)

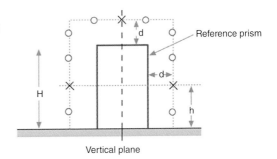

Vertical plane

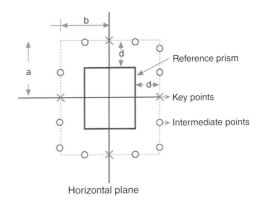

Horizontal plane

d(m)	d(m)
≥ 0.25	1
< 0.25	4L ≤ d ≤ 1
	d > 0.25

L : Max linear dimension of machine
h : H/2 (<0.25 m)
X : Key measurement points
O : Intermediate points at 1 m from key points

ISO 3744:1995, describing the engineering method with an accuracy of grade 2, does not require a special laboratory environment and applies the measurements in the free-field or in open air conditions with the source located over a reflecting plane both outdoors and in large rooms [27a]. The standard introduces "a reference box which is a hypothetical surface defined by the smallest right parallelpiped that just encloses the source under test." The surface on which the measurements are conducted is described as hemispherical or rectangular surfaces with the size depending on the dimensions of the machines and the relations to the nearby reflective walls, the subdivisions of the surface, the positions of the key microphones, and the measurement paths are given in the standard (Figures 5.11a–5.11f). Calculation of the average sound pressure levels needs to take into

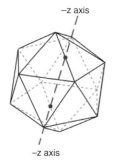

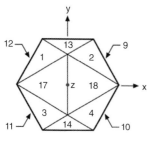

Positions of midpoints of sectors			
Sector numbers	Position		
	x/r	y/r	z/r
1–8	± 0.577	± 0.577	± 0.577
9–12	± 0.934	± 0.357	0
13–16	0	± 0.934	± 0.357
17–20	± 0.357	0	± 0.934

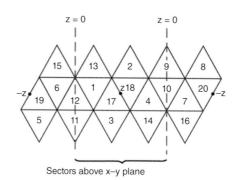

Top view of icosahedron

Sectors above x–y plane

Figure 5.10c Sound power measurement on the multi-façade prism [14].

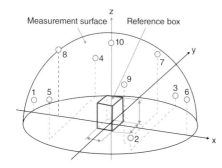

Figure 5.11a Key microphone positions on the hemispherical measurement surface for a broadband noise source: located on a reflective plane (ISO 3744:1995) [27a].
Note: Figures 11a–11f are reproduced with the permission of the International Organization for Standardization, ISO. This standard can be obtained from any ISO Member and from the website of the ISO Central Secretariat at the following address: www.iso.org. Copyright remains with ISO.

account the corrections for background noise and reflected sounds in the room to be determined by room absorption:

$$\overline{L_{pf}} = \overline{L'_p} - K_1 - K_2 \tag{5.14}$$

$\overline{L_{pf}}$: Average surface sound pressure level that can be given as A-weighted or frequency band, dB
$\overline{L'_p}$: Average A-weighted or frequency band measured noise level (not corrected)
K_1: Correction for background noise, dB
K_2: Correction for reflected sound, dB

$$K_1 = -10 \lg \left(1 - 10^{-0.1\Delta L}\right) \text{ dB} \tag{5.15}$$

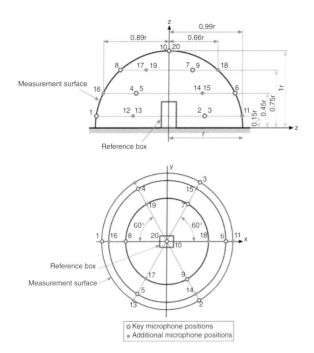

Figure 5.11b Microphone positions on the hemispherical measurement surface for a broadband noise source: side and top views (ISO 3744:1995) [27a].

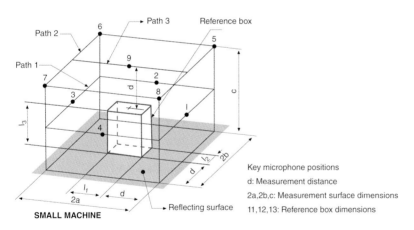

Figure 5.11c Example of a measurement surface for a small machine (ISO 3744:1995) [27a].

$$\Delta L = \overline{L'_p} - \overline{L''_p} \tag{5.16}$$

$\overline{L'_p}$: Average A-weighted or frequency band measured noise level (not corrected)

$\overline{L''_p}$: Average A-weighted or frequency band background noise level

$$K_2 = 10 \lg [1 + 4(S/A)] \quad \text{dB} \tag{5.17}$$

S: Area of the measurement surface, m^2

A: Equivalent sound absorption area of the room, m^2

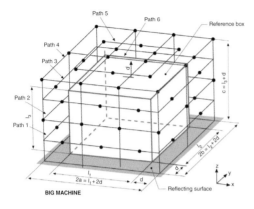

Figure 5.11d Example of measurement surface for a large machine (ISO 3744:1995) [27a].

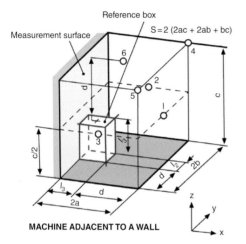

Figure 5.11e Parallelepiped measurement surface with six microphone positions for floor standing noise sources, adjacent to one reflecting plane (ISO 3744:1995) [27a].

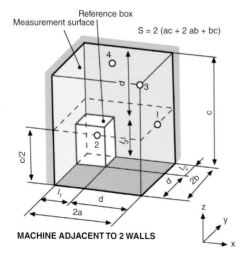

Figure 5.11f Parallelepiped measurement surface with six microphone positions for floor standing noise sources, adjacent to two reflecting planes (ISO 3744:1995) [27a].

Different methods recommended for *S/A* in the standard are the approximate method, the reverberation method, or the two-surface method.

Calculation of sound power level:

$$L_W = \overline{L_{pf}} + 10 \lg \left(\frac{S}{S_0} \right) \text{ dB} \tag{5.18}$$

S: Area of the measurement surface, m^2
S_0: 1 m^2

ISO 3744:2010 (confirmed in 2015) covers more details also defining the cylindrical measurement surface and the surface single event, time-integrated, sound pressure level determinations as described in the reverberation room method (see above) [27b].

ISO 3745:2012 describes a precision method with an accuracy grade of 1, for sound power measurements in anechoic and semi-anechoic rooms for noise sources radiating continuous and intermittent noises [28]. The sound power levels and the sound energy levels are calculated, based on the surface time-averaged sound pressure levels and the surface single event, time-integrated, sound pressure levels for respective source types (as given in the reverberation room method). Absolute acoustic criteria for the background noise levels in the test room are specified as 6 and 10 dB lower than the noise level of the source according to the frequency band. Correction of the background noise is the same as in ISO 3741:2010 (Eq 5.3) [25b].

The geometrical features of the spherical and hemispherical measurement surfaces to be positioned on the acoustic center of the source under test are described in the standard. The array of fixed microphone positions, with the number increased, is given on both the spherical measurement surface (in free field) and the cylindrical measurement surface, as partly shown in Figures 5.12a and 5.12b. The segments with equal areas on the measurement surface are considered while obtaining the surface time-averaged sound pressure levels. Also, measurement of directivity of the test sound source is explained (Section 5.4.1).

For the power level calculation at each one-third octave band in the range of 100–10 000 Hz, the following relationship is given in Eq (5.19):

$$L_w = \overline{L_p} + 10 \lg \frac{S_1}{S_0} + C_1 + C_2 + C_3 \text{ dB} \tag{5.19}$$

$\overline{L_p}$: Surface time-averaged sound pressure level, dB (after being corrected according to background noise and averaged for the number of microphones)

S_1: Area of the spherical measurement surface (if the hemispherical measurement surface is used, then its surface is taken as S_2)

C_1: Reference quantity correction dB, as a function of the characteristic acoustic impedance of the air under the meteorological conditions at the time and place of the measurements

C_2: Acoustic radiation impedance correction, dB, to change the actual sound power relevant for the meteorological conditions at the time and place of the measurement into the sound power under the reference meteorological conditions

C_3: Frequency-dependent correction for air absorption, dB [28]

Guidance on the measurement uncertainties is given in the standard.

At present, computer programs are available to calculate the sound powers by using an interface which is easy to use (Figures 5.13a and 5.13b). The outputs can be obtained in terms of different quantities and presentation techniques (Figures 5.13c) as will be described in Section 5.11.3 [33].

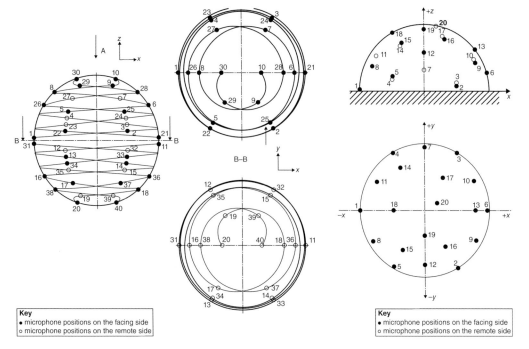

Key
• microphone positions on the facing side
○ microphone positions on the remote side

Key
• microphone positions on the facing side
○ microphone positions on the remote side

Figure 5.12a   The microphone arrays for anechoic room measurements using spherical and hemispherical measurement surfaces (ISO 3745:2012) [28].
Note: The same descriptions are also valid for ISO 3744:2010 for in-situ measurements [27b].
Note: Figures 5.12a and 5.12b are reproduced with the permission of the International Organization for Standardization, ISO. This standard can be obtained from any ISO Member and from the website of the ISO Central Secretariat at the following address: www.iso.org. Copyright remains with ISO.

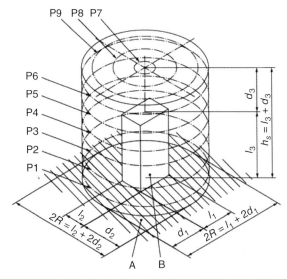

Key

A reflecting plane	d_3 measurement distance (height)	l_3 reference box height
B reference box	h_S height of the measurement surface	P1 to P6 side microphone paths
d_1 measurement distance (length)	l_1 reference box length	P7 to P9 top microphone paths
d_2 measurement distance (with)	l_2 reference box width	$2R$ measurement surface diameter

Figure 5.12b   The microphone arrays for anechoic room measurements using cylindrical measurement surface (ISO 3745:2012) [28].
Note: Same descriptions are also valid for ISO 3744:2010 for in-situ measurements [27b].

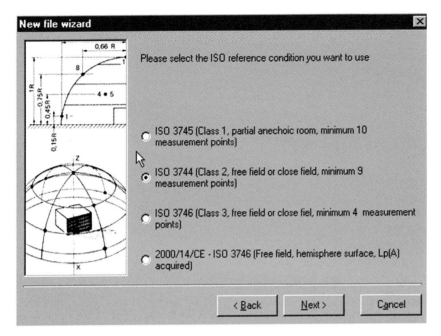

Figure 5.13a Definition of the geometrical parameters of the measurement surface, according to the standard method in the calculation software (01 dB-Metravib) earlier version.

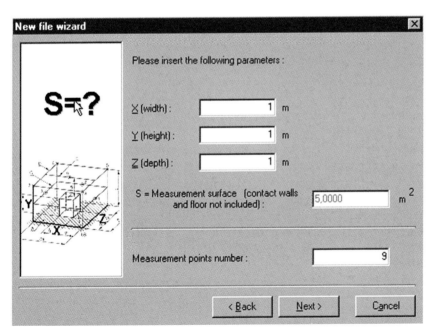

Figure 5.13b Insertion of the geometrical parameters of the measurement surface, according to ISO 3744, into the calculation software (01 dB-Metravib) earlier version.

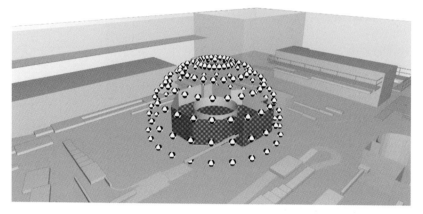

Figure 5.13c Combination of measurements (i.e. measured levels according to ISO 3744) and machine modeling for assessment of noise emission in workplaces. (*Source:* Datakustik.)

5.3.3 Measurement of Sound Intensity

Sound intensity is measured for various purposes, e.g. to determine the total acoustic power of a sound source, to investigate the partial contributions to the total sound radiation from secondary sources, and to apply noise control measures mostly for complex equipment and machinery. However, measurement of sound intensity is more complicated than measurement of sound pressure. Defined as sound energy in a unit area, sound intensity is a vector quantity and can be determined by two types of measurements as given in Section 1.2.2 [34–37].

1) Indirect determination: Using the same techniques as in sound power determination based on sound pressure measurement, a computation procedure is applied.
2) Direct determination: Using special intensity microphones and analyzers, intensity is measured directly.

The basic principle of intensity measurements is to measure the sound pressure (p) by means of two microphones spaced close to each other as a special device called a microphone probe. Particle velocity (u) is calculated based on the measured sound pressures at any instant of time as follows [34–37]:

$$u_r(t) = -\frac{1}{\rho}\int_{-\infty}^{t} \frac{p_2(\tau) - p_1(\tau)}{\Delta_r} d\tau \tag{5.20}$$

$u_r(t)$: Particle velocity at distance r from the source
p_1, p_2: Sound pressures measured by microphone 1 and 2 respectively, N/m^2
τ: Time variable
Δ_r: Distance between microphones, m
ρ: Density of medium, kg/m^3

Instantaneous sound intensity is obtained as follows:

$$I = \frac{1}{2\rho\Delta_r}\left\langle (p_2(\tau) + p_1(\tau))\int_{-\infty}^{t} (p_1(\tau) - p_2(\tau))d\tau \right\rangle_t \ \text{W/m}^2 \tag{5.21}$$

Due to the phase relationship between the sound pressure and particle velocity, the instantaneous active and reactive intensities are presented by the real and imaginary parts of the complex instantaneous intensity. The calculation above can be made by a two-channel FFT analyzer or a sound intensity analyzer (Section 5.11). Using the sound intensity, the sound power (P) and sound power level L_W of the source can be obtained, as given in Box 5.1.

Box 5.1   Determining Sound Power from the Sound Intensity Measurement

Sound power is determined by integrating time-varying products of intensity vectors and surface areas. As explained in the sound power standards, the measurement surface is defined as an imaginary continuous rigid surface enveloping the source under test. Most of the hard objects should remain outside of the measurement surface. The basic formulas are given below:

Instantaneous sound pressure:

$$\vec{I}(t) = p(t)\ \vec{u}(t) \tag{1}$$

Sound intensity:

$$\vec{I} = \lim_{T \to \infty} \frac{1}{T} \int_0^T \vec{I}(t)dt \tag{2}$$

$\vec{I}$: Time averaged sound intensity

Sound intensity level normal to the surface;

$$L_{I_n} = 10 \lg \frac{|I_n|}{I_0} \quad \text{dB} \tag{3}$$

$|I_n|$: Sound intensity of the normal component to the surface

$$I_0 = 10^{-12} \quad W/m^2$$

Partial sound power on a surface segment, S_i:

$$P_i = \overline{I_{n_i}}.S_i \tag{4}$$

$$\overline{I_{n_i}} = \vec{I}.\vec{n} \tag{5}$$

$\vec{n}$: Component normal to the surface
i: Index for the partial surface (segment)
S_i: Surface area of segment, i
n: Total number of the partial surfaces on the measurement surface

Sound power and power level:

$$P = \sum_{i=1}^{N} P_i \tag{6}$$

$$L_W = 10 \lg \frac{|P|}{P_0} \quad \text{dB} \tag{7}$$

P: Sound power, W
P_0: Reference Sound Power 10^{-12} W

Attempts to measure the sound intensity were initiated in the 1930s, however, the acquisition of reliable results was accomplished in the 1970s [36]. The international measurement standard was published in the 1990s.

ISO 9614 (Parts 1–3) (1993–2002, all confirmed in 2018): Different parts of the standard explain the proposed measurement techniques of intensity [38a–c]. In the direct method, the measurements are performed either by a probe microphone positioned consecutively at discrete reference positions all over the source surface or by scanning the entire surface as shown in Figure 5.14. The surface segments, scanning paths, and scanning durations are described in the standard. The uncertainty of the intensity measurements is due to the different geometrical configurations of the measurement surface, the differences between the equipment and measurement techniques, etc. and is determined in the one-third octave frequency bands depending on the limitations of the measuring equipment.

After the measurements are completed, the distribution of intensity levels around the source can be presented as two- or three-dimensional sound intensity maps. Such information is important in the localization of partial noise sources on the entire surface (Section 5.11.3). Some examples regarding source localization are given in Figure 5.15a and Figure 5.15b displays some outputs using different techniques [39]. Figure 5.15c shows an innovative technique called the "expanding scanning method" to characterize the sound energy distribution by using intensity probes and a 3D tracking system. [40]. The three-dimensional surface noise maps of a compressor are given in Figure 5.15d [33].

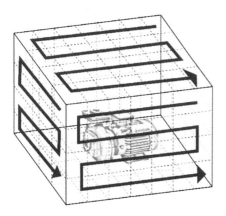

Figure 5.14 Scanning technique in sound intensity measurement for a noise source.

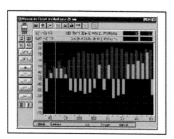
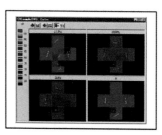
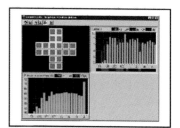

Figure 5.15a Outputs of sound emission measurements (1980) (01 dB-Metravib).

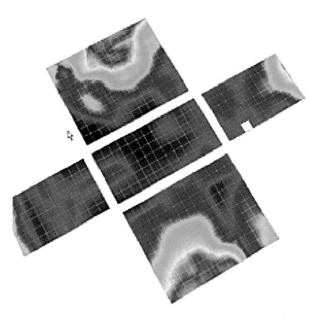

Figure 5.15b Sound mapping and source localization [39].

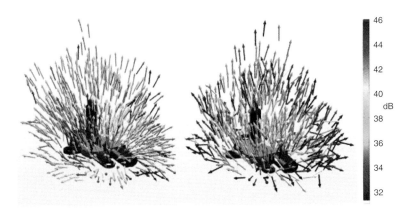

Figure 5.15c A new scanning tool for 3D sound intensity [40]. (*Source:* with permission of Comesana.)

For the sources radiating constant sound and for those with large surfaces, the intensity measurement is useful, as it enables the investigation of sound energy radiated from different parts of the source, so that the necessary precautions can be taken against excessive noise emissions. Also, the direct measurement of intensity is preferred extensively because:

- It can be conducted in a noisy environment under defined background condition (up to a certain extent).
- It does not require an anechoic room.
- It can easily be applied for large machines, AC units, etc.

This technique is also implemented for sound transmission losses of building elements particularly at low frequencies.

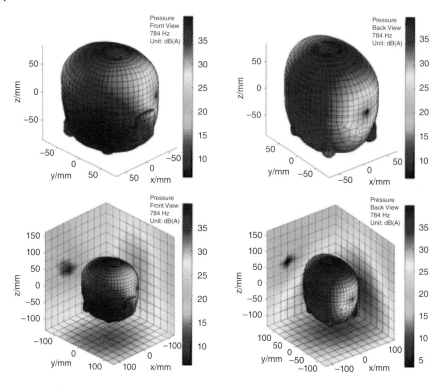

Figure 5.15d Example of surface sound pressure maps obtained from a research study [33]. (*Source:* with permission of Xu and Fagotti.)

5.4 Measurement Methods and Techniques to Acquire Emission Data for Environmental Noise Sources

This section describes the acoustical measurements to obtain emission and immission data for different type of noise sources with the main purpose of assessment and management of environmental noise.

5.4.1 Emission Measurements of Industrial Noise Sources

With respect to environmental noise management, the emission levels of specific noise sources, which can be quantified in terms of sound power, sound energy, and sound intensity, are required for the following purposes:

Product labeling: The emission levels (dB) of various noise sources, i.e. machinery and equipment, are declared in order to inform the public about the quality of the product they purchase, to ensure compliance with the national and international limits and to compare the noises of various products of the same kind. The EC Directive 2000/14/EC and the amending Directive 2005/88 have obliged the EU member states to conduct noise emission measurements for various machines and their parts, especially for equipment used outdoors, such as chain saws, concrete mixers, compressors, etc. The Directives describe the operating conditions for sources during the tests, e.g. working speeds, loads, and product specification and labeling to display the EU certificate for marketing purposes (Figures 5.16a and 5.16b) [41, 42]. The ISO 3744, explained above, is accepted as the basic standard for noise emission measurements in the Directive.

Figure 5.16a Product noise emission label indicating the guaranteed single-number power level as required in the EU Directive [42].

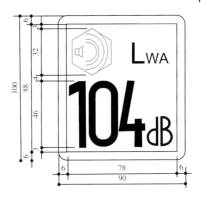

Figure 5.16b Emission measurement in industrial premises. (*Source:* dBKES Eng.)

Noise control at source: Determination of the excessive noise emission levels of specific sources is crucial for the technical measures, according to the source type and operations. Another purpose can be workplace noise assessments for hearing conservation. Directive 2006/42/EC, called the Machinery Directive, published in 2006, aims to protect "essential health and safety requirements relating to the design and construction of machinery" [43]. The Directive describes the health and safety requirements associated with the design and installation of machines so as to reduce the airborne sound emissions to the specified limits at workstations. For example, it is required to display the measured noise levels in workplaces exceeding:

- A-weighted emission sound pressure level of 70 dB(A)
- peak C-weighted instantaneous sound pressure level of 130 dB(C)
- A-weighted sound power emitted by the machine when the A-weighted emission sound pressure level at a workstation exceeds 80 dB(A).

Use of emission data for environmental noise predictions: The emission levels of environmental noise sources, either single or composite, are introduced into the noise prediction models, as sound power or energy levels and source reference levels (see Chapter 4). Noise impact studies need knowledge about the source emission data, which is also important to reveal the significance of source emissions while developing noise action plans (see Chapter 9).

Sound power measurements under defined conditions in laboratories are explained in Section 5.3.2. When the environmental noise sources are concerned, different methodologies have been developed due to the diversity and complexity of noise sources and the sound fields. Source emissions which are independent of environmental effects but dependent on physical, acoustical, and operational properties of noise sources, vary a great deal, necessitating different source-specific measurement techniques. The reference distances at which the sound pressures are measured change with the type, geometry, or size of the source. The basic principle is that the measurement points should be beyond the near field where the sound pressures can fluctuate a great deal and where the sound intensity is not directly related to the average square of sound pressures (to satisfy the inverse-square law) in the far field, before reaching the reverberant field (Chapter 1). Such a field depends on the dimensions of the source, the sound frequency, and the phase of partial surfaces radiating sound. The reference (measurement) distances for a particular source have been specified in different standards given in Table 5.1. During the emission measurements, the mounting, operational, and power conditions (at normal power, maximum power, and idling status) of the sources, such as mechanical and construction equipment, transportation vehicles, and all types of small products, should be reported.

Determination of the directivity pattern of noise sources: The directivity characteristics of sources, as seen in Section 1.3.2, implying variation of sound energy in different directions, are important in environmental noise levels. The directivity pattern displays the sound distribution at the far field of the source in two or three dimensional spaces. It is obtained under the free-field conditions or in a laboratory (an anechoic room) by applying a measurement technique referred to in ISO 3745:2012 (during or before measuring the sound power or intensity) and ultimately the directivity index is calculated at each third octave band by using the difference between the sound pressure level at each microphone position on the measurement surface (spherical or hemi-spherical) and the surface time-averaged sound pressure level [28]:

$$D_{Ii} = L_{pi} - \overline{L_p} \tag{5.22}$$

D_{Ii}: Directivity index for i^{th} microphone position
L_{pi}: Sound pressure level at the ith microphone position, dB
$\overline{L_p}$: Surface time-averaged sound pressure level, dB

Figure 5.17 Directivity chart measured in an anechoic room in 1980 [44].

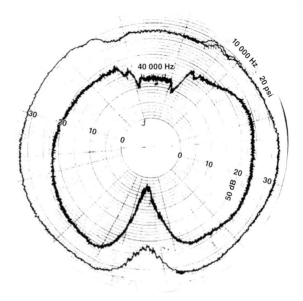

The directivity characteristics of noise sources should be considered in the propagation of noise in the environment, affecting the received noise levels as shown in the prediction models. An example of a directivity chart recorded on a polar sheet by using a graphic level recorder in the 1980s is shown in Figure 5.17 [44].

Source identification and localization: Based on the emission measurements for noise sources, it is possible to identify the individual sound radiating parts of composite noise sources, either small or large, mostly used in industry and transportation. Such analyses can be made for other purposes like predictive maintenance and noise abatement [45–48]. The subject is discussed in Section 5.11.3 in detail.

Table 5.1 gives the recent available ISO standards, as of 2019, regarding emission measurements for different type of noise sources.

Below the specific emission measurement techniques are summarized for various common noise sources.

Emission Measurements of Industrial Noise Sources in the Field

In addition to the laboratory measurements for machinery and equipment/installation to determine the sound power and energy levels, that are explained in Section 5.3.2, field measurements are conducted within the industrial plant or in the open air under the actual conditions. These techniques are rather useful for researchers and environmental planners as well as for acoustical consultants and owners of the plants, when the laboratory data is not available. On the other hand, when treating the entire factory or industrial premise as an extended or composite noise source in environment, a different approach is necessary to obtain the emission levels.

While the measurements are conducted in enclosed spaces, the position of the source in the room is important especially in measuring the narrow bands and discrete tones of sound, since the acoustic power is strongly affected by the room modes.

The measurements in the field, where the machinery, equipment, and installation are operating under normal conditions, are performed to satisfy the specific requirements. However, the effects of specific environmental factors, associated with the measurement area, have to be eliminated by applying some corrections to the measured levels in order to obtain environment-free emission levels for the source under consideration. (As mentioned above, some corrections are also necessary even in the laboratory measurements.)

The basic procedures to determine the sound powers emitted by certain noise sources in the field are described in the ISO standards (Table 5.1). Except for those requiring a special laboratory environment, the standard methods, like the engineering method presented in ISO 3744:2010 (confirmed in 2015), can also be employed outdoors or in large rooms for any type of noise source generating sounds as steady, non-steady, fluctuating, etc. The standard gives the details to measure spectral levels (in one octave and one-third octave bands) and A-weighted sound pressure levels [27b].

ISO 3746:2010 (confirmed in 2015): The standard explains a survey method with accuracy grade 3 and requires no special test environment [29]. It is applicable both indoors and outdoors with one or more sound reflecting planes existing and in connection with the noise source. All types of noise (steady, non-steady, fluctuating, impulsive, isolated bursts of sound energy, broadband or with discrete frequencies, etc.) and all size of sources, such as stationary or slowly moving plant, installation, machine, component or assembly, can be measured using this method. This standard method measures only the A-weighted levels. If the noise is transient, intermittent, and having bursts, the sound energy levels are obtained.

ISO 3747:2010 (confirmed in 2015) describes the comparison method in situ under the reverberation conditions by using the reference source as described in the laboratory measurements [30].

ISO/TS 7849-1:2009 (confirmed in 2017): Sound power of a source (i.e. a machine), radiating airborne sound from vibrations of its solid structures, can be determined by vibration measurement and calculation using the relationship between the sound power and the fixed radiation factor, ε of the machine [49a]. In Part 1 of the technical standard first published in 1987, a survey method is given for the A-weighted sound power levels, by introducing a fixed radiation factor, ε (efficiency of sound radiation). Part 2 describes the engineering method, also presenting a procedure to determine the adequate radiation factor of the vibrating surface and gives the sound power levels in octave bands [49b]. For $\varepsilon = 1$, the max sound power is determined as in Eq (5.23):

$$L_{ws} = \bar{L}_v + 10 \lg \frac{S_s}{S_0} \tag{5.23}$$

L_{ws}: Maximum sound power level, dB
$\bar{L}_v$: Average vibration velocity level on the measurement surface, dB (reference velocity, $v_0 =$: 5×10^{-8} m/s $= 50$ nm/s)
S_s: Vibrating surface area enclosing the machine, m^2
S_0: 1 m^2

The velocity levels of surface vibrations are measured using an accelerator at uniformly or at randomly positioned receiver points on the surfaces of machine components. The method is useful in the cases when the other methods cannot be implemented due to high background noise levels or other factors in the environment.

ISO 11204:1995 (revised in 1997) (confirmed in 2015): The standard proposes a method to determine the emission sound pressure levels at and nearby the workstations in closed or open environments [50a, 50b]. Various positions of a workstation (operator's position) are defined, such as in the plant near the source, in a cabin, or in a partial or total enclosure of large industrial machines. A reference box and reference measurement surface (the workstation may not be on this surface), as in ISO 3744:2010, are used to determine the mean sound pressure level on the measurement surface. The method is applicable for noise sources that are either moving or constant. The measurement duration and technique are explained according to the noise type as steady, non-steady, or impulsive.

The corrections are applied to the measured sound pressure levels; for background noise, for the reflective surfaces other than the floor on which the machine is supported and for the specified position of the workstation or operator, it is called local environmental correction. The standard explains how to obtain these corrections in a way somewhat similar to ISO 3744, by taking into account the acoustic factors of the room.

The microphone positions are described according to the operator's position: seated (fixed) or moving along a specified path. Determination of the significant impulsive components in the total noise output through measurements is explained, defining a descriptor "impulsive noise index" (or impulsiveness) with the relative correction values.

The following units are used in the emission measurements:

- Emission sound pressure, p N/m^2 and emission pressure level, L_{peqT}, dB
- Time-averaged emission sound pressure level, L_{pT}, dB
- Average steady state condition and emission pressure level in T seconds, $L_{pA,eq,T}$, dB. (Energy-averaged in the observation period.)
- Peak emission sound pressure level within a cycle of operation, max instantaneous level, $L_{p,peak}$, dB
- Single event emission sound pressure level, $L_{p,1s}$ (Normalized level for $T_0 = 1$ s.)

The above units can be A- or C-weighted levels or spectral levels. Uncertainties due to the source and environment factors are given in the standard.

In addition to those given above, other standards have been published regarding the measurement of airborne noise emitted from specific products, such as equipment in information technology and telecommunications, internal combustion engines, small air-moving devices (various kinds of fans), gas turbines, etc. ISO 11689:1996 (confirmed in 2017) states a procedure for comparison of noise emission data of various products, applicable to individual machines and to those categorized within a group [32]. Presentation of the measurement results in a test code is explained in ISO 12001:2009 (confirmed in 2018) [31] and the declaration and verification of noise emission values of machines and equipment are described in ISO 4871:1996 (confirmed in 2018) [51]. ISO 6396:2008 and ISO 6393:2008 both aim to provide noise emission data for earth-moving machinery at the operator's position, by applying simulated stationary and dynamic test conditions, respectively.

Emission Measurements of Industrial Premises

For the purposes, such as noise mapping, noise impact assessment, comparison of different sources or only to monitor the noise of a plant, it is essential to determine the total noise emissions of industrial premises covering various activities and machinery in the open air, in closed, or semi-closed buildings. Field measurements are conducted around the plant for this purpose, however, due to a number of factors, such as types and locations of individual noise sources, different type of noises, variation of noise due to different operational conditions, etc. that are explained in Section 3.3, it is rather difficult to evaluate the measured data for determination of the plant emission level.

Attempts to measure the noise emission of industrial premises began in the 1970s, for instance, the guideline prepared by the Building Research Establishment in the UK (BRE) is a pioneering example [52]. Figures 5.18a–5.18c show the arrangement of microphones around the plant according to the "equal angle method." The duration of measurements were simply proposed as 2–5 minutes per microphone position under the condition of wind speed not exceeding 2 m/s at distances less than 25 m from the source.

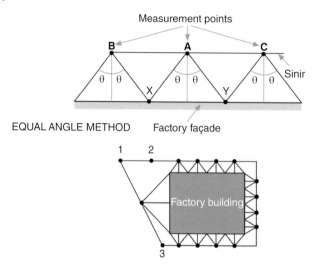

Figure 5.18a Determination of the measurement points around an industrial building, using the equal angle principle (BRE, 1975) [52].

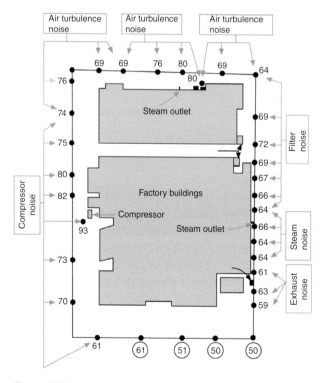

Figure 5.18b Example of measurement points around an industrial building (BRE, 1975) [52].

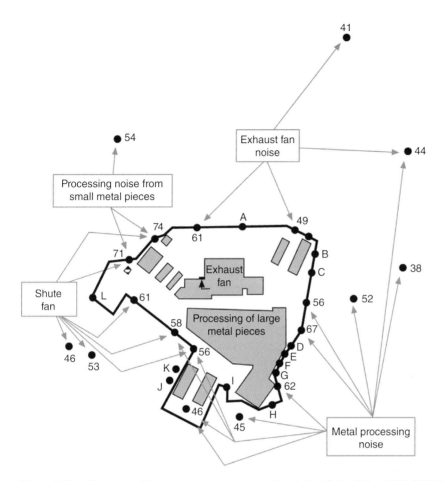

Figure 5.18c Example of Measurement points around an industrial building (BRE, 1975) [52].

The international standard method published for the same purpose is explained below.

ISO 8297:1994 (confirmed in 2018): As in the other engineering methods referred to above, determination of the sound emission of a large multisource industrial plant is based on sound pressure level measurements around the plant [53]. The major assumption is that all the individual sources are integrated as a single point source placed at the geometrical center of the plant. The main dimensions are on the horizontal plane and it emits uniform noise in all horizontal directions. A special measurement contour is designed to accommodate the microphone positions according to a methodology discussed in the standard. The geometry of the contour is dependent on the area of the plant (S_p), the area of measurement (S_m), the distance from the microphone position to the nearest point of the plant or to the boundary of plant d, the average measurement distance, $\bar{d}$, the length of the measurement contour (*l*), and should satisfy some conditions mentioned in the standard (Figure 5.19). The microphone heights are determined based on the characteristic height of the plant, *H*, from the average height of the sources in the plant. The sound pressure levels are measured under the specified weather conditions at the specified positions along the contour within an equal measurement period, *T*, while the sources are operating, and then averaged on the energy basis. The overall emission level of the plant is calculated, by taking into account the effects of

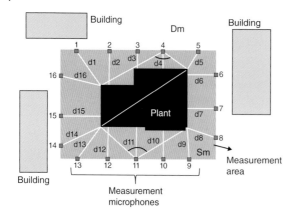

Figure 5.19 An example of implementation of ISO 8297 for a field study to assess the plant noise through measurements.

the physical environmental factors. The corrections given in the standard are related to the measurement surface, proximity, the background noise level, and the atmospheric attenuation:

$$L_w = \overline{L}_p + \Delta L_s + \Delta L_f + \Delta L_M + \Delta L_\alpha \quad \text{dB} \tag{5.24}$$

$$\overline{L}_p = 10 \lg \left[\frac{1}{N} \sum_{i=1}^{N} 10^{0.1 L_{pi}} \right] \quad \text{dB} \tag{5.25}$$

When $L_{p_i} > \overline{L}_p + 5$; $L_{p_i} = \overline{L}_p + 5$

L_w: Sound emission level, dB

$\overline{L}_p$: Average sound pressure level measured at the microphone points on the measurement contour, dB

L_{pi}: Sound pressure level measured at microphone position, i

ΔL_s: Area term according to ISO 3744, dB

ΔL_f: Proximity correction, dB

ΔL_M: Microphone correction (for a directional microphone), dB

ΔL_α: Atmospheric attenuation, dB

The above calculation can give both sound power levels in octave bands between 63–4000 Hz and A-weighted sound power levels.

5.4.2 Emission Measurements of Transportation Noise Sources

Transportation systems generate composite noises contributed by various individual noise sources, e.g. vehicles, trains, aircraft, vessels. Assessment of the total noise of each system, from the standpoint of environmental noise management, requires information about noise emissions of transportation noise sources. Such data are acquired, based on the emission measurements which are involved primarily in noise control engineering and also for comparisons with the regulatory noise limits specified for each type of source (Chapters 7 and 8). Techniques which are employed for noise emission measurements vary with the source type and are described in the specific standards, some of which were published in the 1960s.

A) Emission Measurements of Motor Vehicles

Motor vehicles are composed of secondary noise sources that can be investigated by using different techniques, for example, the distribution of sound pressure levels around a car is shown in Figure 5.20a from an experiment in 1980 and the sound intensity vector field of a car is displayed in Figure 5.20b from another study in 2014 [54a–c]. The emission measurements for motor vehicles are conducted both in the field and in the laboratory according to the specified operating conditions (Figures 5.21a and 5.21b) [54a, 54b]. The measurement procedures have been widely described in the literature for different type of vehicles, such as motorcycles, four-wheel vehicles, trucks, categorizing vehicles with respect to maximum speeds, engine braking conditions (gears), highest "rev/min" at maximum power, stationary vehicles, etc. by taking into account the parameters, such as vehicle mass, axles, engine power, and engine speed [55]. The basic technical standards are outlined below.

ISO 362 and its progress: The preliminary standard ISO R362:1964, "Measurement of noise emitted by vehicles" described the methods both for vehicles moving at full speed and for stationary vehicles, later the recommendation standard was improved by including accelerating vehicles [56a].

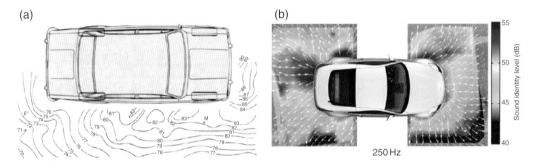

Figure 5.20 (a) A-weighted sound level contours around a car, based on measurements in 1980 (unloaded, stationary automobile at a height of 0.5m for an engine speed of 200 rpm) [54a] (b) Acoustic intensity vector field of a car with a rotational engine speed fixed at 3000rpm. From Comesana thesis, 2014 (Source: with permission of Comesana) [54c].

Figure 5.21a Emission measurement for a vehicle pass-by. (*Source:* photo by A. Güney in 2008.)

Figure 5.21b Exhaust noise emission measurement in a semi-anechoic room. (*Source:* photo: OTAM Laboratory, ITU with permission of A.Guney.)

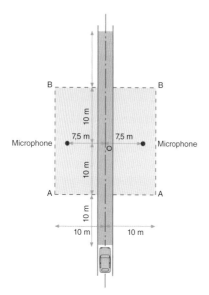

Figure 5.22a Microphone positions during the measurement of pass-by noise of vehicles according to EU Directive 70/157/EEC [56b, 57b].

In the 1998 revision of the standard, it was aimed to measure the highest noise levels from passing-by vehicles under normal driving conditions on dry road surfaces [56b]. Some of the requirements were: The test area should be free of any barrier or a reflective surface at 50 m from the test area. The surface around the microphones should be acoustically hard within a circle of diameter of 10 m and the distance between the microphones to the road axis should be 7.5 m (Figures 5.22a and 5.22b). These conditions are also described in ISO 10844:1994 (revised by ISO 10844:2014), emphasizing the surface characteristics of the test tracks influencing the tire and road noise emission [56c]. Measurements are conducted for the car at a certain initial speed, up to a full acceleration over a distance of 20 m. The recordings are made by using A-weighted levels and the fast response. Figure 5.23 gives the microphone positions during noise emission measurements for moving vehicles, according to the engineering method given in the standard [56b]. Further improved standards take into account the urban driving conditions and the tire/road interaction noise which is dominant at lower speeds (50 km/h).

Current standards: ISO 362-1, 362-2, 362-3, published after 2009, enable applications for all vehicle types, including automatic and manual transmissions, electrical, hybrid, etc. and describe the specific test conditions for each type and the interpretation

Figure 5.22b Conditions for microphone positions during the measurements [57a, 57b].

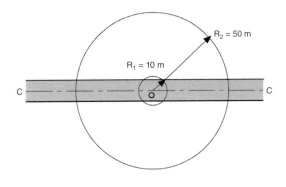

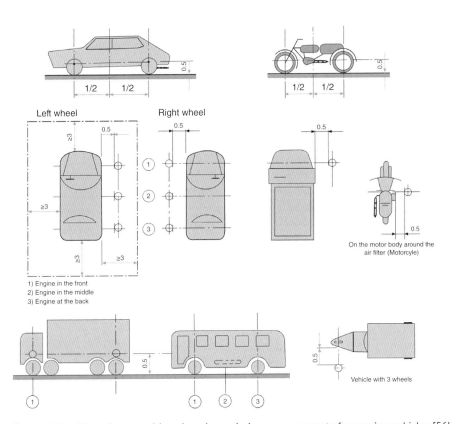

1) Engine in the front
2) Engine in the middle
3) Engine at the back

On the motor body around the air filter (Motorcyle)

Left wheel

Right wheel

Vehicle with 3 wheels

Figure 5.23 Microphone positions in noise emission measurements for passing vehicles [56b, 58].

of the results [57a–c]. Motor vehicles have been categorized for emission measurements in accordance with the EU Directives regarding the declaration of noise limits, as:

- Category L: motor vehicles with fewer than four wheels with different sub-categories. (Sub-categories are for mopeds, for vehicles according to engine cylinder capacity, max speed, gross vehicle mass, unladen mass.)

- Category M: power-driven vehicles having at least four wheels and used for the carriage of passengers. (Vehicles used for the carriage of passengers, sub-categories with number of seats and max mass.)
- Category N: power-driven vehicles having at least four wheels and used for the carriage of goods. (Vehicles used for the carriage of goods, maximum authorized mass.)

EU Directive 70/157/EEC: The European Commission Directive in 1970 describing the measurement conditions for cars, trucks, buses, both stationary and in motion, and instruments, requires an open space of 50 m radius on two sides of the vehicle and a flat concrete-covered surface of 20 m radius at the center of the vehicle. The microphones are to be positioned at 7.5 m from the center of the line and at 1.2 m above the ground (Figure 5.22b) [58]. Three speed conditions were requested (full power, 3/4 power, and 50 km/h). The permissible noise emission levels were declared for different types of vehicles with the defined characteristics.

The EU document has been revised over the years and was last published in 2014 [59]. Conforming to the EU-type approval of all new vehicles, the noise limits are declared for the M and N categories to obtain EU certification. The measurement procedure is explained for motor vehicles both in motion and stationary.

ISO 7188:1994 (withdrawn and revised by ISO 362-1:2007 and ISO 362-2:2001): The standard covers the measurement method based on statistical studies for noise emitted by passenger cars under urban driving conditions. A declaration of the emission level is required as the weighted-average of the test results obtained at full acceleration and at constant speed in open space, for the purposes of official testing, the approval of vehicles, and the manufacturing stage [60].

ISO 16254:2016: This standard, derived from ISO 362-1, describes an engineering method for measuring the sound emission from M and N category motor vehicles under specified test conditions [61], i.e. at a standstill and at low speed operating conditions, over an extensive open space. It is emphasized that the tests providing the objective results should be evaluated with respect to the subjective judgments (i.e. annoyance, perception, etc.) which might differ according to vehicle classes and the actual environmental conditions. The standard also covers an engineering method to measure the external sound generation systems and criteria for their acoustic characteristics.

Procedures in the USA

Noise regulation procedures and the noise emission standards in USA have been the concern, starting in the 1970s, of different institutes, i.e. the Environmental Protection Agency (EPA), and the Society of Automotive Engineers (SAE). The Federal Register, issued in 1976, describes the arrangement of a test area in the field for new medium and heavy trucks as shown in Figure 5.24a [62a, 62b]. Microphones are positioned at 50 ft. (15.2 m) from the centerline of vehicle's path and 1.2 m above the ground with no obstruction element within 1500 angle of view.

The USA FHWA: The procedure issued in 1981 by the Federal Highway Administration (FHWA), requires the test area to be an empty space without any reflective surface within 30 m and keeping the similar microphone positions as described in the Federal Register (Figure 5.24b) [63]. The vehicle should be driven at a constant speed and controlled by a speed measurement system near the microphone position. Sound pressure levels are measured as A-weighted levels by using the fast response. The maximum level during the vehicle pass-by is declared as the pass-by noise with respect to the vehicle category (i.e. automobile, medium truck and heavy truck).

The SAE issued a standard J 986 for passenger cars and trucks in 1986, similar to ISO R-362, however, specifying the track length of 15 m and the microphone positions at 15 m on both sides. The current standard by the SAE J2805 is equivalent to ISO 362-1:2015.

Figure 5.24a Definition of the microphone points for vehicle emission noise measurements according to the SAE (US) method, 1976 [62a].

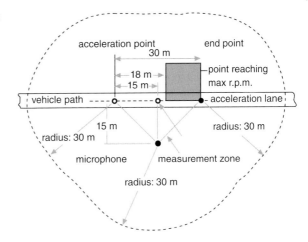

Figure 5.24b Definition of microphone points for motor vehicle noise emission according to the FHWA (US) method, 1981 [63].

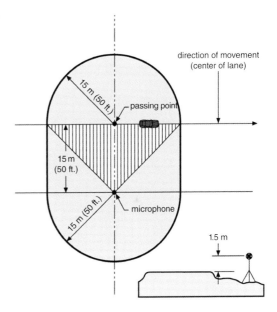

Laboratory Tests

ISO 5130:1982 (withdrawn) and revised by ISO 5130:2007: Based on the earlier standard ISO R-362:1961, the laboratory measurements in an anechoic or semi-anechoic room for stationary vehicles are explained [64a, 64b]. The scope of the standard is to determine the variations of noise emitted from different parts of vehicles in use or after modification in certain components. The revised standard applies to L, M, and N vehicle categories (described above) during official tests. The procedure to measure the exterior sound pressure levels particularly emitted by the engine and the exhaust, is to be repeated under various operating conditions, such as different rpm values, engine speeds, etc. The positions of the microphones according to the exhaust muzzle, angles, and heights are shown in Figure 5.25a. A-weighted maximum sound pressure levels are recorded by taking at least three measurements at each measurement point. Figure 5.25b shows exhaust noise measurements for a motorcycle.

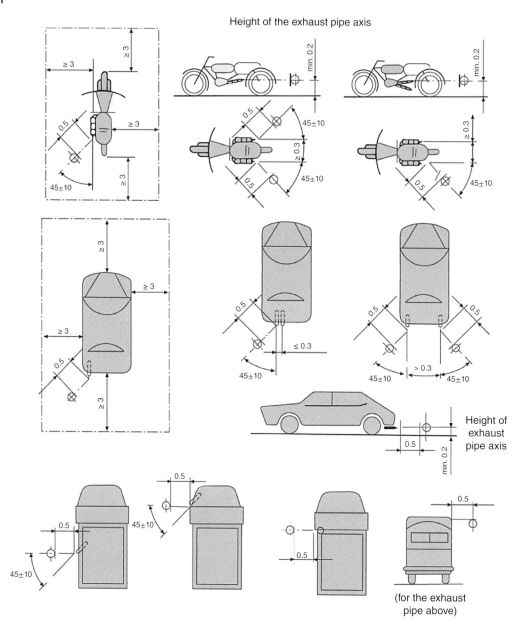

Figure 5.25a Examples of emission measurements for stationary vehicles, e.g. motorcycles, cars and trucks according to ISO standards and EU Directives [64a, 64b].
Note: reproduced with the permission of the International Organization for Standardization, ISO. This standard can be obtained from any ISO Member and from the website of the ISO Central Secretariat at the following address: www.iso.org. Copyright remains with ISO.

ISO 362-3:2016: In the latest standard, the engineering method to measure the noise emission from M and N types of vehicles to be conducted in semi-anechoic rooms is explained [57c]. The requirements for test rooms and test conditions are given, including the calculation of acceleration, the operating conditions, the microphone array, the measurement procedure, the evaluation of the results, the determination of the uncertainty budget, and the calculation of the expanded uncertainty.

Figure 5.25b Exhaust noise measurements for motorcycles.

At present, the numbers of investigations regarding new methodologies to estimate noise from automobiles, have continued, i.e. by combining analytical and experimental approaches [65]. The new standards introducing special techniques to measure noise radiated by different parts of motor vehicles, are published, for example, ISO 13325:2003 (confirmed in 2019) describes the coast-by method for measurement of tire-road sound emission, i.e. while free-rolling with transmission in the neutral position and with the engine and auxiliary systems switched off [66]. Since noise abatement is one of the great concerns in the automotive industry, many indoor tests for tire-road noise have been conducted, some by using advanced visualization techniques [67]. Measurement of low-noise road surfaces is referred to in Section 5.10.1 (under the topic of noise-reducing devices).

Signal System Measurements
EU Directives 70/388/EC and 93/30 EC: In these earlier documents, the measurements are described for the certification of the signal systems of all vehicles including those with two or three wheels, like motorcycles [68, 69]. The measurement conditions regarding the test area, the background noise levels, the monitoring equipment, the operating and the microphone positions are explained. The maximum values of the audible devices are restricted to a maximum 105–118 dB (A) between 1800 and 3550 Hz.

ISO 512:1979 (withdrawn) and ISO 6969:2004 (confirmed in 2019): The measurement procedure for signaling and warning devices of motor vehicles, after being mounted on the vehicle, is presented in these standards. The signals are divided into two categories and the limit values in A-weighted sound pressure levels are given [70, 71]. To check the compatibility with the limit values, two methods are proposed for vehicles without and with the signal systems mounted, applicable in anechoic and semi-anechoic rooms. The measurements are conducted at the microphone positions over a sphere of 5 m diameter where the variation of sound levels remains within 1 dB, ensuring the background levels are 10 dB lower than the signal level. Using Type 1 sound level meter at fast-response, the A-weighted and spectral levels in one-third octave bands are to be recorded, the mounting conditions are described for specific signal equipment.

B) Emission Measurements of Railway Vehicles

As explained in Section 3.2.2, railway noise, composed of rolling noise, traction noise, aerodynamic noise, etc., is emitted by various source components associated with the type of trains, the operation, and the track conditions. Noise emission of trains has become a great concern in Europe involving a lot of data collection, train categorization, and evaluations [72–74]. Noise emission data is important for the prediction of railway noise as an environmental problem and noise impact assessments.

ISO 3095:2005 (revised in 2013): The standard describes the measurement procedures for exterior noise emission levels and noise spectrums of all types of railway vehicles by focusing on rolling noise and minimizing the effect of the track. In the earlier issue of the standard, two types of tests are recommended [75]: Test 1: to check the compatibility of trains with the requirements applicable at the manufacturing stage; and Test 2: for the trains in use or after modifications, aiming at a comparison with trains within the same category.

The requirements are given regarding the test field (free-field with no reflective and obstructive elements), the wind condition (wind speed less than 5 m/s), and the background noise to be less than 10 dBA lower than the signal to be measured. The distance between the source and the microphone can vary between 7.5–50 m depending on the train length. Operating conditions and speeds are described in the standard. The microphones are located in front, at the back, and on the sides as shown in Figures 5.26a–5.26c [75]. Further technical descriptions are given in the latest revised standard and the effect of the tracks is also included. The conditions of the environment, the track, the rail acoustic roughness determination, the vehicles, the measurement configurations, the test procedure for light vehicles, and the determination of measurement uncertainty are explained. The measurement units are:

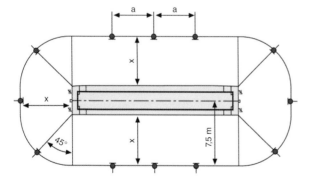

Figure 5.26a Microphone positions in noise emission measurement for stationary trains, 2005 [75]. *Note:* Figures 5.26a and 5.26b are reproduced with the permission of the International Organization for Standardization, ISO. This standard can be obtained from any ISO Member and from the website of the ISO Central Secretariat at the following address: www.iso.org. Copyright remains with ISO.

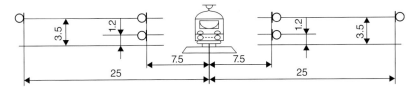

Figure 5.26b Microphone heights for trains with constant speed according to ISO 3095:2005 [75].

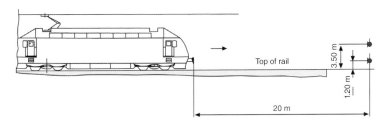

Figure 5.26c Microphone positions for trains with special power units [75].

- The A-weighted sound pressure level as a function of time with fast-time weighting, $L_{pAF}(t)$. The maximum A-weighted sound pressure level determined during the measurement time interval T also using fast-time weighting, L_{pAFmax} (for moving trains).
- The A-weighted equivalent continuous sound pressure level, $L_{pAeq,T}$ (for stationary trains).

Procedures for measurement of noise at platforms and at stopping points, on bridges, and elevated structures are included in the standard. It is necessary to record the track type and wheel roughness, in addition to the other specific parameters contributing to noise emission.

Various studies have been conducted by implementing the standard explained above, to obtain emission data for railway rolling stock in Europe. The results of such measurements have enabled comparisons of emissions according to track and rail types, and the categorization of trains. The EU Commission working group, called "The Railway Noise Working Group WG6," published a report in 2002 introducing the unit of Transit Exposure Level (TEL) expressed in dBA to measure noise emission from railway vehicles. The TEL is determined at 25 m from the rail centerline (horizontally) and at the heights of 3.5 and 5 m above the rail head. The document gave a list of parameters to determine the rail properties, including the definition of a rail bed. The procedures have been developed to measure the rail smoothness, the discrimination of vehicle and rolling noise, and the measurement of brake noise. New measurement units, new equipment to be used, new methods for stationary and moving trains at constant speed, acceleration tests, preparation of the test report, etc. are all explained in the document. The International Union of Railways (UIC), dealing with railway noise since 1998, has also published several documents on railway noise measurements, evaluations, and limits [73b].

In the revised standard (ISO 3095:2013), three types of measurement procedures are described for trains: Stationary noise, starting noise, and pass-by noise to check conformity with the limit values mandated by EU Directive 96/48. The measurements are made on a reference track, however, when the test conditions are not available, the specific procedures to determine the rail roughness and track decay rates are provided. The standard recommends an engineering method with the uncertainty value of ±2 dB and a survey method for accelerating trains. The microphone heights are specified as;

H = 1.2 m and 3.5 m if required at 7.5 m from the rail edge

H = 3.5 m at 25 m from the rail edge.

For uncertainty determinations, at least three measurements are necessary during emission measurements, each at different speeds. When the difference between the measured results exceeds 3 dB, new sets of measurements have to be conducted. The measurement points are shown in Figures 5.26a–5.26c. The following units are to be determined:

- L_{Aeq}: A-weighted equivalent sound pressure levels, with the averaging time T, is selected, so as to cover the total energy during a train pass-by.
- *TEL*: Transit Exposure Level dB(A) to be determined at a horizontal distance of 25 m from the centerline of the nearest track at a height between 3.5 and 5 m above the railhead level. (*TEL* is similar to the sound exposure level, *SEL*, the integration time depends on the length of the train and its velocity.)

$$TEL = 10 \lg \frac{1}{T_p} \int_{t_1}^{t_2} \left(p_A^2(t)/p_0^2 \right) dt \tag{5.26}$$

T: Measurement interval (i.e. one operation cycle), s
T_p: Duration of pass-by, s (see Figure 3.28)

Calculation of the duration of the pass-by:

$$T_p = L/V$$
$$T_p < t_2 - t_1 \tag{5.27}$$

L: Length of train, m
V: Speed of train, m/s

$$TEL = L_{pAeq,T} + 10 \lg \left(\frac{T}{T_0} \right) \tag{5.28}$$

T: Measurement time interval, s
T_0: Reference time, s
$L_{pAeq,\ T}$: A-weighted equivalent continuous sound pressure level in dB

The European Norms have been published for measurements of different parameters in relation to railway vehicle noise emission: EN 15610:2019 describes a direct method for characterizing the surface roughness of the rails affecting rolling noise and the measurement of rail roughness in the one-third octave band spectrum [76]. The measurement procedure associated with the warning devices of high-speed trains is included in EN 15153-2:2013 [77].

C) Emission Measurements of Aircraft

ISO 507:1970: The first standard, already withdrawn, specified the noise emission measurement techniques for aircraft on the ground and in the air and the result was presented as the "Perceived Noise Level," L_{PN} [78]. It was required to conduct the measurements under the three conditions of aircraft in the air: (i) at maximum power during take-off; (ii) at reduced power during climbing; and (iii) at both maximum and minimum powers while landing. The measurement of the directivity pattern for aircraft on the ground was required to draw the noise contours. Based on the measured sound pressure levels at third octave bands, the "Noy" units could be obtained (See table 2.9).

ISO 3891:1978: The standard already withdrawn, defined primarily the noise from a single aircraft operation, and described the determination of "Noise Exposure" for successive operations of flights in a measurement time interval [79]. Certification of aircraft, monitoring noise around airports and propositions regarding land-use planning were also included in the standard.

ICAO Aircraft Certification Method: The International Civil Aviation Organization has declared the standard measurement techniques for the certification of aircraft noise emissions to be expressed in the "Effective Perceived Noise Level" (*EPNL* dB) [80]. The measurements are made under the described reference conditions for take-off and approach noises at the reference measurement points shown in Figures 5.27a and 5.27b. It is necessary to measure the maximum sound pressure levels, to determine the distribution of levels within the frequency range concerned and the variation of noise with time. Frequency spectrums are determined at 24 third octave bands within the intervals of ½ s and afterwards,

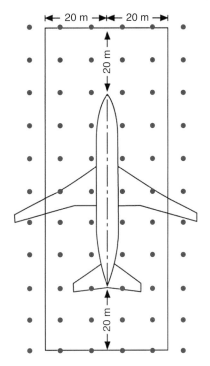

Figure 5.27a Measurement of aircraft noise emission according to ICAO (Annex 16), 1993 [80].

"Perceived Sound Level" (PN) and the "Effective Perceived Levels" (EPNL) are obtained by using the tables provided in the document. The characteristics of the equipment to be used in the measurements (e.g. specific microphones durable under high pressures) are described and the corrections to be applied to the measured results according to flight profiles, the direction angle of sound ray toward the microphone, air absorption, and the atmospheric conditions are explained. The ICAO document (Annex 16) also gives the aircraft noise emission limits in terms of maximum *EPNL*, which vary depending on the number of engines (engine power) and the mass of the aircraft.

The new document (ICAO Environmental Technical Manual), issued in 2004, explains the equivalent measurement procedures as specified in Annex 16, vol. 1 [81]. These reference procedures which are considered cost-effective and time-effective methods in aircraft certification measurements, are displayed in Figures 5.27c and 5.27d for take-off and approach profiles of subsonic jet airplanes.

BS 5331:1976: Sonic booms generated by the supersonic aircraft (Section 1.3.3) was first dealt with by ISO 2249:1973, which provided a basis of physical descriptions and measurements. Later the British standard was developed to describe the measurement procedure and the equipment (both are withdrawn). Both standards recommended using microphones with a dynamic range of 45 dB, which must be durable against pressure changes. Positioning the microphones underground, recording sound pressure signatures, and making evaluations with respect to the effects on humans were described [82].

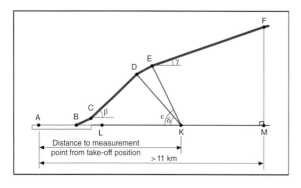

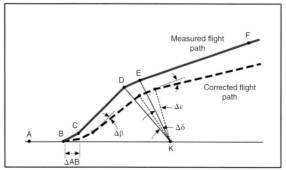

Figure 5.27b Take-off profile and difference between the measured and corrected take-off profiles, ICAO 1993 [80].

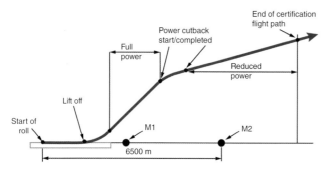

Figure 5.27c The current take-off reference procedure for noise measurements of subsonic jet airplanes, ICAO 2014 [81].

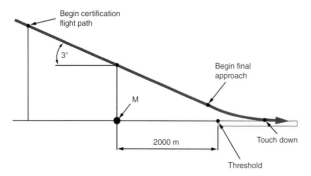

Figure 5.27d The current approach reference procedure for noise measurements of subsonic jet airplanes, ICAO 2014 [81]. (*Source:* with permission of ICAO for all figures.)

D) Emission Measurements of Waterway Vehicles

ISO 2922:2000 (confirmed in 2011+A1:2013): First published in 1975, the standard describes emission measurements for sea vessels in inland waterways, harbors, and ports, except for recreational craft which are discussed in a separate document [66, 83a, 83b]. Specifications for the test site, the operating conditions of the engines, microphone positions, and the test procedure for moving and stationary vessels are explained in the standard.

The measurements are made under the wind speed of less than 5 m/s with the background noise of less than 3 dBA at a distance of 100 m from the vessel and where no barrier or reflective surfaces or other boats exist. The positions of microphones are at 25 m perpendicular to the boat route while traveling parallel to the shore on calm water and at 1.2 m height from the wharf surface (Figure 5.28a). The latest version of the standard gives more detail about the microphone positions since the location of noise sources (funnels, ventilation systems, etc.) varies with the type of ships of different heights. The microphones are arranged at grid points along the ship's length at distances of 1m and 25 m from the ship hull, which is the major noise source (Figure 5.28b) [84, 85]. Sound power levels are calculated from the first group measurements on the first row (on the nearest grid system). The measurements on the second row are used for the derivation of a transfer function (sound propagation model) and for validation of the prediction models [86].

The measurement units are the maximum A-weighted sound pressure level L_{pASmax} and A-weighted sound exposure level, L_{AE} for each passage of single vessels.

ISO 14509-1:2008 (confirmed in 2014): This describes an engineering method for measurement of airborne sounds emitted by powered recreational craft [87].

ISO 2923:1996 (confirmed in 2017): The standard, including ship noise on board, explains the measurements on the deck, in indoor spaces, such as the restaurant, the engine rooms, the control room, the air intake. and the exhaust chutes [88]. Microphone positions in mechanical rooms are selected according to the primary engines. The required measurement conditions for boats traveling at a speed, which is a maximum 95% of the certified speed, are described as: the water should be calm, the route of the vessel should be straightforward, the supplementary engines are operated, and the windows and doors are closed.

Measurements are made to last at least 5 seconds by using fast response, the A-weighted sound pressure levels and the one-third octave or octave band levels starting from 50 Hz are recorded. The corrections according to the background noise are explained and the content of the measurement report is given in the standard.

Mechanical vibrations of ships are included in ISO 20283-5:2016 with an amendment issued in 2014, emphasizing the effects of vibrations on human. The measurements of vibrations caused by specific sources, e.g. propulsion machinery, engine, etc., and structural vibrations are described in the standard [89a].

5.4.3   Emission Measurements of Building Service Equipment

Building service systems, consisting of electrical and mechanical equipment and installations, create noise problems both indoors and outdoors (Section 3.9). Noise emissions of such systems, which are composed of various components, are measured to verify their acoustical quality, and also to obtain EU Certification for the European markets.

Laboratory specifications are included in EN 15657:2017 to measure airborne and structure-borne sounds generated by building service equipment of all kinds [89b]. However, the specific

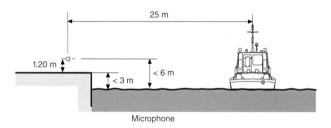

Figure 5.28a Microphone position for measurement of boat noise emission described in ISO 2922:1975 [83a].

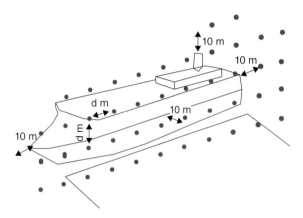

Figure 5.28b Grid points for noise emission measurements of ships [86].

international standard for sanitary systems, including localized noise sources like taps, valves, piping systems and other special appliances, is outlined below.

ISO 3822 (Parts 1-4) (First issued in 1983, current: 2018): Noise generated by water supply and waste water removal systems in buildings is discussed in this standard series, and the measurement techniques in the laboratories to provide noise emission data of the system components, i.e. from water flow through pipework, from appliances and other equipment used in water supply installations, are described [90a–d].

The requirements regarding the operational conditions during tests in terms of the physical parameters of the water flow, i.e. the flow pressure and flow rate (speed) and arrangement and mounting of equipment in the test rooms, the connection of appliances with the test pipe and the corrections to be applied for background noise, are explained in the standard.

An example reverberant test room designed according to the standard is shown in Figure 5.29. A test wall, on which the water pipe is mounted, is constructed in the source room and the noise transmitted through the test wall is measured in octave bands between 125 and 4000 Hz. The acoustic performance of the test device is defined by the "Appliance Spectral Level", L_{apn}, dBA to allow comparisons between the results of different laboratories. The L_{apn} specifies the difference between the noise level measured with a special pipe appliance called INS, which is mounted on the pipe, and the noise level without INS.

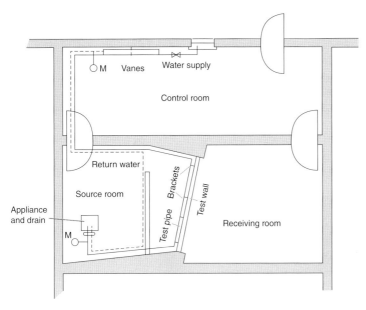

Figure 5.29 An example of a laboratory set-up for measurement of water installation systems described in ISO 3822 [90a, 90b, 90c, 90d].

In addition to the sanitary system equipment, measuring the noise emission of the electrical and mechanical system equipment attached to or installed in buildings, e.g. heating and air conditioning (HVAC) systems or discrete mechanical ventilation, heating and cooling system devices, like boilers, burners, fans, compressors, air supply and exhaust ducts, conditioning grills, diffusers, dampers, vanes, ventilation shafts, unit ventilators, unit refrigerators, electrical motors, other auxiliary service equipment, is discussed in different standards, some of which are given in Table 5.1.

5.5 Measurement Methods and Techniques to Acquire Immision Data for Environmental Noise Sources (Field Measurements and Analysis)

Sound energy emitted by noise sources is received as the sound pressure level (immission level) at a certain point, which is influenced by the physical factors related to the environment (Chapter 2). In practice, the determination of sound pressure levels through measurements is needed to examine the following targets:

- to obtain sound emission data relative to specific noise sources
- to make an assessment of the noise impact on individuals and the community.

The first target can be fulfilled either in a laboratory or in the field by implementing the measurement procedures to determine the sound pressure levels, as explained above. The second target, which is of greater interest in this book, will be dealt with in more detail.

5.5.1   Objectives of the Field Measurements

Assessments of environmental noise require field studies under existing conditions in order to fulfill one or more objectives listed below:

- To quantify noise levels for hearing protection and health in workplaces.
- To determine the magnitude of noise at a certain location in the environment and at a certain time interval.
- To investigate the variation of noise levels under the effect of different physical factors.
- To investigate the effect of a certain physical parameter on noise propagation.
- To compare the noise levels in a noise- sensitive area with the criteria for protection of community health and comfort (Chapter 7).
- To check the compatibility of the noise level of a specific noise source to the noise limits given in regulations and specifications.
- To verify the predicted data and its reliability at reference points.
- To handle a complaint through inspections of the site.
- To determine the appropriateness of an area for a specific purpose of usage and for land-use planning and design.
- To perform a socio-acoustic survey to rate annoyance responses in a specific area (e.g. a neighborhood).
- To derive noise criteria or noise limits to be specified in the national regulations, as a result of dose-response analyses.
- To identify noise sources in the area under concerned.
- To provide reference data for noise mapping and for validation of noise maps (Chapter 6).
- To determine the size of the noise impact by defining the zones exceeding noise limits.
- To develop decisions about noise management for the present and the future, in relation to global noise policy and action plans (Chapters 8 and 9).
- To determine the amount of façade insulation required for buildings which are directly exposed to noise.
- To determine the technical solutions for noise abatement in a specified area or location.
- To quantify the performance of noise control elements in the field, e.g. measurements for noise barriers and to compare the situations with and without a barrier.
- To determine the quiet and noisy façades of buildings (Chapter 9).

Before conducting measurements in the field, various aspects should be questioned, such as what to measure, how to measure it, for which purpose, which equipment to use, at which precision level, etc. Physical conditions of the test site, the variability of the source characteristics, in addition, the knowledge and experience of the personnel performing measurements, all play key roles in pursuing the task.

Different techniques have been discussed in several publications so far, however, the methods recommended in the International Standards are indispensable from the standpoints of measurement accuracy and comparability of results [91–93]. The first ISO standard was ISO 2204, published in 1979 (withdrawn at present), involved with the measurement and valuation of environmental noises [94].

ISO 1996-1:1982 and revised and published as ISO 1996-1:2016: As the prominent international standard regarding the description, measurement, and assessment of environmental noise, ISO 1996 and its different parts have a long past and were revised several times parallel to the increased concern for environmental noise problems [95a–d]. Part 1 gives the definitions and descriptions of the basic quantities and assessment procedures.

ISO 1996-2:2017 (first issued in 1987): This part of the standard includes the determination of environmental noise levels through field measurements, describing the procedures applicable according to the acoustical characteristics of noises and outdoor and indoor environments in brief [95d]. The major aim is to provide a basis for setting environmental noise limits. In addition to the direct measurements, the assessments based on the calculations are also explained. There are great differences between the earlier versions and the latest standard especially regarding the measurement of uncertainty issues.

American Standard ANSI S12.18 (1994 and renewed in 2009): This standard also describes outdoor measurements under the reference conditions, taking into account the effects of ground and meteorological factors. The latest version defines a precision method by using the octave and one-third octave bands, narrow bands, and the A-weighted sound pressure levels, but does not discuss the noise-specific descriptors to be derived from the long-term measurements [96].

A wide range of measurement programs can be implemented to obtain the noise immission levels, from single or multi-point measurements at certain points relatively, in a short period of time, to monitoring the source operations by continuous measurements to obtain long-term data. Single point measurements can be conducted to deal with complaints from a specific noise source or to validate the results of predictions, while simultaneous measurements or multipoint measurements can be performed to provide data for noise mapping (Chapter 6). Defining the objective of measurements properly influences the required precision or the uncertainty degree, which is also dependent on the selection of measurement techniques and the instrumentation, in order to attain the correct results.

5.5.2 Observations in the Field

Prior to the environmental noise measurements, the following factors should be investigated through several site visits:

- Background noises: The residual noises should be at least 3 dB, preferably 5 dB lower than the specific noise. ISO 1996-2:1998 recommends that L_{pASmax} of the noise events of the source under test should be at least 15 dB greater than the average of residual sound [95b, 95d].
- Proximity of the other noise sources to the microphones at site.
- Accessibility to the site and the availability of communication and electrical systems.
- Terrain (topography) characteristics and building obstructions.
- Ease of access to the site and need for permissions.
- Safety conditions for microphone locations, i.e. security for the monitoring station and availability of places for remote control.
- The preferred uncertainty of the measurements.

The following information should be collected at the site where the measurements will be carried out:

Noise source:

- Description of source
- Operational conditions
- Source location and mounting
- Relationship between the source and the physical elements or between the source under investigation and the other sources.

Outdoor acoustic environment:

- Location and direction of source with respect to the measurement point, the distance in between
- Layout plan displaying source and nearby environmental elements (buildings, walls, roads, reflective surfaces, trees, etc.)
- Meteorological conditions (temperature and temperature gradient, relative humidity, atmospheric pressure)
- Winds (direction, speed, and magnitude)
- Other specific factors, such as magnetic fields, mechanical vibrations.

Interior acoustic environment:

- Placement of sources in building
- Size and configuration of indoor space and type of sound field
- Indoor surfaces and acoustic properties
- Building elements transmitting noise
- Indoor atmospheric conditions (temperature, humidity, air pressure).

After the measurement procedure is selected, pilot studies are useful to be able to gain sufficient practice on the methodology and in dealing with unforeseen problems. Such pre-measurements at reference points should provide information about the noise characteristics, directivities, temporal and spectral characteristics in short-term measurements, dynamic characteristics, range of spectra, etc.

In relation to the objectives of the measurement, a time schedule should be prepared also defining the team's work. A report displaying the results of the measurements, the analyses and evaluations should be prepared to cover the above information. The format of the measurement report is normally an annex to the specific technical standard to be implemented in the study (Section 5.5.7).

5.5.3 Measurement Units

The basic acoustic unit to be measured for immission measurements and evaluations is the sound pressure level in decibel, as defined in Section 1.2.2. However, because of the variety in source characteristics and type of impacts, various parameters in quantifying sound exposures have been used in the past and in the present.

ISO 1996-1:2016 gives the basic quantities and assessment procedures in Part 1. Although the symbols have changed in different versions of the standard over the years, with the addition of more descriptive indices, the principal physical measures to describe sounds are preserved. The standard defines the following basic quantities to be used for environmental noises [95a]:

- Equivalent continuous sound pressure level during time interval T; $L_{eq,T}$.
- Sound exposure level during the time integral T, $L_{E,T}$: For a noise event, the levels are recorded until the sound pressure level drops at least 10 dB below the maximum level.

- N Percent exceedance level during the time interval; T, $L_{N,T}$: obtained through measurements of $L_{eq,T}$ (at least in 1 second).
- Maximum time-weighted pressure level, L_{Fmax} and L_{Smax}: measured for a specified number of events of the source.

The descriptors for environmental noises should be chosen according to the objective of the measurement and type of noise sources with their characteristics under different operational and spatial conditions. Formerly, some of the descriptors could be determined by means of calculations from the instantaneous or short-term data, however, they have been delivered directly by advanced instrumentation and analysis techniques for nearly three decades.

The frequency range of measurements is specified as 63–8000 Hz (optionally 50–10 000 Hz) in octave bands in ISO 1996-2:2017, whereas the other standards specify different ranges and 1/3 octave bands. Low frequency measurements generally imply the range 16–200 Hz. Table 5.2 summarizes the international standards for immission measurements.

5.5.4 Overview of the Measurement Methods in the Field

Generally, noise measurements in the field cover the following processes by using the instrumentation explained in Section 5.4:

- *Signal recording and processing:* Conversion of the sound signals into electrical signals, *rms* calculations, calculations of sound power, pressure, intensity and their dB values, display in analogous (in the past) or digital formats.
- *Monitoring:* Instantaneous (real-time), short-, medium-, and long-term.
- *Recording and storing* data and outputs.
- *Analysis* (temporal, spectral, and spatial) and calculation of descriptors.

Development of the Measurement Methods from the Past to the Present

Although some national standards for sound pressure level measurements had been published in the 1960s, there was no international standard for the environmental noise measurements until the 1980s. However, at that time, various scientific documents and guidelines presented information about the measurement techniques, generally categorized into simple and advanced methods. Users were free to select the methodology according to the objective and type of the noise problem. Figure 5.30 displays a scheme for measurement set-up used in the late 1980s [12].

- *Simple method (the observation method):* The method requiring the least time and equipment was employed for simple evaluations of noise by using a single unit, e.g. to check the presence of a noise problem, to propose simple solutions, to apply penalties or to compare different sources in pilot studies. The equipment was portable and easy to control by hand, however, there were some difficulties in reading and handling the data. The portable recorders in the 1970 and the 1980s displayed the levels graphically and the analyses were made manually and the consolidated results were shown in tables.

    The observation method only applied in the existing situation for instantaneous detection of sound signals at certain receiver points, thus the number of receiver positions was restricted. After the 1990s, when the significant effects of certain physical factors on the measurement results were proved, this method was developed so that more accurate evaluations could be made based on the measurements.

Table 5.2 Standards for measurement of immission levels relative to environmental noise sources.

Sources	Content of the standard	ISO standard	Other standard
All sources	Outdoors and indoors	ISO 1996-1, 1996-2 [95a–d]	ANSI S12.18-1994 (R2009) [96] ASTM E 1686-16
Impulsive noise	Single or series of bursts	ISO 10843 [98]	ANSI S12.7 [97]
Traffic noise parameters	Road surface effect	ISO 11819-1, 11819-2 [99a, 99b] ISO 13472-1, 13472-2 [100a, 100b]	—
Railway noise	Inside	ISO 33811 [104]	EN 15892 [105]
Aircraft noise	Outdoor monitoring	ISO 20906 [106]	—
	Interior	ISO 5129 + A1 [107]	—
Waterways and harbor noise	Airborne sound from vessels	ISO 2922 [83a, 83b]	—
	Airborne sound from recreational craft	ISO 14509 [87]	—
	On board	ISO 2923 [88]	—
Noise from shooting ranges	Outdoors (prediction)	ISO 17201-4	—
	Muzzle blast noise	ISO 17201-1 [128a]	—
Industry noise	Outdoors	ISO 1996-1, 1996-2 [95a, 95d]	BS 4142 [112a]
	Workroom noise	ISO 14257 [112b]	–
Construction noise	Outdoors	ISO 1996-1, 1996-2 [95a, 95d]	BS 5228-1 + A1 [114]
Wind turbine noise	Outdoors	ISO 1996-1, 1996-2 [95a, 95d]	IEC 61400-11 [123] ANSI AS 4959 [124]
Building service equipment noise	Indoors	ISO 16032 [135] ISO 10052 [136]	EN 15657 [89b]
Hearing	Audiometric test methods	ISO 8253-1, 8253-2 [146a, 146b]	
	Hearing protectors	ISO 4869-1, 4869-2, 4869-3 [147c]	
	Occupational noise and workers risk	ISO 9612 [148a, 148b]	Directive 2003/10/EC [150]

- *Advanced method:* The expansion of noise pollution after the 1980s necessitated long-term measurements and detailed investigations in the field to obtain more precise results. Different measurements were inevitable due to the variety of noise problems, and the urgent and effective
decisions had to be made for noise control. The technology provided great flexibility in achieving
these objectives with the help of various technical standards published in this period.
- *Precision method:* When detailed descriptions of environmental noises are required, the precision
method is selected, for the purposes of scientific investigations, engineering solutions, strategic
decisions, imposing penalties, analyses in complex noise zones where a number of sources exist

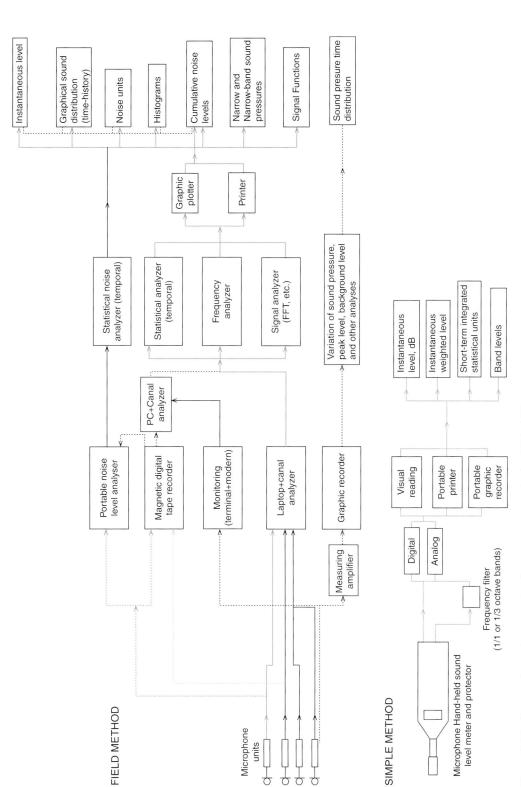

Figure 5.30 Set-ups for environmental noise measurements in the 1980s [12].

and for identification of the sources. Such work requires a preliminary analysis and knowledge about the acoustic environment and the individuals exposed to noise. Continuous noise recordings are acquired in a sufficiently longer period and afterwards, the results are analyzed in laboratories.

The recent measurement technologies are capable of conducting measurements using a number of microphone positions and of implementing instant analyses to speed up the delivery of results. The data stored can be transferred to the laboratory in the required format, for instance, to investigate the effects of different parameters on noise levels.

General Principles and Methodology

The basic principles regarding environmental noise measurements have been widely elucidated in the technical documents and books published so far. The recent International Standards are more comprehensive than in the past with the refined descriptions of the units and procedures to be implemented, in parallel to the advanced measurement techniques targeting accurate results. The following principles are still valid and worth mentioning:

- Correct definition of the noise problem or objective of measurements
- Selection of efficient measurement system
- Correct arrangement of receivers
- Monitoring the environmental factors and their effects on noise levels
- Selection and implementation of most appropriate method
- Conducting measurements according to required precision and accuracy
- Conducting measurements using trained technical personnel and analyses in accredited laboratories.

Implementation of the above principles in environmental noise measurements is outlined below with the updated information according to the current major technical standards.

The objective of the measurement, which should be clearly stated prior to study, is strongly related to the general principles referred to above, in addition, to how to evaluate the results according to expectations.

Outline of Measurement Procedure

The method to be selected for environmental noise measurements should be compatible with the objective, the type of noise source, the temporal and spectral characteristics of noise, and the environmental conditions. The standards published so far cover the comprehensive knowledge about the procedures to be employed for all types of noise measurements based on the earlier experiences [95a–c, 96]. The major standard on environmental noise measurements (ISO 1996-1) designates the sounds which are important for conducting measurements in the field (Figure 5.31) [95a]:

Total sound: Overall sound covering the individual noise sources at a given location at a certain time duration.
Specific sound: Sound produced by a particular noise source.
Residual sound: Implies the background noise when the specific sound is suppressed from the total sound.

Duration of measurements: The time intervals to be considered for conducting measurements are defined in the above-mentioned standards, in addition to BS 4142:2014. Figures 5.32a–5.32c describe the time intervals with respect to noise type:

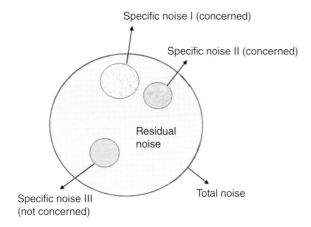

Figure 5.31 Definition of noises in environmental noise measurements.

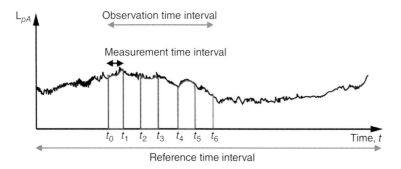

Figure 5.32a Steady (continuous) fluctuating noise and selecting measurement time interval.

- *Measurement time interval:* Duration for which measurements are conducted.
- *Observation time interval:* Duration for which a series of measurements are conducted.
- *Prediction time interval:* Duration for which levels are predicted.

The observation period to obtain the acoustic data depends on the temporal characteristics of the source, the response time of the measuring equipment, and the precision (accuracy) accepted for the measurements. Greater variation of sound level with time requires a longer measurement period. Time intervals for environmental noise measurements are recommended as at least hourly levels or day-time, evening, and night-time levels or total levels in 24 hours might be required. The duration of measurements can be grouped as follows:

- *Short-term measurements:* For the observations lasting one hour or less, relatively simple equipment can be adequate, depending on the units to be measured. The measurements are performed during time intervals under the well-defined emission and meteorological conditions and should cover all the significant variations of noise. For a periodic noise, the measurement time interval should consist of several periods, each representing a part of the complete cycle. For noise having single events, the measurement period is selected covering the duration of total event in order to determine the sound exposure level (Figure 5.33a).

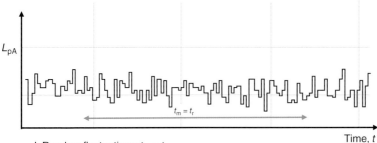

I. Random fluctuations: $t_m = t_r$

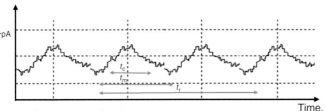

II. Continuous and cyclic: $t_r > t_c$
$t_m < t_r$
(Note: t_m should be selected to cover at least one or more complete cycle)

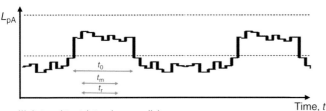

III. Intermittent (steady or cyclic) :
$t_0 < t_r$
$t_m \leq t_0$
(Note: t_m should be selected to obtain a representative value for $L_{Aeq,t\,m}$.
Additional correction for on-time :10lg (t/t_0) is made.)

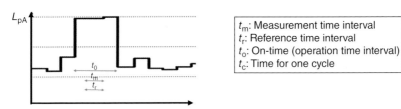

t_m: Measurement time interval
t_r: Reference time interval
t_0: On-time (operation time interval)
t_c: Time for one cycle

IV. Specific noise which is intermittent or cyclic:
$t_m = t_r$. So as to obtain max $L_{Aeq,t\,m}$

Figure 5.32b Selection of measurement time for different types of sources to measure A-weighted noise level
($L_{Aeq,Tm}$) described in BS 4142:2014.
Note: $L_{Aeq,Tm}$ values will be corrected according to the residual noise level in all cases.

Figure 5.32c Impulsive noise and selection of measurement time interval $(t_1-t_0) + (t_3-t_2) + (t_5-t_4)$ according to ISO 10843 [98].

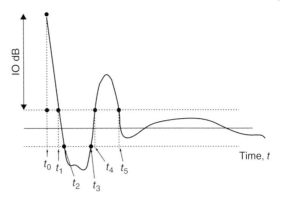

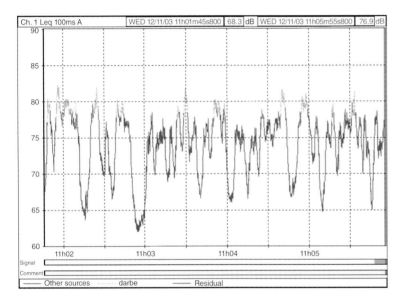

Figure 5.33a Short-term (5 minutes) temporal analysis of environmental noise.

- *Long-term measurements:* Measurements should be of a sufficiently long time, depending on the source operations and must cover all the meteorological conditions, but essentially the favorable situations to be able to derive an average or a representative level in an environment (Figure 5.33b). For detailed evaluations and for use of data in noise maps, the period is at least 24 hours, on certain weekdays or for a complete week in which the measurements are continued. Due to the seasonal factors on noise levels, the measurements are extended throughout a year in order to reduce the uncertainty. ISO 1996-2:2017 recommends that the measurement duration should cover the important emission and propagation conditions as well as possible. Monitoring noise at site is explained in Section 5.5.4.

For the independent measurements or for intermittent sampling, the time-intervals between two consequent measurements are specified according to the noise source, the distance to the source, and the time of day.

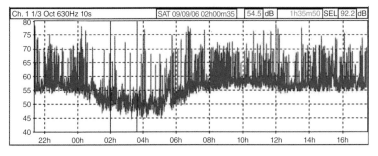

Figure 5.33b Long-term (24 hours) temporal analysis of environmental noise.

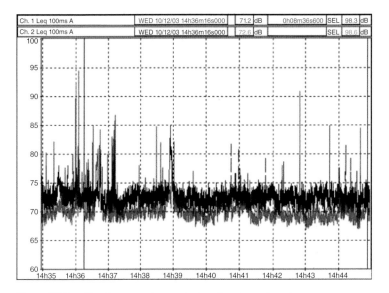

Figure 5.33c Continuous noise analysis using two microphones.

ISO 1996-2:2017 describes a procedure for extrapolation of the measured values to determine the long-term average noise levels from the single short-term measurements [95d].

Measurements according to the temporal characteristics of noise are as follows:

- *Measurement of continuous noises:* The continuous sounds are measured by A-weighted equivalent continuous sound pressure level, L_{Aeq} over a specified time interval with the slow response of the equipment. For fluctuating and intermittent sounds, the maximum A-weighted sound pressure level, L_{Amax} with a time-weighting (fast or slow) is used. It is possible to conduct simultaneous measurements with two (or more) microphones positioned at different locations, the time-history of each signal can compared as shown in Figure 5.33c.
- *Measurement of intermittent noises:* If a noise is steady but intermittent, the durations and intervals are determined at first and after the measurements, then the total levels are calculated by logarithmic summation of the levels obtained at each period. If the difference between the levels exceeds 5 dB, the noise is not accepted as steady. Normally the environmental noises impose such

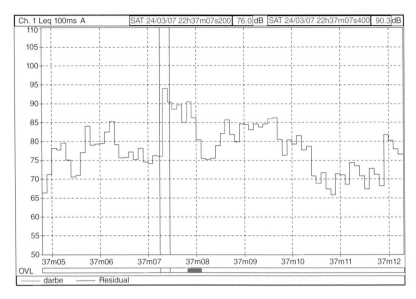

Figure 5.33d Determination of sound exposure level (*SEL*) through event identification from the continuous noise recording.

characteristics. As a general principle, the measurements should be conducted in the reverberant part of the sound field where the noise levels are rather constant (Section 1.3.4).

- *Measurement of events:* Single noise events can be measured by using L_E (*SEL*), L_{eq} and L_{max} with the specified time-weighting (fast or slow) and frequency weighting (A, C, I). The duration of a single event is the total time that the sound pressure level is within 10 dB of its maximum sound pressure level and it depends on the source type, e.g. 10–20s for an aircraft pass-by and less than 1s for gunshot. If there is a series of events, the number of events during the measurement period, the time interval between the events and duration of each event are to be determined. Ultimately the sound exposure levels can be readily obtained through the recorded time-history data (Figure 5.33d)
- *Measurement of impulsive noises:* As defined in Section 2.4, most impulsive sounds are produced during industrial and recreational activities. They can be monitored by special equipment by determining the enveloping curve covering the sound pressure variation (Figure 5.32c). Investigation of impulsive noises, the parameters such as pressure/time pattern (instantaneous sound pressure), rise and decay times, peak/average and *rms* pressures, must be determined. The duration of the impulse is the time in seconds required to reach its peak pressure and return

to zero. For repetitive impulses, the total time of pressure fluctuations (both positive and negative) is measured. The common units which are used for the measurement of impulsive sounds in environmental noise assessments are L_{max} with fast response and peak sound pressure level with C-weighting (L_{Cpeak}), that are explained in the standards [97, 98].

ANSI S12.7-1986 (reaffirmed in 2015 with amendments and changes): This is the first standard giving the methods of impulse noises [97]. It gives the requirements for measurement of time-varying and time-integrated quantities, instrument characteristics and measurement and analysis techniques in detail. For applications of the measurement procedure for impulsive noise, the standard introduces the time parameters such as "A-duration" according to the peak pressure and "B-duration" representing the time interval between impulses.

ISO 10843:1997 (confirmed in 2018): This is the standard aiming to evaluate the workplace noise with respect to hearing conservation and community noise problems, and it describes the measurement of single impulsive sounds and a short series of impulsive sounds emitted from machinery and equipment both indoors or outdoors [98]. Two kinds of measurements are explained: (i) measurement of phase-sensitive parameters, such as peak sound pressure level, L_{peak} and the other parameters characterizing the variation of sound in a time duration; and (ii) measurement of time-integrated quantities, such as frequency weighted sound pressure level, instantaneous sound pressure level, L_p or sound energy level, L_E (instead of time-averaged sound pressure level, L_{eq}).

The measurement conditions, procedures, analysis, and the uncertainty are explained in the standard.

- *Measurements according to spectral characteristics of noise:* It is possible to conduct measurements by recording signals with their spectral content and show them in the same chart (Figure 5.34).

Tonal noise: Noises consisting of dominant tones have been shown to increase the negative impacts on physiological and psychological status (Chapter 7), thus while evaluating noise, the

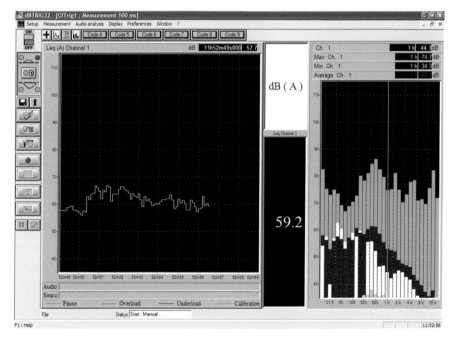

Figure 5.34 Dynamic analyses of temporal and spectral variations of noise.

audibility of tones should be detected. The basic principle during the measurements is to place the microphones so as to be able to receive such discrete tones.

As the simplest application to account for the prominent tones which are present in noise, the earlier version of ISO 1996-1:2003 recommended the adjustment of +3 dB to +6 dB to the measured sound pressure level, to obtain the "Rating Levels."

ISO 1996-2:2017 presents two objective methods for assessing the audibility of dominant tones in noise:

Engineering method (objective method): This method is related to the ability and mechanism of the auditory system and the masking phenomenon. In a critical band around the tone frequency, the difference between the tone and the masking noise, ΔL is determined and compared with a negative masking index. If the difference is greater than the masking index, the tone is audible. The recent ISO 1996-2 gives tonal adjustments, K_t as:

$$\Delta L \leq 2 \, \text{dB}, K_t = 0$$

$$2 \, \text{dB} < \Delta L \leq 9 \, \text{dB}, K_t = 3 \, \text{dB}$$

$$\Delta L > 9 \, \text{dB}; K_t = 6 \, \text{dB}$$

An International Standard dealing with the tone discrimination based on narrow-band analysis (or FFT), is under development at present.

Survey method: Measurements are conducted in the field or laboratory (not applicable in closed rooms due to the modal behavior of the room) and the time-averaged one-thirdoctave band spectrums are analyzed to determine the prominent discrete tone by comparing the sound pressure levels at adjacent bands. When the differences between two neighboring bands remain within the ranges given below, it can be stated that the noise has audible dominant frequency tone:

At low frequency bands (25 – 125 Hz) : 15 dB
At middle frequency bands (160 – 400 Hz) : 8 dB
At high frequency bands (500 – 10 000 Hz) : 5 dB

Measurement of noises with strong low frequencies ($f \leq 200 \, Hz$): Sounds containing low frequencies are important in human health, by increasing annoyance compared to the broadband noises, decreasing the pitch sensation below 60 Hz, and increasing ear pressure. The A-weighting should not be used in the measurements, since it suppresses the effect of low frequencies on the measured levels.

Measurements of such noises in closed spaces are conducted by increasing the number of microphone positions, because of the room resonances at low frequencies. The microphone positions with respect to reflective surfaces and the durations for sampling to extract data from the continuous measurements, in relation to the bandwidth, are explained in ISO 1996-2:2017. During outdoor measurements, the microphones should be on the façade to reduce the effect of background noise, which are dominated at low frequencies. The outdoor measurements for low frequency sounds should go down to 16 Hz in 1/3 octave bands with the microphones to be located at least 16 m from the nearest reflective surface to yield the free-field levels. The standard does not recommend a correction for rating low frequency noises, however, the special descriptors have been developed such as G-weighting or special criteria and limit values can be introduced into the regulations to take the low frequencies into account.

Arrangement of Receivers (Microphones) in the Field

If the measurements are to be performed according to a specification or a technical standard, the arrangement of measurement microphones at a site should comply with that mentioned in the documents. The measurements can be conducted in the open air or in buildings, the layout of the microphones should be shown on a scaled plan with the heights of microphones during pilot studies and the drawing should be acknowledged in the measurement report.

Number of microphones and positions: To be able to obtain the best sample from the noise with the required accuracy, the location of the microphones and the minimum number of microphones (for averaging) are important. Therefore, during the site inspection before taking measurements, sufficient knowledge about the uniformity of the sound field should be obtained. The technical standards discuss this issue in detail. According to ISO 1996-2:2017, the microphones must be positioned to minimize the effect of residual sound.

If the investigations are to be performed to identify the effects of noise on individuals, the microphone height should be positioned at ear level (about 1.1 m for a sitting person) or at the height of the floors of multistory buildings in which the residents live. For noise mapping, the height of the microphone is selected as 4 m $\pm$0.2 m from the ground.

If equal noise contours are the intention, the locations of the microphones, the spacing between them, and the number of microphones, all depend on the irregularity of the sound field and on the required precision of the map. A minimum number should be selected according to the 90% accuracy level. Roughly, if the max dimension of the source is 0.25 m, the reference microphone can be located at 1 m from the source, 2 m from the larger size sources. The microphone height can be taken as half of the source height. The distance between the microphone points should be at least $\lambda/4$ m (1 m at 100 Hz, 2 m at 50 Hz).

Larger machines radiate different sounds from various partial surfaces with different spectrums and vibration modes. For broadband noises, the low frequency sounds are generally variable with time and location, while it is almost uniform at high frequencies. If the wavelength of sound λ is not longer than the source dimensions, it can be considered that the source radiation is somewhat directive. Therefore, a sufficient number of microphone positions should be selected to determine the sound field accurately (e.g. more than 20). If λ is long enough, the sound at low frequencies can be accepted as uniform, radiating equal sound energy in each direction and the directivity pattern is smooth. For the purpose of identifying workplace noise, the microphones should be positioned at each ear of the workers and the measurements should be repeated at these locations without the workers present.

Orientation of the microphones: Since the earlier microphones had directivity characteristics, the orientation problem was important. It is best to receive a flat response when the random sounds are incident to the microphone diaphragm. This situation should be checked through the manufacturer's documents, otherwise, certain corrections should be applied to the recorded levels, especially at high frequencies. When the measurements are made in the free-field, the microphone should be directed toward the direction where the microphone is most sensitive and in the reverberant room, the microphone should receive the random sounds coming from all directions.

This problem has been eliminated with improved microphone designs. However, for hand-held equipment, and if the directivity is not known, the measuring device should be oriented toward the source by keeping it as far as possible from the body.

For indoor measurements by considering the room modes, at least three discrete microphone positions, with distance of 0.7 m in between, are selected or a moving microphone can be used for continuous noise. If low frequencies are dominant, one of the three positions should be in a corner and 0.5 m far from the corner and other walls, and the measurements are conducted in 1/3 octave bands without using moving microphones. The other microphones should be located at least 1 m from the sound transmission element. For larger rooms (i.e. >300 m^3), the numbers of microphones are increased.

Monitoring the Effects of Environmental Factors on Measurements

Environmental conditions, referred to in Section 2.3, should be considered during the measurements. Stationary and transient conditions of the environment affect the measured results and decrease the accuracy of the measurements, therefore, these factors should be observed, measured, or continuously monitored during the measurements and some preventions should be taken if necessary.

If standards require free-field conditions, it is necessary to check the existence of the inverse square law implying that the direct sound is dominant and the sound reduction from a point source is 6 dB per doubling of distance, as explained in Section 1.3.3. Generally, this condition is encountered in anechoic or semi-anechoic rooms or in the open air. If the free-field conditions are not met, while moving the microphone away the source, a series of max and min pressure levels occurs due to the destructive interference between the direct and reflected sounds.

- *Reflections:* Reflections from the nearby objects, whose dimensions are greater than wavelength λ of the sound, and also from the ground, should be considered when setting up the measurement systems. The effects of reflection and diffusion can be minimized while keeping a sufficient distance between the source and the microphones in relation to λ. Also, if the source is at $\lambda/2$ distance from the reflective walls, the reflection problem cannot be ignored. Practically, the inconsistency that can be observed in the received signal, in particular, a maximum increase, implies the effect of reflection. The path difference between the reflected and direct sound rays, which is greater than 3 m, causes an increase of 0.5 dB in the sound level. Since ground reflections produce successive maximums and minimums on the sound levels, the effect of the ground should be determined before measurements. ISO 1996-2:2017 specifies the microphone locations relative to the reflective surfaces in the environment.

- *Meteorological factors:* As explained in Section 2.3.3, the sound pressure levels change significantly with the meteorological factors, so the meteorological conditions should be representative of the noise exposure during the measurements. In the earlier version of ISO 1996-2:2007, only the favorable conditions during sound propagation were recommended for comparison and reproducibility purposes [95d]. The favorable condition occurs when the sound paths are refracted downwards and is defined with the curvature of downward sound ray (i.e. $R < -10$ km or $< +10$ km depending on the distance and height of the source and the receiver positions). The standard gave a meteorological window to determine the favorable conditions in relation to the smallest wind speed, cloud cover, heights, and time of day. The calculation method for R was also presented according to the average gradient of wind speed and temperature and other parameters.

The recent ISO 1996-2:2017 introduces a different methodology in this respect and allows measurements under any meteorological conditions, however, single measurements should be performed under favorable conditions. It is necessary to measure the meteorological parameters during measurements, such as wind speed and direction (to be measured at a height of 10 m), relative humidity, and temperature, in the short and long term and to provide information about the cloud coverage in relation to time of the day and occurrence of precipitation. A matrix meteorological window defining the propagation conditions as favorable, neutral, unfavorable, and very favorable conditions at the shortest distance from the source to the receiver is described (Table 5.3). The radius of curvature R_{cur}, (m) can either be calculated or determined from the meteorological window using the measured wind speed and the horizontal distance between the source and the microphone, D. The negative values indicate an upward condition.

Table 5.3 Determination of the horizontal distance D between source and receiver from the meteorological windows described in ISO 1996-2:2017.

Meteorological windows – description of conditions	Vector wind speed component at 10 m	D/R_{cur}	
		Range	Representative value
M1 Unfavorable	< 1 m/s at day < −1 m/s at night	< −0.04	−0.08
M2 Neutral	1 m/s – 3 m/s	(−0.04) - (0.04)	0.00
M3 Favorable	3 m/s – 6 m/s	(0.04)–(0.12)	0.08
M4 Very favorable	> 6 m/s at day ≥ −1 m/s at night	> 0.12	0.16

Source: adapted from [95d].
The meteorological factors: wind speed, wind direction, relative humidity, temperature (to be measured), the radius of the curvature, R_{cur} (to be calculated)

Since the sound pressure levels change under different meteorological conditions, however, under the below given source-receiver configurations and on soft ground; the levels can be considered as relatively stable:

$$\frac{h_s + h_r}{r} \geq 0.1 \tag{5.29}$$

h_s: Height of source, m
h_r: Height of receiver, m
r: Horizontal distance between source and receiver, m

Uncertainties for the above propagation conditions are given in Section 5.6.2.

The standard requires that the measurement results are presented in a matrix system called an emission window which displays the emission conditions in relation to the meteorological conditions during the measurements. This is important especially for the environmental noises which these two conditions are strongly related to each other, like aircraft noise.

The standard explains the practical measures to be taken against precipitation, magnetic fields, mechanical vibrations, etc.

Other Considerations in Conducting Measurements

The measurements have to be conducted with a required accuracy. Selection of the measurement method, which is described in a technical standard, should be made according to the accuracy grade which is declared in the standard description. Determination of the uncertainty, which is mandatory to be given in the report, varies with the measurement method and objective of the measurement. This subject is explained in Section 5.6.2.

Calibration of equipment and checking the microphone sensitivity frequently are important with respect to the precision of measurements and will be explained in Section 5.11.

The environmental noise measurements should be conducted by well-trained personnel and in accredited laboratories. Sufficient knowledge and experience are important from the standpoint of the reliability of the results. Education and training are discussed in Section 8.5.

5.5.5 Dealing with Background Noise in the Measurements

The measurement results, defined as total sound, contain the sound from the noise source and the residual noise also called the background noise (*BN*) or ambient noise in the previous documents. In order to determine the net level of the specific sound, it is necessary to apply a procedure, if the level of residual noise is known (Section 1.2.2 (Box 1.4)). Different solutions are recommended to eliminate the background noises during the measurements:

1) Selection of the measurement time while the background noise is at its lowest level or exclusion of the measurement intervals when the *BN* is high.
2) Using a directive microphone and orienting toward the source, especially if there is a high frequency component in the noise.
3) Mounting the microphone on the façade to eliminate the noises coming from behind (then the levels will be corrected).
4) Performing statistical analysis on time-history data and eliminating the residual noise if it displays the Gaussian distribution.
5) When the *BN* cannot be detected clearly and when it is not steady, repeating the measurements at different distances until a constant level is obtained.
6) If the *BN* cannot be stopped during the measurements, apply the following procedure:

   Step 1: The total noise from the source (under operation) and other sources is measured, L_{N+S}

   Step 2: The source is stopped and only *BN* is measured. If the source cannot be stopped, the *BN* can be measured by finding a similar environment with the same source and receiver condition, L_N.

   Step 3: The difference between the levels is obtained using logarithmical subtraction or a simple correction is applied to obtain the specific source level, L_S:

$$L_{N+S} - L_N = \Delta L \quad (\Delta L : \text{Difference})$$

   Condition 1: If $\Delta L > 3$ dB(A);
   - Using a simple graphical method, $L_{\text{correction}}$ is found from the equation given in Section 1.2.2 (Box 1.4):

$$L_S = L_{N+S} - L_{\text{correction}}$$

   - Using the logarithmic subtraction operation:

$$L_S = 10 \lg\left(10^{L_{N+S}/10} - 10^{L_N/10}\right) \tag{5.30}$$

   Condition 2: If $\Delta L > 10$ dB(A); $L_{N+S} = L_S$
   Condition 3: If $\Delta L < 3$ dB(A); L_S cannot be determined.
7) If the *BN* levels, which are measured as A- or C-weighted levels, cannot be discriminated from the noise levels of the source but can be detected audibly, the above analysis should be made for the spectral levels (e.g. octave or one-third octave band levels).

5.5.5.1 Comparison with Noise Limits and Evaluations

National regulations contain noise limits that are declared in different formats. In some regulations two approaches are recommended:

- Comparison of the measurement results, after correction, with the limit values directly. Generally, the corrected value representing the specific source level is directly used in such comparisons:

$$L_S < L_{limit}$$

- Comparison according to background noise criteria: Using the uncorrected levels, the combined noise level should not exceed the background noise above a certain level (e.g. about 3–5 dB). This criterion is applied for the A-weighted levels or in the octave band levels of each:

$$L_S < L_N + 5 \text{ (or 3 dB)}$$

The second criterion is declared in some regulations and guidelines as:

$$L_{eq \text{ (BN)}} \text{ dB(A)} + 3 \text{ and } L_{eq \text{ (BN)}} \text{ dB(C)} + 5 \text{ dB}$$

The acoustic characteristics of noise, such as time variations, tonality, impulsiveness, low frequency content, type of source should be considered in comparisons. On the other hand, the limits should be examined with regard to the following points:

- the time (day-time, whole day, or night-time) when the limit value is applicable;
- the time-weighting required in measurements (slow or fast);
- the frequency weighting required in measurements (A- or C-weighting);
- the applicability of the limits for impulsive sounds (I).

Other important aspects are:

- Comparison may be necessary for each frequency band.
- Since generally the limits are not given as spectral values in the regulations, transformation may be necessary by using a normalized average spectrum of the source.
- If the comparison will be performed with the indoor limits, the A-weighted levels can be converted into the indoor noise criteria based on the octave band frequencies, such as *NR, NC* or *NCB* curvatures (see Chapter 7).

In the recent standard ISO 1996-2:2017, the following recommendations are given by substituting L for L_S, L' for L_{N+S} and L_{res} for L_N.

If L_{res} is equal or less than 3 dB below the total noise level, no correction is applied. The results should be reported. When the residual level is more than 3 dB below the total level, then a correction is applied according to the logarithmical subtraction procedure given in Box 5.2.

L is defined as the "corrected sound pressure level", L' is defined as the "measured sound pressure level" and L_{res} is the "residual sound pressure level".

5.5.6 Analyses of the Measured Data

Parallel to the variation of environmental noise levels referred to in Chapter 2, the evaluations of environmental noises are generally accomplished by applying the following analyses:

1) *Temporal analysis:* Chapter 1 gives the descriptions of noise varying in a certain time period (i.e. stationary, interrupted, impact, and impulsive). Measurement of different types of noise necessitates specific measurement techniques as described in Section 5.5.8. A selection of reference times and measurement times is shown in Figures 5.32a–5.32c according to various standards.

Box 5.2   Examples of Evaluation of a Noise Problem by Comparing the Measured and Corrected Source Noise Levels with the Noise Criteria

Practice 1:

$$L_{\text{limit}} = 65\,\text{dB(A)}$$
$$L_{N+S} = 75\,\text{dB(A)}$$
$$L_N = 71\,\text{dB(A)}$$
$$L_S = 10\lg\left(10^{75/10} - 10^{71/10}\right) = 72.8\,\text{dB(A)}$$

a. Since 72.8 > 65 dB(A), there is a noise problem.
b. Since 72.8−71 = 1.8 < 5 dB(A), (a) is not supported and (b) must be checked for each frequency band.

Practice 2:

$$L_{\text{limit}} = 65\,\text{dB(A)}$$
$$L_{N+S} = 75\,\text{dB(A)}$$
$$L_N = 66\,\text{dB(A)}$$
$$L_S = 10\lg\left(10^{75/10} - 10^{66/10}\right) = 74.4\,\text{dB(A)}$$

a. Since 74.4 > 65 dB(A), there is a noise problem.
b. Since 74.4−66 = 8.4 > 5 dB(A), (a) is supported.

Practice 3:

$$L_{\text{limit}} = 65\,\text{dB(A)}$$
$$L_{N+S} = 75\,\text{dB(A)}$$
$$L_N = 72\,\text{dB(A)}$$
$$L_S = 10\lg\left(10^{75/10} - 10^{72/10}\right) = 72\,\text{dB(A)}$$

a. Since 72 > 65 dB(A), there is a noise problem.
b. Since 72−72 = 0 < 5 dB(A). As the source level is equal to the background noise level, the existence of a problem cannot be determined, therefore, (b) must be checked for each frequency band.

Temporal variations are determined by means of time-history diagrams and the statistical analysis of the measured sound levels. The metrics are equivalent sound pressure level, L_{eq} which is time-integrated sound pressure level from short- or long-term measurements, L_E from single noise events and the further parameters such as the exceeded noise level from cumulative distribution analysis, etc. Figures 5.33a and 5.33b, show examples of short-term (5 minutes) and long-term (24 hours) measurements of L_{eq} respectively. From the continuous recordings, the single noise events within a specified duration can be detected through a an algorithm in the software and the L_E (or *SEL*) is calculated as given in Figure 5.33d.

2) *Spectral analysis:* Displaying sound pressure levels or pressure amplitudes at octave or one-third octave bands is inevitable, particularly when the noise spectra contains pure-tone sounds. If the continuous measurements are conducted so as to obtain information about the band levels, the variation of the spectral levels can be investigated dynamically in real time, as shown in Figure 5.34. The spectrum diagrams display instantaneous, average, max/min spectrums and the spectral time-history charts are useful in analysis of dominant tones.

3) *Spatial analysis:* Determination of variations in sound pressure levels at different microphone positions is made through simultaneous measurements using multichannel systems or monitoring terminals. Thus, it is possible to perform comparisons, superpositions, or optimizations of the recorded levels. However, the noise level distribution in an environment including the effects of all the physical factors, can be better investigated through noise maps (Chapter 6).

Managing the Results after Measurements

The current standard ISO 1996-2:2017 states the following processes to apply to the measured levels:

1) Unwanted events with high residual sound levels are eliminated.
2) Measured levels are corrected according to reference microphone values which correspond to the free-field levels (excluding reflections from façades but including reflections from the ground and other objects) according to the microphone locations relative to the reflective surfaces.
3) The measured levels are placed on a specific window (meteorology and operation condition matrix).
4) The measured levels are corrected according to residual (background) noise levels as explained in Section 5.5.5.
5) The levels are corrected according to the reference conditions (including reference traffic and reference atmospheric conditions).
6) $L_{eq,T}$ are calculated for each window as:

$$L_{eq,T} = 10 \lg \frac{\sum_i^N \Delta T_i 10^{0.1 L_{eq,i}}}{\sum_i \Delta T_i} \text{ dB} \tag{5.31}$$

ΔT_i: Duration of each measurement period

Correction for microphone location in relation to reflective surfaces: Since the noise limits given in the regulations are the free-field sound pressure levels, therefore, the measure values obtained in various microphone locations are normally influenced by the reflections from vertical (e.g. façades of buildings) or horizontal surfaces (e.g. the ground) around the microphone. In the informative part of the standard, corrections are proposed for the measured levels according to the microphone locations in relation to the reflective surfaces. The following method is applicable for traffic noise.

Microphone directly on the surface(flush-mounted) : + 6dB

Microphone near reflecting surface : + 3dB

To be able to apply this correction for the extended sources, i.e. road traffic, the required microphone positions have been described with respect to the angle of view of the source from the microphone position [95d]. This configuration ensures that the incident and reflected sounds are equally strong to be able to apply +3 dB correction to the measured level.

Correction to the reference condition: The measurement level is corrected according to the reference condition, by applying the following procedure [95d]:

- The measured results shown in the measurement window with the time periods and with the meteorological parameters, i.e. *t, h, d* (*t*: temperature, *h*: humidity percentage, *d*: measurement distance, *m* from the source) measured during the tests.
- The atmospheric sound attenuation is calculated by using ISO 9613-1 (Chapter 4) for the frequency bands, ΔL_a (*t, h, d*).
- A reference atmospheric sound attenuation, ΔL_a (t_{ref}, h_{ref}, *d*) is determined as described in the standard, e.g. by taking into account the yearly average temperature, or the relative humidity at the same area.
- The correction to apply to the measured levels is determined and the corrected levels are obtained:

$$\Delta L_a = \Delta L_a \left(t_{ref}, h_{ref}, d \right) - \Delta L_a(t, h, d) \quad \text{dB} \tag{5.32}$$

$$L_{eq,ref} = L'_{eq} + \Delta L_a \quad \text{dB} \tag{5.33}$$

L'_{eq}: Measured level

$L_{eq,ref}$: Corrected measured level

The standard uses the corrected measured value to estimate the measurement uncertainty.

Determination of rating noise levels: Evaluation of environmental noises needs some adjustments to be added to the measured or predicted levels with respect to the negative impacts of noises with certain characteristics. The adjustments are declared in the International Standard (ISO 1996-1:2003) according to source type, exposure time, impulsiveness, tonality, duration of the noise events [95a]. Table 5.4 gives the adjustments, with the application notes described in the standard. The adjusted sound pressure levels and sound exposure levels are defined as "Rating Levels" depicted as L_{Req} and L_{RE} respectively and given with a specified time interval. Rating can be applied also to the day, evening, and night level, L_{Rden}, which is defined as "composite whole-day rating level" or to the night-time level, L_{Rn}.

There are differences between the 2003 and 2016 editions of the standard, i.e. aircraft noise adjustment is (5) to (8) and railroad adjustment is (-3) to (-6) in the 2017 edition. Railroad adjustment is applicable for conventional electric passenger trains without causing the vibration problem. The adjustments for impulsive noises are valid when they are audible at the receiver point and if they are not separated from the background noise, no adjustment is applied. The new ISO 1996-1 also covers further notifications on this issue.

5.5.7 Reporting the Measurement Results

The field observations and the measurement results explained above are to be presented in a report. Some of the information to be given in the report is as follows:

Measurement system (instrumentation):

- List of equipment, names, serial number and manufacturer
- Special accessories used during the measurements
- Bandwidth of the analyzer
- Acoustic calibrator type, calibration level, and method of calibration

Table 5.4 Adjustment values applicable to determine the rating units.

Aspect	Specification	Level adjustment, dB
Sound source	Road traffic	0
	Aircraft	(3)–(6)
	Railroad[a]	(-3)–(-6)
	Industry	0
Characteristics of noise	Regular impulsive[b]	5
	High impulsive	12
	High-energy impulsive	Given in the standard
	Prominent tones[c]	(3)–(6)
Time period	Evening	5
	Night	10
	Weekend day-time[d]	5

Source: adapted from Table A.1 of ISO 1996-1:2003. [95a].
[a]No adjustment for long diesel trains and for trains with the speed exceeding 250 km/hr.
[b]Some countries apply objective tests about impulsivity.
[c]The presence of prominent tones should be verified through ISO1996-2.
[d]To be added to L_d as defined by the corresponding authority.

Presentation of the measurement results:

- Name of technicians and observers
- Date, time, and place of measurements
- Type of source, structure, sound emission information, type of noise, location of source, and general layout plan
- Description of operating condition, number of noise events (passing vehicles divided into suitable categories)
- Meteorological conditions (wind, duration/speed, relative humidity, temperature, atmospheric pressure) and location of meteorological measurement equipment
- Sky condition; open, closed, cloudy and the ground condition (dry, wet)
- Methods used to extrapolate the measured values to other conditions
- Result of acoustic calibration
- Weightings used
- Response position of the sound level meter (time-weighting)
- Measurement times, durations and intervals, variability of sound emission
- Measured sound levels in different descriptors L_{eq}, L_{max}, L_E as weighted and spectral (band) levels, N percent exceedance levels
- Background noise at each microphone position (weighted and spectral levels)
- Correction values if applied for microphone, cable, and background noise
- Corrected weighted levels
- Rating levels
- Calculation of uncertainties and the results

5.5.8   Outline of Source-Specific Measurements in the Field

The necessity for assessment of noise impacts in communities has encouraged field studies, by applying the major technical standards explained above. However, some source-specific methodologies are needed in the measurements and evaluations of environmental noises because of the differences in acoustical and operational characteristics of sources and the propagation conditions.

An outline of such dedicated standards currently implemented and the measurement procedures for different sources are given below.

A) Measurement of Traffic Noise

Specific purpose: As the most common noise source in the environment, the major purpose of the immission measurements of traffic noise is to determine the noise impact in urban areas by comparing the measured levels with the scientific noise criteria or the noise limits given in the regulations. Various objectives explained in Section 5.5.1 can also be stated for traffic noise measurements. Measurement of road traffic noise has been undertaken, both in the past and at present, for research and implementation purposes and to acquire data for noise mapping or verification of noise maps. Some of the results derived from the field studies are explained in Section 3.2.1. The noise emission measurements are given in Section 5.4.2.

Type of noise: Continuous and fluctuating noise.

Observations regarding the source in the field: During the measurement of traffic noise, the following information is compiled:

- *Traffic conditions*: Total traffic volumes, numbers of each vehicle category (passenger cars, medium heavy vehicles with 2 axles and heavy vehicles with ≥3 axles), average speeds, heavy vehicle percentages (numbers of pass-by are determined through counting or predicting according to the vehicle categories), type of traffic flow (continuous, interrupted, etc.), stop/start conditions and accelerations due to traffic lights, crossings, roundabouts.
- *Road conditions:* Road gradient, surface cover, dry/wet conditions, quality of surface, number of lanes, widths.

Observations regarding the environment: The factors to be determined are given in Section 5.5.2.

Measurement metrics: Short-term, medium-term, or long-term equivalent continuous sound pressure levels, L_{eq} specified with time (hours), statistics about distribution of sound pressure levels with the standard deviations and the maximum sound pressure level, L_{Amax} for different vehicle categories. L_{Amax} can be directly measured based on a certain number of pass-bys, however, it can be calculated as an arithmetic average, an energy average, or from the exceedance levels. ISO 1996-2:2017 gives a method to determine the energy mean values of L_{Amax} from the arithmetic average of the samples for each vehicle category and the standard deviation. Examples of traffic noise recordings are given in Figures 5.33a and 5.33b.

Measurement duration: The time interval for measurements of L_{eq}, is selected to cover at least three categories of vehicles (passenger cars, medium and heavy vehicles). In the 2007 edition of ISO 1996-2, at least 30 pass-bys of vehicles of two categories (heavy and light) could be selected for L_{eq} and L_{max}.

Measurement uncertainty: Computed as required in ISO 1996-2:2017 (Section 5.6).

Measurement standards to apply: General principles and procedures are described in ISO 1996-1:2016 and ISO 1996-2:2017. Specific standards have been developed to measure the effects of different parameters on tire-road noise only and on total traffic noise emission, such as texture and porosity of road surface: ISO 11819-1:2013 and ISO 11819-2: 2017 [99a, 99b] and sound absorption properties of surfaces: ISO 13472-1:2012 and ISO 13472-2:2013 [100a, 100b].

Evaluation of results: The measured levels to be presented in terms of a selected descriptor (metric), are compared with the limit values given in the national and international regulations or directly used according to the objective of the study, after the statistical and spectral analyses.

Reporting: General principles about information and reporting data are implemented in addition to the requirements of the specific standards.

B) Measurement of Railway Noise

Specific purpose: The major purpose of field measurements of railway noise is to determine the noise impact on buildings near railways, in both urban or suburban areas, and to compare the measured levels with the scientific noise criteria or the noise limits given in the regulations. The other objectives mentioned for traffic noise can also be valid for railway noise measurements (Figures 5.35a and 5.35b) [101, 102].

Train noise measurements are generally conducted to acquire data regarding noise spectrums and noise levels of different types of trains, with great attention given to high speed trains, the effects of various factors on train noise, e.g. track and rail properties, to identify the noise sources. Special emphases have been given to tunnel booms, rail vibrations, rail beds, parked trains, bridges, abatement systems, etc. [103]. Some derivations from these studies are given in Section 3.2.2.

Type of noise: Intermittent noise with individual sound events (pass-bys).

Observations regarding source: During the measurement of train noise, the following information is compiled:

- *Conditions of railway traffic:* The volume of total railway traffic and number of pass-bys per each train category with type of brakes, average speeds, and accelerations near stations, distance to stations, numbers of cars in each train and length of trains (train categories such as high-speed trains, inter-city trains, regional trains, freight and diesel trains).
- *Conditions of railway (track):* Rails (rail unevenness, track roughness), connections, angle of view from the microphone location). Barriers along the railway, cuttings, embankments, railways at grade or elevated structures crossings, railway curvatures, etc. If the measurements are conducted near stations; the platform height, width, surface, station buildings, etc.

Observations regarding the environment: The factors to be determined are given in Section 5.5.2.

Measurement metrics: The equivalent sound pressure levels, L_{eq}, directly measured or calculated from the measured sound energy levels, the sound exposure level, L_E of the train pass-bys, the maximum sound pressure levels, L_{Amax} for each train category (i.e. high-speed, inter-city, regional, freight, and diesel trains) and the average spectrum levels between 50 and 5000 Hz. ISO 1996-2:2017 gives a method to determine the energy mean values of L_{Amax} from the arithmetic average of the samples for each train category and the standard deviation.

Measurement duration: Since train noise varies a great deal with time, the time for measurements is selected according to the objective and uncertainty suggested in ISO 1996-2:2017, e.g. 24 hours continuous for noise mapping by obtaining discrete data for daytime, evening, and night-time levels. In order to determine the above metrics, the measurement duration should include at least 10 train pass-bys and there should be at least 15 minutes or more wait if the required number of trains could not be obtained, whereas the 2007 version of the standard suggested 20 pass-bys for L_{eq} and L_{max}

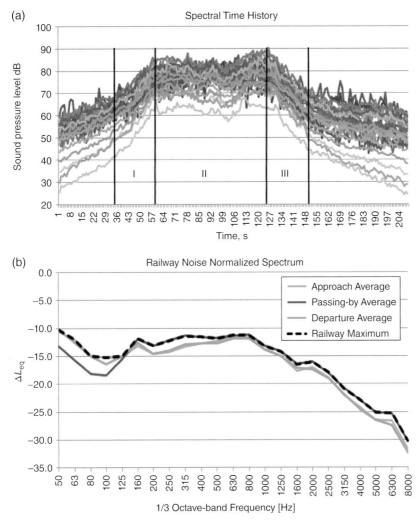

Figure 5.35 (a) Pass-by noises of different trains; (b) the normalized spectrum of railway noise [101, 102].

determinations. The minimum time interval between two independent measurements of trains is given with respect to distance from the railways: at up to 100 m:24 hours for day and night, at 100–300 m:48 hours for day and 72 hours for night, at longer distances: 72 hours for day and night.

Measurement standard to apply: According to ISO 1996-2:2017, L_{eq} directly measured or calculated from the measured L_E values during each train passage. If the measurements are performed because of complaints, the façade levels are measured for at least five different train pass-bys according to the earlier standard (2007). For planning or design purposes or to prepare an Environmental Impact Assessment in an EIA report, the selection of measurement microphones at about 300 m from railway is sufficient. If measurements are to be conducted in maintenance areas, the high pitch rolling sounds are considered in recordings and the rating units are determined by applying corrections for specific tonal noises. Sirens should be analyzed and eliminated from continuous recordings. Measurement standards (e.g. ISO 3095:2013) for noise emitted by railway vehicles, i.e. locomotives, wagons, etc., are explained in Section 5.4.2.

The specific standards including the measurements of interior noise of railway vehicles, including inside the driver's cab have been published [104, 105].

Measurement uncertainty: Computed as required in ISO 1996-2:2017 (Section 5.6).

Evaluation of results: The measured levels described as the required descriptor (metric) are compared with the limit values given in the national and international regulations or directly used according to the objective.

Reporting: General principles on information and reporting data are implemented in addition to the requirements of the specific standards.

C) Measurement of Aircraft and Airport Noise

Specific purpose: The major purpose of immission measurements of aircraft noise is to determine the noise impact in built-up areas near airports, by comparing the measured levels with the noise limits. Investigation and comparison of the different type of aircraft noises and the other objectives mentioned in Section 5.5.1 are also valid reasons for aircraft noise measurements (Figure 5.36) [101, 102]. The noise emission measurements of different aircraft are explained in Section 5.4.2. The results of some experimental studies are given in Section 3.2.3.

Type of noise: Intermittent noise with individual sound events (aircraft flyovers).

Observations regarding source: During the aircraft noise measurements, the following observations are made:

- *Conditions of aircraft traffic:* Number of total flights in day-time, evening, and night-time, and hourly values, numbers per each aircraft category and per operating condition, interval between the two flights, etc.

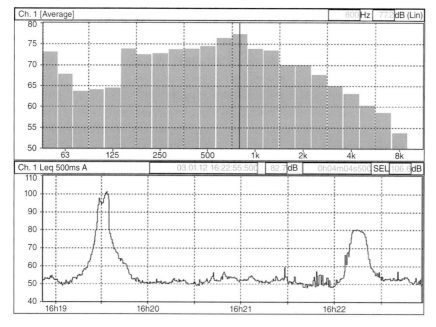

Figure 5.36 Aircraft fly-over noise recording and spectral analysis [101, 102].

- *Conditions at airport:* If measurements are conducted near airports, number of runways and footprints of flight path, total number of landings and take-offs, average flight speeds, flight routes and procedures, distance to microphones.

Observations regarding environment: The factors to be recorded are given in Section 5.5.2.

Measurement metrics: The sound energy levels, L_{AE}, are measured for each aircraft category and the equivalent continuous sound pressure levels L_{Aeq} can be determined for each operation type of flights (traffic pattern: take-off, landing, etc. and the noise propagation conditions). The other units are spectral sound exposure level, L_E, maximum AS-weighted sound pressure level, L_{pASmax} with its time-history, the $N\%$ exceedance level, $L_{p,AS,N,T}$ and the background noise level, $L_{pAS,95}$. The threshold level, $L_{threshold}$ is also defined as an appropriate user-defined sound pressure level to be used for reliable event detection. For busy airports with dense traffic and complex situations, L_{den} or L_{dn} is necessary to be determined for each runway.

During continuous monitoring, it is required to record the equivalent continuous sound pressure level, $L_{p,eq,T}$ (time-averaged sound pressure level) as a series of short time levels each averaged over 1 second, called "one second equivalent continuous sound pressure level," $L_{p,eq,1s}$.

For L_{Amax} measurements, ISO 1996-2:2017 gives a method to determine the energy mean values of L_{Amax} from the arithmetic average of the samples for each aircraft category and the standard deviation.

Measurement duration: Since aircraft noise varies a great deal with time, the time for measurements is selected according to the objective, as 24 hours continuous for noise mapping by obtaining discrete data for daytime, evening, and night-time levels.

Measurement standard to apply: According to ISO 1996-2:2017, L_{eq} is directly measured or calculated from the measured L_E values of aircraft pass-bys. The intervals between the two independent aircraft measurements are recommended as: 24 hours for the distance up to 100 m, 48 hours for distances of 100–300 m, and 72 hours for longer distance from the source [95d].

For noise measurements following complaints, the façade and roof levels of the complainant's building are to be measured for at least five different aircraft operations as landing, take-off or flyover, according to former version of ISO 1996-2:2007, and for L_{pASmax} at least five but preferably 20 events. However, this number in the latest standard is determined according to the required accuracy. The measured levels are compared with the airport limits given in the local regulations. For planning or design purposes or preparation of the Environmental Impact Assessment in the EIA report, it is necessary to obtain noise contours (noise maps) in the area. ISO 20906:2009 + A1:2003 (confirmed in 2014) describes the airport noise monitoring (i.e. operation of permanently installed systems around airports, however, not for validation of noise maps), the requirements for instruments and installation, data recording, reporting data, and uncertainty determination [106]. It also introduces a threshold system of sound event recognition (i.e. aircraft sound events and non-aircraft sound events) in complex noise situations and during uninterrupted measurements. For interior noise of aircraft, the ISO 5129:2001 + A1:2013 (confirmed in 2017) gives test procedures to be applied at crew and passenger locations under steady flight conditions [107].

Measurement uncertainty: Computed as required in ISO 1996-2:2017 and ISO/IEC Guide 98–3:2003 as given in Section 5.6.

Evaluation of results: The measured levels described with the required descriptor (metric) are compared with the limit values given in national and international regulations or directly used according to the objective, including validation of noise maps.

Reporting: General principle on information and reporting data are implemented in addition to the requirements of the specific standards (Section 5.5.7).

D) Measurement of Waterway Noise

Specific purpose: The major purpose of immission measurements of waterway or seaway noise is impact assessment on the residential areas near ports, harbors, naval docks or shipyards, or along seashores or rivers [108, 109]. The other objectives mentioned in Section 5.5.1 can be stated for seaway noise. Some experimental studies and results for waterway noise are given in Section 3.2.4. Noise emission measurements of different vessels are explained in Section 5.4.2.

Type of noise: Intermittent noise with individual sound events for harbors and seashores with low frequency content, stationary, impulsive, and tonal noises at shipyards.

Observations regarding source: During the seaway noise measurements, the following observations are made:

- *Conditions of waterway traffic*: During the waterway or marine noise measurements, traffic conditions should be determined (i.e. types of ships, such as large container ship, cruise ship, etc., their numbers per unit time, volumes per each vessel category, total boat pass-bys, average speeds, accelerations, the passenger terminals nearby, distance to the docks, cruising route, and distance to seashore, etc.).
- *Conditions for harbors, ports, yards, etc.*: Layout in relation to settlements, size of port area, activity type (for loading and unloading goods), topography, and obstructions around

Observations regarding environment: The factors to be recorded are given in Section 5.5.2.

Measurement units: The A-weighted sound energy levels, L_{AE}, measured for each type of vessel or ship, the A-weighted equivalent continuous sound pressure levels, L_{eq}, along with time-history to be directly measured or calculated from the measured L_{AE} for each vessel pass-by to be determined in a certain time interval, the spectral levels of sound exposure, L_E, the A-weighted maximum levels with slow response L_{pASmax} with its time-history, the exceedance level, $N\%$, $L_{pAS,N,T}$, day-time, evening, or night-time level L_{den} or day-night level L_{dn} for large ports carrying dense seaway transportation.

Measurement duration: Since the vessels travel at much lower speeds along rivers and seashores compared to the other transportation systems and the number of pass-bys is also low, implying that waterway noise does not change much with time, thus, the duration of measurements is selected according to the objective and the specific cases. At least five vessel pass-bys in each category should be selected for façade noise insulation purposes. However, it is necessary to measure L_{eq} continuously within 24 hours for the noise assessments in nearby areas.

Measurement standards: ISO 1996-2:2017 is the basic standard for measurements for waterway noise impact assessments, since there is not a specific standard for vessel traffic at the moment. Noise levels at open air manufacturing and maintenance areas (shipyards) can be measured as industrial areas, by paying attention to dynamic range of noise and frequency characteristics, i.e. high-pitched metallic sounds and high volume of impact noise.

Measurements on board: ISO 2923:1996 (confirmed in 2017) describes the procedures for measurements inside the machinery and accommodation spaces, such as engine rooms, cargo spaces, passenger cabins, restaurants, lounges, offices, etc., with respect to the crew's health and performance and, the passengers' comfort, speech intelligibility, and alarm detectability [110]. It is also aimed to compare the results of various vessels and to determine the noise abatement measures. Mechanical noise from air-conditioning units should be measured by applying the equipment noise standard as explained in Section 5.4.1. In addition to airborne sounds, the structure-borne sounds which are transmitted easily through the lightweight metal structures into the noise-sensitive spaces and the mechanical vibrations have to be measured.

Measurements for assessment of risks for underwater life specific standards are applied using different measurement systems as described in various documents [111].

Measurement uncertainty: Computed as required in the specific measurement standards, such as ISO 1996-2017 for the immission measurements (Section 5.6). ISO 2923 ensures that the standard deviation of reproducibility of the A-weighted equivalent continuous sound pressure level, L_{eq} is equal or less than 1.5 dB.

Evaluation of results: The measured overall levels and spectral levels described as the required descriptor (metric) are compared with the limit values given in national and international regulations or directly used according to the objective.

Reporting: General principles on information and reporting data are implemented in addition to the requirements of the specific standards.

E) Measurement of Industrial Noise

Specific purpose: Measurements are performed near industrial premises or in the surrounding areas for environmental impact assessment, the EIA report, or for noise control purposes, e.g. for urban planning. The immission measurements can be conducted either for an entire industrial activity area, for a building, or for a single piece of equipment operated in the open air, e.g. generators, cooling towers, air handling units, chillers or roof-top equipment and installations. Measurements are performed to provide a certificate of permission or simply to obtain noise levels at a certain reference point following complaints. The other objectives mentioned in Section 5.5.1 can be stated also for industrial noise measurements. Measurements of noise emissions of the industrial noise sources and around the premises are explained in Section 5.4.1.

Type of noise: Complex noises with continuous, intermittent, and impulsive characteristics generated during different operation periods. Noise spectrums generally dominate at low frequencies, however, they contain tonal components at high frequencies depending on the type of machinery and equipment, which varies a great deal in industry.

Observations regarding source: During the industrial noise measurements, the following observations are made:

- *Conditions of individual sources*: Type of noise source(s) and activities (number of sources of each type), source locations, acoustic characteristics of noise (airborne or structure-borne sounds, type of temporal variations), tonality, impulsiveness and low frequency components, times of source operations during day and intervals between operations/modes, structure and mounting characteristics.
- *Conditions of industrial site*: Description of the industrial area or building, size and length of buildings, obstructions, nearby buildings, permission of the industrial activity, type of manufacturing, if there are complaints; location (distance to source), and position of the building of complaints, structure, transmission paths of noise to the complainant's residence, construction type of industrial building/size, exits and entrances to the site, heavy vehicle density, openings on the façades of industrial building (doors).

Observations regarding environment: The factors to be recorded are given in Section 5.5.2.

Measurement metrics: The A-weighted equivalent continuous sound pressure levels, L_{Aeq}, directly measured or calculated from the measured L_{AE} for each machine and operations. Spectral levels must be obtained because some corrections, such as tonality and impulsiveness, are applied to the measured results for rating levels, while evaluating the noise impact. The maximum A-weighted noise levels, L_{Amax} and L_{Aeq}, are measured within the periods in which the noise is reasonably consistent.

The interval between two independent industrial plant measurements should be 48 hours at 100–300 m and 72 hours at longer distances from the source (ISO 1996-2:2017).

Measurement duration: Measurement time can be decided with the pilot measurements in a sufficiently long period depending on the type of noise (continuous, impulsive, intermittent) and the distance from the source. The frequency spectrum should be recorded. Obtaining a time-history at least 24 hours prior to the measurements is beneficial for information about the time intervals and operation modes of the industrial activity.

Measurement standards: ISO 1996-2:2017 is the basic standard for measurements of industrial noise impact assessments. Spectral analysis is important for tonal components. To investigate the spectral variations, spectral time-histories should be obtained. Low-frequency sound sources in industrial processes are measured as recommended in the standard. For L_{Amax} measurements, ISO 1996-2:2017 gives a method to determine the energy mean values of L_{Amax} from the arithmetic average of the samples for each operating condition to be divided into classes and the standard deviation.

BS 4142:2014: The British Standard, published in 2014, is more informative about the measurements and evaluations of mechanical noises generated from the following sources [112a]:

- industrial and manufacturing processes
- fixed installations comprising mechanical and electrical plant and equipment
- loading and unloading of goods and materials at industrial and/or commercial premises
- mobile plants and vehicles, such as forklift trucks, or noise from train or ship movements around industrial and/or commercial sites.

The standard describes the methods to measure industrial and mechanical noises indoors and outdoors, enabling assessment of noise impact on people living near the industrial premises and surveillance of the noise conditions upon complaints.

ISO 14257:2001 (confirmed in 2017): The standard aims to determine the spatial distribution of noise levels in workrooms from the measured data and to derive statistical distribution curves, mainly for acoustical qualification and noise control purposes [112b]. Two descriptors are used: excess of sound pressure level compared to a free-field, and the decay of sound pressure level with distance. The standard is used in the calculation of noise immission levels in workrooms.

Measurement uncertainty: Computed as required in ISO 1996-2:2017 (Section 5.6).

Evaluation of results: The measured levels presented with the required metrics are compared with the limit values given in the national and international regulations or directly used according to the objective. The limits for industrial areas are relatively lower than those for residential or mixed areas. The results can also be used for planning purposes, e.g. a tampon belt to be designed around industrial zones, in addition to the necessary measures to be taken for the buildings in the vicinity.

Reporting: General principles on information and reporting data are implemented in addition to the requirements of the specific standards used. For the evaluation of a complaint from an industrial building or operation, the process explained in Chapter 8 is applied.

F) Measurement of Construction Noise

Specific purpose: Since construction noises are generated by temporary and dynamic noise sources, however, which cause great annoyance to the nearby buildings due to the high noise levels, the main purpose is to monitor the noise levels to discover whether they exceed the limits mandated in the local or governmental regulations. Any of the objectives given in Section 5.5.1. can be valid to conduct the field measurements of construction noise generated by building and road construction sites (Figure 5.37) [101, 102]. The studies and some results are given in Section 3.5.

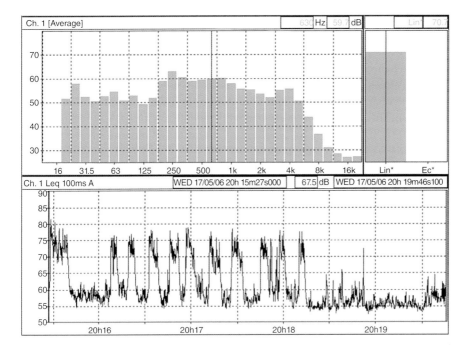

Figure 5.37 Construction site noise recording and spectral analysis [101, 102].

Type of noise: Intermittent noises comprising the individual sound events from construction activities and the continuous noises from the equipment either stationary or moving along an axis. Construction noises, which are composed of sounds generated by different types of machines/equipment operating simultaneously or separately in a restricted area, display a great variability in the spectral content and in time-history characteristics. Low frequencies are dominant within the spectrum of diesel engines, hammering and drilling activities, in addition to the strong tonal components.

Observations regarding source: During construction noise measurements, the following information is obtained regarding the conditions of the source and the operations: Types and numbers of equipment, especially those with a diesel engine, type of work and layout of equipment, operation cycle of each piece of equipment per day and night – if permitted – time-schedule of noisy operations, volume of heavy vehicle transportation, size and configuration of construction site, earth berms, and other material obstructions, reflected surfaces, etc.

Observations regarding environment: The situational factors to be recorded during the measurements are given in Section 5.5.2.

Measurement metrics: Equivalent continuous noise level L_{Aeq} (hourly, maximum hours, or daily), the sound energy levels L_{AE} for each interrupted or intermittent noise, L_{pAFmax} and C- weighted levels, L_{CE} for impact noises and the statistical levels of L_{10} and L_{90} in at least five minutes. Particularly the descriptors should be compatible with the limits in regulations. Rating levels after corrections according to tonality and impulsiveness of noises are determined for impact assessments. Some studies have proposed a new index called the "Construction Noise Index" by classifying the characteristics of construction noises on a scale between 0 and 1000 [113].

Measurement duration: Because of the great temporal variation of noise levels according to the activities, continuous monitoring at the perimeter of the construction site and on the façades of nearby buildings is required in many regulations to inspect the exceedance of the permissible levels of noise.

Measurement standards: Immission measurements are performed by taking into account the type of noises, operation times, according to ISO 1996-2:2017 as described for industrial noises. The dedicated standard on construction noise measurements, namely, BS 5228–1:2009 + A1:2014 has been oriented to predict and control noise and vibration from construction sites, to be used as a guideline [114]. The emission standard for construction equipment, namely, earth-moving machinery, is referred to in Section 5.4.1.

Measurement uncertainty: Computed as required in the specific measurement standards, such as ISO 1996-2:2017 (Section 5.6.)

Evaluation of results: The measured levels presented, using the required descriptor, are compared with the limit values given in national and international regulations or directly used according to the objective.

Reporting: General principles of information and reporting data are implemented in addition to the requirements of the specific standards (Section 5.5.7). The report includes all the information regarding the site and measurements, since the result may be subject to penalties in certain cases, since construction activities are largely restricted in the national regulations.

G) Measurement of Wind Turbine Noise

Specific purpose: Wind farms are a kind of industrial noise sources positioned at about 80–95 m from the ground and distributed in wide areas. The field measurements are conducted mainly for environmental noise assessment in nearby settlements and for inspections upon complaints, in addition to the specific purposes, such as to establish criteria applicable for land-use planning or site-selection and to investigate the effects of wind direction and speed on the propagation of noise, etc. (Figures 5.38a and 5.38b) [115–122]. Some results of those studies are given in Section 3.4.

Type of noise: Intermittent noise varying with wind strength and direction with low amplitude, modulated broadband sound with low frequencies, tonal noise, infrasound from generators.

Observations regarding source: During the wind farm noise measurements, the information regarding the manufacturing and operational characteristics of source, are obtained, e.g. height of tower (center hub and blades), power systems and generators, turning blades, other operational characteristics, number of wind turbines, size of area.

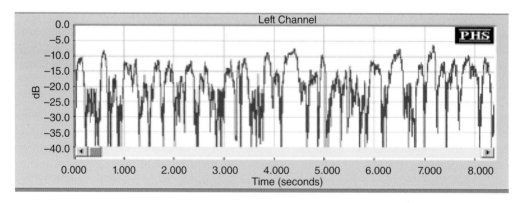

Figure 5.38a   Wind farm noise measurements: a time-history plot [122]. (*Source:* with permission of Noise Measurement Services.)

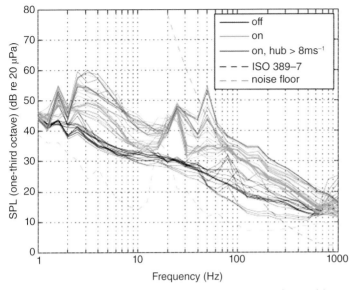

(a) Outside third-octave - 24/7 to 1/8 (shutdown, 24/7 to 26/7).

Figure 5.38b Outdoor noise measurement results for wind farm noise: spectral analysis for different operational conditions [116]. (*Source:* with permission of K. Hansen.)

Observations regarding environment: The factors to be recorded are given in Section 5.5.2. The measurement location is selected as the nearest distance to the wind turbines from the settlement. A weather station is needed near the measurement point at the similar height. The sound pressure levels and the other characteristics of background noises should be observed to apply corrections to the measurement results. The audibility of wind turbine noise is checked by means of audio recordings.

Measurement metrics: A-weighted continuous sound pressure level, L_{Aeq} (total sound levels) and one-third octave band spectra, narrowband spectra, are determined. Different metrics can be used according to the regulations, e.g. L_r, L_{den}, L_{dn}, L_{A50}, L_{A10}, and a specific metric L_{PALF} for low frequency noise indoors [121] (L_r: A-weighted sound pressure level at all wind speeds used in Germany).

Measurement duration: Continuous (24 hours) at least during a week to be repeated to reduce the uncertainty.

Measurement standards: The basic standard is ISO 1996-2:2017 [95d]. Since the source is higher than 90 m in most cases, the effect of air absorption and wind turbulence on the received signals is considered. Measurements need to pay attention to the wind turbulence noise, that can be dominant under the strong winds with the speeds above >5 m/s. (In this situation, the recordings would be useless.) The height of the microphone can be 1.5 m to be directed so as to receive direct sound [117].

IEC 61400-11:2012 standard uses the noise emission measurements to characterize wind turbines with respect to a range of wind speeds and the wind directions, rather than the noise assessment studies and it describes the procedure to compare different wind turbines, including a data reduction procedure, microphone locations,and positions also referring to the tonality, audibility, and masking with background noise [123].

Among the national standards, ANSI AS 4959-2010 is a specific standard for the measurement of wind turbine noise in Australia, based on US/ANSI. At least 10-minute L_{Aeq} measurements are required at a sensitive receiver position under downwind conditions with a direction tolerance of $\pm 45°$. The background noise measurements are made and expressed in L_{A90} with the addition of a minimum 1.5 dB which is applied for the expected difference between wind farm L_{Aeq} and L_{A90} [124a]. As another example, the New Zealand Standard, NZS 6808:2010, expects measurements of at least 10 days period under different wind directions and times of the day, however, for compliance measurement with the planning permits, the downwind sector and $90°$ analysis are needed [124b].

Measurement uncertainty: Computed as required in ISO 1996-2:2017 for industrial noises (Section 5.6). The IEC standard also gives the uncertainty components.

Evaluation of results: The data processing for the recorded signals with a reasonable duration of time, e.g. 10–15 minutes, should comprise a time-history of modulation at dominant frequencies, statistical analysis, tonality, and audibility analyses. The effect of winds and masking with background noise should be investigated with audio recordings. Since the source is far away from the receivers and the noise levels are relatively low, the background noise can be dominant in most situations and it might be rather difficult to distinguish the wind turbine noise in the recorded sound signals. The criteria based on the difference between measured total noise level and the background noise level, $L_{Aeq,T} - L_{A90,T} < 3$ dB(A), can be implemented. In order to eliminate the sounds from birds and rain, the one-third octave band time-history diagrams are also observed.

Reporting: General principles of information and reporting data are implemented in addition to the requirements of the specific standards (Section 5.5.7).

H) Measurement of Shooting Range Noise

Specific purpose: Firearms generating noise from high pressure combustion gases, high energy impulsive noises, for surveillance of the legal requirements, assessment of annoyance caused by shooting ranges or outdoor facilities [125–127]. Some results of experimental studies are given in Section 3.7.

Observations regarding source: During the shooting noise measurements, the following observations are made: size and area, open and closed facilities, topography, numbers of shots fired, types of firearms.

Observations in the field: The physical elements to be recorded are given in Section 5.5.2.

Measurement units: According to ISO 1996-2:2017, sound energy levels, L_E to describe the isolated events, the L_{eq} measured over the time period T_n (e.g. 1 hour), the rating level, L_{RE}. The A-weighted levels with I or F time-weighting are not sufficient for shooting noise, therefore, L_{pAF} or $L_{eq,50ms}$ can be implemented. The onset rate of L_p is determined from the slope (dB/s of the straight line). The correction is for high impulsive noise and penalties according to regulations (e.g. for a small caliber firearm, a penalty of 12 dB).

Measurement standards: The main standard for high energy impulsive sounds is ISO 1996-1 and 2. The specific standard ISO 17201-1:2005 (revised by 2018), is used to measure the energy of the muzzle blast in Part 1 in-situ and the calculation model for immission levels is given in Parts 2 and 3 [128a, 128b]. The standard is applicable for weapons with calibers of less than 20 mm used in civil shooting ranges or explosives of less than 50 g TNT. The standard recommends measuring instantaneous sound pressure and peak sound pressure at least in octave bands in the time-domain waveform which occurs in a very short time (milliseconds) (Figure 5.39). The applicable measurement distance is defined and the corrections are given according to ground reflections. Various

Figure 5.39 Example measurement of high-energy impulse sound (a shot from a rifle) indicating the time-domain wave-form [126]. (*Source:* with permission of G. Flamme.)

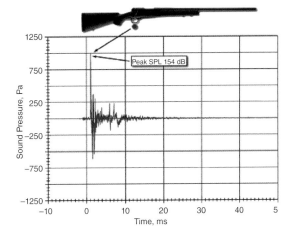

techniques have been implemented in practice for shooting noise measurement, e.g. by using a specific impulsivity meter for gunshot detection and a real-time monitoring system [126, 127, 129, 130].

Measurement duration: Duration should cover all the short-term transient impulses for different type of weapons. Measurements should be repeated under different wind conditions and especially at night-times.

Measurement uncertainty: Computed as required in the specific measurement standard, such as GUM (Section 5.6).

Evaluation of results: The measured levels presented by using the required descriptor are compared with the limit values given in national and international regulations or directly used according to the objective.

Reporting: General principles of information and reporting data are implemented in addition to the requirements of the specific standards. The report should include all the information regarding the site and measurements, since the result may be subject to penalties (Section 5.5.7).

I) Measurement of Entertainment/Recreational Noise

Specific purpose: The measurements are performed to determine the high-level sounds of live music or playback from amplified music systems and to protect people inside and outside the premises, i.e. people living in nearby buildings, the audience (clients), and workers inside. The entertainment premises, some of which are temporary or seasonally operating, display a great variety regarding their locations: they can be in the open air or in semi-closed or entirely closed spaces, like music clubs, discos, restaurants, cafés, amusement centers, open-air concert arenas, even in the city centers, or other recreational and sport activities, festivals, etc. (so-called MOVIDA) and operating under a license received from the local authorities by ensuring certain noise limits. One of the main purposes of noise measurements in and outside such working places, is for surveillance of the target level frequently and to handle the complaints especially in the summer season [131–133]. Some guidelines have been published on measuring recreational noise generated from snowmobiles, motor racing, and other sporting activities [134].

Observations regarding source: During measurements of entertainment noise, the following observations are made:

- *Conditions of entertainment premises in open-air:* Size of premises, distance to the nearest buildings, especially to noise-sensitive buildings, operating times (starting and termination), time schedule of the high-level music, types and numbers of loudspeakers, subwoofers with their powers, layout plan with heights above the floor, position of stage, type of music (impulsive, continuous, or mixture), permissions from local authorities and regulatory restrictions, numbers of similar premises located in the area (if more than one, same observations for each place).

- *Conditions of entertainment premises in buildings of mixed use:* Size of space, numbers in the audience, the construction of the building, the sound insulation of the building elements and the absorptive surfaces inside, windows and other openings on the façades, neighboring spaces (i.e. dwellings), direct and flanking sound transmission paths, in addition to the aforementioned properties of the sound reinforcement systems.

Observations in the field: The factors to be recorded are given in Section 5.5.2. All the physical environmental conditions should be observed especially from the standpoint of their increasing effect on sound levels, i.e. wind directions, absorptive, and reflective properties of the ground, reflective surfaces, natural and artificial barriers, etc.

Measurement metrics: The equivalent sound pressure levels, L_{Aeq}, maximum sound pressure levels measured by fast response, L_{pAFmax}, the C-weighted levels, L_{CE} for the impact noises and the statistical levels of L_{10} and L_{90} in at least 15 minutes by giving the histogram displaying the sound level distribution. The corrections for background noise, tonality and impulsiveness are applied to the measured results for rating levels while evaluating the noise impact. The requirement of specific noise metrics for measurements in amusement parks is discussed [134].

Measurement standards: Outdoor measurements and continuous monitoring are made by applying ISO 1996-2:2017 for outdoors (on the façades of the nearest noise-sensitive buildings) and in the open area outside the entertainment premises. Since there is not a specific standard for this purpose, different techniques are employed in various field investigations [131–133]. A noise map around the premises might be essential to evaluate the impact of especially low frequency noises on the wider area.

Measurements inside the premises are conducted for the prevention of hearing loss on the dance floor and nearby audience sitting areas, therefore, the locations of the microphones are selected accordingly. For inspection following a complaint, the noise levels are measured on the façade of the complainant's building. For the premises within buildings, considering the sound intrusion to the other spaces (rooms), the measurements are conducted according to the above standard applying corrections for the background noise and acoustic characteristics of the sounds. The noise hazard risk to workers is measured the same way as workplace noise. Specific equipment, called intelligent warning devices, is used in most of the premises displaying L_{eq} values in every 15 or 60 minutes or the maximum levels continuously. In some countries, the organizers of festivals or open-air concerts guarantee certain levels such as L_{eq} 100 dB(A) at 60 minutes and surveillance is actively conducted by the microphones mounted on the façades of the nearest building.

Measurement duration: This depends on the time when the amplification systems are operating, and the entire period of the amplified music. If the measurements cannot be realized for such a long time, a time duration can be chosen at least at the beginning and near the termination. When the type of music is known, a sufficient duration can be decided to capture the highest levels. Continuous monitoring is important, as explained in Section 5.4.

Measurement uncertainty: The standard deviations and uncertainties for the indoor and outdoor measurements are computed as required in ISO 1996-2:2017 or applying GUM as given in Section 5.6. However, the uncertainty in entertainment places is rather high due to the variation of music types and volumes according to the DJ's decisions.

Evaluations: The measured levels to be presented in terms of the selected descriptor are compared with the criteria for health and comfort and with the limit values given in local regulations (codes) or directly used according to the objective of the study, after further analyses.

Reporting: General principles of information and reporting data are implemented in addition to the requirements of the specific standard (Section 5.5.7). The reports should include all the information regarding the site and measurements, since the result may be subject to penalties.

J) Measurement of Service Noise in Buildings

Specific purpose: Noises from service equipment, namely, water installations (including water taps, mixing valves of different types, shower cabin, bath tub, water closet, pumps, and pressurizing equipment), mechanical ventilation, heating and cooling service equipment, lifts, rubbish chute, boilers, blowers, pumps and other auxiliary service equipment, motor-driven car park door and other types of building service equipment, are included in building acoustic standards and regulations (Figure 5.40). The purposes of the measurements taken inside the occupied rooms, e.g. noise-sensitive rooms, are to determine the indoor noise levels generated by these kinds of equipment according to the building regulations, and to ensure the guaranteed levels by manufacturers in order to protect residents' health and comfort, especially to prevent sleep disturbance and bad work performance (Chapter 7). The measurements can also be conducted upon complaints.

Observations regarding source: The following information on the noise sources and operations should be obtained: Type of equipment, its location and mounting properties, noise radiating components, maintenance conditions, operating cycles, physical contact with building elements, installation of water pipes and ducts, existence of cabin and boxes and efficiency of sound insulation materials of enclosures, components causing vibration problems, etc. The mechanical rooms need special attention regarding their placement in the building, the layout of equipment and types, the openings outside, the building elements, and direct and flanking sound transmission paths in buildings.

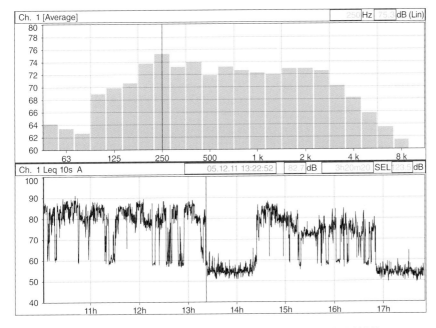

Figure 5.40 Service noise measurements in buildings (pumping noise) [101].

Observations in the field: The factors to be recorded are given in Section 5.5.2.

Measurement metrics: A-weighted sound pressure level, L_A calculated from the octave band values in the frequency range of 63–8000 Hz, C-weighted sound pressure level, L_C calculated from octave band values in the frequency range 31.5–8000 Hz, maximum sound pressure levels, $L_{S\,max}$ and $L_{F\,max}$ in octave bands measured with time weighting "S" and "F," sound exposure level, L_E, maximum sound pressure levels in octave bands, $L_{S\,max,\,nT}$ and $L_{F\,max,\,nT}$ standardized to reverberation time of 0.5 seconds, equivalent continuous sound pressure level in octave bands, L_{eq}, and its standardized and normalized levels taking into account the indoor acoustic conditions.

Measurement standards: The basic standard to measure indoor noise levels is ISO 1996-2:2017 [95d]. The specific standards and recommended measurement procedures are outlined below.

ISO 10052:2004 (confirmed in 2008): This standard explains a survey method to measure noise from domestic service equipment by requiring measurement of only A- or C-weighted sound pressure levels [136]. These are determined as energy-averaged values of the two fixed microphone positions in the receiving room. However, the measurements are simplified by using hand-held sound level instrument by applying manual scanning, as described in the standard. The measuring frequency range is 125–2000 Hz, to calculate the A- or C-weighting. The standard defines the "Service equipment sound pressure level" which is the average of the levels measured in the corner position in the room and in the reverberant field of the room, by applying the weighting of 1/3 and 2/3 respectively. The normalized service equipment sound pressure level, $L_{XY,n}$ is obtained by the reverberation time and the reference absorption area in the receiving room:

$$L_{XY,n} = L_{XY} - k - 10\lg \frac{A_0 T_0}{0.16V} \quad \text{dB} \tag{5.34}$$

L_{XY}: Service equipment sound pressure level calculated as energy-average sound pressure level measured in the room

A_0: Reference absorption area in the receiving room $=10\,\text{m}^2$

V: Volume of the receiving room, m^3

T_0: Reference reverberation time for typical size of rooms $= 0.5$ second

k: Reverberation index calculated from the arithmetic average of the measured reverberation times of 500 Hz, 1 kHz and 2 kHz as Eq (5.35):

$$k = 10\lg \frac{1}{3}\left(\frac{T_{500} + T_{1000} + T_{2000}}{T_0}\right) \text{dB} \tag{5.35}$$

ISO16032: 2004 (confirmed in 2015) [135]: The test procedure is described as an engineering method in the standard. Operating conditions of the equipment during measurements are defined in the standard for each type of equipment. The three microphone positions are selected as one place in the corner of the room. The requirements for the number of measurements and averaging the sound pressure levels are explained. The equipment sound pressure levels are presented as the normalized sound pressure levels L_n and the standardized sound pressure levels L_{nT} according to the reverberation time and the equivalent sound absorption in the receiving room respectively:

$$L_{nT} = L - 10\lg \frac{T}{T_0}\,\text{dB} \tag{5.36}$$

$$L_n = L - 10\lg \frac{A_0 T}{0.16V}\,\text{dB} \tag{5.37}$$

L: Measured indoor level corrected according to the background noise, dB

T: Reverberation time of the room, s (at each octave band frequency)

Other parameters are given above.

In Eqs (5.36) and (5.37), "*L*" corresponds to $L_{S\,max}$, $L_{F\,max}$ or L_{eq} (denoted as $L_{eq,n}$ and $L_{eq,\,nT}$) at each octave band.

The instrumentation required for service noise measurement in the field is an octave band real-time frequency analyzer which is capable of measuring the above units and the microphones and filters meeting class 1 requirements, as will be explained in Section 5.11.2 Standard deviations associated with the reproducibility for octave bands and A- and C-weighted levels, are given in the standard (e.g. maximum 1.5 dB at low frequencies is allowed).

Reporting: General principles of information and reporting data are implemented in addition to the requirements of the specific standard (Section 5.5.7).

5.6 Errors in Measurements and Determination of Uncertainty

The measured emission and immission data obtained either in the field or in the laboratory should be investigated by considering the following aspects:

- whether the same values can be obtained when the measurements are repeated by different technicians or at another point, in another time for the same source under the same operating conditions or in the same laboratory. If there are differences, a decision about which measurement result is correct is made.
- how the measured value is representative of the correct value in the real situation.
- whether the differences are reasonable.

Differences between the results strongly are dependent on:

- different measuring equipment
- different operators
- different methodology
- different time and meteorological conditions
- different determination of residual noise.

5.6.1 Accuracy of the Measurements

Accuracy is defined as "degree of closeness of measurements of a quantity to that quantity's true value" [137]. Accuracy covers all implications regarding: (i) accuracy of measurement methods; (ii) accuracy of results; and (iii) accuracy in applications, and all can be determined based on the standard procedures. The terminology used for the descriptive parameters in relation to accuracy is given in various statistical documents. ISO 5725-1:1994 (confirmed in 2018), including the accuracy of measurements, introduces two concepts: trueness and precision to describe the accuracy of the measurement methods and results [137].

"Trueness" is defined as "the closeness of the mean of a set of measurement results to the actual (true) value." The true value can be a reference value accepted as true, a certified value, or an expected value. Trueness analysis is mostly employed to compare the test results and the theoretical results in scientific studies. "Precision" is defined as "the closeness of agreement between the independent test results" [137].

When the tests are repeated for identical conditions and for the same source, the results may differ due to uncontrolled factors such as the operator (technician), the instrument used by the technician, its calibration, environmental factors, time between the measurements, etc. If the same operator and the same equipment are used in the measurements, the variability of the results are smaller than those obtained by different operators using different instruments. This issue has been discussed in the literature and some guidelines have been prepared [138].

Precision is the statistical term to describe the variability and is determined under two conditions defined as "repeatability" and "reproducibility."

Repeatability conditions: "Conditions where independent test results are obtained with the same method on identical test items in the same laboratory by the same operator using the same equipment within short intervals of time."

Reproducibility conditions: "Conditions where test results are obtained with the same method on identical test items in different laboratories with different operators using different equipment."

Precision depends on the distribution of random errors and is expressed by the calculation of the standard deviation of the test results.

Standard Deviation

As the most common feature in statistic, standard deviation is defined as a measure to quantify the magnitude of variation of a data set (i.e. measurement results) from its mean value (or dispersion of the distribution of the test results). The lower deviation means that the measured values are close to the mean value and the higher deviation implies that the measured values spread over a wider range and are further away from the mean value. The standard deviation is used commonly to provide confidence in the measurement results and is calculated as shown in Eqs (5.38) and (5.39) by assuming that the data is a population with normal distribution:

$$\sigma = \sqrt{\frac{1}{N-1}\sum_{i=1}^{N}(x_i - \bar{x})^2} \tag{5.38}$$

$$\bar{x} = \frac{1}{N}\sum_{i=1}^{N}x_i = \frac{x_1 + x_2 + x_3 + \dots + x_N}{N} \tag{5.39}$$

σ: Standard deviation of the data set (i.e. measurement results)
N: Number of measurements
x_i: Value obtained at each measurement
$\bar{x}$: Average of the measurement results (i.e. mean value, also symbolized as μ)

In a normal distribution, any data remaining in the range between $\mu \pm 2\sigma$ on both sides of the mean value μ, can be the representative value of the total population with the coverage probability of 95%. Also called the "confidence interval," it presents the percentage that the interval covers the true value of the parameter concerned (e.g. the noise level at a point).

The standard deviation of test results is obtained under both repeatability and reproducibility conditions.

5.6.2 Determination of Uncertainty in Different Standards

It is crucial to determine the reliability of the test results by predicting the accuracy levels in relation to the defined parameters, such as the number of samples, the standard deviation, the measurement uncertainty, probability, etc. Uncertainty is defined as the acceptable limits of the measurement errors to be declared with the confidence interval when reporting the results of measurements. The standard procedures to determine the measurement uncertainty are outlined below.

ISO 5725-1:1994 (confirmed in 2018): Giving the general principles of the assessment of accuracy mentioned above, the standard describes the basic method for the determination of the precision of a standard measurement method, including repeatability, reproducibility, and trueness of a standard measurement method [137].

ISO/IEC Guide 98-3: 2008 (GUM: 1995 reissued in 2008): The guideline, first published by International Standards Organization (ISO) and the International Electrotechnical Commission (IEC) in 1995, presents general principles for evaluating and expressing measurement uncertainty according to the accuracy levels required. It is applicable in quality control and quality assurance of measurements, in basic research and development in science and engineering, for calibrations of instruments, in the development of standards and comparisons of measured results, etc. [139a, 139b]. Based on theoretical and statistical approaches, the implementation of the standard in practice has been widely discussed in the literature [140–144]. Briefly explained, when the distribution curve of the measured values is known, the following process is applied to determine the uncertainty:

1) Establishing a relationship f between the measurand Y (implying the output quantity) and the input quantities which are affective in Y:

$$Y = f(X_1, X_2, ...X_N) \qquad (5.40)$$

Function f should cover all the factors contributing to the measurement uncertainty, i.e. corrections applied to the measured value.

X_i: Probability distribution for each quantity whose total is N, which is obtained by a series of observations and the kth observed value of X_i, is denoted by $X_{i,k}$. Strict estimation of X_i is symbolized by x_i.

2) Calculation of the predicted value of X_i, (which is denoted by x_i as the best estimate in the same group) based on either statistical analysis of the observations or by other methods.

3) Computation of standard deviation or what is commonly called "standard uncertainty" $u(x_i)$ for each x_i. The two types of evaluation are explained for inputs of the observed series are similar to those explained in the standard.

4) Obtaining covariance from the correlation of inputs, x_i

5) Calculation of the real measured result by inserting the input value x_i obtained in step 2, into the function X_i, then, y_i values within the measurement results of Y are found by using a functional relationship. The best estimate of the output Y, (which is denoted as y) is obtained by using the best estimates of inputs, x_i:

$$y = f(x_1, x_2, ...x_N) \qquad (5.41)$$

The estimate y can be obtained by the arithmetic average of n independent determination of Y_k of Y according to the formulation given in the standard.

6) Calculation of the composite uncertainty, $u_c(y)$ corresponding to "y," which is the measurement result. In this calculation, the forecast of inputs and the related covariance values are used.

7) Computation of the expanded uncertainty, U, by taking into account a great portion of all the distribution values that can define the "y" (measurement result) in a reasonable way:

$$U = ku_c(y) \qquad (5.42a)$$

k: Coverage factor typically between 2 and 3 and selected according to the confidence level.

8) Reporting the best estimate of the measurement result: "y" is given as composite uncertainty $u_c(y)$ and /or with expanded uncertainty, U.

The measurement uncertainty is identified with a "chosen coverage probability" which is normally selected as 95% (i.e. coverage factor, $k = 2$). This implies that the true result remains in the range of $L \pm 2u$ with the coverage probability of 95%.

Jonasson explains the above procedure in a simple manner [22]: The standard uncertainty is determined for each source of error, u_j. Each error to be given with a sensitivity coefficient c_j depending on its probability distribution. Then the combined standard uncertainty is calculated as:

$$U(L_W) = \sqrt{\sum_1^n (c_i u_j)^2} \qquad (5.42b)$$

$c_j = 1$ for normal distribution.

Determination of uncertainty was required in all the earlier measurement standards published under different topics and the calculation methods were given according to the type of noise source and the measurement procedure. Some examples of uncertainty declarations from the standards are referred to below.

Determination of uncertainty in ISO 8297:1994 (confirmed in 2018): As explained in Section 5.4.1, this method is implemented for large industrial plants with multiple noise sources [53]. The measurement uncertainty is mainly due to the factors, such as (a) variation of sound pressure levels at different microphone positions, which are arranged on the measurement contour according to the size and layout of the plant; (b) non-homogeneous distribution of sound sources within the plant; and (c) varying meteorological conditions during the measurement. The uncertainty of the sound pressure levels, measured as time-averaged levels (L_{eq}) and corrected according to background noises, is given in Table 5.5 with respect to a geometrical parameter; $\bar{d}/\sqrt{S_p}$ within the 95% confidence level [53]. As can be seen from Table 5.5, if the size of the plant is small and the measurement contour is near the industrial activity, the measurement uncertainty is smaller compared to that of a large plant with the contour which is far from the plant.

Table 5.5 Uncertainty required according to the noise prediction of a plant based on the measurements as described in the ISO standard [53].

Geometrical parameter	Range of uncertainty values, dB
0.05	(+3.0)–(−3.5)
0.1	(+2.5)–(−2.5)
0.2	(+2.5)–(−2.5)
0.5	(+1.5)–(−2.0)

The geometrical parameter: Ratio of Average measurement distance to the square root of the plant area: $\bar{d}/\sqrt{S_p}$.

Table 5.6 Standard deviations of the differences between the measured sound power levels according to ISO 3744 [27a] (at 95% confidence level).

Octave bands	1/3 octave bands	Standard deviations of level differences, σ_R
Hz	Hz	dB
63	50–80	51^a
125	100–160	3
250	200–315	2
500–4000	400–5000	1,5
8000	6300–10000	2,5
A-weighted		$1,52^b$

[a]Normally for the outdoor measurements there is no classification at the frequencies.
[b]Applicable for the source with flat spectrum in the range of 100 Hz–10 000 Hz.

Determination of uncertainty in ISO 3744: In the earlier version of ISO 3744 (published in 1995), which described the engineering method for emission measurement of industrial noise sources in-situ, the uncertainty is determined only by the standard deviation between the measurement levels at each frequency [27a]. Table 5.6 gives the acceptable standard deviations of sound power of a source at octave and one-third octave bands and for A-weighted levels with a 95% confidence level.

In the calculation of σ_R (reproducibility) by using Eq (5.38), the x_i values correspond to the measured levels, L_{pi} without applying any correction, at one or more microphone positions on the measurement surface of the source under test.

It should be noted that the later published ISO 3740 series for sound power measurements has declared the three classes of measurement uncertainties: Grades 1, 2, and 3, according to the Precision, Engineering, and Survey methods. Grade 1 has a standard deviation of reproducibility <1 dB, Grade 2 <2 dB, and Grade 3 <3 dB, respectively.

In the recent ISO 3744:2010, the uncertainty of the measured sound power levels is determined by a combination of two types of uncertainties [27b]:

- Uncertainty due to instability of the operating and mounting conditions of the test source, σ_{omc} (standard deviation of repeatability) that can be obtained from repeated tests by readjusting the machine mounting and operating conditions with the same source and operation.
- Uncertainty due to difference in laboratories and different technicians, σ_{RO} (standard deviation of reproducibility) to be obtained from the round robin tests between the laboratories and to be declared as machinery-specific values.

The total standard deviation is obtained:

$$\sigma_{tot} = \sqrt{\sigma_{RO}^2 + \sigma_{omc}^2} \tag{5.43}$$

The expanded measurement uncertainty, U is calculated as:

$$U = k\sigma_{tot} \tag{5.44}$$

k: coverage factor $= 2$ (for 95% confidence)

As a result, for sound power level of the source, the true value lies within the range $(L_W - U)$ to $(L_W + U)$. Determinations of σ_{omc} and σ_{RO} are given in the standard. The rationale of both uncertainties is due to the diversities between the following conditions [27b]:

- Laboratory conditions or differences between the indoor environment.
- The geometry of the indoor space.
- The acoustic properties of indoor surfaces.
- If measurements are outside, the atmospheric conditions.
- Measurement techniques (measurement surface and dimensions), measurement methodology (microphone positions and locations), differences between the equipment.
- Reflections from the nearby environment; the positions of reflective elements.
- Background noise levels and computation techniques.
- Environmental corrections if applied.
- Disturbances around the source (other objects, people, etc.), depending on the source dimensions (their effect is significant, particularly below 250 Hz).

The standard gives the same values which are displayed in Table 5.6, for the standard deviation of reproducibility of the method for measurement of sound power levels and sound energy levels, by remarking that these values could be smaller when using:

- the same measurement system, same method, different laboratory/same field
- the same source, same mounting, same operation, different laboratory
- the same source, same method, same laboratory.

If there are groups of machines, the number of samples for each machine is determined according to ISO 7574-1 to 4 [145a–d]. The total standard deviation and the standard deviations for each machine are calculated and presented in the report.

ISO 12001:2009: This general standard for machinery gives a procedure to determine the measurement uncertainty with the accuracy grade 1 [31].The uncertainties for repeatability (σ_{omc}) and reproducibility (σ_{RO}) are described for certain situations.

Determination of uncertainty in ISO 1996-2:2007: Uncertainty for the immission measurements requires more complex calculations, because of the number of factors influencing the magnitude of errors. Hence, the "composite standard uncertainty" and the "expanded measurement uncertainty" were defined in this earlier version of the standard referred to in [95d] for the environmental noise levels, as shown in Table 5.7.

The information about X, Y, Z values was given in the standard for different type of sources. The Z value is simply the difference between the total measured level (including the background level) and the noise level of the source under concern.

ISO 1996-2:2017: Determination of uncertainty for environmental noise measurements needs to take into account the combined uncertainty which consists of the uncertainties given below [95d]:

- Sound source emission and measurement time duration, u_{sou}.
- Meteorological conditions, u_{met}, to be separately determined for each measurement period (day, evening, night).
- Source location to be given as the distance from the source to microphone, u_{loc}.
- Residual (background) noise, u_{res}.
- Measurement method and instrumentation, $u(L')$.

Applying the basic principles referred to in the GUM: (i) all possible sources of uncertainty are identified; (ii) the inherent uncertainties are quantified separately; (iii) a table which is called the

Table 5.7 Uncertainty table given in ISO 1996-2:2007 (Table 1 Overview of the measurement uncertainty for L_{Aeq}) with permission of ISO/TS).

Standard uncertainty of repeatability dB (due to instrumentation[a])	Standard uncertainty due to operational conditions[b] dB	Standard uncertainty due to weather and ground conditions[c] dB	Standard uncertainty due to residual sound[d] dB	Combined standard uncertainty, σ_t dB	Expanded measurement uncertainty, dB
1.0	X	Y	Z	$\sqrt{1.0^2 + X^2 + Y^2 + Z^2}$	$\pm 2.0\sigma_t$

[a]For IEC 61672-1:2002 class 1 instrumentation. If other instrumentation (IEC 61672-1:2002 class 2 or IEC 60651:2001/ IEC 60804:2000 type 1 sound level meters) or directional microphones are used, the value will be larger.
[b]To be determined from at least three, and preferably five, measurements under repeatability conditions (the same measurement procedure, the same instruments, the same operator, the same place) and at a position where variations in meteorological conditions have little influence on the results. For long-term measurements, more measurements are required to determine the repeatability standard deviation. For road-traffic noise, some guidance on the value of X is given in Section 6.2. (of the standard)
[c]The value varies depending upon the measurement distance and the prevailing meteorological conditions. A method using a simplified meteorological window is provided in Annex A (in this case, $Y = 1$ m). For long-term measurements, it is necessary to deal with different weather categories separately and then combined together. For short-term measurement, variations in ground conditions are small. However, for long-term measurements, these variations can add considerably to the measurement uncertainty.
[d]The value varies depending on the difference between measured total values and the residual sound.

"uncertainty budget" is prepared. The other methods including a modeling approach are described in the standard.

The standard uncertainty of the measurements, u_k within the measurement window k is obtained as Eq (5.45):

$$u_k = 10\lg\left(10^{0.1L_k} + S_k\right) - L_k \tag{5.45}$$

L_k: The energy-averaged sound levels in window k and obtained as:

$$L_k = 10\lg\left(\frac{1}{n_k}\sum_{i=1}^{n_k} 10^{0.1L_i}\right) \quad \text{dB} \tag{5.46}$$

L_i: The measured value representing one independent measurement within window k.
n_k: Number of measurements in each window, k

S_k^2 is the standard deviation given in the basic equation (Eq 5.38), however, by applying energy-averaging of the sound pressure levels within each measurement window, as Eq (5.47):

$$S_k^2 = \left(\frac{1}{n_k-1}\sum_{i=1}^{n_k}\left(10^{0.1L_i} - 10^{0.1L_k}\right)^2\right) \tag{5.47}$$

If the divergence between the Li values, is small, the u_k is calculated:

$$u_k = \sqrt{\sum_{i=1}^{n_k}\frac{(L_i - L_k)^2}{n_k-1}} \tag{5.48a}$$

The standard also gives the determination of the standard uncertainty and sensitivity coefficients for mixed noise conditions and for the residual sound, u_L.

Expanded uncertainty is a combined uncertainty covering the uncertainties due to the sound level meter and the microphone location. It is obtained by the square root of the summation of the squares of standard uncertainties of each factor by taking into account the corrected sound pressure levels.

The standard gives the calculation of the measurement uncertainties separately for road traffic, railway traffic, aircraft. and industrial noises in terms of both L_{eq} and L_{max} measurement units. Uncertainty for traffic noise measurement is calculated by using the number of vehicle pass-bys, n, and the traffic parameter c (for mixed traffic, heavy vehicles, and passenger cars). Uncertainties for railway and aircraft noise measurements are calculated by using the number of train or aircraft pass-bys, n, and the parameter c changing with the conditions of sampling method in the measurements. The determination of uncertainty in industrial plant measurements needs to take into account the operational conditions of the premise [95d]. The source-specific uncertainties of the measurements, u_{sou} are given in Eq (5.48b):

For road traffic:

$$u_{sou} \cong \frac{c}{\sqrt{n}} \tag{5.48b}$$

$c = 10$ for mixed traffic
$c = 5$ for dominant heavy traffic
$c = 2.5$ for dominant light traffic
n: number of pass-bys

For railway traffic:

$$u_{sou} \cong \frac{c}{\sqrt{n}} \tag{5.48b}$$

$c = 10$ (if different operation conditions are not regarded)
$c = 5$ (if different train classes and number of pass-bys are measured)
n: number of sound events (i.e. train pass-bys)

For aircraft traffic:

$$u_{sou} \cong \frac{c}{\sqrt{n}} \tag{5.48b}$$

$c = 3$ for take-offs of jet aircrafts
$c = 4$ for take-offs of other aircrafts
$c = 2$ for landings of jet aircrafts
$c = 3$ for landings of other aircrafts
n: number of sound events (i.e. flights)

For industrial plants:

$$u_{sou} \cong \sqrt{\sum_{i=1}^{n} \frac{\left(L_{mi} - \overline{L_m}\right)^2}{n-1}} \tag{5.48c}$$

L_{mi}: The measured value representing a typical cycle operation, i
$\overline{L_m}$: Arithmetical mean of L_{mi} values
n: total number of independent measurements sound events

5.7 Occupational Noise Measurements

Noise in the workplace has been a great concern globally due to the more stringent health and safety requirements by various international authorities, including the European Agency for Safety and Health at Work and the World Health Organization. The objectives of the measurements relative to workplace noise, which is also called occupational noise, are outlined below:

1) Audiometric tests to determine the noise-induced hearing problems (Chapter 7).
2) Workplace measurements to determine the noise impact and to check the exposure limit values (daily noise exposure levels).
3) Measurements for noise control purposes in the working environment.

5.7.1 Hearing Tests

Assessment of hearing risk and hearing damage caused by noise is included in the field of audiometry. The relative International Standard ISO 8253 deals with the test procedures, particularly in Part 2 [146a, 146b].

ISO 8253-2:2009 (confirmed in 2020) explains the audiometric test methods to be performed in a special laboratory environment, i.e. a listening room with a low background noise. The procedure to determine the threshold of hearing is applied by reference to pure-tone sounds, frequency modulated tones, and narrowband signals, in the range of 125–8000 Hz that can be extended to 20–16 000Hz, to the subjects by means of one or more loudspeakers or by earphones [146b]. The standard describes the measurement procedures, calibrations of instrument, signal presentation, and ambient sound pressure levels in test rooms. The audiometers used for hearing tests have been standardized since 1975 with some technical improvements at present. Some examples of conducting measurements are shown in Figures 5.41a and 5.41b. The results of each test are presented in a standard format called an audiogram which describes the hearing of a person in terms of the lowest signal level which he or she hears at various frequencies. Some examples of audial test reports are shown in Figures 5.42a and 5.42b. At present, the software packages are available to produce hearing assessment reports after transferring data from audiometers to computer.

Figure 5.41a Audiology tests in Japan (1992).

Figure 5.41b Audiology test in NPL, in the UK (1990).

Tests for hearing protectors: Performance tests for hearing protectors which are used to attenuate noise levels at the ears of people exposed to high and intense noises, such as earplugs, earmuffs, or helmets, are conducted by applying the standard methods, e.g. ISO 4869 describes the objective and subjective procedures and instrumentation [147a–c]. ISO 4869-1:2018 presents a subjective method to be conducted in a laboratory to measure the sound attenuation levels of such devices at the threshold level of hearing, i.e. low sound pressure levels [147a].

Measurements of noise-dose: Occupational noise is described by a quantity called the noise-dose which is obtained based on the noise level measurements. The descriptor is the A-weighted sound pressure level normalized to an 8-hour working day and given as a percentage (% Dose).

ISO 1999:1997 (revised in 2013): The former standard gave the definition of noise-dose in terms of the Normalized Equivalent Continuous Sound Level, L_{Aeq8}, to be calculated for a nominal working day of 8 hours [148a, 148b, 149]:

$$L_{Aeq8} = 10\lg_{10}\left[\frac{1}{8}\int_0^T 10^{L_{pA(t)}/10}\ \mathrm{dt}\right] \tag{5.49}$$

$L_{pA(t)}$: A-weighted time-averaged sound pressure level
T: Time between 08:00–16:00 or 8 working hours referred to in most of the national regulations

For a working day less or more than 8 hours, the exposure duration is normalized according to 8 hours. If the sound levels are composed of sound pressure levels at different locations of the worker, the calculation is made as Eq (5.50):

$$L_{Aeq8} = 10\lg_{10}\frac{1}{8}\left[t_1 10^{L_{A1}/10} + t_2 10^{L_{A2}/10} + ... + t_m 10^{L_{Am}/10}\right] \tag{5.50}$$

L_{Ai}: Measured A-weighted sound level at location (workstation), i
t_i: Duration of time at each location

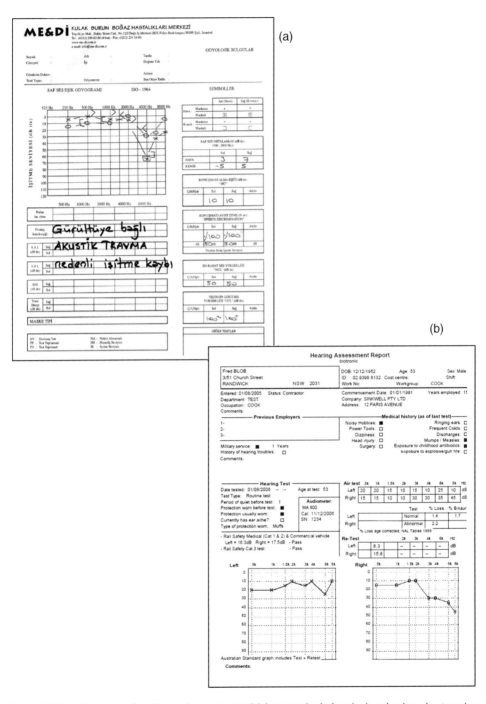

Figure 5.42a Examples of audiometric test reports (a) A report depicting the hearing loss due to noise-related acoustical trauma. (*Source:* provided by M.Ömür, otolaryngologist, 2012) (b) Example of a hearing assessment report produced by a software and the audiograms (*Source:* http://www.biotronic.com.au/ScreenW.htm, 2020).

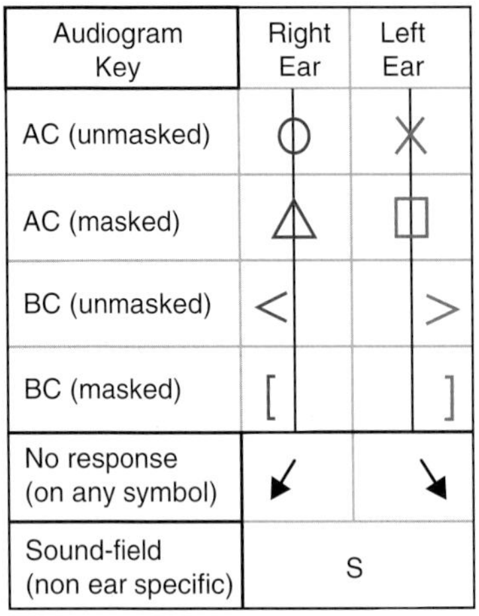

Audiogram Key	Right Ear	Left Ear
AC (unmasked)	⊖	✕
AC (masked)	△	☐
BC (unmasked)	<	>
BC (masked)	[	]
No response (on any symbol)	↙	↘
Sound-field (non ear specific)	S	

Figure 5.42b Standardized audiogram key used in test reports. AC: Air Condition threshold, BC: Bone condition threshold.

The details regarding noise measurements at workplaces by using specific acoustic parameters and the evaluation of results were explained in the standard, including microphone positions, selection of sample times, duration, frequency analysis to be applied according to the noise characteristics.

The later versions of the standard have required the sound exposure level to be measured, by considering the intermittent characteristics of occupational noises in most industrial workplaces. The relationship between the two descriptors is:

$$L_{\text{EX,8h}} = L_{p\text{Aeq},T_e} + 10 \lg\left(\frac{T_e}{T_0}\right) \tag{5.51}$$

$L_{\text{EX,8h}}$: A-weighted noise exposure level normalized to a nominal 8 hours working day
$L_{p\text{Aeq},Te}$: A-weighted equivalent continuous sound pressure level for T_e
T_e: Effective duration of the working day in hours
T_0: Reference duration (8 hours)

ISO 9612:2009 (confirmed in 2014) describes an engineering method for measuring workers' exposure to noise in workplaces by using the frequency-weighted noise levels as follows [148b]:

- A-weighted time-averaged sound pressure level, $L_{p,\text{A},T}$
- A-weighted noise exposure level normalized to an 8-hour working day (daily noise exposure level), $L_{\text{EX,8h}}$
- C-weighted peak sound pressure level, $L_{p\text{Cpeak}}$

The standard presents different measurement methods for more detailed investigations of noise exposures and hearing damage problems at work and to implement noise control measures. The methods are specified as: task-based measurement, job-based measurement, and full-day measurement. The measuring process requires the observation of work activities (full-day work analysis) and analysis of the noise exposure conditions. Estimation of the measurement uncertainty is given in the standard.

5.7.2 Assessment of Noise-Induced Hearing Loss in Workplaces

The earlier issue of ISO 9612:1997 (later revised in 2009 and confirmed in 2014), described the expected variation in threshold of hearing based on the relationship between the A-weighted noise levels and the duration of noise exposure within a working week (40 hours working times) [148a]. The percentage of the workers, whose threshold of hearing is increased about 25 dB or more at the average of 500, 1000 and 2000 Hz, was accepted as a descriptor of noise-induced hearing-loss in the workplace under investigation.

In fact, the assessment of possible hearing losses in workplaces had been a problem since the 1970s, attempting to protect workers from so-called occupational noise. ISO 1999 was issued in 1975 and has been revised over the years up to 2018 [149]. The standard presents a calculation model based on the relationship between noise levels and long-term hearing losses derived from the scientific and experimental studies.

ISO 1999:2013 (confirmed in 2018): The standard estimates noise-induced hearing loss. The threshold level of a worker is calculated at each frequency as in Eq (5.52):

$$H' = H + N + HN/120 \tag{5.52}$$

H': Permanent hearing threshold level for a specified fraction of a population according to the standard, dB (*HTLAN*: Hearing Threshold Level associated with age and noise)

H: Hearing threshold level of associated with age (*HTLA* for a specific section of people)

N: Threshold shift due to actual or possible noise (*NIPTS*: Noise Induced Permanent Threshold Shift)

N is estimated in relation to the measurement of A-weighted continuous equivalent sound pressure levels in 8 hours, $L_{Aeq,8h}$ in the older standard. However, in the latest version of the standard, the measured or calculated A-weighted sound exposure levels, $L_{EA,8h}$ are required for the calculation of N. The estimations derived from the scientific studies are also presented in the standard.

The abbreviations of the parameters employed in the latest standard are shown in the parentheses in Eq (5.52) [148b].

The regulations regarding workers' health and noise specify the limit values based on the determination of the total noise levels during a workday or within the exposed times (Chapter 7). Since noise-dose assessments need to take into account the different exposures of workers during their daily occupational activities, depending on the work schedule and type of task, at different positions with different durations, the continuous measurement of noise exposure is necessary by means of a portable device called a dosimeter which takes measurements and computations of noise-dose (Section 5.11.2) (Figures 5.43a and 5.43b). Such assessments are beneficial to apply noise reduction programs and exposure control for personnel. Also, they provide information on the excessive noise problems in workplaces to take the necessary precautions which are mandated in many countries.

Directive 2003 /10/EC: The European Council has issued a document which specifies the labor noise exposure limits within 8 hours working days in terms of the time-averaged exposures as L_{EX} (8 hours) as recommended in ISO 1999 [150]. The purpose of this Directive is "to protect employees from potential risk from noise at workplace conditions, i.e. industrial places, premises with amplified music, etc." and to apply individual protection and noise control at workplaces. The assessments are mandated in workplaces within the health surveillance of workers.

Measurements in workplaces for noise control: Noise generated by industrial noise sources, like mechanical equipment and installation, not only affects the workers but also the nearby residents outside the premises, by penetrating outside through doors, windows, openings. The measurements related to this problem are explained in Section 5.5.

ISO 11690 (1-3):1996 (confirmed in 2011) provides information about measurements at work and recommendations regarding the design of noise control measures for the attention of inspectors, managers, trade unions, owners of the workplaces, etc. [151a, 151b]. Three parts of the standard

Figure 5.43a Dosimeters used by workers. (*Source:* OSHA Technical Manual, 2005, https://www.osha.gov/dts/osta/otm/new_noise)

Figure 5.43b Measurements with dosimeters in workplaces. (*Source:* photo by C. Kurra in 2008.)

provide information about noise control strategies, measures and sound propagation, and noise prediction in workrooms.

5.8 Scale Model Measurements

Prior to the era of computer simulations facilitating studies on sound propagation in environment, scale model experiments were carried out in the 1970s to the 1980s especially for traffic noise investigations. Despite their size limitations and not being applicable in complex situations, they were useful in the studies regarding the effects of physical elements in built-up areas and for validation of the mathematical models. Acoustic measurements in the prototype of the physical environment concerned also enabled design scenarios for buildings and transportation systems prior to planning, as regards noise control. As they were more economical compared to field measurements and due to the possibility of controlling the physical and acoustical parameters in laboratory, scale model studies were quite popular in those days [152–168]. Some investigations using the scale models, are outlined in Table 5.8. Scale modeling techniques have even been implemented in room acoustics design and analysis of sound distribution in enclosures. However, there are some geometrical and acoustical factors to be considered in performing these experiments.

Table 5.8 Some of scale model experiments conducted between 1964 and 1999.

Name of the investigator/ year/ref.	Sound source model	Scale	Objective	Institution
Maekawa, 1964–1966 [152, 153]	Horn loudspeaker	—	To measure diffraction from noise screens, to investigate the effect of angle of incidence on sound attenuation behind the barrier and sound distribution in the shadow zone	Kobe University, Japan
Ringheim, 1971 [154]	Aerodynamic jet source	20:1	To describe the attenuation in the shadow zone of the screen and influence of geometrical parameters	Akustik Laboratorium, Norway
Delany, 1972 [155], 1978 [163]	Miniature air-jet source	30:1	To investigate traffic noise propagation for motorways, to obtain systematic prediction data, noise penetration through a gap between blocks, validation for different road/housing configurations	National Physical Laboratory, UK
Rapin, 1968, 1972 [156, 157]	Lines of bronze bells	20:1	Motorway noise in tunnels and cuttings	CSTB, France
Ivey and Russel, 1973 [158]	Electrical spark source	64:1	To search for validity of the proposed model for finite and thick barriers and to examine the effects of some geometrical variables	Smith College, USA
Fujiwara, 1973 [159]	Electrostatic loudspeaker	—	Effect of thickness of the barrier on sound attenuation	Kobe University, Japan
Koyasu, Yamashita, 1973 [160]	Incoherent line source of jet noise made of tracing paper and a brass pipe	40:1	To derive an empirical equation as a correction term in attenuation caused by houses	Kobayasi Inst. of Physical Research, Japan

(*Continued*)

Table 5.8 (Continued)

Name of the investigator/ year/ref.	Sound source model	Scale	Objective	Institution
Mohsen, 1975 [161]	—	—	To examine the validity of scale model studies by applying it in a barrier study and investigate the effect of balconies	University of Sheffield, UK
Lyon, 1976 [162]	Electrical spark source	80:1 and 64:1	To study the effects of noise barriers around a runway extension in airport	Massachusetts Institute of Technology, USA
Kurra, 1976 [164]	Pneumatic air-jet source (Delany type)	20:1	To validate the computerized model regarding computation of finite size barriers (buildings) performance model for traffic noise	Istanbul Technical University, Turkey
Kurra, Crocker 1981 [165]	Improved Miniature jet source (moved on a traverse by a motor)	20:1	Investigation of sound propagation in built-up environments, analysis in combination of a computer program	Purdue University, Ray Herrick Laboratories, USA
Jones, Stredulinsky, Vermeulen, 1980 [166]	Hartman whistles	80:1	To obtain sound propagation in the residential area and to compare the results of field studies	University of Calgary, Canada
Jacobs, Nijs, Willigenburg, 1980 [167]	Continuous pneumatic jet source	100:1	To obtain a transfer function for a model of a built-up situation and to validate the results	Delft University of Technology, The Netherlands
Kurra, 1999 [168] (Figures 5.48a and 5.48b)	Recorded signals of transportation noise after applying a transfer function generated by flat LSs on the wall (assumed as façade)	1:1	Psycho-acoustic study on differences between the annoyances from three types of transportation noise sources in the laboratory	Kobe University, Japan

Scale factor: A simulation study needs to decide on a scale which is related to the space availability in the laboratory, which is an anechoic or semi-anechoic room. The size of the area to be modeled, the equipment set-up, and the other simulation principles are related to the scale of the prototype.

Basic principles: The model environment should represent the real environment with its sound field characteristics as much as possible. In addition to the geometric similarity, the kinematic, dynamic, and thermal similarities are important and all the physical factors affecting noise propagation should be modeled with a certain degree of approximation. Sound propagation in relation to the wavelength can be simulated considering the speed of sound and the wavelength, the dimensional ratio of the real and the prototype noise sources, size of the surfaces and objects [152, 153]:

$$\frac{T_m V_m}{L_m} = \frac{T_p V_p}{L_p} \tag{5.53}$$

T_m and T_p: Time parameter in the original and prototype model
V_m and V_p: Speed parameter in the original and prototype
L_m and L_p: Geometrical dimension in the original and prototype

Speed of sound is equal in the model to real life, however, the sound frequency in the model should be increased as the product of the scale factor and the original frequency concerned. That means the measurement frequencies can be very high, up to 20 kHz, to represent the original sound at 1000 Hz, if the scale factor of 20:1 is chosen. Therefore, a specific model source, equipment, and analysis system are needed for the experiment.

Modeling source and noise: Dynamic characteristics of source, spectral properties, temporal variation of levels, type of source (steady or moving point source, line or plane sources) and their properties should be known for acoustic simulation. Sound propagation conditions for the original source should be adequately simulated for accuracy of the model and, for moving vehicles, the real speeds should be considered.

Back to history, the modeling studies in the past used the actual sounds recorded in the field. These signals were put directly into the model. Figure 5.44 presents an experimental set-up used in a scale model study in 1975 [18]. Later, various prototype sources could be used in the studies, such as electrical discharge sources, electroacoustic sources (small loudspeakers), aerodynamic sources (air jets), mechanical sources (bells, balls in a tube, etc.). The types of model sources used

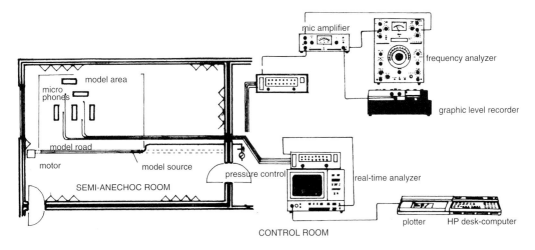

Figure 5.44 .Experimental set-up in a scale model experiment in the laboratory (from a study in 1975) [18].
Note: B&K instrumentation is schematically displayed as shown in the B&K documents in the 1970s.

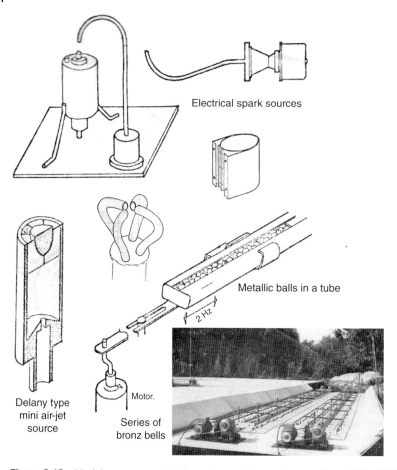

Figure 5.45 Model sources used in the scale model experiments from 1960–1980 [165].

in the former experiments are included in Table 5.8. The model sources giving constant signals (miniature jet sources), impulsive sources (sparkling devices), etc. could be selected according to the characteristics of the original source. Some of the sources used in the studies in the 1960 to the 1980s are shown in Figure 5.45 [165]. The important properties of the model sources are: a flat response in the frequency range of the relevant broad-band noise between 20 and 80 kHz, sound level of about 80 dB at high frequencies, a sufficient sound-radiating capacity within the entire modeling area (signal/noise ratio), similar directivity pattern to the real source (generally non-directional), the size as small as possible not to disturb the sound field, enabling continuous and repeatable sound during measurements. Among those, air-jets have been accepted as the most suitable models because of their relatively small dimensions, omni-directionality, ease of manufacture, high frequency content, stability, and sufficiently high sound outputs. When more than one point source is needed in the model, the phase characteristics of source emissions should be random in order to minimize the interference [162]. A motorway can be simulated as a line source with sets of point sources or a moving point source can be used as a single vehicle passing-by.

Sound absorptions of the surfaces: The ground over which the sound propagation is investigated, the building surfaces, roads, and other physical elements, should be simulated in the model by finding specific materials equivalent to real life. For this reason, the absorption coefficients of the original surfaces should be multiplied by the scale factor to obtain the material absorptions in the model. This requires a database for the sample materials based on measurements. The building surfaces in reality are approximated as 0.1 [160]. For the reflective surfaces of road and pavements, any hard material or aluminum plate is appropriate in the model. Since the absorption properties of soft ground and vegetation are defined by the impedance and flow resistivity, soft wooden boards covered by nylon sheets or textile-like materials are found suitable for modeling.

Air absorption, temperature, and wind gradients are difficult to simulate, thus homogeneous propagation conditions are generally assumed in the model. The air absorption problem, which is important at higher model frequencies, is overcome by calculating a correction factor proposed in the earlier documents [163].

Measurement systems: Sensitivity of the microphones should be constant and sufficiently high throughout the model frequency range. The microphone dimensions should be much smaller compared to the wavelength of the model sound signals, i.e. 1/8″ condenser microphones are suitable for such high frequencies. Filter characteristics should be capable of analyzing the model frequency bands, the filter response should be flat throughout the frequency range of interest with limited distortion and with a large dynamic range. Figures 5.46a–5.51 display some of the experimental studies conducted in laboratories [164, 165, 168].

Consequently, the scale model experiments facilitate investigations on noise propagation problems and noise control measures for transportation routes at the planning stage or the layout of buildings along the existing routes, the design of barriers, etc. Models can be used in cases where mathematical propagation models are not available and the extensive field measurements are not feasible. It is possible to derive empirical relationships between noise levels and different variables

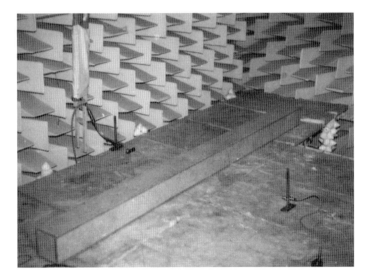

Figure 5.46a Laboratory measurement with a barrier-building model, I.T.U. Turkey, 1975 [164].

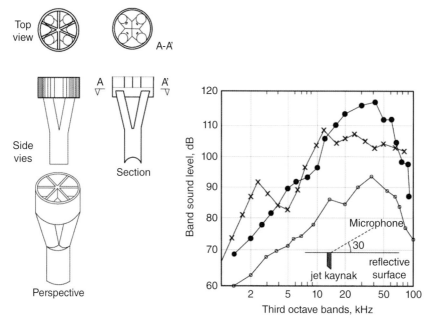

Figure 5.46b The aerodynamic jet source model and its acoustic characteristics used in the study in Figure 5.46a [164].

Figure 5.46c Equipment set-up used in the study in Figure 5.46a [164].

by eliminating the effects of some dynamic factors, such as meteorological factors. However, such studies are not common at present because of advanced computer modeling techniques which are more economical and less time-consuming.

Figure 5.47   Laboratory study for traffic noise propagation in built-up areas, Purdue University, USA, 1980 [165].

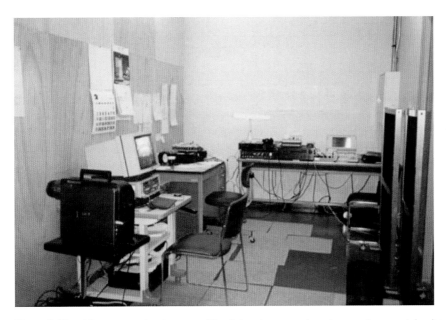

Figure 5.48a   Measurement systems used in a laboratory experiment on environmental noise, control room in Sekisui Lab, Japan, 1998) [168].

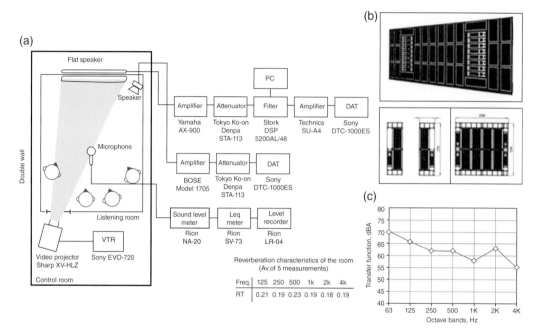

Figure 5.48b Experimental set-up in the laboratory study in Figure 48a, Sekisui Lab, Japan, 1998. (a) Surface-type loudspeakers used in the test room; (b) Transfer function applied for sound signals due to façade transmission loss [168].

Figure 5.49 Scale model measurement in CSTB, France in 1980. (*Source:* photo: S. Kurra.)

Figure 5.50   Scale model measurement in CSTB, France in 1980. (*Source:* photo: S. Kurra.).

Figure 5.51   A laboratory experiment using a scale model, in the UK in 2003 [11]. (*Source:* with permission of B. Berry.)

5.9   Sound Insulation Measurements

When the term "environment" is considered in a large context from the standpoint of noise impact on people, the indoor environment is an indispensable part of the noise problems due to the strong interrelation between outdoor and indoor environments. It is rather difficult to evaluate environmental noise problems without considering buildings, since socio-acoustical studies on community response to noises aim to investigate the responses of people living in residential buildings with different acoustic conditions. These subjective responses naturally include the effect of interior environment. On the other hand, most bothersome noises are generated by indoor sources and it has been found that the neighborhood noise is the second highest disturbing noise source in urban environments due to insufficiency of sound insulation (Chapter 7). Experience has shown that acousticians, consultants, or environmental engineers, who are experts in noise management and control, have to conduct acoustic measurements inside buildings. Therefore, some basic information on building acoustics is given in this section of the chapter.

5.9.1 Basic Knowledge about Sound Transmission

The indoor acoustical problems, covering the phenomenon of sound transmission through building structures and the physical properties of rooms/indoor spaces, are involved in "building acoustics," which is one of the preliminary fields of acoustics and, a well-documented research and implementation area for more than 60 years [169–172]. A brief knowledge of building acoustics associated to noise control, should cover the general principles of sound transmission through constructional elements, i.e. façades, inner walls, doors, windows, floors, roof, etc. and the two types of sounds as:

- airborne sounds which are generated as sound waves in the air around the source;
- structure-borne sounds that originate in the building structure and propagate through it as a vibration, mostly caused by exciting forces of impact and driving forces of vibration.

Transmission of sound from one room to another, from outdoor to indoor, or vice versa, needs analysis of normal (direct) transmission and flanking (indirect) transmission, which need different techniques in measurement and evaluations. Performance of building elements regarding sound insulation is determined by theoretical or empirical calculation methods and by measurements applying special techniques either in the laboratories or in the field. The International Standards relative to measurement of sound insulation are outlined in Table 5.9.

Descriptors for Measurements of Sound Insulation
The sound transmission coefficient of a building element for airborne sounds is defined as the ratio of the energy incident on the surface (W_i) to the energy transmitted through it (W_t). The ratio, which is expressed as a logarithmic value, is called "sound transmission loss" in dB. Since it is difficult to determine the sound transmission loss directly from the measurements, except by using the intensity technique, hence "sound reduction index" is more commonly used in the technical standards. This value is obtained by sound pressure measurements both in the laboratory and in the field by taking into account the effects of acoustic conditions of the receiver room:

$$R = L_1 - L_2 + 10 \lg (S/A) \quad \text{dB} \tag{5.54}$$

R: Sound reduction index, dB
L_1: Energy-average sound pressure level in the source room, dB
L_2: Energy-average sound pressure level in the receiver room, dB
S: Surface area of the test element (i.e. wall), m^2
A: Equivalent sound absorption area of the receiver room, m^2

The equivalent sound absorption of the receiver room can be determined as Eq (5.55):

$$A = 0.161 \, V/T \tag{5.55}$$

V: Volume of the room, m^3
T: Reverberation time of the receiver room, s

Table 5.9 Primary ISO standards in relation to the building acoustics measurements.

	Description	ISO standard	
		Laboratory measurements	**Field measurements**
Airborne sound insulation	Airborne sound insulation measurements for building elements	ASTM E90-09 [173]	
	Airborne sound insulation measurements for building elements	ISO 10140-1, 10140-2 [174a, 174b]	ISO 16283-1 (engineering method) [180a] ISO 10052 (survey method) [136]
	Façade measurements	—	ISO 16283-3 [180c]
	Measurement for flanking transmission for airborne sound	ISO 10848-2, 10848-3 [178a, 178b]	
	Sound insulation performance of cabins	ISO 11957 [179]	ISO 11957 [179]
	Measurement of sound insulation by using intensity technique	ISO 15186-1 [183a, 183c]	ISO 15186-2 [183b]
Impact or structure-borne sound insulation	Impact sound insulation measurements	ISO 10140-3 [174c]	ISO 16283-2 (engineering method) [180b] ISO 10052 (survey method) [136]
	Measurement for flanking transmission for impact sound	ISO 10848-2, 10848-3 [178 a, 178b]	
Room acoustics	Reverberation measurements for performance spaces, ordinary rooms and offices	—	ISO 3382-1, 3382-2, 3382-3[191a, 191b 191,c]
	Sound absorption measurements in reverberation room	ISO 354 [192]	

The reverberation time can be calculated or measured as will be described in Section 5.9.6. Since sound reduction (the acoustic behavior of building elements) is frequency-dependent, therefore, the sound reduction values are to be obtained at each octave or 1/3 octave bands between 100 Hz and 5000 Hz. The sound energy radiated by flanking elements into the room is measured by the "apparent sound reduction index" R'. Various sound pressure-based insulation descriptors are recommended in the recent standards, taking into account the flanking transmission in the field measurements:

- Normalized level difference, D_n:

$$D_n = L_1 - L_2 + 10 \lg \left(\frac{A_0}{A} \right) \quad \text{dB} \tag{5.56}$$

A: Equivalent absorption area in the receiving room, m^2
$L_1 - L_2$: Level difference, D
A_0: Reference equivalent absorption area in the receiving room, m^2 (for dwellings $=10\text{m}^2$)

- Standardized level difference, D_{nT}:

$$D_{nT} = D + 10 \lg \left(\frac{T}{T_0}\right) \quad \text{dB} \tag{5.57}$$

T: Reverberation time in the receiving room, s
T_0: Reference reverberation time for dwellings $= 0.5\,s$

- Standardized impact sound pressure level, L'_{nT} (including flanking transmission)

$$L'_{nT} = L_i - 10 \lg \left(\frac{T}{T_0}\right) \quad \text{dB} \tag{5.58}$$

L_i: Impact sound pressure level in the receiver room, dB (as energy- average level)

- Normalized impact sound pressure level, L'_n:

$$L'_n = L_i + 10 \lg \left(\frac{A}{A_0}\right) \quad \text{dB} \tag{5.59}$$

- Standardized level difference (for facades), $D_{2m,n,T}$:

$$D_{2m,n,T} = D_{2m} + 10 \lg \left(\frac{T}{T_0}\right) \quad \text{dB} \tag{5.60}$$

D_{2m}: Level difference between the sound pressure level at 2m from the facade and indoor sound pressure level

L_1, L_2 and L_i are the energy-average sound pressure levels of the measured levels in different microphone (and source) positions.

General principles for sound insulation measurements are outlined as:

a) For airborne sounds (walls and floors): (Figure 5.52a and 5.52b)

- Two adjacent rooms with a test partition in between.
- Emitting sound signal on one side of the test element.
- Receiving sound on the other side (which is called the receiving room).
- The difference between the two levels is corrected according to the background noise and indoor acoustics to obtain the standardized or normalized levels.

(a)　　　　　　　　　　　　　(b)

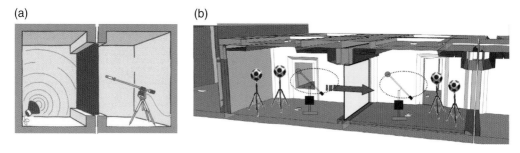

Figure 5.52 Sound insulation measurements in laboratory: (a) simple method in the 1980s; (b) advanced measurement method according to ISO 16283-1:2014 and ISO 10140-2 (From presentation of Belgium Building Research Institute in Cost TU0901 training school, Jan 2013).

Figure 5.53 Impact noise measurements according to ISO 16283-2:2015 and ISO 10140-3:2010.

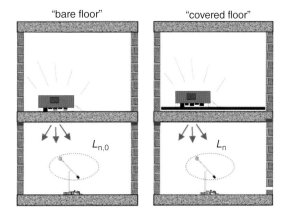

b) For impact sounds (for floors): (Figure 5.53)

- Two rooms, one on top of the other.
- Giving impact sound by a standard machine placed on the floor of the source room.
- Receiving sound in the receiving room below.
- The measured sound level is corrected according to the background noise and indoor acoustics to obtain the standardized or normalized levels.

Sound insulation measurement techniques have improved considerably with time and the latest standards describe the procedures for more accurate results, by increasing the numbers of microphones and source positions in the test area, low frequency corrections, and defining various rating descriptors.

5.9.2 Laboratory Measurement for Sound Insulation

The standard test methods for airborne and structure-borne sounds in a laboratory were first declared in the American Standard, ASTM E90–09 (2016), originally approved in 1955 by the ASTM Committee E33 on Building and Environmental Acoustics [173]. The International Standards Organization published, ISO 10140 series consisting four parts, in 2010, after revisions of the earlier ISO 140 series in the 1990s [174a–d].

Requirements for Laboratories

Two coupled reverberant rooms of unequal size and dimensions located next to each other, or one on top of another, are the primary requirements. The rooms should be constructed with a complete free-standing technique (so-called interrupted construction) to eliminate the sound paths coming from the side elements. The size of the rooms, especially of the receiver room, is selected according to the minimum measurement frequency and both rooms should satisfy the diffusivity condition (diffuse field) as described in the above standards [174a, 174d] (Figure 5.52b). The smaller room is used as source room where a model source is located and the second room is the receiver room where the test specimen is placed in the standard-size opening on the partition between two rooms. Figure 5.54 displays the equipment set-up for airborne sound insulation measurement in the laboratory applied in an earlier experiment [18]. Impact noises are measured in the laboratory, placing the test sample on the floor between the two coupled rooms, one on top of the other. The upper room is the source room where the reference source, called the standard tapping machine, is placed in the source room (Figure 5.53).

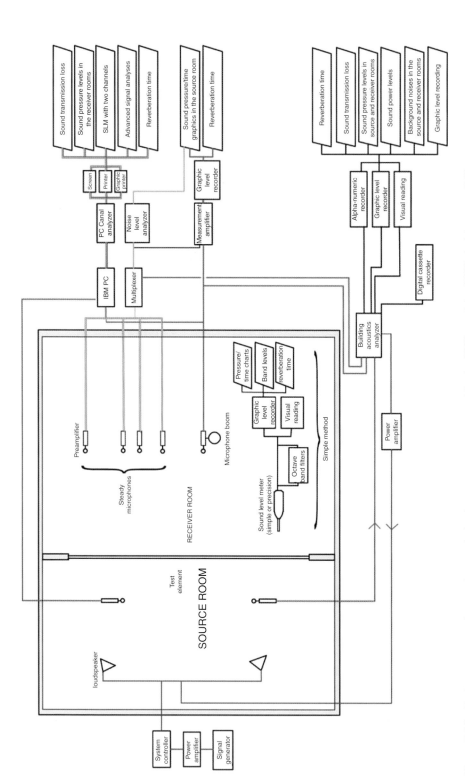

Figure 5.54 Equipment set-up in sound transmission loss measurements used in the 1980s and the 1990s in laboratory [18].

Normally measurements are made in one-third octave bands in the range of 100–3150 Hz. However, this issue is debated widely at scientific levels and the new standard recommends low frequencies down to 50 Hz [175]. The measurements can be made at low frequencies, i.e. 0.50 Hz if the laboratory meets certain principles considering the room modes which have a great influence, especially in relation to the wavelength of sound and the volume of the room (i.e. if 50–100 m³). The receiver microphones are arranged so as to eliminate the room effect while obtaining the space-average result. The construction and mounting detail of the specimen are important since it has been proved that even small details can influence the measured results.

Use of these laboratories has increased throughout the world due to heavy demands by the building sector for sound insulation measurements. As probably the oldest laboratory to measure the sound transmission properties of building elements, the Riverbank Acoustics laboratories (RAL) was founded in 1918 by Wallace Clement Sabine in the USA as referred in Section 5.1 Figure 5.2 [10, 176]. The layout of the test rooms is shown in Figure 5.55a, b [177]. Four independent rooms, constructed by heavy walls and floors and completely separate from each other, are still in active use. For airborne sound measurements, two test openings are available for walls (11.69 m²) and doors/windows (2.96 m²). For impact sound measurements of floors, the opening is 25 m². The measurements are conducted with the aid of computers and using automatic adjustment for room climate (photographs are given in Figures 5.55c–5.55e). An output of a test measurement is given in Figure 5.55f.

Uncertainty of measurements and reproducibility through comparisons between various laboratories (round robin tests) are described in the standards (Section 5.6).

ISO 10140-4:2010 (confirmed in 2017): Part 4 of the standard gives information about the general measurement procedures for sound insulation of building elements and requirements, in addition, the specific measurements, e.g. reverberation time, low frequency measurements, and the measurement of physical parameters, i.e. the loss factor and the determination of radiated sound power by velocity measurement [174d].

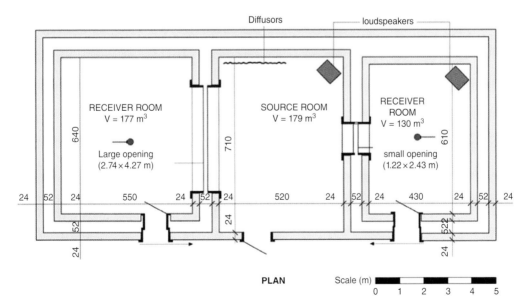

Figure 5.55a Plan of an example laboratory construction for sound transmission loss measurements in Riverbank Acoustical Laboratories, USA [177].

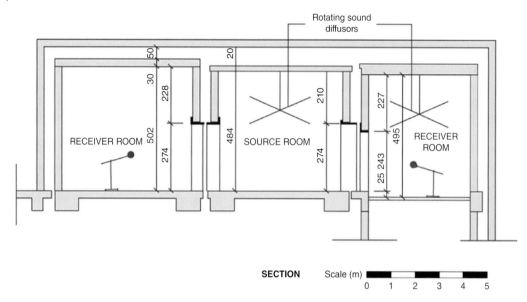

Figure 5.55b Cross-section of an example laboratory construction for sound transmission loss measurements in Riverbank Acoustical Laboratories USA [177].

Figure 5.55c Construction of the test specimens in the laboratory [177]. (*Source:* photo: S. Kurra, 1999.)

ISO 10140-5:2010 + A1:2014 (confirmed in 2016) and published as European Norm: Part of the standard describes general guidelines regarding specific quantities to be used, requirements for the laboratory facility in terms of size, constructional details of the opening, mounting the test element and requirements for equipment [174e]. The amendment of the standard includes the sound of rainfall.

Measurement of airborne sound insulation: ISO 10140-2:2010 (confirmed in 2016) describes the procedure for airborne sound measurement in the laboratory, main quantities, data processing and reporting [174b]. Simply to determine the sound reduction index of an example element in

Figure 5.55d   Insertion of the experimental walls in the openings ([177]. (*Source:* photo S. Kurra, 1999.)

Figure 5.55e   Laboratory control room for the above experiment from the study in Figure 5.55b [177]. (*Source:* photo S. Kurra, 1999.)

the laboratory, sound pressure levels are measured in the source and receiver rooms simultaneously by activating the model source, then according to the descriptors selected (Table 5.10), the calculations are conducted by also taking into account the reverberation time or absorption of the receiver room.

Energy average sound level in source and receiver rooms is calculated by Eq (5.6):

$$L = 10 \lg \frac{1}{n} \left(\sum_{j=1}^{n} 10^{L_j/10} \right) \quad \text{dB} \tag{5.61}$$

$L_1, L_2, ...L_n$: Measured sound pressure levels at n microphone position, dB

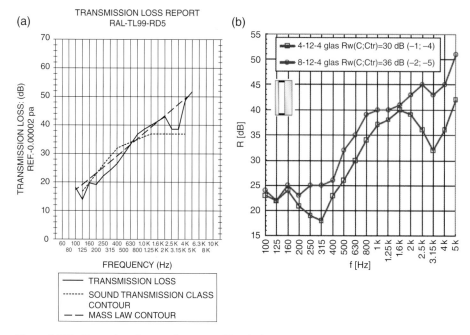

Figure 5.55f Examples of reports from sound insulation measurements in laboratories. (*Sources:* Left: S. Kurra, 1999 [177], Right: B. Ingelaere, COST TU0901 Training School 21–25 January 2013).

Table 5.10 Overview of ISO 717 descriptors for evaluation of sound insulation in buildings [187].

ISO 717:2013 descriptors for evaluation of field sound insulation	Airborne sound insulation between rooms (ISO 717-1)[b]	Airborne sound insulation of façades[a] (ISO 717-1)[b]		Impact sound insulation between rooms (ISO 717-2)[b]
Basic descriptors (single-number quantities)	R'_w $D_{n,w}$ $D_{nT,w}$	R'_w $D_{n,w}$ $D_{nT,w}$		$L'_{n,w}$ $L'_{nT,w}$
Spectrum adaptation terms (listed according to intended main applications)	None C $C_{50-3150}$ $C_{100-5000}$ $C_{50-5000}$	None C $C_{50-3150}$ $C_{100-5000}$ $C_{50-5000}$	C_{tr} $C_{tr,50-3150}$ $C_{tr,100-5000}$ $C_{tr,50-5000}$	None C_I $C_{I,50-2500}$
Total number of descriptors	$3 \times 5 = 15$	$3 \times 9 = 27$		$2 \times 3 = 6$

[a]For façades, the complete indices for R'_w, $D_{n,w}$, $D_{nT,w}$ are found in ISO 717.
[b]For simplicity, only 1/3 octave quantities and C-terms are included in the table, although some countries allow 1/1 octave measurements for field check.

Correction for background noise is made in the same way to the emission noise measurements shown above.

Measurement of impact sound insulation: ISO 10140-3:2010/Amd 1:2015 (also published as European Norm) defines the procedure for impact sound measurements, by using one of the two types of sound sources; a standard tapping machine for lightweight and heavyweight floors, and a heavy/soft impact source (e.g. a rubber ball) for low frequency impact sounds

[174c]. The receiver microphones are fixed or continuously moving, the different source locations are selected on the floors. Calculation of the impact sound pressure level L_{IFmax} is performed by logarithmical averaging of all the microphones and source positions after correction for background noise.

Measuring the effect of flanking transmission on both impact and airborne sound insulation is described in specific ISO standards [178a, 178b]. The sound insulation performances of small products like doors, windows, small technical elements, lining materials, shutters, and cabins, which are required for marketing purposes, can be measured by applying certain standards [174a, 179].

5.9.3 Field Measurement for Sound Insulation

For specific buildings and elements, the measurements are conducted in the real conditions in situ. However, there are some concerns that need attention in the field, for example, the measurements inevitably include the sounds coming by flanking transmission.

Airborne sound insulation measurement in the field: The earliest standard on field measurements of sound insulation, published in 1975 (ISO 140 Series), was applicable in the diffuse fields, however, the revised ISO 16283-1:2014 gives an engineering method, enabling measurements in the non-diffuse rooms [180a]. The measurement process yields the energy-averaged sound pressure levels and low frequency averaging levels by using the measured corner levels L_{Corner}, after correction according to background noise levels.

The frequency range is 100–3150 Hz 1/3 octaves, however, low frequencies at 50, 63 and 80 Hz and also low and high frequencies up to 5000 Hz are considered. The averaging times are defined in relation to the sound frequency range. The standard explains manually scanning microphones, fixed or moving microphones, and how to determine the positions of loudspeakers and microphones according to the floor area of the receiver room. The number of microphones and loudspeakers is determined according to the room volume and floor area.

Impact sound insulation measurement in the field: ISO 16283-2:2018 describes the procedure to measure impact sound insulation in the field by using a tapping machine and a rubber ball [180b]. In addition to the standardized and normalized impact sound pressure levels, by taking into account the flanking transmission for the sounds in the receiver room, the additional sound pressure levels are defined as:

- energy-averaged impact sound pressure level in a room, L_i
- corner impact sound pressure level in a room: $L_{i,Corner}$ (measured for the low frequency range: 50, 63 and 80 Hz)
- low frequency energy-averaged impact sound pressure level, $L_{i,LF}$
- energy-averaged maximum impact sound pressure level $L_{i,Fmax}$
- standardized and normalized maximum impact sound pressure level, $L'_{i,Fmax,VT}$ for the rubber ball.

The correction according to the background noise is required to apply to the measured levels. The standard describes the positions of the tapping machine and the number of its positions on the test floor. The measurements can be made by using several fixed microphones, a continuous moving microphone or a manually scanning microphone. At each time, the energy-averaged sound pressure levels obtained in the receiver room are calculated. The procedures for the low frequencies are described by recording the signals from the microphones positioned in the corners of the room.

Façade insulation measurements: The former standard, entitled ISO 140-5:1998 (withdrawn and revised as ISO 16283-3:2016), determined the level differences for external elements of buildings simply by using the L_{eq} levels due to the time-varying characteristics of most of the external noises:

$$D_{tr,2m,nT} = L_{eq1,2m} - L_{eq2} + 10\lg(T/T_0) \quad \text{dB} \tag{5.62}$$

$D_{tr,\,2m,nT}$: Standardized level difference at 2 m in front of the façade, dB
$L_{eq1,2m}$: Outdoor equivalent sound level at 2 m from the façade, dB
L_{eq2}: Equivalent sound level in the receiver room, dB
T: Reverberation time in the receiver room, s
T_0: Reference reverberation time in the receiver room = 0.5 seconds

The measurement procedure given in the standard is applicable for the rooms with volumes of 10 m³–250 m³ (furnished or unfurnished) at 50–5000 Hz. Figures 5.56a and 5.56b display the earlier equipment used for façade noise measurements, namely, level difference and room parameters [12]. A noise level analyzer could be used to detect the L_{eq} levels of sound for indoors and outdoors. When the background noise is low, it is recommended to use the loudspeakers as the outdoor noise source. The recent measurement systems are shown in Figure 5.57.

The measured sound reduction index or normalized values for airborne and impact sound insulation values in octave and one-third octave bands are presented by using single rating units as described in Section 5.9.5.

ISO 16283-3:2016: In the latest standard, the procedures are described for the façade elements (windows, doors, etc.) and the entire façade in the frequency range from 50–5000 Hz [180c]. Two methods are recommended regarding the noise source used in the measurements.

Global loudspeaker method: To obtain the level difference, $D_{ls,2m}$ by using a loudspeaker position defined in the standard and measured at 2 m from the façade.

$$D_{ls,2m} = -10\lg\frac{1}{n}\left(\sum_{i=1}^{n}10^{-Di/10}\right) \quad \text{dB} \tag{5.63}$$

n: Number of source positions
D_i: Level difference for each source receiver combination

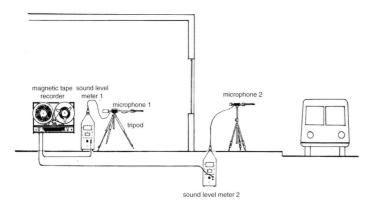

Figure 5.56a Equipment set-up from an earlier study by S. Kurra in 1980 for sound insulation measurement of façades by using traffic noise [12].
Note: B&K instrumentation is schematically displayed as shown in the B&K documents in the 1970s

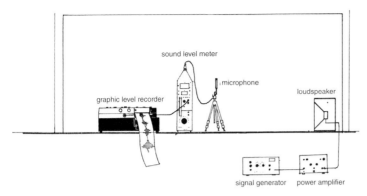

Figure 5.56b Earlier equipment used by S. Kurra in 1981 for reverberation-time measurements [12]. *Note:* B&K instrumentation is schematically displayed as shown in the Brüel & Kjaer document in the 1970s [16].)

(a)   (b)

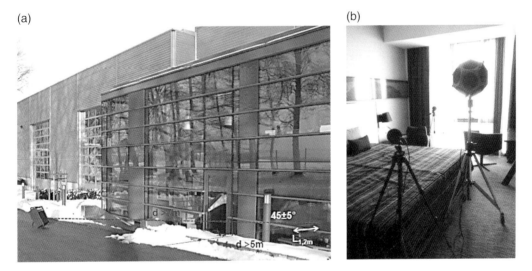

Figure 5.57 Examples of current field measurements: (a) Façade sound insulation measurements by using loudspeaker method (*Source:* COST TU0901 Training School, Jan. 2013, Laboratory of Acoustics, BBRI Brussels). (b) Reverberation time measurement in a hotel room (S. Kurra, 2012).

Global road traffic noise method: Eq (5.62) is applied to determine $D_{tr, 2m,nT}$ as a level difference based on the outdoor sound pressure levels measured at 2 m from the façade. The standard requires certain conditions for road traffic and façade geometry, depending on the angle of view, and describes the microphone positions in order to obtain the energy-averaged sound pressure level in the receiver room, L_2. Low-frequency energy-averaged sound pressure level, $L_{2,LF}$, is also introduced to determine the sound insulation at 50, 63 and 80 Hz by measuring the corner sound pressure levels, $L_{2,Corner}$.

For discrete noise events, Single Event Level, L_E and its normalized and standardized levels are inserted in Eq (5.62). All the other principles given in Part 2 of the standard for airborne sound measurements of partitions are also applicable in façade insulation measurements.

Various experimental studies have been conducted on façade insulation, some emphasizing the measurement uncertainty [181, 182].

5.9.4 Sound Intensity Technique for Sound Insulation Measurements

During the field studies when a high background noise exists and when the sound transmission occurs other than by the direct path through the text element, the so-called flanking paths, it is difficult to inspect the accurate insulation performance of a building element. In that case, the sound intensity technique provides a better option for investigations. Besides, since the pressure methods are restricted with low frequencies below 100 Hz, even below 200 Hz, the lower frequencies can be measured by the intensity technique. Furthermore, the smaller components of the building elements, such as windows, doors, etc., can be readily measured and their influence on the total wall performance can be determined by means of the intensity technique. Sound transmission loss based on the measured intensity level on the wall is obtained by using the following relationship [34]:

$$R = L_{1,s} - 7.5 - (L_{In} + 10 \lg (S_m/S)) + 10 \lg (1 + (S_{b2}\lambda)/8 \, V_2) \text{dB} \tag{5.64}$$

$L_{1,s}$: Average sound pressure level measured at 1m from the wall in the source room, dB
L_{In}: Intensity level measured in the receiver room on the room side of the element, dB
S_m: Surface area of the element, m^2
S_{b2}: Area of the surfaces surrounding the receiver room, m^2
V_2: Volume of the receiver room, m^3
λ: Wavelength of sound, m

The International Standard ISO 15186 describes sound intensity measurements both in the laboratory and in the field [183a, 183b, 183c]. Part 1 and Part 3 both deal with the laboratory measurements for only direct transmission and Part 3 gives the measurements at the low frequencies, ensuring a better reproducibility compared to other methods.

ISO 15186-2:2003 (confirmed in 2013): describes the in-situ (field) procedure to determine the sound insulation of walls, floors, windows, and small elements, also providing the effect of flanking transmission on the measurements.

Concerns about the intensity measurements to be applied in the determination of sound transmission loss of panels were initiated in 1980s and some pioneering studies were published (Section 5.3.3). An example of an earlier investigation (1987), regarding intensity distribution over a window transmitting sound, is shown in Figure 5.58a [184]. Nowadays the new intensity techniques, such as using a three-

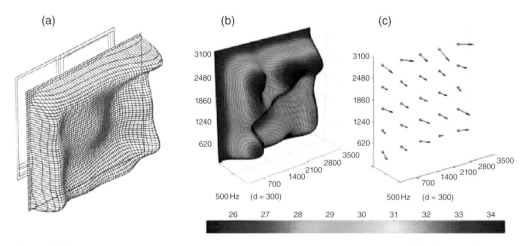

Figure 5.58 Sound transmission measurement by using the intensity technique. (a) Distribution of normal intensity over a window transmitting sound in 2 kHz (1/3 octave band), 1987 [184]; (b) three-dimensional sound intensity measurement results; (c) distribution of intensity vectors in 500 Hz, 2017 [185]. (*Source:* (a) provided by H. Tachibana.)

dimensional vector spectrum, have been developed to facilitate investigation of the weaker points on the surface of building elements, (e.g. a window) as displayed in Figures 5.58b and 5.58c [185].

5.9.5   Evaluation of the Measured Sound Insulation

In regulations and standards, single number insulation rating has been proposed for airborne and structure-borne sounds. These rating units are well correlated with the subjective evaluation of sound insulation and are obtained based on two standard normalized reference spectrums as representative of transportation and living noises. Comparison of the measured sound level index in octave and one-third octave bands with the reference spectrum and applying a curve-shifting method gives the weighted unit in dB which corresponds to the level at 500 Hz on the spectrum diagram. This procedure is described in ISO 717 Parts 1, 2 and 3 for airborne and impact sounds and façade insulation respectively [186a–c]. Five octave bands between 125 and 2000 Hz and 16 one-third octave bands between 100 and 3150 Hz are recommended for use, however, certain adjustments for low and high frequencies and for different source characteristics are also applicable as symbolized by "*C*" with an index denoted for the frequency range. The weighting procedure can be applied to all the insulation units given above, as a result, various rating descriptors are defined. An overview of the descriptors is given in Table 5.10 [187].

The single number ratings facilitate the evaluation of the sound insulation performance of buildings and are useful to manufacturers, engineers, architects, constructors, etc. However, the vast variety of rating descriptors referred to in different standards causes some difficulties at the international level for comparisons, thus, the compatibility between the indices used in different countries in Europe has been achieved at least for residential buildings, by means of the new standard ISO/DIS 19488 [175, 187].

ASTME413-87:1999 (renewed as ASTME413-16): A different single number rating has been implemented in the USA, called the "Sound Transmission Class," to be applied to the one-third octave band levels obtained in the laboratory (STC) and the field tests (FSTC) [188]. The standard also calculates the actual sound insulation in the field by defining the Noise Insulation Class (NIC) and its normalized level (NNIC).

The sound reduction index, determined through measurements of building elements, is used for comparisons with the required quantities declared in the regulations and building specifications. However, the required performance for buildings against noise can also be developed according to the type of noise source, environmental conditions, and usage of building or space. Considering the use in practice, an approach was proposed in a former study for the external elements exposed to road traffic, railways, and aircraft noises, in order to satisfy different indoor noise limits [189].

The sound insulation performances of building elements with various construction properties and materials, which are measured in the laboratories, are collected in a database of the computation programs for assessment of their field performance, as shown in Figure 5.59 [190a, 190b].

5.9.6   Measurements of Room Acoustics Parameters

Two factors influencing indoor sound pressure levels are important in the calculation-based measurements:

1) reverberation time
2) indoor absorption.

Reverberation time is the primary objective parameter representing room acoustical characteristics, which is simply defined as the time required for 60 dB decay of the sound level after the sound source is stopped. The longer the reverberation time, the higher the noise levels in the room. It is

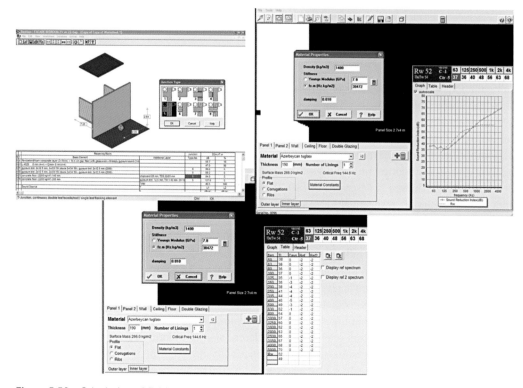

Figure 5.59 Calculation of field sound insulation performance of building elements by means of software (Bastian and Insul) [190a, 190b].

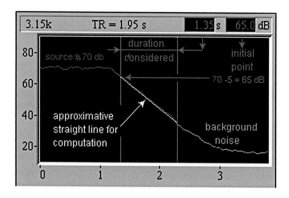

Figure 5.60a Recording of the decay curve in reverberation time measurement. (*Source:* F. Bonfil in 2000.)

important not only in noise control but also in the acoustical design of rooms from the points of view of intelligibility and listening conditions, as in conference or concert halls. Reverberation time and the room absorption are the parameters inversely related to each other, e.g. low reverberation time and high absorption imply a reduction of noise in the spaces concerned. Figure 5.60a shows an earlier measurement of reverberation time in the 1980s.

Measurement of room acoustic parameters is conducted according to ISO 3382, which was first published in 1975 and revised several times until 2015 [191a, 191b, 191c]. The technique described in ISO 3382-2:2008 (confirmed in 2016) is used for reverberation time measurements in ordinary rooms. The advanced computer techniques provide the sound decay curves and the calculations at each frequency band either in octave or one-third octave bands, as shown in Figure 5.60b. They also

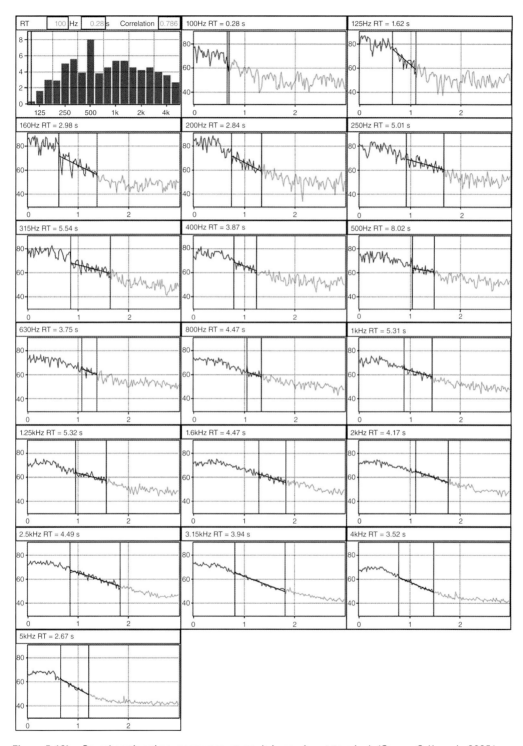

Figure 5.60b Reverberation time measurement result in an airport terminal. (*Source:* C. Kurra in 2005.)

enable the sophisticated analyses and permit various room acoustics parameters to be obtained, which are important in room acoustics.

Another subject in building acoustics is measuring the sound absorption properties of surface materials in the laboratory. This subject has a long history, i.e. 1922 with C. Sabine and 1934 with V.L. Chrisler. In addition to several US standards like ASTM C423, the International Standard ISO 354:2003 (confirmed in 2015) describes the procedure of measurement of sound absorption coefficients in a reverberation room [192]. Such measurements which are mandatory for certification of product are also important in preparing a database which should be ready for acoustical designers to use (Figure 5.61).

Absorption coefficients are measured in the range 250–4000 Hz in the laboratory and the results are converted into single-number rating values with a reference sound absorption curve given in ISO 11654:1997 (confirmed in 2008), to obtain the weighted-sound absorption coefficient, α_w [193].

There are other methods to determine the absorption coefficients of materials, e.g. the impedance tube method, described in ISO 10534-1:1996 (confirmed in 2016), by using the standing wave method (Kundt's tube) and in ISO 10534-2:1998 (confirmed in 2015) by using the transfer matrix method [194]. Some dedicated standards include the measurements of sound absorption properties of outdoor surfaces, e.g. the road surface absorption tests in situ (under the direct sound field) by using flush microphones and a sound source generating a test signal on the road surface [100a, 100b]. The innovative methods for sound absorption measurements can be found in recent literature [195, 196].

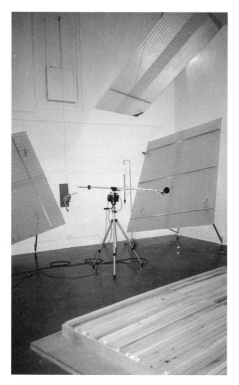

Figure 5.61 Sound absorption measurements in the laboratory in Riverbank Laboratories, USA, 1999.

5.10 Performance Measurements for Noise Abatement Devices

Technical solutions for noise control at source and environmental levels should be certified based on the measurements in laboratories and in the field. These measurements are conducted not only for marketing purposes but also for the design and manufacturing of noise-reducing devices (NRD), such as automotive exhaust silencers and mufflers, specific ducts and duct linings, low-noise road surface materials, low-noise machinery and equipment, cabins, etc. Due to their greater variability in characteristics, specific measurement procedures are applied to determine the acoustic performance given in the technical standards. The results of measurements are declared with the uncertainty values [197–206d]. Noise barrier performance measurements are dealt with in more detail below.

Table 5.11 outlines the international standards those which are widely implemented at present.

5.10.1 Measurements of Silencers

The mechanical noises generated by machinery, generators, fans, compressors or turbines, particularly noise from the air intake and exhaust systems, can be controlled by using sound attenuators, including mufflers, silencers, or acoustic louvers apart from the active noise control systems.

Table 5.11 Technical standards for acoustic performance measurements for noise control devices.

Purpose	Description	ISO standard	
		Laboratory	Situ (field)
Noise control in industry, machines	Silencers	ISO 7235 [198] ISO 11691 [199]	ISO 11820 [200]
	Noise control in machine design	—	ISO 11688-1 [201a] ISO 11688-2 [201b]
	Low-noise workplaces	ISO 11690 [151a, 151b]	—
	Acoustic insulation for pipes, flanges and vanes	ISO 15665 [202]	—
	Noise control by enclosures and cabins	EN ISO 15667 [203]	—
	Noise control in open plant	—	EN ISO 15664 [204]
Noise control elements for road and railway traffic noise	Measurement of barrier performance	—	EN 16272-6 (for railway track noise) [215] ISO 10847 [218] EN 1793-1-6 [219a–f]
	Measurement of removable screen	—	EN ISO 11821 [221]
	Effect of road surface on traffic noise	—	ISO 11819-1, 11819-4 [206a, 206b, 206c]
Interior noise control	Noise control in offices and workrooms	—	ISO 17624 [222]

The test methods are applicable both in the field and in the laboratory and describe the mounting and operating conditions. The general principles of silencer design are explained in ISO 14163:1998 (reviewed and confirmed in 2017) with the information regarding the attenuation of sound transmitted through ventilation openings, the reduction of airflow noise, air-intake and outlet noises from fans, compressors, etc. [197].

ISO 7235:2013 (confirmed in 2018) deals with the performance measurements for ducted silencers which are also used in air-conditioning systems. The method presented in the standard includes the measurements without and with airflow passing through a duct, and defines the effective measurement parameters, such as insertion loss, pressure loss. The arrangement of the laboratory, the layout, and the installation of ducts which need special attention to achieve comparison to the real situation, the configuration of the air inlet and outlet sides, airflow conditions, etc. are described in the standard [198]. The measurements are conducted between 50 Hz and 10 kHz in one-third octave bands (if required, 25 Hz–10 kHz) and the result is expressed in terms of both insertion loss and transmission loss. A laboratory survey method for silencers without air flow is given in ISO 11691:1995 (confirmed in 2010) [199] (Figure 5.62). The measurement for silencer performance under plant operating conditions can be performed in situ according to ISO 11820:1996 (confirmed in 2017) [200].

5.10.2   Measurement of Low Noise Surfaces

The effects of various parameters on tire-road noise are widely discussed in the automotive industry as well as in environmental noise control. Experimental studies are continuing by using the different measurement techniques particularly to improve the tire structure and in the search for low noise pavements [205]. Some of the results are explained in Section 3.2.1.

ISO 11819-1 to 4 (confirmed in 2017): As the main standard on the measurement of the generation of tire-road noise, the standard describes two procedures: the statistical pass-by method in Part 1, and the close-proximity method in Part 2. Part 3 specifies the reference tires for cars and heavy vehicles to be used in the method presented in Part 2 [206a–d] The standard aims to compare

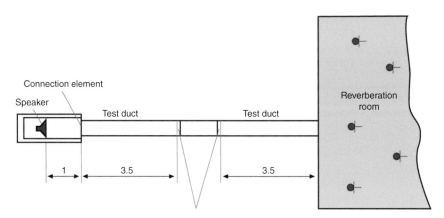

Figure 5.62   Measurement of performance of a silencer according to ISO 11691 [199].
Note: reproduced with the permission of the International Organization for Standardization, ISO. This standard can be obtained from any ISO Member and from the website of the ISO Central Secretariat at the following address: www.iso.org. Copyright remains with ISO.

different road surfaces by taking into account the road categories with respect to traffic speeds in urban, suburban and rural areas and the vehicle categories according to the number of axles.

5.10.3 Measurements of Barrier Performance

Transportation noises are generated from road traffic and railways but can be effectively reduced by properly designed noise barriers along the roads and railway lines. Such barriers provide sound attenuation, depending on various physical and geometrical factors explained in Chapters 1 and 2. As one of the most common practices in environmental noise control, vast numbers of scientific and technical documents and research reports can be found in the literature since the 1970s on the measurement of barrier performance, both in the field and in laboratories [207–217]. The field measurements after the 1990s were standardized, however, by using simple techniques relative to the recent more sophisticated methods. The laboratory measurements in the past were conducted by using scale models in semi-anechoic rooms with hard floors, mostly for investigation of the effects of barriers on outdoor sound propagation and to acquire some empirical relationships (Section 5.8). Figures 5.46a and 5.47 display some experiments conducted in the 1980s for investigation on buildings to be used as noise barriers [164, 165].

Currently, manufacturers need to specify the performances of various types of barriers accurately for marketing purposes. Other purposes of such measurements are for comparison of different barrier constructions and surface materials, to check the validity of the calculation models, or to investigate the applicability of some innovative techniques, such as the "Maximum Length Sequences" (MLS) method (Section 2.2.3). Figures 5.63a–5.63f show some experiments conducted in situ in the past and present. Declaration of the barrier performance (DoP) with the CE marking is mandatory in Europe, according to the standard procedures.

Determination of barrier performance comprises the following measurements:

1) Sound diffraction in relation to the barrier geometry (size, form, and cap detail).
2) Sound insulation of barrier construction.
3) Sound absorption/reflection of barrier surfaces.

Figure 5.63a   Measurement of field performance of an experimental noise barrier in Japan. (*Source:* unknown.)

Figure 5.63b   Field measurement of an experimental noise barrier in the UK. (*Source:* photo: S. Kurra, 1985.)

Figure 5.63c   Measurement for determination of sound reflection index of a barrier surface in-situ, 2017 [217]. (*Source:* with permission of A. Buytaert.)

Figure 5.63d   Example of the measurement of sound reflection in a diffuse sound field according to EN 1793-5 (arrangement of loudspeaker and microphone array). Impulse responses acquisition: near the example and in the free field, 2015 [216a]. (*Source:* with permission of Clairbois and Garai.)

Figure 5.63e Example of the measurement of sound insulation (direct sound field) according to EN 1793-6. Impulse response acquisition: near the example and in the free field, 2015 [216a]. (*Source:* with permission of Clairbois and Garai.)

Figure 5.63f Example of the measurement of sound diffraction in the direct sound field according to EN 1793-6 on a reference wall (reflective, absorptive or in-situ barriers), 2015 [216a]. (*Source:* with permission of Clairbois and Garai.)

Field Measurements of Barrier Performance

a) *Measurement of sound attenuation of barriers in situ:* ISO 10847:1997 (confirmed in 2013) describes the determination of the outdoor barrier insertion loss implying the noise level difference between the "before barrier" and "after barrier" situations." A procedure is given to determine the "before barrier" situation based on the predictions or measurements to be performed at equivalent sites [218]. The standard describes the microphone positions, source conditions, and requirements for the environment including meteorological conditions, and enables comparisons of different barriers constructed on the same site. The descriptors to be used are the A-weighted equivalent continuous sound pressure levels, the A-weighted sound exposure levels or octave and one-third octave band sound pressure levels, and the maximum sound pressure levels.

The *EN 1793 series,* first published in 1997 and renewed in 2016 with six parts, explains the objective measurements of the intrinsic acoustical properties of barriers in situ in the same way as they are installed along the roads [219a–f]. By means of this standard, which is mandatory in Europe, the measurements in the field have been extended also to provide a database for different types of barriers [216, 217]. The field measurements require free-field conditions with low background noise and are conducted under the homogeneous and favorable conditions. The noise levels are measured in octave or one-third octave bands or as A-weighted sound pressure levels. The general procedure of barrier performance measurements is outlined below:

1) Before the barrier is constructed, the sound pressure levels are measured at the reference receiver position in situ.
2) After the barrier is erected at the site, similar measurements are repeated.
3) The difference between the measured sound pressure levels obtained in the two situations indicates the barrier effect at the reference point.

A barrier design, to protect a greater area from noise, needs to repeat the above procedure at a number of receiver points, thus, to facilitate the process, noise maps based on the calculations by using a prediction model, are useful to determine the "before" and "after" cases. However, the field measurements are important for the purpose of validity of the results.

b) *Measurement of airborne sound insulation of barriers in situ: EN 1793-6:2018:* In-situ measurement of airborne sound insulation in direct sound field conditions is explained in this part of the standard [219f]. The standard introduces the "insertion loss of barrier" as the indicator of barrier performance, which is defined as the spectral attenuation with and without barrier situations. When the measurements in "without barrier situations" cannot be performed, the levels are calculated by using a traffic noise prediction model (Chapter 4).

c) *Measurement of diffraction of added devices on the barriers in situ: EN 1793-4:2015:* Additional sound diffraction, provided by the devices installed on the top edge of the barriers, is included in this part of the standard, which aims to help the design process [219d]. A test method to be applied in situ is presented to determine the effect of such devices with a minimum length of 10 m, which are installed on the reference barriers that can be selected as reflective, absorptive, or as an in-situ test wall.

By using the microphones positioned at the reference points near the top edge of the reference barriers, sound pressure levels are measured in one-third octave bands (100 Hz–5 kHz) with and without the device. The standard introduces the sound diffraction index, DI_j as the descriptor of sound diffraction to be computed in one-third octave frequency bands, denoted by *j*. The performance of the added device is expressed as the "sound diffraction index difference" ΔDI, defined as "the difference between the results of sound diffraction tests on the same reference wall with and without the added device on the top" and can also be presented as single number rating values.

The reference height of the test construction, the microphone positions and the sound source characteristics, which is a loudspeaker generating the MLS test signal (impulse response), the data processing using a temporal window, the measuring procedure, the process of the received signal to obtain the diffraction index, measurement uncertainty and the test report are described in detail. The annex of the standard gives also the indoor measurements by eliminating the effect of reverberation.

d) *Measurement of surface reflections of barriers in situ: EN 1793-5:2016/AC:2018:* As one of the intrinsic characteristics of barriers, the reflections from the barrier surfaces are included in this part of the standard and the measurement in situ, under the direct sound field, is explained for the purpose of qualification of the products [219e].

The descriptor of reflections in front of the barrier in the direct field is defined as the "reflection index, *RI*," contrary to the absorption coefficient used in the laboratory measurements in the diffuse field, according to EN 1793-1. The results can also be given as a single number rating of sound reflection, to be obtained as:

$$DL_{RI} = 10 \lg \frac{\sum_{i=m}^{18} RI_i 10^{0.1\,L_i}}{\sum_{i=m}^{18} 10^{0.1L_i}} \tag{5.65}$$

DL_{RI}: Single number rating of sound reflection
m: 4 (number of the 200 Hz one-third octave band)
L_i: Relative A-weighted sound pressure levels of the normalized traffic spectrum, dB (EN 1793-3)
RI_j: Reflection index as a function of frequency in one-third octave bands

The requirements regarding the meteorological conditions (e.g. wind speeds [5 m/s] and air temperature) are given. The reverberant condition of the barriers near the roads is specified as the percentage of open space in the envelope to be less than or equal to 25%, and the different shapes of the envelope formed by the barrier across the road, such as partial covers on one side or on both sides of road, deep trenches, and tall barriers with geometrical properties, are presented as examples in the standard.

The measurement system components are explained in detail: Microphone reference positions in front of the barrier (at the roadside), are given separately for non-flat surfaced and flat- homogeneous surfaced barriers. The size of the maximum sampled area is given in relation to the shortest distances of LS and microphones. The reference circle with its center at the point of incidence is defined to determine the reflection index.

The sampling according to the specified angles of incidence between 50 and 130° is described in the earlier version, however, the recent standard (2018) requires rotating loudspeaker/microphone assembly to be used during the measurements. The results obtained at each rotation of the microphone are averaged and the correction factors for sound source directivity, gain mismatch, sampling rate, etc. are applied. Conditions of low frequency limits are described to determine the reflection index, which is dependent on the size and shape of the barriers.

The overall impulse response to be measured includes a direct and a reflected component in addition to the unwanted components called parasitic elements. Therefore, a signal subtraction technique is described to extract the reflected component from others by using an Adrienne temporal window.

The standard also gives the measurement of the in-situ sound insulation index with the microphones positioned at the rear side. Since both the transmitted wave through the barrier and the diffracted waves from the top edge of the barrier, arrive at the back of the barrier, in order to measure only the transmitted wave in the direct field (free-field), a technique based on the analysis of the power spectra of both waves, by applying corrections according to the path-length differences is presented. The result is shown by the descriptor of "sound insulation index," SI_j to be computed.

Laboratory Measurements of Barrier Performance

a) *Measurement of sound absorption in laboratory: EN 1793-1:2016* explains the measurement of the sound absorption coefficient under diffuse sound field conditions (i.e. reverberant rooms) according to ISO 354:2003 [219a, 192]. From the spectral values of the measured absorption coefficients at 100 Hz–5 kHz at one-third octave bands, the single number rating units are obtained as given in Eq (5.66):

$$Da = -10 \lg \left| 1 - \frac{\sum\limits_{i=1}^{18} \alpha_{Si} 10^{0.1L_i}}{\sum\limits_{i=1}^{18} 10^{0.1L_i}} \right| \qquad (5.66)$$

Da: Sound absorption performance of barrier surface on the source side, dB) (If the result >1, it is taken as 0.99.)

i: Each of the 18 one-third octave bands

α_{Si}: Sound absorption coefficient for the ith band

L_i: Normalized A-weighted traffic noise spectrum value at the ith band

For use in the market, the absorption performance of the barrier surfaces, is classified as:

Da	Sound absorption class of barrier surface
<4	A1
4–7	A2
8–11	A3
>11	A4

Since Da does not emphasize the low frequencies in the traffic noise spectrum, therefore, selection of surface material should be made by considering the spectral values, especially in the environment types where the multi-reflections are present.

b) *Measurement of sound insulation performance of barrier in laboratory:* As explained in Chapter 2, in order to provide an efficient attenuation of barriers used along a roadside, they should possess a sufficient sound transmission loss against the airborne sounds, in brief, the transmitted sound through the barrier should be much less than the sound reached behind it through diffractions. In order to check the sound insulation of barriers through measurements in laboratory; the EN ISO 10140-3 is implemented, as explained in Section 5.9.2 [174b]. The barrier sample to be inserted in the test opening in the laboratory should be constructed together with all the constructional and structural elements and components, e.g. supports, panels, connections, etc. and the measurements are performed between 100 and 5000 Hz at one-third octave bands.

EN 1793-2:2018: This part of the standard describes the airborne sound insulation of barriers under diffuse sound field conditions in laboratory [219b]. Based on the spectral levels to be measured as summarized above, the "single number sound insulation rating value" is calculated as:

$$DL_R = -10 \lg \left| \frac{\sum\limits_{i=1}^{18} 10^{0.1L_i} 10^{-0.1R_i}}{\sum\limits_{i=1}^{18} 10^{0.1L_i}} \right| \text{ dB} \qquad (5.67)$$

DL_R: Single number rating of airborne sound insulation performance of barrier expressed as a difference of A-weighted sound pressure levels, dB

R_i: Sound reduction index in the ith one-third octave band, dB

L_i: Normalized A-weighted sound pressure level of traffic noise in the ith one-third octave band defined in EN 1793-3, dB [219c].

The airborne insulation performance of barriers is categorized as follows:

DL_R	Barrier insulation category
< 15	B1
15–24	B2
25–34	B3
>34	B4

The content of the reports for all different measurements are given in the standard.

Measurement of Other Properties of Barriers

Barriers to be constructed for road traffic noise should possess other non-acoustic characteristics, such as endurance against meteorological conditions (temperature, humidity, exhaust gases, etc.), static durability against strikes (impact or harsh attacks), etc. EN 1794-1 and EN 1794-2 give the measurements of these properties [220a, 220b].

Indoor Noise Screens

Besides roadside barriers, the movable or fixed screens which are widely used indoors for the protection of workers from noise or to provide privacy in open-space offices are commercialized in many countries (Section 2.A.3). ISO 11821:1997 (confirmed in 2018) also adopted as the European Norm, describes the measurements of portable screens to determine the sound attenuation performance by using actual and artificial sound sources in situ [221]. The general effectiveness of screens, in terms of the acoustical and operational requirements, are included in ISO 17624:2004 (confirmed in 2017) [222]. The standard provides technical information covering all type of screens used in offices and workrooms and outlines the performance tests as a guideline for manufacturers, suppliers, and users.

5.11 Instrumentation for Noise Measurements

Instrumentation is a special field of acoustics in relation to electro-acoustics which is included in a great number of published documents [223–230], in addition to those given throughout this chapter. Generally, the measurement system is composed of different functions, i.e. receiving sound signals, data acquisition and storage, post-processing, and continuous monitoring if required. The earliest equipment used for noise measurements were rather simple devices capable of short-term observations and analysis by connecting to other equipment working as calibrated sets. However, nowadays, developments in electronics, computers, and microcomputers have enabled the integration of many functions in a single piece of equipment making measurements a great deal easier to achieve.

5.11.1 Overview of the Measurement Systems from Past to Present

The historical background of instrumentation so far used for environmental noise measurements and analyses, is outlined below, based on the literature:

Phase 1 (Initial Phase)

Sound pressure levels were used to measure in the early periods of the 1960s by using a portable sound level meter with the screen displaying the voltage of signals and dB levels on the analogue screen, averaging by eye and writing the results by hand. For the steady noises, the spectral levels could be obtained by using the portable filters mounted on the equipment and by manually selecting the band. At this time, the graphic level recorders were transported to the field to receive graphics on paper format. Later in this period magnetic tape recorders were used to record sound signals in the field and to analyze them in the laboratory. Some of these audio recorders included two channels in addition to a speech track, enabling simultaneous measurements. By means of aural recordings, the spectral analyses for time-varying noises were possible in the laboratory by connecting the magnetic tape to the real-time frequency analyzers, however, the signal levels were increased with microphone amplifiers.

Phase 2 (Mobile Laboratory)

Temporal analysis to obtain statistical noise descriptors, cumulative distributions, etc. was achieved with the "Statistical Distribution Analyzer" and later with more developed equipment called the "Noise Analyzer," including a memory card. The noise analyzer was programmable so that the statistical analyses could be made within a designed time period. If necessary, the magnetic tape recorders could be connected to the analyzer to obtain the results in the laboratory.

As seen, the measurements and analyses could be made by using sets of equipment connected to each other, however, the connections were available only for a specific brand of equipment. On the other hand, it was difficult to carry those pieces of equipment, which are rather expensive and heavyweight, into the field, to conduct long-term measurements or leave them safely in the field, eventually mini-vans were used as a mobile laboratory equipped with all the measurement set-ups and also carried the technicians.

Phase 3: Computer-Aided Measurements

In this period, measurements began to be processed by using software according to the purpose of measurements. One of the earliest software programs developed in the 1980s was "Canal Analyzer" operated by the DOS system on the earlier personal computers. The signals could be analyzed by connecting the magnetic tape recorder to the computer, to perform spectral and temporal analyses, however, after certain tests. Later, parallel to the development of portable computers, a small measuring unit was sufficient to achieve all the measurements and analyses required in the field.

The numbers of the microphones were increased by means of additional units, such as front end or multiplexer, enabling simultaneous measurements by five or six microphones. In the later phase, the Windows operating system facilitated the measurements enormously with increasing computer capacities. Ultimately, the size and efficiency of equipment and additional accessories have been minimized.

Phase 4: Monitoring from a Distance

In this latest period, by means of protected microphones with accessories and noise monitoring terminals, instantaneous data is transferred to the laboratory for analysis or for observation of the real-time levels. Once the satellite systems, like Global Positioning System (GPS) and Geographical Information System (GIS), were introduced, the data regarding the noise sources and the environment could be accurately obtained, analyzed, managed and presented. These advanced techniques provide information temporarily or continuously for inspections of disturbance from certain noise sources, such as construction or airport activities.

Phase 5: Innovative Techniques

Combining vibration and acoustic measurements and the investigation of acoustic wave transmission in structures by measuring vibration velocity and acceleration, are the subjects still being examined. Measurement of the airborne acoustic field can also be made by using optical methods, i.e. laser techniques. In a new method developed by NPL, the particle velocities are directly measured by means of two laser beams intersecting at a point and the acoustic pressure is calculated [231].

In the twenty-first century, scientists, technicians, and manufacturers will likely continue spending much time, effort, and money in search of the best measurement techniques applicable in the specific fields of interest to acquire accurate results, quickly, economically and through the least complicated means, such as the source identifications for industrial noises described in Sections 5.4.1 and 5.11.2. Ultimately it can be expected that the new techniques will make significant changes in the context of the future measurement standards.

5.11.2 Basic Instrumentation for Acoustic and Noise Measurements From the Past to the Present

Generally, the basic equipment and accessories, used in the past to the present, to obtain field measurements consist of the following systems:

1) Sound receiving (data acquisition): Microphones (sound pressure and intensity microphones), preamplifiers, calibrators, signal amplifiers.
2) Sound pressure measurement: Simple or precision sound level meters, integrating sound level meters and measuring amplifiers.
3) Sound intensity measurement: Intensity probes (to measure sound pressure and particle velocity), calibrators, and intensity analyzers.
4) Frequency analysis: Filters (octave band or one-third octave bands), narrowband filters, FFT systems, real-time analyzers.
5) Statistical analysis: Time distribution analyzer, noise level analyzer.
6) Recording and storage: Graphic level recorders, printers, magnetic tape recorders, noise monitoring terminals.
7) Printing: Graphic level recorders and portable printers.

Figures 5.64a and 5.64b display sample equipment used from the 1970s to the present. Those important in noise measurements are summarized below.

Microphones

As a kind of transducer to measure sounds, microphones are signal-receiving devices which convert sound pressures into electrical signals to enable direct analysis [5]. Of the main three types of microphones, the condenser microphones have been widely used since the 1920s for their uniform sensitivity and low distortion, thus giving correct results. A condenser microphone is composed of a thin diaphragm and an insulated backplate (with holes for air damping) which is separated from the diaphragm by a narrow airgap. The system operates as a capacitor (Figures 5.65a and 5.65b). An external polarizing voltage is applied between the diaphragm and the backplate, with both acting as electrodes of a condenser (polarized condenser microphone). The diaphragm is made of extremely light material, and vibrates with the acoustic pressure, so the pressure changes are transmitted to the air in the gap, causing a change in the capacitance which is converted into a voltage (electrical signal) within the preamplifier, which is a separate unit connected to the circuit. The pre-polarized

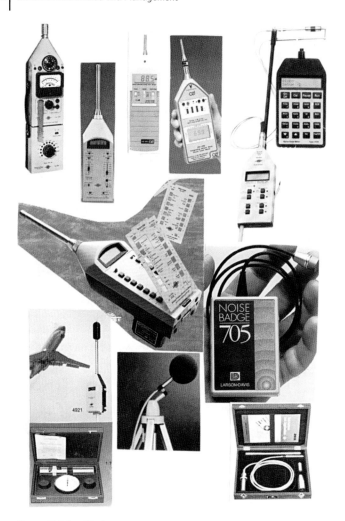

Figure 5.64a Various sound level meters, pistonphone, and dosimeters used in the past and at present.

microphones (electric microphones) emerged in the 1970s, and possess a thin layer of electrically charged material on the backplate without requiring a polarizing voltage. Piezoelectric microphones have a stiff diaphragm connected to a piezoelectric crystal or ceramic element and are used for moderate or less-sensitive measurements. The output voltage of a microphone diaphragm is dependent on its size. The microphones are specified with the following characteristics:

Sensitivity: The sensitivity of a microphone implies the ratio of output voltage to the sound pressure, expressed as mV/Pa or dB with respect to a reference level (dB re 1 V). Sensitivities of typical microphones change between 1 and 50 mV/Pa depending on the size of the microphone diameter, for example, a 1″ microphone has sensitivity of 50 mV/Pa. Small microphones are less sensitive to low pressures. Various sizes of microphones with varying sensitivities are produced as 1/8″, 1/4″, 1/2″, and 1″. Specifications of design, production, and calibration of microphones are described in IEC 61094 [232a–d].

Dynamic range: Defined as the change of output levels between upper and lower limits –above its inherent noise – the dynamic range of microphones is about 100–120 dB, roughly similar to the

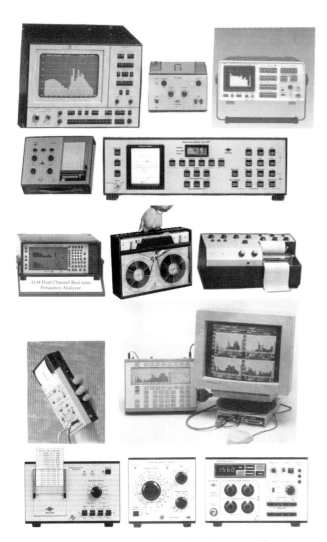

Figure 5.64b Various recorders and analyzers used in the past.

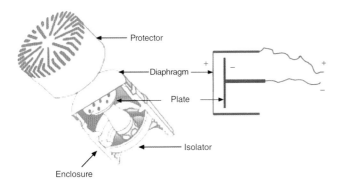

Figure 5.65a Condenser microphone components.

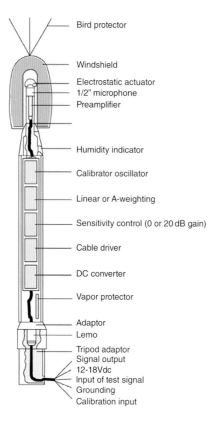

Bird protector

Windshield

Electrostatic actuator
1/2" microphone
Preamplifier

Humidity indicator

Calibrator oscillator

Linear or A-weighting

Sensitivity control (0 or 20 dB gain)

Cable driver

DC converter

Vapor protector

Adaptor
Lemo

Tripod adaptor
Signal output
12-18Vdc
Input of test signal
Grounding
Calibration input

Figure 5.65b A condenser microphone with additional protection against environmental factors. (*Source:* Adapted from G.R.A.S.)

response of the human ear, however, depending on their size. For example; 1″ and 1/2″ diameter microphones have the greater dynamic range about 150 dB. Small microphones, 1/4″ and 1/8″ diameters, can measure sounds without distortion within the dynamic ranges of 140 dB and 100 dB respectively.

Frequency response: Two types of microphones are defined according to the frequency response characteristics as:

- Free-field microphones with flat response for sounds at normal incidence in free-field and used outdoors
- Diffuse-field microphones with flat response for random incidence sounds in diffuse field and normally used indoors (Figure 5.66).

Variation in the sensitivity according to the frequency of sound is specified as the range in which the deviation of output signal is within ±2 dB. An ideal microphone should have a uniform response at the relevant frequency bands. As the wavelengths are small at high frequencies, the unwanted divergencies due to diffractions might occur depending on the size of the microphone, i.e. for the wavelength equal to the diaphragm radius. Some microphones are capable of measurements below 20 Hz (infrasonic) or above 20 kHz (ultrasonic sounds) or only within the hearing range about 20 Hz–20 kHz. Small diameter microphones have frequency responses in a wider range. Approximately, a 1″ microphone covers a range of 2Hz–10kHz and 1/2″; 4 Hz–20 kHz and 1/8″; 6 Hz–150 kHz [5]. The frequency response of microphones is displayed on the calibration

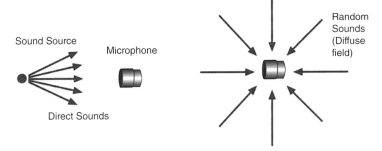

Figure 5.66   Free-field and diffuse-field microphones.

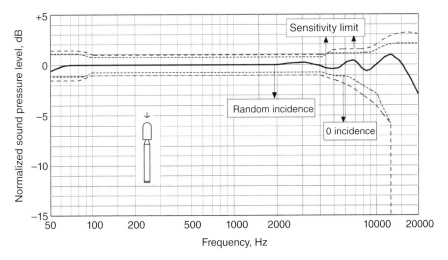

Figure 5.67a   Frequency characteristics of a microphone (calibration chart).

charts provided by the manufacturer. Figure 5.67a shows a typical frequency response chart of a 1″ microphone.

Directivity: The directivity characteristics of a microphone imply the variation of sensitivity with the angle of incidence of sound waves. Directivity changes also with the frequency and the effect is more emphasized at high frequencies. At low frequencies, the frequency response of a microphone is independent of the angle of incidence [93]. The average response of microphones in the diffuse field, where the sound waves are coming from all directions, are measured in the reverberation rooms and is presented on a polar diagram (Figure 5.67b). The microphone directivity is taken into consideration practically in environmental noise measurements, e.g. if the direction of the sound incidence is known, the free-field microphone is directed to the source and if the sound incidence is random, the diffuse field microphone should be chosen. In a diffuse field, orientation of the microphone axis upwards reduces the errors [16].

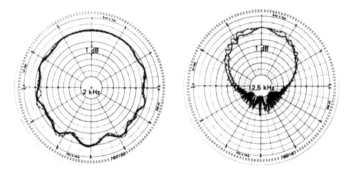

Figure 5.67b Directivity charts of a microphone at different frequencies (polar diagram).

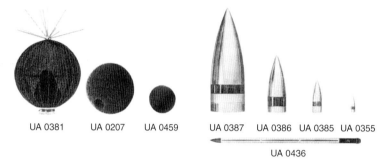

UA 0381   UA 0207   UA 0459   UA 0387   UA 0386   UA 0385   UA 0355

UA 0436

Figure 5.68 Wind and rain shields.

Durability: Most modern microphones have long-term stability and endurance under various environmental conditions, such as temperature and humidity changes and dust. Special accessories are used to protect microphones; a windshield to reduce wind noise at high wind speeds, a noise cone and a rain shield for protection from humidity and rain, a dehumidifier to eliminate humidity and a turbulence screen for the microphone positioned inside the duct, the adaptors for cable connections and the resilient adaptors for mounting microphones (Figure 5.68). Modern microphones are normally attached to a random incidence corrector (microphone cartridge or protection grid) for use in the diffuse field (Figure 5.65a).

As a summary, the following principles play an important role in the selection of a microphone for noise measurements:

- The microphone sensitivity should not change with time or physical environmental factors, if necessary, corrections should be applicable.
- Frequency characteristics with directions (directional characteristics) should be constant.
- Size of microphones is selected according to the purpose of measurement. A free-field condenser microphones with 1″ or ½″ diameter, can be selected for outdoor noises. For diffuse fields, ½″ or ¼″ microphones and for small holes like duct measurements and for the scale-model measurements, 1/8″ microphones are needed.
- Mounting possibilities: The microphones can be mounted on a surface by an additional layer or specific types can be used to measure the surface pressure (Figures 5.69a and 5.69b).

Various types of microphones are available on the market equipped with preamplifiers and other accessories, like hemisphere kits for power measurements, etc. as shown in Figures 5.70a and 5.70b.

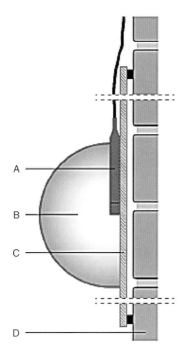

Key

A   Microphone

B   Windscreen

C   Mounting plate

D   Wall or reflecting surface

Figure 5.69a   Mounting microphone on the wall [95d].

Note: reproduced with the permission of the International Organization for Standardization, ISO. This standard can be obtained from any ISO Member and from the website of the ISO Central Secretariat at the following address: www.iso.org. Copyright remains with ISO.

Figure 5.69b   Surface microphones for measurements of surface pressure.

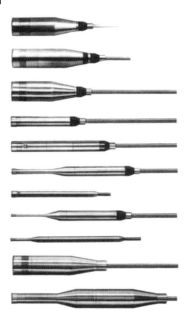

Figure 5.70a Various microphones with preamplifiers (G.R.A.S).

G.R.A.S.
Sound and Vibration

Figure 5.70b Various types of modern
microphones and accessories (G.R.AS.)

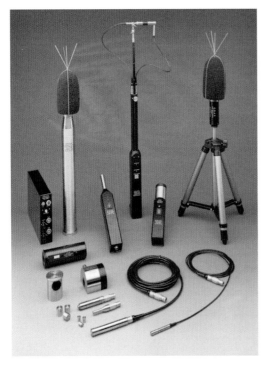

The advances in electroacoustics enable the multi-channel measurements simultaneously, e.g. 120 channels or more, by organizing the microphones or probes as two-dimensional surface arrays. Recently, three-dimensional spherical arrays, called an acoustic camera, have been produced and are able to receive more signals in precision measurements, for source localization and acoustic holography. Different types of arrays can be chosen according to the purpose, e.g. grid arrays for scanned measurements moving manually or automatically from one position to another, arm-wheel arrays, sliced-wheel arrays and hand-held arrays for real-time holography mapping. Arrays provide simultaneous measurements with many numbers of microphones which can be increased as desired to determine the hot spots (most prominent source parts) on noise sources (Figure 5.71).

Sound intensity microphones: As explained in Section 5.3.3, the intensity probe, consisting of the coupled microphones, is used for sound intensity measurements. Two microphones mounted face-to-face with a distance of 6, 12 or 50 mm, which are compatible in phase and amplitude, can measure sound pressure and particle velocity, overcoming the phase-mismatch problem (Figures 5.72a and 5.72b). The sound pressure gradient between the microphones and the particle velocity component in the axis between the microphones is calculated in one-third octave bands [37]. The intensity probe is connected to measuring equipment or a computerized system (Figure 5.72c). Recent technology has provided different designs of intensity kits and multi-microphone intensity probes since the 1990s.

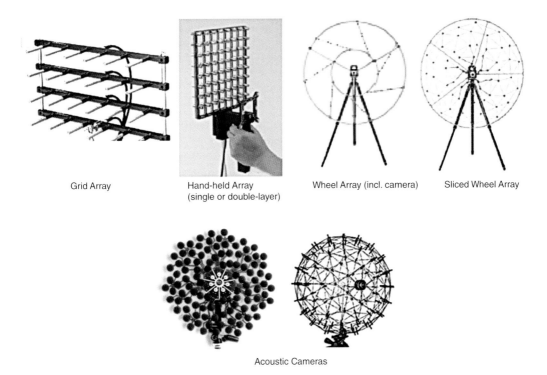

Grid Array   Hand-held Array (single or double-layer)   Wheel Array (incl. camera)   Sliced Wheel Array

Acoustic Cameras

Figure 5.71 Microphone arrays used from the 1990s to date: Grid arrays, spherical arrays (acoustic camera) for hologram measurements and 3D arrays for beamforming measurements.

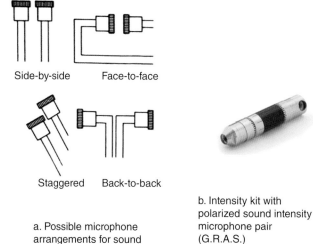

Side-by-side   Face-to-face

Staggered   Back-to-back

a. Possible microphone arrangements for sound intensity measurements (Fahy,1977 Crocker,1993)

b. Intensity kit with polarized sound intensity microphone pair (G.R.A.S.)

c. Connecting intensity probe to computer through an acquisition device (01dB Areva)

Figure 5.72   Sound intensity transducers and computer connections; (a) possible microphone arrangements for sound intensity measurements; (b) intensity kit with polarized sound intensity microphone pair (G.R.A.S.); (c) connecting intensity probe to computer through an acquisition device (01dB Areva).

Sound Level Meters

The basic equipment to measure and report the sound pressure levels is a sound level meter, also called a sonometer in the 1980s. The specifications regarding their design are described in IEC 61672:2013 based on the earlier IEC651:1979 [233, 234a–c]. The performance characteristics of sound level meters are as follows:

Accuracy: Sound level meters have been classified in the standards as:

Type 0: The standard reference size which is used in the laboratories and for calibration of all the equipment (accuracy for random incidence sounds can be: ±0.7 dB).

Type 1: A precision sound level meter which is used for sensitive measurements both in the laboratory and in the field (accuracy for random incidence sounds can be: ±1 dB).

Type 2: General purpose sound level meter which is used in the field for recording and spectral analysis of noises (accuracy for random incidence sounds can be: ±1.5 dB).

Type 3: Survey sound level meter with less sensitivity, which is used for preliminary measurements and observations in the field.

Type 1 (sometimes called Class 1) instruments has lower tolerance limits than Type 2.

Frequency response: The accuracy of the measured signals in relation to the frequency range is declared by the manufacturers as the free-field and the diffuse-field responses, and not only for the microphones, but also for the other elements of the measurement system, e.g. preamplifiers, filters, etc.

Dynamic range: The difference between the maximum input level and the self-noise of the sound level meter implies the dynamic characteristics as explained for the microphones above.

Response time: The signals are received by using the techniques called time-weighting characteristics of the sound level meter (Figure 5.73). These are:

Fast response: One signal received per 100, which is the equivalent response of the human ear up to 200 ms

Figure 5.73 Comparison between fast and slow response in signal receiving.

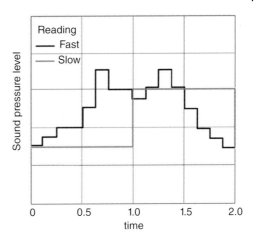

Slow response: One signal received in one second which is used if the noise is steady and continuous. This response is suitable to determine the average levels.

Impulse response: The sampling time is as small as 35 ms for the intense noises with fast rise and decay. This response is also used to measure the peak noise levels within a short duration such as 50 μs.

Frequency-weighting networks: The weighted levels, according to the selected frequency weighting network, are obtained by using electronic filters built into the sound level meters (Section 2.2). The selected weighting (A, C, D, L_{in}) is displayed on the digital screen.

Combining the microphone, preamplifier, attenuator, filter and the reading window showing the sound pressure level Lp, this device is still widely used in environmental measurements and analyses. Filters are used with band pass or A-weighting. A block diagram of a digital sound level meter is given in Figure 5.74 [235].

Advances in sound level meters: Sound level meters have shown a great improvement from past to present. Figures 5.75a–k display various samples from different manufacturers.

Conventional sound level meter: The first devices, some called sonometers, were manufactured to measure the pure-tone sounds for audiological measurements. A type of galvanometer, they could give a constant or slow response and instantaneous sound levels could be displayed with 1 dB accuracy, though hardly readable due to analog display, they could give graphical records. An example of this device, used by NPL in the UK, is shown in Figure 5.75a. In later years, more advanced types were produced as shown in Figures 5.75b–f.

Integrating-averaging sound level meter: Over the years, the sound level meters were designed and manufactured to provide the following features (Figures 5.75g–k):

measurement of non-steady noises (fluctuating)
extension of duration of measurement
increasing the numbers of samples
energy summation and time-averaging (measurement of time-average sound level L_{eq} and event noise level, L_E)

The first products had a dynamic range of 60, 80 and 100 dB depending on the brand and were used for all type of environmental noises. Starting in the 1970s, improvements were seen, such as liquid crystal display, greater dynamic range of 25–140 dB, and longer periods (between 60 seconds to 8 hours). When the microprocessor emerged in the 1980s, the memory of the sound level meters

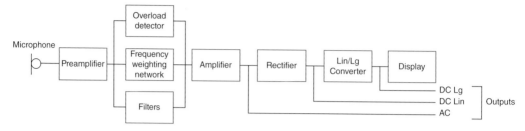

Figure 5.74 Block diagram of a sound level meter. (*Source:* adapted from [235].)

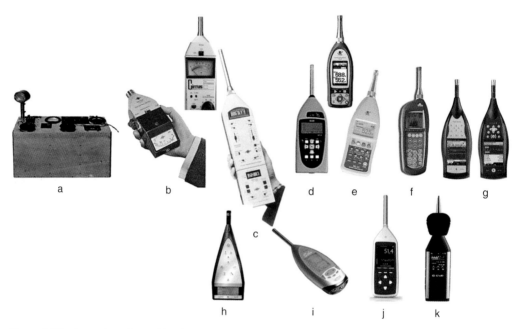

Figure 5.75 Examples of sound level meters from the past to date: (a) NPL Objective Noise Meter; (b) and (c) Brüel & Kjaer (above: old analogue meter by Cirrus); (d) SIP 95 (01dB-Metravib); (e) Rion (above: Rion); (f) Norsonic; (g) Brüel & Kjaer; (h) Brüel & Kjaer; (i) Solo by 01dB; (j) Cirrus; (k) 01dB-Areva.

was increased with the multifunctional characteristics. These advances have facilitated longer-term measurements and the digital storage of average levels. Detecting impulse noises with an abrupt pressure change, e.g. within less than 1 second, permits measuring and analysis of industrial noises which contain impact and impulsive characteristics. Different modules can be added to these sound level meters, e.g. for analyses of room acoustics and loudness parameters and advanced statistical analyses.

Computerized sound level meters: By means of digital signal processing, these systems can translate the signals to digitals, store them, and can make computations as required. After the basic signal is received, it is possible to obtain the required results instantly. After the development of computer technology and micro-processors, two types of measurement systems were manufactured and are still in use:

Computer attached systems: Various functions, such as signal receiving, signal processing, editing, and data organization, report writing and giving the digital outputs of weighted levels, spectral diagrams, time-history diagrams and tables, etc. are combined in a data acquisition unit and a data

Figure 5.76   A computer-based noise measurement.

processing software. Noise source identification can also be made by means of a special data proc-essor. They consist of the following components (Figure 5.76):

- Microphone and preamplifiers: (explained above).
- DC conditioning unit: Adapting the electrical signal for the analysis, since they are fed by elec-trical power.
- Processor unit: Digital data acquisition.
- Laptop: Visual reading as real-time and able to calculate the analytical calculations and store the results.
- Software: Detecting the measurement type, processing according to the objective.

The interactive software called a dedicated tool, embedded in the measurement system can be purchased separately as modules for different measurements and operations as required. The micro-information technology provides for organization of measurements, signal receiving and optimization data process. Software programs are prepared according to the technical standards, specifying the measurement methodology. Recent software is able to make precision analyses of sounds depending on their signal processing and computation performances (Section 5.11.3).

The new systems are more efficient with the increased storage capacity for digital signals, with enlarged dynamic range, longer recording time, and a decrease in the processing time and improved visual functions. Some of them have the capability to make audio recordings. As they are light-weight and portable, field measurements can be conducted unattended or controlled from a dis-tance (Figures 5.77a and 5.77b).

Advanced "one-meter type" measurement system: The new generation of instruments is the latest advance, combining all the functions in one piece of equipment, implying a return to the "one sound level meter" period [13]. By means of advanced microprocessors, they have great capacity for programming for computations, making sophisticated analysis, data communication, and stor-ing. They measure, display, and store the one-third octave range of 12.5 Hz to 20 kHz, all the new and older metrics, a user-defined event setting, detection with event/start/stop triggers with min-imum threshold exceedance.

At present, the numbers of manufacturers producing such advanced equipment competitively, have been increased by developing user-friendly, aesthetic, and colored screens facilitating usage and readings.

(a) (b)

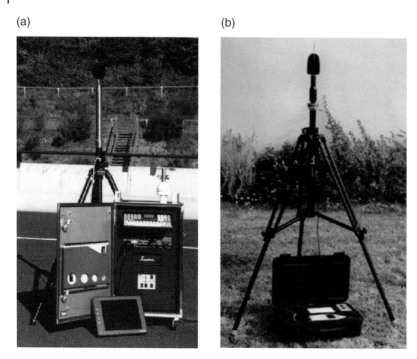

Figure 5.77 (a, b) Noise monitoring equipment used for long-term measurements in the field.

Figure 5.78 Multichannel measurement systems (Orchestra 01 dB-Metravib) and NETDB12 01 dB-Metravib-12 channels (for frequency analyzing, power measurement, modal analyses, and sound quality measurements).

Long-term monitoring systems: In the initial phase of noise measurements, the data collected in a long period of time, i.e. 24 hours or 48 hours at stationary microphone positions automatically or manually, could be transferred to the main computer via modem and telephone networks for analyses. In the next phase, the computerized smart systems are connected to each other or other systems, such as radar or GPS, so that the information, once transferred, could be converted to different formats by means of modern telecommunication technology. Figures 5.78a and 5.78b show a multichannel measurement and data collecting systems that can be remote-controlled, by means of Bluetooth technology. A number of technical documents can be found in this topic [236–239].

Use of monitoring for the acquisition of environmental noise data and to bring the results to the attention of public, has been mandated by legislations and standards globally (i.e. 2002/49/EC in Europe) [236]. The characteristics of the current monitoring systems in the market, are summarized below:

Enabling short-term, long-term, and permanent monitoring.

Using different software and hardware.

Including large number of metrics.

Easier data management.

Embedding large amounts of memory in instrumentation.

Relatively inexpensive and with wireless connection capability.

Capable of collecting wide data range and storing automatically for future processing.

Able to measure the new metrics.

Supplying an extreme database, and presenting outputs graphically and in tabular format.

Providing kits to apply corrections according to the atmospheric conditions.

Well-protected from the meteorological conditions in cabinets and microphones with safety protectors.

Details of the computer-based data acquisition system and network-based data control are given in Section 5.11.3.

Other Supplementary Equipment and Systems

Calibrators: Calibration is necessary to ensure the accuracy of the microphones and the sound level meters by checking their sound pressure responses. For this reason, a uniform sound pressure is applied over the diaphragm of the microphone [5, 223, 230, 241]. In practice, a reference pure-tone signal, whose level and frequency are known, is given on the microphone diaphragm, generated by a small sound source called a calibrator, however, by providing a good seal between the microphone and calibrator. The calibration signal is a plane-wave with 94 dB re 20 μPa at 1000 Hz or 124 dB re 20 μPa at 250 Hz. The signal level, displayed on the sound level meter or on the computer by using the calibration mode of the measurement software, is checked to see whether there is a difference between the true value. The difference – even one tenth – implies the value is out of tolerance and the equipment is corrected up to ±0.2 dB. Bigger differences imply the possibility of a destroyed microphone diaphragm. Such a check should be made prior to and frequently during the measurements. Performance requirements for calibrators are described in IEC 60942:2017 (first issued in 1988) for Class 1 and Class 2 equipment. Calibrator selection should be compatible with the sound level meter [234b, 234d].

Various types of calibrators have been manufactured according to the specification of IEC 60942:2017 and some examples are found on the market, shown in Figure 5.79 [241]. The portable calibrator, also called a pistonphone, with a small battery-powered electric motor is purchased as a piece of separate equipment – with the adaptors to be used on different size of microphones and with a barometer to make corrections according to the atmospheric pressure. For sound intensity probes, a special coupler attached to the pistonphone is used as sound intensity calibrator, adaptable to ½″ and ¼″ microphones.

Total sensitivity of the microphone and the equipment should be checked in the accredited laboratories within the range of hearing. This time-consuming procedure should be made every five years to comply with the technical standards.

Analyzers and recorders used in the past: In the earlier period, separate bits of equipment were used to connect to the sound level meters or the receiver system to perform the time-domain and frequency-domain noise analyses which have been explained scientifically in various publications [242, 243].

Frequency analyzers: To measure the frequency spectra of the noises, portable band-filters, frequency analyzers, real-time analyzers, computer-aided frequency analyzing (operated by the DOS

Figure 5.79 Microphone calibrators (pistonphones) used in the 1970's to date.

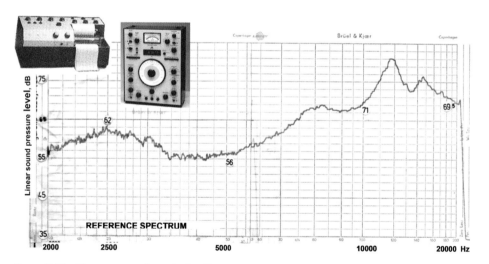

Figure 5.80 The spectrum chart obtained by using frequency analyzer (Brüel & Kjaer) (S. Kurra, 1982); below: real-time frequency analyzer (Brüel & Kjaer).

system), recently built-in analyzers and advanced software, have been used in the past to the present. All were specified in the international standards [244a–c]. Early examples of frequency analyzers are displayed in Figure 5.80.

Statistical analyzers: For time-varying noises, simple statistics like distribution of noise levels with time, histograms and cumulative curves to determine the statistical noise units, were produced by certain equipment in the 1970s to the 1980s. The oldest one is B&K's "Statistical Distribution

Analyzer" followed by the "Noise Level Analyzer" in the 1980s, which has a memory capacity for stored data and was capable of input, timing, and output programming (Figures 5.81a and 5.81b).

Data recording and storage systems: Before the computer era, the results were obtained by using thermal graphic printers as paper output, such as alphanumerical histograms or tables (Figure 5.82). Later magnetic tape recorders were carried to the site for audio recording over the tapes which were rolled on a wheel or a cassette, and could be analyzed in the laboratory (Figure 5.83).

Signal generators and test sound sources: Various tests in environmental and building acoustics need the test signals generated by standard sources, such as for the barrier performance tests or sound insulation measurements. The standard signals are white, pink, and pure-tone sounds as

(a)

(b)

Figure 5.81 Statistical noise level analyzers: (a) Statistical Distribution Analyzer Type 4420 in the 1980s; (b) Noise Level Analyzer Type 4426 in the 1990s, (Brüel & Kjaer).

Figure 5.82 Graphic level recorder (Brüel & Kjaer Type:2307) and series of recorders in a moving laboratory. (*Source:* photo: S. Kurra, TRRL, 1997.)

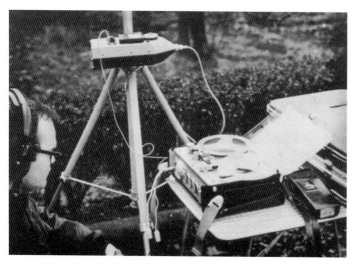

Figure 5.83 Magnetic Tape Recorder (Nagra) used in the field in the 1970's. (*Source:* photo: S. Kurra, TRRL, 1975.)

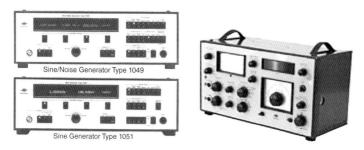

Figure 5.84 Signal generators (Brüel & Kjaer) used in the 1980s.

explained in Chapter 1. The signal generators are composed of a sound-emitting device which is an omni-directional loudspeaker and an amplifier Figure 5.84. Examples of sound sources are shown in Figure 5.85.

One of the compact pieces of equipment, from the early 1980s, was the "Building Acoustics Analyzer" from Brüel & Kjaer, which consisted of a signal generator, amplifier, and signal receiving system. Including sufficient memory capacity, it was used for reverberation time and sound insulation measurements with dual channel microphone connections for simultaneous measurements.

Back in the 1970s, the environmental noise measurements could be performed by using sets of equipment connected to each other in the field as shown in Figure 5.86 [12]. Owing to the above-mentioned developments in measurement and analyzing equipment and set-up, a vast number of investigations on environmental noise have been carried out so far, leaving a great deal of experience and valuable knowledge for later generations. It has been shown that even with simple instrumentation, significant findings could be derived to achieve particular goals.

Dosimeters: Assessment of the hearing risk of those people exposed to high noise levels or intense noise is measured by a portable small bit of equipment called a dosimeter (Section 5.6). Mounted on the shoulder with the microphones at the ear position, the noise-dose of a dynamic or stationary worker is determined through measurement of sound energy during an 8-hour

Figure 5.85 Sound sources used in emission measurements and building acoustics tests, since the 1970s to the present.

working day. The exposure time can be adjusted according to regulations regarding work hours. Dosimeters consist of a microphone, a frequency weighting meter, an rms processor, a logarithmic summation processor, a descriptor for time, and the threshold value. They give A-weighted sound levels and the calculated cumulative noise-dose based on continuous measurements. Dosimeters can be used to detect noise in a constant position in a workplace, so if a worker has to do his work in different work situations under different noise exposures, individual dosimeters are recommended. Different types of dosimeters are given in Figure 5.87.

5.11.3 Computer-Based Data Acquisition System and Network-Based Data Control

Long-term noise monitoring uses the computer-based measurement system and remote control. The collection of information from microphones, processing, and storing, to be able to use in various applications, is called data acquisition to be used in real-time monitoring, control, post-processing, and analyses. Advances in computer technologies have made it possible to use the computer-based instruments to realize this process through network data acquisition systems at present [240, 241]. Figure 5.88 gives the components of this system, comprising microphones for measuring sound, signal conditioning, measurement, and analysis using software (to be conditioned for the measurement device, by amplification, filtering, etc.), connection to computer by interface, then to internet connection for remote tests and control and ultimately sharing the data with public through a web interface. Connections can be provided with the noise mapping software, as described in Chapter 6.

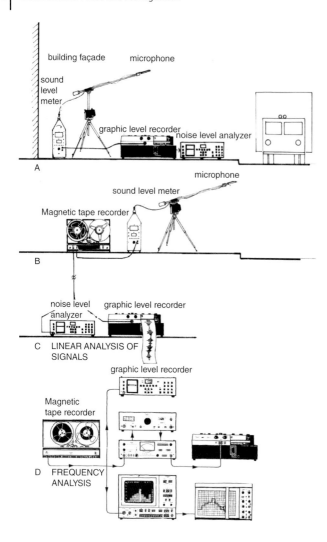

Figure 5.86 Equipment set-ups for environmental noise measurements in the 1970s [12].
(*Note:* B&K instrumentation is schematically displayed as in the B&K documents in the 1970s, e.g. [16].)

The main element is the monitoring terminal which is set in airports or near airports motorways/ railways, construction sites, industrial premises and entertainment activities with purposes of inspection, research and development studies, for noise mapping, for investigations on various factors, for checking the validity of the prediction models, in addition to a general assessment of community noise impact. Examples of such systems produced by different manufacturers are shown in Figure 5.89 [237–239]. Figure 5.90 gives the principles of how the terminal works.

- *Signal processing:* Defined as "The science of extracting relevant information from measured data signals representing a phenomenon of interest" [241], signal processing is used to investigate the industrial noise sources as well as environmental noises to describe the noise conditions based on the data recorded. Digital instruments and computation systems are required for analysis of digital data at different phases of the process, e.g. converting analog signals to digital, computations of rms values, one-third octave band levels, Fourier transformations, and various other functions which are used in noise problems, like the frequency-domain function, the spectral-density function, cross-correlation, etc.

Figure 5.87 Various dosimeters on the market.

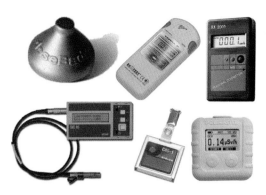

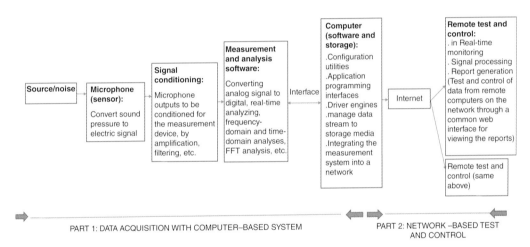

PART 1: DATA ACQUISITION WITH COMPUTER–BASED SYSTEM

PART 2: NETWORK –BASED TEST AND CONTROL

Figure 5.88 Basic components of the computer-based noise measurement system and remote control. (*Source:* adapted from [240, 241].)

Figures 5.91a–5.91c display some outputs from dedicated software analyzing the results from the environmental noise measurements. The last figure shows the dynamic analyses of two channel measurement on the real-time monitoring window. Figures 5.92a and 5.92b give the outcomes of signal processing of road traffic noise data recorded in a short time by using different techniques.

- *Source localization:* Identification of the sound-radiating parts of industrial noise sources and visualization of the sound field through acoustic source maps are accomplished by using the source localization techniques, also called diagnostic tools for noise control problems. They are used in determining the magnitude of noise emission of each part, in the selection of the dominant source, optimization of the noise emission of a source, evaluations, comparison with noise limits and in the application of noise control measures. Different algorithms and different metrics are employed in various source identification techniques, all using arrays of microphones in the measurements.

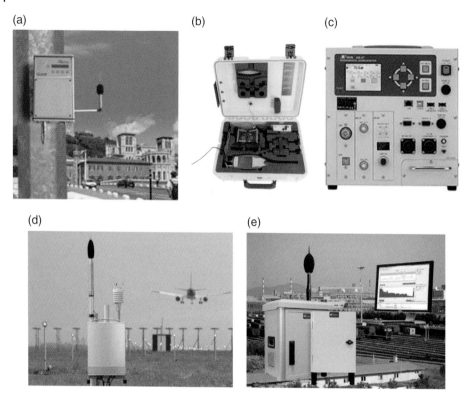

Figure 5.89 Various environmental noise monitoring terminals: (a) portable noise monitoring terminal (B&K) [237]; (b) and (c) Rion [239]; (d) permanent noise monitoring terminal (B&K) [237]; (e) Norsonic [238].

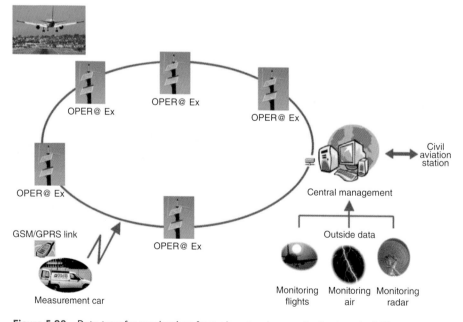

Figure 5.90 Data transfer mechanism from airport noise monitoring terminal (Opera system and 01 dB-Areva).

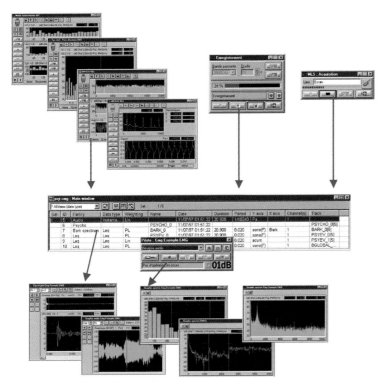

Figure 5.91a Signal analyses using computer software (Symphony system and dBFA-01dB-Areva).

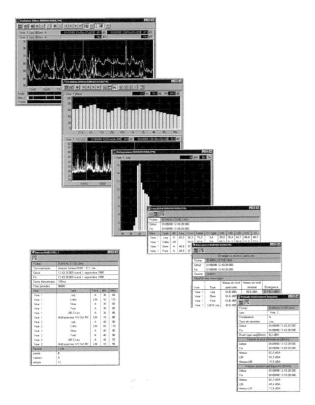

Figure 5.91b Statistical and spectral analyses of environmental noises using computer-based system (Symphonie and dBENV-01dB-Areva).

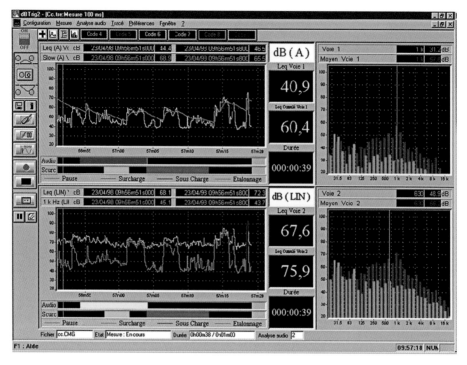

Figure 5.91c Dynamic analysis of noise signals using dual channel. Left: Time-history diagrams, middle: instantaneous noise levels, right: maximum and minimum spectrums (Symphonie system and dBTrait-01 dB-Areva).

- *Planar nearfield acoustic holography*: The holography technique, that was developed in the 1940s as optical holography, was adapted for acoustics in 1980, called near-field acoustic holography [33, 45–47]. Acoustic holography, as briefly referred to in Section 5.11, is applied to the smaller sources at smaller distances. The acoustic power and spectral content are determined through a single test on a two-dimensional hypothetical surface by using microphone arrays with different spacings according to the wavelength of sound. The measured data obtained on a plane parallel to the measurement plane, at near field of the source, is then used in the calculations based on spatial Fourier transformation, explained in Section 2.2.2. Acoustic holography enables investigations of a sound field with the parameters of sound pressure, intensity, and particle velocity as vectors on the spherical or cylindrical sound fields. By means of special software, the sound field is visualized in the near-field and the sound pressure distribution in the far field of the source under test can also be estimated. At present, the studies on noise source localization in industry and transportation noise sources (i.e. interior noise in aircraft and cars) employ acoustical holography for observation of sound transmission paths, ranking of secondary sources, etc. Some applications are shown in Figures 5.93a and 5.93b [45].
- *Beamforming technique:* As one of the tools for the source localization technique, beamforming (BF) is used to estimate the position and intensity of non-coherent sub-sources emerging from the large noise sources (e.g. construction machinery, cars, trains, airplanes, and large machinery, etc.) at far distances, in modal analysis for source locations and for acoustic source mapping. The

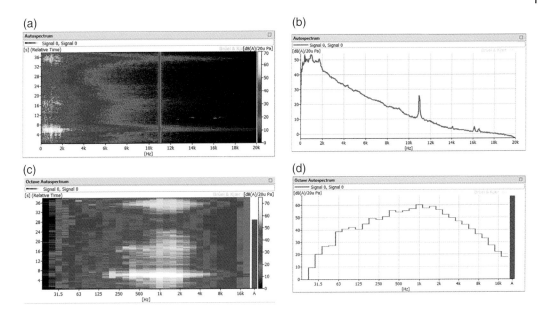

Figure 5.92a Signal analyses of road traffic noise using advanced software (*Source:* courtesy of A. Akgul, 2017.) (a) FFT analysis and autospectrum contours of road noise signal (sound pressure, Pa, frequency [Hz] and time; (b) frequency spectra (sound pressure level, dB vs frequency, Hz); (c) octave autospectrum (A-weighted sound pressure levels, octave band frequencies, Hz and time, s); (d)octave-band spectra (A-weighted sound pressure level, dB vs octave bands, Hz).

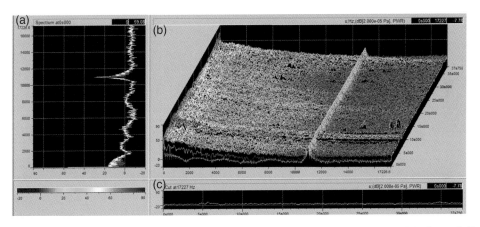

Figure 5.92b Signal analyses of road traffic noise using advanced software (Waterfall diagram) (*Source:* courtesy of S. Kurra, 2017.) (a) spectrum at a certain instant of time; (b) waterfall diagram (time, frequency [linear] and sound pressure level, dB); (c) spectral time-history.

technique, using 2D microphone arrays, which were spatially distributed on a plane of different geometries, was used in aeroacoustics applications in 1976 for the complex sources to identify sub-source positions [48]. The microphone arrays can be fixed, moving, or scanning. The calculation is performed by means of special software for summation of all the microphone signals by applying a delay

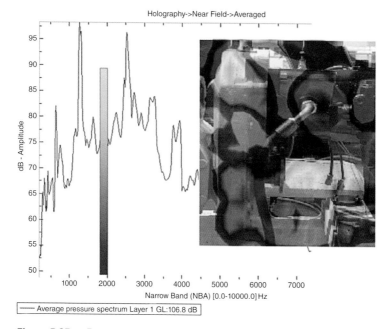

Figure 5.93a Example of acoustic hologram in the near-field of a car (01 dB-Metravib). (*Source:* with permission of D. Croix, Areva, 01dB-Metravib.)

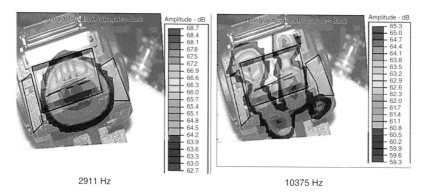

Figure 5.93b Examples of acoustic holograms for seat belt retractor [45]. (*Source:* with permission of D. Croix, Areva, 01dB-Metravib.)

in relation to virtual and real source positions. It uses the deconvolution approach, since the multiple sources make it difficult to separate the sources in the acoustical source maps, especially when the source is composed of non-coherent sub-sources. The sub-sources can be investigated in terms of position, frequency content, and sound power radiation, and ranking of sub-sources is possible. There are other techniques using deconvolution, however, beamforming is preferred due to its simplicity and speed. Figure 5.94 displays an example of output using the beamforming technique.

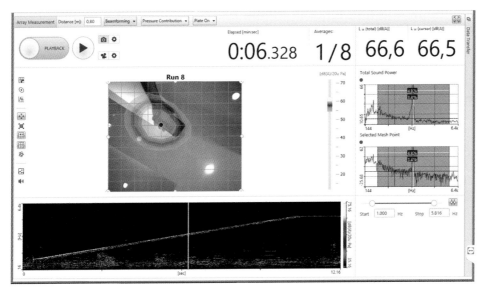

Figure 5.94 Output of software for an example beamforming measurement and analysis of sound field (acoustic source map) (Brüel & Kjaer). (*Source:* courtesy of A. Akgul.)

5.11.4 Purchasing Noise Measurement Equipment

Due to the great variability of the instrumentation, they should be selected in accordance with the objectives of the measurement, the standard method employed, and the budget. Manufacture of the electro-acoustic equipment has become a sector with a growing number of producers and their representatives competing in the market. Thus, it is necessary to consider the following factors in the selection of an efficient measuring system:

- Possible type of work conditions (outdoors and indoors)
- Acoustic parameters and units to be used
- Necessary analyses to be made
- Measurement durations
- Measurement techniques (i.e. sampling duration)
- Noise source and type of noise
- Noise limits and descriptors to be used
- Equipment type and technical properties, auxiliary devices
- Specific standards

The instrumentation should be compatible with the requirements in the technical standards describing the measurement methodology. For existing equipment, the memory capacities, dynamic range, and range of frequency should be checked especially for long-term measurements.

For new equipment, the information regarding dynamic range, loading conditions, and frequency range can be found in the technical documents provided by the manufacturer. Besides, the effect of the extension cables on the signal level, the calibration process of the equipment, the analyzing techniques, especially for the complex sounds, and connections to other devices, permit adjustments, if necessary, to obtain the maximum signal to noise ratio. The characteristics of the filter, the microphone capacity, the display properties, etc. should be determined.

The preferred sound level meter should possess at least a statistical analysis capability, a high capacity of storage memory, greater dynamic range (up to 120 dB), the lowest measurement level of 20 dB(A) and should be Class 1 or 2 equipment. Earphones should be connected to AC output. It should be capable of frequency analysis of at least the hearing range both in octave band and one-third octave bands and giving all the weighted levels. Average values and the statistical descriptors should be displayed instantly. Source identification should be made by means of exceedance level, noise events, and numbers and durations by histograms and cumulative diagrams.

Microphones and preamplifiers to be used in outdoor environments should meet the requirements for frequency and directivity responses. Protection from wind, humidity, rain and birds, should be provided by the essential accessories. For instance, a case or cabinet durable against environmental factors, spare battery, lemo 7-pin cable and other accessories are vital when making field measurements and it is important to be prepared for all unexpected situations. Elimination of background noise caused by other sources is possible by using earphones, and one can also listen to the source signals only.

Other considerations when purchasing new equipment, are:

availability of maintenance and repairing service
calibration charts and provision of yearly calibration certificates
possibility of satellite connections
seminars for training
possibility of upgrading with low cost
company references
the cost of the equipment

Before starting taking measurements, preliminary checks are essential regarding capacity and duration of energy source (battery, portable generator, etc.) and the possibility of taking power from other sources, the possibility of leaving the equipment safely unattended in the field, the sufficient length of the extension cables for measurements at distance, the possibility of raising microphones to higher positions. The equipment should be transported safely from one place to another.

5.12 Conclusion

Investigations on environmental noises by means of acoustic measurements are crucial to determine the magnitude of the noise problem to be declared in terms of noise impact assessment. The acoustic measurements, in simple or complicated situations, require a certain investment, time, trained personnel and experience, also an accredited laboratory according to EN/ISO/IEC/17025 [245]. The objective of the measurements, how to use and evaluate the data, should be clarified prior to the experiment. Generally, the type of noise sources, noise characteristics, and the physical environment where measurements are conducted, are the basic factors in implementations.

Due to the variability of noise sources and operations, measurements require different methodologies, as described in various technical standards. Developments in electro-acoustics and computer technology and in measurement equipment reveal a great revolution, starting from heavy and bulky equipment that could be transported via a mobile laboratory, to the simple hand-held equipment or a laptop conducting all measurements and complex analyses, occasionally by using interactive monitoring systems controlled out of the field.

In the selection of the most efficient system, reasonable decisions are also made by keeping up with the new technologies and new designs of the measuring instrumentation. The accuracy and

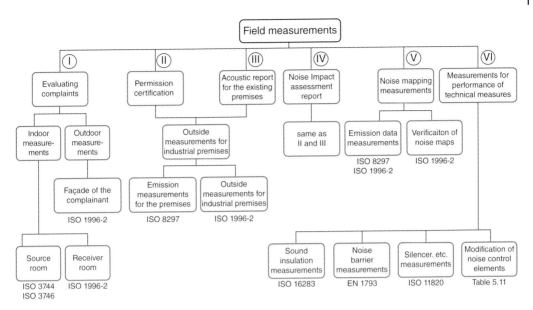

Figure 5.95 Noise measurements to be conducted in noise impact assessments.

precision of the results are other issues of concern and should be evidenced by repeatable measurements and by uncertainty calculations.

Environmental noise measurements have become mandatory through the national regulations and international commitments, such as European Directives. The types of measurements by applying different standards according to the objectives are summarized in Figure 5.95.

This chapter gives an overview of the rather sophisticated field of acoustic measurements which are directly or indirectly related to environmental noise assessments. Methods implemented for the field measurements of noise are emphasized, in addition, source-inherent measurements are acknowledged to acquire emission data for most common noise sources, which are important in noise management, as will be explained in Chapters 8 and 9.

References

1 Berry, B. (1998). Standards for a quieter world: some acoustical reflections from the UK National Physical Laboratory. *Noise News International* 6 (2): 74–83.

2 WHO (2001). Noise: Acoustic measurement, World Health Organization Regional Office for Europe, No. 37.

3 Brüel, P.V., Pope, J., and Zaver, H.K. (1998). Introduction. In: *Handbook of Acoustics* (ed. M.J. Crocker), 1311–1317. Wiley.

4 Ampel, F.J. and Uzzle, T. (1993). The history of audio and sound measurement, Audio Engineering Society, 94th Convention. Berlin. http://hps4000.com/pages/special/sound_history.pdf.

5 Crocker, M.J. (2007). General introduction to noise and vibration transducers, measuring equipment, measurements, signal acquisition and processing. In: *Handbook of Noise and Vibration Control* (ed. M.J. Crocker), 417–434. Wiley.

6 Harris, C.M. (ed.) (1979). *Handbook of Noise Control*. McGraw-Hill.

7 Fahy, F. (2001). *Foundations of Engineering Acoustics.* Academic Press.

8 Marks, E. and Florentine, M. (2011). Measurement of loudness, Chapter 2. In: *Loudness*, Springer Handbook of Auditory Research, vol. 37 (eds. R.R. Fay and A.N. Popper), 17. Springer.

9 Stevens, S.S. (1955). Measurement of loudness. *Journal of Acoustical Society of America* 27: 815–829.

10 Kopec, J.W. (1997). *The Sabines at Riverbank: Their Role in the Science of Architectural Acoustics.* Acoustical Society of America, Peninsula Publishing.

11 Berry, B. (2006). History bits: conference notes in Bahçeşehir University, Istanbul.

12 Kurra, S., Aksugür, N., and Arık, A. (1981). Determination of environmental noise in Istanbul and proposal for noise control criteria. Project Report, TÜBITAK MAG 524/A.

13 Turner, M. (2015). How 21st century technology can improve sound level measurement, Euronoise 2015, Maastricht, The Netherlands, paper no. 0000890.

14 Beranek, L.L. (1971). The measurement of power levels and directivity patterns of noise sources, Chapter 6. In: *Noise and Vibration Control* (ed. L.L. Beranek). McGraw-Hill.

15 (a) Broch, J.T. (1971). *Application of the Brüel & Kjaer Measuring Systems to Acoustic Noise Measurements.* Brüel & Kjaer reprinted June 1975.
(b) Ginn, K.B. (1978). *Architectural Acoustics.* Brüel & Kjaer.

16 Hassell, J.R. and Zaveri, K. (1979). *Acoustic Noise Measurements.* Brüel & Kjaer.

17 Beranek, L. (1986). *Acoustics.* American Institute of Physics for the Acoustical Society of America.

18 Kurra, S., Tamer, N., and Altay, A. (1998). The report on the establishment of the Noise Pollution Research Unit: Analysis and evaluation of noise sources, supported by TUBITAK, Project no: KTCAG No. 22, Istanbul Technical University, Turkey.

19 (a) American National Standards Institute, Inc. (1971). ANSI S1.13-1971, American National Standard methods for the measurement of sound pressure levels (partial revision of S1.2-1962);
(b) American National Standards Institute, Inc. (2005). ANSI S1.13-2005 (Reaffirmation with all amendments and changes in 2010), Measurement of sound pressure levels in air.

20 ISO (2000). ISO 3740:2000 (revised by ISO 3740:2019) Acoustics: Determination of sound power levels of noise sources. Guidelines for the use of basic standards.

21 Peterson, E.C. (1995). An overview of standards for sound power determination, application notes. https://www.bksv.com/doc/BO0416.pdf.

22 Jonasson, H.G. (2007). Determination of sound power level and emission sound pressure level. In: *Handbook of Noise and Vibration Control* (ed. M. Crocker), 526–533. Wiley.

23 ANSI (1972). ANSI S1. 21-1972, American National Standard methods for the determination of sound power levels of small sources in reverberation rooms (Revision of Section 3.5 of S1. 2-1962).

24 ASTM (1997). ASTM E 1124-97, Standard test method for field measurement of sound power level by the two surface method, republished in 2016.

25 (a) ISO (1999). ISO 3741:1999 (corr.: 2001), Acoustics: Determination of sound power levels of noise sources using sound pressure. Precision methods for reverberation rooms;
(b) ISO (2010). ISO 3741:2010 (confirmed in 2015), Acoustics: Determination of sound power levels and sound energy levels of noise sources using sound pressure. Precision methods for reverberation test rooms.

26 (a) ISO (1994). ISO 3743-1:1994 (revised in 2010 and confirmed in 2015) Acoustics: Determination of sound power levels of noise sources using sound pressure. Engineering methods for small, movable sources in reverberant fields. Part 1: Comparison method for hard walled test rooms;
(b) ISO (1994). ISO 3743-2:1994 (revised in 2018) Acoustics: Determination of sound power levels of noise sources using sound pressure. Engineering methods for small, movable sources in reverberant fields. Part 2: Methods for special reverberation test rooms.

27 (a) ISO (1994). ISO 3744:1994 Acoustics: Determination of sound power levels of noise sources using sound pressure. Engineering methods in an essentially free field over a reflecting plane;

(b) ISO (2010). ISO 3744:2010 (confirmed in 2015) Acoustics: Determination of sound power levels and sound energy levels using sound pressure. Engineering methods for an essentially free-field over a reflecting plane.

28 ISO (2012). ISO 3745:2012 (amd 1:2017) Acoustics: Determination of sound power levels and sound energy levels of noise sources using sound pressure. Precision methods for anechoic rooms and semi-anechoic rooms.

29 ISO (1995). ISO 3746:1995 (revised in 2010 and confirmed in 2015) Acoustics: Determination of sound power levels and sound energy levels of noise sources using sound pressure. Survey method using an enveloping measurement surface over a reflecting plane.

30 ISO (2010). ISO 3747:2010 (confirmed in 2015) Acoustics: Determination of sound power levels and sound energy levels of noise sources using sound pressure. Engineering/survey methods for use in situ in a reverberant environment.

31 ISO (2009). ISO 12001:2009 (confirmed in 2017) Acoustics: Noise emitted by machinery and equipment. Rules for the drafting and presentation of a noise test code.

32 ISO (1996). ISO 11689:1996 (confirmed in 2017) Acoustics: Procedure for the comparison of noise-emission data for machinery and equipment.

33 Xu, J., Fagotti, F., and Jiang, W. (2012). Nearfield acoustical holography research and application on compressor and refrigerator, International Compressor Engineering Conference at Purdue University, Mechanical Engineering (16–19 July)

34 Crocker, M.J., Raju, P.K., and Forssen, B. (1981). Measurement of transmission loss of panels by the direct determination of transmitted acoustic intensity. *Noise Control Engineering Journal, vol.* 17, 6–11.

35 Fahy, F.J. (1997). International standards for the determination of sound power levels of sources using sound intensity measurement: an exposition. *Applied Acoustics* 50: 97–109.

36 Fahy, F.J. (2002). *Sound Intensity*, 2e. CRC Press.

37 Jacobsen, F. (2007). Sound intensity measurements. In: *Handbook of Noise and Vibration Control* (ed. M.J. Crocker), 534–548. Wiley.

38 (a) ISO (1993). ISO 9614-1:1993 (confirmed in 2018) Acoustics: Determination of sound power levels of noise sources using intensity. Part 1: measurement at discrete points;
(b) ISO (1996). ISO 9614-2:1996 (confirmed in 2018) Acoustics: Determination of sound power levels of noise sources using intensity. Part 2: measurement by scanning;
(c) ISO (2002). ISO 9614-3:2002 (confirmed in 2018) Acoustics: Determination of sound power levels of noise sources using intensity: Part 3: precision method for measurement by scanning.

39 OROS Group (2018). Sound intensity, Dec. 14. http://www.oros.com/3906-sound-intensity.htms.

40 Comesana, D.F., Steltenpool, S., Korbasiewics, M.K. et al. (2015). Direct acoustic vector field mapping: new scanning tools for measuring 3D sound intensity in 3D space, Euronoise 2015, Maastricht, The Netherlands.

41 EU (2000). Directive 2000/14/EC of the European Parliament and of the Council of 8 May 2000 on the approximation of the laws of the Member States relating to the noise emission in the environment by equipment for use outdoors.

42 EU (2005). Directive 2005/88/EC of the European Parliament and of the Council of 14 December 2005 amending Directive 2000/14/EC on the approximation of the laws of the Member States relating to the noise emission in the environment by equipment for use outdoors.

43 EU (2006). Directive 2006/42/EC of the European Parliament and of the Council of 17 May 2006 on machinery, and amending Directive 95/16/EC.

44 Crocker, M.J. and Kurra, S. (1981). Measurement of noise levels and directivity of air pumps, consulting report for Gas Manufacturing Corporation, Purdue University, R.W. Herrick Laboratories, Department of Mechanical Engineering.

45 de la Croix, D.V. and Pischedda, P. (2005). How can nearfield acoustical holography help in efficiently improving the acoustic quality of small size products? Report for AREVA, 01dB-Metravib.

46 Ginn, K.B. and Haddad, K. (2012). Noise source identification techniques: simple to advanced applications, *Proceedings of Acoustics*, Nantes, France.

47 Williams, E.G. (2007). Use of near-field acoustical holography in noise and vibration measurements. In: *Handbook of Noise and Vibration Control* (ed. M.J. Crocker), 598–611. Wiley.

48 Padois, T. and Berry, A. (2017). Two- and three-dimensional sound source localization with beamforming and several deconvolution techniques. *Acta Acustica United with Acustica* 103: 392–400.

49 (a) ISO (2009). ISO/TS 7849-1:2009 (confirmed in 2017) Acoustics: Determination of airborne sound power levels emitted by machinery using vibration measurement: Part 1: survey method using a fixed radiation factor;
(b) ISO (2009). ISO/TS 7849-2:2009 (confirmed in 2017) Acoustics: Determination of airborne sound power levels emitted by machinery using vibration measurement. Part 2: Engineering method including determination of the adequate radiation factor.

50 (a) ISO (1995). ISO 11204:1995 (revised and corrected in 1997) (Technical Memorandum) Acoustics: Noise emitted by machinery and equipment: measurement of emission sound pressure levels at a work station and at other specified positions. Method requiring environmental corrections;
(b) ISO (2010). ISO 11204:2010 (confirmed in 2015) Acoustics: Noise emitted by machinery and equipment: determination of emission sound pressure levels at a work station and at other specified positions applying accurate environmental corrections.

51 ISO (1996). ISO 4871:1996 (confirmed in 2018) Acoustics: Declaration and verification of noise emission values for machinery and equipment.

52 Jenkins, M.P., Salvidge, A.C., and Utley, W.A. (1976). Noise levels at boundaries of factories and commercial premises. BRE current paper no. 43/76.

53 ISO (1994). ISO 8297:1994 (confirmed in 2018) Acoustics: Determination of sound power levels of multisource industrial plants for evaluation of sound pressure levels in the environment. Engineering method.

54 (a) Augusztinovicz, F. and Buna, B. (1980). An investigation of the close proximity vehicle noise survey method. *Noise Control Engineering* 14 (2): 79–87.
(b) Morrison, D. (1987). Road vehicle noise emission legislation, Chapter 9. In: *Transportation Noise Reference Book* (ed. P. Nelson). Butterworths.
(c) Comesana, D.F. (2014). Scan-based sound visualisation methods using sound pressure and particle velocity, Thesis for: Doctor of Philosophy.

55 Sandberg, U. (2001). Noise emissions of road vehicles: effect of regulations. Final report 01-1 by the I-INCE Working Party on noise emissions of road vehicles.

56 (a) ISO (1964). ISO R 362:1964 (withdrawn) Measurement of noise emitted by vehicles;
(b) ISO (1998). ISO 362:1998 (withdrawn) Acoustics: Measurement of noise emitted by accelerating road vehicles. Engineering method;
(c) ISO (2014). ISO 10844:2014 Acoustics: Specification of test tracks for measuring noise emitted by road vehicles and their tyres.

57 (a) ISO (2015). ISO 362-1:2015 Measurement of noise emitted by accelerating road vehicles. Engineering method. Part 1: M and N categories;
(b) ISO (2009). ISO 362-2:2009 (confirmed in 2020) Measurement of noise emitted by accelerating road vehicles. Engineering method. Part 2: L category;
(c) ISO (2016). ISO 362-3:2016 Measurement of noise emitted by accelerating road vehicles. Engineering method. Part 3: indoor testing M and N categories.

58 EEC (1970). Directive 70/157/EEC, Council Directive of 6 February 1970 on the approximation of the laws of the Member States relating to the permissible sound level and the exhaust system of motor vehicles.

59 EU (2014). Regulation EU/2014 No. 540 of the European Parliament and of the Council of 16 April 2014 on the sound level of motor vehicles and of replacement silencing systems, and amending Directive 2007/46/EC and repealing Directive 70/157/EEC.

60 ISO (1994). ISO 7188:1994 (withdrawn and revised by ISO 362-1:2007 and ISO 362-2: 2008) Acoustics: Measurement of noise emitted by passenger cars under conditions representative of urban driving.

61 ISO (2016). ISO 16254:2016 Acoustics: Measurement of sound emitted by road vehicles of category M and N at standstill and low speed operation. Engineering method.

62 (a) Anon (1976). Transportation noise emission controls: medium and heavy trucks noise. *Federal Register* 41 (72).
(b) EPA, Office on Noise Abatement and Control (1976). Medium and heavy trucks noise emission standards.

63 FHWA (1981). Sound procedures for measuring highway noise: final report.

64 (a) ISO (1982). ISO 5130:1982 Acoustics: Measurement of noise emitted by stationary road vehicles: survey method (withdrawn);
(b) ISO (2007). ISO 5130:2007 +A1:2012 (confirmed in 2019) Acoustics: Measurements of sound pressure level emitted by stationary road vehicles.

65 Ibarra, D., Ramirez-Mendoza, R., and Lopez, E. (2016). A new approach for estimating noise emission of automotive vehicles. *Acta Acustica United with Acustica* 102: 930–937.

66 ISO (2003). ISO 13325:2003 (confirmed in 2019) Coast-by methods for measurement of tyre-to-road sound emission.

67 Sandberg, U. and Ejsmont, J.A. (2007). Tire/road noise: generation, measurement, and abatement. In: *Handbook of Noise and Vibration Control* (ed. M.J. Crocker), 1054–1071. Wiley.

68 EC (1970). Council Directive 70/388/EC of 27 July 1970 on the approximation of the laws of the Member States relating to audible warning devices for motor vehicles.

69 EC (1993). Council Directive 93/30/EEC of 14 June 1993 on audible warning devices for two- or three-wheel motor vehicles.

70 ISO (1979). ISO 512:1979 (withdrawn) Road vehicles: sound signaling devices on motor vehicles acoustic standards and technical specifications.

71 ISO (2004). ISO 6969:2004 (confirmed 2019) Road vehicles: sound signaling devices. Tests after mounting on vehicle.

72 TNI (1999). Railway applications: Acoustics: Measurement of noise emitted by railway vehicles, draft standard.

73 (a) EU (2014). Commission Regulation (EU) No. 1304/of 26 November 2014 on the technical specification for interoperability relating to the subsystem 'rolling stock noise' amending Decision 2008/232/EC and repealing Decision 2011/229/EU;
(b) UIC (2002). Noise creation limits for railways: background information from UIC Subcommission on Noise and Vibration, 01.10.02

74 Lutzenberger, S., Gutmann, C., and Miller, B.B.M. (2013). Noise emission of European railway cars and their noise reduction potential: data collection, evaluation and examples of Best-Practice railway cars. Environmental research of the Federal Ministry of the Environment, Nature Conservation and Nuclear Safety, 12.

75 ISO (2005). ISO 3095:2005 (revised by ISO 3095:2013) Acoustics: Railway applications. Measurement of noise emitted by railbound vehicles.

76 EN (2009). EN 15610:2019 Railway applications: noise emission: rail roughness measurement related to rolling noise generation.

77 EN (2013). EN 15153-2:2013 Railway applications: external visible and audible warning devices for high speed trains. Part 2-Warning horns.

78 ISO (1970). ISO R 597:1970 (withdrawn) Procedure for describing aircraft noise around an airport.

79 ISO (1978). ISO 3891:1978 (withdrawn in 1999) Acoustics: Procedure for describing aircraft noise heard on the ground.

80 ICAO (1993). Aircraft noise certification: Annex 16, vol. 1, 3e.

81 Böttcher, J. (2004). Noise Certification Workshop Session 2: Aircraft noise certification. Annex 16, vol. 1 and equivalent procedures, Montreal, Canada.

82 BS (1976). BS 5331:1976 (withdrawn).

83 (a) ISO (1975). ISO 2922:1975 (withdrawn) (confirmed in 2011) Acoustics: Measurement of noise emitted by vessels on inland water and harbours;
(b) ISO (2000). ISO 2922:2000 +A1:2013 Acoustics: Measurement of airborne sound emitted by vessels on inland waterways and harbours.

84 Badino, A., Borelli, D., Gaggero, T. et al. (2012). Control of airborne noise emissions from ships, International Conference on Advances and Challenges in Marine Noise and Vibration 21, MARNAV 2012, Glasgow, Scotland (5–7 September).

85 Di Bella, A. (2014). Evaluation methods of external airborne noise emissions of moored cruise ships: an overview, ICSV 21, Beijing, China (13–17 July).

86 Rizzuto, E. and Soares, C.G. (eds.) (2011). *Sustainable Maritime Transportation and Exploitation of Sea Resources*. CRC Press.

87 ISO (2008). ISO 14509-1:2008 (confirmed in 2014) Small craft. Airborne sound emitted by powered recreational craft. Part 1: Pass-by measurement procedures and Part 3 (2009 and confirmed in 2014). Sound assessment using calculation and measurement procedures.

88 ISO (1996). ISO 2923:1996 (confirmed in 2017) Acoustics: Measurement of noise on board vessels.

89 (a) ISO (2016). ISO 20283-5:2016 Mechanical vibration: measurement of vibration on ships. Part 5: Guidelines for measurement, evaluation and reporting of vibration with regard to habitability on passenger and merchant ships.
(b) EN (2017). EN 15657:2017 Acoustic properties of building elements and of buildings. Laboratory measurement of structure-borne sound from building service equipment for all installation condition.

90 (a) ISO (1999). ISO 3822-1:1999 (confirmed in 2016) Acoustics: Laboratory tests on noise emission from appliances and equipment used in water supply installations. Part 1: Method of measurement, (ISO 3822-1:1999/Amd 1:2008 Measurement uncertainty);
(b) ISO (1995). ISO 3822-2:1995 (confirmed in 2016) Part 2: Mounting and operating conditions for draw-off taps and mixing valves;
(c) ISO (1997). ISO 3822-3:1997 (revised by ISO 3822-3:2018) Part 3. Mounting and operating conditions for in-line valves and appliances;
(d) ISO (1997). ISO 3822-4:1997 (confirmed in 2013) Part 4: Mounting and operating conditions for special appliances.

91 Bies, D.A. and Hansen, C.H. (eds.) (2002). *Engineering Noise Control, Theory and Practice*, 2e. Spon Press.

92 Beranek, B. (1998). *Acoustical Measurements*, 2e. Acoustical Society of America.

93 Crocker, M.J. (ed.) (2007). *Handbook of Noise and Vibration Control*. Wiley.

94 ISO (1979). ISO 2204:1979 (withdrawn 11.01.79) Guide to international standards on the measurement of airborne acoustical noise and evaluation of its effects on human beings.

95 (a) ISO (1982). ISO 1996-1:1982 (revised and published as ISO 1996-1:2016) Acoustics: Description, measurement and assessment of environmental noise. Part 1: Basic quantities and assessment procedures;
(b) ISO (1987). ISO 1996-2:1987/Amd 1998 (withdrawn) Acoustics: Description and measurement of environmental noise. Part 2: Acquisition of data pertinent to land use;

(c) ISO (1987). ISO 1996-3:1987 (withdrawn) Acoustics: Description and measurement of environmental noise. Part 3: Application to noise limits;

(d) ISO (2017). ISO 1996-2:2017 Acoustics: Description, measurement and assessment of environmental noise. Part 2: Determination of environmental noise levels.

96 ANSI (1994). ANSI S 12.18-1994 (R2009) Outdoor measurement of sound pressure level.

97 ANSI (1986). ANSI S12.7-1986 (reaffirmed in 2015 with amendments and changes) Methods for measurements of impulse noise.

98 ISO (1997). ISO 10843:1997 (confirmed in 2018) Acoustics: Methods for the description and physical measurement of single impulses or series of impulses.

99 (a) ISO (1997). EN ISO 11819-1:1997 (confirmed in 2013) Acoustics: Method for measuring the influence of road surfaces on traffic noise. Part 1: Statistical pass-by method;

(b) ISO (2017). ISO 11819-2:2017 Acoustics: Method for measuring the influence of road surfaces on traffic noise. Part 2. The close-proximity method.

100 (a) ISO (2002). ISO 13472-1:2002 (confirmed in 2012) Acoustics. Measurement of sound absorption properties of road surfaces in situ. Part 1: Extended surface method;

(b) ISO (2010). ISO 13472-2:2010 (confirmed in 2013) Acoustics: Measurement of sound absorption properties of road surfaces in situ. Part 2: Spot method for reflective surfaces.

101 Kurra, S. (2012). Derivation of reference spectrums for transportation noise sources to be used in rating sound insulation, Euronoise 2012, Prague, the Czech Republic (10–13 June).

102 Kurra, S. (2013). Source-specific sound insulation descriptors for transportation noise and proposal for insulation classes, paper no. 1250, Inter-Noise 2013, Innsbruck, Austria (15–18 September).

103 Yanliang, L., Lanhua, L., Xiaoan, G. et al. (2015). The experimental studies on high-speed railway noise field vertical distribution and propagation characteristic, Euronoise 2015, Maastricht, The Netherlands.

104 ISO (2005). ISO 3381:2005 (confirmed in 2013) Railway applications. Acoustics: Measurement of noise inside railbound vehicles.

105 EN (2011). EN 15892:2011 Railway applications. Noise emission. Measurement of noise inside drivers' cabs.

106 ISO (2009). ISO 20906:2009 +A1:2013 (confirmed in 2020) Acoustics: Unattended monitoring of aircraft sound in the vicinity of airports.

107 ISO (2001). ISO 5129:2001 +A1:2013 (confirmed in 2017) Acoustics: Measurement of sound pressure levels in the interior of aircraft during flight.

108 Witte, R. (2015). Regulation of noise from moored ships in ports, Euronoise 2015, Maastricht, The Netherlands.

109 Witte, R. (2010). Noise from moored ships, Inter-Noise 2010, Lisbon, Portugal.

110 ISO (1996). ISO 2923:1996 (confirmed in 2011) Acoustics: Measurement of noise on board vessels.

111 NPL (2014). Good practice guide for underwater noise measurement, NPL No. 133.

112 (a) BS (2014). BS 4142:2014+A1:2019 Methods for rating and assessing industrial and commercial sound;

(b) ISO (2001). ISO 14257:2001 (confirmed in 2017) Acoustics: Measurement and parametric description of spatial sound distribution curves in workrooms for evaluation of their acoustical performance.

113 Van der Maarl, W. and de Beer, E. (2015). Construction and urban noise: automatic assessment of noise monitoring results, Euronoise 2015, Maastricht, The Netherlands.

114 BS (2009). BS 5228-1:2009 +A1:2014 Code of practice for noise and vibration control on construction and open sites.

115 Anon (2013). A good practice guide to the application of ETSU-R-97 for the assessment and rating of wind turbine noise, Institute of Acoustics, May.

116 Hansen, K., Zajamsek, B., and Hansen, C. (2014). Comparison of the noise levels measured in the vicinity of a wind farm for shutdown and operational conditions, Inter-Noise 2014, Melbourne, Australia (16–19 November).

117 Fauville, B. and Moiny, F. (2015). Detection of wind turbine noise in immission measurements, Euronoise 2015, Maastricht, The Netherlands.

118 Pedersen, E., van den Berg, F., Bakker, R. et al. (2009). Response to noise from modern wind farms in the Netherlands. *Journal of the Acoustic Society of America* 126: 634–643.

119 Møller, H. and Pedersen, C.S. (2011). Low-frequency noise from large wind turbines. *Journal of the Acoustic Society of America* 129: 3727–3744.

120 Evans, T. and Cooper, J. (2012). Comparison of predicted and measured wind farm noise levels and implications for assessments of new wind farms. *Acoustics Australia* 40 (1): 28–36.

121 Koppen, E. and Fowler, K. (2015). International legislation for wind turbine noise, Euronoise 2015, Maastricht, The Netherlands.

122 Noise Measurement Services Pty Ltd (2011). Wind Farm Noise Guideline, March.

123 IEC (2012). IEC 61400-11:2012 Wind turbines: Part 11: Acoustic noise measurement techniques.

124 (a) ANSI (2010). ANSI AS 4959-2010 Acoustics: Measurement, prediction and assessment of noise from wind turbine generators;
(b) NZS (2010). NZS 6808:2010 Acoustics: Wind farm noise.

125 Vos, J. (1995). A review of research on the annoyance caused by impulse sounds produced by small firearms, *Proceedings of Inter-Noise 1995*.

126 Rasmussen, P., Flamme, G., Stewart, M. et al. (2009). Measuring recreational firearm noise. *Sound & Vibration* August: 14–18.

127 Brueck, S., Kardous, C.A., Oza, A. et al. (2014). Measurement of exposure to impulsive noise at indoor and outdoor firing ranges during tactical training exercises, Health Hazard Evaluation Program, Report No. 2013-0124-3208, US Dept of Health and Human Services, Centers for Disease Control and Prevention, National Institute for Occupational Safety and Health.

128 (a) ISO (2005). ISO 17201-1:2005 (revised by ISO 17201-1:2018) Acoustics: Noise from shooting ranges. Part 1: Determination of muzzle blast by measurement;
(b) ISO (2006). ISO 17201-2:2006 (confirmed in 2013) Acoustics: Noise from shooting ranges. Part 2: Estimation of muzzle blast and projectile sound by calculation.

129 Pedersen, T.H. (2002). Nordest Method Proposal: Impulsive method for the measurement of prominence of impulsive sounds and for adjustment of LA_{eq}. Final report.

130 Witsel, A.C. and Moiny, F. (2015). Smart sound meter for shooting noise monitoring, Inter-Noise 2015.

131 Clark, W. (1991). Noise exposure from leisure activities: a review. *Journal of Sound and Vibration* 90: 175–181.

132 Kok, M. (2015). Sound level measurements at dance festivals in Belgium, Inter-Noise 2015.

133 I-INCE (2012). Outdoor Recreational Noise, vol. 1: A Review of Noise in National Parks and Motor Sport Activities, I-INCE publication: 12-1, Final Report of the I-INCE Technical Study Group on Outdoors Recreational Noise (TSG 1).

134 Institute of Acoustics (2003). Good Practice Guide on the Control of Noise from Pubs and Clubs.

135 ISO (2004). ISO 16032:2004 (confirmed in 2015) Acoustics: Measurement of sound pressure level from service equipment in buildings. Engineering method.

136 ISO (2004). ISO 10052:2004 (confirmed in 2008) Acoustics: Field measurements of airborne and impact sound insulation and of service equipment sound. Survey method.

137 ISO (1994). ISO 5725-1 (confirmed in 2018) Accuracy (trueness and precision) of measurement methods and results. Part 1: General principles and definitions.

138 Craven, N.J. and Kerry, G. (2007). A good practice guide on the sources and magnitude of uncertainty arising in the practical measurement of environmental noise, DTI Project: 2.2.1: National Measurement System Programme for Acoustic Metrology, University of Salford, UK.

139 (a) ISO/IEC (2003). Guide to the expression of uncertainty in measurement (GUM). Part 1: Basic and general terms (VIM), Part 2. Vocabulary of legal metrology (VLM), Part 3: Evaluation of measurement data: Guide to the expression of uncertainty in measurement (GUM);
(b) ISO/IEC (2008). ISO/IEC Guide 98-3: (JCGM/WG1/100) (GUM: 1995). Part 3: Guide to the expression of uncertainty in measurement (with Suppl. 2: 2011).

140 Taylor, B.N. and Kuyatt, C.E. (1994). Guidelines for evaluating and expressing the uncertainty of NIST measurement results, NIST Technical Note 1297 (TN 1297).

141 Fidell, S. and Schomer, P. (2007). Uncertainties in measuring aircraft noise and predicting community response to it. *Noise Control Engineering Journal* 55 (1): 82–88.

142 Cox, M. (2003). Guide to the Expression of Uncertainty in Measurement (GUM) and its supplemental guides, National Physical Laboratory. http://www.accreditation.jp/council/image/1_5.pdf.

143 Cox, M. and Harris, P. (2003). Up a GUM tree? Try the Full Monte!, www.npl.co.uk/ scientific_software/tutorials/uncertainties/up_a_gum_tree.pdf.

144 Grabe, M. (2001). Estimation of measurement uncertainty: an alternative to the ISO Guide. *Metrologia* 38: 97–106.

145 (a) ISO (1985). ISO 7574-1:1985 (confirmed in 2012) Acoustics: Statistical methods for determining and verifying stated noise emission values of machinery and equipment. Part 1: General considerations and definitions;
(b) ISO (1985). ISO 7574-2:1985 (confirmed in 2012) Acoustics: Statistical methods for determining and verifying stated noise emission values of machinery and equipment. Part 2: Methods for stated values for individual machines;
(c) ISO (1985). ISO 7574-3:1985 (confirmed in 2012) Acoustics: Statistical methods for determining and verifying stated noise emission values of machinery and equipment. Part 3: Simple (transition) method for stated values for batches of machines;
(d) ISO (1985). ISO 7574-4:1985 (confirmed in 2012) Acoustics: Statistical methods for determining and verifying stated noise emission values of machinery and equipment. Part 4: Methods for stated values for batches of machines.

146 (a) ISO (2010). ISO 8253-1:2010 (confirmed in 2015) Acoustics: Audiometric test methods. Part 1: Basic pure tone air and bone conduction audiometry;
(b) ISO (2009). ISO 8253-2:2009 (confirmed in 2020) Acoustics: Audiometric test methods. Part 2: Sound field audiometry with pure tone and narrow-band test signals.

147 (a) ISO (1990). ISO 4869-1:1990 (revised as ISO 4869-1:2018) Acoustics: Hearing protectors. Part 1: Subjective method for the measurement of sound attenuation;
(b) ISO (1994). ISO 4869-2:1994 (confirmed in 2018) Acoustics: Hearing protectors. Part 2: Estimation of effective A-weighted sound pressure levels when hearing protectors are worn;
(c) ISO (2007). ISO 4869-3:2007 (confirmed in 2017) Acoustics: Hearing protectors. Part 3: Simplified method for the measurement of insertion loss of ear-muff type protectors for quality inspection purposes.

148 (a) ISO (1997). ISO 9612:1997 Acoustics: Guidelines for the measurement and assessment of exposure to noise in a working environment;
(b) ISO (2009). ISO 9612:2009 (confirmed in 2014) Acoustics: Determination of occupational noise exposure. Engineering method.

149 ISO (2013). ISO 1999:2013 (confirmed in 2018) Acoustics: Estimation of noise-induced hearing loss.

150 EU (2003). Directive 2003/10/EC of the European Parliament and of the Council of 6 Feb. 2003, on the minimum health and safety requirements regarding the exposure of workers to the risks arising from physical agents (noise).

151 (a) ISO (1996). ISO 11690-1:1996 (confirmed in 2011) Acoustics: Recommended practice for the design of low-noise workplaces containing machinery. Part 1: Noise control strategies. Part 2: Noise control measures;
(b) ISO (1997). ISO/TR 11690-3:1997 (confirmed in 2009) Acoustics: Recommended practice for the design of low-noise workplaces containing machinery. Part 3: Sound propagation and noise prediction in workrooms.

152 Maekawa, Z. (1965). Noise reduction by screens: Memoirs of the Faculty of Engineering, no. 11.

153 Maekawa, Z. (1965). Noise reduction by screens: Memoirs of the Faculty of Engineering, no. 12.

154 Ringheim, M. (1976). An experimental investigation of the attenuation produced by noise screens, Report no. LBA 461, Akustik Laboratorium, Trondheim, NTH.

155 Delany, M.E., Rennie, A.J., and Collins, K.M. (1972). Scale model investigations of traffic noise propagation, NPL Acoustics Report, AC 58.

156 Rapin, J.M. (1968). Mise au point et premier application d'une méthode d'étude sur modele réduit de la propagation des bruits de traffic routier, Cahier du CSTB No. 93, Cahier 810.

157 Rapin, J.M., Roland, J., and de Tricaud, P. (1972). Réalisation de maquettes destinées à simuler l'environment sonore du milieu urbain, CSTB Final report.

158 Ivey, E. and Russel, G.A. (1977). Acoustical scale model study of the attenuation of sound by wide barriers. *Journal of the Acoustical Society of America* 62 (3): 601–606.

159 Fujiwara, K., Ando, Y., and Maekawa, Z. (1973). Attenuation of a spherical sound wave diffracted by a thick plate. *Acustica* 28: 341–347.

160 Koyasu, M. and Yamashita, M. (1973). Scale model experiments on noise reduction by acoustic barrier of a straight line source. *Applied Acoustics* 6: 233–242.

161 Mohsen, E. (1975). Use of the technique of acoustic scale modelling for traffic noise research, BS 20, Department of Building Science, University of Sheffield.

162 Lyon, R.H. (1976). Environmental noise and acoustic modelling. *Technological Review* 78 (5).

163 Delany, M.E., Rennie, A.J., and Collins, K.M. (1978). A scale model technique for investigating traffic noise propagation. *Journal of Sound and Vibration* 56 (3): 325–340.

164 Kurra, S. (1980). A computer model for predicting sound attenuation by barrier buildings. *Journal of Applied Acoustics* 3: 331–355.

165 Kurra, S. and Crocker, M.J. (1981). A scale model experiment on sound propagation at building sites. Research report, Department of Mechanical Engineering, Purdue University, USA.

166 Jones, H.W., Stredulinsky, D.C., and Vermeulen, P.J. (1980). An experimental and theoretical study of the modelling of road traffic noise and its transmission in the urban environment. *Applied Acoustics* 13: 251–265.

167 Jacobs, L.J.M., Nijs, L., and Willigenburg, J.J. (1980). A computer model to predict traffic noise in urban situations under free-flow and traffic light conditions. *Journal of Sound and Vibration* 72 (A): 623–537.

168 Kurra, S., Morimoto, M., and Maekawa, Z. (1999). Transportation noise annoyance: a simulated environment study for road, railway and aircraft noises. Part 1:overall annoyance. *Journal of Sound and Vibration* 220 (2): 251–278.

169 Maekawa, Z., Rindel, J.H., and Lord, P. (2011). *Environmental and Architectural Acoustics*, 2e. Spon Press.

170 Hopkins, C. (2007). *Sound Insulation*. Butterworth-Heinemann.

171 Rindel, J.H. (2018). *Sound Insulation in Buildings*. CRC Press.

172 Warnock, A.C.C. and Fasold, W. (1998). Sound insulation: airborne and impact. In: *Handbook of Acoustics* (ed. M. Crocker), 953–984. Wiley.

173 ASTM (2009). ASTM E90-09, superseded and published as ASTM E90- 09 (2016) Standard test method for laboratory measurement of airborne sound transmission loss of building partitions and elements.

174 (a) ISO (2010). ISO 10140-1:2010 + Amd 2:2014 (revised by ISO 10140-1:2016) Acoustics: Laboratory measurement of sound insulation of building elements. Part 1: Application rules for specific products;
(b) ISO (2010). ISO 10140-2:2010 (confirmed in 2016) Acoustics: Laboratory measurement of sound insulation of building elements. Part 2: Measurement of airborne sound insulation;
(c) ISO (2010). (ISO 10140-3:2010 + Amd 1:2015 (confirmed in 2016) Acoustics: Laboratory measurement of sound insulation of building elements. Part 3: Measurement of impact sound insulation;
(d) ISO (2010). ISO 10140-4:2010 (confirmed in 2017) Acoustics: Laboratory measurement of sound insulation of building elements. Part 4: Measurement procedures and requirements;
(e) ISO (2010). ISO 10140-5:2010 +AMD 1:2014 (confirmed in 2016) Acoustics: Laboratory measurement of sound insulation of building elements. Part 5: Requirements for test facilities and equipment.

175 ISO (2019). ISO/DIS 19488 (deleted) and ISO/AWI TS 19488 (under development) Acoustics: Acoustic classification of dwellings.

176 Alion Science's Riverbank Acoustical Laboratories. https://www.alionscience.com/riverbank-acoustical-laboratories.

177 Kurra, S. and Arditi, D. (2001). Determination of sound transmission loss of multilayered elements. Part 2: an experimental study. *International Journal of Acoustics: Acta Acustica* 87 (5): 592–604.

178 (a) ISO (2006). ISO 10848-2:2006 (revised as ISO 10848-2:2017) Acoustics: Laboratory measurement of the flanking transmission of airborne and impact sound between adjoining rooms. Part 2: Application to light elements when the junction has a small influence;
(b) ISO (2006). ISO 10848-3:2006 (revised as ISO 10848-2:2017) Acoustics: Laboratory measurement of the flanking transmission of airborne and impact sound between adjoining rooms. Part 3: Application to light elements when the junction has a substantial influence.

179 ISO (1996). ISO 11957:1996 (confirmed in 2017) Acoustics: Determination of sound insulation performance of cabins. Laboratory and in situ measurements.

180 (a) ISO (2014). ISO 16283-1:2014 Acoustics: Field measurement of sound insulation in buildings and of building elements. Part 1: Airborne sound insulation;
(b) ISO (2015). ISO 16283-2:2015 (revised as ISO 16283-2: 2018) Acoustics: Field measurement of sound insulation in buildings and of building elements. Part 2: Impact sound insulation;
(c) ISO (2016). ISO 16283-3:2016 Acoustics: Field measurement of sound insulation in buildings and of building elements. Part 3: Façade sound insulation.

181 Scrosati, C., Scamoni, F., Asdrubali, F. et al. (2015). Uncertainty of façade sound insulation measurements obtained by a round robin test: the influence of the low frequencies extension, ICSV 22, Florence, Italy (July).

182 Berardi, U., Cirillo, E., and Martellotta, F. (2010). Measuring sound insulation of building façades: interference effects and reproducibility, Inter-Noise 2010, Lisbon, Portugal (June).

183 (a) ISO (2000). ISO 15186-1:2000 (confirmed in 2016) Acoustics: Measurement of sound insulation in buildings and of building elements using sound intensity, Part 1: Laboratory measurements;
(b) ISO (2003). ISO 15186-2:2003 (confirmed in 2013) Acoustics: Measurement of sound insulation in buildings and of building elements using sound intensity. Part 2: Field measurements;
(c) ISO (2002). ISO 15186-3:2002 (confirmed in 2013) Acoustics: Measurement of sound insulation in buildings and of building elements using sound intensity. Part 3: Laboratory measurements at low frequencies.

184 Tachibana, H. (1987). Application of sound intensity technique to architectural acoustics, *Proceedings of the 2nd Symposium on Acoustic Intensity* (in Japanese).

185 Hong-wei, W., Long, Z., and Ya-jie, Y. (2017). Application of three-dimensional sound intensity method in sound insulation measurement of building components, Inter-Noise 2017, Hong Kong.

186 (a) ISO (2013). ISO 717-1:2013 (confirmed in 2018) Acoustics: Rating of sound insulation in buildings and of building elements. Part 1: Airborne sound insulation;
(b) ISO (2013). ISO 717-2:2013 (under review) Acoustics: Rating of sound insulation in buildings and of building elements. Part 2: Impact sound insulation;
(c) ISO (1982). ISO 717-3:1982 (withdrawn) Acoustics: Rating of sound insulation in buildings and of building elements. Part 3: Airborne sound insulation of facade elements and facades (revised by ISO 717-1:1996).

187 Rasmussen, B. and Machimbarrena, M. (eds) (2014). Building acoustics throughout Europe, vol. 1: Towards a common framework in building acoustics throughout Europe, COST Action TU0901: Integrating and Harmonizing Sound Insulation Aspects in Sustainable Urban Housing Constructions, COST Office.

188 ASTM (1999). ASTM E413-87 (revised as ASTM E 413-16) Classification for rating sound insulation.

189 Kurra, S. and Tamer, N. (1993). Rating criteria for facade insulation against transportation noise sources. *Applied Acoustics* 40 (3): 213–237.

190 (a) Bastian The Building Acoustics Planning System, Datakustik. http://www.datakustik.com/en/products/bastian;
(b) Insul. Innovation software. http://www.marshallday.com/innovation/software.

191 (a) ISO (2009). ISO 3382-1:2009 (confirmed in 2015) Acoustics: Measurement of room acoustic parameters. Part 1: Performance spaces;
(b) ISO (2008). ISO 3382-2:2008 (confirmed in 2016). Acoustics: Measurement of room acoustic parameters. Part 2: Reverberation time in ordinary rooms;
(c) ISO (2012). ISO 3382-3:2012 (confirmed in 2017) Acoustics: Measurement of room acoustic parameters. Part 3: Open plan offices.

192 ISO (2003). ISO 354:2003 (confirmed in 2015) Acoustics. Measurement of sound absorption in a reverberation room.

193 ISO (1997). ISO 11654:1997 (confirmed in 2018) Acoustics: Sound absorbers for use in buildings. Rating of sound absorption.

194 ISO (1998). ISO 19534-2:1998 (confirmed in 2015) Acoustics: Determination of sound absorption coefficient and impedance tubes. Part 2: Transfer function method.

195 Garai, M. (1993). Measurement of the sound-absorption coefficient in situ: The reflection method using periodic pseudo-random sequences of maximum length. *Applied Acoustics* 39 (1–2): 119–139.

196 Cuenca, J. and De Ryck, L. (2015). In-situ sound absorption of ground surfaces: Innovative processing and characterization methods, Euronoise 2015, Maastricht, The Netherlands (1–3 June).

197 ISO (1998). ISO 14163:1998 (confirmed 2017) Acoustics: Guidelines for noise control by silencers.

198 ISO (2003). ISO 7235: 2003 (confirmed in 2018) Acoustics: Measurement procedures for ducted silencers and air terminal units: insertion loss, flow noise and total pressure loss.

199 ISO (1995). ISO 11691:1995 (confirmed in 2010) Acoustics: Measurement of insertion loss of ducted silencers without flow: laboratory survey method.

200 ISO (1996). ISO 11820:1996 (confirmed in 2017) Acoustics:Testing of silencers in situ.

201 (a) ISO (1995). ISO/TR 11688-1: 1995 (confirmed in 2009) Acoustics: Recommended practice for the design of low-noise machinery and equipment. Part 1: Planning;
(b) ISO (1998). ISO/TR 11688-2:1998 (confirmed in 2009) Acoustics: Recommended practice for the design of low-noise machinery and equipment. Part 2: Introduction to the physics of low-noise design.

202 ISO (2003). ISO 15665:2003 (confirmed in 2013) Acoustics: Acoustic insulation for pipes, valves and flanges.

203 ISO (2000). ISO 15667: 2000 (confirmed in 2015) Acoustics: Guidelines for noise control by enclosures and cabins.

204 ISO (2001). ISO 15664:2001 (confirmed in 2017) Acoustics: Noise control design procedures for open plant.

205 Yoon, S.C., Yoo, Y.J., Woo, J.H. et al. (2017). Management of low noise pavements in Korea, Inter-Noise 2017, Hong Kong (27–30 August).

206 (a) ISO (1997). ISO 11819-1: 1997 (confirmed in 2013) Acoustics: Measurement of the influence of road surfaces on traffic noise. Part 1: Statistical pass-by method;
(b) ISO (2017). ISO 11819-2:2017 Acoustics: Measurement of the influence of road surfaces on traffic noise. Part 2: The close-proximity method;
(c) ISO (2017). ISO/TS 11819-3:2017 Acoustics: Measurement of the influence of road surfaces on traffic noise. Part 3: Reference tyres;
(d) ISO/PAS (2013). ISO/PAS 11819-4:2013 (confirmed in 2017) Acoustics: Method for measuring the influence of road surfaces on traffic noise. Part 4: SPB method using backing board.

207 Watts, G.R., Surgand, M., and Morgan, P.A. (2002). Assessment of noise barrier diffraction Using an in-situ measurement technique, *Proceeding of the Institute of Acoustics*, Spring Conference, 24.2.

208 Watts, G.R. and Godfrey, N.S. (1999). Effects on roadside noise levels of sound absorptive materials in noise barriers. *Applied Acoustics* 58 (4): 385–402.

209 Watts, G.R., Crombie, D.H., and Hothersall, D.C. (1994). Acoustic performance of new designs of traffic noise barriers: full scale tests. *Journal of Sound and Vibration* 177 (3): 289–305.

210 Watts, G.R. (1996). Acoustic performance of a multiple edge noise barrier profile at motorway sites. *Applied Acoustics* 47 (1): 47–66.

211 Guidorzi, P. and Garai, M. (2013). Advancements in sound reflection and airborne sound insulation measurement on noise barriers. *Open Journal of Acoustics* 33 (2A): 25–38.

212 Guidorzi, P. and Garai, M. (2013). Impulse responses measured with MLS or Swept-sine signals: a comparison between the two methods applied to noise barriers measurements, Proceedings of the Audio Engineering Society Convention, Rome, Italy, paper no. 8914.

213 Duhamel, D., Sergent, P., Hua, C. et al. (1998). Measurement of active control efficiency around noise barriers. *Applied Acoustics* 55 (3): 217–241.

214 EN (2014). EN 16272-3-2:2014 Railway applications. Track: Noise barriers and related devices acting on airborne sound propagation. Test method for determining the acoustic performance. Normalized railway noise spectrum and single number ratings for direct field applications.

215 EN (2014). EN 16272-6:2014 Railway applications. Track: Noise barriers and related devices acting on airborne sound propagation. Test method for determining the acoustic performance. Intrinsic characteristics. In situ values of airborne sound insulation under direct sound field conditions.

216 (a) Clairbois, J.P. (2015). Noise barriers and standards for mitigating noise, CEDR Conference on Road Traffic Noise, Hamburg, Germany (8–9 September);
(b) Houtave, P., Clairbois, J.P., Vanhooreweder, B. et al. (2016). Acoustic performances of roadside barriers: Wide scale on-site measurements in Flanders, Inter-Noise 2016, Hamburg, Germany.
(c) Garai, M., Guidorzi, P. (2015). "Sound reflection measurements on noise barriers in critical conditions". *Building and Environment* 94 (2): 752–763.
(d) Garai, M., Schoen, E., Behler, G., Bragado, B., Chudalla, M., Conter, M., Defrance, J., Demizieux, P., Glorieux, C., Guidorzi, P. (2014). "Repeatability and reproducibility of in situ measurements of sound reflection and airborne sound insulation index of noise barriers". *Acta Acustica united with Acustica* 100, 1186–1201.

217 Buytaert, A., Vanhooreweder, B., Clairbois, J.P. et al. (2017). In-situ measurements according to EN 1793-5 and EN 1793-6: first results and impressions, Inter-Noise 2017, Hong Kong (27–30 August).

218 ISO (1997). ISO 10847:1997 (confirmed in 2018) Acoustics: In-situ determination of insertion loss of outdoor noise barriers of all types.

219 (a) EN (2017). EN 1793-1:2017 (first published in 1997) Road traffic noise reducing devices: test method for determining the acoustic performance. Part 1: Intrinsic characteristics of sound absorption under diffuse sound field conditions;
(b) EN (2018). EN 1793-2:2018 Road traffic noise reducing devices: test method for determining the acoustic performance: Part 2: Intrinsic characteristics of airborne sound insulation under diffuse sound field conditions;
(c) EN (1998). EN 1793-3:1998 Road traffic noise reducing devices: test method for determining the acoustic performance. Part 3: Normalized traffic noise spectrum;
(d) EN (2015). EN 1793-4:2015 Road traffic noise reducing devices: test method for determining the acoustic performance. Part 4: Intrinsic characteristics: in situ values of sound diffraction;
(e) EN (2016). EN 1793-5:2016/AC:2018 Road traffic noise reducing devices. Test method for determining the acoustic performance. Part 5: Intrinsic characteristics: in situ values of sound reflection under direct sound field conditions;
(f) EN (2018). EN 1793-6:2018 Road traffic noise reducing devices:test method for determining the acoustic performance. Part 6: Intrinsic characteristics: in situ values of airborne sound insulation under direct sound field conditions.

220 (a) EN (2018). EN 1794-1:2018 Road traffic noise reducing devices: non-acoustic performance. Part 1-Mechanical performance and stability requirements;
(b) EN (2011). EN 1794-2:2011 Road traffic noise reducing devices: non-acoustic performance: Part 2-General safety and environmental requirements.

221 ISO (1997). ISO 11821:1997 (confirmed in 2018) Acoustics: Measurement of the in situ sound attenuation of a removable screen.

222 ISO (2004). ISO 17624:2004 (confirmed in 2017) Acoustics:Guidelines for noise control in offices and workrooms by means of acoustical screens.

223 Crocker, M.J. (ed.) (1998). Acoustical measurement and instrumentation, Part XV. In: *Handbook of Acoustics*. Wiley.

224 Kuehn, J. (2010). Noise measurement. In: *Instrumentation Reference Book*, 4e (ed. W. Boyes), 593–614. Elsevier.

225 Dyer, S.A. (2004). *Wiley Survey of Instrumentation and Measurement*, e-book. Wiley.

226 Vorlander, M. (2013). Acoustic measurements. In: *Handbook of Engineering Acoustics* (eds. G. Müller and M. Möser), 23–53. Springer.

227 Nakra, B.C. and Chaudhry, K.K. (2009). Acoustic measurements. In: *Instrumentation:Measurement and Analysis*, 3e (eds. B.C. Nakra and K.K. Chaudhry), 333–349. McGraw-Hill Education Limited.

228 Bies, D.A. and Hansen, C.H. (2003). Instrumentation for noise measurement and analysis. In: *Engineering Noise Control: Theory and Practice*, 3e (eds. D.A. Bies and C.H. Hansen), 92–122. CRC Press.

229 Hixson, E.L. and Busch-Vishniac, I.J. (2007). Transducer principles. In: *Handbook of Acoustics* (ed. M.J. Crocker), 1383–1393. Wiley.

230 Hixson, E.L. and Busch-Vishniac, I.J. (2007). Types of microphones. In: *Handbook of Acoustics* (ed. M.J. Crocker), 1417–1429. Wiley.

231 NPL (2012). A new generation of acoustic measurements. www.npl.co.uk/news/a-new-generation-of-acoustic-measurements.

232 (a) IEC (2000). IEC 61094-1:2000 Measurement microphones: Part 1: Specifications for laboratory standard microphones;
(b) IEC (2009). IEC 61094-2:2009 Electroacoustics: Measurement microphones. Part 2: Primary method for pressure calibration of laboratory standard microphones by the reciprocity technique;
(c) IEC (2016). IEC 61094-3:2016 Electroacoustics: Measurement microphones. Part 3: Primary method for free-field calibration of laboratory standard microphones by the reciprocity technique;
(d) IEC (1995). IEC 61094-4:1995 Electroacoustics: Measurement microphones. Part 4: Specifications for working standard microphones;
(e) IEC (2016). IEC 61094-5:2016 Electroacoustics: Measurement microphones. Part 5: Methods for pressure calibration of working standard microphones by comparison.

233 IEC (1979).IEC 651:1979 Sound level meters (replaced by IEC 61672-1:2002 and IEC 61672-1:2003).

234 (a) IEC (2013). IEC 61672-1:2013 Electroacoustics: Sound level meters. Part 1: Specifications;
(b)IEC (2013). IEC 61672-2:2013 Electroacoustics: Sound level meters: Part 2: Pattern evaluation tests;
(c) IEC (2013). IEC 61672-3:2013 Electroacoustics: Sound level meters. Part 3: Periodic test;
(d) IEC (2017). IEC 60942:2017 Electroacoustics: Sound calibrators.

235 Malchaire, J. (2001). Sound measuring instruments. In: *Occupational Exposure to Noise* (eds. B. Goelzer, C.H. Hansen and G.A. Sehrndt), 125–140. WHO http://www.who.int/occupational_health/publications/noise6.pdf.

236 National Academy of Engineering (2010). Appendix E, Modern instrumentation for environmental noise measurement. In: *Technology for a Quieter America*. The National Academies Press https://doi.org/10.17226/12928.

237 https://www.bksv.com/en/products/environment-management/noise-and-vibration-monitoring-terminals/portable-noise-monitoring-terminal.

238 http://www.norsonic.com/en/applications/environmental_noise_monitoring.

239 https://www.environmental-expert.com/products/rion-model-na-37b-class-1-aircraft-noise-monitor-411936.

240 Li, Z. and Crocker, M.J. (2007). Equipment for data acquisition. In: *Handbook of Noise and Vibration Control* (ed. M.J. Crocker), 486–492. Wiley.

241 Piersol, A.G. (2007). Signal processing. In: *Handbook of Noise and Vibration Control* (ed. M.J. Crocker), 493–500. Wiley.

242 Randall, R.B. (1987). *Frequency Analysis*, revised edition. Copenhagen: Brüel & Kjaer.

243 Randall, R.B. (2007). Noise and vibration data analysis. In: *Handbook of Noise and Vibration Control* (ed. M.J. Crocker), 549–564. Wiley.

244 (a) IEC (2014). IEC 61260-1:2014 Electroacoustics: Octave-band and fractional-octave-band filters. Part 1: Specifications;
(b) IEC (2016). IEC 61260-2:2016 Electroacoustics: Octave-band and fractional-octave-band filters. Part 2: Pattern-evaluation tests;
(c) IEC (2016). IEC 61260-3:2016 Electroacoustics: Octave-band and fractional-octave-band filters. Part 2: Periodic tests.

245 BS (2005). EN/ISO/IEC/17025 (revised by ISO/IEC 17025:2017) General requirements for the competence of testing and calibration laboratories.

6

Noise Mapping

Scope

After being recognized as a type of environmental pollution, environmental noise has become an issue of concern globally and field studies have increased to reveal the size of the areas and the populations exposed to excessive noise from various sources. Noise conditions prevailing in urban settlements, or even in rural areas, are determined by means of field measurements and prediction models. However, the results are presented to the public and the authorities by means of visual tools, as it is important to turn their attention to the noise problems and to increase their awareness. One of these visual techniques, called noise maps, demonstrate the distribution of sound levels virtually on a map and rate the noise impact according to a simple criterion. A noise map has a number of benefits in developing noise management strategies and in urban planning (see Chapter 9). Noise mapping has been practiced for two decades in Europe and is accepted as the initial stage of the technical and political decision-making process to control environmental noise. In this chapter, the characteristics of noise maps, the principles of their preparation and evaluation are included with emphasis on accuracy issues. The techniques for noise mapping implemented in the past and their history up to the present are explained with examples.

6.1 Descriptions and Objectives of Noise Mapping

Defined as the "establishment of a digital and visual model of the physical environment and noise sources," noise maps are used in various fields and serve various purposes [1]. Basically, noise maps are graphical representations of noise level distribution in the environment, over a layout plan or regional plan incorporated into a geographic map. Since the radiation of sound from a source is dealt with as a space-related phenomenon, it is quite reasonable to illustrate noise levels according to certain categories in two- or three-dimensional spaces.

Generally, the objectives of noise maps are as follows [1–4]:

- To determine the acoustic conditions in terms of the average noise levels in different sections of the environment.
- To inform the public about the magnitude of noise and the risks to public health.
- To analyze the effects of the source and the environment-related factors on noise levels.

Environmental Noise and Management: Overview from Past to Present, First Edition. Selma Kurra.
© 2021 John Wiley & Sons Ltd. Published 2021 by John Wiley & Sons Ltd.

- To check whether the noise levels in a specific urban area exceed the noise limits given in the regulations.
- To assess the boundaries of areas where noise levels exceed the noise limits (noise zoning).
- To display the percentage of the population exposed to a certain range of noise levels.
- To determine the negative effects of major and specific sources in the environment (such as road traffic, railways, airports or industrial areas, wind turbines, entertainment centers, etc.).
- To evaluate the mixed source situations by overlapping the source-specific maps and to compare the individual sources and impacts.
- To determine the effects of new transportation systems during the planning and modification stages.
- To determine the façade noise levels to decide on efficient sound insulation.
- To prepare the environmental impact assessment (EIA) reports on noise pollution caused by a specific noise source.
- To evaluate the performance of a noise-mitigating measure by comparing the "before and after" maps.
- To confirm the results of the prediction models by comparing the calculated results with those measured.
- To provide a basis for urban planning, land-use decisions, and layout design of buildings.
- To determine the priority areas (hot spots) for action plans.
- To enable comparisons between noise pollution and other types of pollution, such as air pollution, to be used for strategic decisions on the pollutant areas.
- To develop action plans according to the noise control strategies and policies.
- To develop future scenarios, comparing different measures that might guide the optimization studies, for example, to compare the possible outcomes of the effects of different planning alternatives.
- To compare the noise impacts in different regions.
- To use the map data in a cost–benefit analysis for action planning.

To accomplish any of the objectives above, precision in assessing noise exposure is important, especially when planning new roads, railways, industrial facilities, and airports near urban settlements, since the decisions are closely related to the noise limits and economic aspects. Ultimately, noise maps have become a common tool for the assessment of environmental noise and a new business sector all over the world.

The data on a noise map are presented mainly by the following:

1) Noise levels at single or more receiver points, of which the coordinates are specified.
2) Points with equal noise levels depicted in contour lines, called noise exposure contours.
3) Distribution of noise impact rating scores.

6.2 Strategic Noise Maps

Directive 2002/49/EC (END), first issued by the European Commission in 2002, defines a strategic noise map as: "a map designed for the global assessment of noise exposure in a given area due to different noise sources or for overall predictions for such an area" [5a]. The European member states adapted this regulation in their national regulations, as will be discussed in Chapter 8. Two types of noise maps are recommended in the Directive:

- *Environmental noise maps* or simply "noise maps" to determine the noise conditions due to a specific noise source either in the existing or future situations.
- *Strategic noise maps* to display the overall noise exposure of more than one major source existing in the environment (mixed noise condition).

Strategic noise maps rating the impact of overall human noise both in urban settlements and in rural areas have been included in END with the declared objectives as follows:

1) To provide noise data in the areas (sites) around main highways, railways, airports, and industrial areas.
2) To determine the strategic noise levels to be able to develop a European Noise Policy.
3) To provide data regarding noise levels for use by community, local, national and international decision-makers.
4) To provide data for dose/response relationships.
5) To develop an action plan
6) To identify hot spot definitions and distributions to be based on proposed Noise Scores [6].

The major aim of strategic maps is to support the national noise policies of the European countries, assisting in the preparation and revision of urban planning, for comparison of different types of noise sources and for comparison with other types of pollution [7].

A strategic map is generally the static representation of a noise field and facilitates the investigation of the total effect of noise corresponding to the actual case. Depending on the graphical scale selected, some maps cover a large area and present rough approximations, whereas others provide more details for small-scale local urban areas. The algorithms of noise mapping have been given in the standards and design guidelines and recent computer technology has provided different presentation techniques to ease the interpretations.

Noise maps display the average noise levels in an environment – such as yearly averages – therefore it is inevitable one will find some deviations between the actual (i.e. measured) and map noise levels at the same receiver point. Also, strategic noise maps have certain approximations within the limits of acceptable errors of estimation. The accuracy of noise maps and their reliability should be checked during the process of development and should be declared in the final report, as will be explained in Section 6.6.

6.3 Noise Mapping Techniques

According to the afore-mentioned objectives, different presentation techniques have been used in noise mapping since their inception. At present, the general principles of the preparation of noise maps are established in the European Document, WG-AEN [8].

6.3.1 Development of Noise Mapping

First attempts at noise mapping began in the 1970s for traffic noise by using the measured data and drawing the noise contours manually. After the emergence of the simple prediction models based on the traffic parameters, the preliminary noise contours were displayed in some design guides published in the 1980s (see Chapter 4). The results were set out in the design guides not only as the distribution of noise levels in neighborhood areas, but also the noise reductions by the physical elements, such as barriers. An example appeared in the design guide published by the Centre Scientifique et Technique du Bâtiment (CSTB) in France in 1980 (Figure 6.1) [9]. A simple noise map around an airport (1980) is given in Figure 6.2 depicting only the NEF 40 contour which was accepted as the noise control criterion for new housing development [10].

Calculations for numbers of grid points could be made by the computer programs mainly using the codes written in Fortran IV language with the aid of earlier computers operated by DOS system (in the pre-Windows era).

In the late 1970s, the concerns in urban planning and land-use decisions were on the rise in many countries, the computerized studies were expanded gradually. As an example, the study conducted in 1978 to assess traffic noise levels behind buildings can be referred [11, 12]. The results of the study enabled the calculation of noise contours and were later confirmed by a scale model experiment in

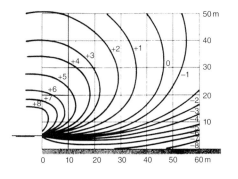

ROAD PLATFORM			Type of ground
Width	Number of lane	Height from ground	Natural reflective surface
12 m	1×2	(+7,5)–(+2,5)	

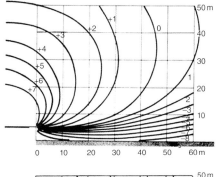

ROAD PLATFORM			Type of ground
Width	Number of lane	Height from ground	Natural reflective surface
27 m	2×2	(+7,5)–(+2,5)	

ROAD PLATFORM			Type of ground
Width	Number of lane	Height from ground	Natural reflective surface
34 m	2×4	(+7,5)–(+2,5)	

Figure 6.1 Contours displaying effect of road height and change in noise levels, CSTB [9].

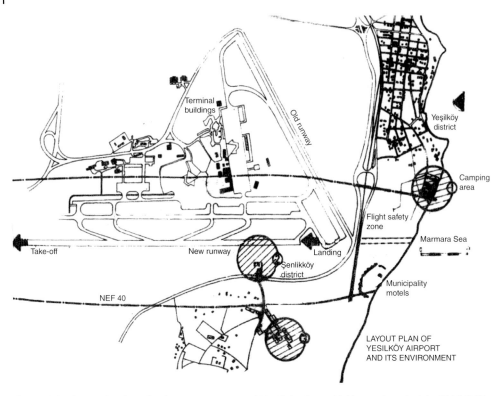

Figure 6.2 Determination of noise contours around Yeşilköy (Ataturk) Airport, Istanbul, in 1980 [10].

the laboratory in 1982 (Figure 6.3) [13]. In another study in 1980, a computer program, written in Fortran IV, was prepared to determine noise contours displaying the required insulation performances of buildings located in mixed-source environments (Figures 6.4a and 6.4b) [14].

One of the mapping techniques that was used in Japan, in 1982, is shown in Figure 6.5. The dots with varying sizes, according to the categories of noise levels obtained from the field measurements, were displayed in a grid system [15]. From 1985 until the 1990s, the noise maps representing the noise conditions along roads also were able to categorize roads with respect to noise levels. Two examples of such maps are given in Figures 6.6 and 6.7 for European cities.

One of the first prototypes of recent noise maps was prepared for the city of Düsseldorf in 1978, shown in Figure 6.8 [16]. A German design guide, published in 1983, displayed the noise contours around various building layouts to aid architectural planning and design with respect to traffic noise (Figures 6.9a and 6.9b) [17]. Another example, which is given in Figure 6.10, was prepared by employing an earlier computer program called TSØY developed in Norway in 1995 [18, 19].

Noise mapping for industrial premises have shown great advances over the years, particularly in modeling industrial noise sources. Drawing simple contour lines around one or two factory buildings as seen in Figure 6.11 and later grading noise zones by using colors as given in Figures 6.12 and 6.13, were then transformed into the precision noise maps of large industrial areas by modeling all indoor and outdoor equipment (Figure 6.14) [20–22] (Section 6.3.5).

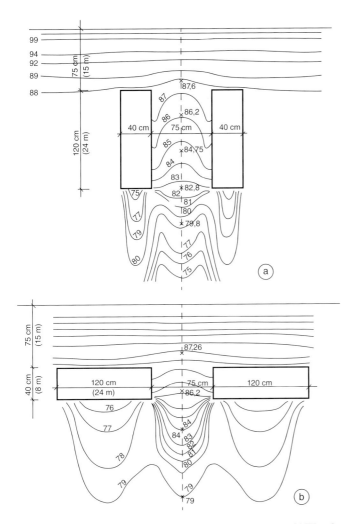

Figure 6.3 Noise contours obtained by calculations with SATT software [13].

Noise mapping in the vicinity of airports, which is a rather complicated task relative to the other noise sources, began in the 1970s, then spread by means of computer programs, such as Nortim in Norway 1995 and INM by the Federal Aviation Administration (FAA) in the USA, was first issued in 1978, then revised in 2015. Recently aircraft noise models have been greatly improved, as aircraft noise maps are now mandatory in many countries [23, 24].

Following the introduction of speed-up processors and the increased capacities of computers, various presentation techniques have been developed. Categorization of noise levels and the bands between the noise classes are emphasized by using colors, which had been standardized in 1987 with ISO 1996-3 (see Chapter 5) [25]. Relating the noise level to the impact of noise, blue/black represents the highest levels, red is the indicator of intense annoyance, green represents serenity and quietness, corresponding to the lowest levels of noise.

The recent software also provides the building evaluations, i.e. vertical noise contours at site, noise patches on the façades of buildings representing the floor levels or the maximum noise levels on the roofs.

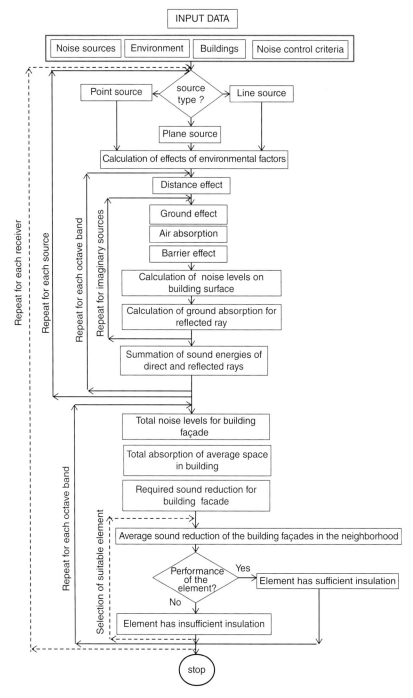

Figure 6.4a Flow chart for calculation of noise contours and the required sound insulation by using GURULT, software in 1980 [14].

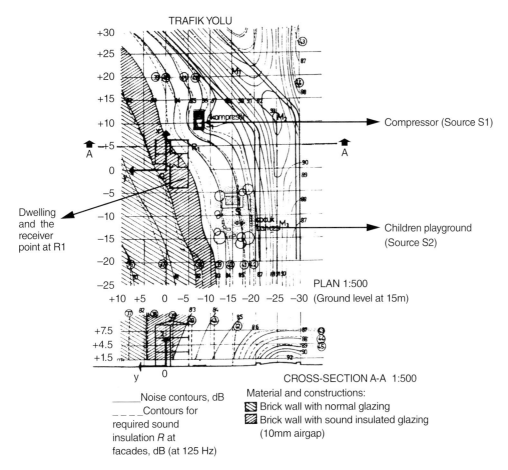

Figure 6.4b Example of manually prepared noise contours in a hypothetical built-up area computed by GURULT software, 1980 [14].

Algorithms of the software improved immensely in the late 1990s and the early 2000s, so that the precise description of the land topography and source data has been accomplished. While only flat ground could be modeled 15 years ago, recently geometry of land can be described in three dimensions by means of the visualization and animation techniques. Geographic Information Systems (GIS) have facilitated the definition of land, roads, and buildings by utilizing information from satellites. Currently it is mandatory to use GIS complying with Directive 49/EC, aiming to produce comparable noise maps. In order to facilitate comparisons, collaboration among the various countries has been established regarding collecting, organization, and evaluation of data in addition to the presentation of noise levels. The uncertainty assessments of noise maps have become important to increase reliability and the accuracy of noise maps, as will be explained below.

When large-scale noise maps (e.g. covering thousands of kilometers of main roads outside agglomerations) were required to aid the regional and urban planning decisions in action plans, more advanced techniques were introduced in the 2000s for noise mapping. It is possible to export the three-dimensional noise model into Google Earth and to integrate the study site with the complete urban area. When the model size is too large to include in one project file, it is possible to share the computation task among connected computers.

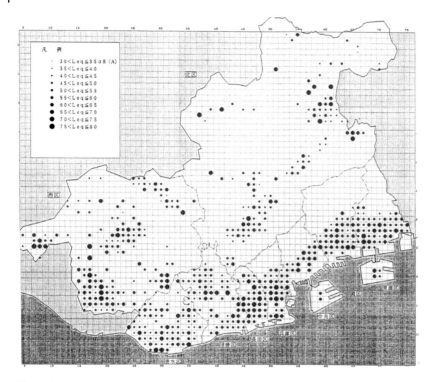

Figure 6.5 Example of an early noise map prepared in Japan (1991) [15].

6.3.2 Preliminary Determinations for Noise Mapping

At the initial stage of a noise mapping study, some decisions and clarifications regarding the following subjects are necessary:

1) Size of the area (km^2) concerned and its perimeters.
2) Methods applicable in noise mapping (measurement and prediction or combined).
3) Positions of the reference receiver points where the reference noise levels are measured and introduced as input to the model.
4) Graphical scale of noise map which depends on the objective and the size of the land area, e.g. 1:1000 or 1:5000 for overall layout plans or 1:20000, 1:50000 or 1:100000 for regional noise maps, 1:5000 or 1:1000 for detailed analysis in partial urban sectors, and 1:500 for a specific noise-sensitive building, such as a hospital or school, can be selected. The German Standard (DIN 45687) recommends:

"The usual scale is 1:5 000 down to 1:10 000 for the sound immission maps. It may be useful to choose a scale of 1:1000, 1:500 or larger for the neighborhoods where sound immissions are distributed over a small area or, when the efficiency of noise control measures at the source or along the propagation path is to be investigated to allow the preparation of development plans or for noise maps in the vicinity of noise-sensitive buildings or noise sources [26]".

5) Heights of grid points on the horizontal noise maps are selected according to the standards (e.g. 4 m suggested by END). For vertical noise maps, the cross-sections are placed considering the topography, height of buildings, noise-sensitive buildings or a particular piece of land. Vertical

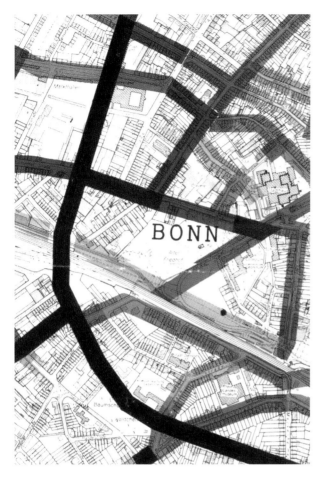

Figure 6.6 Earlier display of city traffic noise: Road noise map in Bonn, Germany, 1985.

noise maps are useful to check the precision of the land model before starting the calculations. Positions (locations and heights) of the single receiver points at which the noise levels will be calculated and displayed on the map.

6) Some of the other issues to consider prior to noise mapping, are as follows:

1) Project budget and available resources.
2) Performance of the software to be employed.
3) Capacity of the computers.
4) Adequacy of the available measurement systems.
5) Quality of available data regarding the source and the environment.
6) Total period of the project (time allowed and the schedule).
7) Technical personnel and their experience in mapping practice and general knowledge of environmental noise.
8) Accuracy of the meteorological data.
9) Details and quality of the geographic maps and layout plans to be provided from the local authorities.

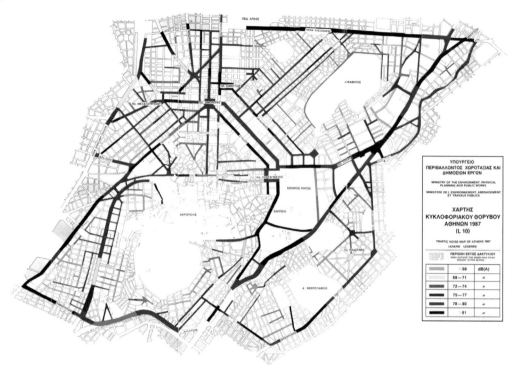

Figure 6.7 Earlier display of city traffic noise: Road noise map in Athens, Greece, 1990.

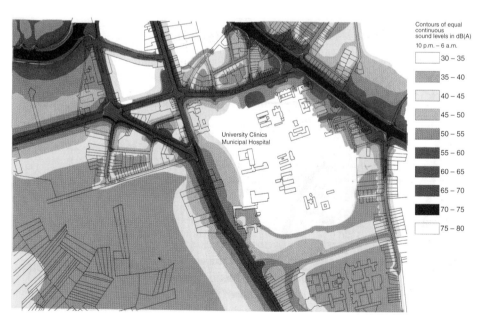

Figure 6.8 Area noise map prepared for Düsseldorf, Germany, in 1978 [16]. (*Source:* Permission granted by VDA.)

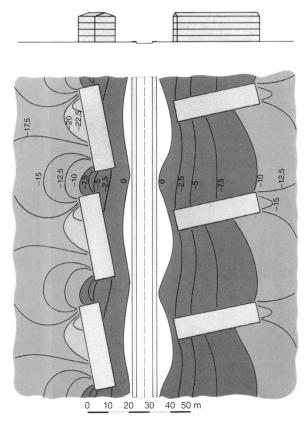

Figure 6.9a A detail noise map from a German design guide for road traffic noise (Plan), 1983 [17].

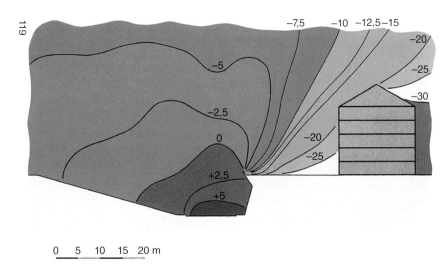

Figure 6.9b A detail noise map from a German design guide for road traffic noise (Cross-section), 1983 [17].

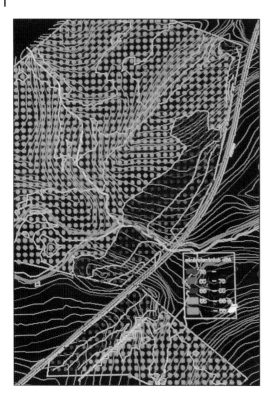

Figure 6.10 Earlier noise map prepared by employing TSØY software, in Norway, 1995 [18]. (*Source:* Permission granted by T. Giestland, Sintef.)

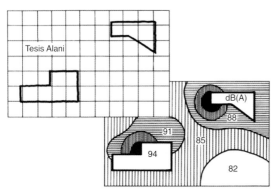

Figure 6.11 Display of noise zones for an industrial plant. (*Source:* from B&K Manual [20]., Permission granted.)

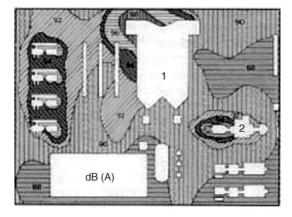

Figure 6.12 An example noise map in an industrial area. (*Source:* from B&K in the 1990s [21]. Permission granted.)

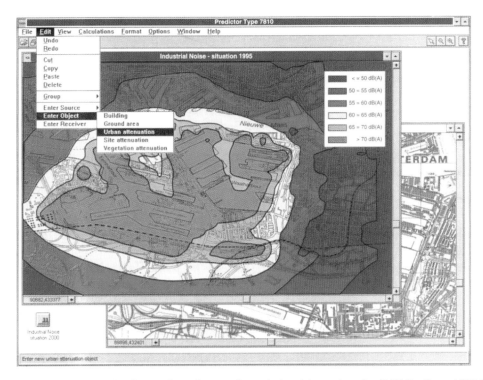

Figure 6.13 Example of an earlier noise map for an industrial area by using B&K Predictor, 1993 [21]. (*Source:* Permission granted by B&K.)

Figure 6.14 An example noise contour map for industrial premises located in an urban area, using Cadna-A, Datakustik [22].

10) Availability of the Geo GIS system and updating the data (data verification).
11) Present standards, regulations, and specifications for noise mapping and evaluation of results.
12) Required accuracy of the noise map in terms of uncertainty.

6.3.3 Measurements for Noise Mapping

Even though the noise contours cannot be constructed solely based on the measured data except in very small neighborhoods, because of the great variation of noise levels occurring with time, however, actual measurement data regarding noise sources and particular reference points in the study area are needed as input data for noise mapping. Measurements can be performed for the following reasons:

1) *Measurements of emission levels:* In order to obtain the source emissions to use as input data into the calculation model, the standardized field measurements can be conducted if laboratory data is not available (Section 5.4). If data regarding emission levels are provided from other documents, they should be checked to discover whether they represent the latest information about noise sources.
2) *Immission measurements:* Single-point or multi-point measurements for sound pressure levels can be conducted to provide the reference levels by applying ISO 1996-2 [27] (Section 5.5.1). Multi-channel recording systems are rather efficient in performing simultaneous measurements at various points although they are limited to distances. For time-varying sources like motorway traffic, continuous recordings are necessary to obtain day, evening, and night-time L_{eq}, dB(A) levels. Spectral information for typical sources or average spectrums is also useful after the effects of background noises have been eliminated. The measured noise levels at different reference receivers are then introduced into the software manually or automatically. Instant data transfer to the laboratory where the mapping is undertaken was made by means of a modem with telephone system in the past, by wireless communication (Bluetooth technology) and by greatly advanced techniques at present, in monitoring the real-time noise levels and to update the noise maps called "interactive maps" (Section 5.11.2).
3) *Measurements for validation of noise maps:* Field measurements are made to check the compatibility between the calculated results and the actual levels at certain receiver positions (Section 6.5.4). The measurements should be performed for longer periods under the same conditions of meteorology and source operations.

The important aspects to be considered during the measurements are:

- The consistency of source configuration and operational characteristics should be continuously observed, e.g. if road noise is to be measured, the volume and speed of traffic and heavy vehicle percentage should also be recorded during the measurement period.
- Variation of meteorological conditions (wind direction, cloudy/sunny hours, temperatures and humidity) should be monitored and statistically analyzed. Measurements should be performed under homogeneous and favorable (downwind) conditions according to ISO 1996-2 (see Chapter 5) [27].

Noise measurements for noise mapping are time-consuming studies with a rather high cost. Furthermore, the uncertainties in the measured results may be high due to the variations of noise levels and environmental factors. Hence, noise mapping by employing a reliable prediction model with controllable parameters is quite feasible for any size of complex source and environmental situation.

6.3.4   Computation Models and Input Parameters for Noise Mapping

Source-specific noise prediction models were explained in Chapter 4. Selection of a model requires the validity of the calculated results (Section 6.5). In Europe, the first issue of Directive 2002/49/EC (END) recommends the following interim methods to be used for noise mapping [28]:

- NMPB (XP S 31–133), the French calculation method (for road traffic noise)
- ISO 9613-2 (for industrial noise)
- ECAC International Civil Aviation Organization method (for aircraft and airport noise)
- RMRS (for railway noise)
- Harmonoise (for all kind of environmental noise)
- Cnossos-EU (for all kind of environmental noise) (adopted by the European Union, mandatory as of December 31, 2018) [5a].

The Directive allows the use of the adopted national methods when the compatibility of the results with those of interim methods has been proved. Generally, all these models consist of three parts: (i) source emission model; (ii) a model for noise propagation; and (iii) a model for impact assessment at certain receiver points within the map area (see Chapter 4).The results are presented by using the descriptors of L_{day}, L_{night}, L_{den} as recommended by END.

Some of the national methods used in European countries are:

- TemaNord 1996:525 in the Nordic countries except Denmark;
- Austrian standard RVS 3.02 Lärmschutz (1997) in Austria;
- NF S 31–133:2011 (NMPB 2008: for roads, railways, and industry) in France;
- RLS-90. Richtlinien für den Lärmschutz an Straßen, in Germany;
- Reken-en Meetvoorschrift Wegverkeerslawaai 2012, specifying a basic method (Standaard Rekenmethode I) and an advanced method (Standaard Rekenmethode II) in The Netherlands;
- CRTN-88 in the United Kingdom.

In the USA, for the computation of traffic noise. the FHWA-TNM 1998 has been the standard model since 1998 and the computerized version based on the earlier program called Stamina and Optima (Section 4.2). In Japan, road traffic noise prediction model ASJ RTN-Model 2013 is the national model.

Some of the national models for railway noise are:

- Nord 2000. New Nordic Prediction Method for Rail Traffic Noise, in the Nordic countries except Denmark;
- Berechnung des Beurteilungspegels für Schienenwege (Schall 03) (2014) in Germany;
- Calculation of Railway Noise (CRN), 1995 in the United Kingdom.

The national models used in the countries for other kinds of noise sources are given in chapter 4. The EU member states have started to integrate the CNOSSOS-EU calculation method for road, railway, aircraft, and industrial noise according to EU Directive 2015/996, into their regulations, since the use of CNOSSOS-EU will come into effect for the 2022 noise mapping round [5b].

Noise mapping requires the following input data to be collected as accurately as possible.

Data Regarding Noise Sources

In order to calculate the acoustic emission levels of environmental noise sources, the numbers of parameters have to be specified, as explained in Chapters 2–4. The commercial software packages generally include global libraries containing the reference acoustic emissions for individual noise sources, vehicles, trains, etc., based on the measurements, some as grouped data according to certain classes, or operation characteristics. However, the emission databases should be checked to find out whether they represent the actual noise sources and if the category system is applicable or not. Most of the environmental noise sources are complex in nature, thus the emission levels should be calculated by specifying the following parameters in the actual conditions:

For motorway traffic noise:

- *Traffic speeds* (average speed or speeds for light and heavy vehicles separately that can be measured via a speedometer or a Doppler radar operating with radio waves). Speed limits can also be accepted by some software.
- *Average traffic volumes per day, evening, and night:* Numbers of light and heavy vehicles for each lane and their variations daily, weekly, seasonally, can be determined at certain intervals by using an automatic traffic counter, video recordings, or a manual counter. A suitable model can be used to estimate the traffic volumes of different roads to be based on short periods of counting. WG-AEN (2007) recommends approximation functions to obtain the long-term traffic flow of different types of roads by implementing some conversion factors for day, evening, and night-times [8].
- *Heavy vehicle percentage:* Ratio of number of heavy vehicles to total volume per each of the day-time periods.
- *Characteristics of traffic flow:* Free, interrupted, accelerating, decelerating traffic, halts due to stop signs and crossings, specified with their distance to the receiver points.
- *Road geometry:* Total length of road, number of lanes, widths of road shoulder and center strip, road gradient, curvatures and road profile remaining in the mapping area, however, with a certain extension to outside the boundaries. Road alignment should be checked by using the GIS tools also to solve possible errors in geographical and local maps.
- *Type of road surface* with absorbing properties, heights of parapets along road, etc.

When the road geometry is supplied through the geographic and transportation maps, the information should be compared with the real situations.

For aircraft noise:

- *Type of aircraft and categories* according to take-off weight, powers, etc.
- *Number of flights for each aircraft type or category:* In addition to total aircraft traffic, the numbers of landing, take-offs, and flyovers, with respect to hour, day and yearly averages (sometimes weekend and seasonal values) are provided.
- *No of runways, routes and flight procedures* implying change in speed and flight profile. *Ground operations:* Initial tests before take-off, maintenance, and repairs in the open air or in semi-closed buildings, the machines used in the tests, etc.

For railway noise:

- *Train types and categorization* in terms of length of trains, brake types, heights of source, etc.
- *Percentage of trains with disc brakes.*
- *Train schedule:* Volume of trains implying numbers of trains in day and night, at weekends
- *Average and maximum train pass-by speeds*, acceleration and deceleration of speed.
- *Railway characteristics:* Ballast and traverses, tracks, type of connections of rails (welded, or bolted), bridges with their locations, constructions and materials, at-grade crossings, tunnels, curvatures with their radius and shapes, etc.

- *Geometry of tracks*, heights from surrounding land, cuttings, embankments, earth-berms, and other obstacles.
- *Vertical reflective surfaces*, heights and distances to rails.
- *Locations of train stations.*

For industrial noise:

- *Layout and geometry* of entire area covered by the industrial premises.
- *Configuration* of each individual noise source.
- *Locations and positions of individual noise sources* and installations with the coordinates and physical relations with the industrial building.
- *Types of individual sources* and their operational characteristics (static or dynamic, operation modes).
- *Acoustic characteristics* of noise sources to be determined by measurements or from product catalogs in terms of:
  sound powers (emissions) or power levels;
  measured sound pressure levels at receiver points after corrected;
  directivity pattern of sound radiation (directivity indexes at octave bands);
  average or maximum spectral levels (octave band levels).
- *Structures of industrial buildings* which are connected or attached to the sources. *Radiating characteristics* of structures and sound insulation values of building elements.

Data in Relation to Physical Environment

The parameters in relation to sound propagation paths throughout the mapping area are summarized below (see Chapters 2 and 4):

- *Topographical conditions*, including lines of faults, embankments, etc. to be updated essentially in urban areas.
- *Types of ground:* Ground cover, reflective and absorptive parts of ground and their size and shape.
- *Building locations and heights:* Number of floors, height from ground or above a reference surface, façade configuration, balconies, etc.
- *Meteorological factors:* Wind speeds, temperature and humidity in terms of daily and yearly averages or long-term statistics, in accordance with software requirements.
- *Natural and man-made barriers and other obstacles:* Locations and geometries; height, size and shapes, acoustic characteristics of surfaces (e.g. sound absorption, scattering, etc.).

Demographic Data

- Categories of buildings in the mapping area.
- Number of residential buildings and noise-sensitive buildings.
- Number of residents in each building, number of dwellings in multifamily buildings, and average number of residents per dwelling, etc.

Receiver Points and Grid System

Specific receiver points and the grid system, assuming that the multiple receiver points are positioned at grid points, should be determined for calculations by considering mainly;

- the required precision of the map
- the topography and variations of heights
- the layout plan and density of the buildings.

The mapping area is divided into the sectors whose size can vary according to the design of the grid system, e.g. rectangular or triangular framing (Figures 6.15a and 6.15b) [29a]. Spacing of the grid points, which are related to the size and scale of the noise map, mainly depends on the precision of the map, e.g. if the mapping area is thousands of km^2, the minimum spacing can be about 10 m. Some programs implement dynamic grid size, i.e. the points representing the receivers are denser with narrow spacing near buildings and behind barriers, whereas they are sparse in the areas with less density of buildings and with few physical elements. The grid points remaining under the buildings and noise sources, are eliminated and the nearest point to the

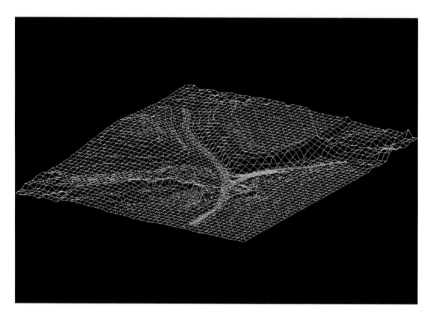

Figure 6.15a 3D view of the rectangular grid system using Mithra software.1998 [29a]. (*Source:* Permission granted by J. Defrance.)

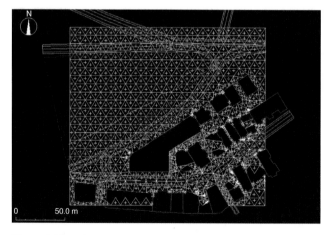

Figure 6.15b Plan view of the triangular grid system using Mithra software, 1998 [29a]. (*Source:* Permission granted by J. Defrance.)

building is selected automatically by the software. When the vertical noise maps are required, the grid system can be selected with larger spacing by also considering the maximum height of land and buildings. The height of the grid points is decided on purpose, however, the standard height required by EU Directive is 4 m above ground level for comparisons.

In order to obtain the outdoor noise levels for the noise impact assessments, it is required to determine the number of people on different floors, the receiver points are distributed on the building façades at each floor level with appropriate spacing to be defined in the software.

Computer-Aided Noise Mapping

In the 1970s, the model procedures were rather simple so the noise exposure contour maps could be made using hand-held calculators and presented as simple drawings for small-scale areas. After the comprehensive prediction models were developed by taking into account the increased number of physical factors in the environment, the calculations to be reiterated at each grid point to obtain the octave band sound pressure levels, as well as the statistical analyses to be performed, required a longer time for the computation process and the higher capacity of computers. The initial computer programs were developed by government departments, universities, and other scientific institutes in many countries and their use was restricted. Nowadays, however, the mapping software is commercially available, thanks to the advances in computer technology. National and international standards and guidelines were published dealing with software development and the producers have made a constant effort to improve the software performance on the competitive market [30].

The properties of the software to be employed in a mapping project should be investigated with respect to the required data quality and data organization, the applicability of the standard calculation models, user friendliness, the requirements for the capacity of the computers, etc. Most of the software comprises calculation modules specific to the noise source, that can be purchased separately on request. Correct use of programs depends on well understanding of the program algorithm and how applicable the features are, by following the manuals, and this may also require special training. The interface of the computation module should be designed to be easily comprehensible by all software users.

6.3.5   Strategies of emission calculations

Many mapping programs require detailed acoustic power information associated with noise sources which can be discrete or in combination, located outside or inside buildings that might transfer noise to the nearby environment. Emission levels, which are introduced as A-weighted levels or octave band levels into the mapping model, can be obtained by different methods depending on the source type. The emission measurements to be conducted in the field and laboratory applying the technical standards are explained in Section 6.3.3 and Section 5.4.1, especially for machinery, equipment, installations, and for vehicles. Such emission data might be acquired readily from the product documents or from the software database. Emission levels of some sources can also be calculated based on the technical parameters. The source emissions are inserted into the model in terms of the sound power level in octave bands or the sound pressure level measured at a reference point for omnidirectional sources in octave bands covering a wide range, i.e. 31.5–8000 Hz [31]. For directional sources, the directivity pattern should be defined.

Transportation Noise Sources

The emissions for road traffic noise are calculated for segmented road lines (including reflections from the road surface). The road gradient can be calculated from the terrain model or can be specifically defined in the model. Recent software provides modeling for bridges, viaducts, etc. The models apply calculations using the traffic/road parameters also given in Section 4.2. The sound power level of a road is calculated for 1 m length of road (*PWL'*: Length-related *PWL*) by using the database for emission levels of vehicles in classes according to the number of vehicles in each class.

In the railway models, each train, comprising locomotive and wagons, is assumed to be a finite-length source and the total sound power is calculated for each train group and corrected according to the railway characteristics. The databases of most mapping programs provide emissions of various train classes.

The aircraft noise models for the calculation of noise emissions in octave bands assume that the aircraft are moving point sources, and the emissions during take-off or landing are calculated for each aircraft group according to flight path profiles and operational procedures (see Section 4.4). However, there are some differences between the models with respect to aircraft groups and segmentation principles. For mapping purposes, the individual emissions of aircraft can be obtained through the International Civil Aviation Organization (ICAO) certificates (see Section 5.4.2).

Industrial Noise Sources

The total acoustic energy of an industrial activity, either in the open air or in a building, is obtained through calculations and acoustic measurements.

• *Determination of sound emissions of individual sources:* For industrial noise sources, the product data sheets from the manufacturers' catalogs might provide the noise emission data available from laboratory measurements, some as A-weighted sound pressure levels, others in octave band levels.

For mapping purposes, the acoustic emission measurements can be made in the field by using the technical standards ISO 3744 and ISO 3746 by applying on-site corrections [32, 33] (Section 5.4.1). Recent software is able to predict the emission data of the individual sources from the known technical parameters, such as electrical power, speeds, revolutions per time or cooling capacity. Furthermore, the simulation of a composite source which radiates noise from different parts of its structure, such as motors, pumps, piping and cooling towers, can be made by means of the advanced techniques, to estimate the overall source emission.

The contributions of all the individual source emissions are considered in the calculation of the immission levels of the entire industrial area or building (generally called an industrial plant). Thus, it is possible to investigate the effects of different sectors of the plant and of the separately positioned machinery and equipment. Such an approach facilitates the noise control implementations, e.g. changing equipment layout, etc. according to different scenarios.

• *Determination of overall sound emissions of industrial buildings:* Emissions of industrial activity at lower level from the ground are represented by the area-related sound power (*PWL''*, dB per km^2). This unit is useful when comparing the noise maps of different plants. However, for an industrial building or for a building complex in which the noise sources can be both inside and outside, the external elements are considered as discrete area sources and their emissions are then combined to find the total emission of the building. The models can be of two situations: (i) sources inside the industrial building, and (ii) sources attached to industrial buildings.

a) *Sources inside industrial building:* Computation of total indoor noise first needs insertion of the acoustic powers specifying the source heights and locations. The indoor acoustic characteristics should be determined by measuring the impulse responses of room. The recent models enable the modeling of complex machines and apply ray tracing techniques to calculate indoor levels. Sound pressure level distribution can be displayed as indoor noise maps by means of specific computer programs or supplementary modules of mapping software [34–37]. Some examples are given in Figures 6.16 and 6.17. Generally, workplace noise maps are useful for occupational noise assessments and for protecting other parts of buildings from excessive noise. In the second step, a transfer function model is established to assess the total indoor noise transmitting outside and propagating in the nearby environment. Basic calculation procedures take account of the sound reduction properties of external building elements (façades and roof), as given in Boxes 6.1 and 6.2 [31].

b) *Sources attached to industrial buildings:* Information about all the individual noise sources located on the roof, or connected to the external walls of the industrial building, like AC ducts, is provided and simulated in the physical model similar to outdoor noise sources. Some examples displaying the 2D and 3D simulations are shown in Figures 6.18a and 6.18b.

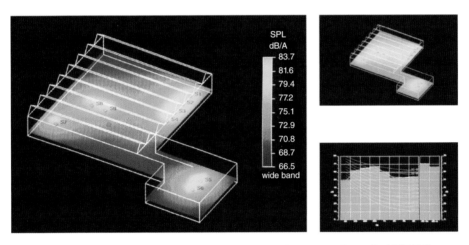

Figure 6.16   Indoor noise mapping in a workplace, using Raynoise software in 1985 [34].

Figure 6.17   Indoor noise mapping in a workplace and definition of noise sources, using Odeon software [35]. (*Source:* C.L. Christensen, Odon A/S.)

Box 6.1   Computation of Noise Emission Level of Industrial Premises When the Noise Sources Are Inside the Building

1) The emission levels are obtained as follows:
   a) Emissions can be determined based on the measurements of sound pressures after being normalized according to reflections, diffractions, etc. Sound power levels are presented in octave bands.
   b) Product data for noise sources can be used if available. However, differences from the documented levels due to the age of the sources, the maintenance conditions, the connections, the mounting types, the silencers, or the auxiliary systems, should be considered and checked through the site measurements.   All these data are inserted into the software interface. An example is shown below [30].

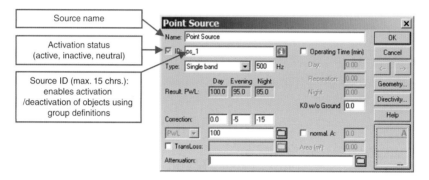

The acoustic power of an omnidirectional source can be calculated if a measured sound pressure level at a certain distance is provided (L_i) after applying the area and nearfield corrections when necessary. If a source radiates sound within only a part of the sphere, the percentage of the area is taken into account in the calculations.

2) Logarithmic summation of all the industrial source emissions (L_w) in the workplace or plant is applied.
3) Total indoor noise level is calculated by taking the parameters of surface absorption/reflection, simply as below:

$L_i = L_{wt} - 10 \lg A + 6$ dB

L_i: Total indoor noise level, dB

L_{wt}: Total sound power level inside the workplace, dB

$A = S\alpha$

S: Total indoor surface area, m^2

α: Sound absorption coefficient of indoor surfaces (assuming equal for each surface)

When the surfaces have different absorption coefficients with different materials, the total areas of each one with the absorption coefficients in octave bands have to be introduced in the above formula to find the $S\alpha$ values.

4) The workspace may comprise at least four enveloping walls and a roof which are defined as horizontal and vertical plane sources in the model. The sound reduction values (R) of these elements are determined by using one of the following approaches:
   a) Selection material R from the software database.
   b) R is measured applying standards (ISO 10140-2; in the laboratory, ISO 16283-3 at the site) [38, 39] (Section 5.9).
   c) Using the R values of similar elements.
   d) Calculation of R of each different element transmitting sound outside, applying EN 12354 calculation model at octave bands by taking into account the flanking transmissions [40].
   e) Calculation of R of composite elements including windows, doors, and other openings (Box 6.2).
5) All external building elements are assumed to be individual plane sources in noise mapping.

Box 6.2   Calculation of Sound Reductions of External Elements of Industrial Buildings

The sound transmission phenomenon is included in the field of building acoustics. Reduction of sound through building elements can be calculated simply by using the basic relationships given in Eq (1):

$$\tau = 10^{-R_i/10}$$

$$R = 10 \lg \left(\frac{1}{\tau}\right) \quad \text{dB} \tag{1}$$

τ: Sound transmission coefficient (unitless)
R: Sound reduction of element, dB

When there are components on the surface (composite element), the area of each component and its sound insulation property is considered separately to obtain the average sound transmission coefficient:

$$\bar{\tau} = \frac{\tau_1 S_1 + \tau_2 S_2 + \ldots + \tau_i S_i \ldots}{S}$$

$$R_{total} = 10 \lg \left(\frac{1}{\bar{\tau}}\right) \quad \text{dB} \tag{2}$$

τ_i: Sound transmission coefficient of each component (unitless) $(\tau_1, \tau_2,\ldots)$
S: Total wall area, m^2
$\bar{\tau}$: Average sound transmission coefficient of composite element (unitless)
$S_{1,\,2,\,3,\ldots}$: Areas of different components such as door, massive part, window, opening, etc., m^2

When more than one industrial activity or building exists in the area, the emission levels of each activity or building are summed (on an energy basis) to obtain the sound emission of the entire area in developing the large-scale noise maps.

Determination of noise emissions of industrial premises by measurements: Noise emission data regarding entire industrial area for noise mapping can be provided by acoustical measurements

Figure 6.18a   Simulation of individual noise sources in an industrial area, 3D view, using Cadna-A [31]. (*Source:* Permission granted by Datakustik.)

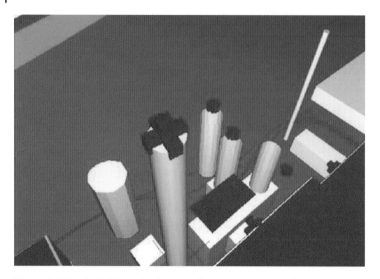

Figure 6.18b Definition of point and plane sources: a detail from the industrial premises given in Figure 6.14 [31].

over an imaginary contour around the area, as explained in Section 5.4.1. ISO 8297 (1994) describes a procedure to predict the average sound power level by applying ISO 1996-2 to measure the sound pressure levels [27, 41] (Chapter 4). The industrial area or building is considered to be a radiating box and the emission is determined by converting all the individual noise sources to a single source with the acoustic power representing the whole plant. Before the measurements, it is necessary to determine the operating conditions of all the sources. If the noise is not consistent, the emission classes in which the source emission is reasonably stable are identified according to the temporal variation of noise.

There might be several premises in the industrial area concerned for noise mapping. In that case, the average area in km^2 per premise is found from the ratio of total area to the number of premises. First, sound emission level per km^2 (PWL'') is determined, second, the total emission level is calculated by taking into account the size of the area, operational hours, etc.

6.3.6 Strategies of immission calculations

The propagation of sound generated by environmental noise sources and the prediction models are explained in Chapters 2 and 4 respectively. For applications of these models in noise mapping, various algorithms have been employed in the computations.

Point source models: In this approach, all the environmental noise sources are assumed as point sources or multiples of stationary and moving sources, as explained in Chapters 1 and 4 (Figure 6.19a and 6.19b). The transmission paths between sources and receiver points are determined and the sound energy transferred by this path is calculated by a selected propagation model taking account of the reflections, diffractions, etc. Geometrically linear sources, such as motorways, railways, etc., are divided into finite-length segments and the calculations are made for the mid-point of each segment, then the total effect is computed by logarithmic summation (Figure 6.19c) [29a, 42]. Plane sources (area sources), like stadiums, are also divided into sectors and the calculations are conducted for the geometrical center of each sector by applying a sound propagation model, then the noise levels

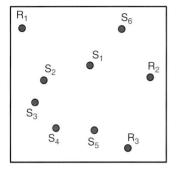

Figure 6.19a Sources and receiver points distributed in enclosure.

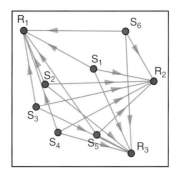

Figure 6.19b Identification of sound paths from all sources to all receivers.

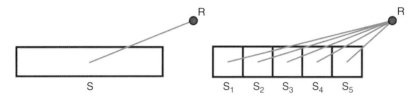

Figure 6.19c Conversion of a line source into an array of point sources.

are summed on an energy basis to find the total sound level at a receiver point (Figure 6.20). Single and multiple reflections from the vertical surfaces (e.g. façades) are taken into account for each point source (Figure 6.21).

Line source models: In the earlier noise mapping programs, the linear sources such as motorways, railways, etc. were directly included as line sources by using the basic equations given in Chapter 4. While modeling the line sources, the sound rays were assumed to be coming from the receiver to the source (inverse-ray analysis). Figure 6.22 shows a single sound path consisting of the reflected and diffracted paths of 1, 2, 3, and 4 [29a].

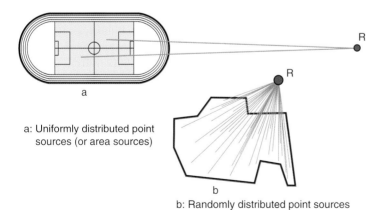

a: Uniformly distributed point sources (or area sources)

b: Randomly distributed point sources

Figure 6.20 Conversion of a plane source into the point sources. (*Source:* adapted from [42].) (a) Linearly distributed sources; (b) randomly distributed sources. (*Source:* Permission granted by J.B. Chene, CSTB.)

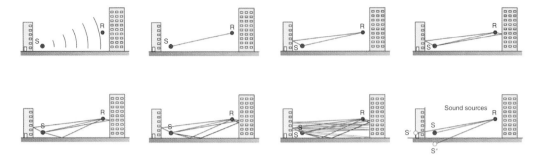

Figure 6.21 Single and multiple reflected rays between buildings and the imaginary sources. (*Source:* adapted from [42].) (Permission granted by J.B. Chene, CSTB.)

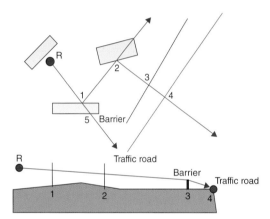

Figure 6.22 Analysis of inverse ray for a line source.

Calculation of immission levels at receivers: The recent computer programs apply basically three calculation strategies to calculate noise levels at grid points or at single receivers [43]:

1) Ray tracing: This deterministic approach is employed in most of the mapping software by searching for all the possible sound rays arriving at a receiver point from a source point (Figure 6.23a). Extended linear sources are subdivided by applying the projection method and Figure 6.24 shows the definition of a sector over a line source. Contributions from all paths to the sound pressure level at the receiver, i.e. not only from the direct ray but also from reflected and diffracted rays due to the surfaces and other objects between the receiver and the noise source, are calculated.

2) Angle scanning: This approach, also called the "constant angle step method," applies a procedure to search for the sources contributing to the receiver level by dividing the two-dimensional angle of 360° around the receiver point into equal angle sectors. Then the axis of each sector is used as a search ray for noise sources and the objects existing on the ray. The point sources inside a sector are projected to the axis. For the line sources, such as roads, the intersection point between the search ray and the axis of the road, is taken as a point source and, after the calculations are performed for each point, the sound pressure levels are combined (Figure 6.23b).

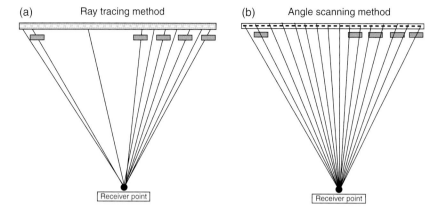

Figure 6.23 Different calculation strategies for line source and definition of sectors contributing sound at the receiver point [43]. (a) Ray tracing method; (b) angle scanning method.

Figure 6.24 Definition of a sector on a line source.

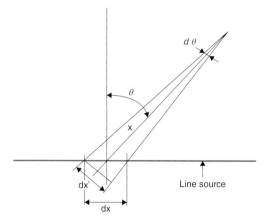

3) Random search: In this less common strategy, a control volume around the receiver is defined. The source is assumed as emitting sound rays which are randomly distributed. The number of sound rays intersecting the control volume is determined and the total acoustic energy is obtained by combining the individual energies of the counted rays.

6.3.7 Insertion of Data into the Software

Basic topographical data which are vital for noise mapping can be drawn from various sources:

a) Digital models of terrain
b) Topographical maps with height contours, in steps of less than 5 m
c) Heights of drain manholes (e.g. from sewage plans)
d) Elevation heights of streets or road profiles (e.g. from road construction plans)
e) Specific altitude measurements at local points (spot heights)
f) Aerial photographs (Google Earth).

Information regarding the site concerned and the source/receiver geometry are introduced into the computer program following the procedure described in the software manuals. Since noise mapping has been around for about four decades, different techniques have been used to insert the data into the software:

1) *Direct input on the screen (free-drawing):* The earlier software, developed by scientific or government institutions in the late 1970s, necessitated manual input directly into the computer screen since they were capable of only small-scale maps. When the geometrical characteristics of the physical elements, sources, receiver points (locations, dimensions) were known, they could be drawn on the computer screen by a mouse according to a linear coordinate system. The heights were inserted into the tables displayed on the interface.

2) *Use of a digitizing tablet:* In the 1980s, when a paper copy (mainly a Xerox copy) of the area plan comprising the topographical contours, roads, buildings, etc. was available, digitizers were used to obtain the digital maps. Coordination between the computer screen and the linear coordinate system of the paper copy, attached to the digitizer's electronic table, was established by means of a special software and a scaling technique. Each element or level point was transferred from the table to the software by using the digitizer's cursor (Figure 6.25). The points were defined with their heights according to a reference level. Finally, the electronic map was obtained and stored for mapping calculations.

3) *Transfer from different files in different formats:* After the 2000s, the geographic maps, regional and local plans and building layout plans could be provided readily in digitized format also with the heights of the objects which are automatically transferred into the mapping software. However, the compatibility of the drawing files with the software requirements had to be searched. The unnecessary points and shapes or auxiliary lines, which may not be important in noise mapping, should be removed to save memory and for visual clarification (Figure 6.26a). Cuttings, embankments, different ground types, and all the surfaces should be simplified, however, but keep the important properties.

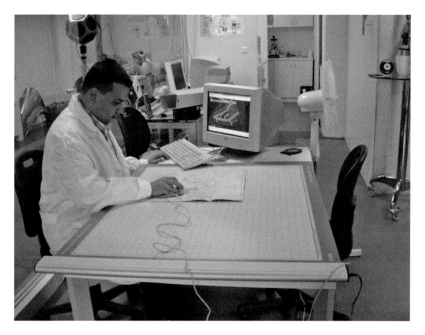

Figure 6.25 Using the digital tablet (GTCO Roll-up II Digitizer) to transfer geometrical data into the computer. (*Source:* Bahcesehir University Technology Center, Butech, Istanbul, 2002.)

Figure 6.26a An example layout plan of an urban area to be examined prior to noise mapping.

   Digital maps including geographical information, that are provided by the local authorities, can be imported into the software directly or after transformation into the file type which is required by the software. Most of the mapping software programs accept the drawing file formats, such as ".dxf" and ".dwg" from AutoCad, ".shp" and "ASCII" from Arcview, Atlas Gris, etc. The geodesic design files; ".dxf" and ".shp," bitmaps from Google Earth, etc., can be transformed for use in noise mapping. While importing the drawing data, the dimensional units used in the original map and in the map imported, should be controlled.

4) *Using satellite systems for noise mapping:* As well as producing data, updating, storing, analyzing and data management, the techniques developed in the 2000s, like Geographic Information System (GIS) and Global Positioning System (GPS), have been a great advance in monitoring the stationary and dynamic points in the area (e.g. noise sources), via satellite systems. GIS gives the geometrical data of the terrain from laser scans or from other sources, transferring this to the

Figure 6.26b   Integration of the noise map with the Google Earth satellite and aerial imagery.

Figure 6.27   Example of data input tool (Cadna-A software) [31].

computer with the aid of various software such as GeoSamba, whereas the GPS are used to trace the dynamic sources. Google Earth provides land-use and building information through photographs and three-dimensional visual materials. All the information obtained from these systems can be integrated, to establish an accurate basis for the physical model (Figure 6.26b). Some software programs transform data for online noise mapping with the required information, (like ArcGis-Pro) which is coupled with noise mapping programs [29b].

Although the noise maps give the average noise levels at the site concerned, however, the recent systems enable real-time monitoring to be shown on the noise maps: the actual data provided by the noise monitoring systems are transferred to the noise map which is updated continuously, so that the changes in contour lines or in noise levels at specific points can be observed (Chapter 5 and Section 6.3.9).

Defining the physical elements in the mapping software: All the elements affecting sound propagation in the environment can be defined as the design parameters by identifying their characteristics in the software. The descriptions are facilitated by means of a toolbox viewed on the interface comprising the following elements (Figure 6.27):

- topographic level contours (combination of the points with the same heights)
- roads and railways

- industrial sources (point and plane sources)
- sound screens
- walls (barriers)
- earth-berms (trapezoidal shape)
- cuttings and faults
- green areas (vegetation) and reflected surfaces
- bridges
- specific receiver points.

The physical elements in the mapping area which are defined with their three-dimensional coordinates are displayed on the map as open or closed polygons. The height parameter of the vertical elements is given as absolute or relative heights above ground. The terrain model and buildings can be visualized in 3D, which is useful for checking the realistic appearance (Figures 6.28–6.30). Barriers are defined with their vertical dimensions, shapes, and their distances to source within an error of less than 1 m. The barrier surface on both sides is defined with

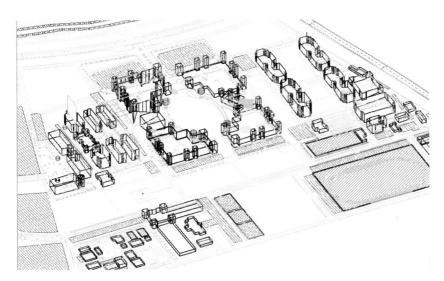

Figure 6.28   Axonometric view of site with wireframe technique, using Cadna-A.

Figure 6.29   3D-view of site topography and buildings [29a]. (*Source:* from earlier version of Mithra software.) (Permission granted by J.B. Chene, CSTB.)

Figure 6.30 Detailed modeling of complicated urban configuration, using Cadna-A software [31]. (*Source:* Permission granted by Probst, Datakustik.)

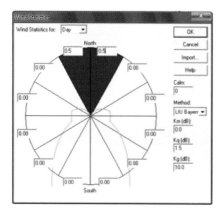

Figure 6.31 Insertion of meteorological data into the model. The wind diagram shows the percentage of time when the wind is blowing from a certain direction [31].

the absorption coefficients given in octave bands. Apartment blocks and individual buildings, parks, bridges, and other urban configurations, can be viewed slightly under the noise contours on the final map without disturbing the presentation of noise contours. While entering the building data, it is possible to assign the number of residents as described by the software. Insertion of the meteorological data into the model is possible through the wind speeds in each direction as percentages as shown in Figure 6.31. Since it is important to construct the model with sufficient detail, the accuracy of the physical properties should be inspected through vertical cross-sections.

6.3.8 Performing Operations

Software programs for noise mapping improved a great deal in the 2000s and the modeling has become more realistic and easier to apply. Calculation times have been remarkably reduced, thus, completion of a map becomes much faster. The new features could be introduced like diffractions around the edges, elevated roads, bridges, detailed description of surface absorption characteristics. Data importing has become more efficient.

The German National Standard DIN 45687 describes the required properties of the software packages. As one of the criteria, the "user friendliness" is associated with the design of graphical interface with the Windows operating system [26]. The graphical features, including 3D viewing of the overall site and sound rays, are helpful to ensure the precision of the map.

Different calculation models can be incorporated in the programs to allow comparisons of the results by repeating the calculations, for example, by means of "orthogonal programming." Complex theoretical methods are integrated to solve the wave equations or to simulate the sound propagation in all kinds of environment.

Data organization is managed readily, such as sorting and eliminating data, categorizing, subgrouping, numbering, etc. The building blocks of similar types can be grouped according to their geometrical and functional properties. Particular receiver points and the reference points at which the field measurements will be made, can be identified.

The new generation software can perform calculations according to the desired strategies, such as correct segmentation of linear sources, different propagation conditions, or changing the reflection order, defining the search radius as a configuration parameter. The search radius is the distance from the receiver point at which all sources and objects will be taken into account while calculating the sound level, for example, by ignoring the sources away from this distance.

The operation time needed to complete the calculations can be estimated at the beginning. Errors are calculated based on the statistical evaluations and the maximum errors (e.g. 0.5 dB) can be controlled by the intelligent methods.

For the large- scale noise mapping over vast areas as required in the EU Directive, the new acceleration techniques have been developed to facilitate and speed up calculations, for instance, "Automated Shared Processing" (ASP) for the data presented in the computer's Random Access Memory (RAM) and "Program Controlled Segmented Processing" (PCSP) [43]. The entire mapping area is divided into sectors with rectangular frames in order to share the task with several computers. The main computer is programmed to organize the whole process and to integrate the results by means of modern processors with multithreading features (dual, quad-core, etc.) and a supporting software requiring advanced computers, e.g. a 64-bit version computer. Ultimately, the number of receiver points for simultaneous calculations is increased, the size of the project that can be undertaken is extended and dealing with the problematic industrial noises and high order reflections becomes easier.

6.3.9 Managing the Outputs

Obtaining noise exposure contours (isophones) and presentations: After calculations at the grid points, the points with equal noise levels are determined through interpolations. These points are then combined to construct the equal noise lines (noise exposure contours) on both the horizontal and vertical cross-sections, with intervals of 1, 2, 5 dB(A). Figure 6.32 displays the calculated noise levels at grid points and the contour lines. The points are generally compressed in the vertical sections according to the z ratio to be selected (Figure 6.33).

The quality of the map has a great influence on transferring information to the public. Color is one of the visual factors affecting the interpretation of the results derived from noise maps. Coloring the areas is made in compliance with the noise level intervals and is depicted as a legend on the noise map. The color scale (implying classes and class sizes) was standardized by ISO 1996-3:1987, however, this was not included in the later issues of the standard [25]. The importance of color selection, when considering the physiological and psychological reactions of people, has been emphasized by some researchers and options to increase clarity and comprehensibility of noise maps through different visual variables, such as sharpness, value, color, texture, etc. have

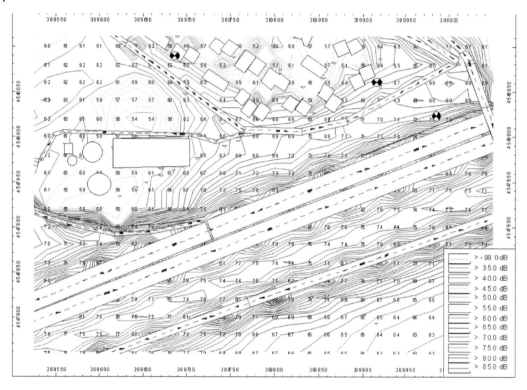

Figure 6.32 Grid levels and the contour lines on the noise map. (*Source:* Butech, BAU Istanbul, 2004.)

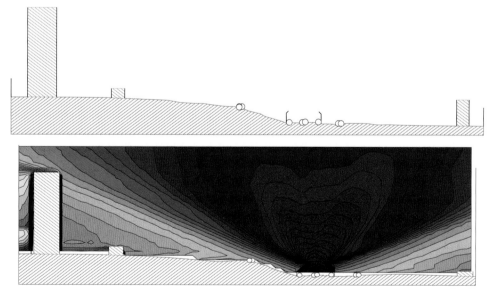

Figure 6.33 Cross-section of road and vertical noise map, using Cadna-A. (*Source:* Butech, BAU Istanbul, 2006.)

been discussed. The debate continues for better use of color in terms of hue, value, and saturation, based on human perception theories [44].

Recently many software packages are available on the market, some with powerful display options. They can provide the visual materials, such as elevation of the physical terrain model in 3D (isometric, axonometric, parallel line, etc.) and video animations (Figures 6.34 and 6.35). Animation scenarios can be made by adjusting the camera positions, the directions on the horizontal and vertical planes, moving the speed, rotation speed, resolution, etc. and by flying over the site or moving parallel to the ground, the noise levels and their variations can be detected.

Figure 6.34 3-D noise map by Cadna-A [31]. (*Source:* Permission granted by Probst, Datakustik.)

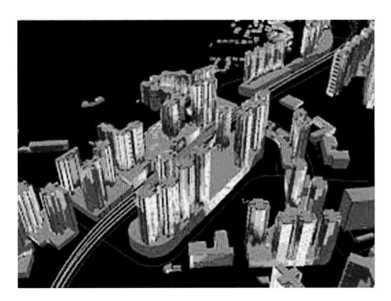

Figure 6.35 3-D noise map for Hong Kong [45].

Superposition of different kind of maps and optimization of barrier design are other features provided by many of the recent software packages.

Dynamic noise mapping: Currently real-time noise mapping is widely used in Europe and the USA, using the technique explained in Section 5.11.3. This technique requires a localized noise monitoring network that collects noise data continuously and transmits it to a data center where the noise model is operated [46–49]. The data management system works through a dedicated software, some of which enable observations on the noise maps so that real-time monitoring and change in noise conditions can be observed instantly or with certain intervals, e.g. the average hourly levels (L_{eq}). Dynamic noise maps are useful for noise-sensitive areas and buildings, especially in the vicinity of airports, highways, and other major noise sources or to monitor excessive construction noises and entertainment noises. Such informative maps are presented to the authorities to take precautions or apply incentives to comply with the regulations, and to the community, to attract their attention to the problem.

The attempts to develop different approaches to real-time noise mapping are continuing in Europe, to achieve a low-cost monitoring by simplification of the systems, to develop an algorithm to eliminate anomalous noise events, to establish a web-based GIS software application and reporting for public information. The "Life Dynamap" project aims to facilitate and speed up noise mapping due to the necessity of updating and submitting them to the EU every five years. Automatic monitoring systems have already been established in the pilot areas [48a-d].

Statistical Analyses of the Results Derived from Noise Maps

1) *Emission levels:* Outputs of emission calculations for various noise sources are tabulated for different statistical analyses according to the purpose. For traffic noise, sound energies radiated from all the segments are displayed, so that the contribution of each segment to the total energy and dominant parts can be observed. An example regarding statistical analysis of road traffic parameters is given in Table 6.1. The result tables for a railway, for a point source (mechanical equipment) and for a plane source (parking lot), are shown in Table 6.2. The emission data at each octave-band are collected in the local libraries of the software, facilitating the decisions on noise control.

2) *Immission levels:* The results of calculations at specific receiver points or at grid points can be evaluated for different purposes, e.g. the variation of noise generated by a dynamic source at a certain receiver point can be illustrated graphically. Figure 6.36 shows a time-history of the noise levels during a vehicle pass-by with a constant speed. Recent programs enable the auralization of the pass-by noise of motor vehicles and trains including the Doppler effect. Calculated sound pressure levels for each source and receiver are tabulated with their IDs facilitating statistical analysis such as the dispersion of receivers exceeding certain noise levels in the mapping area or the percentage of receivers with noise levels remaining in predefined noise classes.

   The calculated results at receiver points corresponding to noise levels on façades are tabulated with their identities (Table 6.3). The recent software include the operations as called building evaluation and the calculated noise levels on the receiver points are indicated on plans, elevations, and as three-dimensional views (Figures 6.37a, 6.37b, and 6.38).

3) *Geometrical ray analyses:* It is possible to observe sound rays with their energy changing with direction. The reflected and diffracted rays from surfaces and obstacles can also be traced on the mapping area. Such detailed analyses, that could be conducted even in the earlier software, are of importance in the search for the path carrying the maximum sound energy to the receiver, in order to take measures for noise control or to eliminate the paths with negligible contribution to total sound energy (Figures 6.39a and 6.39b).

Table 6.1 Output table displaying the calculated A-weighted emission levels, *PWL'(L_{n'w})* for unit meters of different roads and the input traffic parameters obtained using Cadna-A [31].

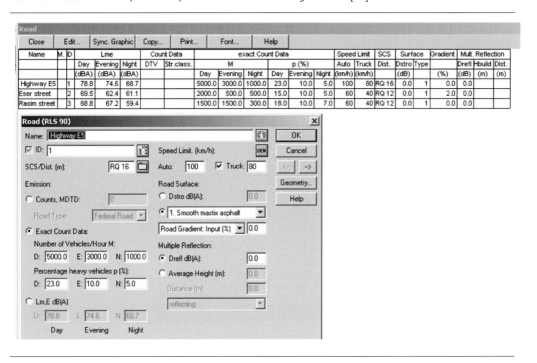

4) *Checking the limit values:* The noise limits and standard values can be entered directly to the software or defined in relation to land use, so that the calculated noise levels at particular receiver points can be compared. When the limits are exceeded, a warning sign appears on the map, such as a red spot or receiver point sign turning red. This technique facilitates the instant detection of a problem, also, it is useful for the observation of efficiency of a noise measure. Some software can provide so-called "conflict maps" displaying the magnitude of excessive noise and deviations from a noise limit [31].

5) *Analysis of indoor noise levels:* Some models yield the levels of noise transmitted from outside into the noise-sensitive spaces through external building elements (this subject has been explained in Section 6.3.5 for interior noises in industrial buildings). If the indoor noise criteria are defined, the acoustic conditions of indoor spaces can be evaluated from the point of health and comfort of those using the spaces and the necessity of measures to be determined. The supplementary software which calculates the building acoustics parameters like, sound reduction of walls, windows, roofs, etc., can be employed to export the calculated data into the mapping program, or can be operated simultaneously to obtain the indoor noise levels [50, 51] (Figure 6.40). For specific constructions, i.e. multilayered or composite elements, various software programs, are also available. Sound insulation values of external and internal elements through direct and flanking transmission paths are modeled according to ISO EN 12354 [40]. Ultimately when the sound insulation performance of the external element is improved, the effect can easily be observed on the indoor noise maps, similar to those given in Figures 6.4a and 6.4b. Such applications facilitate the evaluation of the noise conditions with respect to the indoor acoustic parameters.

Table 6.2 Analysis of the emission levels for other types of sources and the parameters involved (example of outputs obtained using Cadna-A [31]).

Finite-line source (Railway):

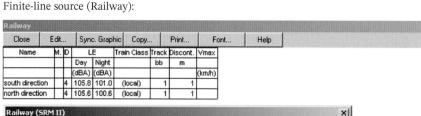

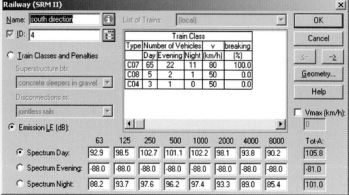

Point source (Compressor):

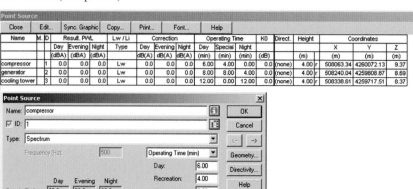

6) *Design and optimization of noise barrier:* As explained in Chapter 2, the effect of the barrier is determined by means of an immission model and the effects of the geometrical and acoustical parameters including the top profile are taken into account. The recent software enables the optimum design of sound barriers to be used for noise control, e.g. along a road. The optimization process requires initial decisions about location and the length of the barrier to be designed. If required, an updated noise map is prepared. The optimum height is determined by the following process:

Table 6.2 (Continued)

Area source (Parking lot):

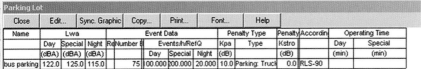

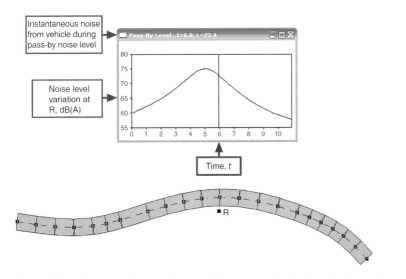

Figure 6.36 Example of a time-history of the computed noise levels at a receiver point during a car pass-by, using Mithra (CSTB) in the 1980's [29a].

Table 6.3 Example of the calculation results of different floors of a building by using the earlier version of Mithra software, in 1998 [29a].

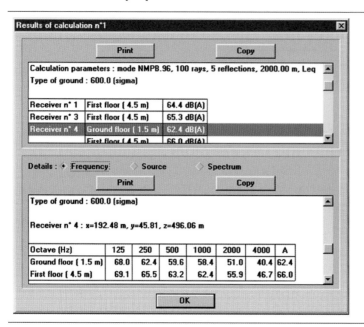

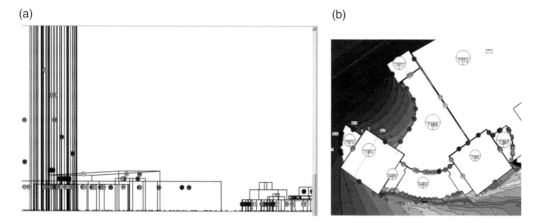

(a)   (b)

Figure 6.37 Example of building evaluation displayed on elevation (a) and plan (b). Circles show the receiver points on the façades, using Cadna-A.

a) Standard values (i.e. noise limits) are defined for daytime and night-time with respect to the land-use.
b) The total barrier length is divided into segments of 1–5 m lengths.
c) Sound levels are calculated behind the barrier at a certain receiver position and checked to see whether the limits have been exceeded or not, as explained above. If yes, the model increases the height of each segment gradually, such as by 1 m each time and repeats the calculations.

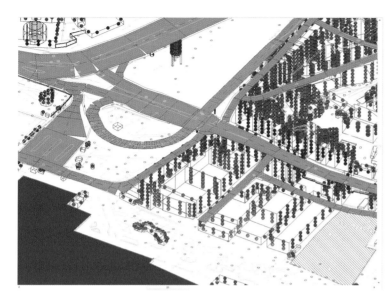

Figure 6.38 Determination of noise levels at different reciever points on the façades of buildings [31]. (*Source:* prepared in Butech training programs in 2008, using Cadna-A.)

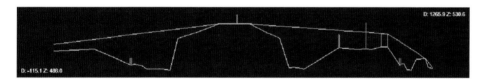

Figure 6.39a Ray analysis on the vertical plane, using Mithra, 1998 [29a]. (*Source:* Permission granted by J.B. Chene, CSTB.)

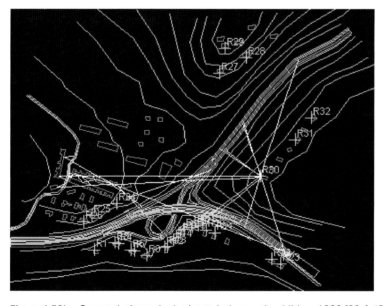

Figure 6.39b Ray analysis on the horizontal plane, using Mithra, 1998 [29a]. (*Source:* Permission granted by J.B. Chene, CSTB.)

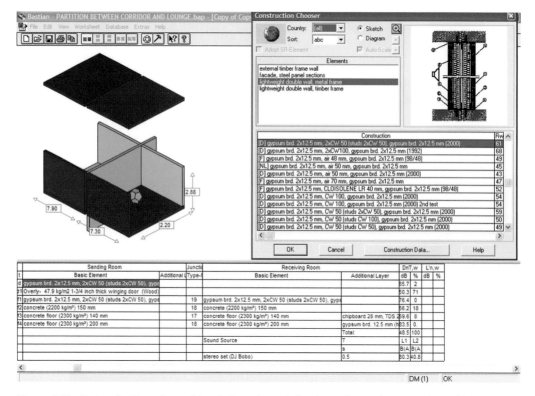

Figure 6.40 Determination of sound insulations through Bastian software in connection with Cadna-A [50].

d) In the end, the optimum height for each segment is determined along the total length of the barrier and the barrier profile is displayed as shown in Figures 6.41a and 6.41b. The total surface area and the cost of construction based on pre-estimated value per unit area, m², are calculated.

e) The barrier impact area can be shown on a new noise map, the so-called "after barrier map."

7) *Demographic analyses:* END requires the population density, i.e. the number of people per building and calculation of number of dwellings exposed to noise levels according to a classification system. This information can be provided by most of the mapping software based on the statistical analysis of the results (Figure 6.42). While determining the exposure levels which also represent the noise levels in rooms with open windows, the average or maximum sound pressure levels, which are calculated on the façades, are taken into account according to a strategy. Since the façade levels are about 2.5–3 dB(A) higher than the calculated levels due to the façade reflections, the exposure levels are corrected accordingly.

WG-AEN 2007 recommends submitting the numbers of people living in specially insulated dwellings, although not mandated in END [8]. In addition, the dwellings whose windows of noise-sensitive rooms are not facing the noise source should be counted. Outputs relative to such statistical data should be included in the report.

8) *Cost analysis:* Most of the software programs enable a cost valuation by converting the noise levels into monetary values, for example, estimation of the reduced rental or sale prices in noisy areas. A procedure applicable in cost analysis is given below:

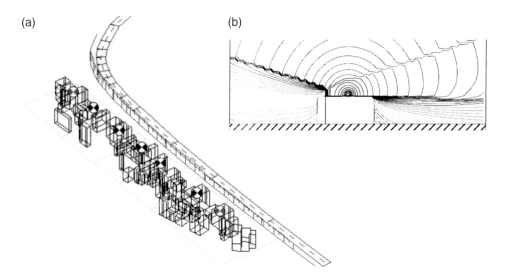

Figure 6.41 Barrier optimization process: (a) 3D view of a barrier profile and the receiver points to check the criteria; (b) noise contours on the cross-section with barrier along an elevated road, using Cadna-A [30]. (*Source:* Permission granted by Probst, Datakustik.)

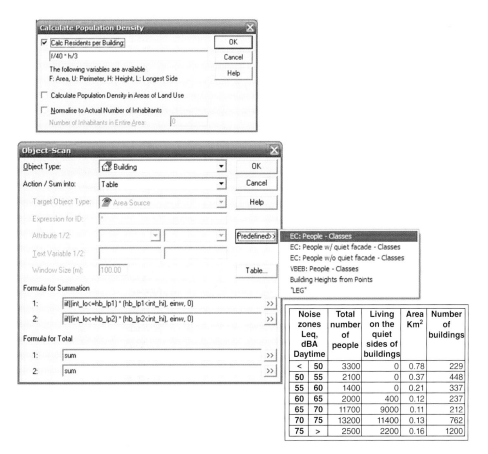

Noise zones Leq, dBA Daytime		Total number of people	Living on the quiet sides of buildings	Area Km²	Number of buildings
<	50	3300	0	0.78	229
50	55	2100	0	0.37	448
55	60	1400	0	0.21	337
60	65	2000	400	0.12	237
65	70	11700	9000	0.11	212
70	75	13200	11400	0.13	762
75	>	2500	2200	0.16	1200

Figure 6.42 Demographic analyses as a result of noise mapping (an example using Cadna-A).

a) The area with respect to noise level category is specified with its size and perimeter on the noise map. The number of existing buildings and the percentage of rental dwellings are determined automatically since this information is given while inserting the building data into the model.

b) The ratios of the floor/ground area (area of single housing unit in the building to total floor area) and the ratio of rental units are determined.

c) Calculations are made by taking into account the decrease in rent or sale prices due to noise. For example, it can be assumed that a noise level increase by 1 dB causes a decrease of 1% in the rental rate. This value varies with the countries and/or land uses (Section 9.5).

d) Ultimately, the average economic losses due to noise can be obtained and displayed in a separate map.

9) *Comparison with other kinds of pollution:* Recent software can perform analyses for both noise and air pollution and the predictions of their critical values are presented in discrete maps. Since traffic noise levels are higher in densely populated areas or along highways, there is also a serious air pollution problem due to exhaust fumes and various particles. Integration of such maps facilitates the general evaluation of the polluted areas in developing environmental policies and providing reliable information for action plans (Chapter 9).

10) *Preparation and use of supplementary maps:* Some information required in the evaluation of the results and interpretations of the maps can be obtained with the aid of the supplementary maps in combination with the noise exposure maps. Such maps that might contain information should be at the same scale of the noise maps if they are intended for the development of noise policies. Some examples of these are as follows:

a) Density and population maps

b) Maps displaying the density of highway, railway, airport, and industrial premises

c) Maps for density and distribution of on-going construction in urban areas

d) Maps covering other noise sources

e) Maps separately prepared for different meteorological conditions

f) Maps prepared for seasonal activities (e.g. summer times)

g) Maps exhibiting land uses and building types

h) Maps indicating noise-sensitive or non-sensitive buildings and areas

i) Maps showing required insulation for buildings

j) Maps for hot spots and quiet areas

k) Soundscape maps [52]

l) Maps showing the distribution of the standard deviation or uncertainty [53]

m) Contour maps of the Tranquillity Rating Prediction Tool (TRAPT) [54]

n) Noise map showing noise levels only in a selected frequency band

o) Future maps for global noise pollution

6.4 Evaluation of Noise Maps

6.4.1 Identifying Noise Zones

A noise zone is defined as the area between the noise classes representing the long-term average values, for example, a piece of land between two specified contour lines, e.g. 60–70 dB(A). Different categories and class intervals have been selected in the past up to the present for noise zoning. A classification scheme should cover the clear principles to be used for acoustical planning and urban development policy [25, 55].

- ISO 1996-2:1987 recommended noise zones with contour boundaries in multiples of 5 dB. However, the zone width of 10 dB could be used to report the measurement results on the map, in terms of long-term average sound level, $L_{A,eq;LT}$ and long-term average rating values $L_{A,r;LT}$ [25]:

$L_{A,eq;LT}$ and $L_{A,r;LT}$ dB(A)
<45
45–55
55–65
65–75
75–85

- In a former regulation issued in the UK in 1994, four categories regarding noise exposure were defined for road traffic noise in terms of the average noise levels to be implemented in urban planning (L_{day}: 07:00–23:00), L_{night}: 23:00–07:00) [56]:

	L_{day} dB(A)	L_{night} dB(A)
Category A	<55	<45
Category B	55–63	45–57
Category C	63–72	57–66
Category D	>72	>66

- Identification of noise zones in the designated urban/rural area or in a selected region or for a specific source can be made in simple groups of noise level ranges. The Green Paper, issued by the European Commission in 1996, categorized the noise zones with respect to L_{eq}, dB(A) [57]:

A. Very noisy (Black zone):	$L_{eq} > 65$ dB(A)
B. Moderately noisy (Gray zone):	$L_{eq} = 55$–65 dB(A)
C. Less noisy (White zone):	$L_{eq} < 55$ dB(A)

An example noise map displaying the zones according to above three-category system is shown in Figure 6.43. However the above mentioned simplifed zoning system are not commonly used in the legislations at present.

- In recent practice, the designation of noise classes is made with respect to noise levels in L_{den}. END and the European Acoustic Association (EAA) recommend the following categories to provide compatibility in the evaluation of noise impacts in European countries [28, 58]:

L_{den} dB(A)
>75
70–74
65–69
60–64
55–59
50–54
45–49
<45

The END does not require submission of the last two category areas.

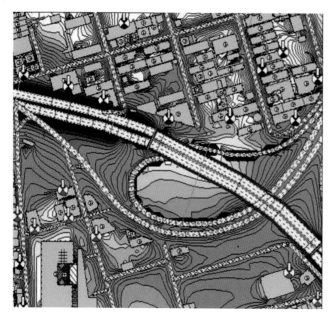

Figure 6.43 Displaying the black, gray, and white zones in a noise map complying with the Green Paper, 1996, using Cadna-A. (*Source:* BAU student assignment, Istanbul 2002.)

6.4.2 Rating Noise Impact and Noise Scores

In order to achieve the objectives of noise mapping (Section 6.1), the evaluation of overall noise impact can be made by using an appropriate single-number descriptor, that will be used also for comparisons of different regions and different planning alternatives. Derivation of such an exposure indicator was attempted in the 1950s, however, studies increased in recent years along with the discussions to reach a consensus over the best indicator to be adopted for rating noise impact. The descriptors for rating noise impact that were proposed in the past and modified at present are outlined below. They all take into account the noise levels and number of people exposed.

Noise Impact Index

The first index was developed in 1977 by Von Gierke in the USA when the concerns on noise impact statements emerged for the guidelines [59, 60]. The index was defined based on a demographic parameter "total weighted population" to be calculated as the product of "weighted impact function" and the total population in the area:

$$TWP = \sum W_i P_i \tag{6.1}$$

TWP: Total weighted population to be calculated based on the day-night average noise level, L_{dn}
W_i: Annoyance weighting factor for i^{th} noise level category (derived from a noise impact function)
    (Chapter 7)
P_i: Number of people associated with the ith noise level category

In earlier documents, W_i were given in the tables in relation to noise groups. The relative impact of noise in a particular environment is defined as the Noise Impact Index:

$$NII = \frac{TWP}{\sum P_i} \tag{6.2}$$

NII: Noise Impact Index

$\sum P_i$: Total population in the area

NNI was recommended for comparisons of noise impacts in different environments.

Noise Impact Value

The relationship between the population exposed to a certain level of noise and the impact is defined as a single number descriptor in Germany [31]:

$$LB = \sum N_i \times U_i \tag{6.3}$$

LB: Noise impact value

N_i: Number of people subject to excess level U_i

U_i: Excess level, dB(A)

i: Noise classes

The above process is implemented in noise mapping programs, simply given as follows:

a) The area is divided into zones identified with their boundaries and size in km^2.
b) The noise limit values accepted by local or national governments are declared either as a single value or in a range.
c) The number of buildings are determined in each zone.
d) The population density is obtained from the noise maps, which is defined as the number of residents living in each building (based on prior approximations: person per m^2).
e) The noise level (in classes) to which each building is exposed and the number of people exposed to each noise class are obtained.
f) The noise impact value is calculated for each zone.
g) Noise impact values at all designated zones are displayed on the noise maps.

Potential Road Noise Exposure

Developed by a group of researchers in 2012, the Potential Road Noise Exposure (PNEI) is based on L_{den} that can be acquired from the noise maps [61]. A section of the urban layout is defined as a census block which is the smallest sub-municipal division with 1,500 inhabitants on average [62]. The indicator is obtained as shown in Eq (6.4):

$$PNEI_{Lden} = 10 \lg \left(\frac{1}{nhab_{tot}} \sum_{bat=1}^{N} nhab_{bat} \cdot 10^{(L_{denbat}/10)} \right) \tag{6.4}$$

$PNEI_{Lden}$: Average potential noise exposure indicator for a census block

L_{denbat}: Average L_{den} on the façades of a building (i.e. energetic mean)

N: Number of buildings in the census block

$nhab_{bat}$: Number of inhabitants in each building

$nhab_{tot}$: Total inhabitants in a census block

The range of variation of the *PNEI* (maximum and minimum values) can be assessed based on the standard deviation of L_{den} to be calculated through a mapping software.

Rating Systems Developed by Miedama on the Basis of Noise Maps

Emphasizing the importance of a rating system to be acquired from the noise map, Miedama declared, in his report in 2007, that such a system should describe the noise impact in terms of some descriptive properties, such as non-quiet area, quiet area, acoustic climate, hot spots, etc. [63–67].

1) *Non-quiet area:* Miedama and Vos in 2003 proposed a simple indicator giving the percentage of the area where $L_{den} > 50$ dB (A) (i.e. the ratio against the total area) to describe the ambient

acoustical quality [63]. This metric, defined as AREA$_{50}$, is incorporated with the noise score approach later developed for night-time noise impact.

2) *Noise climate (or acoustic climate):* The acoustic climate is determined based on the relationships between the levels and impacts of noise that have been derived from the field surveys. As discussed in Section 7.3, the "dose-response relationships," introduced in the noise climate concept, are based on two negative impacts of noise in residential areas: "annoyance" and "sleep disturbance." For assessment of these two impacts separately, Miedama developed the source-specific polynomial equations for transportation noise sources (road traffic, railway, and aircraft) and some stationary noise sources (industrial premises, shunting yards, and seasonal sources) [63–67]. Ultimately, noise climate is defined in two forms:

- overall noise climate in relation to the percentage of highly annoyed people (*HA%*);
- night-time noise climate in relation to the percentage of highly sleep-disturbed people (*HSD%*).

Miedama has proposed a procedure to describe noise climates on the basis of noise maps in 2003. The dose-response relationships (as polynomial equations given in Section 7.3.2), prepared for *HA%* and *HSD%*, were then translated into the annoyance measure (*A%*) corresponding to the noise scores [63]. The noise levels in these equations are given as L_{den} and L_{night} on the façades of buildings to be obtained from the noise maps, however, after applying corrections for the other factors which were quantified as:

- the buildings with above average extra sound insulation;
- the buildings with a quiet façade;
- the ambient sounds other than those concerned.

The noise climate concept is applied to derive the noise scores given below.

Noise Scores

The noise score is defined as "a single number representing the social weightings of an unwanted situation due to high noise exposures" [68]. Different approaches can be implemented to determine the noise scores in local areas or in designated parts of larger areas. By assuming a simple linear relationship between the noise levels and the noise impacts (in terms of noise annoyance), the noise score is determined as Eq (6.5):

$$NS = \sum_i n_i \cdot (L_i - L_R) \tag{6.5}$$

L_i: Noise level in certain classes (noise bands), i
L_R: A reference noise level or a limit value
n_i: Number of people exposed to L_i

While implementing the above noise score concept, L_i can be given in terms of $L_{den,i}$ and $L_{night,i}$, dB corresponding to the noise level on the façade of buildings at 4 m above ground and taken as the value on the most exposed façade (the building-related noise score). Accordingly, the limit value L_R should be specified in both descriptors. The n_i is estimated as the number of people living in buildings that are exposed to each of the noise bands.

The noise score has been accepted by the EAA as the common descriptor of noise impact [69]. Probst has suggested the area-related noise scores in noise mapping should be calculated for two noise conditions, as $L_{den} < 65$ dB(A) and $L_{den} > 65$ dB(A) [70].

Hot Spots

The determination of unacceptable exposures and non-quiet areas based on noise maps has been widely discussed at present in terms of hot spots. Identification of hot spots can be made considering various features, for example, places with high noise levels, places where the noise levels exceed a defined limit, places with high population numbers and with high noise levels, or places with a high density of annoyed people. Analysis of hot spots may also require estimations of the numbers of the annoyed, highly annoyed or highly sleep-disturbed people based on the dose-response relationships given separately for road, rail, traffic, and industrial noise sources.

Miedama (2007) described the hot spots within residential areas in relation to two factors: the weighted number of individuals (n_L) unacceptably exposed to noise above a limit L, and the weighted noise level to be obtained from noise maps. He proposed two weighting functions for L_{den} [66, 67]:

- Linear function:

$$W(L_{den}) = 1 + a(L_{den} - L) \qquad (6.6)$$

  L_{den}: Day, evening and night level acquired from noise map
  L: The limit value in L_{den}, dB(A)
  a: Parameter to be defined by the local authority

- Exponential function:

$$W(L_{den}) = 10^{a(L_{den} - L)} \qquad (6.7)$$

In both formulas, $a > 0$; $W(L_{den}) = 0$ when the noise level is below the limit.

Detection of Noise Scores and Hot Spots on Noise Maps

In order to detect the hot spots on the noise maps, based on area-related noise scores, a gliding frame method is recommended [68]:

1) A noise map is prepared for larger areas with the calculated noise levels displayed on the grid points in L_{den} (and L_{night}) and on the façades. The building-related noise scores are determined for each building or housing unit (dwelling).
2) A window frame (polygon) with size (e.g. 100×100 m) is constructed and placed on the map so that the first grid point matches its center point. (Figure 6.44).
3) The noise scores of all the buildings within the frame are calculated and summed to find the area-related scores in the framed area.
4) The obtained sum is divided by the window area and then multiplied by a reference area (e.g. 1000 m^2) to find the area-related noise score (NS/per unit area). This value is assigned for the grid point.
5) The frame is moved to the second grid point, as shown in Figure 6.44, and the same procedure is repeated. By sliding the frame from one point to the next, the calculated noise scores are assigned for all the grid points in the map area.
6) Distribution of the noise scores/1000 m^2 can be displayed on a new map called "area-related noise score maps."
7) A certain limit value for noise score (symbolized as NS_{limit}) can be specified and the points with the NS values exceeding the limit can be marked by using a red color on the NS map, which are declared as hot spots.

Distribution of the total noise scores are displayed on the noise map by using the three colors of red, yellow and green, corresponding to the percentage of areas with noise scores of greater than

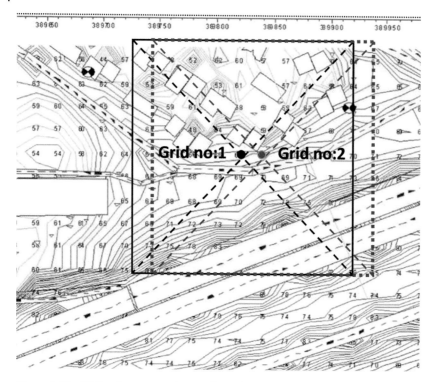

Figure 6.44 Gliding frame procedure to determine the noise scores at grid points.

90%, 10–90% and less than 10%, respectively [70]. The red area is declared a "hot spot" representing the place where the area-related noise score is higher than 90% of all values in the entire area. The green area is designated as the "quiet area."

Since the hot spots are determined from the strategic noise maps and they indicate not only the high noise levels but also the numbers of people exposed, they represent the unacceptable exposure areas also called high priority areas, where noise protection is strongly recommended or mandatory in national legislations (Figure 6.45) [71a,b]. Ultimately, detection of hot spots on noise maps allows the implementation of relevant noise measures in action plans and enables comparisons of different solutions that can be displayed on the re-calculated noise maps.

6.5 Uncertainty and Validity of Noise Maps

The accuracy of noise maps, which is simply defined as "truth," indicates the quality of the assessment of noise levels in the environment. Accuracy is related to the purpose of the map, i.e. regional or local maps aiming to determine the global impact or to prepare action plans or implement a technical measure in a particular area. Uncertainty, which is a quantified aspect (or numerical presentation) of accuracy, implies the accuracy of calculated noise levels. Accurate predictions of environmental noise depend on the quality of various factors, associated with the acoustical modeling, input data, the characterization of the area, data processing, output organization, etc. Estimation of the uncertainty of a noise map needs to analyze all the partial uncertainties in order to establish a criterion for the total uncertainty.

Figure 6.45   Displaying the hot spots on a noise map, using Soundplan [71b].

At present, accuracy of noise maps has emerged as a major concern in Europe, where the noise mapping studies which are mandatory through the EU Directives, have already been completed in many countries after the second round of noise maps along with the action plans and some valuable experience has been gained.

6.5.1   Sources of Uncertainty in Noise Mapping

A number of aspects of uncertainty in noise maps are discussed by the experts. De Muer and Botteldooren remarked: "Uncertain information can be characterized by the partial knowledge of the true value of a statement. Imprecise information is linked to approximate information or not exact information" [72]. Uncertainty regarding the results obtained from noise maps is determined under the following topics. The measurement uncertainty is discussed in Section 5.6.

Input Data Uncertainty
Craven has emphasized that "if all the relevant input data is known with certainty and precision, a correct calculated result can be obtained even if it differs from the measured result" [73]. Accurate information derived from the noise maps is based strongly on the quality of the input data or data sets regarding the geometrical model of the site and the acoustic properties of sources, for instance, the emission level related to the operational conditions, etc.

As it is clear that some input data may have higher uncertainty than others for various reasons, the question that should be answered is: "If the input data is not absolutely correct, by how many decibels could the calculated noise level vary from the correct result?" [7]. The raw data is generally collected from various sources which may not be explicitly provided for noise mapping, therefore the data will need to be processed, simplified, combined with other data sets and digitized, before introducing it into the calculation model. For instance, when the general purpose GIS data is converted for use in noise mapping, the degree of accuracy which is necessary for simplification should be determined. Investigations have revealed that simplification of data can cause a level of uncertainty of 3–5 dB in many common situations. In traffic noise maps, using the default traffic volumes instead of the direct counts has increased the deviations of the calculated noise levels in L_{den} at individual receiver positions, between 8.8 and 11.7 dB (95%CI) [53]. As stated by some researchers: "The output can only be as good as the input."

Model Uncertainty

The quality of the model affects the accuracy of the noise map. Kragh stated that "the uncertainty of a predicted noise level is an interval in which the true value lies" [74]. Therefore, the success of a model in representing the real world should be checked by making accuracy assessments.

Uncertainty of the calculated results in terms of the noise levels and the population exposure assessments can be determined by comparing the outputs of other models that might give different values for the same situation. This fact is due to the discrepancies between the model algorithms or the conditions taken into account in the model, such as meteorological factors. It has been shown that the discrepancy could be as high as 4-6 dB when the interim and the national methods are employed as recommended by END [75].

The effects of different aspects on the accuracy of the model have been discussed in the literature, for example, it has been pointed out that the complexity of the calculation model, by increasing the number of physical factors involved in the algorithm, caused higher uncertainty and decreased the precision and transparency [76]. The uncertainty of the calculated noise level at a single point against the real value of the calculation could be influenced by different calculation strategies (i.e. the angle scanning or ray tracing), the interpolation techniques or the order of reflections selected. Researchers have shown that the different approaches in the modeling of certain features, such as accelerated traffic, crossings or tunnel entrances, could cause overestimation or underestimation regarding the consequences [76]. Clear description and proper understanding of the model features to be employed in noise mapping are important for the correct implementation and interpretation of results.

Uncertainty in the Execution (Running) of the Software

The algorithms of the computer programs managing the operational phase differ from each other, hence, the calculated results from different software packages may vary even though the same prediction model is employed. The uncertainty due to the software structure is associated with the following factors:

- *Calculation time:* For the completion of the task and the speed of operation.
- *Grid design:* Shape and size of the grid defining the number of receiver points and spacing. A finer set of grid points is essential for precision.

- *Height of grid points:* The standard height of 4 m may not be appropriate for areas with high rise buildings.
- *Segmentation of linear sources:* By constant view angle or equidistant approaches.
- *Interpolating techniques:* Applied to generate noise contours after calculations (the interpolated result does not represent the real value which is calculated for the same point).

Uncertainty due to calculation time and speed: An acceptable calculation time in relation to uncertainty can be determined according to DIN 45687 [26]. Software developers want to improve the quality of their products and to increase the efficiency of the operation by decreasing the running time. Reducing the uncertainty is also associated with the balance between the calculation times for each physical element [6].

In recent software packages, the calculation time has been reduced significantly by introducing different acceleration techniques, some of which are given below:

- Eliminating some of the physical factors (e.g. meteorological factors in building agglomerations) by assuming that they have a relatively small influence in strategic noise mapping.
- Applying a proper search radius which influences the duration of calculation. It was found that changing the search radius from 2000 to 500 m reduced the calculation time about 97%, however, it increased the overall uncertainty 4.25 dB up to 10.7 dB in L_{den} at specific receiver points [77].
- Restricting reflections (i.e. the order of reflections) within a specified distance between the source and the receiver.
- Specifying the maximum error (e.g. 0.5 dB).
- Employing approaches such as the "Parabolic Equation" (PE), the "Fast Field Program" (FFP), etc.
- Using task-sharing techniques, as explained in Section 6.3.8.

Uncertainties associated with the calculated results can be investigated by comparing the results with and without application of the acceleration procedures referred to above.

Exposure assessment uncertainty depends on the assessment of the number of buildings and the estimation of the population associated with the size of the search radius, in addition to the differences in determination of the façade levels. The influences of all these factors on the total uncertainty have been quantified in the document published by Defra in the UK [78].

6.5.2 Determination of Uncertainty

Investigation of the overall accuracy after noise mapping is important, especially for large projects and for the areas with complex noise sources, such as industrial activities. DIN 45687, dealing with this problem, presents a method on how to evaluate the accuracy of a noise map through a quality assurance system. The Standard describes a calculation procedure to achieve the best accuracy desired and the calculation of maximum error [26].

The Directive END, 2002, made accuracy evaluations of strategic maps mandatory and published a document entitled WG-AEN (also called GPG) in 2007 to determine the input data uncertainty [8, 28]. GPG has later provided a series of toolkits for a qualified accuracy statement in relation to the assessment of noise.

Generally, the uncertainty calculations are based on standard deviations. The errors, assumptions, and the uncertainties are presented in decibels, in line with the calculated results (Section 5.6). A simple scheme of the uncertainties to be determined in noise mapping and in the verification of noise maps is given in Figure 6.46a and an example of an uncertainty map is given in Figure 6.46b.

Since the type and accuracy of the input data for the noise model vary greatly, thus influencing the global uncertainty, the sensitivity analysis (SA) is performed, by dividing the uncertainties according to the inputs of the model. At present, a global SA framework to be integrated with the Cnossos noise prediction model has been developed to facilitate comparisons between the calculated exposures [79a].

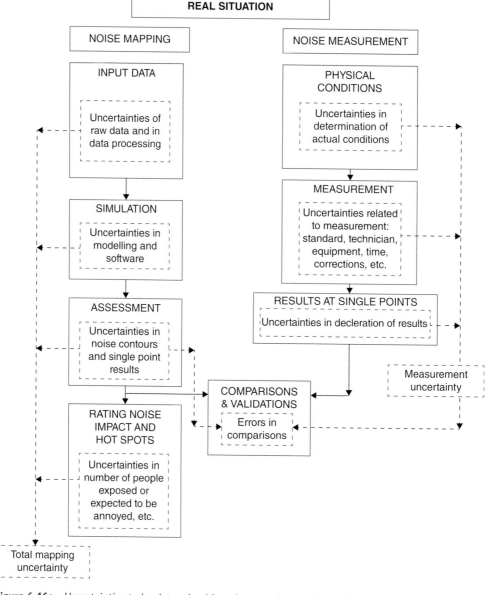

Figure 6.46a Uncertainties to be determined in noise mapping and in verification of noise maps.

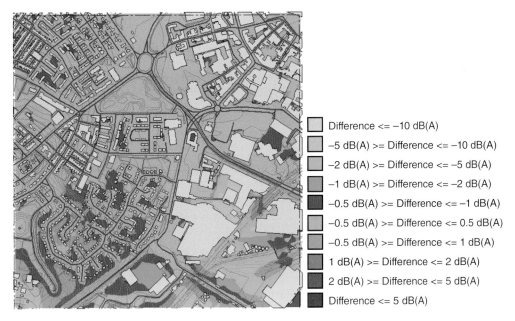

Figure 6.46b An example of an uncertainty map [53]. (*Source:* Permission granted by S. Shilton.)

Development of Toolkits

Based on preliminary recommendations by GPG, Defra, in the UK, conducted a research project to determine the potential factors in the acoustic accuracy of the calculated noise levels [78]. In this study, the aim was to quantify the accuracy statements in relation to the assessment of noise and to provide a guideline for estimating the uncertainty of the input data according to the assigned quality levels. Ultimately, the influence of the quality of input data upon the overall uncertainty of the calculated results was revealed by establishing the relationships between the quality of the input data and the quality of the calculated data in assessing noise impact from road and railway noises.

In this study, a wide range of input data was collected, categorized, and analyzed to show how each data set affected the overall uncertainty of the map model by employing the uncertainty propagation analysis (UA) in addition to SA [79b–81]. The order of importance of some input parameters was presented, e.g. for the calculation of noise emission of road traffic noise. Briefly, the estimation of uncertainties in the model inputs in relation to model outputs was accomplished. The consequences of the project were converted into the toolkits, later adopted in WG-AEN Version 2 in 2007 [8].

Benefits of toolkits:

- Toolkits contain uncertainty levels for the input data declared, as decibels.
- The toolkits are guidelines to determine the acoustic accuracy of the calculated noise levels.
- The toolkits indicate the quality of the input data used for the assessment of noise levels for road traffic and railway noises through noise maps.
- Accuracy of input data can be determined through the toolkits, according to how the input data is obtained for the noise mapping.
- Toolkits give knowledge about the potential uncertainty when the input data is inaccurate and insufficient.

- Toolkits guide the necessary steps, especially while conducting large-scale mapping projects.
- Toolkits provide a complete understanding and evaluation of input uncertainties with a statement of acoustic accuracy.

As stated in WG-AEN, "Toolkits are helpful to promote further investigations into the various technical aspects affecting the accuracy of the results" [8]. In the second version of the document, the quantified accuracy statements presented in the tables are indicated by symbols using different colors. Various symbols are used in the toolkits, to rate the data in terms of complexity, accuracy, and cost within the ranges given below (Table 6.4):

- *Complexity of data:* From simple to sophisticated. The color code of complexity changes from light blue to dark blue (blue code gradation).
- *Accuracy level:* From low to high.
- *Accuracy code:* Between (>5 dB) and (<0.5 dB). Color code of accuracy changes from yellow to green.
- *Cost code:* From inexpensive to expensive. Color code of cost changes from ochre to brown (red code gradation).

The titles of the toolkits associated with input parameters are given below:

Toolkit 1: Area to be mapped (agglomeration, major road or railway, major airport)
Toolkit 2: Road traffic flow
Toolkit 3: Average road traffic speed
Toolkit 4: Composition of road traffic
Toolkit 5: Road surface type
Toolkit 6: Speed fluctuations at road junctions
Toolkit 7: Road gradient
Toolkit 8: Sound power level of trams and light-rail vehicles
Toolkit 9: Train (or tram speed)
Toolkit 10: Sound power level of industrial sources
Toolkit 11: Ground elevation close to the source
Toolkit 12: Assignment of population data to residential buildings
Toolkit 13: Cutting and embankments
Toolkit 14: Barrier heights near roads
Toolkit 15: Building heights

Table 6.4 Quantified accuracy implications in WG-AEN (2007) [8].

Colour code to rate Tools					
complexity	colour code	accuracy	colour code	cost	colour code
simple	△	low	> 5 dB	inexpensive	△
-		-	4 dB	-	
-	◇	-	3 dB	-	◇
-	⬠	-	2 dB	-	⬠
-		-	1 dB	-	
sophisticated	⬡	high	< 0.5 dB	expensive	⬡

Toolkit 16: Sound absorption coefficients a_r for buildings and barriers

Toolkit 17: Occurrence of favorable sound propagation conditions

Toolkit 18: Humidity and temperature

Toolkit 19: Assignment of population data to residential buildings

Toolkit 20: Determination of the number of dwelling units per residential building and the population per dwelling unit

Toolkit 21: Assignment of noise levels to residents in dwellings in multi-occupied buildings

6.5.3 Required Accuracy

The level of uncertainty to obtain a sufficiently accurate noise map should be specified in terms of the accuracy levels for each input data. Decisions about the required accuracy prior to the noise mapping are important also for the data suppliers in delivering the appropriate input data and in processing the data for mapping projects.

In the abovementioned study conducted by Defra, the required accuracy percentages and the ranges were determined after sensitivity testing for the railway traffic parameters (e.g. for train speed, flow, and source height) [81]. Eventually a proposal has been presented for a dataset specification which is suitable for the purpose of noise mapping and accepted in the WG-AEN as accuracy recommendations. The five accuracy groups for the map outputs are defined in terms of the potential errors in outputs depending on the input data quality [8]:

Group A: Very detailed input data (suitable for detailed calculation and for validation).

Group B: Each element produces a potential output error; less than 1 dB (best standard recommended by END).

Group C: Each element produces a potential output error; less than 2 dB

Group D: Each element produces a potential output error; less than 5 dB (pass standard).

Group E. Data is not confident and it is recommended the data quality be improved.

WG-AEN has pointed out that the input data should not be unnecessarily complex (or detailed). For example; if Group C (i.e. <2 dB) is chosen or calculated for each input dataset, the total uncertainty of the model results becomes of the order of 5 dB.

6.5.4 Validation of Noise Maps

One of the shortcomings of noise maps, except for the accuracy, is the validity to be confirmed through different techniques. As defined, "Validity of relationships (i.e. prediction of noise levels) means that there is no systematic error or bias in their estimation, whereas the uncertainty implies random errors that can be described by the confidence intervals" [82]. The random error might be caused because of the wrong displacement of the noise contours.

The validity of the calculated noise levels in noise maps can be investigated through the following studies:

1) Comparisons are made between the calculated and the measured results. The field measurements are conducted by using the technical standards as explained in Section 5.5 and in Section 6.3.3 [27]. Discrepancies are inevitable between the calculated and the actual results which are caused by a number of factors, however, they should remain within the acceptable limits based on the statistical analyses. Since the noise maps give long-term average levels while the measured values represent the actual conditions, the discrepancies that might occur are due to the differences between the source operations or between the meteorological conditions

during measurements. Therefore, some corrections are necessary to apply to the measured levels when comparing the results. This process should be repeated for various receiver locations or in different environments and under different conditions to determine the uncertainty. After the results are corrected, the preparation of a calibrated noise map is useful to display the validity. For calibration purposes, some techniques, such as "reverse engineering," have been recommended for the emission models [75].

The inaccuracy related to field measurements is another issue to be considered, according to the measurement standard, i.e. ISO 1996-2. It has been argued that if the repeated site measurements for the same source on different days result in 5 dB(A) difference, it can be accepted as reasonable, considering the meteorological and other differences.

As a verification process, the level differences between the calculated and measured data at each measurement location can be statistically analyzed to investigate the average level differences, and also the standard deviations by considering the measurement and model uncertainties [83].

2) The calculated results can be compared with the laboratory experiments applying the simulation techniques explained in Section 5.8. However, the result from a scale model study represents the stable conditions, mostly neglecting the wind effect, and the accuracy is dependent mainly on the success of the simulation.

3) The results obtained from a map can be compared with those calculated for a specific receiver point by using a mathematical model with the same numerical values of the parameters involved. However, it should be considered that the theoretical models might have certain assumptions and boundary conditions that make the comparisons rather difficult or the number of parameters might be limited.

4) Results from a noise map using a specific computer program can be compared by employing the other programs or software packages with the uncertainty similar to the model.

6.6   Reporting the Results of Noise Maps

Commercial mapping software programs provide a plot designer to present the calculation results to the community and responsible authorities to attract their attention as much as possible. The information should be given in graphical and written formats with sufficient detail. Generally, the technical reports, either paper-based reporting or electronic submissions, should contain the following information:

1) Aim of noise mapping.
2) Area (geographical description, land use, description of settlement with size, location, type of buildings, number of people living, etc.).
3) Characteristics of noise sources:
   - layout
   - list of sources
   - acoustic data (emission)
   - reference for the acoustic data obtained.

4) Physical environmental data (meteorology parameters to be given as statistical values and ground conditions).
5) Height of grid points and specific receiver characteristics.
6) Time for mapping and noise descriptors.

7) Map scale and grid size.
8) Outputs:
   - source emission list
   - noise levels at specific receivers (list)
   - areas (km^2) where noise levels exceed certain levels (i.e. noise zones)
   - number of buildings and population in each zone
   - façade levels for specific buildings or all buildings
   - horizontal and vertical maps with 3D presentations (enriched by additional informative maps, such as maps for seasons, favorable and unfavorable conditions, etc.)
   - statistics about the quiet façades
   - numbers and locations of the buildings with higher insulation
   - uncertainty value calculated for the map
   - results of the validity study.

9) Comparisons with the standard levels (i.e. noise limits)
10) Problematic areas: Hot spots and locations where noise control measures are necessary (to be indicated also in action plans)
11) Other information:
   - calculation models used for predictions
   - if the buildings are insulated, type of constructions.

Electronic Reporting for Noise Maps

Environmental Noise Directive (END) Annex 6 sets out the data to be sent to the European Commission. Aiming to provide uniform data submission and distribute the information among the EU countries, the Electronic Noise Data Reporting Mechanism (ENDRM), was developed in 2007 and updated in 2012 [84a]. It established several communication systems, like a delivery process called Reportnet, the Infrastructure for Spatial Information for Europe (INSPIRE), and the Shared Environmental Information System (SEIS), to store information in electronic databases and to share the environmental data. The reporting mechanism includes strategic noise maps, action plans, national limit values for major roads, railways, airports, and agglomerations, noise control programs in the past and current action plans (Chapter 9). ENDRM requests information for major sources which are obligatory in the Directive and the submission of noise maps on which only the 55 and 65 dB(A) L_{den} contours are displayed.

The document covers the END reporting requirements, a model for data handling, descriptions of table formats, visualization of the reports, and the relations from national databases. The templates are prepared in Microsoft Excel and Word, which are commonly used in Europe. The quality assessment (QA) rules are described to check the individual reports. The data submitted by the EU Member States is managed by the European Environment Agency (EEA) using the Reportnet system. GIS information is obligatory in order to combine all the information throughout Europe. The technique has provided a great development for noise mapping [84b].

The impact assessments to be submitted to the EU are the percentage of the population exposed per noise zone. The EEA collects the results of all the countries in a graphical format displaying the percentage of the population exposed to noise and the agglomerations in a country (cities) [58].

At present, the updated maps, the so-called "interactive maps" or the "intelligible noise maps," can be viewed on the webpages prepared by some countries. For example, Defra, which is the responsible authority in the UK, has prepared intelligible noise maps for different regions of

London, also providing the possibility of zooming in to the required parts of the maps to observe the noise level variations in detail (www.services.defra.gov.uk) (Figure 6.48).

6.7 Examples of Noise Mapping Practices

Nowadays noise mapping has become a business and field of expertise. Many projects have been continuing throughout the world, especially in Europe, after the EU Directive obliged the member countries to finalize noise mapping for every settlement whose population exceeds 250 000, by 2009, and later 150 000, in the second round. Most of the mapping projects and action plans (as explained in Chapter 9) are submitted to the EU Commission and the outcomes are debated also on scientific platforms while the third round was due at the end of 2017. Some examples of noise mapping projects from various countries are available on the webpages, which are open to the public, or, through special contacts, will be outlined below to give information about the visual techniques used in the past few years and at present.

6.7.1 Large-Scale Maps (Regional Maps)

Urban or regional noise maps covering transportation noise mainly are prepared for noise pollution at present or also to develop future noise management strategies. Some examples are shown in the figures according to the countries:

United Kingdom: Interactive noise map and overall traffic noise map for London (Figures 6.47 and 6.48) [85]

USA: New York City noise map (Figure 6.49)[86]

Germany: Noise map for Munich city traffic (Figure 6.50a), for Bayern traffic, 2017 (Figure 6.50b) [87], and noise map for major railways (Figure 6.51) [71a];

France: Noise map for Paris (Figure 6.52) [88];

Italy: Flight pattern map for Milan airport (Figure 6.53);

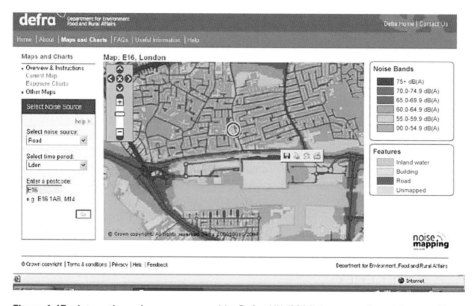

Figure 6.47 Interactive noise map prepared by Defra, UK (2016) (www.services.defra.gov.uk).

Figure 6.48   Greater London traffic noise map [85].

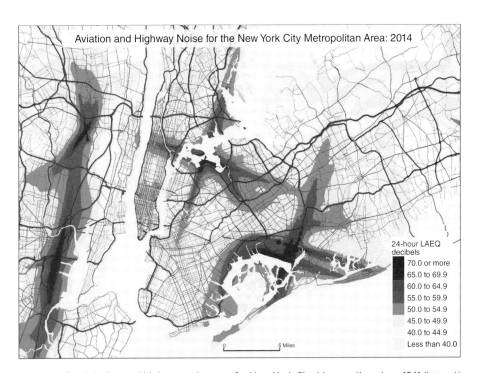

Figure 6.49   Aviation and highway noise map for New York City Metropolitan Area [86] (https://www.bts.gov/newsroom/national-transportation-noise-map)

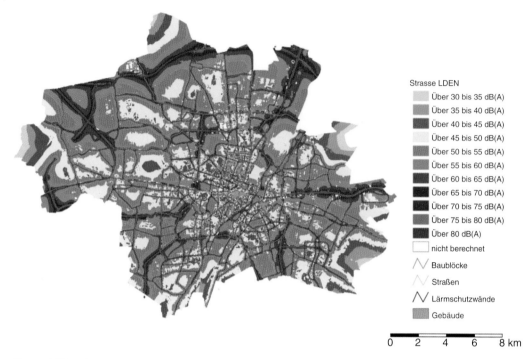

Figure 6.50a A large-scale noise map for L_{day}: München, Germany. (Permission granted by D. Holzmann, Referat für Gesundheit und Umwelt, Landeshauptstadt, München.

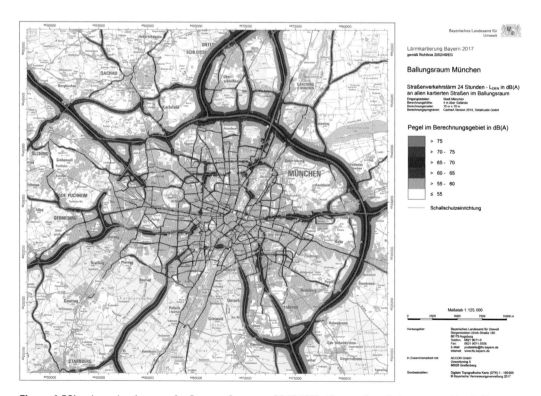

Figure 6.50b A road noise map for Bayern, Germany, 2017 [87]. (*Source:* Permission granted by S. Bauer, Bayerisches Landesamt für Umwelt.)

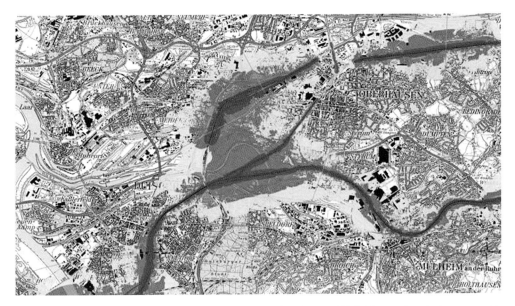

Figure 6.51 Railway noise map for German Eisenbahn-Bundesamt [71a]. (*Source:* Permission granted by Soundplan.)

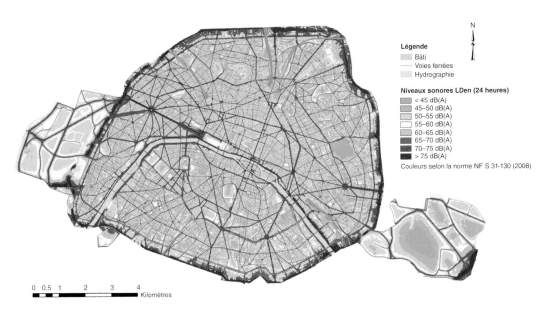

Figure 6.52 Noise map for Paris, France [88]. (*Source:* Mairie de Paris.)

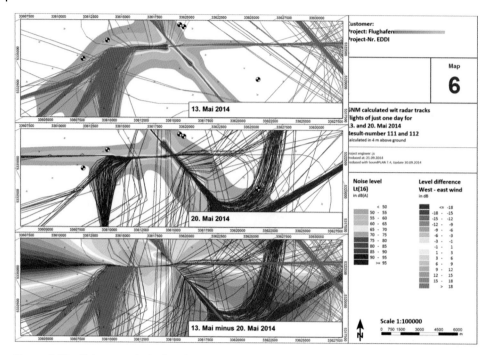

Figure 6.53 Noise maps including the flight patterns in Milan airport calculated with radar tracks. (*Source:* provided by Soundplan.)

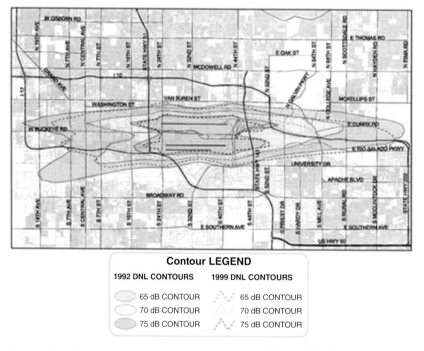

Figure 6.54 Airport noise map prepared for Phoenix, USA, in 1999 [89].

Figure 6.55 Airport noise map for Chicago O'Hare airport, USA, displaying the changes in various years [90]

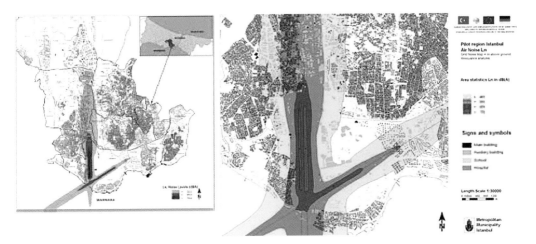

Figure 6.56 Presentations of pilot noise map for Ataturk airport, Istanbul, 2014 [91].

(a)   (b)

Figure 6.57 (a and b) Photorealistic 3D noise model for Hong Kong, 2011 [45].

the USA: Airport noise maps (Figures 6.54 and 6.55) [89, 90];
Turkey: Ataturk Airport pilot noise map (Figure 6.56) [91];
China: Hong Kong noise map (Figures 6.57a and 6.57b) [8, 45].

6.7.2 Small-Scale Maps (Local Maps)

For local urban environments, the more detailed noise maps are prepared for noise control decisions in layout planning or building design. Some examples are given in Figures 6.58–6.66.

Highway noise at different residential areas Figures 6.58 a and b for a Trans-European Motorway connection road (2000) [92];
Trans-European Motorway noise in a residential area with high-rise buildings (Figures 6.59 a–e) (2014);
Local noise maps for urban traffic noise (urban noise for traffic): for Buca Izmir Municipality (2008) (Figures 6.60a–b);
Local noise maps for urban traffic noise for the Beşiktaş area, Istanbul (2008) (Figures 6.61a–b);
Detail noise maps around a school complex in Istanbul (2000) (Figures 6.62a–b) [92];
An example of a detail noise map around buildings (Figure 6.63) [71a];
A detail noise map for equipment noise on the terrace of a shopping mall (Figure 6.64).

(a)

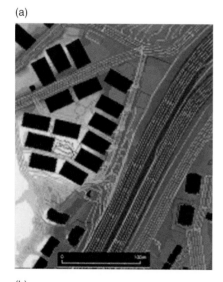

(b)

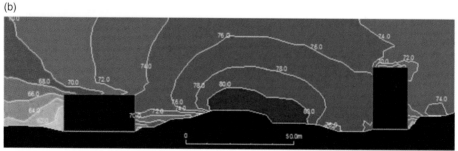

Figure 6.58   (a and b) Horizontal and vertical noise maps in the same area as Figure 6.58 after a barrier design. Note: h = 2 m. [92]

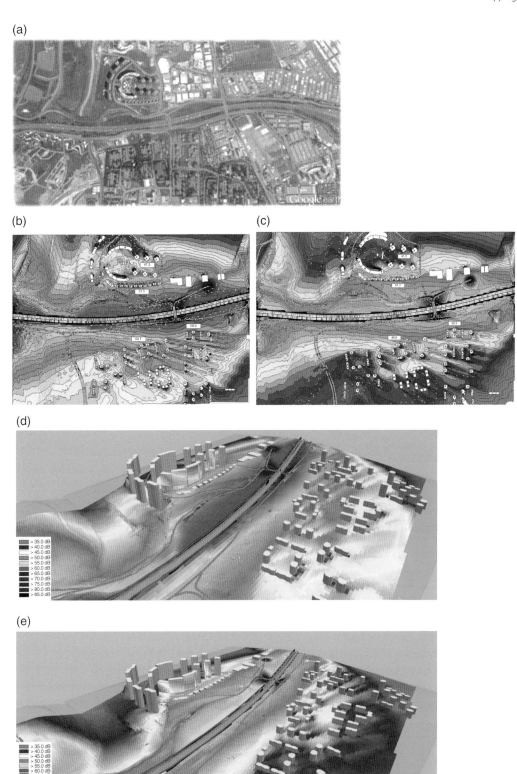

Figure 6.59 (a–e) A noise map project for a residential area near the Trans-European Motorway in Istanbul, 2014. (a) Google Earth map (b) L_{den} (c) L_{night} (d) 3-D view of building and land evaluations L_{den} and (e) L_{night}. (*Source:* dBKES Eng.)

(a)
(b)

(c)

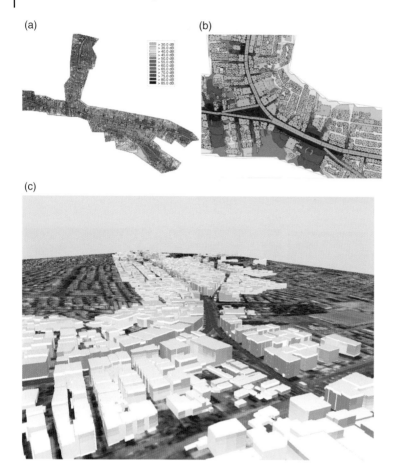

Figure 6.60 (a–c) Noise map prepared for Buca, Izmir, with different scales. (a) and (b) horizontal maps; (c) 3D view with building evaluations. (*Source:* prepared by Butech, Bahcesehir University, 2008.)

6.7.3 Noise Mapping for Specific Environmental Noise Sources

Strategic noise maps may cover sources other than those required in the Directive and WG-AEN, such as wind farms, harbors, shooting ranges, mixed-used areas, etc. Some of them, like recreational activities that can be performed in open areas and in semi-open facilities, may need different approaches in source characterization and modeling with use of specific noise indexes considering on–off times, seasons, etc. It is widely accepted that these sources should also be included in the strategic maps of agglomerations.

Examples of the noise maps prepared for wind farms are presented in Figures 6.65a and b [93, 94]. In addition, the noise generation from seaports containing various chain operations with the heavy equipment and traffic (also railway traffic) is modeled as in Figure 6.66 [95]. Lastly, the noise map depicting shooting range noise contours in and around a Police Academy campus, is presented in Figure 6.67.

6.7.4 Recent advances in noise mapping

Following the demands on noise impact studies and legislative obligations in many countries, noise mapping has become widespread worldwide, in line with the remarkable improvements, some of which are mentioned in this chapter.

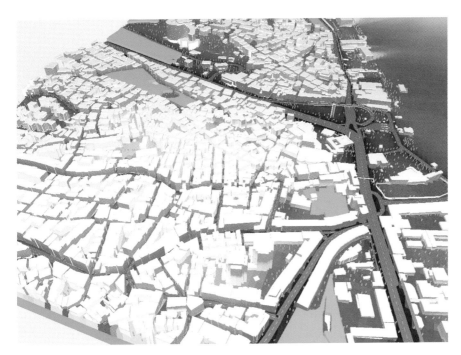

Figure 6.61a   3-D view of Beşiktaş, Istanbul, 2008. (*Source:* prepared by Butech, Bahcesehir University 2008.)

Figure 6.61b   Horizontal noise map of the same area given in Figure 6.63a. (*Source:* prepared by Butech, Bahcesehir University 2008.)

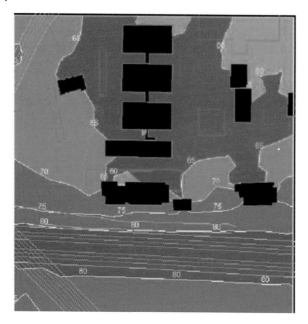

Figure 6.62a Detail of horizontal map for the school buildings along a highway, by using Mithra 2000 [92].

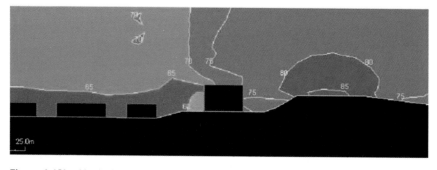

Figure 6.62b Vertical map for the same buildings as in (a) [92].

As regards the technical features; 3D noise mapping is improved to make more accurate assessment of noise impact also with the aid of Google mapping techniques integrated with 3D-viewers. The calculation efficiency is maximized and new interpolation processes are employed to reduce the calculation time without jeopardizing the accuracy of the results. The software tools are increased in modeling the environment and describing the source parameters correctly. Editing graphics, reporting, multitask operations are facilitated and quality of graphics is much improved.

It has been possible tracing noise levels even in local communities within countries, through interactive maps displayed on the websites. As an example; the comprehensive noise maps at state and county levels, have been created by the U.S. Bureau of Transportation Statistics BTS, for public attention, of which a segment is given in Figure 6.68 [86]. So-called The National Transportation Noise map is also incorporated with the BTS Geospatial Data Catalog to supply information. Besides, the U.S. Federal Aviation Agency, FAA website provide links for airport noise and land use information including Noise Exposure Maps, to assist the policy makers [96].

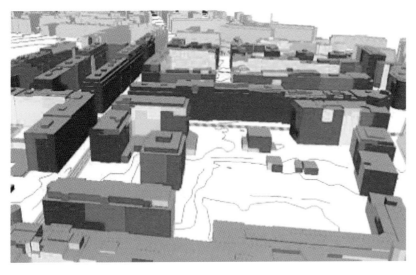

Figure 6.63 Figure 6.63 An example of an advanced noise mapping in 3D, using MithraSIG [29b]. (*Source:* Permission granted by CSTB.)

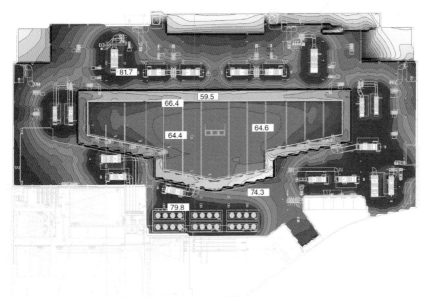

Text Single	Text interval	Color	Line
>= 30.0	... < 35.0		
>= 35.0	35.0 <= ... < 40.0		
>= 40.0	40.0 <= ... < 45.0		
>= 45.0	45.0 <= ... < 50.0		
>= 50.0	50.0 <= ... < 55.0		
>= 55.0	55.0 <= ... < 60.0		
>= 60.0	60.0 <= ... < 65.0		
>= 65.0	65.0 <= ... < 70.0		
>= 70.0	70.0 <= ... < 75.0		
>= 75.0	75.0 <= ... < 80.0		
>= 80.0	80.0 <= ...		

Figure 6.64 A noise map on the terrace roof of a shopping mall accommodating various HVAC equipment, with a skylight of 2 m height in the middle By using CADNA-A) (*Source:* Kurra, 2016).

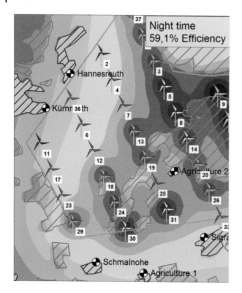

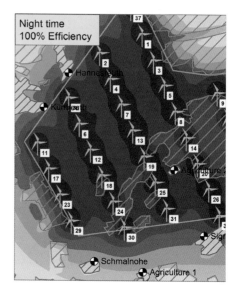

Figure 6.65a An example of wind farm noise mapping, night-time, working at 59% efficiency [93]. (*Source:* Soundplan.)

Figure 6.65b An example of wind farm noise mapping, night-time, working at 100% efficiency [94]. (*Source:* Soundplan.)

Figure 6.66 Seaport noise map: Port of Livorno, Italy (including all sources in L_{den}) [95]. (*Source:* Permission granted by T. Van Breemen, Project NoMEPorts.)

In Europe, as a result of END, the noise fact sheets have been published by EEA based on the third noise mapping round. It is expected that the use of Cnossos model to harmonize future noise mapping assessments, will make easier to compare data across the countries and to take common actions towards noise control. The Dynamap project (i.e. application of real-time noise mapping) as mentioned in Section 6.3.9, is at the final step "where the accuracy, reliability and sustainability of the system are being assessed" through various implementations in Europe [97]. New techniques are experimented in order to detect and eliminate anomalous noise events during monitoring [98].

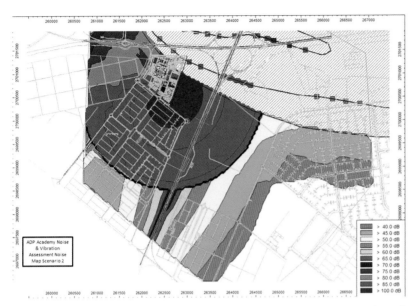

Figure 6.67 Shooting range noise contours during practice in the open air, from the assessment report for ADP Police Academy-NV (*Source:* S.Kurra, Hawk Technology, UAE, 2014)

Figure 6.68 Interactive noise map for the USA [86] (*Source:* the Bureau of Transportation Statistics, BTO)

One of the popular issues to facilitate noise mapping, is the use of audio recordings of mobile devices, i.e. smart phones as a low- cost and efficient method. For example, in Belgium, the mobile technology is implemented during collection of environmental noise measurement data for noise mapping, through "participatory sensing campaigns" [99]. Another study has provided the evidence of achieving same accuracy as the standard noise mapping techniques with the aid of intensive calibration tests [100].

In France, acquisition of the hourly noise maps based on measurements through smart phones, and a data assimilation method combining the simulation maps and measurements, are introduced in the case studies. The analysis maps are derived to reduce errors [101] and the spatial interpolation technique for the measured data is developed [102].

In Korea, the studies on real-time noise monitoring and collecting data via mobile devices aiming sustainable method in noise mapping, are undertaken by claiming "to overcome the shortcomings of existing station-based noise monitoring systems in urban areas" [103]. However, issues regarding accuracy of smart phone sound sensors, standardization etc. are still discussed. In China, a calibration system for the smart phones, along with the technique for data transmission and the interpolation method to improve precision of such noise maps, have been developed [104].

Currently, a wide range of new publications, specific journals, books, guidelines and the articles have emerged regarding noise mapping and monitoring practices, some concerning the efficiency of noise management systems on the basis of noise maps.

6.8   Conclusion

Noise maps are defined as "Robust visual tools for sound which is heard only" [2]. In general, they are developed to investigate the noise pollution in magnitude; size, and extension, especially in dense urban areas, aiming to protect the quiet environments. The outcomes of noise mapping are exploited for various purposes. They give information about temporal and spatial variation of sound pressure levels under certain physical environmental conditions and play an important role in land use and urban planning. It is possible to determine the unacceptable exposure areas defined as "hot spots" by applying various techniques and specified criteria. The quiet areas should be preserved or newly created to keep the human right for living in healthy environments. Based on the noise maps, the need for sound insulation for buildings can be determined and the applicable solutions for noise control can be integrated into regulations, the development of noise policy, and action plans (Chapters 8 and 9).

Promotion of comparable noise maps depends on compliance with the technical standards and the establishment of a data management system. The EU Commission pays a great deal of attention to the collection and categorization of data for noise mapping, particularly source emission data to be used in computation models and the conversion of existing databases to the new models proposed in the Directive. The quality of noise maps has always been a great concern and, at present, the basis of accuracy valuation has been structured under the concept of uncertainty.

Preparation of a noise map, including collecting and organizing input data, conducting computations for a specific noise source or many of the sources existing in an area and proper presentation of results in terms of the noise level distribution within regional or local maps using an appropriate scale, interpretations to derive solutions efficiently for action plans, definitely require certain experience, skilled and trained personnel, sufficient budget and time. There are further issues to be considered in noise mapping:

- knowledge of the prediction model prior to noise mapping;
- knowledge of the software, i.e. input requirements, variables, restrictions, descriptors, calculation parameters, capacity, and organization of the database, and efficient design of the interface in relation to the algorithm;
- the validity and applicability of the calculation models and the computer programs available in global markets under different names, to be proven in the country's own conditions.

Although the EU Directive mandates noise mapping only for the major noise sources (roads, railways, aircraft, and industry) and for the agglomerations defined in size, other temporary and stationary noise sources can also be taken into account in the strategic noise maps, e.g. seaway traffic with harbors, wind turbines, amusement centers, shooting ranges, temporary constructions, etc. Also, seasonal noise maps are important in touristic areas in providing the forecasts regarding the increased surface and air traffic.

Finally, as Probst stated, "Noise mapping is a never-ending war" [1].

References

1 Probst, W. and Huber, B. (2003). The sound power level of cities. *Sound and Vibration* 37 (5): 14–17.

2 Freeborn, P. (2003). Constructing a noise map. *Acoustics Bulletin* Nov./Dec (4).

3 Wing, L.C., Kwan, L.C., and Kwong, T.M. (2003). *Visualization of Complex Noise Environment by Virtual Reality Technologies*. Environmental Protection Department www.science.gov.hk/paper/EPD_CWLaw.pdf.

4 Probst, W. and Petz, M. (2007). Noise mapping, hot spot detection and action planning: an approach developed in the frame of the EC-Project Quiet City, Inter-Noise 2007, Istanbul, Turkey, paper no. 214.

5 (a) EU (2015). Directive EU 2015/996 of 19 May 2015 establishing common noise assessment methods according to Directive 2002/49/EC of the European Parliament and of Council; (b) Kok, A. (2019). Refining the CNOSSOS-EU calculation method for environmental noise, Inter-Noise 2019, Madrid (16–19 June).

6 Probst, W. and Bernd, H. (2007). How to evaluate the accuracy of a noise map? 19th International Congress on Acoustics, Madrid (2–7 September).

7 Olny, X., Vincent, B., Carra, S. et al. (2015). From regional strategic maps to miscroscopic scale models: multi-scales approaches to improve the assessment of the exposure to pollutants associated with transportation, Euronoise 2015, Maastricht, The Netherlands.

8 Anon (2007). Good Practice Guide for Strategic Noise Mapping and the Production of Associated data on Noise Exposure, Version 2, 13.01.2007, European Commission Working Group, Assessment of Exposure to Noise (WG-AEN). www.europa.eu.int/comm/environment/noise/home.htm.

9 CETUR (1980). Guide du Bruit des Transports Terrestres Prevision des Niveaux Sonores, Ministère de l'Environnement et du Cadre de Vie, CETUR.

10 Kurra, S. and Yılmaz, S. (1985). Airport noise control in design of settlements and Yeşilköy airport. Technical Bulletin, Turkish Scientific and Technical Research Establishment (TUBITAK), Building Research Center, No. 21.

11 Kurra, S. (1978). A computer model for determination of criteria units against traffic noise while using buildings as barriers. PhD thesis. Istanbul Technical University.

12 Kurra, S. (1980). A computer model for predicting sound attenuation by barrier buildings. *Journal of Applied Acoustics* 13: 331–355.

13 Kurra, S. and Crocker, M. (1982). A scale model experiment of noise propagation at building sites: report of a research project conducted in Purdue University, Ray Herrick Laboratory, February.

14 Kurra, S. (1980). A computer model for rating performance of building external elements against environmental noises, *Proceedings of Science Congress*, organized by Turkish Scientific and Technical Research Establishment (TUBITAK).

15 Tsukiyama, A., Takagi, N., Matsui, M. et al. (1991). Analysis of environmental noise in Nagano City: a study in the fluctuation for 7, preprint of National Conference of Architectural Institute of Japan.

16 Verband der Automobilindustrie (1978). Urban traffic noise. E. V. VDA.

17 Anon (1983). Dienstanweisung betreffend Lärmschutz an Bundesstrassen, Bundesministerium für Bauten und Technik.

18 Storeheir, S.A., Hunstadt, K., and Bornes, V. (1995). TSTØY calculation and analysis of road traffic noise in a digital terrain model, Norwegian Public Road Administration, SINTEF.

19 NOISELAB (1993). *Noise Laboratory in PC, Delta Acoustics and Vibration*. Danish Acoustical Institute.

20 Brüel & Kjaer (1980). B&K course material, Brüel & Kjaer.

21 http://www.bksv.com/products/environmentalnoisemanagement/noisemanagementsolutions/noisepredictionsoftware/7810predictor.aspx.

22 http://www.datakustik.com/en/products/cadnaa/modeling-and-calculation/presentationof-results/horizontal-noise-maps.

23 Anon (1995). NORTIM Norwegian topography integrated model, SINTEF Delab, May.

24 FAA (2015). Integrated Noise Model, INM version 7.0 2015: Aviation Environmental Design Tool (AEDT), May. https://www.faa.gov/about/office_org/headquarters_offices/apl/research/models/inm_model.

25 ISO (1987). ISO 1996-2:1987 Acoustics: Description and measurement of environmental noise. Part 3: Acquisition of data pertinent to land use.

26 DIN (2006). DIN 45687:2006–05 Acoustics: Software products for the calculation of the sound propagation outdoors. Quality requirements and test conditions (English title), 2006–05.

27 ISO (2007). ISO 1996-2:2007 Acoustics: Description, measurement and assessment of environmental noise. Part 2: Determination of environmental noise levels.

28 EU (2002). Directive 2002/49/EC of the European Parliament and of the Council of 25 June 2002, Relating to the Assessment and Management of Environmental Noise (The Environmental Noise Directive – END).

29 (a) Mithra (1998).Version 4.0, The prediction of outdoor acoustics, User's manual, 01dB-Stell, France;
(b) CSTB Mithra Sig 4.0 The Acoustic mapping software. Centre Scientifique et Technique du Bâtiment (CSTB). http://editions.cstb.fr/Products/Mithra-EN.html.

30 Probst, W. (2008). Multithreading, parallel computing and 64-bit mapping software: advanced techniques for large-scale noise mapping. http://www.Paper_AdvancedNoiseMappingSoftware_Probst_DAGA08.pdf.

31 Cadna-A (2008). State-of-the-art noise prediction software Reference Manual Release, 3.8, Datakustik GMBH.

32 ISO (1994). ISO 3744:1994 Acoustics: Determination of sound power levels of noise sources using sound pressure: engineering methods in an essentially free field over a reflecting plane.

33 ISO (1995). ISO 3746:1995 Acoustics: Determination of sound power levels of noise sources using sound pressure: comparison method in situ.

34 RAYNOISE (1985). *System for Geometrical Acoustics*. Numerical Integration Technologies.

35 ODEON (2008). Version 9.1, User Manual, Industrial, Auditorium and Combined Editions.

36 Cadna/SAK (2016). Computer program for calculation of noise levels from emission data of the machines. http://www.datakustik.de.

37 Navcon Engineering Network (1994). Theory and applications of indoor noise model (VDI 3760), SoundPlan Tech Note, Indoor Factory Noise.

38 ISO (2010). ISO 10140-2:2010 Acoustics: Laboratory measurement of sound insulation of building elements. Part 2: Measurement of airborne sound insulation.

39 ISO (2016). ISO 16283-3:2016 Acoustics: Field measurement of sound insulation in buildings and of building elements. Part 3: Façade sound insulation.

40 ISO (2005). EN 12354 (ISO 157124:2005) Building acoustics: Estimation of acoustic performance of buildings from the performance of elements. Part 4: Transmission of indoor sound to the outside.

41 ISO (1994). ISO 8297:1994 Acoustics: Determination of sound power levels of multisource industrial plants for evaluation of sound pressure levels in the environment. Engineering method.

42 Gabillet, Y. (1990). Un méthode inverse de recherche de rayons, le logiciel, MITHRA, CSTB, Cahier 2444, octobre.

43 Probst, W. and Rabe, I. (2008). Techniques to accelerate noise mapping calculations for large areas and cities, ICSV 15, Korea.

44 Wenninger, B. (2015). A color scheme for the presentation of sound immision in maps: requirements and principles for design, Euronoise 2015, Maastricht, The Netherlands.

45 Law, C.W., Lee, C.K., Lui, A.S.W. et al. (2011). Advancement of three-dimensional noise mapping in Hong Kong. *Applied Acoustics* 72: 534–543.

46 Wei, W., Botteldooren, D., and Van Renterghem, T. (2014). Monitoring sound exposure by real time measurement and dynamic noise map, Forum Acusticum, Krakow, Poland (7–12 September).

47 Schweizer, I., Barti, R., Shulz, A. et al. (2011). Noise map: real time participatory noise maps. http://msr-waypoint.com/en-us/um/redmond/events/phonesense2011/papers/NoiseMap.pdf.

48 (a) Belluci, P., Peruzzi, L., and Zambon, G. (2016). Life Dynamap: an overview of the project after two years working, Inter-Noise 2016, Hamburg, Germany;
(b) Bellucci, P., and Peruzzi, L. (2019). Life Dynamap: accuracy, reliability and sustainability of dynamic noise maps, Inter-Noise 2019, Madrid (16–19 June);
(c) Zambon, G., Roman, H.E., Smiraglia, M. et al. (2018). Monitoring and prediction of traffic noise in large urban areas. Review. *Applied Sciences (Switzerland)* 8 (2).
(d) Zambon, G., Benocci, R., Bisceglie, A. et al. (2017). The Life Dynamap project: towards a procedure for dynamic noise mapping in urban areas. *Applied Acoustics* 124: 52–60.

49 Bisceglie, A., Benocci, R., Edwardo, H. et al. (2017). Dynamap Project: Procedure for noise mapping updating in urban area, Inter-Noise 2017, Hong Kong (27–30 August).

50 Datakustik (2018). Bastian: The Building Acoustics Planning System. http://www.datakustik.com/en/products/bastian.

51 Insul (2018). Predicting sound insulation. http://www.insul.co.nz.

52 Rodriguez-Manzo, F.E. and Vargas, E.G. (2015). Moving towards the visualization of the urban sonic space through soundscape mapping, ICSV22, Florence, Italy, paper no. 62.

53 Shilton, S.J., Stimac, A., and Nota, R. (2009). Equivalence within noise mapping projects, Euronoise 2009, Edinburgh.

54 McKenzie, F., Odell, S., and Hewlett, D. (2017).Tranquility: an overview, Technical Information Note 01, Jan. 2017. Landscape Institute Information Note.

55 Nurzynski, J. (2016). How to exploit effectively noise maps: a proposal for acoustic categorization of residential areas, ICSV23 Congress, Athens, Greece, paper no. 860 (1–14 July).

56 Kang, J. (2007). *Urban Sound Environment*. Taylor & Francis.

57 Anon (1996). The Green Paper on Future Noise Policy, COM(96) 540, European Commission, November.

58 EEA (2014). Noise in Europe, Report 10/2014. European Environment Agency.

59 Von Gierke, H.E. (1977). *Guidelines for Preparing Environmental Impact Statement on Noise*. National Research Council.

60 Von Gierke, H.E., Yaniv, S.L., and Blackwood, L.B. (1979). Environmental impact statements. Chapter 45. In: *Handbook of Noise Control*, 2e (ed. C.M. Harris). McGraw-Hill.

61 Bigota, A., Boutinb, C., Davida, A. et al. (2012). Construction of an average indicator of potential noise exposure and its sensitivity analysis in Marseilles city (France), Acoustic, Nantes, France (23–27 April).

62 EC (2002). EC-WG/2: Position paper on dose-response relationships between transportation noise and annoyance, Luxemburg.

63 Miedama, H.M.E., and Vos, H. (2003). Comparison of noise quality indicators, *Proceedings of 8th International Congress on Noise as a Public Health Problem*, ICBEN.

64 Miedama, H.M.E. (2004). Relationship between exposure multiple noise sources and noise annoyance. *Journal of the Acoustical Society of America* 116 (2): 949–957.

65 Miedema, H.M.E. and Vos, H. (2004). Noise annoyance from stationary sources: relationships with exposure metric day-evening-night level (DENL) and their confidence intervals. *Journal of the Acoustical Society of America* 116: 334–343.

66 Miedama, H.M.E. and Borst, H. (2007). Rating environmental noise on the basis of noise maps, TNO report 2007-D-R0010/B, Delft, The Netherlands.

67 Miedama, H.M.E. and Borst, H.C. (2007). Rating environmental noise on the basis of noise maps within the framework of the EU Environmental Noise Directive, Inter-Noise 2007, Istanbul, Turkey.

68 Probst, W. (2012). From noise maps to critical hot spots: priorities in action plans. In: *Noise Mapping in the EU: Models and Procedures* (ed. G. Licitra), 361–368. CRC Press.

69 EAA (2010). Good practice guide on noise exposure and potential health effects, EAA Tech Report No. 11.

70 Probst, W. and Petz, M. (2007). Quiet City: a European project to support cities in noise mitigation, ICSV14, Cairns, Australia (9–12 July).

71 (a) http://www.soundplan.com/info1.htm;
(b) http://www.soundplan.eu/english/soundplan-acoustics/picture-book.

72 De Muer, P. and Botteldooren, D. (2003). Uncertainty in noise mapping: comparing probabilistic and a fuzzy set approach, IFSA 2003, LNAI 2715.

73 Craven, N.J. and Kerry, G. (2007). A good practice guide on the sources and magnitude of uncertainty arising in the practical measurement of environmental Noise, University of Salford DTI Project 2.2.1, National Measurement System Programme for Acoustical Metrology.

74 Kragh, J. (2001). News and needs in outdoor noise prediction, Inter-Noise 2001, The Hague, The Netherlands.

75 Manvell, D., Aflalo, E., and Stapelfeld, H. (2007). Reverse engineering: guidelines and practical issues of combining noise measurements and calculations, Inter-Noise 2007, paper no. 068, Istanbul, Turkey.

76 Van Leeuwen, H.J.A., and Van Banda, S.E.H. (2015). Noise mapping state of the art: is it just simple as it looks? Euronoise 2015, Maastricht, The Netherlands.

77 Probst, W. (2012). Uncertainty and quality assurance in simulation software. In: *Noise Mapping in the EU: Models and Procedures* (ed. L. Gaetano), 181–211. CRC Press.

78 DEFRA (The Department for Environment, Food and Rural Affairs) (2005). Research project NANR 93: WG-AEN's Good Practice Guide and the implications for acoustic accuracy. Final Report: Data Accuracy Guidelines for CRTN, HAL 3188.3/9/2.

79 (a) Aumond, P., Can, A., Mallet, V. et al. (2019). Global sensitivity analysis of a noise mapping model based on open-source software, Inter-Noise 2019, Madrid (16–19 June);
(b) Shilton, S.J., Van Leeuwen, J.J.A., Nota, R. et al. (2005). Accuracy implications of using the WG-AEN Good Practice Guide Toolkits, Forum Acusticum 2005, Budapest, Hungary.

80 Shilton, S., Van Leeuwen, H., and Nota, R. (2005). Error propagation analysis of XPS31–133 and CRTN to help develop a noise mapping data standard, Forum Acusticum 2005, Budapest, Hungary.

81 Shilton, S., Van Leeuwen, H., Nota, R. et al. (2007). Accuracy implications of using the WG-AEN Good Practice Guide Toolkits for railways, Inter-Noise 2007, Istanbul, Turkey (28–31 August).

82 EU (2002). Position paper on dose-response relationships between transportation noise and annoyance: the EU's future noise policy, WG2-20.

83 Shilton, S., Stimee, A., and Grilo, A. (2017). Verification of noise mapping using stratified assessment windows, Inter-Noise 2017, Hong Kong (27–30 August).

84 (a) EEA (2012). Electronic noise data reporting mechanism: A handbook for delivery of data in accordance with Directive 2002/49/EC, EEA Technical report No. 9/2012;
(b) Bocher, E., Guillaume, G., Picaut, J. et al. (2019). Noise modelling: an open source GIS-based tool to produce environmental noise maps, Inter-Noise 2019, Madrid (16–18 June).

85 http://www.londonnoisemap.com.

86 https://www.bts.gov/newsroom/national-transportation-noise-map. and also; https://www.bts.gov/about-BTS.

87 https://www.umweltatlas.bayern.de/mapapps/resources/apps/lfu_laerm_ftz/index.html?lang=de.

88 http://parisisinvisible.blogspot.com.tr/2011/10/noise-maps-of-paris.html.

89 http://www.mapwatch.com/gallery/airport-noise-map.shtml.

90 http://mapsontheweb.zoom-maps.com/post/72577485752/chicago-ohare-airport-noise-in-surrounding-area.

91 Ozkurt, N., Sarı, D., Akdağ, A. et al. (2014). Modeling of noise pollution and estimated human exposure around Istanbul Ataturk airport in Turkey. *Science of the Total Environment* 482–483: 486–492.

92 Kurra, S., and Bayazıt, N. (2000). Results of a pilot study about teacher's annoyance relative to noise exposure in three schools, International Symposium on Noise Control and Acoustics for Educational Buildings, Turkish Acoustical Society, Istanbul, Turkey, May.

93 https://i0.wp.com/ontario-wind-resistance.org/wp-content/uploads/2011/01/wind-farm-noise-map.gif.

94 http://www.soundplan.com/info5.11.jpg.

95 Anon (2008). *Good Practice Guide on Port Area Noise Mapping and Management*. Amsterdam, The Netherlands: NoMEPorts.

96 https://www.faa.gov/airports/environmental/airport_noise/noise_exposure_maps/.

97 Benocci, R., Bellucci, P., Peruzzi, L, Bisceglie, A., Angelini, F., Confalonieri, C., and Zambon, G. (2019), Dynamic Noise Mapping in the Suburban Area of Rome (Italy), *Environments, 6* (7), 79.

98 Socoró, J.C., Alías, F., and Alsina-Pagès, R.M., (2017). An Anomalous Noise Events Detector for Dynamic Road Traffic Noise Mapping in Real-Life Urban and Suburban Environments. *Sensors 17*, 2323.

99 D'Hondt E., Stevens M., and Jacobs A. (2013). Participatory noise mapping works! An evaluation of participatory sensing as an alternative to standard techniques for environmental monitoring, *Pervasive and Mobile Computing*, Vol. 9, Issue 5, 681–694.

100 Aumond, P., Lavandier, C., Ribeiro, C., Boix, E. G., Kambona, K., D'Hondt, E., and Delaitre, P. (2017). A study of the accuracy of mobile technology for measuring urban noise pollution in large scale participatory sensing campaigns," *Appl. Acoust.* 117, 219–226.

101 Ventura, R., Mallet, V., and Issarny V. (2018). "Assimilation of mobile phone measurements for noise mapping of a neighborhood", *The Journal of the Acoustical Society of America* 144, 1279.

102 Aumond, P., Can, A., Mallet, V., De Coensel, B., Ribeiro, C., Botteldooren, D., and Lavandier, C. (2018). "Kriging-based spatial interpolation from measurements for sound level mapping in urban areas," *J. Acoust. Soc. Am.* 143(5), 2847–2857.

103 Shim, E., Kim, D., Woo, H., and Cho, Y. (2016). Designing a Sustainable Noise Mapping System Based on Citizen Scientists Smartphone Sensor Data. PLoS ONE 11(9): e0161835.

104 Jinbo, Z., Hao X., Shuo L., and Yanyou Q. (2016). Mapping Urban Environmental Noise Using Smartphones, *Sensors*, 16 (10), 1692.

7

Effects of Noise and Noise Control Criteria

Scope

Noise pollution is one of the main environmental problems causing risks to human health, which is defined as "the situation of being complete wellness from the standpoints of physical, emotional (psychological) and social aspects." In the early years of the last century, it had been assumed that noise created problems solely in the hearing sensation, however, its effect on health and well-being has nowadays been realized through scientific investigations. Research is continuing in medical sciences on the physiological impacts of noise which leads to chronic pathological changes. The psychological effects and the effects on human performance have been explained more unambiguously.

The physical definition of noise is given in Chapter 2. Other descriptions can be found in the literature by emphasizing the subjective effects of noise, such as "artificially produced unwanted sounds whose quality is deteriorated and whose quantity is unacceptably high." Since individual preferences are associated with the unwanted aspects of noise, the effect of noise on psychological and neuro-vegetative systems may conditionally differ in line with this fact. Obviously, the most consistent impact of noise is on the hearing mechanism.

Two groups of people are distinctly concerned with noise protection:

- Those who are directly exposed to noise sources: Industrial workers, operators of heavy vehicles or construction machinery, those professionally involved in high level amplified music. etc.
- Those who are indirectly exposed to a noise source: People living or working near motorways, railways, airports, industrial or construction areas, generally those in the noisy environment.

In the first group; the effect of noise is more intense although the number of people exposed is limited, whereas in the second group, the impact is extended to larger areas and the numbers of people affected are greater. It is known that noise is present, whether people are aware of it or not, thus, noise resembles "a sneaky disease".

Noise impacts on people's health and well-being are investigated under the topics of physical (hearing), psychological, physiological and performance impacts. This chapter discusses the general aspects of noise impacts outlining the basic knowledge of human hearing and perception. The concept of noise control criteria and principles are introduced under the measures of noise management.

Environmental Noise and Management: Overview from Past to Present, First Edition. Selma Kurra.
© 2021 John Wiley & Sons Ltd. Published 2021 by John Wiley & Sons Ltd.

7.1   The Concept of Hearing, Perception, and Loudness

Hearing is a physical phenomenon in the fields of otolaryngology, the medical science dealing with ear, nose and throat diseases, and, audiology which is a branch of science studying hearing, balance, and related disorders. Although this subject is beyond the scope of this book, some background information is given below to emphasize the importance of hearing protection from noise.

7.1.1   Anatomy of the Hearing System

The hearing mechanism has been explained by numbers of scientists since the 1930s. Stevens (1938), Barany (1938), Bekesy (1941, 1951, 1953, 1954), Davis (1947, 1951, 1953), and Tasaki-Davies (1952) and well outlined in their books and other documents [1–5]. The ear which is the hearing organ consists of three main components: the outer ear, the middle ear, and the inner ear (Figure 7.1a).

The pinna, acting as a sound collector, and the ear canal of 2.5–3 cm in length constitute the outer ear. The eardrum (or tympanic membrane), of about 7 mm in diameter, divides the two parts of the ear at the end of the ear canal. In the middle ear, which is an air-filled cavity, there are ossicles which are tiny bones called a smalleus, incus, and stapes, all connecting the outer ear to the inner ear, and the eustachian tubes act as a balance element for the air pressures in the outer canal and in the middle ear cavity. The inner ear is composed of the semicircular canals which are balancing

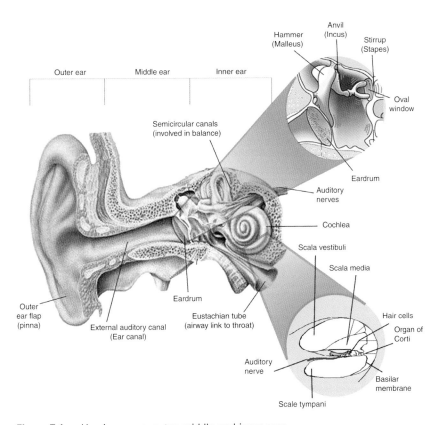

Figure 7.1a   Hearing organ: outer, middle and inner ears.

elements, the round and oval windows both connected to the cochlea and the auditory nerve fibers. The cochlea, which is in the spiral shape of 30–35 mm in length, is divided into three compartments, and includes the sensory organ of the corti in which are found the sensory cells and the supporting structures embedded on a basilar membrane. The sensory organ contains about 20 000 hair-like cells. All of these organs play a role in the detection of sounds.

7.1.2 The Hearing Mechanism

Sound, which is funneled into the ear by the pinna, sets the eardrum into vibratory motion and is transferred to the middle ear through the ossicles. The sound waves are transmitted to the oval window, which is connected to the ossicles, and pass into the cochlea which is filled by a water-like liquid (Figure 7.1a). The pressure waves propagate rapidly within a few micro-seconds in the fluid of the cochlea and in the basilar membrane. The hair cells which amplify sound waves, are located in order along the basilar membrane and are named as outer and inner hair cells according to the central axis of the spiral. They move periodically with the flowing of the fluid within the cochlea and the acoustic vibrations are converted into electrical signals that are sent to the brain through the auditory nerves and perceived as sound. Hair cells are grouped according to their various lengths and each group is associated with a different frequency of sound, i.e. the hair cells closest to the middle ear are shorter and sensitive to high frequency sounds whereas those located farthest away, are sensitive to low frequencies (Figure 7.1b).

Variation and amplification of the pressure in different parts of the ear, related to acoustic input impedance on the eardrum, transfer functions of the middle ear, normal modes of the outer ear, inter-aural time differences, directionality characteristics, have all been explained comprehensively in the scientific literature. The transformation of sound pressure in the ear has been investigated through measurements conducted in real ears or by using artificial ears, and physical models have been created [5].

Auditory Perception

The auditory system is involved in two different scientific branches [6]: (i) Auditory physiology, mostly studying non-humans, deals with "the construction, organization and function of individual elements and subsystems at the various stages in the auditory nervous system;" and (ii) psychophysics or psychoacoustics reveals how human listeners sense and perceive auditory stimulus and measures the limits and the performance with the effects of interrelated parameters, simply "to provide a functional description of hearing process."

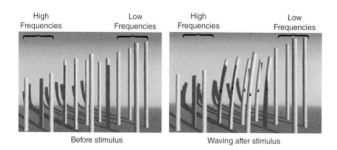

Figure 7.1b Hair cells and their frequency responses. (*Source:* Architectural Acoustics education software by J.P. Cowan for Acentech, 1999, specially provided by Acentech.)

Auditory perception is dependent on age, personal sensitivity, the functional status of the inner ear, and the general health of the individual. For a normal hearing person, the auditory system has a great dynamic range based on the measurements by Robinson and Dadson [10] who defined the threshold of hearing as "the least sound pressure needed to make a tone audible" [1]. The threshold is low at frequencies near 3000–4000 Hz and this implies that the human ear is most sensitive in this range. The upper limit of the auditory area, called the threshold of discomfort, is rather independent of frequency and the 120 dB boundary is the pain threshold. The auditory area and the approximate range of environmental sounds are shown in a simplified chart given in Figure 7.2. Section 7.5.1 gives more detail on perception related to soundscape.

Loudness
Loudness is a subjective intensity identifying the perceived magnitude of sound. In psychoacoustics, sometimes it is described as the volume of sound. The adjective "loud" is applied to an auditory experience in a scale ranging from "nearly audible" to "intolerably loud" [6].

Due to its nature, the human ear cannot perceive (judge) sound with its real physical properties, but by comparing it with a reference sound whose characteristics are known, e.g. the minimum audible sound. Perception of sound is explained by the loudness concept which depends on the non-linear frequency sensitivity of the human ear [7–14]. Some sounds, although they have different frequency tones and different pressure levels, stimulate the same sense of loudness on the ear. However, the sounds which are equal in loudness may not impose the same annoyance degree. The sounds perceived equally are displayed on a diagram as a set of curves so that each point on the curves represents the free-field sound pressure level of a single frequency tone. The diagram covers all the sounds in the full range of frequency that an ear can detect. They were derived from the loudness judgment tests applied to a group of observers listening to the given sounds and comparing them with a reference signal of 1000 Hz tone (Figure 7.3). The curves, called the equal loudness level, were first obtained by Fletcher-Munson in 1933, aiming to find a sensation unit to describe sounds. The chart was revised by Robinson and Dadson [10] and the curves which were somewhat different were named generically "equal

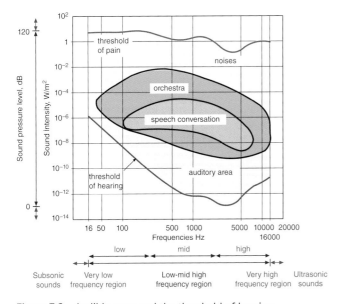

Figure 7.2 Audible zone and the threshold of hearing.

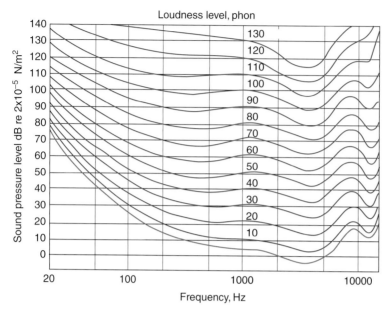

Figure 7.3 Equal loudness curves [10]. (*Source:* © IOP Publishing. Reproduced with permission. All rights reserved.)

loudness contours." Zwicker developed a calculation method to estimate the loudness impression using psychoacoustic principles [14,15].

As can be seen in Figure 7.3, the spaces between the equal loudness contours are about 10 dB but their shape changes into a complex curvature above 1000 Hz and they tend to bunch at low frequencies, all rising rapidly when the sound pressure level is increased. Measurement of loudness by sound level meters is difficult unless they are equipped with the same auditory characteristics of the ear, such as special band pass filters. Loudness has been calculated by using different procedures developed by various scientists, such as Fletcher, Stevens, King, etc., called the equal tone method, the corrected equal tone method, and the equal noise method [9–15]. The models for loudness estimation have been standardized in ISO 532 and ANSI S1.11-1986, based on the one-third octave band spectrum [16–18]. However, whether these models are good enough to measure the complex sounds and the amplitude modulated waveforms is still being debated and new techniques are developed [19–22].

Two types of units are used in measurements: (i) Loudness: The unit is the sone which is defined as the loudness of a 1 kHz tone as a function of the sound pressure level. It has been revealed that loudness in sones doubles with every 10 dB increase in level, describing a power function. 1 kHz tone at 40 dB sound pressure level (SPL) is equal to 1 sone; and (ii) loudness level: The unit is the phon which is defined as the sound pressure level at which a 1 kHz tone is judged as loud as another tone. The basic unit of the loudness level is the decibel but this is depicted as phon to distinguish it from the physical definition of dB.

The relationship between the sone and the phon is given as:

$$\text{Lg } N = 0.03 L_N\text{-}1.2$$

N: Loudness, sones

L_N: Loudness level, phons

A loudness of 1 sone has been arbitrarily selected to correspond to a loudness level of 40 phons.

As mentioned in Section 2.2.4, the frequency weighting networks that are applied to the measured noise levels as adjustment factors are widely used in noise descriptors, e.g. the A-weighted noise level has been derived from the equal loudness contours. The increase of 3 dB in the A-weighted level is perceived as a slight change in loudness, however, a 5 dB(A) increase is clearly perceptible. A 10 dB(A) increase in noise level implies a twofold increase in the loudness sensation. For a reference level of 60 dB(A), the level of 70 dB(A) corresponds to two times louder, 80 dB(A) equals four times louder. and 90 dB(A) is perceived as eight times louder.

Effect of Duration of Sound on Perception

The human ear has the ability to perceive sounds as short as 30 ms. Impulsive and intensive sounds, such as from hammers, blasts, and sonic booms, are heard as discrete noise events of short duration. Generally, sound pressure changes abruptly within one second and determines the impulsiveness characteristics. Perception of impulsiveness is dependent on the duration of the event, the rising and decreasing rate of the sound pressure and the number of impulses.

Pitch

Pitch is defined as a subjective attribute associated with the repetition rate of the waveform of sound. Although it is described as a function of the frequency of sound, pitch is also related to sound pressure level and the composition of sound. Pitch perception is an aspect of auditory sensation that was found to be roughly proportional to the number of the nerve fibers. The pitch cannot be measured directly but only by comparison with other sounds as loudness assessments. The judgments for pitch are ordered on a scale ranging from "low" to "high" or on a musical scale. It is stated that variations in pitch resemble the sense of melody, thus it is involved in musical acoustics as a descriptor of musical sounds [21]. The magnitude of the pitch sensation is determined by using a unit called a "mel."

The Masking Effect

When a noise is so loud that it prevents a person from hearing another sound, this is commonly known as masking. Masking of pure tones by noise depends on its sound pressure spectrum and temporal differences of the two signals. In a steady state noise, there is a band in the vicinity of the tone concerned called the critical band, suppress the audibility of the tone as shown in diagram as a function of frequency (Figure 7.4). The unit of critical band is the "bark." The second

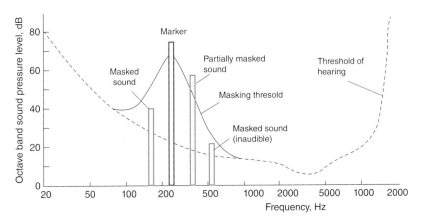

Figure 7.4 Spectral masking effect and masking threshold.

masking behavior is temporal masking (phase effect) due to the duration of masker. Auditory masking theories have been developed by Zwicker, Fastl and others [15, 22a,b]. Fastl investigated this complicated phenomenon in terms of temporal effect, masker duration, masking of pure-tones by broadband noises and impulses, fluctuation strength of broadband noises, short-term masking etc. However masking is simply explained as: if two sounds at same level present simultaneously, the lower frequency tends to mask the higher frequency sound. Above and below the band very little masking occurs, depending on the level of noise. For instance, a low frequency pure tone must be at least 14–18 dB higher than the sound pressure level of noise to avoid the masking effect. The derivation of the spectral shape of the masking noise is important in "active" noise control. Masking effect in the human auditory system is also a great concern in speech intelligibility problems.

Localization of Sound

The orientation of the ears on both sides of the head, and the nature of the pinna enhancing sounds, enable humans to localize sounds on a horizontal plane. The direction of a sound arriving at both ears is determined at a different horizontal orientation because of the differences in sound pressure levels and in times of arrival at each ear.

Directionality Characteristics of Hearing

The directionality of hearing implies the directions of high sensitivity in the human ear depending on frequencies. At 4 kHz, high sensitivity is obtained between 0–80° of an azimute angle and - 40° to +80° of the angle of elevation

Sensitivity of Hearing

The auditory system acts as a limited resolution frequency analyzer in which complex sounds are broken down into their sinusoidal frequency components. Human sensitivity to certain noises is described by using various subjective descriptors, shown on a scale with upper and lower boundaries. Some of these attributes are given below:

1) Threshold scale: Low–high
2) Loudness scale: Soft–hard
3) Tonal quality: Simple–complex
4) Dynamic quality: Steady–fluctuating
5) Directivity scale: Directive–non-directive
6) Acceptability scale: Acceptable–unacceptable
7) Perception quality: Perceivable–imperceptible

Subjective scales are important in the evaluation of sound quality and soundscape, as will be explained in Section 7.5.1.

7.2 Effects of Noise on Human Health, Comfort, and Work Performance

The effects of noise on people have been of great concern since the 1970s and numbers of books and scientific documents were published, dealing with noise as a factor of intrusiveness, and the physical variables with the threshold levels were described in detail [23–27]. The adverse effects of noise are defined by the World Health Organization (WHO) as: "change in the morphology and physiology of an organism that results in impairment of functional capacity to compensate for additional stress," or "include any temporary or long-term lowering of the physical, psychological or social functioning of humans or human organs". Noise can cause "increases in the susceptibility of an organism to the harmful effects of other environmental influences" [24]. The Directive 2002/49/EC issued by European Council gives a simpler description of the harmful effects; "Negative effects on human health" [28].

The adverse effects of noise that can be stable (consistent), dynamic, impulsive, fluctuating, etc., have been scientifically investigated under the following topics (Figure 7.5).

1) Physical effects (pathological effects: Hearing impairment and noise-induced hearing loss, threshold shift).
2) Physiological effects (change in body activities, increase in blood pressure, metabolism change, increase in breath rate, speeding up the heart impulses, startle and defense reactions, sleep disturbance, etc.).
3) Psychological effects (behavior changes, anger, disturbance, general dissatisfaction, headaches, fatigue, depression, etc.).
4) Performance effects (reduction in intelligibility of conversation, inability to hear radio or TV, decrease in concentration, blocking the activities, interference with task performance causing decrease in work efficiency or productivity, etc.).

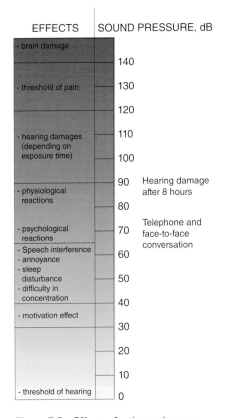

Figure 7.5 Effects of noise on humans.

In the earlier literature in the 1970s, the dependence of the negative effects of noise in relation to noise levels were categorized as follows [27]: (The B-weighting, which is not currently used, represents the loudness level of 60 phons.)

1) Degree: $L = 30\,\text{dB(A)}–65\,\text{dB(B)}$ Annoyance, anger, nervousness, deterioration of concentration and sleep disturbance.
2) Degree: $L = 65–90\,\text{dB(B)}$ Physiological responses, increase in blood pressure, heart rate and breath speed increase, reduction in pressure in the brain liquid, sudden reflexes.
3) Degree: $L = 90–120\,\text{dB(B)}$ Increasing physiological responses, headaches, etc.
4) Degree: $L > 120\,\text{dB(B)}$ Permanent hazard and loss of balance.
5) Degree: $L > 140\,\text{dB(B)}$ Brain damage.

At present, more detailed and accurate categorizations regarding adverse effects have been developed as will be given below. Scientists have explained that when the physical impact due to noise exposure occurs, the human body acts as a passive system; the physiological responses cause variations in the indicators associated with the central nervous system, the psychological responses determine the individual's acceptance or rejection of the sound, and the performance effects emerge as the interference between noise and activity.

Over the past two decades, the health impacts of exposure to environmental noise have been better understood based on the evidence, and a new guideline emphasizing the following health outcomes was published in 2018 by the WHO Regional Office for the European Region [29a–c]:

- sleep disturbance
- annoyance
- cognitive impairment
- mental health and well-being
- cardiovascular diseases
- hearing impairment and tinnitus
- adverse birth outcomes.

The guideline aims to make assessments on the relationships between noise exposure and adverse health effects and to quantify the risks increasing with increased noise level [29a, 29c].

The systematic reviews on the recent findings presented in the guidelines are relevant to quality of life, mental health, and well-being and the effectiveness of interventions in reducing noise exposure and negative health impacts. The guideline covers new recommendations for different noise sources, namely, transportation (road traffic, railways, and aircraft) noise, wind turbine noise, and leisure noise (Section 7.7.2).

7.2.1 Physical Effects: Noise-Induced Hearing Loss

As a result of exposure to high-level noises over a sufficient amount of time, the harmful effects emerge related to hearing, such as threshold shift, known as noise-induced hearing loss, tinnitus, or acoustic trauma (after an impulsive noise or blast sound like shooting). These can be either temporary or permanent. Generally, environmental noise is an increasing risk factor for hearing impairment [30, 31].

Temporary and Permanent Hearing Loss

When the tiny hair-like cells in the inner ear are over-stimulated by sound energies that are too high, the hair cells connected to auditory nerves lose their flexibility, resulting in reduced hearing (Section 7.1.2). It may emerge suddenly after exposure to intense sounds and it takes a few minutes or a few hours to recover, called as temporary hearing loss or auditory fatigue. Continuous exposure to high-level noise results in destruction of the hair cells, eventually leading to permanent hearing loss.

Noise-induced hearing loss and impairment (NIHL) implies an increase in the threshold of hearing. Deformation of the hair cells primarily starts at the higher frequency range of 3000–6000 Hz with the greatest effect at 4000 Hz. With increasing noise level and extended duration of exposure, deformation eventually spreads toward the lower frequencies down to 500, 1000 and 2000 Hz. The rise of the hearing threshold, as shown is Figure 7.6, requires a louder sound affecting hearing compared to a normal ear. At this stage the losses may not be detected since it does not obstruct speech conversation, however, the telephone and doorbell ringing are heard as weak sounds, but later the negative impact extends to the range of speech frequencies. When the loss occurs at mid-frequencies, the person has difficulty in understanding speech in noisy environments, since the frequency range of human speech covers 500–4000 Hz (Figure 7.6). Figure 7.7 shows an audiogram of a person's hearing loss in both ears based on the tests (Section 5. 7.1). The chart depicts the degrees of hearing losses such as mild, moderate, severe, and profound hearing losses in relation to hearing level [31].

As is known, the normal speech level is 55–65 dB at 90–180 cm away from the speaker, the hearing level in this position is 35–45 dB as the average level in 500–1000–2000 Hz. If there are hearing losses about this range, conversation is negatively affected. Noise-induced hearing losses emerge cumulatively in years, for example, a research study of mine workers, who had worked for 1–5

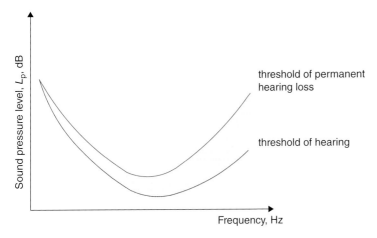

Figure 7.6 Rise of threshold of hearing.

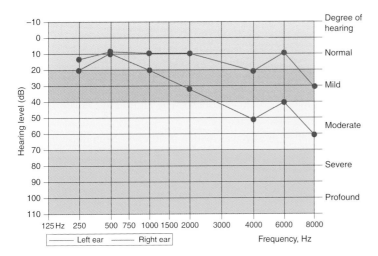

Figure 7.7 An example of an audiogram of a person with noise-induced hearing loss in octave bands.

years, showed that the hearing loss is 22 dB at 4000 Hz and is 38 dB in those who had worked 6–10 years. Figure 7.8 displays the hearing loss of a 25-year-old carpenter who has the same hearing ability as a 50-year-old person who has not been exposed to noise, as detailed in the National Institute of Occupational Safety and Health (NIOSH) (US) statistics [32]. Permanent loss cannot be cured since it is related to the auditory nerves. It has been proved that there was no effect below L_{Aeq}, 8h = 75dB or L_{Aeq}, 24h = 70 dB for occupational noises.

Hearing loss increases 1.5 dB per each dB of noise level when the level increases from 99 dB to 119 dB, and, 119 dB is accepted as the critical level for cochlea deformation. As a result, 120 dB is defined as the "threshold of pain."

ISO 1999: 2013 (confirmed in 2018) has made it possible to make assessments of the variation in threshold levels of workers who are exposed to 84 dB, depending on the duration of the exposure (up to 40 years) [33]. Hearing loss is measured by using audiometric test methods, according to ISO 8253:2010 (Section 5.7.1).

Hearing damage occurs not only with the noise level but also is complicated by other reasons, such as individual differences and illnesses. The main factors causing hearing impairment in

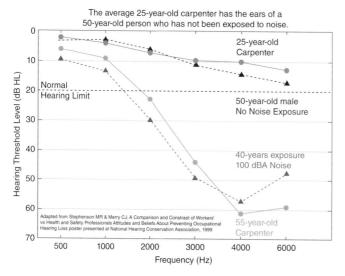

Figure 7.8 Chart depicting how a 25-year-old carpenter who does not protect his hearing has the hearing of a 50-year-old [32].

workplaces are noise levels in L_{Aeq},8h, the number of noise-exposed years, and individual susceptibility. Impulsive and high intensity noises, such as a rifle shooting noise, create an increasing risk of noise-induced impairment especially for L_{Aeq}, 24h >80 dB(A). The cumulative effect of noise exposure in daily life contributes to hearing loss.

Deterioration of hearing due to age is called presbycusis, which is different from noise-induced loss and occurs in both ears at about the same level. However, in some cases, e.g. when the worker is wearing one-sided earmuffs, like those working in telephone centrals and reservation offices, one-sided hearing loss can be seen. Hearing risk criteria will be explained in Section 7.7.1.

Hearing loss is not only a workplace disease (illness) or an accidental occurrence but it is also a social event, affecting social intercourse, causing behavioral disorders and isolation of the individual in the community or in working life. Those with hearing difficulty can feel lonely, in working life or as children in playing or in school life, leading to unhappiness.

The quantitative data regarding noise-induced hearing loss in different countries are obtained based on surveys. According to a report published in 2010 by the U.S. Department of Health and Human Service, approximately 15% of Americans between the ages of 20 and 69 – or 26 million Americans – have a hearing loss that may have been caused by exposure to noise at work or in leisure activities [34]. The UK Social Security Department issued a report based on a long-term survey and indicated that deafness rates caused by workplace noise were still high, and over 15 000 people per year suffered from NIHL, although it is lower now than in the past.

Musicians and those using personal audio players are at risk o hearing loss based on the scientific studies conducted by using an artificial head model and microphone positions near the ear drum [35a].

Tinnitus

When the hair cells are over-stimulated by sound energies that are too high, a loss of flexibility and damage occurs and the hair cells keep acting for a time after the sound is gone. This fact, caused by high-level industrial noise along with other reasons, such as an ear infection, allergy, side effects of medicine, etc., is perceived as a ringing, buzzing or hissing noise in both ears, either persistent or intermittent. Treated as a symptom not as a disease, it is difficult to measure since it is usually a subjective aspect [1, 5].

Acoustic Trauma

Acute acoustic trauma can emerge under the effects of short duration of exposure or exposure once or more times to excessive levels of sound. This is damage of the inner ear physiology and deterioration of the cochlea and the corti organ. Sudden blasts and explosions may cause acoustic trauma by damaging the tympanic membrane.

7.2.2 Physiological Effects of Noise

Numerous studies have been carried out so far and various reports have provided substantial evidence regarding the adverse effects of noise on the human physiology [2, 17–26]. Industrial workers, people living near airports, industrial premises or motorways, or particularly susceptible individuals are at risk of permanent effects associated with a high level of noise and the duration of exposure. The magnitude and duration are partly dependent on the individual or personal factors, health, lifestyle, and environmental conditions.

Sudden unexpected noise evokes startle responses that can cause accidents at work. It has been revealed that workers after exposure for 5–30 years have developed an increased risk of hypertension. Cardiovascular effects may emerge after long-term exposure of L_{Aeq},24h of 65–70 dB(A). With regard to mental illness, environmental noises intensify the development of neurosis and increase the use of drugs, e.g. tranquilizers and sleeping pills.

Physiological responses are explained as reactions of the human body automatically and unconsciously against continuous, impulsive, and intensive high-level sounds. In general, L_{Aeq} (A-weighted equivalent sound pressure level) and L_{Amax} are recommended as predicting short-term or instantaneous health effects. The WHO report discussed the noise-induced physiological effects in the mid-term and the long term, based on the results found by monitoring the physiological parameters through electroencephalograms [24]. Studies investigating the interrelation between noise and cardiovascular diseases yielded the following symptoms:

- high blood pressure
- high heart rate
- increase in cholesterol
- increase in adrenaline
- increased breath rate
- stretched muscles
- reflexes (startles) causing work accidents
- metabolism changes
- fatigue
- increased gastric secretion
- sleep disturbance

Figures 7.9a and 7.9b display the physiological effects of noise based on Babish, 2002 and the modified version in the EEA Report, 2014 [35b and 35c]. The importance of noise on stress has been emphasized in scientific studies concluded that noise was one of the reasons for stress among other environmental impacts [24]. Figure 7.9c gives a reaction scheme displaying the relationship between noise exposure, stress, and cardiovascular diseases [35b]. The cumulative effects of noise

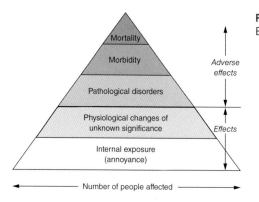

Figure 7.9a Effects of noise on health and discomfort, Babish 2002 [35b].

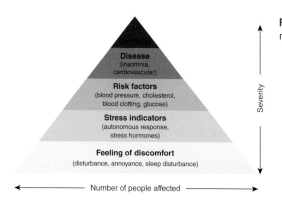

Figure 7.9b Modified diagram presented in EEA report, 2014 on effects of noise on health [35c].

may cause ulcers, asthma, or headaches, as reported. Noise affects the immune system and decreases the resistance of the body to infections. People may complain about the weight in their head and heart pulses. A recent study (2018) revealed more evidence that the transportation noise level and noise annoyance could affect the respiratory system of adults and both might independently exacerbate asthma [36a].

Unborn babies are not protected from noise and the reactions of the mother are transferred to them, as declared by earlier studies conducted in the 1980s in Japan. The relationship between low birth weight and noise exposure has been the focus of various investigations published in the early 2000s, and although the evidence is not consistent, the adverse birth outcomes have been included in the systematic review in the WHO guidelines in 2018 as one of the important health outcomes [29c, 36b, 36c]. However, the subject is not yet included in the estimation of burden of disease in the WHO report published in 2010 [37].

Despite all the evidence, the effects of noise as a source of physical and mental disease cannot be easily determined because of various underlying factors which require detailed investigations and source-specific studies. Regarding aircraft noise effects, research studies have proved that cardio and circulation problems are more observed among elderly people living under low-level flights. It has been shown that with the noise of subsonic aircraft with high speeds (i.e. with a pressure increasing 30 dB in a very short time, like 0.4 second) [38], the blood pressure increases considerably, also causing a rise in heart rate, e.g. 11–15 impulses per minute. For a single flight, the heart rate increases to 21 per minute within 3–9 minutes. Ultimately, $L_{\max} = 115$ dB(A) and the noise

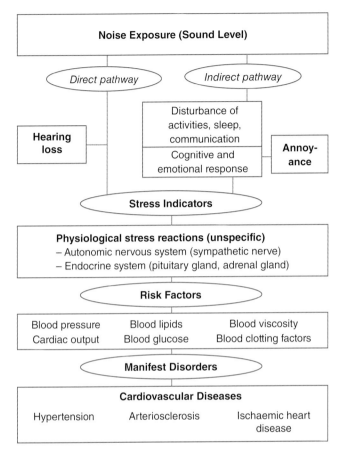

Figure 7.9c   Noise effects reaction scheme. (*Source:* adapted from [35b].)

level increase of 30 dB within 0.5 second, have been proposed as heart risk criteria (at 75 m below the aircraft with max speed of 420 km/h). The factors affecting heart risk are the height of the flight, the content of acoustic energy (0.8–2 kHz or broadband), flight speed, the level difference within 0.5 second, the distribution of average peak levels, the max level, and the number of flights. Non-regular flights were found to increase the risk to health of a certain section of the community.

Investigations highlighted that the high-level noise increases the status of illness and has serious impacts on the physiological processes of vulnerable individuals such as patients, children, students, and elderly people [39]. In recent years, the Quality of Life Index has been defined to include a set of symptoms due to environmental noises, such as fatigue, sleep complaints, stomachaches, dizziness, and headaches [40].

Effect of Noise on Sleep

The adverse effects of noise disturbing sleep have been an extensive research field for many years [24, 41–48]. A relaxed and undisturbed sleep is definitely essential for humans as chronic sleep disturbance can cause various disorders, namely, insomnia, day-time fatigue, and somnolence. Quantitative and qualitative effects of noise are investigated in three phases of sleep: pre-sleep, during sleep, and after sleep.

Pre-sleep phase: The most important effect in the pre-sleep phase is difficulty in falling asleep or lengthening the time to fall asleep. In the earlier study conducted in the 1980's France, a good relationship was found between the noise levels and sleep difficulty resulted in higher consumption of sleeping pills [47, 48a]. A recent study supports the stronger correlation between the high traffic noise levels and the usage of psychotropic medication (antidepressants) [7.48b].

During sleep: As stated in the WHO guideline for night noise, "uninterrupted sleep is a prerequisite for good physiological and mental functioning." Different effects emerging during sleep are tested in sleep laboratories from the point of view of different parameters, such as awakening and alterations of sleep stages or depth. Disorders in different phases of sleep, particularly in the REM phase, are emphasized during sleep. A change in the heart rate, cardiac arrhythmia, increased blood pressure, a change in vegetative functions, a change in respiration, finger pulse amplitude, and increased body movements, etc. have been evidenced so far.

People complain more about sudden awakening from noise rather than having difficulty in falling asleep due to noise. It is known that noise in the early mornings was more disturbing than noise at the beginning of night. Children are seen to be less sensitive to night noise and the threshold of awakening is higher by age. Vulnerable people, such as pregnant women and sick people, are at risk of sleep disturbance.

Field and laboratory studies on noise-induced sleep disturbance and the effect on sleep quality intensified in the 1990s by pioneering researchers, such as Vallet and Vernet, Öhrström, Ollerhead, Fidell et al., Pearsons, Griefahn et al., Passchier-Vermeer, etc., as compiled in the WHO Report authored by Berglund, Lindvall and Schwela [24].

Griefhan has listed the factors to be considered while determining the critical values: if interruption of sleep in about a minimum 4 minutes (time between awaking and falling asleep), remembering this situation next day, subjective sleep quality, psychological mood, and performance [43]. A good correlation was found between sleep quality and disturbance in terms of efficiency, wellness situation, and extroversion.

For the average 40-year-old, the relationship between maximum noise levels, number of noise events (for aircraft noise). and awakening statistics is shown in Figure 7.10 [44b].

The WHO Night Noise Guideline, 2009 presents two charts for observed sleep disturbances due to the effects of road traffic noise (RTN) and aircraft noise, given in Figures 7.11a and 7.11b respectively [49]. Aircraft noise causes less awakenings due to the reduced number of events at night compared to train noise events.

After sleep effects: The noise effects the following morning are: reduced perceived sleep quality, change in psychological situation, depressed mood or low well-being, increased fatigue, the sense of

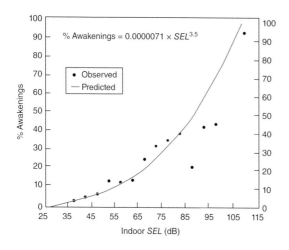

Figure 7.10 Sleep disturbance curve proposed by Finegold et al., 1992 Percentage of awakenings vs. indoor A-weighted sound exposure level (*SEL*) [44b].

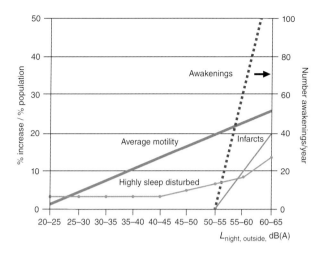

Figure 7.11a Effects of road traffic noise at night. (*Source:* WHO Report NNG [49].)

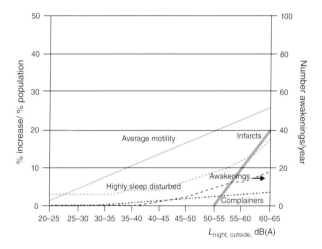

Figure 7.11b Effects of aircraft noise at night. (*Source:* WHO Report NNG [49].)

not being relaxed and rested, headache and neurotic stomach disorders, and decreased performance [41–46].

Development of Criteria against Sleep Disturbance

The development of criteria needs sufficient evidence based on the collected data and the evaluation data and must find substantial evidence, mostly derived from dose-response analysis. Studies have revealed that the physiological responses increase and the sleep disturbance starts at $L_{eq} = 37$–40 dB(A) for traffic noise. Even a weak noise of 32 dB(A) can cause awakening in a room with closed windows [47].

When the background noise is lower, the acceptable noise level outside of buildings should be less than 45 dB(A) for intermittent noises [7.36]. Vallet et al., in 1983, recommended $L_{eq}=37$dB(A) and $L_{pmax}=45$ dB(A) as the threshold levels above which the sleep quality is impaired inside bedroom [7.37].

Considering indoor levels, $L_{Aeq,indoors} < 30$ dB(A) is proposed as a limit for continuous background noise and $L_{Aeq,\,indoors} < 45$ dB(A) for individual noise events. The WHO, 2009 (Night Noise Guidelines for Europe) has depicted $L_{night,outside} = 40$ dB(A) as threshold of night-noise exposure for outdoors, based on scientific evidence, and has acknowledged that if this target is not applicable, a stepwise approach can be adopted as the interim target [23, 24, 49].

For intermittent noises, the criteria have been developed in terms of average exposure at night, characteristics and time pattern of noise, number of noise events per night or peak levels of individual noise events. The criteria recommended for intermittent noise cover a larger range (45–68 dB (A)) since the effect is also dependent on individual factors. Repetition of 10 times of a noise of 60 dB (A) at night-time is a kind of threshold affecting the quality of sleep [47].

In addition to various sleep disorder measures, motility is accepted as a descriptor which shows acceleration of the body movements in the successive time intervals (i.e. 15 seconds). The European Commission Position Papers, published in 2002 and 2004, recommend using a percentage of sleep disturbance and a percentage of high sleep disturbance to describe the night-time noise climate [50a, 50b]. The latter document in 2004 gives the noise-induced motility during sleep in relation to noise level at night based on the research by Miedama et al. in TNO (Section 7.3.2).

7.2.3 Effects of Noise on Human Performance

Since human tolerance to different noise types and noise levels varies a great deal, the effect of noise on performance is rather difficult to define quantitatively. However, numbers of investigations on task performance and interference with noise have been published so far. In research conducted as early as in 1955, the acoustic comfort region was defined in relation to reverberation time and background noise, and the type of noise which is more disturbing has been investigated [26, 50–55]. Disruption of activities at home and at work due to noise has been involved in the field surveys and specifically examined through laboratory experiments [25, 56, 57]. The effect of noise on the efficiency of work in terms of performance and interference with activities is described below:

- Interference with reading and comprehension.
- Interference with conversation (face-to-face speech and on the phone).
- Interference with daily activities such as resting, working, etc.
- Interference with task performance and reduced efficiency.
- Difficulty in focusing on task or concentration.
- Accidents or delay in completion of work due to startles.

Hall, Taylor, and Birnie, in 1985, published data regarding activity interference as a function of event noise levels in terms of interruption of indoor and outdoor speech, difficulty getting to sleep and awakening for aircraft, road traffic, and train noises in Canada. Their aim was to predict the probability of annoyance as a function of the combination of activity interferences. The model has been tested by field studies [56].

Berry, Porter and Flindell compiled the results of various surveys indicating the effect of noise on various activities and presented two charts for day-time and night-time at home and outside the home in comparison with the noise exposure distribution [58] (Figures 7.12a and 7.12b).

Kurra et al. investigated the variation in the annoyance during reading and listening in a laboratory experiment and the annoyances are compared in the chart given in Figure (7.13) [59].

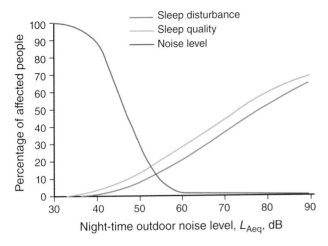

Figure 7.12a Comparison of noise impacts on various activities at night [58]. (*Source:* Permission granted by Berry.)

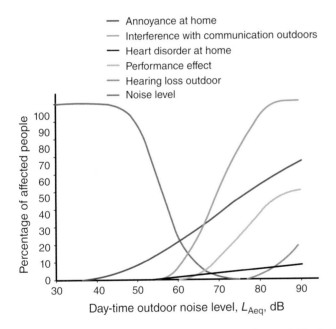

Figure 7.12b Comparison of noise impacts on various activities in day-time [58]. (*Source:* Permission granted by Berry.)

Influence of Noise on Work Performance

Noise can adversely affect performance of cognitive tasks, i.e. reading, studying, learning, problem solving, memorizing, etc. [24]. Although simple tasks may not be influenced by noise in the short term, even increasing the motivation to work, however, cognitive performance substantially deteriorates for more complex tasks. Sudden and intense noises act as a distracting stimulus and intermittent noise events have a disruptive effect as a result of startle responses. Noises with high noise

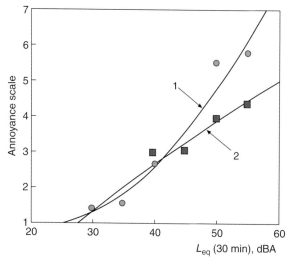

● Listening-1 y = 2.7606 − 0.18001x + 4.3786e−3x² r^2 = 0.940
■ Reading-2 y = −3.0980 + 0.16031x − 4.2857e−4x² r^2 = 0.951

Figure 7.13 Relationship between noise levels and annoyance (average group scores) during reading and listening tasks obtained in the laboratory [59].

levels, particularly intermittent and impulsive noises, can affect task performance in workplaces. Effects of noise on quality of work (e.g. delay, speed and accuracy of task) have been evidenced in the research studies.

Completion of task in due time: During focused work requiring attention, noise causes disturbance, disperses attention, lengthening the perception time and the time needed for refocusing on task, and involves spending more effort on reconcentration, lengthening the time for completion of the work, and introducing fatigue.

Accuracy of work: A decrease in work accuracy especially under impulsive sounds and an increase in the number of errors in the results have been observed and it was found that impulsive noises increase errors by about 5% [60].

Work accidents: The startles due to sudden intermittent noises can cause work accidents. Also, high-level noises can mask warning signals, thus preventing attention to the approaching danger.

Experts underline that it is difficult to obtain evidence on the influence of noise on work performance, depends on the type of task. In some cases a minimum level of sound intruding into the room makes it impossible to perform the task, such as in recording studios. Impulsive sounds may cause discontinuity temporarily or periodically, such as interruption in visual perception and in receiving information (e.g. due to eye blinks). The effect of noise on task performance also depends on the frequency of sound. Above 2000 Hz, the negative effect is greater and a sudden decrease in performance can be observed. The rhythm (period) of intensive sounds is also important in task performance. If the duration of the sound is 1 second and the intervals of interruptions (or repetitions) occur within 1 second, the decrease in task performance is found to be less than that of a continuous sound [60].

The reactions of people while doing the same task may differ considerably. Some people may concentrate more on their task under a moderate level of noise, instead of silence. Others may have reactions in the initial phase of task within ½ hour [51]. In noisy working places, the employer's efforts and good will may prevent negative reactions against noise for a short duration of exposures.

Investigations have aimed to establish descriptors or weighting units to assess the reactions of people under different noise conditions. The apparent reactions with respect to type of source are summarized as:

1) Aircraft and train noises interfere more with speech communication.
2) Continuous road traffic noise (RTN) affects resting and sleep.
3) The effect of train noise on work performance is greater but the general annoyance is less.
4) Dose-response relationships differ for train and aircraft noise.
5) Interference with activity varies with the source type.
6) With train and heavy vehicle noises, speech interference becomes profound, parallel to general annoyance, which rises with the increasing number of events, but decreases after a certain number. However, the increase in annoyance with number of flights is consistent.
7) While conducting a task, the percentage of highly annoyed people (*HA%*) decreases with the number of passing trains and aircraft or remains constant.

Interference with Speech Communication

Some of the consequences derived from the field and laboratory studies on the effect of noise on speech interference are as follows:

- *Difficulty in listening and comprehension:* This fact emerges when the low frequencies mask the high frequency components within speech.
- *Interrupted communication:* High-level sounds, repetitive, and interrupted noise events cause stop/start situations during conversation which is very disturbing for the speaker and listener.
- *Necessity to talk aloud:* High-level continuous noises according to the signal-to-noise-ratio cause extreme fatigue in the speaker, who spends more energy in raising their voice.
- *Disruption of social relations:* People talk less in a noisy environment or talk only about the important things, even talk to themselves.
- *Interference with telephone conversations:* At the noise level above 70 dB, it becomes difficult to talk on the phone and the conversation becomes not comprehensible, even impossible.
- *Difficulty in listening to radio, TV, and music:* Getting pleasure from music by comprehending its real characteristics or listening to other pleasant sounds is minimized under noisy conditions.
- *Ultimately social life is negatively affected* with noise intrusion in an environment.

Speech interference is the result of the masking process causing difficulty in understanding. For a complete sentence intelligibility, the signal-to-noise ratio should be at least 15 dB(A) in the frequency range of 1000–6000 Hz where most of the acoustic energy in speech is concentrated. The distance between the speaker and listener and the reverberation time (*RT*) of the room are other effective factors. It is accepted as a criterion that the *RT* should be <0.6 second in small rooms.

A chart presented in 1979 by Webster, on the relationship between the A-weighted speech levels and the distance for speech comfort under a noisy environment, is given in Figure 7.14 [53]. Generally, speech is understood when the background noise level is 35 dB(A) corresponding to the intelligibility of 100%. Speech interference starts at 55 dB(A) and conversation can be possible at 65 dB(A) only by spending a high speech effort. In some investigations this level was found as 50 dB(A). Laboratory experiments provide information particularly on speech comprehension of

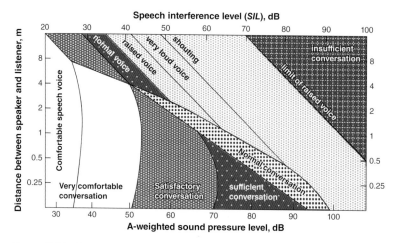

Figure 7.14 Determination of permissible distance between talker and listener at various voice levels [53].

those with normal and impaired hearing [54]. Speech interference is evaluated by using specific units like preferred speech interference level (*PSIL*) and speech interference level (*SIL*) [55, 60–62].

Speech intelligibility is also an important issue in room acoustics. The background noise and the reverberation time associated with the room size are factors that play a role in satisfactory speech intelligibility. For existing situations, intelligibility tests are helpful in the evaluation of noise and room effects on listening conditions. In these experiments which are standardized by the International Standards Organization (ISO) and the American National Standards Institute (ANSI), a certain number of phonetically balanced phrases are read in the speaker's position to the subjects and the number of correct answers yields the corresponding articulation index [63, 64].

Negative Effects on Reading and Learning

Cognitive activities are negatively affected by the intrusion of noise for all people, however, significantly for children. There are two types of adverse effects of noise in schools:

- *Effect on teaching*: Because of the interference between the teacher's talk and the background noise, the teachers' raised voice in order to be heard, causes excess fatigue, stress, and has stress-related effects on them.
- *Effect on students' concentration:* Interruption of class because of intruding noise into the classroom either from outside or inside, distracts the attention of students from the subject and destroys their concentration. Studies in relation to cognitive memory and task which needs attention, thinking, and learning are negatively affected by noise or by music and songs played in the background. The destructive impact of noise is significant in learning language, the ability to read, comprehension, and in general, throughout the mental progress based on the scientific evidence [24]. Section 7.3.5 explains some of the field studies on noise in schools.

7.2.4 Psychological Effects of Noise

Subjective Evaluation of Noise

Before the 1970s except very few study, the effects of noises generally were dealt with as an issue related only to hearing and the unwanted noises. When attention to environmental noises grew, the

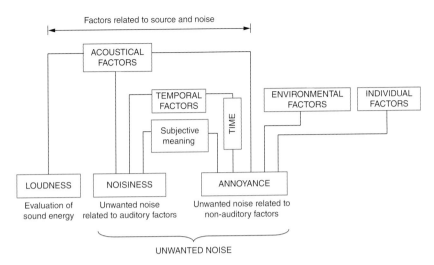

Figure 7.15 Subjective evaluation of noise. (*Source:* adapted from [65]. Permission granted by Rice.)

research on the psychological effects of environmental noises intensified between 1980 and 1990, to elaborate the subjective responses under various noise situations and to generalize the consequences for an entire community. In those days, the subjective judgments of noise were investigated in terms of three aspects: loudness, noisiness, and annoyance (disturbance) (Figure 7.15) [65–69].

Loudness of sound has been defined as an unbiased measure of sound. However, the physical, auditory, and cognitive factors play an important role in loudness perception, as explained above. For example, in a concert, a person can perceive only the sound of a special instrument according to own attention. When noise is concerned, loudness is a perceptual aspect of noise that can be changed, e.g. by turning the volume down. Noisiness is defined as a quality of noise which can be distinguished from other sounds although both have an equal loudness level, i.e. the sound of a jack hammer may be judged as less noisy than that of a motorcycle [70]. Disturbance (annoyance), defined as "feeling of displeasure" or "a degree of community noise annoyance," is a nuisance aspect emerging after a longer period of exposure time that can be determined by means of field surveys. Since it is a more complex feature compared to the others, it will be explained in detail in Section 7.3.1. Table 7.1 gives the three types of subjective evaluations and the relevant factors.

The investigations have enabled comparisons of the above-mentioned subjective responses through psychoacoustic tests. Some of them aimed to find the best descriptor for environmental noises and the specific factors influencing subjective evaluations. For example, Kuwano, Namba, and Fastl aimed to find out the effects of social and cultural differences on annoyance, loudness and noisiness by conducting a cross-cultural study between Japan and Germany [68]. As a result, the differences were emphasized, as Germans were more sensitive to noise level changes (noisiness) while the Japanese were more sensitive to the sound energy level (loudness) [71, 72]. The perceived noisiness alone was investigated in various studies for specific noise sources, such as a motorcycle noisiness study conducted in England [73].

Relationships between the three subjective descriptors were sought also in field studies, e.g. on neighborhood noise by asking the respondents the questions below [69]:

Loudness: How loud is the neighboring noise coming into your home? (verbal scale: High, medium, low)

Table 7.1 Subjective evaluation of noise and effective factors.

Evaluation	Concept	Basic factors	
Loudness	Perception	Acoustic factors	Sound intensity (sound energy) Sound frequency
		Individual factors	Hearing physiology Health Age
Noisiness	Unwantedness due to auditory factors	Acoustic factors	Sound pressure level Time Spectrum (tonal content) Temporal variation of sound (steady, intermittent, impulse, etc.)
		Individual factors	Adaptation Meaning of sound
Annoyance	Unwantedness due to auditory and non-auditory factors	Acoustic factors	Type of noise source Sound pressure level Duration of exposure Tonal components Temporal variation of sound (steady, intermittent, impulse, etc.) Background sound level
		Environmental factors	Land use (residential, mixed, etc.) Urban-suburban Type of building Physiography and climate Transportation facilities Amenity Parks and green areas New or existing sites
		Individual factors	Physiological Psychological Social and educational status Economic status
		Other factors	Meaning of sound Personality and likeness Expectations Motivation Adaptation Activity and task type Relations with source (neighborhood communication, fear, economic dependence, etc.) Culture and traditions
		Temporal factors	Day-time, evening, and night-time Seasons Weekday and weekend, holidays

Noisiness: Do you feel that this environment is noisy? (6-degree numerical scale)

Annoyance: Does the noise coming from your neighbor disturb you, at what level? (4° scale from high to low)

Behavior against noise: What do you think (or do) about noise? (open-ended)

The above study has revealed that the relationship between loudness and annoyance was weak and two sounds at equal loudness level could impose different degrees of annoyance whereas the loudness judgment was well correlated with noisiness. Noisiness and annoyance are more subjective aspects than loudness. However, the differences in judgment of noisiness and annoyance are greater than the differences in judgment of loudness [68].

In Japan, the "Perceived Level" has been introduced as the best noisiness descriptor according to the results of psychoacoustic tests performed on the insulation performance of walls [74]. Izumi declared that the "loudness judgment" can be used for environmental noise rating based on his laboratory experiments, comparing the other three descriptors: loudness, noisiness, and annoyance [75].

Sound Quality

Starting in the 1990s, the subjective impressions of various noise sources were evaluated within a new concept of sound quality. Relating the physical characteristics of sounds (objective parameters) and the various subjective attributes, such as sharpness, tonality, roughness, fluctuation, strength, etc., can be defined quantitatively based on spectral and statistical analyses. Such parameters describing the quality of sound are well correlated to the desirability or the acceptability or general perceptibility of sounds. However, their weights on annoyance are still being debated.

In sound quality evaluations, listening tests are performed to determine the relevant characteristics in a particular sound context by applying different methods. The subjects are given the short-term noise signals recorded by an artificial head (binaural recordings) and semantic differential techniques are implemented in evaluations. The subjects declare their preferences or annoyance in relation to the subjective parameters depicted on a set of scales, including two opposing terms such as pleasant/unpleasant, soft/sharp, loud/soft, powerful/weak, etc. The results are evaluated through different analysis techniques, e.g. multifactor analysis of variance. Another method is using a multidimensional test performed through paired comparisons using various verbal comments (e.g. which is the loudest sound?) [76, 77].

The sound quality of noise sources is a different branch in acoustics which has been of interest to various researchers in the past, and still is enthusiastically undertaken, mostly because of the demands from manufacturers. At present engineers work on improving the sound quality of the noise sources, i.e. powerful engines generating high noise levels as in the car manufacturing industry.

In the pioneering studies dealing with subjective evaluations conducted by Namba and Kuwano in Japan in the 1980s, the semantic differential techniques were widely implemented. As an example of their findings, the semantic profiles of annoyance from neighbourhood noise are shown in Figure 7.16, derived from a cross-cultural study between Japan, England, and Germany [71]. In another study, the subjective descriptions taken for each car type, enabled the averaged profile and later the cluster analysis was applied to the grouped response data for sound quality identification of the sounds [78]. A further experiment is given in Figure 7.17 to evaluate the sounds in a forest [76].

The sound quality studies have led to a new era in environmental acoustics called soundscape design (Section 7.5.1).

Psychological Effects

Scientific research has proved the various psychological impacts on people exposed to noise, especially long-term effects. It is generally accepted that noise-induced stress upsets the balance of the human psychological situation also, depending on the non-auditory variables. If the noise is high

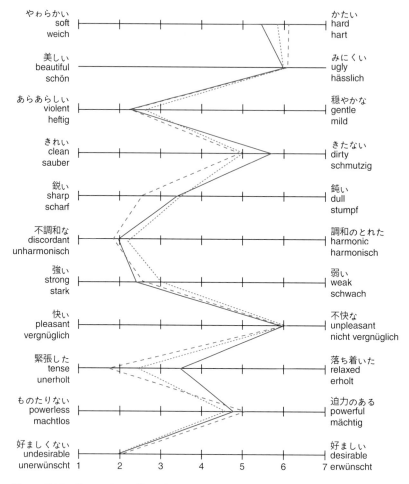

Figure 7.16 Semantic profiles of annoyance derived from a cross-cultural study, on neighborhood noise 1985 [71]. (*Source:* Permission granted by Namba, Kuwano, and Schick.)

enough and the source is unknown or if the stress is sufficiently high, the following indications of stress-related disorders can be observed in different scales and dimensions [25, 66, 79–85]:

- Negative feelings such as disappointment, dissatisfaction, desperation, anxiety, discomfort, depression, and remaining silent.
- Negative impacts that can be transformed into excess reactions and behaviors (sudden and uncontrollable anger, losing one's temper, or a tendency toward crime) and tolerances vary enormously.
- Increasing aggressive behavior (at above 80 dB(A)).
- Orienting anger externally.
- Introversion.
- Orienting anger to oneself (introversion).
- Using sedatives.
- Decreasing tolerance.
- Unwilling to help others.

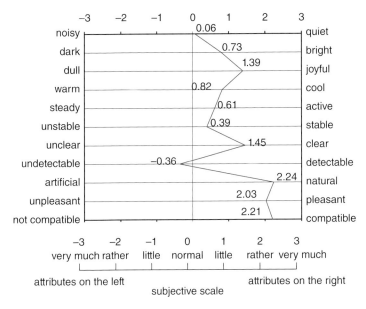

Figure 7.17 Example of semantic differential study for sounds in a forest [76]. (*Source:* Permission granted by Badelt Verlag.)

- Other responses (frequently going to doctor, closing windows, dividing sleep into portions, and spending less time outside, staying at home, writing complaints, etc.).

The research studies have emphasized the relationship between noise and mental health [86]. Recently the effects of noise on children's hyperactive symptoms have been reported and the restorative function of quiet and green areas was confirmed [40].

Researchers investigated the personal and situational variables playing a role in psychological impacts and found some attitudes were statistically significant. For example, Fields determined the four non-auditory variable, which were significant through social surveys: fear of danger from the noise source, belief in the need for noise protection, belief in the importance of noise source, annoyance with the non-noise impacts of the noise source (or non-acoustic properties of the noise source).

Different descriptors have been developed to assess the subjective effects of environmental noises, such as Psycho-Social Wellbeing Index (*PSW*) that was recommended by Öhrström and Rylander for use in studies regarding noise-related depression and annoyance [66].

As is known, workers after exposure to a high level of occupational noises at work can recover psychologically in the quietness of their home, as a protection from the more serious impact of noise. The magnitude and duration of the above referred negative effects may last long or short periods, depending on the basic acoustic factors and sensitivity to noise, but also non-acoustic factors apply, such as living conditions, environmental characteristics, temporal factors, etc. [60, 68, 83].

For 30 years or so, there have been numbers of scientific studies on the relationships between noise and depression and noise and cognitive failures, that were found to be statistically significant.

The effects of environmental noises on the community require various intervening variables to be dealt with. Therefore, investigation of subjective effects of noise should be made with the collaboration of experts from different disciplines and through field studies in the working groups.

Sensitivity to Noise

People respond differently to noises at the same level. The difference between the attitudes of individuals has been explained by sensitivity differences and its effect on annoyance has been investigated by numbers of scientists since the 1970s, i.e. Langdon and Buller [81, 82], Scholes (1981), Fields [83], Stansfeld [84], Fidell and Pearson [87], Van Kamp et al. [88], Jobs [89], Miedama and Vos [90], and many other researchers. Sensitivity is a parameter indicating the variance of individual responses against noise. Noise-sensitive (*NS*) people are likely to be sensitive to other aspects of the environment as discussed by Broadbent, 1972; Weinstein, 1978; Thomas and Jones, 1982; Stansfeld et al., 1985. It is confirmed those who are more sensitive to noise complain more about their environment, since the interrelation between noise sensitivity and neuroticism (psychiatric symptoms) has been proved scientifically, however, the physiological indicators have not been investigated. A scientific study in 2017 has investigated the correlations between *NS* and physical illness and revealed that "individuals with high *NS* were more likely to develop diabetes and hyperlipidemia, compared to those individuals with low *NS*, even at a similar level of noise exposure" [91a]. Luz found that approximately one in five people were acutely sensitive to moderately loud noise and presented the "Weinstein Noise Sensitivity Index" as a descriptor for determining noise sensitivity [91a]. Another study declares that the percentage of people who are extremely noise-sensitive varies between 12–15% depending on culture and climate as well as a higher level of annoyance and sleep disturbance [86–91b].

The role of noise sensitivity is accepted as one of the non-acoustical modifiers of noise and reactions (dose-response) relationships.

7.2.5 Adaptation to Noise

Generally, it is accepted that the people who responded as "not annoyed" by noise could have adapted to noise. Adaptation to noise due to individual characteristics is still debated at the scientific level [92, 93]. Compatibility to a noisy environment implies a delay in reactions. The level and extent of adaptation to noise and the type of sounds which are acceptable or habitual are the issues to be investigated.

Numbers of researchers consider that the habituation to noise is not correct (true) even if it is thought to be so, but the biological reactions cannot be denied [25]. The criteria for habituation have been established in experimental studies by measuring the reaction time against noises while doing a task. Namba and Kuwano declared 30 seconds of reaction time as the criterion and explained the individual and physical factors affecting the adaptation criteria. As a result, it was found that some people are less sensitive to certain sounds [92]. Although adaptation could be observed in field studies in the initial phase of a task, however, the differences in the quality of work were apparent.

A certain amount of habituation against the controllable human voices has been reported in the studies where the factors influencing habituation were investigated by using white noise, speech conversation, vocal and instrumental music. Adaptation to environmental noises was involved in the railway noise study by Walker (Chapter 7) and people's approach to railway noise in England was found to be positive since they had been used to railway noises for a long time, about 80 years. However, this finding is not confirmed in the new communities.

Investigations revealed that people even after five years of exposure may not indicate habituation against traffic noise. However, sleep disturbance among those living near roads with busy traffic, may indicate a consistent effect because of adaptation, but long-term exposure increased the consumption of medicine and fatigue in some sections of the community [35, 36]. Adaptation to aircraft noises near airports cannot be in question because of fear of danger.

7.3   Noise Impact Studies and Dose-Response Relations

A socio-acoustic survey is defined as a survey "in which noise-induced annoyance is assessed and values of measured or calculated noise metrics are attributed to the subjects' residential environment." Generally, assessment of the health impacts is based on an exposure-effect (dose-response) relationship, which should be derived through scientific studies. In noise impact assessments; the exposure is the noise level presented using a descriptor and the response implies different health outcomes, such as sleep disturbance, cardiovascular diseases, total health loss or annoyance to be quantified according to numerical scales. There are several uncertainties in generalizing the exposure relationships and in comparisons of international assessments [94].

Miedama and Rijckevorsel in the TNO, the Netherlands studied the dose (as metric) and response (non-metric) relationships in 1988. Based on the analyses of a significant number of survey results in a data matrix and by introducing coding functions, they provided a framework on the assessment of the relationship between the objective and subjective variables [95].

The WHO has defined the exposure-response relation as the "relationship between specified sound levels and health impacts" [24]. EC Directive 49 describes the same concept, defining the "Dose-effect relation" as "the relationship between the value of a noise indicator and harmful effect" [28].

Attempts at describing the dose-response relation has a long past, since the 1970s in the USA and in Europe at the same time. Since 2002, its importance in community noise assessments has been emphasized extensively in the EU member states which are obliged to develop their individual dose-impact assessments.

Noise impact assessment studies are conducted by applying a methodology comprised of finding sufficient evidence regarding noise impact, the assessment of exposure distribution, the selection of the exposure response functions, the calculation of the proportion of cases in a population exposed to noise, and eventually the calculation of the total burden of disease and costs.

7.3.1   Annoyance and Related Factors

Descriptors for Noise Community Noise Impact

When paying greater attention to community noise and its adverse effects was suggested in the 1970s, the urgent need for a common indicator to quantify the overall reactions was profound and various numerical indicators (or metrics or indexes) were acquired, based on the statistical analysis. The scientific platforms debated whether such a descriptor would be of importance with respect to the following tasks:

- describing the dimension of noise pollution;
- determining community reactions to noise;
- defining the quantified criteria for the design of an acceptable environment, i.e. town planning, etc.

Different terminology has been used to describe the psycho-social effects of community noise, e.g. causing nuisance, disturbance, dissatisfaction, bother, displeasure, discomfort, irritation, etc. Since the 1990s, "annoyance sensation" has become the common terminology in many European countries to indicate the overall negative impact from noise.

Annoyance was defined by the WHO in 1999, as a "feeling of displeasure associated with any agent or as determined condition, known or believed by an individual or a group to adversely affect them" [24]. Another WHO report on noise effects and morbidity, issued in 2004, describes annoyance as "a central effect of noise" and defines it as "a feeling of discomfort which is related to adverse influencing of an individual or a group by any substances or circumstances" [96]. Based on recent scientific evidence, the revision of the WHO report was completed in 2018 by emphasizing the attitudinal and annoyance responses in "steady state" and "change situations" implying sudden changes in noise exposures [29a,b,c].

The FICON Report in the USA, published in 1992, declared that annoyance is a "summary measure of the general adverse reaction of people to noise," and that "the percentage of the area population characterized as 'highly annoyed' by long-term exposure to noise" could be the preferred measure of annoyance [97]. UK regulations define annoyance as a dissatisfaction "in an individual and community imposed in relation to a factor or situation which is believed to impose a negative effect" [84].

The word annoyance has become the standard metric to measure adverse reactions to community noise in English, then it is translated into other languages based on an experimental study [98]. Non-English-speaking countries have conducted surveys applying a common methodology to select the best wording for expressions regarding various degrees of annoyance. Eventually the results were reflected in the international standard, ISO/TS 15666:2003 (confirmed in 2013), based on statistical analyses [99]. The standard defines noise-induced annoyance as "one person's individual adverse reaction to noise" and acknowledged that the reaction might be referenced with different wording in English, e.g. dissatisfaction, bother, disturbance, in addition to annoyance.

Sounds causing annoyance can be grouped with respect to various aspects. For example, a sound can stimulate annoyance under the following conditions:

- If a sound intrudes into an environment without a necessity.
- If it generates an unreasonable fear.
- If noise is not appropriate to the activity.
- If the person believes that the noise is produced by a person due to anti-social behavior.

In general, sounds triggering a high level of annoyance are identified as:

- Sounds with higher levels compared to background noise.
- Sounds containing components with high and very low thresholds (which are more disturbing than those having a balanced spectrum; like screams).
- Sounds whose type and direction are unknown.
- Sounds whose intensity and rhythm cannot be detected.
- Sounds whose levels are close to the perceptibility threshold.
- Sounds interrupting a task.

In fact, an environment with complete quietness (or 0 dB sound level) does not exist, even in nature. The lowest level achieved in acoustic laboratories is about 10 dB. On the other hand, it is accepted that a low level of noise below 30 dB imposes a motivation and stimulant sensation during simple tasks or in some activities. Also, there are various high intensity sounds, i.e. warnings, sirens, and alarms, which are useful in social life. The decision on whether a noise is acceptable or not should be based on the psycho-acoustic and socio-acoustic tests in a laboratory or the field.

Annoyance is related to various factors, as given in Table 7.1. The level (or degree) of annoyance in a community is determined by field studies, since it varies with a great range of situational factors. Generally, 50 dB(A) is accepted as the starting point of serious annoyance. The level of annoyance is measured and evaluated by various methods, as will be explained below.

Variables Affecting Annoyance

Variation of annoyance from noise has been widely investigated in terms of the acoustical and non-acoustical factors. Jobs gave an extensive review of the factors of influence in noise exposure and reaction [100, 101].

a) ***Acoustical factors:*** Annoyance depends on various acoustical attributes of the noise source, such as sound pressure level, temporal variations of noise, peak levels of noise events, dominant tonal components, the shape of the spectrum and background noise levels. Dose-response relations should cover all these factors by using an appropriate noise descriptor which is correlated well with the subjective reactions. For example, in Sweden, L_{max} levels were found to be well correlated with the percentage of people annoyed by traffic noise [102].

Low frequencies have been shown to be a factor increasing annoyance and the relationship between annoyance and low frequency noise is high, even though the A-weighted noise level is low. Also infra sounds below 20 Hz can be heard if they are sufficiently loud. Since low frequency noise radiation is omnidirectional, it is difficult to detect the source location, especially in closed spaces. Necessity of a different criterion for low frequency noise generated by heating, ventilation, and air conditioning (HVAC), boilers, etc. in buildings, or industrial sources from outside, is still discussed, although some correction factors are recommended in the international standards for low frequency sounds. Indoor sound measurements are made using $L_{pA,LF}$ down to 50 Hz and the insulation rating units are selected accordingly [103].

b) ***Source-related factors:*** The manufacturing and operational characteristics of the noise source (Chapter 3) have a direct or indirect effect on annoyance. When traffic noise is concerned, as the most common environmental noise, road and traffic conditions affecting the total noise level have to be taken into consideration in the impact assessments. A number of studies found good relationships between some of these indirect factors and annoyance. Impulsive and intermittent noises depending on the number of noise events (e.g. train or aircraft pass-bys) and the duration of each event, impose greater annoyance than steady consistent noises [104].

The effects of number of aircraft movements were investigated based on the classification into high-rate of change (HRC) airports and low-rate of change (LRC) airports. It was found that the prevalence of highly annoyed residents in relation to CTL value (noise level at which the 50% of the exposed population is highly annoyed) was higher in the HRC airports [105].

c) ***Environmental factors:*** Physical characteristics of the environment such as meteorological factors, green areas, noise barriers, etc., as explained in Section 2.3, have direct effects on increasing or decreasing noise levels and are indirectly related to the annoyance of people living in these environments [106–109]. The characteristics of built-up areas, the layout of buildings, the building configuration, the orientation of façades with respect to the noise sources, the age of buildings, the sound insulation properties of building façades, are all associated with dissatisfaction from noise in the environment. The social and cultural characteristics of the environment, i.e. social life attractiveness, transport facilities, easy access to parks and green areas, etc. have also proved to be effective factors against annoyance. It was reported that outdoor vegetation seen from the window of a living room facing on to a road, acts as an annoyance-reducing factor at high noise levels (L_{den} = 65–85 dB(A) at the most exposed façade) [108]. Some studies focused on the effects of noise source visibility on annoyance [110] and found that the effect of visibility was dependent on noise levels in the laboratory experiment; e.g. at 65 dB(A), the annoyance from traffic noise was higher when the source was visualized [111]

d) ***Temporal and situational factors:*** Annoyance varies a great deal during day-time and night-time, when very low levels of annoyance are required for sleeping. Noises disturb people more at weekends and on holiday rather than on weekdays. In the summertime, people are more sensitive to noise while relaxing, whereas, on the contrary, social activity noises reach their peak, therefore seasonal differences should be taken into account in organizing field studies.

Annoyance is related to situations called the "steady-state" and the "change situation," implying a sudden increase or decrease in noise levels, for instance, under new infrastructures, such as an increase in flight routes, the responses differ from the steady-state situation. The rate of change depending on the situation is defined as low-rate change or high-rate change [29, 112]. It was shown that a change in people's reactions to noise is associated to their knowledge about future change in the noise source and noise exposure, for example, due to the extension of an airport and flight routes [113].

e) ***Individual factors:*** The role of personal variables on subjective responses and sensitivity has been explained in Section 7.2.4 [25, 83, 113–116]. The various moderating factors such as the diversity of people with regard to gender, age, health, socio-economic status, education, meaning of noise, economic dependence on the noise source, activity during exposure to noise, type of task, sensitivity to the environment in general and specifically to noise, previous experiences, existence of a liked sound together with a noise, visual appearance of the noise source, last exposure, habituation to noise, ownership or tenancy, sense of desperately living with noise, confidence in the authorities, efficiency of the complaints system, etc., all may influence the personal strategy of coping with noise. Although substantial relationships have been achieved for some of these parameters and annoyance, the correlations of many others were not satisfactory because of insufficient data.

Meaning of sound: If the type of source or type of sound has a special meaning for a person exposed to noise or if he or she has a special interest in it, the annoyance degree can be reduced. The degree of annoyance is higher from low-level intelligible speech, rather than unintelligible noises at the same level. The numbers of complaints are higher for fast motorways rather than nearby street traffic. High-level concert sounds affect people differently depending on the type of music, giving displeasure or pleasure.

There are also concerns about the combined effects of different physical environmental factors acting as stressors on the human perception of noise [109]. A number of scientific studies have investigated the discrete effects of these factors.

Schomer and Wagner, in 1996, discussed the contribution of noticeability of environmental sounds to noise annoyance and proposed taking this variable into account to predict human response to noise [114].

7.3.2 Methodology for Noise Impact Studies

Socio-acoustic field studies aim to determine the community noise impacts under real conditions; i.e. in the living environment and, if available, to search for the effects of some parameters with an actual range on the overall responses.

Studies relative to the quantification of the negative impact of noise were conducted as early as the 1970s based on "Community Noise Impact Analysis" and are widely implemented at present under the title of socio-acoustic field surveys. A socio-acoustic field survey is defined in ISO/TS 15666:2003 as a social survey in which noise-induced annoyance is assessed and the values of measured or calculated noise metrics are attributed to the subjects' residential environment [99].

Field surveys are associated with local built-up areas in urban or suburban areas, however, if the findings are to be generalized nationwide, different types of environments need to be selected.

Noise impact studies, to reveal the actual situation, are conducted through social surveys whose results can be used in different ways, by applying certain methodologies, some of which are standardized, others are derived from earlier experiences considering cost, time, manpower, etc. If the outputs are to be generalized nationwide and to be compared with those of other regions in a country or of other countries, standardized techniques must be implemented.

As explained in Section 7.3.3, laboratory experiments can be performed to investigate variation in annoyance responses with a designed range of variables that can be controlled or to search for the effect of a specific variable.

Objectives and Methodology

Defining the main objective of noise impact studies, followed by a good strategy and plan to achieve it, is vital in conducting community impact assessments. There may be secondary purposes (such as activity disturbance) to be derived during data evaluation. However, the works to be completed within a certain time, a budget and according to availability or the feasibility concerns, may prevent all the objectives from being fulfilled. Studies may be organized according to a specific type of source or oriented to the general dissatisfaction of a community with the aim of generalizing the results throughout a region, or limited to a local area, or for a group of people under a special condition, such as the temporary impact of summer amusement activities. Therefore, the strategy of the study should be determined according to the objective.

Field studies are conducted for one or more of the following purposes:

1) To determine the degrees of negative impacts and behaviors under excessive noise conditions.
2) To investigate the effects of certain source-related or environmental factors on annoyance assessments (for scientific reasons only).
3) To investigate complaints from individuals or a community.
4) To check the opinions or acceptability of a new source.
5) To obtain noise criteria (acceptable maximum levels) for various noise sources and propose limit values to incorporate as regulatory values.
6) To develop assessment methods for noise impacts.
7) To develop noise control strategies and prepare action plans which are mandatory in the legislation.
8) To verify the efficiency of (or satisfaction with) a noise abatement measure when applied and quantify satisfaction (through surveys before and after), e.g. a noise barrier.
9) To check the existing noise limits achieve satisfaction and provide validity or modify the values based on the tolerance or complaints.
10) To compare the assessments in different regions or in different countries.

According to Fields, the basic purpose in noise impact studies is to determine the shape of the dose-response relationship (derived from regression equations) in order to provide the equivalency of dBs of the impacts and report the probability range [115]. The main target and usability of the results should be specified after a decision-making process involving the collaboration of various stakeholders.

The results can be of extensive or specific use in practice. Governmental or local authorities conduct field surveys reluctantly to enhance people's satisfaction with their environment. Nowadays, noise impact assessments are mandatory in Europe.

Since the environmental noise levels are increasing for various reasons and the community's socio-economic and cultural structures are changing, especially in cities, the analyses regarding noise impact in a certain area may give different results over the years. Therefore, field studies are undertaken every three to five years to investigate the extension of noise pollution with respect to the differences in annoyance degrees. Evaluations need to be made also based on the noise maps presenting recent data on the increase in numbers of people exposed to a certain noise level (Section 6.4.2).

In order to accomplish the overall study target safely, accurately, smoothly, in due time without interruption, economically, by eliminating possible errors, a proper definition and organization of the tasks depend on the following aspects in the initial phase:

- identifying objectives and sub-objectives;
- decision made about how to use the results;
- decision about the size of the study.

There are four strategies in noise impact analyses:

1) Conducting pilot studies.
2) Determination of acoustic conditions (through field measurements and noise maps).
3) Conducting social surveys.
4) Statistical evaluations of results and derivations to obtain sufficient evidence for the initial hypothesis.

Phase 1. Pilot studies

The following tasks are important in developing a strategic plan for conducting the field survey:

- Collecting information about acoustical and environmental conditions: noise sources, monitoring changes, etc.
- Providing information about the social structure of the community: average cultural, educational, and socio-economic status, etc.
- Obtaining demographic data: number of buildings, population density, etc.

The above information which is necessary for sampling design, categorizing data, and uncertainty computations, can be acquired through the preliminary studies comprising various tasks:

- Determination of sample size (explained below).
- Preparation of questionnaire and pre-tests.
- Pilot noise measurements and determination of strategy in noise sampling at site (i.e. receiver positions).
- Decision of the methods to execute each task.
- Organizing the work teams, defining their duties and training personnel.
- Preparation of measurement equipment, technical standards, etc.
- Compiling a work schedule, taking into account meteorology, source operations, holidays and other factors which possibly interrupt work and cause delay, etc.

Sampling: In order to assess the noise impact from a specific noise source in an environment, a sufficient sample size should be determined to represent the environment, building groups, and respondents of different physical, psychological, and social status. Thus, the number of residential units and number of people in the field survey are determined.

Prior to the study, it is strongly recommended to carry out a pilot survey on a small group of people living in the environment to examine their reactions to the survey in general and to the questions specifically. The experience obtained from this survey is useful while conducting the real study.

It is necessary to define the objective of the study clearly, thus, the research should be appropriately designed, the samples should be selected adequately, and the scientifically proved techniques should be implemented in the analyses. Different sampling strategies can be developed after discussion. By means of the pilot noise measurements, the areas can be categorized as low, mid, and high noise levels. The samples should be distributed in a wide range of noise levels with a great standard deviation. The people to be surveyed should be selected from those living under normal conditions continuously or frequently.

Grouping the sources: The typology of noise sources in the area to be surveyed should be determined by;

- Categorizing roads: Types of roads in urban and inter-suburban transportation are grouped, such as highways, motorways, secondary roads, arterial roads, slow speed roads, city streets or main roads, collecting roads, arterial roads, service roads, inner roads, etc., also by using the local transportation authority classification regarding traffic volume and road configuration.
- Categorizing railways: Types of railway lines and trains should be categorized with respect to speed and configuration, i.e. at-grade, elevated, depressed, etc.
- Categorizing airports and air traffic: International, national, touristic and military airports and adjacent areas under take-off or landing routes or parallel to the runways.

Selection of building site and categorizing buildings

- Decision for land use: Urban, suburban or rural, pure residential, administrative or office areas, health, educational districts, industrial or mixed areas.
- Typology of built-up area: Topographic situation, layout of buildings, type of buildings; high-rise apartment blocks, single houses in low density areas, four-to-five floor suburban or urban buildings, etc.
- Building use: All-year use buildings, summer houses, weekend houses.
- Building configuration: Orientation to the noise sources, façade insulation, quiet façades, typical layout of indoor spaces, construction typology
- Socio-demographic, cultural, economic, educational levels of settlement, average age of people, ownership or tenancy ratios, year of occupation. and daily usage time of house. etc.

Selection of subjects (residents): Various sampling methods are implemented in the social studies [117]. One of the important issues is balanced (or equal) distribution of a number of subjects within the entire range of noise levels. For example, if a survey is targeted for 1500 people, then one-third of the subjects should be exposed to the noise level below 60 dB(A), one-third to between 60–68 dB(A), the rest to above 68 dB(A) [116].

Selection of the subjects can be made by implementing one of the following statistical methods:

- Probability method: Selection of subjects randomly.
- Equal probability method: Selection of all individuals with equal probability.
- Non-equal probability: Certain characteristics which are more probable to be selected.
- Stratified sampling (biased selection method): Selection is made according to a pre-decision such as subjects older than 18 years old and having normal hearing ability.

Following the determination of objectives and methodology, as a simple method, the buildings which are close to each other and having similar characteristics, can be randomly selected. A completely random sampling of buildings does not give scientific results in noise impact studies.

There should be compatibility between the objective of the study and the selection of the building site. If there are different objectives, discrete sampling should be made for each objective. Different sampling strategies can be implemented, according to the type of respondents, land-use or source type to be able to compare the results, if necessary.

In the selection of the respondents (i.e. subjects), some pre-decisions are essential regarding the age classes or ranges, numbers of males and females, general health status and specifically hearing health or other issues. Random selection of the subjects can be made by using the Kish-Grid method. For this purpose, the addresses and the people living in these addresses are listed and the subjects to be interviewed are determined by producing random numbers.

Size of samples: The number of respondents to be interviewed should represent the entire community, considering the diversity of individuals. At this stage, the cluster sampling technique is applicable. In each cluster, the numbers of individuals giving the same response are specified as a design criterion to determine the average sample size. A greater cluster design increases the accuracy but also increases the cost of the study. For decisions and implementation of the sampling strategies explained above, it is necessary to collaborate with a statistician who is expert in environmental impact studies.

Phase 2. Obtaining data regarding acoustic conditions through measurements

The actual noise levels in the environment or at the building site under concern should be determined through noise measurements and should be presented by using noise metrics compatible to the noise source and its acoustic characteristics. If noise maps are available, the noise levels can be acquired from the noise contours, however, remembering that the noise levels on the maps are the calculated yearly average values.

Primarily the reference points at the site are selected based on the pilot measurements. It is reasonable to conduct acoustic measurements at the same time as the interviews according to a proper schedule. If the physical parameters vary a great deal, for example, the number of buildings, complex building configurations or meteorological conditions, the number of measurement points are increased. During the measurements, the ISO 1996-2 should be implemented to determine the façade levels which should be reported with uncertainty values [118]. The duration of measurements, microphone locations, obtaining traffic data, etc. are all explained in the standard (Section 5.5.4.). An economic sampling strategy for noise data can be developed by calculating the accuracy limits [119].

Phase 3. Conducting the survey

To assess the noise impacts on people living in noise environments, social surveys should be designed appropriately and be compatible with the objectives of the study. Development of the questionnaire needs to take into account the following aspects:

Preparation of the questionnaire: The questions should be organized so as to obtain opinions of the respondents regarding the following aspects:

1) Overall environmental problems, existence of noise problem and given importance, by asking about:
   a) General dissatisfaction with the environment
   b) Determination of indoor and outdoor problems
   c) General annoyance from environmental noise sources
   d) Evaluation of dissatisfaction from each source or a specific source, according to the annoyance scale
2) Investigation of noise impact with respect to independent parameters:
   a) Dependence of annoyance on individual factors, such as age, gender, socio- economic status, education, etc.
   b) Dependence of annoyance on the total daily noise dose (including noise at work)

c) Variation annoyance with time

d) Variation of annoyance with psychological status and sensitivity to noise

e) Dependence on annoyance on activity and task type (working, sleeping, listening to TV or radio, conversation, relaxing, etc.)

f) Variation of annoyance according to sound insulation of façades and window conditions (open or closed)

3) Observation of individual reactions to noise with open-ended questions, such as:

a) Behavior against noise sources and noise-makers by asking about the economic dependence on the noise source, meaning of noise, etc.

b) Opinions about applicable measures, reporting complaints, etc.

ISO/TS 15666, issued in 2003, divides the questions mainly into three groups: (i) direct rating questions which are accepted as the primary measure of the dose-response relationships; (ii) indirect questions which are useful to ascertain the underlying impact of noise on people with open-ended questions; and (iii) questions geared toward comparing the specified noise to some other noise or situation [99].

It is important to conduct pilot tests for a group of people outside the study, to be able to check the applicability of the questionnaire, the order of the questions, the way of asking, comparison of the wording. The feedback of the pilot survey can be used for revision of the questionnaire if necessary and the efficiency of the analyzing methods can be evaluated. Based on the pilot survey results, some behavioral responses can be transferred to the questionnaire to clarify the ambiguity.

Placement of control questions is a necessity in social surveys for the sake of reliability in the responses. For example, the general annoyance question can be repeated at the end of the survey using different wording to check the consistency of the response and to see the bias if it exists [119, 120]. If it is desired to find the effect of noise on certain activity, similar questions should be repeated for each activity type using the same annoyance scale. The questions should be simple and comprehensible since the answers are influenced by the way of asking. Questions to be prepared, should be clear, intelligible, have sufficient length and include also some open-ended behavioral questions. Such questions are worth asking to learn about behavioral reactions, such as shutting windows in a noisy environment, raising the TV or radio volume, not using balconies, individual complaints, public petitions or even moving home, etc. Based on the field studies conducted in the past and recent ones, various supplementary data about noise impacts has been collected so far, e.g. anger or tendency to violence, decrease in social relations, differentiation in daily moods, reduced performance during some tasks, decrease in success in schools, negative influence on short- term memory, prolonging decision time, and reluctance in bodily reactions, increasingly consuming medicines, increasing accidents at work or at home, increased number of hospital trips, and how willing to pay to get rid of noise. These findings are revealed by asking questions like "Are you thinking of moving somewhere else within two years?" or "How much can you pay to modify the situation without moving from this site?". The results can be evaluated within the economic valuation of the noise impact (Chapter 9).

Selection of the annoyance (dissatisfaction) scale: In noise impact studies, assessment of annoyance is rated by means of verbal and numerical rating scales. Selection of the scales and the associated scores is an important issue in scientific research [121]. A linear scale consisting of 5–10 points marked on a line with equal intervals is the widely applied technique. The annoyance scale should be compatible with the purpose of the study, easy to use, and should give sufficient information about annoyance. Response descriptors can be used on the scale by choosing the words appropriately.

The same rating scale should be applied for each activity and performance. For precise evaluations, it is recommended to use both verbal and numerical responses or combined versions. The annoyance scale has a great influence on the overall assessment to obtain the group median scores and the percentage of highly annoyed people, etc. For example, when a group median level on a five-point scale is three or above, it implies the need for a noise abatement measure against the noise concerned.

Various questionnaire forms and different annoyance scales have been used in the past, however, there have been several efforts to standardize the annoyance surveys, which is important in the comparison of the different survey results [122, 123]. ISO/TS 15666, describes the specification for assessing the degree of annoyance by recommending two main annoyance rating questions by using verbal and numerical scales (Box 7.1 gives the verbal question and the numerical question with the introduction) [99].

Box 7.1   Questions Related to Annoyance in the Survey

Question in verbal rating scale:

"Thinking about the last (12 months or so), when you are here at home, how much does noise from (noise source) bother, disturb or annoy you?"

CARD·QV

NOT·AT·ALL

SLIGHTLY

MODERATELY

VERY

EXTREMELY

Question in numerical rating scale:

Introduction:

"This uses a 0-to-10 opinion scale for how much (noise source) bothers, disturbs or annoys you when you are at home. If you are not at all annoyed, choose 0; if you are extremely annoyed choose, 10; if you are somewhere in between, choose a number between 0 and 10."

Question:

"Thinking about the last (12 months or so), what number from 0 to 10 best shows how much you are bothered, disturbed, or annoyed by (noise source)?"

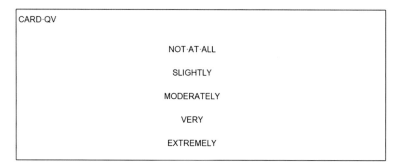

CARD QN										
NOT AT ALL									EXTREMELY	
0	1	2	3	4	5	6	7	8	9	10

Execution of the survey at the building site: Surveys are conducted by applying the techniques, as in a face-to-face interview, an interview on the phone, by mail, or by internet. Direct contact is more appropriate if possible, to receive more accurate results and to increase the efficiency of the survey. However, some drawbacks can be mentioned, such as difficulty in arranging appointments, finding trained personnel, increasing the cost of the study. Subjects' judgments are received by showing the pre-prepared cards (i.e. show cards) containing both annoyance scales and options of the words to be selected, according to the above standard.

The standard gives further specifications regarding the execution of the questionnaire: by stating that the respondents should not be eliminated on the basis of some previous questions, such as hearing or not, etc.

Phase 4. Analyses and evaluation of data

The basic statistical analyses on the output data comprise the following tests with the aid of a statistical program based on scientific methods (e.g. SPSS).

- *T*-tests on all the answers,
- Descriptive tests to obtain percentages of similar answers, statistical distributions, group median scores, average group scores, percentage of highly annoyed people *HA%*, percentage of people exposed to noise in all the noise categories, demographic profiles of respondents, etc.
- Correlation and regression analyses (to make annoyance prediction model)
- Partial least squares (PLS)
- Variance analysis and multifactor analysis of variance
- Cluster analyses
- Principle component analysis (PCA)
- Multilayer correlations
- Factor analysis

Correlations between dose-response (namely, noise levels and annoyance) indicate the strength of the relationship between noise and response. Further analyses can be implemented to search for interrelations between the noise and the different type of impacts or noise and other related parameters, such as demographic factors. (For example, dependence of annoyance on the sensitivity of people.) The results are evaluated based on correlation coefficients, significance levels, and other statistical evidences to be reported.

Derivation of dose-response curvature: Based on the analytical and, sometimes, the descriptive analyses which are widely implemented in health research studies, the relationship between noise levels and responses are sought. Annoyance which is described in terms of various descriptors, such as the percentage of annoyed people derived from the overall annoyance questions, such as disturbed or not, the percentage of highly annoyed people (those marking the annoyance scores on the upper part of the scale), group median scores, group average scores, etc., to be computed from data, are displayed in the scatter diagrams of the total data [124]. Prediction of the probability distribution, either exhibiting Gaussian and non-Gaussian distribution, is helpful in determining the qualitative assessments of noise impact. When the relation is found to be statistically significant with the confidence intervals (i.e. probability percentages), the regression equations are acquired and presented as the dose-response functions along with the polynomial variables and confidence level. The curve fitting is displayed in the diagrams where the noise levels are depicted in the abscissa and the impact values are on the ordinate.

The benefits and use of dose-response curves are summarized below:

- If the sample size is sufficient, the results can be used to derive a prediction model for noise impact and generalize to entire community.
- They can be used to obtain noise criteria, implying the acceptable highest levels by a certain percentage of people annoyed or highly annoyed.
- The weight of the other factors in noise impact can be determined, which are important in noise management strategies, policies and action plans (see Chapter 9).
- They are used for quantitative risk assessment, which is mandatory in the EU.
- They are essential for the derivation of guidelines for the public health noise policy of governments.

The basic behavior of the generic relationship curvature between the noise and annoyance which represents an exponential function is given in Figure 7.18.

While determining the criterion values, generally the upper part of the percentages of the annoyance scale, the group median scores or the group averages of all the numerical values are taken into account. It was found that the correlations are higher when the average group responses are used instead of individual responses.

By means of the regression analyses, variation in different reactions to different noise sources can be determined and the different formulations can be acquired as prediction models.

Organization of outputs of the field survey: The criteria to be used in the design of noise control schemes in action plans are based on the regression functions defining the maximum noise levels, which are acceptable to a certain percentage of people. However noise criteria should be confirmed from the standpoints of applicability, cost, and performance, as explained in Chapter 9, then can be transformed into the regulatory limits. The criterion values can be differentiated according to the noise source, noise conditions, environment type, time of the day, etc. based on the evidences (Section 7.7).

It is important to have strong evidence for the findings. Accuracy of the results is associated with the annoyance scale, the noise descriptor, and the best control of other variables. It has been indicated that twice the standard deviation of noise levels increases the accuracy four times [125].

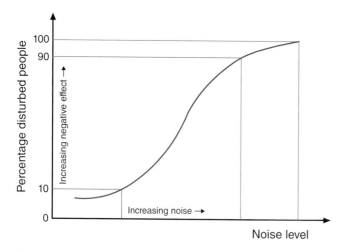

Figure 7.18 Generic relationship between noise levels and annoyance.

Further elaborations of the results: The consequences of the study should be presented in a suitable format for the attention of the public and the authorities. Relations between noise and responses should be clearly interpreted by presenting the noise impacts in terms of corresponding decibel values. The results of the variance analysis and the standard deviations of group responses should be indicated as dB(A) correspondence within certain intervals. For example, a reduction in annoyance degrees can be expressed as dB/equivalent effect = −0.21 (with 85% probability).

While elaborating the results of the study, it should be realized that the annoyance is not limited to an impact on a specific activity, but is extensively dependent on other factors regardless of noise, as explained above. Different annoyance reactions can be obtained at the same L_{Aeq} level at different building sites, a reason for this situation might be the significant differentiation between the L_{Aeq} and L_{max} values in such environments. Therefore, appropriate noise descriptors are selected at the sites where intermittent noises or sound events are dominant, like L_{AE}.

In the field surveys, whether the initial hypotheses are proved or not at the end of the study, the results should be stated clearly in the reports and scientific papers. According to ISO 15666:2003 (confirmed in 2013 and under review at present), the results of the field studies on noise impacts should be submitted with the minimum information depicted in Table 7.2 [99].

Table 7.2 Core information for reporting the results of socio-acoustical surveys based on ISO 15666:2003 (2013) [99].

Study area	Item	Topic	Required information
Overall design	1	Survey date	Year and month of survey
	2	Site location	Country and city of study sites
	3	Site selection	Any important and unusual characteristics of study period and sites Map or description of study site, building locations relative to the noise source
	4	Site size	Rationale for site selection Site selection and exclusion criteria
	5	Purpose of study	Number of study sites Number of respondents per site Goals of the original study
Social survey sample	6	Sample selection	Method for selection of sample respondents (probability, judgmental, etc.) Respondent exclusion criteria (age, gender, length of residence, etc.)
	7	Sample size and quality	Response rate Number of non-responses and reasons
Social survey data collection	8	Survey methods	Method (face-to-face, telephone, internet, etc.)
	9	Questionnaire wording	Exact wording based on pilot survey (including answer alternatives)
	10	Precision of sample estimate	Number of responses for main analyses
Acoustical conditions	11	Noise source	Type of primary noise sources (aircraft, road traffic, etc. Type of noise source operations that are included or excluded Protocols to define the noise source (e.g. minimum level, operations, days of week)

(Continued)

Table 7.2 (Continued)

Study area	Item	Topic	Required information
	12	Noise metrics	Give the complete description of any noise metric reported, according to ISO 1996-1, ISO 1996-2, ISO 19996-3 or ISO 3891 (if applicable):
			Provide $L_{Aeq,24h}$, L_{dn} and L_{den} (or L_{Aeq} by time period) for all locations or
			Provide conversion rule(s) to estimate $L_{Aeq,24h}$, L_{dn} and L_{den} under the specific conditions from the study's preferred metric
			Discuss the adequacy of the conversion rule(s)
			Provide impulse and/or tone corrections
	13	Time period	Hours of day represented by noise metric
			Period (months, years) represented by noise metric
	14	Estimation/ measurement procedure	Estimation approach (Modeling, measurement during sampled periods, etc.)
	15	Reference position	Nominal position relative to noise source and reflecting surfaces
			Present exposure (or give conversion rule) for noisiest façade, specifying whether reflections from the façade are taken into account or not
	16	Precision of noise estimate	Best information available on precision of noise exposure estimates
Basic dose - response analysis	17	Dose/response relationship	Tabulation of frequency of annoyance ratings for each category of noise exposure

Overview of Field Surveys for Different Environmental Noise Sources

Some of the field studies that have been conducted in the past up to the present and those that presented the dose-response functions in the literature, are summarized below with respect to source and noise type.

A) *Noise annoyance from traffic noise*

Studies aiming to describe the dose-response relationships for road traffic noise (RTN) have been extensive, starting in the 1970s, and 1975 has probably been the most efficient year. One of the preliminary studies related to traffic noise impact was carried out by Griffiths and Langdon in the Building Research Station (BRS), later the Building Research Establishment (BRE) in the UK, by using 1200 people in 14 sites in London in 1968–1977 [80–82]. The effects of various parameters like sensitivity to noise, seasonal variations, etc. were also investigated in the study. Correlation between the noise level and the individual dissatisfaction scores was found to be rather low, however, it was better for the average group scores ($r = 0.60$) and even higher for *TNI* ($r = 0.88$) [82]. In the later study conducted by Langdon with 3000 people in 53 regions in the UK, he found that the product moment correlation coefficient, by using L_{10},24h and the median dissatisfaction score, on a 1–7-point scale was 0.521. In this study, the effect of noise on sleep disturbance was examined and 50% of people were found to have sleep disturbance at $L_{10} = 78$ dB(A) under the open-windows condition whereas 60% of people had sleep disturbance at $L_{10} = 52$ dB(A). The relationship is indicated in the chart given in Figure 7.19 [80–82].

Griffiths and Raw declared that the group median scores were the best indicator in the field surveys. The relationship between the highly annoyed people and the L_{dn} was also found to be good, as shown in Figure 7.20.

A number of investigations were conducted in the Scandinavian countries in 1976, 1988, and so on, and the dose-response relationships were obtained using different noise descriptors [102, 126–129]. The correlation coefficient between the group median scores and L_{eq} and L_{90}, was rather good, $r = 0.81$. However, in Sweden, a weak relation between dissatisfaction and L_{eq} was found but

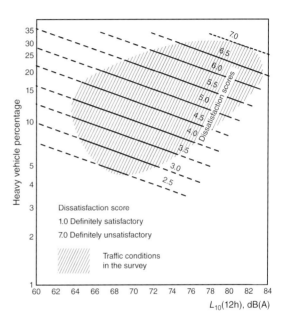

Figure 7.19 Annoyance from traffic noise in terms of heavy vehicle percentage and noise levels (Langdon, BRS, 1968) [80, 81].

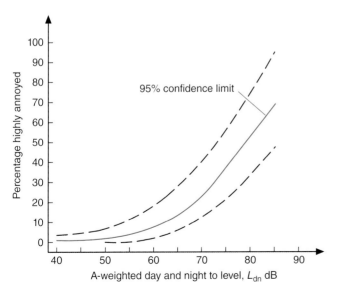

Figure 7.20 Relationship between traffic noise levels, L_{dn} and highly annoyed people, (Langdon, BRS, 1977) [82].

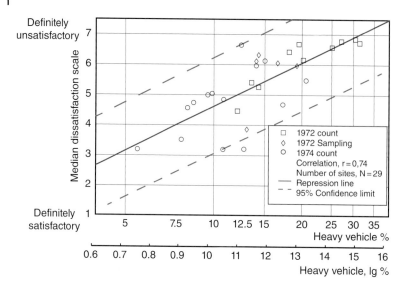

Figure 7.21 Median dissatisfaction score in relation to heavy vehicle percentage of traffic (based on experimental study) [81].

the correlation with L_{max} was higher [102a, 102b]. The assessment model, developed by Bjorkman and Rylander in the 1980s, revealed 1800 vehicle pass-bys in 24 h corresponds to 40% of the highly annoyed people ($r = 0.93$) [102, 104]. A relationship diagram is given in Figure 7.21 as a function of number of heavy vehicles which was found as the prime factor on annoyance [81] (the figure to be decided).

The estimations of dose-response relationships for annoyance of transportation noise have also been established by Schultz [130], Fidell et al. [131], Miedama and Vos [136], and Miedama and Oudshoorn [138]. Some of those studies have constituted the scientific bases of the assessment models [103, 106, 107].

There are other field studies aimed at revealing the performance of abatement measures relating to the change in community annoyance. Conducting such a study, Brown stated that a change in annoyance cannot be expected when a noise measure reduces noise level less than 3 dB. A higher reduction provided by the abatement would have a greater effect in lowering the degrees of annoyance [128].

Evaluations of community noise in the USA were based on the "Complaints" until 1970 at the federal level and later the descriptor called "Community Noise Rating" (*CNR*) that was taken as a measure for community reaction to noise in 1974, based on the percentage of complaints. *CNR* was developed as early as 1955, and was obtained by applying a certain methodology to the estimation of the noise level category from the idealized spectral shapes of noise sources obtained in the laboratory through loudness tests. However, *CNR* was a descriptive measure, not a numerical value. The relationship between complaints and the percentage of *HA%* were determined and transformed into the dose-response diagrams for aircraft noise, as shown in Figure 7.22 [129]. US Federal Interagency Committee on Noise, FICON, in 1992, adopted a function between L_{dn} and percentage of highly annoyed people as a dosage effect relationship (Figure 7.23) [97].

Combination of the dose-response functions for traffic noise: When the field surveys, especially for traffic noise, were expanded widely in various countries and conducted by independent institutions, comparisons of the dose-response functions have been attempted. However, it has been

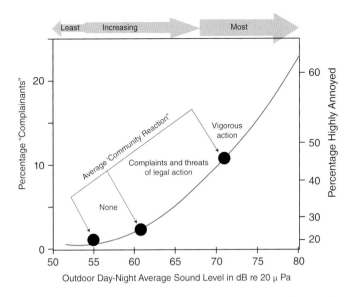

Figure 7.22 *CNR* methodology on assessment of community reaction to aircraft noise exposure based on complaints, 1974 [129].

Figure 7.23 Fitting function on the data points of Schultz and USAF adopted by FICON, 1992 [97].

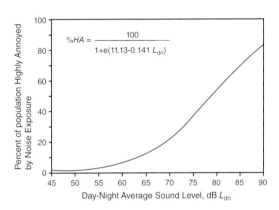

difficult to compare the findings of the various field studies because of different annoyance scales, different noise descriptors and different methodologies. Thus, some transformations have been essential in compiling the data.

Schultz (1978) combined the numbers of dose-response curvatures obtained in different countries for all types of noises, including street traffic, expressways, aircraft, and railroads by using a translation technique on 11 original clustering surveys. The chart, given in Figure 7.24, was prepared using the noise level L_{dn} and *HA%*. He stated that such comparisons would be easier if median values and *HA%* were used [130]. The reason for the scattered data in Schultz's chart was due to differences in background noise levels, façade insulation properties, size of the survey, sampling techniques, and the type of activities performed in the individual surveys.

Applied least square curve fitting procedure on the data sets in 1988 and 1991 were used by Fidell et al. to develop a statistical model, including L_{dn} and *HA%*, and they gave a third-order polynomial

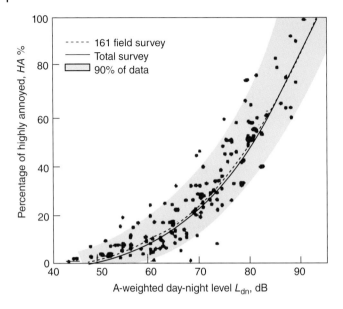

Figure 7.24 Combination of dose-response relationships obtained by a number of field surveys. (*Source:* prepared by Schultz, 1978 [130].)

function to describe the informal approximation on noise and responses, rather than a relation derived from regression analysis [131, 132]:

$$HA\% = 0.8533L_{dn} - 0.0401L_{dn}^2 + 10.00047L_{dn}^3 \tag{7.1}$$

HA%: Percentage of highly annoyed is determined as the upper 27% of the annoyance scale (about 2/7 of the scale)

Emphasizing the great dispersion in the Schultz diagram, Schomer [133], in the USA, proposed certain correction values to be added to the measured L_{dn} values according to different situations and for use in comparisons of the survey results [133]. He revised the reference values, which the Environmental Protection Agency (EPA) developed in 1973, as given in Table 7.3.

Fields compiled about 521 field surveys conducted within 60 years up to 2001, and discussed whether the differences in communities play an important role in noise annoyance [134]. He made a diagram for comparison of the results using a seven-point annoyance scale (Figure 7.25). When all the data from 26 different surveys were taken into account, the standard deviation was found to be 7 dB. This study revealed that the effect of demographic variables, such as gender, sensitivity, dependency on the source, owner/tenant, background noise, etc., had a negligible effect on annoyance. The research also showed that the annoyance could be present also at very low noise levels, rapidly increasing with the increasing noise levels, but there was no a cut-off point on the curvature [134]. Fields also showed that the rejection of annoyance (null annoyance), in some situations, could be related to the visibility of the noise source and the steep rise of annoyance curve was dependent not only on the noise levels but also other variables. He emphasized that the noise was the greatest environmental problem in a building site.

Figure 7.26 shows the combined results of the field surveys that were conducted between 1979–1985 in Istanbul and the linear regression equations were also given [135].

Table 7.3 Correction factors for impulsive noises to be added to measured L_{dn} to obtain normalized L_{dn} [133].

Type of factor	Type of correction	Description	Addition (Correction value)
Acoustical-physical	Characteristics of sound	Highly impulsive sound	+12
		Regular impulsive sound	+5
		High-energy impulsive sounds	ANSI -S 1.29
		Prominent discrete tones	+5
	Rattling sounds	Detectable rattles	+10
Acoustical – physiological	Time period in L_{dn} calculation	Usage in certain times a year	Calculate L_{dn} during these times
		All year long	Calculate yearly average
		Night	+10
		Day-time at weekend	+5
	Community relations	If good and continuous relationship between the noise-maker or if responsible	−5
Acoustical-psychological	Settlement	Rural areas	+10
		All other areas	0
	Prior experience	Not previously experienced	+5

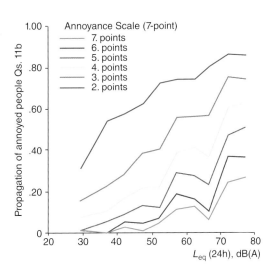

Figure 7.25 Comparison of dose-response relationships for traffic noise (L_{eq}, 24 h and people of annoyed % on 7-point scale) Fields, 1980 [134]. (*Source:* Permission granted by Fields.)

Miedama et al., in 1999 and 2007, combining the results of 26 studies and 19 172 observations, derived the following prime function for traffic noise by using L_{den} [136–139]: (Figure 7.27a)

$$\%HA_{road} = 9.868 \times 10^{-4} \, (L_{den} - 42)^3 - 1.436 \times 10^{-2} \, (L_{den} - 42)^2 + 0.5118 \, (L_{den} - 42) \quad (7.2)$$

$\%HA_{road}$: percentage of highly annoyed.

The above relationship has been included in the EC Position paper, 2002 and ISO 1996-1:2015, as the estimated prevalence of a population highly annoyed as a function of adjusted L_{den} within the 95% prediction interval [50a, 140].

B) *Noise annoyance from railway noise*

One of the basic research studies on railway noise impact was conducted by Fields and Walker in 1981–1982 [141]. In their investigation, made of 1453 people at 75 zones, it was found about 2% of the population in the UK (170 000 people) were exposed to a noise level above 65 dB(A). A good

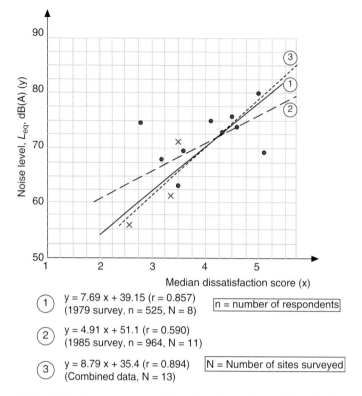

1. $y = 7.69 x + 39.15 (r = 0.857)$
(1979 survey, n = 525, N = 8)

n = number of respondents

2. $y = 4.91 x + 51.1 (r = 0.590)$
(1985 survey, n = 964, N = 11)

3. $y = 8.79 x + 35.4 (r = 0.894)$
(Combined data, N = 13)

N = Number of sites surveyed

Figure 7.26 Dose-response relationship obtained for traffic noise in Istanbul (compiled data between 1979–1985) [135].

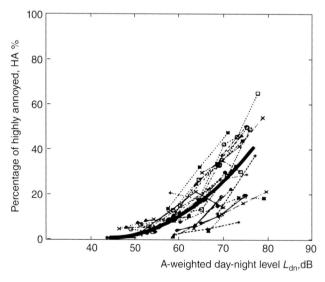

Figure 7.27a Combination of dose-response relationships for traffic noise (compiled from 26 studies, 19172 subjects). (*Source:* Miedama 1965–1992 [136, 138].)

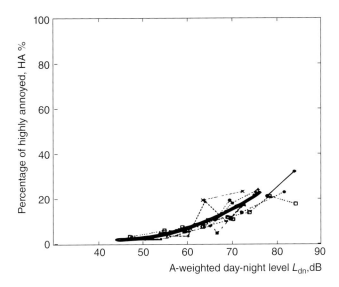

Figure 7.27b Combination of dose-response relationships for railway noise (8 studies, subjects: 7632). (*Source:* Miedama, 1965–1992 [136, 138].)

correlation was obtained between L_{eq} 24h and the annoyance degrees on a 10-point scale. The study also revealed the significant effect of train types on dissatisfaction, as the electric trains caused about 10 dB less equivalent annoyance compared to other types of trains.

Sleep disturbance due to railway noise was investigated in a study conducted in Germany in the 1990s and the effects of various parameters on sleep disturbance were investigated, particularly sudden awakenings in quiet and noisy environments [43, 142]. In Japan, where transportation is widely dependent on railways, a number of field surveys were conducted in the 1990s also with the purpose of deriving the noise criteria for railway noise control [143].

Miedama gave a polynomial equation, based on 7632 observations and eight studies, to assess the percentage of highly annoyed people from railway noise, in relation to L_{den}: (Figure 7.27b) [136, 137]:

$$\%HA_{rail} = 7.239 \times 10^{-4} \left(L_{den} - 42\right)^3 - 7.851 \times 10^{-3} \left(L_{den} - 42\right)^2 + 0.1695 \left(L_{den} - 42\right) \tag{7.3}$$

$\%HA_{rail}$: Percentage of highly annoyed

The above relationship has also been included in the EC position paper 2002 and ISO 1996-1:2015 [50a, 140].

C) *Noise annoyance from aircraft noise*

Fastl investigated the relationship between the loudness of aircraft noise and sound quality in 1990 and stated that parallel to the substantial decrease in aircraft noise emissions in years, the loudness of aircraft decreased by half and the radiated sound tended to have softer characteristics. He also established the relationship between the loudness of aircraft and L_{eq} and examined the time function of loudness [144, 145].

Among the aircraft noise impact studies, Kryter discerned a relationship between the percentage of highly annoyed people and the noise levels in L_{dn} based on his study in 1985 [25].

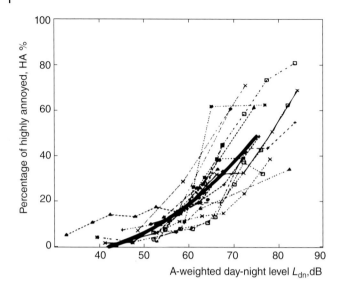

Figure 7.27c Combination of dose-response relationships for aircraft noise (19 studies, subjects: 27081). (*Source:* Miedama, 1965–1992 [136, 138].)

Miedama gave a polynomial equation, based on 27 081 observations, in 19 studies, to assess the percentage of highly annoyed people from railway noise in relation to L_{den} [136, 138]: (Figure 7.27c)

$$\%HA_{air} = -9.199 \times 10^{-5} \left(L_{den} - 42\right)^3 + 3.932 \times 10^{-2} \left(L_{den} - 42\right)^2 + 0.2939 \left(L_{den} - 42\right) \tag{7.4}$$

HA_{air}: percentage of highly annoyed

The above relationship has also been included in the EC position paper 2002 and ISO 1996-1:2015 [50a, 140].

D) *Comparison of annoyance against different types of transportation noises*

Some researchers have focused on the differences between people's dissatisfaction with basically three types of transportation noises [142, 146–152]. One of the pioneering studies on this topic is the London Noise Survey, performed in 1978, in which the Parkin chart was derived, enabling a comparison between road traffic and aircraft noise annoyances (Figure 7.28). It was revealed that community reaction to motor vehicle noise was much higher than that of aircraft noise at the same level, despite the expectation otherwise, likely because of the belief that the noisiness of aircraft noise was greater [126].

Differences between railway noise and motorway noise annoyances have been investigated in Germany, France, the Netherlands, and the UK as displayed by the combined charts. The results were interpreted as showing that 20% of people were annoyed by motorway noise while 2–4% of people were disturbed by railway noise, emphasizing that, generally, people have more tolerance of trains. Statistical analyses (variance analysis) have proved that the railway noise exposure should have 4–5 dB(A) higher noise levels compared to traffic noise, giving an equal level of annoyance, however, at higher noise levels, the impact tends to be opposite, i.e. annoyance from traffic noise increases [146]. Similarly, Fields and Walker [141] revealed that the railway noise in Europe

Figure 7.28 Comparison of dose-response relationships for traffic and aircraft noises, by Parkin, 1978 [126].

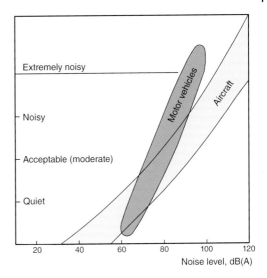

Figure 7.29 Comparison of annoyance functions obtained for road, railway and aircraft noises (Number of studies for aircraft noise: 173, railway noise: 170, traffic noise: 170). (*Source:* Fidell et al. 1991 [131].)

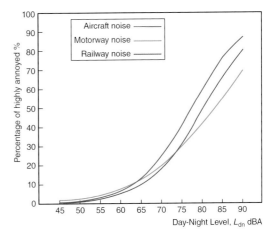

disturbed people less than motorways and the disturbance was increasing slowly with increasing noise levels [141].

Fidell et al. combined data for each of the aircraft, traffic and railway noises and produced a chart (Figure 7.29) in 1991 [131].

Fields and Walker, in 1982, found that a 10 dB(A) difference between the railway and road traffic noise annoyances at the same degree of annoyance [142].

Izumi, in 1990, compared the impacts of road traffic and train noises based on a laboratory experiment and found that traffic noise impact displayed a steeper increase with noise level below 55 dB (A) and, above this threshold, the increase in traffic noise annoyance became slower whereas the train noise exhibits a consistent increase. However, when the HA% was taken into account, the increase of annoyance with noise level was faster for traffic noise [147].

The dose-response relationships obtained separately for aircraft, road traffic, and railway noises by Kurra et al., using the field and laboratory data, were compared and the combined effect was investigated in 1997 (Figures 7.30a and 7.30b) [153, 154]. The regression curves were derived through probit analysis taking into account the higher annoyance scores. The result of the

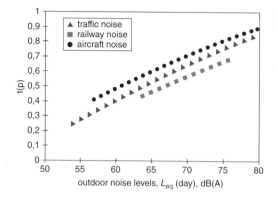

Figure 7.30a Comparison of noise dose-general annoyance functions for road, aircraft and railway noises (field data) [153].
Note: Probit analysis on the respondents who declared higher annoyance.

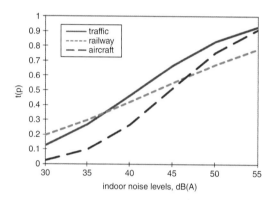

Figure 7.30b Comparison of noise dose-general annoyance functions for road, aircraft and railway noises (experimental data) [153].
Note: Probit analysis on the respondents who declared higher annoyance.

comparative study revealed that traffic noise was the major source of annoyance based on the laboratory experiment. Railway and aircraft noises displayed somewhat similar annoyance degrees at $L_{eq,\ indoor} = 45$ dB(A) where 67% of the subjects were disturbed by traffic noise, 55.2% by railway noise and 51.5% by aircraft noise. Above this level, aircraft noise rises sharply over the railway noise impact. Comparison between traffic and railway noises resulted in a similar impact below 35 dB(A) and, above this level, traffic noise proved to be the dominant source of annoyance. When the field data using the facade noise levels were analyzed, the noise impacts from the three sources were somewhat parallel to each other, stressing aircraft noise as the dominant source of annoyance.

Miedama et al. applied a methodology to compare the results of different studies and converted the different annoyance descriptors to a 0–100 point scale. Consequently, they derived Eqs (7.2–7.4) between L_{den} and $HA\%$ (within the range of $L_{den} = 42$–75 dB(A)) [136, 137]. Based on the studies by Miedama et al., the European position paper in 2002, presented the polynomial equations which were organized with respect to the percentage of annoyed people % A after transformation, as given in Eqs (7.5–7.7). It was recommended using % A in rating noise impact (noise score) by acquiring data from the noise maps [50a, 138].

Aircraft noise:

$$\%A = 8.588 \times 10^{-6}(L_{den} - 37)^3 + 1.777 \times 10^{-2}(L_{den} - 37)^2 + 1.221(L_{den} - 37) \tag{7.5}$$

Road traffic noise:

$$\%A = 1.795 \times 10^{-4}(L_{den} - 37)^3 + 2.110 \times 10^{-2}(L_{den} - 37)^2 + 0.5353(L_{den} - 37) \qquad (7.6)$$

Railway noise:

$$\%A = 4.538 \times 10^{-4}(L_{den} - 37)^3 + 9.482 \times 10^{-3}(L_{den} - 37)^2 + 0.2129(L_{den} - 37) \qquad (7.7)$$

Night-time annoyances have been investigated with respect to noise level and sleep disturbance for road traffic, railway, and aircraft noises, in terms of motility (movement during sleep). The mean motility is given in Eq. 7.8 as a function of nighttime noise levels indoors ($L_{night,inside}$) and the age of respondent. On the other hand, the sets of the polynomial approximation in Eqs (7.9–7.17) are recommended by the EU position paper, 2004, based on Miedama's work in TNO, (The Netherlands Organisation for Applied Scientific Research) [50b, 137]:

$$\text{Mean motility} = 0.0587 + 0.000192*L_{night,\,inside} - 0.00133*age + 0.0000148*age^2 \qquad (7.8)$$

For road traffic:

$$\%HSD = 20.8 - 1.05*L_{night} + 0.01486*\left(L_{night}\right)^2 \qquad (7.9)$$

$$\%SD = 13.8 - 0.85*L_{night} + 0.01670*\left(L_{night}\right)^2 \qquad (7.10)$$

$$\%LSD = -8.4 + 0.16*L_{night} + 0.01081*\left(L_{night}\right)^2 \qquad (7.11)$$

For aircraft:

$$\%HSD = 18.147 - 0.956*L_{night} + 0.01482*\left(L_{night}\right)^2 \qquad (7.12)$$

$$\%SD = 13.714 - 0.807*L_{night} + 0.01555*\left(L_{night}\right)^2 \qquad (7.13)$$

$$\%LSD = 4.465 - 0.411*L_{night} + 0.01395**\left(L_{night}\right)^2 \qquad (7.14)$$

For railways:

$$\%HSD = 11.3 - 0.55*L_{night} + 0.00759*\left(L_{night}\right)^2 \qquad (7.15)$$

$$\%SD = 12.5 - 0.66*L_{night} + 0.01121*\left(L_{night}\right)^2 \qquad (7.16)$$

$$\%LSD = 4.7 - 0.31*L_{night} + 0.01125*\left(L_{night}\right)^2 \qquad (7.17)$$

%HSD: percentage of high sleep disturbance
%SD: percentage of sleep disturbance
%LSD: percentage of less sleep disturbance

Figure 7.31 shows the *HSD*% curves with 95% confidence intervals for three types of noise source. The above relationships can be applicable in the range $40 \leq L_{night} \leq 70$ dB(A). Below 40 dB(A) the percentage of highly sleep-disturbed is assumed to be zero.

E) *Effect of impulsiveness and tonality characteristics of noise on annoyance*

Evaluation and rating of noise necessitate information about the acoustic characteristics of noises. Some experimental studies were conducted to define the perceptional noticeability of noises [155]. Based on these investigations, different noise descriptors have been developed by taking into consideration the temporal and spectral characteristics of various types of noises [62]. For example,

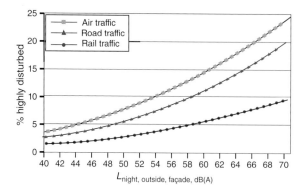

Figure 7.31 High sleep disturbance (*HSD* %) by three types of noise (2005) [165].

variation of noise levels in time was taken into account in the derivation of an earlier noise descriptor (1969–1971), called the Noise Pollution Level, defined as $L_{NP} = L_{eq} + 2.56\sigma$ (σ: standard deviation of noise levels).

Rice (1983 ISVR, in the UK) investigated both the effect of tonal components and the effect of sudden change in sound pressure levels, on disturbance, through laboratory and field studies. He revealed that the impulsive sounds were more annoying depending on the rise in sound pressure levels with time, in comparison to traffic noise. Rice's study revealed that source-specific noise impacts, determined by the laboratory and field studies, were higher compared to the combined annoyance of impulsive and steady noises (Figures 7.32a and 7.32b). Another study by Vos indicated that annoyance from traffic noise was more steady (consistent) at low noise levels, but the response to impulsive noise increases sharply with L_{eq} level [156].

Berry and Porter (2005) conducted several research studies in the National Physical Laboratory, in the UK, on impulsiveness of noises, aiming to find correction factors applicable for rating environmental noises when they have an impulsive nature. It was proposed that when the sudden decrease is about 10 dB per second, the sound is defined as impulsive. Reviewing various methodologies, they proposed an objective penalty for tonal noises applicable to the A-weighted levels [157–160]. Their result has been taken into consideration in the ISO 1996:1971.

Berry also developed an experimental method to search for the detectability of tonality to obtain corrections for rating environmental noises [161]. He proposed 0–5 dB corrections based on a good correlation when the 0–5 point scale was used (if no tonality is detected, the penalty is 0dBA, or if a tonality is detected clearly, the penalty is 5 dB(A)) [160].

Schomer also worked on the impulsiveness of environmental noises and proposed the use of C-weighted *SEL* for impulsive noises and recommended 6 dB addition to C-weighted L_{dn} for conversion to A-weighted L_{dn} level [162, 163].

F) *Noise annoyance from industrial noise*

Relationships between exposure to noise from stationary sources (shunting yards, seasonal industry and other industries) were investigated by Miedama using L_{den} and the curve-fitting functions were derived as follows [50]:

$$\text{Industry}: \%HA = 36.307 - 1.886\,L_{den} + 0.02523\,L_{den}{}^2 \tag{7.18}$$

$$\text{Shunting}: \%HA = 16.980 - 1.367\,L_{den} + 0.02980\,L_{den}{}^2 \tag{7.19}$$

$$\text{Seasonal}: \%HA = 18.123 - 0.887\,L_{den} + 0.01091\,L_{den}{}^2 \tag{7.20}$$

The range of L_{den}: 35–65 dB(A).

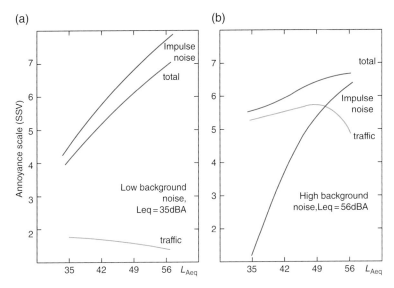

(a) (b)

Figure 7.32 Comparison of dose-response functions for traffic noise and impact noise (Rice, 1983): (a). With low background noise, (b). With high background noise [189]. (*Source:* Permission granted by Rice.)

Nowadays, various new noise sources have appeared in neighborhoods (Chapter 3), such as sounds generated by electronic toys, concerts, fireworks, roller coasters, computer games, etc., in addition to the recreational activities from shooting ranges, jet-skis, water games, amusement parks, shouts, bells, announcement systems for warning or religious purposes, parades, children's playgrounds, etc. Many of them are reported as disturbing noise sources in the specific noise surveys.

The report, published by the EU in 2005, indicates the impacts of other sources is causing serious annoyance in three EU countries: Germany, the UK, and the Netherlands (Figure 7.33a) [165].

G) *Noise annoyance from entertainment noise*

In recent years, amplified music and recreational music have been considered a source of nuisance, particularly at night-time (Section 3.6). The most important negative effect of high-level music is the risk of the threshold shift, for example, in pop music concerts. The electronically amplified music which is enjoyed by many people, was found to be a great source of disturbance in certain situations, depending on various factors, such as the level of sound (outdoor concerts), exposure time and time duration, bass content, interruptions, impulsiveness, content, meaning of the music, and psycho-acoustic factors. Some of these parameters were found to be well correlated with the annoyance. It was recommended that the correction factors should apply to the measured total background noise levels for comparisons with the limit values, by also taking into account seasonal differences [166, 167]. The effects of these sounds, called Movida, have started to be of great concern nowadays and various surveys have reported this as a highly annoying source, even on boundaries of the urban areas. There are precautions, such as strict noise noise limits for maximum levels, restrictions as regards loudspeakers' use and locations, inspections, etc., that can be taken by the owners and local administrations [119, 168–170].

H) *Noise annoyance from construction noise*

Although construction activities and their machine operations are temporary, the negative effects of construction noise have been investigated since the 1970s [171–173] (Section 3.5). Due to the loud noise, impulsive characteristics, containing pure tones, being uncontrollable, occurring

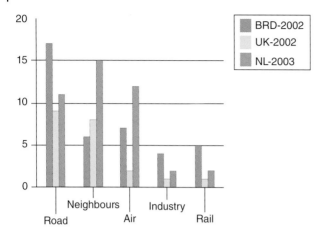

Figure 7.33a Impacts of other sources causing serious annoyance in three EU countries, 2005 [165].

unexpectedly and at undesirable times, they affect – like the other types of noises – the quality of life and are accepted as a contributory factor in health deradation [172]. A field study, conducted in 2000, in a student campus, revealed that the study-related activities were negatively affected by the building construction noise nearby, such as more error in work, interruption of thought, being distracted, etc. [173]. In another study (2010) aiming to manage construction noise in the city, the potential noise impact matrix was developed. The elements of the matrix were noise levels converted into the type of activity, the duration of the activity, and distance to sensitive buildings [174].

I) *Noise annoyance from wind turbine noise (WTN)*

Field surveys, regarding wind turbines and their effects on the living environment in the vicinity, have expanded siginificantly at present, because of the increasing number of wind farms all over the world. However, it is difficult to make assessments about annoyance due to the masking of background noise levels and the influence of other aspects related to the environment as well as personal factors. The acoustical characteristics of WTN, generated by various parts of their structure, are described using terms such as swishing (or whooshing), thumbing, whistling, pulsating/throbbing, resounding, low frequency, scratching/squeaking, tonal, and lapping (Section 3.4). In a study conducted in Sweden in 2004, it was found that swishing was the most annoying character of wind turbines, resulting in the percentage of highly annoyed (*HA%*) of 33% with $r = 0.718$, as the nighttime impact [175]. The same study yielded a dose-response relationship which was found to be statistically significant. It was also revealed that a higher proportion of people, especially those spending time outdoors, was annoyed by WTN, compared to those annoyed by RTN (road traffic noise) at the same indoor noise level, below 40 dB(A) [176]. The other study in the Netherlands (2009) yielded similar findings by revealing that the reasons for high annoyance were "swishing" from wind turbines, temporal variations of noise, continuity of noise at night, and visual impact [177]. A study in New Zealand (2011) reported that the health-related quality of life of individuals (HRQOL by using the WHO scale), was significantly lower in terms of overall quality of life, physical quality of life and environmental quality of life, compared to the results obtained for those living in similar environments without wind turbines. People living in the proximity of turbines up to 2 km declared they had decreased sleep quality and a less restful environment [178]. The two studies were conducted in Japan compiling the results from the social surveys between 2010–2012 on community response to wind turbine noise, and the dose-response relationships were presented

Figure 7.33b Annoyance from wind turbine noise: relationship between L_{Aeq} and annoyance derived from a study in Japan [180]. (*Source:* Permission granted by Kuwano.)

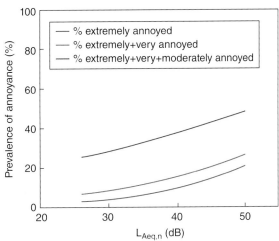

depicting the variation of annoyance with noise level and with distance, to be used as noise control criteria [179a, 179b, 180] (Figure 7.33b). The results provided sufficient evidence that people were annoyed by wind turbine noise more than by RTN with the difference of 6–9 dB in L_{dn} at the same percentage of highly annoyed people. This fact was explained by the visual disturbance, low background noise, and temporal characteristics of wind turbine noise. Also, in the past investigations on wind turbine noise-induced annoyance in Canada, the recent study conducted in 2016 for a larger area evidenced that the long-term wind turbine noise levels (L_{eq}) (WTN) could be as high as 46 dB (A) outdoors. Based on the 1238 interviews for those living between 0.25–11.2 km from the wind turbines, the multiple regression models were constructed by taking into account all the effects of factors on annoyance [181a]. It was revealed that community response to wind turbine noise was found to be statistically related to the A-weighted WTN levels, reporting that the number of "very" or "extremely annoyed" increased up to 13.7% in the areas with the background levels below 30 dB (A). The laboratory listening tests also support this finding; declaring that the same $HA\%$ corresponds to 3–5 dB(A) lower L_{Aeq} of WTN than RTN after visual factors are excluded [181b].

Overviews of the recent studies regarding wind turbine noise annoyance are also given and the findings are evaluated extensively in some published documents [181b, 181c]. The same judgment that WTN causes higher annoyance than other environmental noises is explained as due to: (i) rhythmic characteristics (amplitude modulation on top of the existing annoyance), may be leading the structural vibrations of houses, and perceived indirectly; and (ii) low frequency sounds (≤ 200 Hz) and infrasound that can be heard sometimes contrary to the hearing theories and causing inaudible health hazards, since "the levels of infrasound are comparable to the level of internal body sounds and pressure variations at the ear while walking" [181b].

However, the debates are continuing about whether the infrasounds are causing Vibroacoustic Desease (VAD) and the Wind Turbine Syndrome (WTS), which cover sleep disturbance, headache, tinnitus, ear pressure, vertigo, nausea, etc. which are not scientifically proven [181c]. There is insufficient evidence of the direct health effects of WTN based on the field studies and not a clear correlation found between annoyance and self-reported sleep disturbance. The field studies emphasize the relevance of the visual aspects in annoyance, such as destroying the landscape, shadow occurrence, blinking lights due to moving blades, and sensitivity to noise, however, the search for the influences of non-acoustical variables on annoyance could not give significant correlations because of the complexity of the phenomenon. Some say it might be possible to find the lower reaction of some people benefitting from the WT, according to the on-going conflicts between environmentalists and manufacturers.

J) *Noise annoyance from neighborhood noise*

Although attention paid to neighborhood noise was rather scarce in the past, the surveys have focused on investigating satisfaction with sound insulation in residential buildings. The WHO LARES report, in 2004, reveals the pathological effects of neighbor noise and significantly elevates relative risks in various health problems [96]. It was confirmed that the percentage of people moderately bothered by general neighborhood noise was 35% and strongly bothered was 12.4% whereas annoyance from traffic was 14.4%.

A sample survey carried out in 1997 with 1242 household in the Netherlands revealed the following types of noises from neighbors were heard at night [182]:

Contact noise: 22%
Noise from sanitary fittings, central heating, etc.: 19%
Noise from radio, TV, and hi-fi: 12%
Do-it-yourself noises: 8%
Pets: 6%

The Danish Health and Morbidity Survey, in 2003, revealed that the disturbance from neighbor noise was 32.7% while disturbance from road traffic was 15%, although it was difficult to compare the results due to the differences in methodologies [183a]. In a further study, the survey data acquired in 2010 and 2013 has been analyzed to investigate the relationship between neighbor and traffic noise annoyance and health outcomes [183b]. In conclusion, neighbor annoyance of the adult population living in multi-storey housing has been proved to be associated with poor mental health and high levels of perceived stress. A recent study compiling the survey results in Denmark has revealed that neighborhood noise annoyance was found to be higher than the annoyance from traffic noise and caused fatigue and sleeping problems [183c]. The other negative impacts stressed in the study are: neighbor noises such as voices/shouts, dogs, radio or TV, music, parties, footsteps, children jumping, doors banging, washing machines etc., and these disturb one's own activities at home, i.e. sleeping, using every room in the house, listening to TV and radio, reading, resting, and having a conversation. Above all, the lack of privacy at home provokes annoyance and restrains activities.

So far, there is no model to assess neighborhood noise annoyance, since the noise sources and their acoustic quality diverge significantly.

K) *Noise annoyance from mixed noise sources*

In the field studies generally, the response of people to a specific noise source has been investigated and the source-dependent annoyance degrees have been determined, as explained above. However, the evaluations are more difficult when there is more than one source in an environment [65, 184–186]. Also, it is not easy to develop limits for the total noise in such a combined exposure environment, including noise sources other than what is specifically concerned. In scientific studies this problem is solved by focusing on the target noise and taking the others as background noises.

Researchers have turned their attention to assessment of the combined effects on people living in the mixed environments. Normally traffic noise exists in combination with either railway or aircraft noises or industrial noises. Fields declared that if a dominant noise source exists, then people's response is not affected by the other noises or the effect of other noises would be very low. Miedama and Vos revealed that at low levels of noise, the aircraft noise was more annoying in comparison with road traffic noise existing at the same time [136]. Izumi conducted the laboratory tests in 1990 and evaluated the separate and combined effect of train and road noises. He found that the source-specific noise was prominent compared to the total annoyance [75, 147].

Great efforts have been spent in development a complex matrix for the assessment of the combined noise impact [147, 148, 184–186]. Flindell in 1983 compared the effects of interrupted and continuous noises and proposed a correction factor in relation to exposure duration [148]. Taylor proposed the assessment models in Eqs (7.21–7.26) to evaluate the total effect of aircraft and traffic noises [184]:

1) Energy summation model:

$$A = f(L_T) \tag{7.21}$$

$$L_T = 10 \lg \sum_{i=1}^{n} 10^{Li/10} \tag{7.22}$$

L_i: Noise level, L_{eq} from specific source, i
L_T: Total noise level
A: Annoyance response to the combined sources
n : number of sources in the environment

2) Independent effect model:

$$A = f_1(L_1) + f_2(L_2) + \ldots + f_n(L_n) \tag{7.23}$$

$L_1, L_2, L_3, \ldots L_n$: L_{eq} values for distinct sources
$f_1(L_1), \ldots \ldots f_n(L_n)$: Functions determined for each source separately

3) Energy difference model:

$$A = f_1(L_T) - f_2(|L_1 - L_2|) \tag{7.24}$$

4) Response summation model:

$$A = f \left\{ L_t + \sum_{i=1}^{n} D_i 10^{(L_i - L_T/10)} \right\} \tag{7.25}$$

D_i: Effective level of source, i

5) Summation and inhibition model:

$$A = f(L_T + E) \tag{7.26}$$

E: Correction factor for the summation and inhibition effects among sources. A chart is given for correction E.

Taylor declared that the best model was the energy difference model. On the other hand, dealt with the combined source situations, Rice proposed the dominant source model, in 1986 [65].

To assess the combined effect of various noise sources, the annoyance equivalent model has been developed. Briefly, the procedure is to transform the noise from the individual sources into the equally annoying sound level of a reference source (which is road traffic), and then to sum these levels to obtain the total level L_T. The assessment of the total noise level and the corresponding percentage highly annoyed *(HA%)* for combined exposures can be computed as follows.

HA% is given as a function of L_{den} for air, road and rail traffic noises in Eqs (7.27–7.30). recommended in the EU position paper, 2002 [50a]. Assessment of the total noise level and the corresponding *HA%* in the mixed-noise environments is made by applying the procedure as suggested by Miedama [139]:

1) Determination of L_{den} for traffic, train, and aircraft noises as $L_{den,road}$, $L_{den,train}$, and $L_{den,air}$
2) Calculation of *HA%* for railways and aircraft noises using Eqs (7.3) and (7.4).
3) Calculation of equally annoying road traffic noise (RTN) levels for railway and aircraft noises (*i*: either rail or train):

$$\begin{cases} 48.85 + 168.9\, F\,(\%HA_i) - \dfrac{0.8843}{F(\%HA_i)} & \text{for } L_{den,i} > 42 \\ re(L_{den,i}) = L_{den,i} & \text{for } L_{den,i} \le 42 \end{cases} \tag{7.27}$$

Where

$$F(x) = \left(-2.374 \times 10^{-4} + 1.05 \times 10^{-4}x + \sqrt{2 \times 10^{-7} - 5 \times 10^{-8}x + 1.11 \times 10^{-8}x^2} \right)^{1/3} \tag{7.28}$$

4) Calculation of the total noise level:

$$L_{den,T} = 10\lg\left(10^{0.1 \times re(L_{den,air})} + 10^{0.1 \times L_{den,road}} + 10^{0.1\, x\, re(L_{den,rail})} \right) \tag{7.29}$$

5) Calculation of *HA%* for the combined, multiple sources

$$\%HA\% = 9.868 \times 10^4 (L_{den,T} - 42)^3 - 1.436 \times 10^{-2}(L_{den,T} - 42)^2 + 0.511810^4(L_{den,T} - 42) \tag{7.30}$$

Similar to above, the night-time sleep disturbances at highly, moderate and less disturbance levels have been presented under the combined exposure.

7.3.3 Laboratory Experiments on Annoyance

Investigations of annoyance assessments can also be performed in the acoustical laboratories under controlled conditions. These studies need careful design and implementation according to the strategies adopted [75, 154, 156, 184, 187–194]. There are important aspects to be considered in the simulation of noise and the environment and in the selection of the subjects. Generally, it is known that annoyance increases when people are tired, due to increasing fatigue. Parallel to that, it was found that the subjects were more concentrated in the laboratory experiments and more careful in answering questions, thus resulting in higher annoyance degrees compared to the real condition. Therefore, the subjects in the laboratory, are given some simple tasks to distract their attention rather than focusing only on listening to noise.

Physical Properties of the Laboratory
Acoustical laboratories include listening rooms whose physical properties are defined in Chapter 5 and mainly are used for subjective tests like loudness and noisiness rating, comparisons between the sound sources, emission tests for loudspeakers, etc., for investigations on the psycho-acoustic quantities explained above, different spatial characteristics of sound fields, or location of sources, etc. [191]. The laboratory room should be constructed in compliance with the sound insulation requirements and eliminating floor vibrations. The room dimensions and surface materials should be selected so as to avoid the room modes, as explained in Chapter 2. When these rooms are used for the judgment of loudness and noisiness, the test sounds, recorded binaurally using an artificial head, are given to the listener (subject) through earphones with a short duration and at intervals. The subject fills in the questionnaire form during the experiment. If such a laboratory has to be used

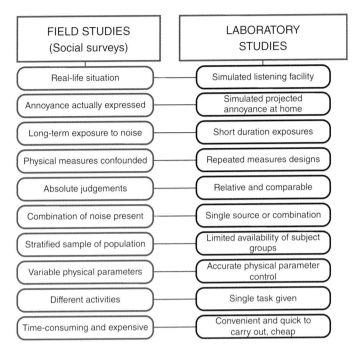

FIELD STUDIES (Social surveys)	LABORATORY STUDIES
Real-life situation	Simulated listening facility
Annoyance actually expressed	Simulated projected annoyance at home
Long-term exposure to noise	Short duration exposures
Physical measures confounded	Repeated measures designs
Absolute judgements	Relative and comparable
Combination of noise present	Single source or combination
Stratified sample of population	Limited availability of subject groups
Variable physical parameters	Accurate physical parameter control
Different activities	Single task given
Time-consuming and expensive	Convenient and quick to carry out, cheap

Figure 7.34 Comparison between field studies and laboratory experiments. (*Source:* adapted from [65]. Permission granted by Rice.)

for annoyance judgments, the sound diffusion and reverberation time of the room are important as they are effective in the sound signals given into the room through a sound emitter. The volume of the room should be $V > 30\,m^3$ to avoid uneven frequency response and the reverberation time should be about 0.5 second.

Location of the loudspeakers is determined so as to provide sufficient diffusion in the room. The correct annoyance judgments can be obtained if the real conditions are simulated adequately for the long-term annoyance on listeners.

Evaluations in terms of annoyance degrees from a specific source in laboratory are expected to equal the real life conditions. Thus, various physical environmental, and non-acoustic factors should be taken into account. For example, a stable temperature degree in the test room is important because increasing the temperature causes somewhat higher annoyance. Comparison of the laboratory and field experiments (benefits and disadvantages) are given in Figure 7.34 based on the earlier discussions highlighted by Rice in 1984 [65].

The Method to Be Implemented in the Laboratory for Annoyance
Deriving the results, which are close to those of real conditions in the field, depends on the success of the simulation techniques. The design and execution of a laboratory test have been explained in a number of publications in the past [154, 188]. A summary is given below:

a) *Simulation of environment:* The listening room which will be used for annoyance evaluations should be decorated like a normal living room or bedroom or an office environment and, prior to the experiment, the subjects should be asked to imagine this room as their own living room. The noise signals are given by the sound emitters hidden behind the walls or curtains. For visual simulation of an outdoor noise source, a false window is helpful, even playing video clips shot in the real environment. Figure 7.35 shows some experiments conducted in Japan and the UK in the past.

b) *Acoustic simulation:* The noise samples to be used in the experiment have to be recorded in the field under the real conditions by using sensitive equipment without losing the original characteristics of noise. The test signals should be processed according to the purpose of the study and the acoustical characteristics of the noise sources that can be controlled, such as total sound levels, peak levels, number of noise events, time intervals, spectral and temporal characteristics, mixture of different sounds, etc. A transfer function is determined to apply to the test signals according to the sound insulation characteristics of the external walls and windows, so as to correspond to the indoor sounds. The background noise in the laboratory, e.g. from an air-conditioning (AC) system or lighting equipment or others, should be eliminated. If required, a specific background noise can be mixed with the spectrum of the indoor sound. As an example, the equipment set-up used in a simulated environment laboratory experiment has been given in Chapter 5, (Figure 5.48b) [59, 154].

Categorizations of the parameter values are necessary according to the design of the experiment. To avoid bias or adaptation of the subjects, the sound samples are given into the room in a random order to be designed prior to the experiment.

c) *Duration of the experiment:* In order to get their opinions of annoyance, subjects should be exposed to the noise samples in a period of time which is sufficiently long. Generally, in the psychoacoustic experiments, e.g. for noisiness and loudness judgments, very short durations of speech signals are used, since the short-term memory is required for perception and remembering the sounds containing words. Sometimes a TV or a computer screen is helpful. The subjects are allowed to repeat the signals until they make their decisions properly, especially in the

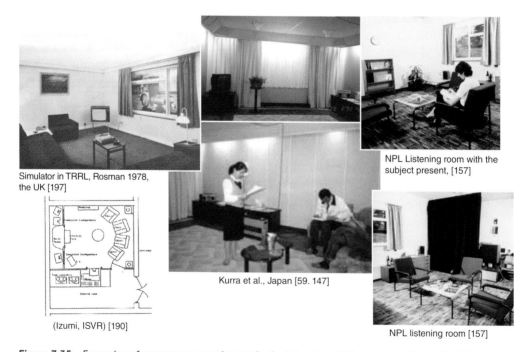

Simulator in TRRL, Rosman 1978, the UK [197]

(Izumi, ISVR) [190]

Kurra et al., Japan [59. 147]

NPL Listening room with the subject present, [157]

NPL listening room [157]

Figure 7.35 Examples of annoyance experiments in the laboratory (listening room) [compiled from 154, 157, 187, 190, 197].

comparative tests. However in annoyance evaluations, the longer period implies higher annoyance and the researchers recommend half an hour period as sufficient duration of exposure for annoyance responses, also considering the feasibility of the study [156]. To eliminate focusing too much on listening noise, the subjects are given simple tasks or mental work to perform during the test, like playing cards, solving puzzles, reading a book or magazine, listening to the news, or watching TV. The experiment should be designed to take into account the total period of study, the number of subjects, technical facilities, the project budget, etc.

Figures 7.36 and 7.37 give the details of the procedure that can be applied in such experiment [187,188].

d) *Selection of the subjects and organization of the sessions:* At the beginning of the experiment, the subjects take auditory tests to ensure that they have normal hearing. The total number of subjects (voluntary or paid) and equal distribution according to the individual characteristics, such as gender, age, socio-economic, and educational levels, etc., are initially decided as in the field

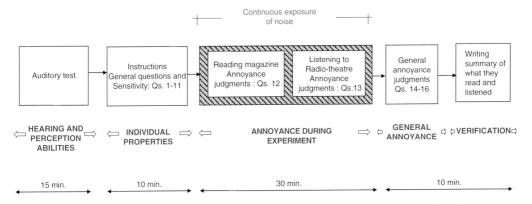

Figure 7.36 Test procedure applied to the subjects during the laboratory experiment [187, 188].

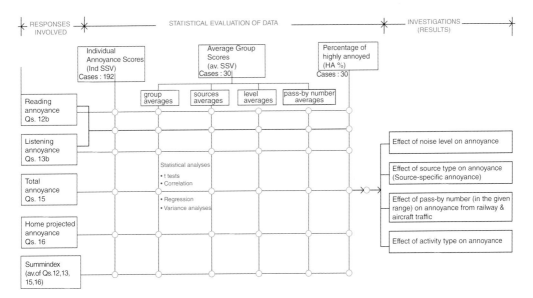

Figure 7.37 Evaluation of experimental data [187, 188].

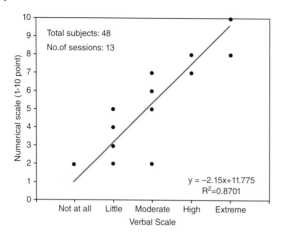

Figure 7.38 Relationship between verbal and numerical scales of annoyance [154].

studies explained in Section 7.3.2. In the design of the experiment, the number of sample noises, a list of the subjects with their identification number, number of sessions, number of subjects to be surveyed at each session, duration of sessions, order of the sound signals to be given into the room, order of sessions (can be designed by using the Latin square method) and other related issues are decided and the sessions are scheduled [194].

e) *Preparation of the questionnaire form:* The reliability and accuracy of the test are governed by an appropriate inquiry technique somewhat similar to the field studies. The surveys are conducted by using a questionnaire form with well-organized questions, including both verbal and numerical scales for annoyance judgments. Figure 7.38 presents the correspondence of both scales derived from a laboratory experiment using 48 subjects in different test sessions [154]. The questionnaire form to be prepared is associated with the scenario of the experiment and is short and comprehensible. Application of pilot tests is necessary, as in the field studies. If the results are intended to be compared with the field results, the similar form of questions and evaluation techniques should be used.

f) *Execution of the tests:* It is also important to assist the subjects so that they feel as if they are at home. At the beginning of the experiment the trained surveyors explain the purpose of the test and briefly give instructions about the procedure. Warning systems like blinking a light can be used to indicate termination of the sessions. Technicians should be available to solve technical problems in the control room. A sufficient number of pilot experiments should be performed prior to the study.

g) *Evaluation of the results:* The response data is analyzed to determine the effects of noise on the subjects, namely, annoyance degrees, similar to the field studies explained in Section 7.3.2.

Overview of the Laboratory Studies on Annoyance from Environmental Noises

The laboratory experiments regarding assessment of annoyance were the focus of great attention in the 1980s (Chapter 5), however, they are less used at present, mostly they are oriented toward evaluation of satisfaction from insulation performances of building elements [195, 196]. Among the premier studies conducted in the laboratory for annoyance research, the following studies are worth mentioning:

In research conducted in 1990, the aim was to compare the annoyances from construction and traffic noises. The noise samples were categorized as L_{eq} (indoor) = 40, 50, 60 and 70 dB(A) with the background levels changing as 0, 10, 20, 30 dB(A). Number of subjects was 48 subjects and the 7-point annoyance scale was used in the experiment. Duration of each session was six minutes [164].

Another study (2006) aimed to investigate the annoyance degrees from entertainment noise and to derive a criterion for amplified music. Various types of musical sounds, consisting of different spectrums and time-varying impulses, were used, like rock music, sports events playing on TV, karaoke music (vocal music). It was found that the C-weighted L_{eq} is recommended as the best noise descriptor for amplified music which is found to be well-correlated with annoyance, whereas the A-weighted L_{eq} is recommended for sports and vocal music [169]. Eventually, L_{Aeq} (five minutes) = 34 dB as indoor level with the closed windows and $L_{Aeq} = 47.5$ dB(A) for outdoors were proposed as the acceptable limits.

Traffic noise has been the most investigated noise in the laboratory experiments. Among the earliest studies, the one conducted in TRRL in the UK [197], and those by Rice and Izumi, are the leading experiments by introducing the techniques for laboratory simulations and yielding valuable findings about the annoyance from transportation noises [189, 190].

In the study performed in Sweden by Öhrström, the noise signals recorded in the field were grouped into two levels as 45 and 55 dB(A) and the responses of subjects were received during various tasks. A linear annoyance scale of 100 mm was used for assessment of annoyance degrees. The duration of each test was 10 minutes [192].

An experimental study conducted by Kurra et al. in 1991–1992 in collaboration with Kobe University in Japan was designed with the following objectives [154, 187, 188]:

a) Effect of source type (three types of transportation noise sources) on annoyance: Road vehicles, trains and aircraft.
b) Effect of activity type (reading and listening) on annoyance.
c) Effect of individual factors on annoyance responses.
d) Comparison of the field and laboratory data to search for differences between annoyance responses.

The 1–7-point annoyance scale was used in the annoyance evaluations. Ultimately overall annoyance and task-specific/source-specific annoyances, and the effects of various factors on annoyance in the simulated home environment were evaluated and the significance of each factor was statistically analyzed. Some of the results are displayed in Figure 7.30b.

Öhrström, in 2007, in Sweden, conducted several laboratory studies to discover annoyance assessments from road and railway noises, in addition to the mixed noise sources. The findings of these studies and criteria recommendations were included in the WHO documents later and established the basis for soundscape evaluations in the later years [192].

7.3.4 Comparison of Field and Laboratory Study Results on Annoyance Responses

When the surveying techniques and the noise samples recorded under real conditions are used in the laboratory experiments, the results can be compared with the measured results for validation purposes. As an example, the dose-response functions derived from the study explained above were compared with the earlier field data [153]. The comparisons were made based on the results after probit analyses performed on the cumulative values of the responses. Figures 7.30a and 7.30b compare three types of noise sources. The group median annoyance scores correspond to those of 50% of subjects giving scores of four and above in the annoyance evaluations. Variance analysis indicated differences between the laboratory and field studies. Effect of source type on annoyance in the laboratory was not found to be significant (confidence level of $p < 0.005$ in laboratory with $r = 0.1806$), whereas it was significant in the field (p < = 0.0086). This situation can be explained as due to the reasons given above.

7.3.5 Investigations on the Effects of Noise in Schools

Swedish scientists conducted various studies, in 2004–2005, on the effect of aircraft and traffic noise in schools [151]. In a project supported by the EU, 2800 students aged 9–10 years old were surveyed in countries in Europe and the intelligibility test were applied [152]. It was found that the relationship between the cognitive memory remembering knowledge was well correlated with aircraft noise and the report emphasized the excessive impacts of aircraft and road traffic noise on children's cognition and particularly reading.

Various other surveys on the effects of noise on the performance of both teachers and students were published. Another comparison study proved that the aircraft noise was a more disturbing noise compared to road traffic noise. In addition, it was remarked that the social support of the parents could be a protection factor for the students and their stress could decrease in the quiet home environment.

7.3.6 Noise Impact Assessment in General

Environmental impact assessment (EIA) is defined as "the possible effects of an action (activity) in the physical or social environment of humans or in nature – wildlife – in the short or long term" [198]. EIA also covers noise pollution and declares the assessment techniques based on analysis and synthesis of the data. Rating of noise impact in terms of different descriptors has been concerned since the 1970s, as outlined in Section 6.4.2. Based on the field surveys explained above, the consequences can be generalized to global assessments of noise pollution as a part of the EIA. In an earlier study (1994); Lambert and Vallet established the basic strategies with the aid of GIS technology as one of the first applications at that time, and later, in 2000, they estimaed that 80 million people in the European member states could be exposed to noise levels above 65 dB(A) [199, 200]. The updated information on the noise impact assessments, which are important in noise management and policy statements, will be discussed in Chapter 9.

Since noise maps are taken as the basis for impact assessments at present, the demographic data and associated noise levels are obtained through noise maps within the noise classes, i.e. people living in buildings remained in certain successive noise contour areas. The noise scores can be derived based on the dose-response relationships given above and using the demographic data. Ultimately the hot-spots implying the priority areas can readily be determined for any size of area concerned. The details of such studies are given in Section 6.4.2.

7.4 Community Reactions to Environmental Noise

According to sociologists, a community implies groups of people living in a restricted geographical area who possess a common interest and the same culture, with similar activities, comprising neighborhoods and families and individuals of various ages. Communities might have several problems but health is the most important issue. Noise pollution is a community's problem since it is directly related to a community's health and welfare. Consequently, if the level of exposure reaches a certain level (degree), the community reactions inevitably emerge. The reactions are revealed as different behavioral actions – as explained above – depending on various factors [81–84, 115].

In earlier reports, types of reactions against noise and noise-makers in a community were simply declared by relating the noise levels as below:

Noise level, dB(A)	Reactions
<55	No reactions
55–65	Acceptable but disturbance for sensitive people
60–65	Behavioral responses like closing windows
>65	Severe reactions

Reactions to noise are differentiated according to the socio-cultural and economic status of the community. For example, it has been shown that the complaints from environmental noise increase in regions where the level of socio-economic status of the population is high. Contrarily, the complaints were shown to be higher in the areas where low-income groups were living, likely because of the weakness in coping with noise as in other problems.

Investigations on the types of community reactions first started in 1973, undertaken by the Environmental Protection Agency (EPA) in the USA, and the relationship was established between the complaints and noise levels [129]. The sociological reactions of a community were grouped as:

4: No complaint
3: Frequent complaints
2: Writing petitions to local authorities
1: Serious obstructions

In 1974, the EPA declared the following reaction levels against aircraft noise, according to the Assessment of Reactions Method [201] as shown in Figure 7.39:

I) No individual complaint
II) Frequent complaints with some neighbors
III) Individual complaints against noise-makers
IV) Extension of complaints
V) Group organization for the complaints
VI) Organized group actions (demonstrations
VII) Increased group actions and legal actions?

Figure 7.39 Assessment of people's reactions to noise, 1974 [201].

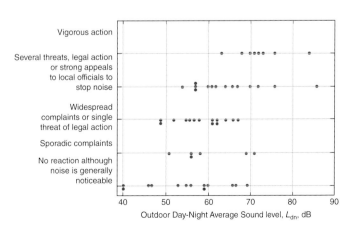

These types of reactions in relation to *CNR* for aircraft noise were displayed by Fidell in 1979 (see Figure 7.22) [129, 202]. However, he suggested some corrections by taking account of various factors affecting reactions, such as background noise, type of neighborhood location (quiet, rural, suburban, in city apartments, near industry or heavy industry, mixed types areas), climate, day-time or night-time, type of noise (continuous, intermittent, tonal, etc.).

In the WHO report on Nighttime Noise Guideline [49], the complaints from airport noise at night-time were investigated in Europe and a number of complaints were analyzed around the airports like Manchester, in 2003, and Amsterdam, in 2005. In the first study, by selecting a unit of 1 complaint per 1000 aircraft traffic in the airport, 10 complaints were found in the day-time and up to 80 at night-time. Regarding the increase of complaints about noise level, one complaint was found on average at 70 *PNL* dB and two complaints at 114 *PNL* dB. For Amsterdam Schipol Airport, based on the relationship between complaints and L_{Aeq}, the threshold for day-time and night-time complaints was suggested as around L_{den} 45 dB and L_{night} 35 dB. Above these levels the percentage of complaints increased to about 7% of the population.

Noise complaints for other types of noise sources are involved in the administrative and legal stages as given in Section 8.4.5.

7.5   The Soundscape Concept, Objectives, and Implementations

The urban environment where people live, relax, gather socially, do various activities or just move from one place to another, can be investigated acoustically in terms of two aspects [203]:

1) The acoustic environment, which is determined by the physical measurements using physical parameters, such as sound pressure levels, the spectral and temporal pattern of noises.
2) The acoustic environment, which is perceived and understood by people in context, determined by means of specific evaluation techniques or psychological experiments.

Both acoustic environments exist simultaneously, while the former requires physical instruments and objective assessment techniques, the latter is called the soundscape, which needs scaling methods for human perception. Soundscape is a descriptor of the environment as well as the other descriptors describing physical attributes of the environment.

The soundscape was called acoustic ecology initially, and it is the acoustic environment whose characteristics are described by people associated with their subjective perception as a result of the action and interaction of natural and individual factors. Schaffer introduced the concept of the soundscape in 1969 as an ecological approach against noise pollution in natural sound environments [204]. Thomson defined a soundscape as "like a landscape, a soundscape is simultaneously physical environment and a way of perceiving that environment; it is both a world and a culture constructed to make sense of that world." He declared that the objective of soundscape as "a way of perception of the physical environment," could be developed "to establish a meaningful world" [205].

Ecological concern about the sonic environment may be oriented either to a specific noise source in an exposed area for the purpose of noise control or modification of source (direct acoustic ecology), or to the environment with a multitude of sources (diffuse acoustic ecology). The soundscape is the aural equivalent of the landscape, hence the analogy of landscape policy, landscape planning, and landscape management can be constituted for soundscape policy, soundscape management, and soundscape planning to be pursued by the competent public authorities.

The physical environment often refers to an outdoor area, space or location, either man-made or natural, comprising various other properties that might influence the sonic environment. The soundscape covers the temporal and spatial variation of sounds in the physical environment, generated by various dynamic sources [206a, 206b, 207]. Since these audible and perceived sounds vary in time (i.e. in the long or short term or even instantly) and in space parallel to variations of the physical environment, the soundscape has a permanently varying nature. In such an environment, although the satisfactory sounds dominate at certain times, some disturbing sounds may exist at other times. "Soundscape is an emphasis on the way the acoustic environment is perceived and understood by the individual or by society" and exists through human perception of the acoustical environment, but always within the context of a particular time, place, and activity [206a, 206b, 207].

According to the new standard ISO 12913-1:2014, "Soundscape is an acoustic environment as perceived or experienced or understood by a person or people, in context" [208]. It is accepted as a perceptual construct distant from physical phenomenon and exists through human perception of the acoustic environment.

Several methodological procedures have been proposed in the literature to evaluate and categorize soundscapes.

7.5.1 Sound Quality and Soundscape

Although discussions among experts, on various aspects of soundscape and on the related terminology, have been continuing, the following concepts are important in the understanding of a soundscape.

Sound Quality

Sound quality is a perceptual reaction indicating how acceptable the sound is. As briefly referred to in Section 7.2.4, it is defined as "a concept of audible suitability of a product when compared with a person's expectations", widely applied in audio and manufacturing engineering. The subject is associated with the cognitive and emotional situation of a person. The cognitive processes create an emotional effect (cognitive response), such as satisfaction or dissatisfaction, to be measured by using a response measuring method. Like the sounds of products, characteristics of urban sounds are also described by the sound quality associated with the human/user sensation. There are various modifiers in sound quality evaluations and one of the determinants is the meaning attributed to sounds.

Context

Context is defined as the relationships between a person, an activity and a place, in space and time influencing soundscape. Sounds and the acoustic environment are the major determinants in context. Figure 7.40a outlines the process of soundscape evaluation associated with context and the affective variables, referred to in the standard [208, 209].

Acoustic Environment

The acoustic environment implies the sound at the receiver coming from all sound sources and modified by the environment due to meteorological and physical factors, such as absorption, reflection, diffraction, etc. The acoustic environment can be actual or simulated, outdoors or indoors. Perceiving, understanding, and experiencing an acoustic environment depend upon complex relationships between various aspects [208, 209].

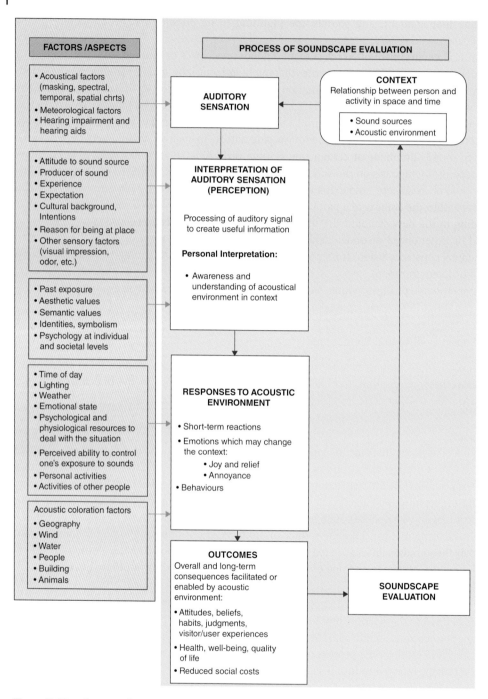

Figure 7.40a Process of soundscape evaluation and relevant factors. (*Source:* based on [208] and [209].)

Sounds

Human assessment of a soundscape critically depends on distinguishing between different sounds such as mechanical sounds and those produced from natural sources, etc. In some documents the sounds are simply divided into two groups as generated by nature and by human activity (road traffic, birds, voices, footsteps, etc., and the distribution of sound sources in space and time is concerned. Natural sounds and other sounds heard in the rural environment create a continuous flow of non-visible but highly attention-demanding sounds. The invasion of non-visual sounds with a great variety in character provides sense impressions.

ISO/TS 12913-2:2018 on data collection and reporting requirements, covers the taxonomy of the acoustic environments and sound sources to be used in soundscape studies. The outdoor sound sources are given in categories, such as urban, rural, wilderness, and underwater acoustic environments and focus on urban environment. The sounds are divided as generated or not by human activity. The latter contains nature (wildlife, wind, water, thunder, earth/ice movements) and domestic animals [210].

Auditory Sensation

Auditory sensation is a function of neurological processes and is influenced by the acoustic environment and non-acoustical factors, such as meteorological factors and their variations, hearing impairment etc. Physical descriptors that correlate with auditory sensation are the acoustical parameters of exposure, such as sound level, frequency, and temporal variations [208].

Human Perception

As referred to briefly in Section 7.1.2, perception is the interpretation of auditory sensation by the processing of the auditory signal to create useful information. The factors which influence the perception of environmental sounds are listed by some researchers as follows:

1) Personal factors: Activity, personal goals, traits, psychological state of mind.
2) Environmental factors: Other people, landscape and architecture, lighting, source location and time period and the moment.
3) Auditory factors: Loudness, spectral content, temporal variation, information content, meteorological factors: temperature and humidity.
4) Other senses: Sight and smell.

Figure 7.40 displays these factors in different groups according to different processes of soundscape evaluation.

The perceptual cues for discriminating sounds, are categorized into six groups:

- Homogeneity
- Details
- Sharpness
- High frequency sounds
- Localization
- Loudness

Based on psychoacoustics and sound quality theory, the relation between sound and its perception can be established. For example, a single sound event has higher priority in perception. Irregular sounds are more annoying, whereas the sounds which can be identified and localized are apparently less annoying than those which evoke confusion [211, 212].

Researchers have succeeded in linking the acoustic and psychoacoustic parameters to the subjective judgment of environmental sound by considering its dynamic properties. Coensel, Botteldooren, and Muer attempted to find similarity between musical sounds and soundscapes in rural and urban environments, in terms of the long-term variations in level and pitch of different kinds of music [211]. They found that all music types exhibit the same $1/f$ behavior both for level and pitch variations and, the same behavior appears in many soundscapes within the frequency interval 0.2 Hz–5 Hz, corresponding to a few seconds. It was proposed that this finding can be used to categorize the sounds according to certain time intervals. However, for longer time intervals, the loudness slope in most of the rural soundscapes is steeper than $1/f$. "In fact, a music comprising a flat and balanced spectrum, is perceived as 'chaotic' and non-comprehensible, whereas a spectrum with sudden drops is much more predictable and thus boring" [212].

In an acoustic environment, the acoustic and visual components interact in human perception. Studies reveal that auditory perception improves when sound is accompanied by the related visual display [213]. Soundscape is experienced by keeping sounds in memory and joining auditory information together with visual information enhances the efficiency of memorizing.

Identification of Acoustic Environment

The prominence of the sounds and potentially the ratio of certain sound types in the acoustic environment play an important role in soundscape evaluations [214]. Similarly, the areas of high acoustic quality are identified as wanted or unwanted or liked or disliked, in particular contexts, not just by the levels of sound. The attributes with respect to identification of place and its importance, are outlined below:

- *Liveliness, naturalness:* relaxation, solitude, well-being.
- *Nature appreciation, nostalgic attachment:* safety, tranquility, sense of excitement.
- *Peacefulness:* satisfaction, uniqueness, providing clarity and conveys safety.
- *Place attachment:* sense of control, variety, providing information.
- *Hearing sound marks:* a soundscape that has unique cultural or natural characteristics.

Preference

Soundscape quality is expressed in terms of human appreciation or preference but cannot be determined by physical measures. The preference for the soundscape and the context are determined in terms of the attributes, such as acceptability, appropriateness, clarity, comfort, communication, enjoyment, excitement, happiness, harmony [215, 216]. Some researchers have suggested that the percentage of time when the attended sound source exists would be a better predictor for preference. Sounds of preference are also involved in the field of building acoustics for the design of music halls, but not used in environmental noise control.

Noise Control Versus Soundscape

Sound is conceived as a waste product in the environmental noise approach, to be reduced and managed. It focuses on sounds of discomfort and adverse health effects. On the contrary, in the soundscape approach, sound is perceived as a resource and the focus is on sounds of preference. Brown stated:

"The soundscape field regards sound largely as a resource-with the same management intent as in other scarce sources, such as water, air and soil: rational utilization and protection and enhancement where appropriate. The resource management has a particular focus on the usefulness of a resource to humans and its contribution to the quality of life for both present and future generations" [215].

The acoustic environment measured by physical parameters relies on community noise surveys for assessment. A soundscape is a positive approach emphasizing the wanted sounds and applying listening tests.

Environmental noise management includes measures to integrate all sounds at a receptor and focuses on reducing levels. In the soundscape approach, the outdoor acoustic environment is a resource whose diversity is to be managed and enhanced, as complementary to noise management.

The soundscape approach requires a differentiation between sound sources with the aid of sound source identification and human judgment, targeting the wanted sounds and not being masked by unwanted sounds. The measures should minimize the negative impacts by changing the meaning and perception of the acoustic environment.

Benefits of Soundscape

Brown emphasized the "beneficial use" of soundscape to make decision-makers aware of the inadequacy of applying a single criterion (such as reducing annoyance) in the management of outdoor noise. The concept of beneficial use covers [215]:

- the wilderness experience,
- restoration of health and well being
- respite, relaxation
- enjoyment or excitement
- enhancement of culture
- safety and security
- wildlife habitat protection
- diversity in the outdoor acoustics and sustainability

END, Directive 2002/49/EC requires the identification of quiet areas in agglomerations and in open country, based on noise maps [28]. Actually, this task is accomplished based on noise level criteria, such as 40–55 dB(A), as described in Chapter 9. Nowadays, it is accepted that the notion of quiet cannot be limited to only noise levels, but also must be complemented by the definition of high quality soundscapes. The concept of "natural quiet" used for parks, gardens, or forests, is defined as the absence of mechanical noises, but containing the sounds of nature, such as wind, streams, and wildlife. Natural quiet implies that man-made sounds are absent, not audible above natural sounds and audible for a smaller percentage of time.

As stated, "Soundscape has become a major tool in facilitating people's involvement in decision processes about acoustical environments" [209].

7.5.2 Methodology of Soundscape Research and Design

A good soundscape design needs a proper methodology for implementation. The new standard ISO/TS 12913-2:2018 aims to construct a systematized framework for perceptual data collection and to analyze it to allow comparisons of studies, to provide design tools, and to improve perceived sonic environments [209].

Procedure for Soundscape Assessment

Although a standard methodology on soundscape research has not been issued at present, the initial step of soundscape design in a study area is the assessment of soundscape by applying the following process:

Collecting information: In urban areas, the soundscape is composed of background sounds and individual sound events. The background sound mainly determines the overall feeling of quietness, which is the basic quality of the soundscape [206a]. Quality assessment of urban quiet areas has to address the quality of both background and sound events, which are perceived with recognition, accentuation or disturbance and association [216–218a]. Collection of sound data –meaningful knowledge about the area – can be acquired by organizing sound walks in the urban study area and open interviews and with the help of local experts living in the area [209], and proper acoustic measurements [218b]. During sound walks, the tape recorders can be given to the people (subjects) and they are asked to describe spontaneously their impressions of places through which they travel. A sound walk should be led by a moderator and the interview with a certain questionnaire are both described in the ISO/TS 12913-2:2018 [210] (Figure 7.40b). The questions are grouped as: (i) source identification (do you hear or not, at what levels?); (ii) perceived affective quality of sounds; (iii) assessment of surrounding environment; and (iv) appropriateness of the sound environment, to be answered according to given verbal scales. The subjects are requested to rank the sounds at the end of the experiment. The binaural measurements using an artificial head is explained at in the stationary condition and with the time intervals, so as to cover typical sources and sound events.

The next step is the classification of urban sound events recorded in real-life conditions and the preparation of a database, applying a method, such as neural network-based models [219, 220], or inventory methods, such as a "machine learning-based aggregation scheme" proposed for complex and dynamic urban sounds by combining their local and global acoustic features [221a].

Information about the noise impact in the study area (e.g. annoyance from major sources) can be obtained through the noise maps and field surveys.

Figure 7.40b Sound walk and sound recording spots in the old city, Diyarbakır (Organized by Turkish Acoustical Society, 2019).

The analysis of data to be collected, according to ISO/TS 12913 Part 2, is explained in Part 3 of the standard ISO/DTS I12913-3:2019 which is under development [221b]. Briefly, the draft standard describes the techniques for the analysis of quantitative data through statistical methods (significance tests, correlations, etc.) and analysis of qualitative data through transcriptions, text analysis methods, etc. and reporting, as well as the analysis of binaural data to characterize the acoustic environment at the receiver.

Identification of sounds in the field and laboratory: To evaluate the identity of sounds which are perceived varies with the effective factors:

A) Uniqueness or singularity of local sounds in relation to those of other cities.
B) Informative condition or the extent of the activity in place and spatial form.
C) Delightfulness of the sounds (i.e. the quality of sounds which caused them to be liked or disliked).
D) Correlation between the visual and auditory perception (whether supported by visual activity).
E) Categorization of the types of sounds according to preferences of people (or like/dislike).

Laboratory experiments can provide information about the preferences and qualities of sounds that are given to subjects as prepared settings according to certain categories. Evaluations are made by using psychoacoustic techniques. Various subjective testing methods, e.g. classic ones, such as the semantic differential method (SD), categorical scaling (CS), etc., were applied in the earlier studies by giving sets of sound signals of different types (low, mid and high frequencies) to the subjects. At the end, the sounds attracting attention, giving information, or culturally preferred are accepted as "cultural," natural sounds, for example, light wind, water and bird sounds, soft and warm human voices (like a whistle) have been found to be the sounds "liked" by the subjects. Human voices and sounds of steps in narrow alleys around houses are also good sounds because they indicate human existence. The time of exposure is important (as in the annoyance assessments). The judgments in long-term exposures (e.g. 15 minutes) are important in addition to instant or short-term judgments (e.g. in 0.5 second). Generally, individuals combine the short perceptions by remembering past experiences.

Ultimately, some sounds maybe evaluated as lack of uniqueness and information while others give pleasure, therefore they are preferred (or liked). These are called sonic delights and depend on much more than their physical qualities. As Brown states:

"Low to middle frequency and intense sounds are preferred but the delight is increased when sounds are novel, informative, responsive to personal action, culturally approved, as birds and bells. Quiet is liked but informative places are preferred such as constantly varying soft personal sounds such as footfalls, fragments of conversations, whistling or shuffling. Big, long, cool sounds of a waterfront symbolize tranquility as nature sounds and the warm human sounds are preferred." [222]

Definition of soundscape quality areas: Information about soundscape covers the soundscape quality areas, the so-called sonoscapes, where people perceive or react. These areas can be shown in maps of sonoscapes [223a] or "tranquility maps," as in the UK [223b] to be used with noise abatement plans. Noise maps detailing noise levels generated by sources can be supplemented by soundscape maps emphasizing the areas of natural sounds, iconic sounds as church bells, and other sounds of preference explained above.

A quality assessment of urban quiet areas considering both background noise and sound events needs subjective and objective criteria. De Coensel and Botteldooren have proposed the following criteria in their methodology [212]:

- pleasantness and the presence and number of unusual sound events (determined using a questionnaire survey)
- quality of background measured using indicators for loudness, temporal and spectral content.
- congruence of the area, biological, natural, and landscape value of the environment.

They applied a detailed study of auditory perception of noise sources (i.e. cars and trucks passing), changing peak sharpness, abrupt changes of levels in time, etc., to estimate the variation of subjective annoyance. Then the measures were applied for tuning the sound quality by changing the traffic parameters and the improvement in perception and soundscape was investigated.

Soundscape design and planning: The important features to be considered in design and planning are:

- Soundscapes involve diverse fields of practice, approaches and disciplinary interests, i.e. soundscape design needs collaboration between architects, acoustical engineers, environmental health specialists, psychologists, social scientists, urban developers, etc.
- The urban sonic environment plays a role in the overall assessment of the quality of life.
- It is emphasized that sonic design has real significance for city design because sound has also important emotional and social consequences. The sonic environment may have effects on an entire community's mental health [222]. Therefore, the "design of soundscape alone may be a way of making the city less stressful but more delightful and informative to its users."
- As Kang stated: "The study of soundscape is not only the passive understanding of human acoustic preference but can be placed into the intentional design process comparable to landscape" [224, 225].
- The objective of soundscape design is to provide a high quality of sonic environment in which there is a good match between the sounds that can be heard and the sounds that are expected, these are called "fitting sounds."
- Soundscape design, if the goals are properly defined, may have a really beneficial effect on the community [226].
- An ideal urban sonic environment can be achieved primarily by making the sonic environment as quiet as possible. However, absolute quietness is not necessary and sometimes is not welcome. The main focus is on changing the characteristics of tonality, impulse, and noticeability of sounds to reduce annoyance, particularly for industrial or mechanical noises.

Soundscape design process involves the following steps:

1) Selection of particular places or locations and a particular context (based on noise zonings).
2) Establishment of acoustic objectives (goals).
3) Identification of wanted/unwanted sounds that may influence the objective (through listening walks and laboratory experiments).
4) Management and design.

Selection of area: The principles of selection are defined in various publications as:

a) Sound-controlled and climate-controlled public areas in the center of the city which would ensure quiet, like shopping malls.

b) Quiet area to be decided by zoning according to a noise map, such as urban parks, gardens, waterfronts, recreational areas or areas requiring preservation and reinforcement of sound marks.

c) Areas used by citizens particularly between stressful city trips, like pedestrian ways, alleys, any public areas in cities.

Establishment of goals: One or more benefit of soundscaping, as explained above, can have priority.

Identification of wanted sounds through assessment of actual soundscape: as explained above.

Management and design: A good design can be achieved either:

by masking the unwanted sound by the wanted sound or,

by ensuring the unwanted sound does not mask the wanted sound.

Some of the design tools for soundscape are given below, based on the literature:

- Natural sounds, such as splashing water geysers, boats with horns, bells, other big sounds that can ring periodically from a distance [222].
- Responsive spaces such as alleys or other small hard surfaces can be more attractive by selecting sonic floor materials (i.e. those that squeal, rumble, squish or pop when walked upon would be fun) or sounds made by hidden speakers and lights to make people more attentive (artificial soundscape).
- Novel sounds amplified, distorted, reflected, repeated at the receiver's command.
- Sounds to provide interest and attention.
- Sounds to distract dull or ugly visual settings, e.g. with large animated sculptures, sounds would be attractive.
- Sonic signs: They have advantage of being attention-demanding, less distracting to tasks, evoking images with provocative advertisements.
- Sounds to communicate public information: Chiming clocks to tell the time or sirens to warn or symbolic sounds to inform about the weather, approaching buses or trams or special events like sports meetings, public sounds, police whistles with special character to strengthen the identity of the place.
- Sounds to draw attention to certain parts of the visual scene.
- Sonic signs for children to help them use the city and learn from it.
- Sound from a water fountain to mask by an appropriate flowing technique and fountain design.
- Attracting songbirds.
- Adding greenery in well- arranged spaces [216]

Artificial soundscape: To change actual soundscape by using different installations by help of modern computer simulations [209]. An artificial soundscape is produced with the auralization technique, however, new approaches, such as an artificial neural network based on the psychoacoustical principles have been developed and experienced [219].

Providing realistic, artificial simulation of the preferred sounds and reproduction of the various sound sources and generating the fitting sounds by hidden loudspeakers or creation of the artistic sound installations are some of the current applications in various cities.

7.5.3 Overview of Soundscape Studies

Soundscape studies started in the 1970s with the assessment of sound environments in urban areas and have continued to attract a great deal of attention from the researchers from different disciplines besides acousticians, such as architects, landscape planners, musicologists, sociologists, etc. Schaffer has described soundscape studies as "the middle ground between science, society and the arts" and a "new interdiscipline-acoustic design."

Although numbers of publications can be found in literature, a brief outline of the soundscape studies and practices, in the past and the present, is given below.

As early as the 1970s, Southwork conducted an experiment to assess the soundscape in Boston with blind and deaf people to evaluate the identity of the sounds in Boston and to obtain visual–auditory correlations [222]. He proposed various design tools, some were explained above.

Studies regarding the subjective reactions to noise had been extensively conducted in Japan long before the evolution of the soundscape concept. In a study on the soundscape of the city parks, various natural sounds samples (trees, birds, water), public sounds (social sounds, like children playing, human voices), and artificial sounds (music, announcements) were recorded and given to the subjects in the listening rooms. The acceptable or pleasant sounds components were obtained through semantic differential techniques (profiles) [76]. Based on the results, the parks were divided according to their qualities using the vineyard method and the appropriate soundscape regions were determined. Recently the soundscape projects conducted in Japan since 1996 have aimed to preserve cultural heritage, and were reviewed in terms of success of implementation from the points of view of publicity, preservation and utilization [227].

Great efforts were spent to establish relationships between the acoustic and psychoacoustic parameters in subjective evaluation of city noise, as explained above. A new qualitative method was developed by Paul, Fortekamp, and Genuit in 2004 by emphasizing that subjective evaluation of sounds is highly sensitive to the context and the tested individuals (TI) [228]. The evaluation steps and decision-making process of TIs were investigated. Difficulties were reported in sound evaluations, understanding wording, difficulty in describing daily life acoustic phenomena with a high grade of accuracy and dependence on evaluation (pre-test).

In China, many of the soundscape studies paid attention to Chinese gardens covering idealized miniature landscapes, expressing harmony between man and nature, and to development strategies of soundscape design through a systematic review [229, 232]. Some researchers emphasized the differences between soundscape and listening/hearing in Buddhist temples, which are landmarks of peace and quiet in China, whereas others focused on the influence of soundscape on tourism in historical and ethnic tourist areas, based on long-term studies for heritage protection.

Cross-cultural studies regarding soundscape evaluations are increasing, thus, it will be possible to reach a compromise on a soundscape standard to benefit the whole world [230].

In a novel study on soundscape by Berglund, Nilsson, Pekala, they categorized the environment by using neural networks and developed a predetermination system [231, 232]. In this study the mono-aural and binaural recordings were made in nature and in the city and then used in the laboratory tests. The sounds whose acoustical characteristics were controlled in one-third octave bands and 12 different emotional attributes were selected and described with words such as "relaxing-comforting," "weak or live," "positive-negative," "stressful, hard, intense, disturbing, annoying, noisy," etc. They attempted to find a psychological relationship between the physical environment and data sets regarding the perceived soundscape. One of the aims was to make an assessment about the perceived soundscape based on the acoustic data that can readily be measured. Another was to develop a design tool for town planners, designers, architects and constructors, and material producers in the future. The ultimate goal was to construct a classification system for the perceived soundscapes leading to green labeling for certification of residential soundscape.

In the Swedish research program entitled "soundscape supportive to health," the perceived soundscapes were focused on creating appropriate tools for assessing the characteristics of residential soundscapes [233]. In the study, the recorded samples from various city noises were given to the subjects to listen and they were asked which part of city was represented by which sound, then the sounds were grouped based on the evaluations. Consequently, six categories of sound properties were found to be important with respect to perception: Homogeneity, detail, sharpness, high frequency components, localization, and loudness. The second phase of the research comprised a perceptual evaluation of various noise mitigation measures, such as noise barriers and, a study on future change in the perceived quality of soundscape [234]. It was focused on the effect of a noise barrier and whether the difference in acoustic conditions was noticeable or not, by using the sound signals recorded before and after the barrier construction in the experiment. Eventually, a difference in the perceptual soundscape was observed implying that sound reduction, by any means, could change the quality of the environment. The participants freely described the perceptual cues to discriminate the before and after situations and expressed the after the barrier sounds as monotonous, indistinct and homogeneous.

In another study, the recorded samples from various city noises were given to the subjects and they were asked which part of city was represented by each sound, which were grouped based on the evaluations. Consequently, six categories were found to be important with respect to perception: Homogeneity, detail, sharpness, high frequency components, localization, and loudness [233]. Further investigation in the same study focused on the effect of the noise barrier and whether the difference in acoustic conditions was noticeable or not, by using the sound signals recorded before and after the barrier construction. It was found that the barrier caused a difference in the perceptual soundscape. This implied that sound reduction, by any means, could change the quality of the environment.

Various scientific studies highlight perception and evaluation of acoustic comfort in relation to soundscape, characterizing factors which will lead to further studies [234–239]. It should be acknowledged that soundscape studies attract many researchers and numbers of new publications can be found in literature at present.

7.6 The Concept of Noise Pollution

Pollution is defined as the "introduction of harmful substances or products into the environment" or "introduction of contaminants into the natural environment that cause adverse change" [240a, 240b]. Noise pollution is "a type of energy pollution in which distracting, irritating or damaging sounds are freely audible" [241]. However, unlike light and heat energy, the physical contaminants causing noise pollution are sound waves and are not perceived visually or tactually, but aurally. One can find numbers of definitions of noise pollution in the literature emphasizing the negative impacts on human health and well-being, the deterioration of the natural environment, etc.

Noise pollution is sometimes called acoustic pollution, and has found its way into environmental problems and was first described as a technological residue (waste). It became a great concern in the USA between 1950 and 1970, before the Noise Control Act came into force like other federal laws against air and water pollution (Chapter 8). In Europe, the interest in transportation noise pollution started in the 1960s when the consequences of scientific investigations revealed that environmental noise caused a great risk to human and public health. It was also shown that the tolerance of people decreases in parallel to increasing numbers of the exposed population. The studies gave rise to public awareness extensively through informative publications, guidelines, booklets, attracting community attention to the problem. Nowadays, noise pollution has been legally involved at national levels, also international strategies to minimize its effects have been coordinated in the OECD, United Nations Development Programme, and later in the European Union (Chapters 8 and 9). The reality is that the increase in noise levels is still a threat even in developed countries, the situation is worse in the developing countries with varying degrees due to the following reasons outlined in the 1980's [135, 242]:

- Planning processes lagging behind the existing situation in view of the rapidly growing population in the big cities, resulting in unplanned and irregular settlements or general avoidance of planning.
- Increasing traffic and its extension into quiet areas.
- Increase in construction activities.
- Increase in the volume of transportation with increasing traffic speeds, due to the priority given to motorized traffic rather than mass transit systems.
- Insufficient knowledge of individuals about noise and its impact on community health.
- Usage of noisy household equipment and increase in outdoor mechanical equipment which are sources of complaints.
- The entertainment noise, with high amplified music, emerging as a serious problem and needing immediate precautions.
- Wide use of lightweight building construction with poor insulation.
- Disregarding acoustics in the low cost apartment buildings due to lack of regulation at ocal levels.
- Lack of sufficient acoustic consultancy and professional training for technical solutions.
- Economic restraints obstructing technical slutions since the manufacturers, owners, and constructors are not willing to make an additional investment.
- Difficulty in developing the realistic criteria (noise limits) for noise control.
- Difficulty in noise mapping, at least for critical areas.
- Lack of regulations directly referring to noise issues, inability of sufficient inspections during construction lack of enforcement, and difficulty in applying incentive measures.

Although the significant improvements in the above conditions might have been achieved in some countries over the years, however, any of them contributes to intensify the noise pollution problem. Before an alarming situation is emerged, urging the necessary steps to be taken is crucial particularly with respect to the human right of living in a silent world.

7.7 Noise Criteria and Limits

Generally, a criterion is defined as a tool or a measure accepted to value something or an action in order to reach a judgment. Literally, criteria imply the references expressed either qualitatively or quantitatively to evaluate some variables or results, hence, they are used as basis for comparisons. Different criteria can be developed to evaluate the same situation. However, the general trend is to measure the magnitude of hazard and to protect humans and the environment by specifying the upper limit of a parameter related to the pollutant.

Noise control criteria should be developed by applying a process with different steps. First of all, determination of the permissible conditions quantitatively needs the following observation tests or surveys:

a) Physiological tests for all the physical variables employed.
b) Performance tests aiming to check the task quality, speed, and accuracy level under noise exposure.
c) Subjective tests to measure people's psychological situation, i.e. annoyance.

Noise criteria are declared the highest acceptable (permissible) magnitude of noise under a certain condition in a certain duration of time in addition, they may define/restrict the quality properties. They are important in determining the required noise reduction and to describe the quality and quantity of the measures to be implemented. The criteria for rating noise and its negative effects are included under the following captions:

- hearing risk criteria for individuals and especially workers
- environmental noise criteria for public health and well-being.

7.7.1 Hearing Risk Criteria

Noise-induced hearing loss occurs under certain conditions described in Section 7.2.1. The criteria have been developed based on the degrees of destruction related to noise level, tonality, impulsiveness, or temporal and spatial factors. Also called damage risk criteria for hearing, two groups of criteria have been developed: (i) the criteria for threshold of shift (hearing impairment criteria); and (ii) the criteria for hearing loss in relation to exposure time.

Criterion for threshold of shift (hearing impairment criteria): The degree of hearing loss corresponds to the percentage of the inability of hearing under certain conditions. Hearing impairment criteria (noise-induced permanent threshold shift, NIPTS) determines the situations when the average hearing losses exceed certain threshold (reference) levels, given in dB within a certain frequency range at which the ear is more sensitive [243]. The threshold level and the frequency range considered have changed over the years in different standards.

ISO 1999:1971 and in revised editions suggested the threshold of 25 dB in the frequency range of 500, 1000 and 2000 Hz, accepting 40 hours of working week [244a, 244b]. In the latest publication in 2013 (confirmed in 2018), the frequency range is extended to 10 kHz [245]. The audio tests are described in Section 5.7.1.

In the USA, the Occupational Safety and Health Administration (OSHA), in 1983, accepted a threshold shift as having average losses more than 10 dB at 2000, 3000, 4000 Hz (age corrected), later revised as 25 dB average loss in the same frequency range in 2009. The National Institute for Occupational Safety and Health (NIOSH) (1998) defined the hearing impairment of 15 dB and more as an average between 500–6000 Hz [246] according to age and lifestyle. The Australian Standard (AS1269.4:2014), gives more detailed description of hearing threshold in relation to the five groups of frequency range to be considered as a shift from baseline, e.g. averaged at 3000, 4000 and 6000 Hz $\geq$ 5 dB, averaged at 3000 and 4000 Hz $\geq$ 10 dB, averaged at 6000 Hz $\geq$ 15 dB, etc. (Information provided by Biotronics)

Hearing loss risk is defined as the difference between the threshold levels of a normal ear and the ability of hearing of a person, obtained from the audio tests. The risk criteria varied in countries in the past because the incentives were differentiated. The WHO (1991) defined hearing risk, in relation to the average losses in the defined range of frequency (e.g. up to 25 dB), as corresponding to a risk of 0% and when losses reach to 92 dB, the degree of risk becomes 100%. On average, excess of 30 dB at 1.2 and 4 kHz implies a significant hearing loss or hearing damage. The models for assessment of risk of damage (excess risk of hearing impairment) have been developed with respect to noise level, age, year of exposure. NIOSH made risk estimates for workers in all types of industries, i.e. mining, metals, agriculture, manufacturing, construction, etc. [247].

Criteria for hearing loss in relation to exposure time: The endurance of a person to a certain noise level before hearing impairment emerges is directly related to the exposure time. The noise guidelines/standards, in the past, declared the permissible exposure levels in dB(A) in 8-hour working day (L_{18h}) such as 90, 87, 85 or 80 dB(A). By accepting an exchange rate of either 5 or 3 dB(A) and halving the exposure time presented in hours and minutes, the permissible duration of exposure under the higher levels of noise, could be determined. Table 7.4 gives the standards in the USA and Europe for workers' exposure to noise in workplaces, also called occupational noise and the exposures above these levels are considered hazardous. The criteria vary with countries and years.

NIOSH in 1998 used the limit of 85 dB(A) to define the Risk Exposure Level (*REL*) for eight hours and the increase of 3 dB(A) by halving the exposure time [247]. OSHA (2009) started at 90 dB(A) with the exchange rate of 5 dB(A) [248].

In Europe, the EU Directive 2003/10/EC which is called noise at work regulation, recommends a different system, as displayed in Table 7.4, by stating "minimum requirements for the protection of workers from risks to their health and safety arising or likely to arise from exposure to noise and in particular the risk to hearing" [249]. The exposure levels which imply noise levels at the worker's ears, for continuous and impulsive noises are given by using the following units:

- For continuous noise: L_{EX}(8h), dB(A) (defined as Daily Noise Exposure)

$$L_{EX}(8h) = L_{Aeq,Te} + 10 \lg (T_e/T_0) \tag{7.31}$$

$L_{Aeq,Te}$: A-weighted equivalent continuous sound pressure level for T_e
T_e: Effective duration of the working day in hours
T_e: Reference duration (8 h)
L_{EX} (8h) is also called the "Time-weighted average (*TWA*)" of the noise exposure levels for an 8-hour working day.

- For impulsive noise: the C-weighted peak level L_{peak}, dB(C) or peak pressure, Pa

Table 7.4 Hearing health criteria for occupational noise (at workplace).

Permissible noise exposure according to OSHA 2009 (TWA) and (NIOSH 1998) standards in USA [247, 248]			EU Directive 2003/10/EC [249]*		
L_{max}, dB(A) (measured by slow response)	Duration, hours	Definition	Exposure values for continuous noise L_{EX}, 8h dB(A)	Peak level limit for impulse noise P_{peak} (pascal)	L_{peak}, dBC
90 (85)	8	Exposure limit value	87 (exposure limit)	200 Pa	140
95 (88)	4				
100 (91)	2				
105 (94)	1	Upper exposure action value	85 (upper exposure action value)	140 Pa	137
110 (97)	1/2				
115 (100)	1/4				
120 (103)	1/8	Lower exposure action value	80 (lower exposure action value)	112 Pa	135

* These values are converted to time-equivalent level trade-off 3 dB for halving the time.

In the Directive, two groups of criteria are described: Exposure limits and exposure action values (as upper and lower values). The exposure limits imply the noise levels that cannot be exceeded at all and the workplace owners might be subject to various sanctions. When the action values are exceeded, the employers have to implement special programs, training, or measures that are defined in the standard to protect the workers. In fact, a worker's effective exposure (not the action value) is determined by taking into account the noise reduction provided by the hearing protectors if used. Two conditions are important:

- If the daily noise exposure changes considerably from one weekday to another, the weekly averages are concerned.
- If the daily noise exposure varies (e.g. due to change in the working stations or noise variation because of the source operation), the exposure levels should be corrected according to the length of each work duration. The combined effect is calculated as:

$$D = 100 \left(C_1/T_1 + C_2/T_2 + \dots + C_n/T_n \right) \tag{7.32}$$

D: Daily dose as given as a percentage (for the noise limit of 85 dB(A) and for a person exposed to L_{eq} 85 dB(A) in eight hours, the dose is 100%)
$C_1, C_2, \dots C_n$: Duration of noise exposure at a specified noise level
$T_1, T_2, \dots T_n$: Exposure duration for the noise at this level (Permissible duration will be taken from Table 7.4)

If $D \geq 100$, the exposure exceeds permissible levels.
The daily dose (D) can be converted into L_{EX} (8 h) by using Eq (7.33):

$$L_{EX} \left(8\,h \right) = 10 \lg \left(D/100 \right) + 85 \tag{7.33}$$

The Directive obliges employers to make risk assessments at workplaces and take necessary precautions, obligatory to use hearing protectors under the described conditions. The risk assessment procedures are described in an additional document published as an application of the Directive in 2007 [250].

Various applications of the Directive in practice can be found in literature, such as damage risk criteria for military noise [251] and risk of hearing loss in orchestra musicians [252]. By combining both criteria, ISO 1999: 2013 (confirmed in 2018) described a calculation method for the expected noise-induced permanent threshold shift and the population at risk by using L_{EX}(8 h) and number of years of noise exposure [245]. The standard covers all kinds of noises (steady, intermittent, irregular, fluctuating) below 10 kHz.

7.7.2 Environmental Noise Criteria

With respect to community protection from environmental noises, a different kind of criteria to evaluate different aspects, has been developed:

1) Emission criteria for noise sources.
2) Environmental noise criteria for outdoors and indoors (also called immission criteria).
3) Sound insulation criteria.
4) Performance criteria for sound-reducing devices.

Emission Criteria for Noise Sources
The primary action for noise control is to minimize the source emissions which should be described in terms of sound power levels or sound pressure levels at a reference distance from the source. The

permissible emission levels for common noise sources used in daily life and industry are given in the national and international documents for use of manufacturers and buyers. The proposed emission criteria to be compared with the product emissions which are determined through either laboratory or field tests (Chapter 5), need detailed definitions regarding the source characteristics in categories, by considering the applicability, the market, and economy. Thus the limit values are subject to change over the years according to developments in recent technology and achievements in product performance regarding noise outputs.

The first noise emission criteria for motor vehicles were issued in 1970 by the OECD and the EU. The limits have changed great deal since the 2000s. Tables 7.5 and 7.6 summarize the earlier and recent limits for motor vehicles in Europe [253–255]. The EU Directive 2000/14/EC gives the permissible sound power levels for industrial machinery in relation to their mass and electric power [256], similar to the standards published in the USA for industrial and construction equipment.

Environmental Noise Control Criteria

Determination of the health and comfort criteria for environmental noises is a rather difficult task due to the number of factors. The WHO declared the negative effects of noise and the threshold levels based on the scientific results, as explained above [49]. Table 7.7 gives an overview of the health criteria.

Naturally the international standards do not mandate the environmental noise limits which specify the highest permissible levels for outdoor and indoor noise, because of the difficulties in reaching a compromise among the countries and the inflexibility of different countries' own regulations. In the USA, the EPA identified noise criteria as yearly average noise levels in L_{dn} and L_{Aeq} (24 hours) in 1974 [257]. The U.S. Department of Housing and Urban Development (HUD) issued a

Table 7.5 Noise emission limits for moving road vehicles in ten years perspective according to European Directive (1970) 70/157/EEC, measurements at a distance of 7.5 m from the path of vehicle center line [253].

Vehicle category	Noise level, dB(A)
Vehicles intended for the carriage of passengers and comprising not more than nine seats including the driver's seat	82
Vehicles intended for the carriage of passengers and comprising more than nine seats including the driver's seat, and having a permissible maximum weight not exceeding 3.5 metric tones	84
Vehicles intended for the carriage of goods and having a permissible maximum weight not exceeding 3.5 metric tones	84
Vehicles intended for the carriage of passengers, comprising more than nine seats including the driver's seat, and having a permissible maximum weight exceeding 3.5 metric tones	89
Vehicles intended for the carriage of goods and having a permissible maximum weight exceeding 3.5 metric tones	89
Vehicles intended for the carriage of passengers and comprising more than nine seats including the driver's seat, and having an engine power equal to or exceeding 200 HP DIN.	91
Vehicles intended for the carriage of goods or materials, having an engine power equal to or exceeding 200 HP DIN and a permissible maximum weight exceeding 12 metric tones	91

Table 7.6 Noise emission limits for moving vehicles according to Regulation EU No 540/2014 [255].

Vehicle category	Description P/M: Power to mass ratio, n: Number of seats	Limit values, dB(A)[a]		
		2013 (2016)[b]	2015–2017 (2022)[b]	2020–2022 (2026)[b]
M1 (Passenger vehicles)	≤120 kW/1000 kg	72	70	68
	120 kW/1000 kg < P/M ≤ 160 kW/1000 kg	73	71	69
	160 kW/1000 kg < P/M	75	73	71
	P/M > 200 kW/1000 kg and n ≤ 4	75	74	72
M2	M ≤ 2500 kg	72	70	69
	2500 kg < M ≤ 3500 kg	74	72	71
	3500 kg < M ≤ 5000 kg Rated engine P ≤ 135 kW	75	73	72
	3500 kg < M ≤ 5000 kg Rated engine P > 135 kW	75	74	72
M3	Rated engine P ≤ 150 kW	76	74	73
	150 kW < Rated engine P ≤ 250 kW	78	77	76
	Rated engine P > 250 kW	80	78	77
N1 (for goods)	M ≤ 2500 kg	72	71	69
	2500 kg < M ≤ 3500 kg	74	73	71
N2	Rated engine P ≤ 135 kW	77	75	74
	Rated engine P > 135 kW	78	76	75
N3	Rated engine P ≤ 150 kW	79	77	76
	150 kW < Rated engine P ≤ 250 kW	81	79	77
	Rated engine P > 250 kW	82	81	79

[a]See the original document for the application conditions.
[b]According to 2016 amendments.

regulation in 1979 for outdoor noise levels, specifying that $L_{dn} > 65$ dBA was unacceptable and 65–75 dB(A) required a 5 dB(A) attenuation for new and rehabilitated buildings [258]. In 2000, the WHO accepted 50 dB(A) as the criterion for outdoor noises during day-time.

The new WHO Guideline for the European region (2018) provides recommended exposure levels for environmental noise in order to protect the population's health. It covers new recommendations for different noise sources, namely, transportation noise (road traffic, railway, and aircraft), wind turbine noise and leisure noise, by aiming "to support public health policy that will protect communities from the adverse effects of noise, as well as stimulate further research into the health effects of different types of noise" [29c].

The recommendations given in Table 7.8 are considered to be taken into account in the policy-making process, as explained in Chapter 9.

Factors playing a role in environmental noise criteria: Declaration of the maximum acceptable noise levels for environmental noise control purposes, should consider the following aspects:

1) Source type and specific characteristics (structural, operational and acoustical).
2) Characteristics of environment (land-use).

Table 7.7 Overview of effects of noise on health based on WHO, 2009 [49].

Effect	Description	Indicator	Threshold	Evidence
Biological effects	EEG awakening	$L_{Amax,inside}$	35	
	Body movement (motality)		32	
	Changes in sleep stages, durations, etc.		35	
Sleep quality	Waking-up		42	Sufficient
	Increased average motality		42	
Wellbeing	Use of drugs and sedatives		40	
Medical conditions	Environmental insomnia	$L_{night,outside}$	42	
Well-being	Complaints		35	
	Hypertension		50	
Medical conditions	Myocardial infarction		50	Limited
	Psychic disorders		60	
	Guideline value (NNG)	$L_{night,outside}$	40	–
	Interim target	$L_{night,outside}$	55	

Note: L_{night}: 23:00–07:00 (8 hours): A-weighted long-term average sound level according to ISO 1996-2 (yearly average).

Table 7.8 Overview of the recommendations given in WHO Guidelines for the European region, 2018, based on the systematic review of the health outcomes.

	L_{den}, dB	L_{night}, dB	Rationale	Type of recommendation
Road traffic noise	< 53	< 45	To reduce adverse health effect and adverse effect on sleep	Strong
Railway noise	< 54	< 44	To reduce adverse health effect and adverse effect on sleep	Strong
Aircraft noise	< 45	< 40	To reduce adverse health effect and adverse effect on sleep	Strong
Wind turbine noise	< 45	Not recommended due to lack of quality of evidence	To reduce adverse health effect	conditional
Leisure noise	< 70 $L_{Aeq, 24h}$ the yearly average from all leisure noise sources combined, applying the equal energy principle can be used to derive exposure limits for other time averages.	For single-event and impulse noise exposures, the GDG conditionally recommends following existing guidelines and legal regulations to limit the risk of increases in hearing impairment from leisure noise in both children and adults.	To reduce adverse health effect and to limit the risk of increases in hearing impairment from leisure noise in both children and adults.	Conditional

3) Intended use of buildings and indoor spaces.
4) Background noise levels (if the criteria is for a specific noise source).
5) Time intervals for implementation of the criteria (reference intervals).

The descriptors to rate the environmental noise considering the above factors, can be given as average levels, maximum noise levels, statistical levels, or in other metrics that are also used in the measurements (Section 2.5.2). Two approaches are applicable in the development of environmental noise criteria for outdoors:

- Specific limit values for each condition/source/environment/building/space
- An overall limit value together with the correction factors according to each situation.

Both approaches in practice have been selected in various national regulations. Declaration of a criterion needs to provide proof of its validity by answering the following questions:

- Which operational conditions of source?
- Where exactly are the limits applied (near or on the façades)?
- Which measurement standards are used and under which meteorological conditions (favorable or homogeneous)? (see Chapter 5)
- What are the uncertainties and reasons?

Considering the above factors, criteria development for environmental noise control has become a wide range of interests in the 1980s. A process for determination of traffic noise criteria is given in Figure 7.41.

Transferring the criteria into regulations: Implementation of environmental noise criteria, derived from scientific studies or field surveys, is enforced through regulations/laws or local legislation. However, some difficulties can arise in practice when transferring the scientific criteria into the regulatory limit values, that must be discussed widely in the community:

1) A limit or a noise level accepted by the majority of the population may not be accepted by a considerable number of people in a community. Thus, the limits should be specified to find an optimal level, such as (i) tolerated levels and (ii) acceptable levels. The communities or countries with a lack of economic and technical abilities might adopt the tolerance level, which is lower than the acceptable level.
2) The background noise levels have to be handled appropriately. In fact, the allowable noise limit for a new establishment, constructed in a noisy community with similar premises that already exist, should be much lower than the existing total background level to satisfy the environmental criteria. This requires extra investment for noise control. To be fair, the costs should be shared by all the premises (the difference between the limit and the existing background noise level).
3) In the proposition of a criterion for entertainment places, different objectives, such as protection of clients or workers, should be distinguished. Different types of common sources may require specific solutions, for example, the limitation of noise emerging from misbehavior of individuals is proposed in the British standard (BS 4142), i.e. the maximum levels to be measured outside at 1 m from the windows should be less than 70 dB(A) between 23.00–07:00 hours [259]. Different criteria are suggested for car parks, warehouses, garbage collecting, service vehicles, etc.
4) Proposals of limits for construction or sports activities need to take into account the other restrictions in the regulations regarding the operating times. Construction noise levels are generally very high, but the duration of the activity may be brief, therefore construction noise limits should be balanced with the necessity to continue and complete the work. For sports activities, the criteria should be decided according to the frequency of usage of these spaces.

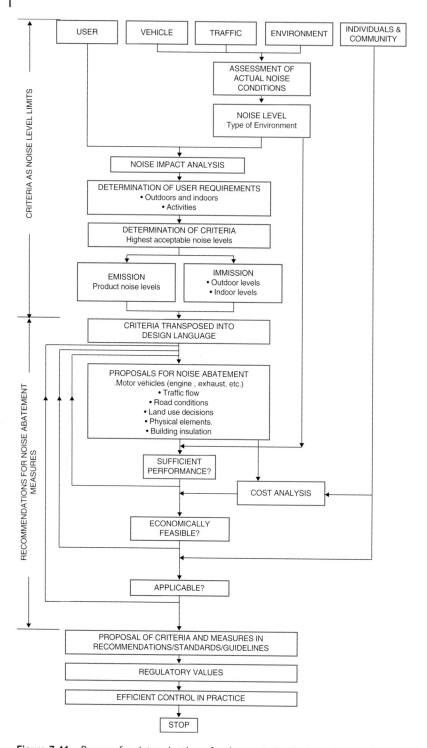

Figure 7.41 Process for determination of noise control criteria against traffic noise (*Source:* Kurra, 1986).

Consequently, the regulatory limits to be developed for community noise should be realistic and applicable to complex situations without causing additional social problems or insoluble situations, and should take into account the socio-cultural, traditional, and economic status of the community.

Further information about regulatory actions against environmental noise has been given in Section 8.4.

7.7.3 Building Acoustics Criteria for Noise Control

The environmental noise criteria aim to achieve the health and comfort of individuals in buildings, hence they cover the allowable (maximum) indoor noise levels and the minimum sound insulation requirements for building constructional elements. Although this issue is involved in the field of building acoustics in detail, some brief information is given.

Criteria for Indoor Spaces

The effects of noise and acoustic comfort are associated with the activities performed indoors and outdoors. Although the approaches are similar, the type of activities and sources with different acoustical characteristics, some of which are unpredictable, vary considerably in buildings.

Units for indoor noise criteria: Although the noise descriptors are the same as outdoor criteria, however, the additional metrics are also used for indoor noises. Allowable noise levels are presented along with the time intervals, as average values L_{eq}, dB(A), maximum levels (L_{max}), peak levels and C-weighted levels, depending on the type of noise. Considering schools, for classrooms and other spaces where listening and speech activities are performed, the acceptable acoustic condition is defined also by using the intelligibility criteria, e.g. the Speech Transmission Index (STI) associated with the signal to noise ratio (speech level/background noise). Indoor noise levels are related to the reverberation time, which is a room acoustic parameter that plays an important role in efficient noise control in closed spaces. These parameters are used also for the spaces where speech privacy is important, such as banks, offices, and other public areas.

The WHO recommended the time-based indoor noise criteria for various spaces in buildings (Table 7.9) [260].

The following considerations need to be taken into account when declaring indoor noise limits for specific cases:

1) The outdoor criteria should be compatible with the indoor criteria under the open or closed window conditions.
2) While proposing the limits to be compared with the actual noise levels (mostly to be measured), it should be kept in mind that bedroom windows half-open will increase the indoor levels by about 10–15 dB(A) and closed windows will decrease the outdoor noise levels by about 25 dB(A).
3) Implementation of criteria for manufacturing areas inside the industrial buildings might be opposed by the owners or managers due to various reasons. However, they cannot avoid the protection of workers from excessive noise, as well as to minimize the transfer of noise to outside.

At present, many countries have declared permissible noise levels in buildings in their national standards and regulations according to the type and usage of indoor space and activity type, conditions of windows (openable or fixed), and time intervals (hours). Some take into account the characteristics of noise source and noise (i.e. steady, intermittent, impulsive, impact, tonal). An example matrix displayed in a national regulation for the new and existing buildings is given in Table 7.10 [261].

Table 7.9 WHO Guidelines values for community noise, 2001 [260].

Environment	Critical effect	L_{eq}, dB(A)	Time base (hours)	L_{Amax} (fast response) dB
Outdoor living area (day-time and evening)	Serious annoyance	55	16	–
	Moderate annoyance	50	16	–
Dwelling indoors	Annoyance, speech interference	50	16	–
	Moderate annoyance	36	16	–
Bedroom (inside)	Sleep disturbance	30	8	45
Bedroom (outside)	(window open)	45	8	60
School classroom (indoors)	Speech intelligibility, disturbance in communication	35	During class	–
Preschool bedrooms (indoor)	Sleep disturbance	30	During sleep	45
School playground	Annoyance	55	Play time	–
Hospital patient room	Sleep disturbance (night)	30	8	40
	Sleep disturbance (day and evening)	30	16	–
Hospital treatment rooms	Interference with rest and recovery	As low as possible		
Industrial, commercial, shopping, indoors and outdoors	Hearing impairment	70	24	110
Ceremonies, festivals, entertainment events	Hearing impairment	100	4	110
Public addresses indoors and outdoors	Hearing impairment	85	1	110
Music and other sounds through headphone/ earphones	Hearing impairment (adapted to free-field value)	85	1	110
Impulse sounds from toys, fireworks and firearms	Hearing impairment (adults)	–	–	140 (peak sound pressure)
	Hearing impairment (children)	–	–	120 (peak sound pressure)
Outdoors in parklands	Disruption of tranquility	Quietness should be preserved and the ratio of intruding noise to natural background noise should be kept low.		

Spectral criteria: As explained in Section 7.2, the spectral characteristics of noise play an important role in noise effects. Although the human perception of low frequency sounds is lower than that of middle frequencies, however when the low frequency components in noise carry considerably high energies, such as noise from heavy vehicles or machinery or amplified music, the disturbance has been proven to increase. The earlier indoor noise criteria, based on the loudness concept, were developed as spectral levels displayed by curvatures, like Noise Criteria, *NC* (1957),

Table 7.10 Example of indoor noise limits given in a sample building regulation [261].

Building type	Indoor space		Time duration Night: 23.00–07.00 Evening: 19.00–23.00 Day: 07.00–19.00	Indoor noise level, L_{Aeq}	
				New buildings	Existing buildings
Houses	Bedrooms		Night	34	38
	Living rooms		24 h	39	43
	Kitchens		24 h	39	43
Educational buildings	Classrooms		Day-evening	39	43
	Special classrooms		Day-evening	44	48
	Administrative rooms		Day-evening	39	43
	Sports hall		Day-evening	49	53
	Reading rooms		Day-evening	39	43
	Circulation areas		Day-evening	49	53
	Pre-school	Playing and eating	Day	44	48
		Bedrooms	Day	34	38
Health buildings and nurseries	Private patient rooms		24 h	34	38
	Hospital Wards		24 h	39	43
	Operating rooms		24 h	39	43
	Treatment rooms		24 h	39	43
	Laboratories		24 h	44	48
	Circulation areas (corridors)		24 h	49	53
Office buildings, (Administrative and private)	Executive offices		Day-evening	44	48
	Open plan offices		Day-evening	44	48
	Meeting rooms		Day-evening	39	43
	Teleconference rooms		Day-evening	34	38
	Resting or relaxing rooms		Day-evening	44	48
	Circulation areas		Day-evening	49	53
	Court rooms		Day	39	43
Hotels	Bedrooms		Night	34	38
	Restaurants		24 h	49	53
	Service areas		24 h	54	58
	Circulation areas		24 h	49	53
Dormitories	Wards		Night	34	38
	Study room		Day-evening	39	43
	Restaurant		24 h	49	53
	Circulation areas		24 h	49	53
Cultural buildings	Theater and conference rooms		24 h	39	43
	Cinema halls		24 h	39	43
	Concert halls		24 h	34	38
	Museums		Day	44	48
	Libraries		24 h	39	43
	Music and TV studios		24 h	29	33
	Circulation areas		24 h	49	53
Commercial buildings	Shops		Day-evening	49	53
	Gallery, atrium, circulation areas		Day-evening	54	58
	Supermarkets		Day-evening	54	58
	Post offices, bank offices		Day-evening	49	53
	Circulation areas		24 h	49	53
Terminal buildings	Waiting lounges		24 h	49	53
	Offices and staff rooms		24 h	44	48
Religious buildings	Praying areas		24 h	39	43

(Continued)

Table 7.10 (Continued)

Building type	Indoor space		Time duration Night: 23.00–07.00 Evening: 19.00–23.00 Day: 07.00–19.00	Indoor noise level, L_{Aeq}	
				New buildings	Existing buildings
Entertainment and sports buildings	Restaurants, cafees, etc.		24 h	49	53
	Bars, live music halls		Night	59	63
	Sports halls	Halls	Day	49	53
		Swimming areas	Day	49	53
Industrial buildings	Laboratories		24 h	54	58
	Precision montage and tests		24 h	49	53
	Control rooms		24 h	59	63
	Staff rooms		24 h	44	48
	Health rooms		24 h	39	43
	Circulation areas		24 h	49	53

Noise Rating Curve *NR* (1973 by ISO), Room Noise criterion, *RC* (1981) and the curves were expressed with single numbers derived from spectral levels [262, 263].

As indoor noise may comprise of various tonal components, such as AC noise, neighborhood noise, office or outdoor traffic noise, thus, it is important to identify the total noise spectrum in the space.

Balanced noise criterion (NCB): Developed by Beranek in 1989 as the improved version of *NC* curves, the *NCB* curves, which are shown in Figure 7.42, include corrections at low frequencies [264, 265]. They are applicable to steady noises in time and for noises having a continuous and balanced spectrum whose tonal components are not greatly emphasized, like AC noise, some human activity, and vehicle noises.

Table 7.11 gives the proposed single number *NCB* units for different space types in buildings. The *NCB* curves to be selected as indoor criteria provide (i) a good speech comprehension and satisfactory speech, and (ii) comfortable listening to vocal music and strings by perceiving the dynamic color of natural music. Due to the shape of the *NCB* curves at the low frequencies (16–63 Hz), the roaring sounds in mechanical noises and the rattling sounds due to mechanical vibrations are taken into account, besides the high frequency tones, like a whistling sound (125–5000 Hz), particularly above 1000 Hz, can be evaluated. In order to check the existence of such sounds in a noise spectrum, a method has been proposed based on Speech Interference Levels (*SIL*) [265].

Beranek suggested the quality control criteria for an acceptable noise could be:

a) The noise levels should not exceed a specific *NCB* value which is accepted at the beginning for a specific room.
b) There should be no excessively low and high frequency components in the spectrum, such as roaring and whistling.
c) The spectrum should be uniform and balanced.

He pointed out that it was possible to improve or correct the imbalance characteristics of a noise spectrum by increasing the other parts in the spectrum, i.e. raising the sound levels of the frequency components at 500–4000 Hz, ultimately eliminating the roaring effect.

Development of an indoor noise criteria for only AC noise, the acceptable level for the mechanical equipment, should be lower than the accepted *NCB* value for that space.

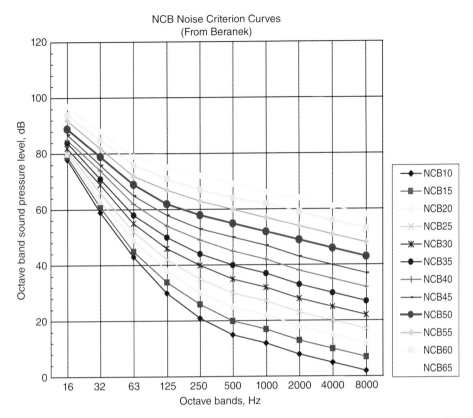

Figure 7.42 Balanced noise control criteria. (*Source:* Beranek [264]. Permission granted by INCE.)

Rating Noise Criteria (RNC): Proposed by Schomer in the USA for very low levels of noise (below the hearing level), *RNC* combines *NCB* and *RC*, which was recommended by ANSI and ASA for fans and for high turbulence-borne sounds [266]. *RNC* has been validated by the psychoacoustics tests by Schomer and Bradley [267].

Criteria for Sound Insulation in Buildings

The principal indicator of noise control in buildings is sound insulation performance of building elements, such as façades, indoor walls, partitions, and floors. The requirements regarding insulation values, are included in the standards and regulations. In building acoustics, the sound reduction levels of existing or designed elements are determined by considering the following factors:

- Magnitude and characteristics of noise which the element is exposed (e.g sound pressure levels on the facades or in the source room);
- noise criteria to be selected for indoor spaces (indoor noise limits);
- sound absorption in the receiver room.

The outdoor and indoor noise levels are measured (as explained in Section 5.5) or predicted by using simulation models (Chapter 4). The actual performance of the existing building elements against airborne and structure-borne sounds are obtained through sound insulation measurements at the site as spectral values or in single number rating units. It can also be calculated

Table 7.11 *NCB* values recommended for various indoor venues [264, 265].

Space	NCB
Recording studios	The lowest curve
Broadcasting studios (Distant microphone records)	10
Concert halls, operas, recital halls	15–18
Small auditoriums	25–30
Large auditoriums, large theaters	20–25
TV studios	15–25
Legitimate theaters	20–25
Bedrooms in houses	25–30
Multifamily residential buildings	30–40
Living rooms	30–40
Classrooms in schools with $<70\,m^2$	35–40
Classrooms in schools with $>70\,m^2$	30–35
Open-plan classrooms in schools	35–40
Rooms in hotels and motels	30–35
Meeting rooms and banquet halls in hotels	25–35
Service areas in hotels	40–50
Offices	25–35
Executive rooms in office buildings	25–35
Large office areas with meeting table	30–35
Large conference rooms	25–30
Small conference rooms	30–35
Secretary areas in offices	40–45
Open space offices	35–40
Machinery and computer areas	40–45
Circulation areas	40–50
Rooms in hospitals	25–30
Wardrooms in hospitals	30–35
Operating rooms	25–35
Laboratories	35–45
Corridors	35–45
Waiting rooms	40–45
Cinemas	30–40
Small churches	30–35
Court rooms	30–35
Libraries	35–40
Restaurants	40–45
Maintenance rooms, light industrial space, kitchens, laundry rooms	45–55
Garages and shops	50–60

by using scientific or standard methods. The descriptors for rating insulation were briefly introduced in Section 5.9.

Simple methods can be implemented to determine the required sound reductions for building elements (e.g. facades and indoor partitions). For example, a study in 1993 presented the sets of diagrams to obtain the weighted sound reductions for facades exposed to different transportation noises. (Figure 7.43).The calculation procedure accounted the differences between the frequency spectrums of road, railway and aircraft noises and applied the *NCB* criteria for different indoor spaces [268]. Nowadays, the environmental noise maps provide façade noise levels also in octave bands, thus, it becomes possible to display the insulation requirements on the façades of by means of so-called sound insulation maps. Such a technique is useful to enhance the variation of

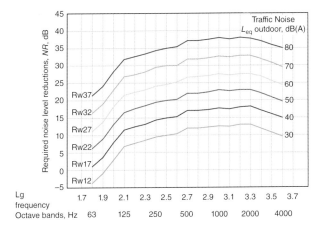

Figure 7.43 Criterion curves for façade noise insulation in terms of weighted noise reductions (*R*w) applicable for different traffic noise levels at façades. [268]) (prepared for indoor noise NC35).

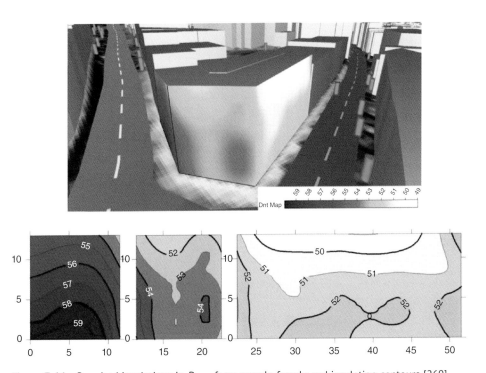

Figure 7.44 Required insulations in $D_{nT,w}$ for a sample façade and insulation contours [269].

insulation degrees throughout the façades of different configurations which are exposed to different noise conditions outdoors (Figure 7.44) [269].

When such detailed information is not available, the general requirements that are given in regulations or building specifications are applied. The proposed quantities regarding sound insulation exhibit a great variation in different countries/regions/communities or even buildings. Development of insulation criteria should also be based on subjective evaluation tests from which the specific noise dose & insulation prerequisite functions can be derived. Field studies are beneficial to learn about the people's satisfaction with acoustic conditions in their houses. For example, in Denmark, 10–28% of the population declared that their insulation was not sufficient at all. It was found that people in Austria and France requested the highest insulation for their homes. Based on such studies Rindel recommended the following criteria for airborne and structure-borne sounds for residential buildings in terms of the insulation descriptors given in Section 5.9.5 [270, 271]:

For new houses: $R'_w > = 60\,\text{dB}$ and $L'_{nw} < = 53\,\text{dB}$ (For the parameters see Section 5.9.1)

In a joint EU project conducted in 2009–2013 (COST TU 0901), a great discrepancy was revealed between the insulation descriptors and the criteria values declared in the regulations of the European member states [272a]. Figures 7.45a and 7.45b display the updated sound insulation requirements in the regulations of 35 countries, after converting the original insulation descriptors to $D_{nT,W}$ for airborne sounds and $L'_{nt,w}$ for impact sound insulation respectively [272a, 272b]. Recently, the acoustic classification of buildings for the purpose of rating the acoustic performance

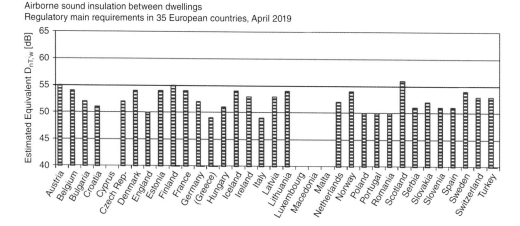

Figure 7.45a Overview of information on airborne sound insulation between dwellings: main requirements in 35 European countries (provided by B. Rasmussen, 2019) [272b, 272c].

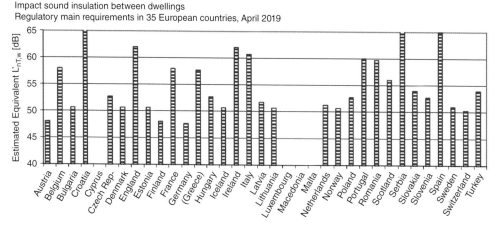

Figure 7.45b Overview of information on impact of sound insulation between dwellings. Main requirements in 35 European countries (provided by B. Rasmussen, 2019) [272b, 272c].

of buildings by using a harmonized system at the international level, has become a matter of concern in the building sector and the draft ISO standard based on the Cost project, has been prepared as ISO/AWI TS 19488 [273].

Criteria for Noise Reduction Devices (NRD)

The requirements regarding the minimum acoustic performance of noise control elements used against noise sources in transportation or in industry, such as road pavements, silencers, noise barriers, cabins etc. are declared in standards or specifications, for the attention of the manufacturers, planners or the users. An example from the past is given in Figure 7.46 displaying the recommendation about geometry of noise barrier along a high speed road (Trans-European Motorways) in 1986. The required heights of barriers are shown, in relation to traffic capacity, road geometry, and type of terrain (flat, rolling, and mountainous) [274].

7.7.4   Implementation of Criteria and Noise Control

Implementation of criteria/requirements, whether declared as recommended values or regulatory limits, needs comparisons with the existing (actual) conditions. The next step is to decide on the noise abatement measures and provide the incentives for implementation.

Evaluation of the Actual Conditions

The decision on whether a noise problem exists or not is simply made by comparing the criteria with the actual conditions that can be determined either by measurements or predictions, applying the standard techniques explained Chapter 4 and 5. Deviations from noise criteria reveal the magnitude of the reduction of excessive noise level to be fulfilled by using the engineering methods, planning strategies, etc. However, the criterion levels or the limit values implying legal enforcement should be checked after several years of application – generally five years – to see whether the criteria have been met or not, and whether people's tolerance and sensitivities to increasing noise levels differ significantly or slightly in time.

If the noise control procedures are satisfactory, the criterion levels can be decreased several dBs, depending on technology, costs, and people's reactions, as well as the rate of change in noise levels. Field surveys are useful to observe the differences in people's reactions and the limits are lowered in years based on the updated dose-response relationships. In addition, the criterion levels

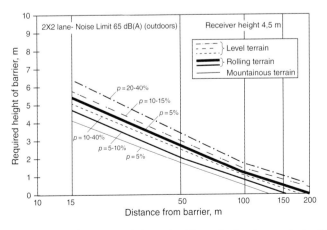

Figure 7.46   Recommended standard for noise barriers against noise from Trans-European Motorways [274] (*p*: Heavy vehicle percentage) .

derived from the noise levels corresponding to a certain percentage of highly annoyed people (*HA* %) can be revised (i.e. by taking a lower *HA%* which corresponds to a limit level to be proposed.). When the actual noise levels and the criterion levels are compared, the results demonstrate the amount of reduction to be provided in noise levels, either outdoors or indoors. Such a study is conducted in a systematic approach.

Concept of Noise Control

Noise control has long been a special topic since the 1950s that has been widely elaborated in a great number of publications from 1970 to the 1990s, and still is important issue at present as a part of noise management system. Some of the outstanding books are given in [275–280]. However, the main principles are summarized below to facilitate an understanding of noise management (see Chapter 9) and its historical background.

Generally, noise control has been defined as reducing the negative effects of noise completely or partly by using the following techniques:

- To reduce the noise levels to an acceptable level.
- To change the acoustic characteristics of noise (sound quality) or reduce the duration of noise.
- To mask the noise with an acceptable or non-attracting sound spectrum.
- To reduce the exposure time (duration) to noise.

The above methods are only the technical and engineering part of noise control. However, there are other aspects to be considered:

- Noise control is a technology to obtain an economic and acceptable environment, compatible with operational aspects. The acceptable environment can serve an individual, a group of people, a community or sometimes even noise-sensitive equipment.
- Noise control is a system problem whose elements are source, environment and user.
- Noise abatement can be conducted within the noise control concept and its variables are performance, cost. and applicability.
- For each case the amount of required noise reduction should be determined mostly based on predictions made at the beginning.
- Noise control processes need comprehensive work due to the temporal variability and diversity of the sources and the environments. Acoustic analyses and inspections through measurements, should be repeated until the satisfactory and sustainable solutions are achieved.
- Noise control engineering under the title of general acoustic engineering nowadays covers the development of technical solutions at different scales, using advanced technology based on basic acoustic theories.
- It is necessary to search for the applicable solutions to fulfill the required noise reduction. In this process, the simple evaluation methods are used and the final decisions about solutions should be checked from the standpoints of cost/benefit analysis, keeping in mind that the cost is directly related to the criteria and limit values.
- In general, noise control is a costly application that requires investment but its benefits cannot be refunded directly. Economic impacts are discussed in Section 9.5. However, if noise control is avoided, dealing with the negative effects would be more expensive.
- Strategies of noise control are to be decided (planned) in coordination with national policies.
- The roles of various responsible authorities should be determined correctly and clearly.

Consequently, noise control can be accomplished in a systematic approach:

1) Determination of sources and analyses (Chapter 3).
2) Examination of the physical environment (Chapter 2).

3) Determination of noise conditions through:
   prediction/assessment methods (Chapter 4)
   field measurements (Chapter 5)
   noise maps (Chapter 6)
4) Establishment of noise criteria and limits (Chapter 7).
5) Evaluation of actual conditions with comparisons with the criteria and limits.
6) Establishment of legal base for enforcements (Chapter 8)
7) Decision about alternative solutions applicable separately or combined by effective management and action planning (Chapter 9).
8) Verification of noise control measures with respect to meeting of the criteria.

The above-mentioned subjects are involved with in the discipline called noise management and action planning (Chapter 9).

7.8 Conclusion

An understanding of human hearing and perception mechanisms leads to better evaluating sounds and the noise impacts that can be expressed as physical, physiological, psychological, and performance effects. Studies have shown that the problem of noise pollution is increasing throughout the world, although great efforts have been made in the last 40–50 years to reduce noise in the environment. The results urge communities to improve the situation through national or international collaborations.

This chapter gives an overview of human behavior in relation to sounds, mainly noise, and presents knowledge on noise impact, by emphasizing the complexity of noise rating parameters, summarizing past and recent scientific studies and discussions, explaining the methodologies applied in pursuing research studies, referring to the development of criteria at the receiver, source, and environment scales, and explaining the implementation for a better world through soundscaping studies aligned with the noise control studies.

The numbers of variables and interrelationships between these and noise conditions in different environments make the assessment of noise impact rather complicated. However, as a widely accepted subjective response, "annoyance" due to environmental noise is investigated in the field surveys. The quantification of noise impact and future forecasts are possible by deriving the dose-response relationships or by applying the standard formulas given in the international standards. Specific type of dose-response relationships organized to express satisfaction from noise measures implemented, e.g. satisfaction from sound insulation in buildings, can be prepared for developing criteria to be transferred into the regulations as enforcements. The legal aspect of the problem will be explained in Chapter 8 and the skillful management in tackling with the environmental noises, are discussed in Chapter 9.

References

1 Davis, H. (1957). The hearing mechanism, Chapter 4. In: *Handbook of Noise Control* (ed. C.M. Harris). McGraw-Hill.

2 Crocker, M.J. (ed.) (1998). *Handbook of Acoustics*. Wiley.

3 Moller, A.R. (2006). *Hearing: Anatomy, Physiology, and Disorders of the Auditory System*, 2e. Academic Press.

4 Musiek, F.E. and Baran, J.A. (2006). *Auditory System: The Anatomy, Physiology, and Clinical Correlates*. Pearson Education.

5 Shaw, E.A.G. (1998). Acoustical characteristics of the outer ear. In: *Handbook of Acoustics* (ed. M. Crocker), 1093–1104. Wiley.

6 Green, D.M. and McFadden, D. (1998). Introduction. In: *Handbook of Acoustics* (ed. M.J. Crocker), 1141–1146. Wiley.

7 Scharf, B. (1998). Loudness. In: *Handbook of Acoustics* (ed. M.J. Crocker), 1181–1196. Wiley.

8 Fletcher, H.F. and Munson, W.A. (1933). Loudness, its definition, measurement, calculation. *Journal of the Acoustical Society of America* 5: 82–108.

9 Fletcher, H. (1940). Auditory patterns. *Reviews of Modern Physics* 12: 47–65.

10 Robinson, D.W. and Dadson, R.S. (1956). A re-determination of the equal loudness relations for pure tones. *British Journal of Applied Physics* 7: 166–181.

11 Stevens, S.S. (1956). Calculation of the loudness of complex noise. *Journal of the Acoustical Society of America* 28: 807–832.

12 Stevens, S.S. (1972). Perceived level of noise by mark VII and decibels (E). *Journal of the Acoustical Society of America* 51: 575–601.

13 Moore, B.C.J., Glasberg, B.R., and Baer, T. (1997). A model for the prediction of thresholds, loudness and partial loudness. *Journal of the Audio Engineering Society* 45: 224–240.

14 Zwicker, E., Flottorp, G., and Stevens, S.S. (1957). Critical bandwidth in loudness summation. *Journal of Acoustical Society of America* 29 (5): 548–517.

15 Zwicker, E. (1958). Über psychologische und methodische Grundlagen der Lautheit. *Acustica* 8: 237–258.

16 ISO (1985). ISO 226:1985 and ISO 226:2003. Acoustics: Normal equal-loudness level contours.

17 (a) ISO (1975). ISO 532:1975 Acoustics: Method for calculating loudness level;
(b) ISO (1975). ISO 532-1:1975 Acoustics: Methods for calculating loudness. Part 1: Zwicker method.

18 ANSI (2012). ANSI S3.4–2007 American National Standard Procedure for the Computation of Loudness of Steady Sounds (Reaffirmation Notice, June 15, 2012).

19 Moore, B.C.J., Vickers, D.A., Baer, T. et al. (1999). Factors affecting the loudness of modulated sounds. *Journal of the Acoustical Society of America* 105: 2757–2772.

20 Gockel, H., Moore, B.C.J., and Patterson, R.D. (2002). Influence of component phase on the loudness of complex tones. *Acta Acustica United with Acustica* 88: 369–377.

21 Moore, B.C.J. (1998). Frequency analysis and pitch perception. In: *Handbook of Acoustics* (ed. M.J. Crocker), 1167–1180. Wiley.

22 (a) Fastl, H. (1975). Loudness and masking patterns of narrow noise bands. *Acustica* 33: 266–271.
(b) Fastl, H., Zwicker, E. (2007). Masking. In: *Psychoacoustics*. Springer, Berlin, Heidelberg pp. 61–110.

23 Berglund, B. and Lindvall, T. (1995). Community noise. *Archives of the Center for Sensory Research* 2 (1).

24 Berglund, B., Lindvall, T., and Schwela, D.H. (eds.) (1999). *WHO Guidelines for Community Noise: A Complete, Authoritative Guide on the Effects of Noise Pollution on Health*. World Health Organization http://www.ruidos.org/Noise/WHO_Noise_guidelines_4.html.

25 Kryter, K.D. (1985). *The Effects of Noise on Man*. Academic Press.

26 Passchier-Vermeer, W. and Passchier, W.F. (2000). Noise exposure and health. *Environmental Health Perspectives* 108 (Suppl. 1): 123–131.

27 Rettinger, M. (1973). *Acoustics Design and Noise Control*. New York: Chemical Publishing Co.

28 EU (2002). Directive 2002/49/EC of 25 June 2002 of the European Parliament and Council relating to the assessment and management of environmental noise.

29 (a) Guski, R. and Schreckenberg, D. (2015). First steps in development of the new WHO Evidence Review on Noise Annoyance, Euronoise 2015, Maastricht, The Netherlands;
(b)Guski, R., Schreckenberg, D., and Schuemer, R. (2017). WHO Environmental Noise Guidelines for the European Region: a systematic review on environmental noise and annoyance. *Review, International Journal of Environmental Research and Public Health* 14: 1539.

(c)WHO (2018). Environmental Noise Guidelines for the European Region, World Health Organization, Who Regional Office for Europe.

30 Benn, T. (2000). How many people suffer hearing problems from noise at work? *Acoustics Bulletin* Sept–Oct: 15–20.

31 Bohne, B.A. and Harding, G.W. (2014). Noise-induced hearing loss. http://oto2.wustl.edu/bbears/noise.htm.

32 The National Institute for Occupational Safety and Health (2013). Facts and statistics. https://www.cdc.gov/niosh/topics/noise/factsstatistics/charts/chart-50yrold.html.

33 ISO (2013). ISO 1999:2013 (Confirmed in 2018) Acoustics: Estimation of noise-induced hearing loss.

34 US Dept. of Health and Human National Service (2010). Instigation of deafness and other commodation disease (N/DCD).

35 (a) Hohman, B.W., Mercierve, V., and Felchlin, I. (1999). Effects on hearing caused by personal cassette players, concerts and discotheques and conclusions for hearing conservation in Switzerland. *Noise Control Engineering Journal* 47 (5): 163–173.
(b) Babish, W. (2002). The noise/stress concept, risk assessment and research needs. *Noise and Health* 4 (16): 1–11.
(c) EEA (2014). EEA Report Noise in Europe 2014, No 10/2014.

36 (a) Eze, I.C., Foraster, M., Schaffner, E. et al. (2018). Transportation noise exposure, noise annoyance and respiratory health in adults: a repeated-measures study. *Environment International* 121: 741–750.
(b)WHO (2010). WHO Report on Burden of Diverse from Environmental Noise (2011). Burden of disease from environmental noise: Practical guidance. Report on a working group meeting, 14–15 October 2010;
(c) WHO (2009). Children and Noise, Children's Health and the Environment, WHO Training Package for the Health Sector. World Health Organization. www.who.int/ceh;
(d) Carvalho, W.B., Pedreira, M.L., and de Aguiar, M.A. (2005). Noise level in a pediatric intensive care unit. *Journal of Pediatrics* 81: 495–498.

37 WHO (2010). Quantifying Burden of Disease from Environmental Noise. Practical Guidance. WHO Report on Working Group meeting, Germany, 14–15 October.

38 Spreng, M. (2017). Effects of noise from military low level flights on humans, Congress on Noise as a Public Health Problem, Proceedings Part 1, Stockholm.

39 Mazer, S.E. (2006). Increase patient safety by creating a quieter hospital environment. *Biomedical Instrumentation and Technology* 40 (2): 145–146.

40 Van Kempen, E., van Kamp, I., Nilsson, M. et al. (2010). The role of annoyance in the relation between transportation noise and children's health and cognition. *Journal of the Acoustical Society of America* 128 (5): 2817.

41 Öhrsröm, E. (1989). Sleep disturbance, psycho-social and medical symptoms: a pilot survey among persons exposed to high levels of road traffic noise. *Journal of Sound and Vibration* 133 (1): 117–128.

42 Griefahn, B. and Gros, E. (1986). Noise and sleep at home: a field study on primary and after-effects. *Journal of Sound and Vibration* 105: 373–383.

43 Griefahn, B. (1990). Critical loads for noise exposure, Inter-Noise 1990.

44 (a)Griefahn, B. (2002). Sleep disturbances related to environmental noise. *Noise & Health* 4 (15): 57–60.
(b) Finegold, L.S., Harris, C.S., and von Gierke, H. (1994). Community annoyance and sleep disturbance: Updated criteria for assessing the impacts of general transportation noise on people. *Noise Control Engineering Journal* 42 (1): 25–30.

45 Öhrström, E., Rylender, R., and Björkman, M. (1988). Effects of nighttime road traffic noise: an overview of laboratory and field studies on noise dose and subjective noise sensitivity. *Journal of Sound and Vibration* 127 (3): 441–448.

46 Vallet, M. (1979). Psychophysiological effects of exposure to aircraft or road traffic noise, *Proceedings of the Institute of Acoustics*, Southampton.

47 Vallet, M., Gagneux, J.M., Blanchet, V. et al. (1983). Long term sleep disturbance due to traffic noise. *Journal of Sound and Vibration* 90 (2): 173–191.

48 (a) Vallet, M., Maurin, M., Page, M.A. et al. (1978). Annoyance from and habituation to road traffic noise from urban expressways. *Journal of Sound and Vibration* 60 (3): 423–440.
(b) Okokon, E.O., Yli-Tuomi, T., Turunen, A.W., Tiittanen, P., Juutilainen and Lanki, J.T. (2018) "Traffic noise, noise annoyance and psychotropic medication use", *Environment International*, Volume 119, 287–294..

49 WHO (2009). WHO Report, Night Noise Guidelines for Europe.

50 (a) EU (2002). Future Noise Policy, EC WG/2 –Dose-effect: Position paper on exposure response relationships between transportation noise and annoyance, Luxembourg;
(b) EU (2004). EC WG/2 position paper on dose-effect relationships for night time noise, European Commission Working Group on Health and Socio-Economic Aspects, 11 November.

51 Purcell, J.B.C. (1955). Acoustics in dwellings. *Architectural Record* Sept.: 229–232.

52 Dalton, B.H. and Behm, D. (2007). Effects of noise and music on human and task performance: a systematic review. *Occupational Ergonomics* 7 (3): 143–152.

53 Webster, J.C. (1979). Effects of noise on speech, Chapter 14. In: *Handbook of Noise Control* (ed. C.M. Harris). McGraw-Hill.

54 Aniansson, G. and Björkman, M. (1983). Traffic noise annoyance and speech intelligibility in persons with normal and persons with impaired hearing. *Journal of Sound and Vibration* 88 (1): 99–106.

55 Klumpp, R.G. and Webster, J.C. (1963). Physical measurements of equally speech-interfering navy noises. *Journal of Acoustical Society of America* 35: 1328–1338.

56 Hall, F.L., Taylor, S.M., and Birnie, S.E. (1985). Activity interference and noise annoyance. *Journal of Sound and Vibration* 103 (2): 237–252.

57 Ahrlin, U. (1988). Activity disturbance caused by different environmental noises. *Journal of Sound and Vibration* 127 (3): 599–603.

58 Berry, B.F. and Porter, N.D. (1999). Health effect-based noise assessment methods. *Acoustics Bulletin* Jan./Feb.: 13–15.

59 Kurra, S. (1993). Evaluation of annoyance against transportation noises with respect to reading and listening activities, Noise & Man 93, Noise as a Public Health Problem, Nice, France (July).

60 Flindell, I.H. and Stallen, P.J.M. (1999). Non-acoustical factors in environmental noise. *Noise & Health* 1 (3): 11–16.

61 Beranek, L.L. (1954). *Acoustics*. McGraw-Hill.

62 Schultz, T.J. (1972). *Community Noise Rating*, 2e. Applied Science Publishers.

63 ISO (1991). ISO/TR 4870:1991. Acoustics: The construction and calibration of speech intelligibility tests.

64 ANSI (1997). ANSI S3.5–1997 (R2007) Methods for the calculation of the Speech Intelligibility Index.

65 Rice, C.G. (1986). Factors affecting the annoyance of combinations of noise sources, *Proceedings of the Institute of Acoustics*.

66 Öhrström, E. and Rylander, R. (1990). Psycho-social effects of traffic noise, Inter-Noise 1990.

67 Berglund, B., Berglund, U., and Lindvall, T. (1975). Scaling loudness, noisiness and annoyance of community noises. *Journal of Acoustical Society of America* 57: 930–934.

68 Kuwano, S., Namba, S., and Fastl, H. (1988). On the judgment of loudness, noisiness and annoyance with actual and artificial noises. *Journal of Sound and Vibration* 127 (3): 457–465.

69 Hiramatsu, K., Takagi, K., and Yamamoto, T. (1988). A rating scale experiment on loudness, noisiness and annoyance. *Journal of Sound and Vibration* 127 (3): 467–473.

70 Kryter, K.D. (1968). Concepts of perceived noisiness, their implementation and application. *Journal of Acoustical Society of America* 43: 344–346.

71 Namba, S., Kuwano, S., and Schick, A. (1985). Cross-cultural survey on noise problems in apartment houses, Inter-Noise 85.

72 Namba, S., Kuwano, S., and Schick, A. (1986). A cross-cultural study on noise problems. *Journal of the Acoustical Society of Japan (E)* 7: 5.

73 Watts, G., Godfrey, N., and Berry, B. (1990). An examination of the relationship between vehicle noise measures and received noisiness, Inter-Noise 1990.

74 Tachibana, H., Hamada, Y., and Sato, F. (1988). Loudness evaluation of sounds transmitted through walls basic experiment with artificial sounds. *Journal of Sound and Vibration* 127 (3): 499–506.

75 Izumi, K. (1988). Annoyance due to mixed source noises: a laboratory study and field survey on the annoyance of road traffic and railroad noise. *Journal of Sound and Vibration* 127 (3): 485–489.

76 Ge, J. and Hokao, K. (2004). Research on the sound environment of urban open space from the viewpoint of soundscape: a case study of Saga Forest park, Japan. *Acta Acustica United with Acustica* 90: 555–563.

77 Parizet, E. and Nosulenko, V.N. (1999). Multidimensional listening test: selection of sound descriptors and design of the experiment. *Noise Control Engineering Journal* 47 (6): 1–6.

78 Hashimoto, T. and Takao, H. (1990). Subjective estimation of running car interior noise, Inter-Noise 1990.

79 Langdon, F.J. (1985). Noise annoyance, Chapter 6. In: *The Noise Handbook* (ed. W. Tempest). Academic Press.

80 Langdon, F.J. and Scholes, W.E. (1968). The Traffic Noise Index: a method of controlling noise nuisance. BRS 38/68 April.

81 Langdon, F.J. (1976). Noise nuisance from road traffic noise in residential areas: Parts 1 and 2. *Journal of Sound and Vibration* 47: 243–263.

82 Langdon, F.J. (1977). The effects of road traffic noise in the residential environment, *Proceedings of the Institute of Acoustics*.

83 Fields, J. (1992). Effect of personal and situational variables on noise annoyance. U.S. Dept. of Transportation, NASA Final Report.

84 Stansfeld, S.A. (1992). Noise, noise sensitivity and psychiatric disorder. *Psychological Medicine* 22: 1–44.

85 Fields, F.M. (1998). Reactions to environmental noise in an ambient noise context in residential areas. *Journal of Acoustical Society of America* 104 (4): 2245–2260.

86 Van Kamp, I. and Davies, H. (2008). Environmental noise and mental health: five-year review and future directives, 9th ICBEN, Zurich, Switzerland.

87 Fidell, S. and Pearson, K. (2003). Sensitivity to prospective transportation noise exposure. *Noise Control Engineering Journal* 51 (2): 106–117.

88 Van Kamp, I., Job, S.R.F., Hatfield, J. et al. (2004). The role of noise sensitivity in the noise-response relations: a comparison of the international airport studies. *Journal of Acoustical Society of America* 116: 3471–3479.

89 Job, R.F.S. (1999). Noise sensitivity as a factor influencing human reaction to noise. *Noise & Health* 1 (3): 57–68.

90 Miedama, H.M.E. and Vos, H. (2003). Noise sensitivity and reactions to noise and other environmental conditions. *Journal of Acoustical Society of America* 113: 1492–1504.

91 (a) Luz, G.A. (2005). Noise sensitivity rating of individuals. *Sound and Vibration* August: 14–17.
(b) Jangho, P., Seockhoon, C., Jiho, L. et al. (2017). Noise sensitivity, rather than noise level, predicts the non-auditory effects of noise in community samples: a population-based survey. *BMC Public Health* 17 (315): 1–9.

92 Namba, S. and Kuwano, S. (1988). Measurement of habituation to noise using the method of continuous judgement by category. *Journal of Sound and Vibration* 127 (3): 507–511.

93 Vallet, M., Favre, B., and Pachiaudi, G. (1978). Annoyance from and habituation to road traffic noise from urban expressways. *Journal of Sound and Vibration* 60 (3): 423–440.

94 Millieu Ltd (2010). RPA report by Millieu Ltd for DG Environment of the European Commission. Final Report on Task 2. Inventory measures for a better control of environmental noise. Social cost of noise.

95 Miedama, H.M.E. and Van Rijckevorsel, J.L.A. (1988). Dose and non-metric response analysis, Congress organized by Society for Multivariate Analysis in the Behavioral Sciences (December).

96 Niemann, H. and Maschke, C. (2004). Final report: Noise effects and morbidity, WHO LARES

97 FICON (1992). Report by the U.S. Federal Interagency Committee on Noise.

98 Guski, R., Schuemer, R., and Suhr Felsher, U. (1999). The concept of noise annoyance: how do international experts see it? *Journal of Sound and Vibration* 223: 513–527.

99 ISO (2003). ISO/TS 15666:2003 (confirmed in 2013) Acoustics: Assessment of noise annoyance by means of social and socio-acoustics surveys

100 Job, R.F.S. (1988). Community response to noise: a review of factors influencing the relationship between noise exposure and reaction. *Journal of Acoustical Society of America* 83: 991–101.

101 Job, R.F.S. (1993). The role of psychological factors in community reaction to noise. In: *Noise as a Public Health Problem*, vol. 3 (ed. M. Vallet), 47–79. Arcueil Cedex, France: INRETS.

102 (a) Rylander, R., Björkman, M., Ahrlin, U. et al. (1986). Dose-response relationship for traffic noise and annoyance. *Archives of Environmental Health* 41 (1): 7–10.
(b) Björkman, M. (1988). Maximum noise levels in road traffic noise. *Journal of Sound and Vibration* 127 (3): 583–587.

103 Roberts, C. (2008). A guideline for the assessment of low frequency noise. *Acoustics Bulletin* 33: 31–36.

104 Björkman, M. and Rylander, R. (1990). Community noise: The importance of noise levels and number of noise events, Inter-Noise 1990.

105 Gjestland, T. and Gelderblom, F.B. (2017). Prevalence of noise-induced annoyance and its dependency on number of aircraft measurements. *Acta Acustica United with Acustica* 103: 28–33.

106 Brown, A.L. and Lam, K.C. (1987). Levels of ambient noise in Hong Kong. *Applied Acoustics* 20: 85–100.

107 Quirt, J.D. (1976). Effects of road traffic noise on a residential community, Building Research Note, 118, National Research Council, Canada.

108 Van Renterghem, T. and Botteldoren, D. (2015). The effect of outdoor vegetation as seen from the dwellings on self-reported noise annoyance, Euronoise 2015, Maastricht, The Netherlands, paper no. 0000251.

109 Stienen, J., Schmidt, T., Paas, B. et al. (2015). Noise as a stress factor on humans in urban environments in summer and winter, Euronoise 2015, Maastricht, The Netherlands, paper no. 000215.

110 Bangjun, Z., Lili, S., and Guoqing, D. (2003). The influence of the visibility of the source on the subjective annoyance due to its noise. *Applied Acoustics* 64 (12): 1205–1215.

111 Xu, J.M., Chou, C.K., and Tang, S.K. (2017). The effects of sound visibility on noise annoyance, Inter-Noise 2017, Hong Kong (27–30 August).

112 Brown, A.L. and van Kamp, I. (2009). Response to a change in transportation noise exposure: a review of evidence of a change effect. *Journal of Acoustical Society of America* 125: 3018–3029.

113 Job, R.F.S., Topple, A., Carter, N.L. et al. (1996). Public reactions to changes in noise levels around Sydney airport, Inter-Noise 96, Liverpool, UK.

114 Schomer, P.D. and Wagner, L.R. (1996). On the contribution of noticeability of environmental sounds to noise annoyance. *Noise Control Engineering Journal* 44 (6): 294–305.

115 Fields, J.M. (1996). Policy related goals for community response. *Noise* 88 (115).

116 Vallet, M. (1996). Caractéristiques et indicateurs de la gêne due au bruit des avions, INRETS no. 29.

117 Kish, L. (1965). *Survey Sampling*. Wiley.

118 ISO (1996). ISO 1996-2 Acoustics: Description, measurement and assessment of environmental noise. Part 2: Determination of environmental noise levels.

119 Brown, A.L. and Lam, K.C. (1987). Urban noise surveys. *Applied Acoustics* 20: 23–29.

120 Lambert, M., Simonnet, F., and Vallet, M. (1984). Patterns of behaviours in dwellings exposed to road traffic noise. *Journal of Sound and Vibration* 92 (2): 159–172.

121 Edwards, A.L. and Kilpatrick, F.D. (1948). Scale analysis and the measurement of social attitudes. *Psychometrika* 13 (2): 99–114.

122 Fields, J.M., Jong, R., Brown, A.L. et al. (ICBEN Team 6) (1997). Guidelines for reporting core information from community noise reaction surveys. *Journal of Sound and Vibration* 206 (5): 685–695.

123 Fields, J.M., Jong, R., Brown, A.L. et al. (ICBEN Team 6) (2001). General-purpose noise reaction questions for community noise surveys: research and recommendation. *Journal of Sound and Vibration* 242 (4): 641–679.

124 Ikuta, A., Ohta, M., and Siddique, N.H. (2005). Prediction of probability distribution for the psychological evaluation of noise in the environment based on fuzzy theory. *International Journal of Acoustics and Vibration* 10 (3): 107–114.

125 Franken, P.A. and Jones, G. (1969). On response to community noise. *Applied Acoustics* 2: 241–246.

126 Parkin, P.H. and Purkis, H.J. (eds.) (1978). *London Noise Survey*. Her Majesty's Stationery Office.

127 Sörensen, S., Rylander, R., and Kajland, A. (1976). The disturbance caused by traffic noise in Stockholm, S.72–1976. Swedish Building Research Report.

128 Brown, A.L. and Bullen, R.B. (2004). Exposure to road traffic noise in Australia. *Acoustics Bulletin* May–June: 22–27.

129 EPA (1973). Public health and welfare criteria for noise: US Environment Protection Agency, Office of Noise Abatement and Control, Report, EPA 550/9-73-002.

130 Schultz, T. (1978). Synthesis of social surveys on noise annoyance. *Journal of Acoustical Society of America* 64 (2): 377–405.

131 Fidell, S., Barber, D.S., and Schultz, T.J. (1991). Updating a dosage effect relationship for the prevalence of annoyance due to general transportation noise. *Journal of Acoustical Society of America* 89: 221–233.

132 Fidell, S., Schultz, T., and Green, D.M. (1988). A theoretical interpretation of the prevalence rate of noise-induced annoyance in residential populations. *Journal of Acoustical Society of America* 84 (6): 2109–2113.

133 Schomer, P.D. (2002). On the normalizing DNL to provide better correlation with response. *Sound and Vibration* 36: 14–23.

134 Fields, J.M. and Walker, J.G. (1980). Reactions to railway noise: a survey near railway lines in Great Britain. Vol. I. ISVR Technical Report 102. Institute of Sound and Vibration, Southampton, England.

135 Kurra, S. (2005). Overview of the community noise studies in Turkey and Introduction to the new regulation conforming to Directive 49/EC, Environmental Noise Control, The 2005 Congress and Exposition on Noise Control Engineering, Rio de Janeiro, Brazil (7–10 August).

136 Miedama, H.M.E. and Vos, H. (1998). Exposure–response relationships for transportation noise. *Journal of Acoustical Society of America* 104 (6): 3432–3445.

137 Berry, B. and Miedama, H.M.E. (2002). Origins and applicability of the EU position paper on dose-response relationships between transportation noise and annoyance, *Proceedings of IOA Meeting on Noise and Health* (October).

138 Miedama, H.M.E. and Oudshoorn, C.G.M. (2001). Annoyance from transportation noise: relationships with exposure metrics DNL and DENL and their confidence limits. *Environmental Health Prospectives* 109 (4): 409–416.

139 Miedama, H.M.E. (2004). Relationship between exposure to multiple noise sources and annoyance. *Journal of Acoustical Society of America* 116 (949).

140 ISO (2016). ISO 1996-1:2016 Acoustics:Description, measurement and assessment of environmental noise. Part 1: Basic quantities and assessment procedures.

141 Fields, J.M. and Walker, J.G. (1982). The response to railway noise in residential areas in Great Britain. *Journal of Sound and Vibration* 85: 177–255.

142 Fields, J.M. and Parker, J.G. (1982). Comparing the relationships between noise level and annoyance in different surveys: a railway noise versus aircraft and road traffic comparison. *Journal of Sound and Vibration* 81 (1): 51–80.

143 Yoshida, T. and Nakamura, S. (1988). Subjective ratings of health status and railway noise. *Journal of Sound and Vibration* 127 (3): 593–598.

144 Fastl, H. (1990). Trading number of operations versus loudness of aircraft, Inter-Noise 90.

145 Fastl, H. (1985). Loudness and annoyance of sounds: subjective evaluation and data from ISO 532 B, *Proceedings of Inter-Noise 85.*

146 De Jong, R.G. (1990). Review of research developments in community response to noise, The Netherlands Institute for Preventive Health Care, TNO, Report. 1990. *Environment International* 16 (44): 515–522.

147 Izumi, K. (1990). A survey on the community response to road traffic noise in the mixed noise environment, Inter-Noise 90.

148 Flindell, I. (1983). Pressure Leq and multiple noise sources: a comparison for railway and road traffic noise. *Journal of Sound and Vibration* 87 (2): 327–330.

149 Vernet, M. (1983). Comparison between train noise and road noise annoyance during sleep. *Journal of Sound and Vibration* 87 (2): 331–335.

150 Knall, V. and Scaumer, R. (1983). The differing annoyance levels of rail and road traffic noise. *Journal of Sound and Vibration* 87 (2): 321–326.

151 Gunnarsson, A.G. and Nilsson, M.E. (2004). An environmental health model for children exposed to aircraft and road traffic noise, *Proceedings of Inter-Noise 2004.*

152 Stansfeld, S.A., Berglund, B., and Clark, C. (eds.) (2005). Aircraft and road traffic noise and children's cognition and health: a cross-national study. *The Lancet* 365: 1942–1949.

153 Kurra, S. and Onur, I. (1997). Significance of the variation of annoyance relationships based on the field and laboratory studies, Inter-Noise 1997, International Institute of Noise Control Engineering, Budapest, Hungary (24–27 August).

154 Kurra, S., Morimoto, M., and Maekawa, Z. (1999). Annoyance from transportation noise sources: a simulated-environment study for road, railway and aircraft noises, part 1: overall annoyance. *Journal of Sound and Vibration* 220 (2): 251–278.

155 Lercher, P. (1996). Environmental noise and health: an integrated research perspective. *Environment International* 22 (1): 117–129.

156 Vos, J. (1997). A re-analysis of the relationship between the results obtained in laboratory and field studies on the annoyance caused by high-energy impulsive sounds. *Noise Control Engineering Journal* 45 (3).

157 Berry, B.F. (1987). The evaluation of impulsive noise, National Physical Laboratory Report Ac111.

158 Berry, B.F. and Porter, N.D. (2004). A critical review and inter-comparison of methods for quantifying tonal and impulsive features in environmental noise, Inter-Noise 2004, Prague, the Czech Republic (22–25 August).

159 Berry, B.F. and Porter, N.D. (2005). A review of methods for quantifying acoustic features in environmental noise. *Proceedings of the Institute of Acoustics* 27 (4).

160 Berry, B.F. and Porter, N.D. (2005). A review of methods for quantifying tonal and impulsive features in environmental noise, Inter-Noise 2005, Rio de Janeiro, Brazil.

161 Berry, B.F. and Porter, N.D. (2006). Raising the tone: results from a workshop on tonal assessment. *Acoustics Bulletin* March–April: 13–15.

162 Schomer, P.D. (1978). Growth function for human response to large amplitude impulse noise. *Journal of Acoustical Society of America* 46 (6): 1627–1632.

163 Schomer, P.D. (1982). A model to describe to impulse noise. *Noise Control Engineering Journal* 18 (1): 5–15.

164 Yano, T. and Kobayashi, A. (1990). Disturbance caused by impulsive, fluctuating and combined noises, Inter-Noise 1990, Goteborg, Sweden.

165 European Commission (2005). Working paper on the effectiveness of noise measures, Working Group Health & Socio-Economic Aspects. http://circa.europa.eu/Public/irc/env/noisedir/library?l=/position_papers/workingpaper_sept2005pdf/_EN_1.0_&a=d.

166 Grimwood, C. and Tinsdall, N. (1995). An investigation into a method for the assessments of disturbance caused by amplified music from neighbours. *Acoustics Bulletin* March–April: 17–20.

167 Fothergill, L. (1993). Assessing nuisance caused by amplified music. *Environmental Health* 101 (09): 287–288.

168 Axelsson, A. (1996). Recreational exposure to noise and its effects. *Noise Control Engineering Journal* 44 (3): 127–134.

169 Griffin, J. and Seller, J. (2006). Amplified music from licensed premises, developing the night time noise offense. *Acoustics Bulletin* 31 (5).

170 Menge, C.W. (1999). Noise from amusement park attractions: sound level data and abatement strategies. *Noise Control Engineering Journal* 47 (5).

171 Large, J.B. and Ludlow, J.E. (1975). Community reaction to construction noise, Inter-Noise 1975, Sendai, Japan.

172 FHWA (2006). Construction Noise Handbook, Final report, FHWA-HEP-06-015, DOT-VNTSC-FHWA-06-02, NTIS No. PB2006–109102.

173 Ng, C.F. (2000). Effects of building construction noise on residents: a quasi-experiment. *Journal of Environmental Psychology* 20 (4): 375–385.

174 Garcia, I., Spuru, I.A., Diez, I. et al. (2010). Protocol to manage construction noise in urban areas: practical case in Bilbao municipality, Euronoise 2010, Maastricht, The Netherlands.

175 Pedersen, E. and Persson, W.K. (2004). Perception and annoyance due to wind turbine noise: a dose-response relationship. *Journal of Acoustical Society of America* 116 (6): 460–470.

176 Pedersen, E. and Persson, W.K. (2007). Wind turbine noise, annoyance and self-reported health and well-being in different living environments. *Occupational and Environmental Medicine* 64 (7): 480–486.

177 Pedersen, E., Van den Berg, F., Bakker, R. et al. (2009). Response to noise from modern wind farms in the Netherlands. *Journal of Acoustical Society of America* 126 (2): 634–643.

178 Shepherd, D., McBride, D., Welch, D. et al. (2011). Evaluating the impact of wind turbine noise on health-related quality of life. *Noise & Health* 13 (54): 333–339.

179 (a) Kuwano, S. (2013). Social survey on community response to wind turbine noise in Japan, Inter-Noise 2013, Innsbruck, Austria (15–18 September);
(b) Kuwano, S., Yano, T., Kageyama, T. et al. (2014). Social survey on community response to wind turbine noise in Japan. *Noise Control Engineering* 62 (6): 503–520.

180 Yano, T., Kuwano, S., Kageyama, T. et al. (2013). Dose-response relationships for wind turbine noise in Japan, Inter-Noise 2013, Innsbruck, Austria (15–18 September).

181 (a) Michaud, D.S., Keith, S.E., Feder, K. et al. (2016). Personal and situational variables associated with wind turbine noise annoyance. *The Journal of Acoustical Society of America* 1390 (3): 1456–1466.
(b) Schaffer, B., Schlittmeier, S.J., Heutschi, K. et al. (2015). Annoyance potential of wind turbine noise compared to road traffic noise, Euronoise 2015, Maastricht, The Netherlands;

(c) van den Berg, F. and van Kamp, I. Health effects related to wind turbine sound. *Acoustics Australia* 46 (1): 31–57. https://doi.org/10.1007/s40857-017-0115-6.

(d) Lotinga, M.J.B., Perkins, R.A., Berry, B. et al. (2017). A review of the human exposure-response to amplitude-modulated wind turbine noise: health effects, influences on community annoyance, methods of control and mitigation, 12th ICBEN Congress on Noise as a Public Health Problem, Zurich, Switzerland (18–22 June).

182 Leidelmeijer, K. and Marsman, G. (1997). Sociale normen van geluidproducerend woongedrag. (Neighbourhood noise and noise from neighbours), Amsterdam: RIGO Research en Advies BV.

183 (a) Rassmussen, B. and Ekholm, O. (2015). Neighbour and traffic noise annoyance at home: prevalence and trends among Danish adults, Euronoise 2015, Maastricht, The Netherlands;

(b) Jensen, H.A.R., Rasmussen, B., and Ekholm, O. (2018). Neighbour and traffic noise annoyance: a nationwide study of associated mental health and perceived stress. *European Journal of Public Health* 01 (10): 1–6.

(c) Rasmussen, B. and Ekholm, O. (2019). Is noise annoyance from neighbours in multi-storey housing associated with fatigue and sleeping problems?, *Proceedings of the 23rd International Congress on Acoustics*, Aachen, Germany (9–13 September).

184 Taylor, S.M. (1982). A comparison of models to predict annoyance reactions to noise from mixed sources. *Journal of Sound and Vibration* 81 (1): 123–138.

185 Burges, M.A. (2004). Criteria for multiple environmental noise sources, *Proceedings of Inter-Noise 2004.*

186 Howarth, H.V.C. and Griffin, M.J. (1990). Subjective response to combined noise and vibration: summation and interaction effects. *Journal of Sound and Vibration* 143 (3): 443–454.

187 Kurra, S., Morimoto, M., and Maekawa, Z. (1999). Transportation noise annoyance: a simulated environment study for road, railway and aircraft noises, Part 1: Overall annoyance. *Journal of Sound and Vibration* 220 (2): 251–278.

188 Kurra, S., Morimoto, M., and Maekawa, Z. (1999). Annoyance from transportation noise sources: a simulated environment study for road, railway and aircraft noises, Part 2 Activity annoyance. *Journal of Sound and Vibration* 220 (2): 279–295.

189 Rice, C.G. (1983). CEC joint research on annoyance due to impulse noise: laboratory studies, *Proceedings of the 4th International Congress on Noise as a Public Health Problem*, Italy.

190 Izumi, K. (1986). On the measurement of annoyance in the laboratory: 5 case studies to validate the simulated environment method, Muroran Institute of Technology, N-85, 10-2.

191 Fels, J. and Klemenz, M. (2003). Annoyance perception of spatially distributed sound sources. *Acta Acustica United with Acustica* 89: 547–555.

192 Öhrström, E. (2007). Listening experiments on effects of road traffic and railway noise occuring seperately and in combination, Inter-Noise 2007, İstanbul, Turkey, paper no. 116.

193 Rice, C.G. (1977). Investigation of the trade-off effects of aircraft noise and number. *Journal of Sound and Vibration* 52 (3): 325–344. (ISVR lab).

194 Sandrock, S., Griefahn, B., Kaczmarek, T. et al. (2008). Experimental studies on annoyance caused by noises from trams and buses. *Journal of Sound and Vibration* 313 (3–5): 908–919.

195 Rychtáriková, M. and Horvat, M. (2013). Developing a methodology for performing listening tests related to building acoustics: toward a common framework in building acoustics throughout Europe, Cost Action TU 0901.

196 Pedersen, T.H., Antunes, S., and Rasmussen, B. (2012). Online listening tests on sound insulation of walls: a feasibility study, Euronoise 2012, Prague, the Czech Republic.

197 Rosman, P.F. (1978). A simulation approach to evaluating environmental effects of roads. Digest of report, TRRL Digest LRB26.

198 Von Gierke, H.E. (1977). *Guidelines for Preparing Environmental Impact Statements on Noise.* National Research Council.

199 Lambert, J. and Vallet, M. (1994). Study related to the propagation of a communication on a future EC Noise Policy, INRETS-LEB Report 9420, prepared for CEC-DG XI Dec. 1994. 69675, Bron, France.

200 Lambert, J. and Vallet, M. (2004). GlpsyNoise (DAMEN), Inter-Noise 2004, Prague, the Czech Republic.

201 EPA (1974). Information on Levels of Environmental Noise Requisite to Protect Public Health and Welfare with an Adequate Margin of Safety, US Environmental Protection Agency (EPA) Report 550/9-74-004.

202 Fidell, S. (1979). Community response to noise, Chapter 36. In: *Handbook of Noise Control* (ed. C.M. Harris). McGraw-Hill.

203 Brown, A.L. (2010). Soundscapes and environmental noise management. *Noise Control Engineering Journal* 58 (5).

204 Schafer, R.M. (1969). *The New Soundscape.* Vienna: Universal.

205 Thompson, E. (2004). *The Soundscape of Modernity: Architectural Acoustics and the Culture of Listening in America, 1900–1933.* Bigger Books.

206 (a) Schafer, R.M. (1977). *The Soundscape-our Sonic Environment and the Tuning of the World.* Rochester, VT: Destiny Books.
(b) Schafer, R.M. (1994). *Our Sonic Environment and Soundscape: The Tuning of the World.* Rochester, VT: Destiny Books.

207 Kihlman, T., Kropp, W., Öhrström, E. et al. (2001). Soundscape support to health. a cross-disciplinary research program, Inter-Noise 2001, The Netherlands.

208 ISO (2014). ISO 12913-1:2014 Acoustics: Soundscape. Part 1 Definition and conceptual framework.

209 Brooks, B. and Schulte-Fortkamp, B. (2016). The soundscape standard, Inter-Noise 2016, Hamburg, Germany.

210 ISO (2010). ISO/TS 12913-2:2018 Acoustics: Soundscape. Part 2. Data collection and reporting requirements.

211 de Coensel, B., Botteldooven, D., and de Muer, T. (2003). Noise in rural and urban soundscapes. *Acta Acustica* 89: 287–295.

212 de Coensel, B. and Botteldooren, D. (2006). The quiet rural soundscape and how to characterize it. *Acta Acustica United with Acustica* 92 (6): 887–897.

213 Broadbent, D.E. (1958). *Perception and Communication.* Pergamon Press.

214 Davies, W., Adams, M., Bruc, N. et al. (2009). The Positive Soundscape Project: a synthesis of results of many disciplines, Inter-Noise 2009, Ottawa, Canada (23–26 August).

215 Brown, A.L., Kang, J., and Gjestland, T. (2011). Towards some standardization in assessing soundscape preference. *Applied Acoustics* 72: 387–393.

216 Brown, A.L. (2012). A review of progress in soundscapes and an approach to soundscape planning. *International Journal of Acoustics and Vibration* 17 (2): 73–81.

217 Maffiolo, V. (1999). De la caracterisation semantique et acoustique de la qualite sonore de l'environnement sonore urbain. Ph.D thesis. University of Maine.

218 (a) Maffiolo, V., Castellengo, M., and Dubois, D. (1999). Qualitative judgments of urban soundscapes, Inter-Noise 99, Fort Lauderdale, FL, USA;
(b) Cameron, H., Smyrnova, J., Brendan, S. et al. (2019). The practicalities of soundscape data collection by systematic approach according to ISO 12913-2, Inter-Noise 2019, Madrid, Spain (16–19 June).

219 Prante, H.U. (2001). Modeling judgements of environmental sounds by means of artificial neural networks. PhD dissertation. Technical University of Berlin, Germany.

220 Piczak, K.J. (2015). Environmental sound classification with convolutional neural networks, IEEE international workshop on machine learning for signal processing, Boston, USA (17–20 September).

221 (a) Ye, J., Kobayashi, T., and Murakawa, M. (2017). Urban sound event classification based on local and global features aggregation. *Applied Acoustics* 117: 246–256.
(b) ISO (2019). ISO/DTS (under development) Acoustics: Soundscape. Part 3: Data analysis.

222 Soutworth, M. (1970). The sonic environment of cities. *Ekistics* 178: 230–239.

223 (a) Klaeboe, R., Engelien, E., and Steines, M. (2004). Mapping neighborhood soundscape quality, Inter-Noise 2004, Prague, the Czech Republic;
(b) Campaign to Protect Rural England Tranquility map: England. www.cpre.org.uk/campaigns/landscape/tranquility.

224 Kang, J. (2007). *Urban Sound Environment*. Taylor & Francis.

225 Kang, J. (2007). A systematic approach towards intentionally planning and designing soundscape in urban open public spaces, Inter-Noise 2007, Istanbul, Turkey (August).

226 Botteldooren, D., de Coensel, B., Van Renterghem, T. et al. (2008). The urban soundscape: a different perspective. In: *Sustainable Mobility in Flanders: The Livable City* (eds. G. Allaert and F. Witlox), 177–204. Ghent University.

227 Daimon, S. and Minoura, K. (2017). Soundscape preservation policy and local society correspondence: A case of 100 soundscapes of Japan, Inter-Noise 2017, Hong Kong (27–30 August).

228 Paul, S., Fortekamp, B.S., and Genuit, K. (2004). Sound quality and soundscape: a new qualitative approach to evaluate target sounds, Inter-Noise 2004, Prague, the Czech Republic

229 Senqi, Y., Hui, X., Huasong, M. et al. (2016). A summary of the spatial construction of soundscape in Chinese gardens, 22nd International Congress on Acoustics, ICA 2016, Buenos Aires, Argentina (5–9 September), paper ICA2016–678.

230 Yu, C.J. and Kang, J. (2007). Soundscape in the sustainable living environment: a cross-cultural comparison between the UK and Taiwan, Inter-Noise 2007, Istanbul, Turkey (August).

231 Berglund, B., Nilsson, M.E., and Pekala, P. (2004). Towards certification of indoor and outdoor soundscape, Inter-Noise 2004, Prague, the Czech Republic (22–25 August).

232 Haykin, S. (1999). *Neural Networks*. Prentice Hall.

233 Kihlman, T., Kropp, W., Öhrström, E. et al. (2001). Soundscape support to health: a cross-disciplinary research program, Inter-Noise 2001, Maastricht, The Netherlands.

234 Brown, A.L. (2014). Soundscape planning as a complement to environmental noise management, Inter-Noise 2014, Melbourne, Australia (16–19 November).

235 Nilssen, M.E., Kaczmarek, T., and Berglund, B. (2004). Perceived soundscape evaluation of noise mitigation methods, Inter-Noise 2004, the 33rd International Congress and Exposition on NCE, Prague, the Czech Republic.

236 Lercher, P., Van Kamp, I., and Von Lindern, E. (2015). Transportation noise and health related quality of life: perception of soundscapes, coping and restoration, Euronoise 2015, Maastricht, The Netherlands.

237 Schulte-Fortkamp, B., Brooks, B.M., and Bray, W.R. (2007). Soundscape: an approach to rely on human perception and expertise in the post-modern community noise era. *Acoustics Today* 3 (1): 7–15.

238 Schulte-Fortkamp, B. and Fiebig, A. (2006). Soundscape analysis in a residential area: an evaluation of noise and in people's mind. *Acta Acustica United with Acustica* 92 (6): 875–880.

239 (a) Zhang, X., Ba, M., and Kang, J. (2017). Correlations between acoustic comfort and characterizing factors in urban open public spaces, Inter-Noise 2017, Hong Kong (27–30 August);
(b) van Kamp, I. and Brown, A.L. (2019). Soundscape and restoration: observations from planning case studies, Inter-Noise 2019, Madrid, Spain (16–19 June);
(c) Pierre, A. and Lavandier, C. (2019). Links between representation of outdoor soundscape, noise annoyance at home and outdoor acoustic measurement, Inter-Noise 2019, Madrid, Spain (16–19 June).

240 (a) Pollution. www.dictionary.com;
(b) Pollution. https://en.wikipedia.org.

241 Goines, L. and Hagler, L. (2007). Noise pollution: a modern plague. *Southern Medical Journal* 100 (3): 287–294.

242 Kurra, S. (1988). Analysis of traffic noise problems in developing countries with reference to a case study in residential areas, the 5th International Congress on Noise as Public Health Problem (Noise 88), Stockholm, Sweden (21–25 August).

243 Passchier-Vermeer, W. (1993). Noise induced hearing loss from daily occupational noise exposure: extrapolation to other exposure patterns and other populations, Noise and Man 93, Noise as a Public Health Problem, *Proceedings of the 6th International Congress*, France.

244 (a) ISO (1971). ISO/R1999:1971 Acoustics: Assessment of occupational noise exposure for hearing conservation processes (withdrawn);
(b) ISO (1990). ISO 1999:1990 Acoustics: Determination of occupational noise exposure and estimation of noise-induced hearing impairment(withdrawn).

245 ISO (2013). ISO 1999:2013 (Confirmed in 2018) Acoustics: Estimation of noise-induced hearing loss.

246 NIOSH (1998). Criteria for a recommended standard: occupational noise exposure. Revised Criteria 1998, U.S. Department of Health and Human Services, National Institute for Occupational Safety and Health (NIOSH).

247 NIOSH (1998). Criteria for recommended standard: Occupational noise exposure revised criteria for 1998, DHHS (NIOSH) Publication No. 98–126.

248 OSHA (2009). Standard No. 1926.52: Occupational Noise Exposure, published by United States Department of Labor, Occupational Safety and health Administration, OSHA. https://www.osha.gov/pls/oshaweb/owadisp.show_document?p_table=STANDARDS&p_id=10625.

249 EU (2003). Directive 2003/10/EC of the European Parliament and of the Council of 6 February 2003 on the minimum health and safety requirements regarding the exposure of workers to the risks arising from physical agents (noise) (Seventeenth Individual Directive within the meaning of Article 16(1) of Directive 89/391/EEC).

250 EU (2007). Non-binding guide to good practice for the application of Directive 2003/10/EC "Noise at work," European Commission Directorate-General for Employment, Social Affairs and Equal Opportunities Unit F.4 December.

251 Buck, K., Hamery, P., and Zimpfer, V. (2010). The European regulation 2003/10/EC and the impact of its application to the military noise exposure, Proceedings of 20th International Congress on Acoustics, ICA 2010, Sydney, Australia (23–27 August).

252 Behar, A., Wong, W., and Kunov, H. (2006). Risk of hearing loss in orchestra musicians: review of the literature. *Medical Problems of Performing Artists* 21 (4): 164–168.

253 EC (1970). European Council Directive of 6 February 1970 on the approximation of the laws of the Member States relating to the permissible sound level and the exhaust system of motor vehicles, (70/157/EEC).

254 EU (2013). EN 2013: Noise Limits for Motor Vehicles, Council of the European Union, November.

255 EU (2014). Regulation (EU) no. 540/2014 of the European Parliament and of the Council of 16 April 2014 on the sound level of motor vehicles and of replacement silencing systems, and amending Directive 2007/46/EC and repealing Directive 70/157/EEC.

256 EU (2000). European Council Directive 2000/14/EC of The European Parliament and of the Council of 8 May 2000, on the approximation of the laws of the Member States relating to the noise emission in the environment by equipment for use outdoors.

257 EPA (1974). Information on levels of environmental noise requisite to protect health and welfare with an adequate margin of safety, EPA report No.550/9-74-004, US Environment Protection Agency.

258 Galloway, J.W. and Schultz, T.J. (1979). Noise assessment guidelines 1979, Report no.HUD-CPD-586, U.S. Department of Housing and Urban Development.

259 BS (2014). BS 4142:2014 (former:1997) Methods for rating and assessing industrial and commercial sound.

260 Schwela, H. (2001). The new World Health Organization guidelines for community noise. *Noise Control Engineering Journal* 49 (4).

261 Bayazit, N., Kurra, S., Özbilen, B.Ş. et al. (2016). New regulation on noise protection for buildings and sound insulation in Turkey, 23rd International Congress on Sound and Vibration, ICSV 23, Athens, Greece (10–14 July).

262 Beranek, L.L. (1957). Revised criteria for noise control in buildings. *Noise Control* 3: 19–27.

263 Blazier, W.E. (1981). Revised noise criteria for application in the acoustical design and rating of HVAC systems. *Noise Control Engineering Journal* 16: 64–73.

264 Beranek, L.L. (1989). Application of NCB noise criterion curves. *Noise Control Engineering Journal* 32: 209–216.

265 Beranek, L.L. (1997). Applications of NCB and RC noise criterion curves for specification and evaluation of noise in buildings. *Noise Control Engineering Journal* 45 (5): 209–216.

266 Schomer, P.D. (2000). Proposed revisions to room noise criteria. *Noise Control Engineering Journal* 48 (3): 85–96.

267 Schomer, P.D. and Bradley, J.S. (2000). A test of proposed revisions to room noise criteria curves. *Noise Control Engineering Journal* 48 (4): 124–129.

268 Kurra, S. and Tamer, N. (1993). A rating criteria for facade insulation. *Journal of Applied Acoustics* 40: 213–237.

269 Kurra, S. and Dal, L. (2012). Sound insulation design by using noise maps. *Journal of Building and Environment* 49: 291–303.

270 Rindel, J.H. (1999). Acoustic quality and sound insulation between dwellings. *Journal of Building Acoustics* 5: 291–301.

271 Rindel, J.H. (2007). Sound insulation of buildings, Inter-Noise 2007, Istanbul, Turkey (28–31 August).

272 (a) COST Action TU 0901 (2014). Integrating and harmonizing sound insulation aspects in sustainable urban housing constructions, building acoustics throughout Europe, Volume 1: Towards a common framework in building acoustics throughout Europe;
(b) Rasmussen, B. (2019). Sound insulation between dwellings: comparison of national requirements in Europe and interaction with acoustic classification schemes, *Proceedings of the 23rd International Congress on Acoustics*, Aachen, Germany (9–13 September).

273 ISO (2019). ISO/AWI TS 19488: Acoustics: Acoustic classification of dwellings (under development).

274 Kurra, S. (1986). Draft standard for TEM Traffic Noise Control, UN-ECE project report, TEM/CO/TEC/WP50 (prepared for TC Directorate of Highways).

275 Harris, C.M. (1979). *Handbook of Noise Control*, 2e. McGraw-Hill.

276 Beranek, L.L. (ed.) (1971). *Noise and Vibration Control*. McGraw-Hill.

277 Crocker, M.J. (1975). *Noise and Noise Control*, vol. 1. CRC Press.

278 Beranek, L.L. and Ver, I.L. (eds.) (1992). *Noise and Vibration Control Engineering: Principles and Applications*. Wiley.

279 Crocker, M.J. (2007). *Handbook of Noise and Vibration Control*. Wiley.

280 Bies, D.A. and Hansen, C.H. (2009). *Engineering Noise Control: Theory and Practice*, 4e. Spon Press.

8

Regulations on Environmental Noise

Scope

The significance of environmental and workplace noise on community health and the need for efficient noise control have been recognized by many countries, since the second half of the twentieth century. The outcomes of scientific investigations and social studies have guided local, national, and international noise management policies which have been legalized as the noise control regulations issued in line with the technical standards and guidelines covering knowledge about noise control techniques. Legislative action implying enforcements is the final step of environmental protection instrument, and, in this chapter, the development of regulations as a part of a noise management concept (see Chapter 9), is explained. State-of-the art in different countries are outlined and framework of regulations applying the principles of better law making agreement recommended by EU, are explained. Importance of training in implementations and, efficient acoustical consultancy have been stressed. Managing the complaints from community and individuals against noise sources and noise makers, are summarized with some practices.

8.1 Legislative Terminology

There are vast differences between countries with respect to the definitions of the operative terminology related to the regulations and the issuing authorities. The vocabulary has been presented in a study (2006) by the Technical Study Groups (TSG) of the International Institute of Noise Control Engineering (I-INCE) to foster greater agreement between countries on the keywords used in legislation [1–8]. The general terms describing the legal documents which were adopted by the TSG 5 for the purpose of developing noise policies are given below [1]:

Act – A statute.
Advisory – A report or recommendation with advice on action to be taken.
Code – A systematic statement of a body of law, especially one given statutory force.
Convention – A general agreement about basic principles or procedures.
Directive – An order issued by a high-level administrative body or official. Note: In Europe, a European Directive is an order issued by the European Union (EU) prescribing that EU Member States shall enact the contents of the Directive in their national legislation, and shall enforce the requirements of the Directive.
Guideline – A recommended way of doing or managing something.

Environmental Noise and Management: Overview from Past to Present, First Edition. Selma Kurra.
© 2021 John Wiley & Sons Ltd. Published 2021 by John Wiley & Sons Ltd.

Harmonization – For a group of nations, the process of modifying their national laws to make them equivalent in all aspects.

Harmonized – A standard for which enforcement is mandatory within a group of nations bound by a treaty.

Law – A written set of principles governing an action or procedure established by a sovereign authority and expected to be observed by all who are subject to that authority.

Legislation – The written enactments of a legislative body that has the power to make laws.

Ordinance – An order, statute, or regulation governing some details regarding procedure or conduct and enforced by a limited authority, such as a municipality.

Policy – A high-level overall plan embracing the general goals and acceptable procedures of a governmental body or other authority regarding a particular subject.

Protocol – A preliminary diplomatic agreement that forms the basis for a final convention or treaty; the records or minutes of a diplomatic conference or congress incorporating the agreements arrived at by the negotiators to amend, clarify, or add to a treaty.

Regulation – A set of rules, ordinances, or laws by which an action, conduct, or procedure is controlled or governed.

Rule – An authoritative regulation governing actions, methods, or procedures.

Specification – A detailed precise requirement or standard; a common reference, method, or quantity, established by an authorized body.

Statute – A law enacted by a legislative body and set forth in a written document.

Treaty – A formal written agreement between two or more nations on a subject of mutual interest, such as peace or trade.

The TSG 5 of I-INCE, in the final working report, 2006, proposed definitions of the technical terms relating to noise policies to be used in legislative documents [1]. Some are given below:

Community noise (also referred to as environmental noise) – Unwanted sound in a non-occupational setting, indoors or outdoors, caused by sources over which an individual has little or no control, including sounds produced by neighbors.

Noise action threshold – The maximum amount of sound that triggers an action as set by a threshold-responsible authority.

Noise control engineering – The selection or design of techniques or materials to control noise and vibrations by means of engineering.

Noise declaration – A statement of the noise emitted by a source.

Noise guideline – Recommendations without mandatory requirement for compliance.

Noise regulation – All kinds of legislative documents issued, enacted, and enforced by the administrative bodies comprising of restrictions on the amount of noise, the duration of noise, and the source of noise with their operations. It is also defined as "legally-imposed requirement on the upper emission limit for a noise source or an upper limit on noise immission or noise exposure" [1]. However, it should also restrict the duration of noise and the time. Some regulations mandate the noise abatement and the administrative processes. A survey conducted by TSG 3 of I-INCE in 2005 on the national regulations throughout the world has revealed diversities between the legislative structures for the enactment and enforcement of noise policies, regulations, and standards [2].

Noise standard – Technical description of the procedure and process for the measurement, assessment or prediction of noise.

The operational systems implemented in legislative work are divided into three groups:

1) Centralized system: Regulations prepared, enacted, and enforced by the national government.
2) Tiered (shared) system: Regulations prepared, enacted by a national government; however, enforcement is made by local governments or municipalities by means of specific regulations, ordinances, building codes, etc.
3) Decentralized system: Regulations, or special requirements enacted and enforced by local governments. A further implementation is a partly centralized or semi-decentralized system combining two systems, for example; the emission requirements (centralized) and the immission requirements (decentralized).

The legislative and administrative authorities, with a shared responsibility of the enaction and enforcement of noise regulations, vary according to countries, as given Box 8.1 [2, 3].

Box 8.1 The Legislative Authorities in Different Countries Based on a Survey (2009) [2]

Legislative documents issued by enacting bodies	Administrative authorities (enforcement bodies)
Act of Parliament Regulation Standards: Noise standard: technical description of the procedure and process for the measurement, assessment, or prediction of noise Guidelines Code of Practice Provision Statute law Public laws By-law Ordinance European Commission Directives Notices Orders Circulars Notifications Building Codes Recommendations	Federal authority (government) Local councils, local government and local authority Governor Environmental Protection Authority and Environmental Protection Agency Government transport and planning agency Government planning organization Parliament/Congress Ministries of Environment, Transport, Labor, Industry and Trade, Interior, Physical or Urban Planning, Local Government and Regional Development, Transport and Communications, Health and Social Affairs Environmental Health Department Planning Department Criminal courts of law Pollution Control Authority Civil Aviation Authority Health and Safety Executive Police

In many countries, besides the national (e.g. federal and state) government's enacting laws, the scientific and technical organizations, i.e. national research institutes, national standards institutes, and non-governmental institutes publish various recommendations, national standards, or guidelines, etc. [4]. National governments are also involved with setting noise exposure limits to control noise at receiver locations and publish emission specifications aimed at reducing noise emission from noise sources. However, this creates confusion when trying to understand whether the documents are obligatory (legally required) or non-obligatory (recommended). Most countries also comply with the standards and recommended practices on different aspects of noise which are issued by the international organizations, such as: the World Health Organization (WHO) of the United Nations, the European Union Council of Parliament, the International Labour Organization (ILO) of the United Nations, the World Trade Organization (WTO), the Organization for Economic Co-operation and Development (OECD), the World Bank Group, the International Organization for Standardization (ISO), the International Civil Aviation of the International Civil Aviation Organization (ICAO), etc. Some of the professional and scientific organizations, like the I-INCE, the International Commission on the Biological Effects of Noise (ICBEN), European Acoustics Association (EAA), etc. also play an important role in establishing consensus between countries in the development of new regulations and rules for implementation practices, publishing guidelines about the noise exposure limits.

There are considerable differences between national regulations on noise, some regulations mandate noise abatements and declare administrative procedures. The major laws or ordinances relevant to noise problems comprise of noise protection in design, manufacturing, and operational phases of noise sources, concerning the noise protection within environment and in buildings by means of recommendations, restrictions, and, in some cases, incentives. Obligations concerning periodic measurements for specific noise sources, noise limits, and exposure times of labor are also dealt with in the noise regulations. Environmental noise limits are subject to change over the years and the initial levels can be reduced 1–2 dB, according to technological developments and the change in community tolerances or reactions to noise (see Chapter 7).

Legislative work is the vital step in the reduction of noise pollution and a major part of action plans against noise. Primarily, the protection of health and well-being, which is a basic human right guaranteed by the national constitutions, is strengthened by the environmental laws which address noise problems. In some countries, the noise regulations, either local or regional, are enacted under the general environmental protection laws, whereas in many countries there are separate noise regulations even issued for specific noise sources.

Noise regulation is accepted as one of the main policy instruments or a tool, as will be described in Chapter 9.

8.2   Overview of Noise Regulations

Declarations of prohibitions on excessive noise have a long history dating back to ancient Greece, in the sixth century BCE [9–13]. Historians believe that the first noise decree, which prohibited metalwork and rooster ownership within cities, was created to help people who wanted to sleep. In 44 BCE, the Roman consul Julius Caesar brought in a rule banning carts at night in the streets of Rome and in suburban areas where houses were located. In the UK, the first noise ban came into force to prohibit carriages running at certain hours at night, due to the excessive noise of their steel wheels rolling across stone pavements in the London streets. Another ban, in 1595, was against neighborhood noise at night, particularly excessively loud human voices.

The "terrible" noise from automobiles when they first came into use in the nineteenth century had to be regulated. In 1957, the Chicago Zoning Ordinance for the first time specified maximum noise levels. In the 1970s, excessive workplace noises were regulated in the laws and by-laws. In Europe, the first aircraft noise law came into force in 1978. Later, the legal dimensions on strengthening enforcement for noise control extended widely throughout Europe; for example, the first noise regulations were decreed in The Netherlands in 1979, in France in 1985, in Spain in 1993, and in Denmark in 1994. The restrictions in housing layout and the mandatory insulation were also described in some of the airport, motorway, and industrial regulations. However, not all of these legislative documents were adequately comprehensive or fully enforceable about noise protection, and they generally addressed rising ambient noise, numerical noise criteria for aircraft and motor vehicles or directives to local governments.

Although great efforts have been made in legislative activities for noise control at present, generally the development of noise regulation is a time-consuming process requiring extensive debates at political and administrative levels to reach a consensus also convincing the scientists and environmentalists. In various countries it is not easy to make the politicians aware of noise problems to be considered in a distinct category from other environmental pollutions. In addition, the enforcements through regulations may not be adequately implemented because of a country's social and economic conditions. The situation is more difficult in developing countries trying to keep up with the legislative procedures of developed countries while struggling with severe environmental problems [14, 15].

From the earlier period of development of noise regulations up to the present, the major topics found in the contexts of regulations have been: occupational noise regulations, emission regulations, and community noise regulations. Some are composed in one regulation structure, but nowadays they are included in separate regulations due to specific interests and the sophistication required for each topic. The regulations set the criteria for certain situations derived from the scientific investigations regarding the effects of noise (see Chapter 7).

A) *Occupational noise regulations:* These regulations aim to protect workers' health and safety against risks caused by noise at work, mainly inside an industrial space. However, these regulations are applicable for open-air industrial establishments like farming, mining, etc. and in music and entertainment sectors. They give provisions on:

- noise exposure limits for workers and exposure times
- noise control programs to reduce noise and to improve workplace safety
- training of workers
- risk assessments (level of risk) and employer's actions
- evaluation of effectiveness of the measures against noise.

The enforcement of occupational noise regulations is carried out by the Ministry of Labor, Labor Departments or Ministry of Health, Health and Safety Executives (HSEs), health surveillance authorities, etc.

B) *Emission regulations:* The maximum permissible noise emission levels based on the criteria referred to in Section 7.7.2 for different noise sources, operating principles and times, noise abatements, etc. are enforced in national, regional, and international regulations, that are grouped as:

- aircraft noise emission regulations
- railroad vehicles noise emission regulations
- motor vehicle noise emission regulations
- product noise emission regulations.

Emission regulations comprise provisions on:

- noise limits
- quiet products
- certification process.
- noise mission labeling
- tests and documentation (see Chapter 5).

Enforcement is carried out by the Ministries of Industry and Trade, Environment and Transportation, and the authorized departments.

C) *Community noise regulations:* The maximum permissible noise immission levels (i.e. noise level at receiver) for different noise sources, environments, types of users are declared based on the criteria and principles explained in Chapter 7, specifying the exposure time, duration, and abatement measures also in urban and transportation planning. They are grouped as:

- construction noise regulations and permits
- road traffic noise regulations
- railway traffic noise regulations
- airport noise regulations
- noise from domestic premises and public spaces.

Immission regulations comprise provisions on:

- noise immission limits
- noise abatements
- notices to be given to the noise-maker or owner of the noise source
- the appeal mechanism
- penalties.

Enforcement is carried out by the Ministries of Environment, Transportation, Interior or Health, the environmental agencies, local governments, urban planning departments, civil aviation departments, etc.

8.2.1 A Survey on Noise Regulation

The I-INCE TSG 3 conducted a comprehensive survey between 1999 and 2005, on the noise regulations and guidelines enforced in various countries regarding the prevention of community noise. The objective of the study was to investigate the efficiency of regulations in achieving noise control policies in different countries [3–5]. The information regarding the nature of regulations, context, numerical criteria, and descriptors used for rating noise, were all collected and a database was organized. Global noise policies were discussed in the follow-up studies, targeting harmonization and the development of a framework in environmental noise regulations [1, 2, 7, 8]. Due to the great discrepancy between the national regulations and between the countries' noise policies, it was decided to seek future harmonization, at least in the basic topics. The major outcomes of the survey are as follows:

1) Most countries have a basic community law for environmental noise control. This law details the responsible authority for the implementation and enforcement of the law, also the administrative structure.
2) There are great differences in the approaches to development of major noise control laws, regulations, ordinates, guidelines, etc.

3) Many countries have organizations or institutes to develop standards and guidelines. However, they publish various documents, which create some confusion as to awhether they are only recommendations or obligatory documents.

4) Some countries include environmental noise regulations at a national level, while others consider it only at the local level.

5) Noise emission limits are generally declared in national documents for specific products (e.g. transportation vehicles or machinery) and are legally required.

6) The immission limits have, in general, a *recommended* status and are incorporated in design guidelines.

7) Those who are following the regulations are listed as contractors, manufacturers, equipment operators management, area planners, industrial facilities, property owners, or exporters.

8.2.2 Noise Regulations in Different Countries

The historical development of the noise abatement legislations in some countries are summarized below, however the information might need to be updated at present.

Noise Regulations in the USA

The USA took a leading role in the world in developing noise regulations, some of which are state noise ordinances. The history of the establishment of noise regulations is as follows [16, 17]:

- The first *US Federal Noise Regulation* under the *Walsh-Healey Act* in 1969: a noise standard published, by defining a maximum noise exposure level of 90 dB(A) within eight hours for government employers [18].
- *National Environmental Policy Act* (NEPA) in 1969 [19].
- *Noise Pollution and Abatement Act* (also called the "Noise Control Act") (NCA) in 1972 [20]: Initiated a federal program regulating noise pollution control (established) mechanism of setting emission standards and a framework of noise regulations. The Environmental Protection Agency (EPA) was assigned to undertake activities to reduce noise pollution, in product noise emissions.
- *Quiet Communities Act of 1978*: The role of the EPA's Office of Noise Abatement and Control (ONAC) was expanded in state and local governments.
- *U.S. Code of Federal Regulations Noise Abatement Programs in 1982*: Shifted responsibilities to state and local governments, raising awareness of the need for noise control [21].
- The EPA was established in 1982; it has continued to support research, public information, and published a structural framework of regulations to be taken as a basis for the development of state and local laws [21]. U.S. Environmental Protection Agency described the content of a model for ordinance on community noise in 1975 [22, 23].
- About half of the US states passed substantial noise control laws or ordinances between 1970 and 1981 [21, 24].

Other codes for specific noise sources and the land uses in the USA are:

- Mines and Mining Equipment Noise (30CFR62.0)
- Codes for Aviation Noise Control, by the FAA
- Codes for workspace noise and Hearing Conservation Instruction by the Occupational Safety and Health Administration (OSHA)
- Legislative and non-legislative rules and management, along with procedural rules for railroad equipment by the Federal Railway Administration (FRA)
- Regulations for highway traffic noise by the Federal Highway Administration (FHWA)

- Requirements on noise limits and noise amounts from military equipment/ships (for hearing conservation) by the Department of Defense (DoD)
- Guidelines for control of noise from mining machinery by National Institute for Occupational Safety and Health, DHHS (NIOSH)
- HUD Guidebook published by U.S. Department of Housing and Development
- Local ordinances for construction and industrial noises
- Local ordinances for housing programs considering planning and zoning against environmental noise.

Noise Regulations in Japan

Japan also initiated enactment process regarding noise as early as 1968. The legislative developments are introduced below [17, 25]:

- *Basic Environmental Law* (revised in 1993): Basic Environment Plan and Quality standards.
- *Environmental Impact Assessment Law, EIA* (1997): First guideline published in 1972 on environmental conservation measures relating to Public Work and the process for EIA initiated in 1974 and submitted to the government in 1982.
- *Noise Regulation Law* (1968), amended in 2000 (regulations regarding specified factories, construction work, road traffic noise, designation of areas and penalties).
- *Environmental Quality Standard for noise*, 1971 (revised in 1998), based on Article 9 of the Basic Law. The noise limits were renewed in 1998 and enforced in 1999.
- *Environmental Quality Standard for Aircraft Noise*, 1973 (revised in 1993) to prevent aircraft noise pollution.
- *Environmental Quality Standards for Shinkansen Superexpress Railway Noise*, 1975, revised 1994.
- The future policy for motor vehicle noise reduction 1979, 1986, 1992.
- *Civil Aeronautics Law* (revised in 1994) concerning gradual restraints of operation of the noisier aircraft.

Noise Regulations in the UK

- *The Control of Noise at Work Regulations* 2005 (the Noise Regulations): These came into force for all industry sectors in Great Britain on April 6, 2006 (except for the music and entertainment sectors which were included on April 6, 2008) [26]. Regulations have provisions for exposure limit values, assessment of the risk to health and safety, control of exposure, hearing protection, maintenance and use of equipment, health surveillance, information, instruction and training, exemptions relating to the Ministry of Defence and other issues.
- Guidance published for above regulations by Health and Safety Executive, HSE [26].
- *Clean Neighbourhoods and Environment Act*: Authorizing the district councils to make all or part of a district an alarm notification area for legal action about noise.
- *Environment Protection Act* of 1997: Sets out zones identifying the acceptable noise levels for daytime and night-time, when the threshold is lowered by 10 dB(A) in comparison to day-time levels.
- *Environmental Protection Noise Regulations*, 2006: In relation to the measures in European Directives relating to the assessment and management of environmental noise.

Noise Regulations in Germany

Two types of legislations on noise exist in Germany: German Federal State Legislations and the EU legislations mainly for noise immissions [27]. Some of the basic national regulations are given below:

- *Federal Immission Control Act of 26, August 1998* and *Technische Anleitung zum Schutz gegen Lärm, TA Lärm* [28a].

- *German Federal Immission Control Act* of June 2005: Noise mapping and action plans to transpose the Directive 49 into German national law.
- *Equipment and Machinery Noise Protection Ordinance*, 2002.
- *GS Mark for Labeling of Noise Emission* of all products covering more extensive equipment than EU Directives.
- *German Equipment and Product Safety Act* (GPSE), 2011: Regarding minimum requirements for interest of manufacturers and suppliers.
- *Traffic Noise Protection Ordinance,199*0: Applies to the construction of new roads and expansion of existing ones and noise protection measures.
- *Traffic Routes Noise Protection Measures Ordinance*, 1997.
- *Air Traffic Noise Act* (Federal Environment Ministry), 2007: Advocating limits and operational rules.

Noise Regulations in France

- *LOI no 92-1444 du 31 décembre 1992 relative à la lutte contre le bruit* is a basic law enforcing noise prevention and implementation on a broad scale, establishing a framework for the development of a number of regulations for the execution of the main Law. Regulations have been issued for each type of source between 1996 and 1999, such as "Environmental Regulation" and "Construction and Housing Regulation," which describe the details and practices of the Articles of the Basic Law. The Law aims to prevent noise, limit the emission and propagation through preventive measures, regulating certain noisy activities, setting noise measures for road and railway transportation-related noise in urban areas, or measures against air traffic noise [28b].

For implementation of the Basic Law, the following decrees were enacted:

- *Decree No. 2006–1099*, August 31, 2006: Involving neighborhood noises, authorizes the Mayors to inspect and control the offenses.
- *Decree No. 98–1143,* December15, 1998: Concerning establishments playing amplified music; describes the operating rules and penalties to protect the hearing of the people by setting the limit of 105 dB(A).
- *Decree No. 2010–1226, October20, 2010*: Related to traffic helicopters in areas of high population density.

Environmental regulation involves the restriction of building around airports, insulation of buildings, assistance with soundproofing, or classification of land transport infrastructure according to their acoustic characteristics and traffic conditions. Directive 2002/49/EC has been adopted in French legislation, including the action plans to abate noise in the environment, to protect quiet areas and "to ensure a better coherence between the various policies in a perspective of sustainable development" in five stages explained comprehensibly in the document.

Noise Regulations in Italy

- *Law No. 447 of October 26, 1995, Frameworklaw on noise pollution*, which came into force in January 1996, is a framework on noise pollution to assess and control noise by assigning specific responsibilities to the municipalities, local authorities, regions, and central administrations [28c]. The government regulations on noise zoning areas have been issued according to the territory classification with regards to noise level limits, as Class I (implying quiet areas) to Class VI (industrial areas): and the noise limits are assigned for each class.
- *Decree of October 31, 1997 of the Ministry of the Environment and the Ministry of Transport* includes airport noise and introduces an indicator L_{AV} to evaluate the airport noise level based on the *SEL* evaluation to be used for noise mapping and planning.

- *Decree of November 18, 1998 on railway noise* to regulate noise by railway transport, establish the test methods for emission levels, set up action plans, and fix planning criteria for urban areas around railways. Grouping the railways, according to the maximum speeds, below or above 200 km/h. and noise limits are assigned for the day-time period and the night-time period. Regulated separately for railways in operation and for new railways, and taking into account the width of zones and with respect to building types in each zone.
- *The Decree of the Ministry of the Environment of November 29, 2000 and the Presidential Decree 142 of March 30, 2004* established the noise limits and the implementation methods of the mitigation measures. Due to the basic Law and the subsequent implementing decrees, on June 28, 2007, the Autostrade per l'Italia prepared noise maps for the entire network and action plans, open to the public, on June 20, 2017, covering the upgrade of its network over the next 15 years in the 14 regions and 706 municipalities concerned [28d].

Noise Regulations in Australia

- The *Environmental Protection (Noise) Regulations 1997* issued under the *Environmental Protection Act 1986* set limits on noise emissions. The major changes were issued in 2013 with various amendments published up to 2014, such as motor sports noise, shooting ranges, entertainment premises, waste collection, and air blast noise. Some of the limits have been revised and provisions to improve clarity and enforceability of the regulation, are introduced [28e].

Noise Regulations in The Netherlands

- *The Environment Management Act (Wm)* regulates also the noise exposure of national roads and railways with noise limits and declares the European obligations for noise maps and action plans.
- *Noise Abatement Act (Wgh)* All environmental aspects of noise are regulated including industrial noise, railway noise, road traffic noise, buildings along roads, construction noise, railway noise, acoustic report, definitions of noise limits, zoning plans, procedure for environmental permits, assessment frame work for road constructions and industrial zoning.
- *Environmental provisions-General Provisions Act (Wabo)-Noise Pollution of The Activities Decree*; declares noise regulations for all activities which are listed in addition to wind turbines and shooting ranges. The decree includes rules for companies to protect the environment and to prevent noise nuisances by them.

Noise regulations in Norway

- *General Municipality By-Law* "prohibits any loud, unreasonable, unnecessary or unusual noise or any noise that annoys, disturbs, injures or endangers the comfort, repose, health, welfare, peace or safety of any person, resident, or property owner within the city".
- *The Environment Directive* has been implemented since 2002.

Noise regulations in Turkey

The legislation issued in the past indirectly referring to noise and bringing restrictions and penalties are: Civil Law (1926), Criminal Law (1926), Public Health Law (1930), Labor Law (1971), Regulation for Labor Health and Work Safety (revised in 1974), Building Law (1975), Regulation for the

Organized-Industrial Zones based on the Building Regulation (1982), Motorways Traffic Law (1983), Ordinance for Manufacturing, Modifying and Assembling of motor vehicles (1985). Those which are directly addressing the environmental noise are:

- *The Environment Law* (1983): defines noise pollution as one of the environmental problems and prohibits noises which are harmful to the physical and psychological health of people.
- *Noise Control By-law* (1986): is the most comprehensive legal document about environmental noises declaring noise limits and rules, also defining the responsibilities of various parties in implementation of the by-law: the central and local governments, municipalities, etc. (Figure 8.1)
- *The Environment Impact Assessment By-law* (1993) (revised in 1997): requires the preparation of impact reports that include noise emission levels and prevention measures for new establishments.
- *Evaluation and Management of Environmental Noise By-Law* (2005): revised and re-issued in 2010 with new amendments in line with EC Directive 49. The acoustical consultancy, preparation of acoustic reports and certification obligations for the inspectors and technical personnel are brought.
- *Protection of Buildings from Noise*, 2018 (will be explained in Section 8.6.1).

Further remarks

Beyond the examples given above which are available at present, almost all countries in the world have their own regulations or other legislative documents, giving priorities to noise-related

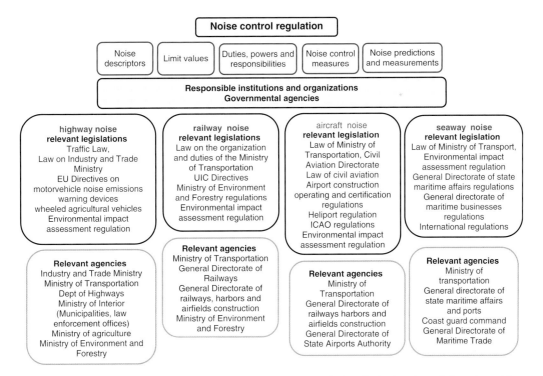

Figure 8.1 An example of operational framework of a noise regulation displaying the collaboration with governmental agencies and other legislations (for only transportation noises).

aspects according to their cultural and social life and stressing the significance of enforcements. For example, in Mexico fighting with the noise problems within the "State Law of Ecological Balance", it has been argued that sanctions should be imposed strictly on businesses like night-clubs and individuals creating excessive noise. The proposal of up to 36 hours of arrest for noisy neighbors, has been approved in the Anti-Noise Law of Jalisco, as of Feb.2020, to reduce "auditory contamination". [https://themazatlanpost.com/2020/02/21/jalisco-approves-anti-noise-law-up-to-36-hours-jail-time-for-violators/].

8.3 Recommendations by International Organizations

The major international bodies paying great attention to environmental noise as health and well-being issues are:

- the ICAO The International Civil Aviation Organization
- the UNECE The United Nations Economic Commission for Europe
- the OECD The Organisation for Economic Co-operation and Development
- the WHO The World Health Organization.

The ICAO, founded in 1944, based on the Chicago Convention to foster the planning and development of international air traffic, described noise certification procedure for aircrafts through noise measurements in 1969 and published Annex 16 of the document entitled "Environmental Protection" after several revisions over the years [29]. The purpose of certification is "to ensure that the latest available noise reduction technology is incorporated in the aircraft design." Almost all countries implement the ICAO certification system due to the increasing volume of air traffic. It has been proved that the requirements of noise certification has, over time, yielded significant noise reductions in communities around airports in terms of noise generated by civil and subsonic airplanes. Annex 16 describes the uniform measurement of aircraft noise levels and the noise certification standards (Section 5.4.2). In years, the Committee on Aviation Environmental Protection (CAEP) updated the technical noise standards, as regards to aircraft type and airworthiness certification, and issued amendments to Annex 16 with supplementary conditions and changes in noise limits. The numbers of documents with the purpose of "to limit or reduce the number of people affected by significant aircraft noise" have been so far published by ICAO, e.g. the manuals about (i) land-use planning and management to provide a guideline for prevention of the impact of aircraft noise in the vicinity of airports, (ii) uniform methodology of assessing noise around airports, (ii) noise charges within the ICAO's policy first developed in 1981 [29b].

UNECE: the Working Party 29 of the United Nations Economic Commission for Europe founded in 1947, aims "to establish regulatory instruments concerning motor vehicles and motor vehicle equipment" through UN Regulations, United Nations Global Technical Regulations and UN Rules. The Sustainable Transport Division organizes forums on harmonization of vehicle regulations and publishes and updates noise emission requirements and test procedures for motor vehicles since 1958.

The OECD: Noise pollution issues were handled in particular by the OECD's Noise Pollution Committee in the 1970s and various data from member countries were collected between 1975 and 1991 [30–31]. The report, published in 2001, revealed the situation on noise pollution in OECD countries, for instance, the increase in transportation routes (per km) was about 40% and the numbers of people exposed to traffic noise above 55 dB(A) and 65 dB(A) increased by 30% and 10%, respectively [32a]. It was stated that even though advanced technologies were proven to have been adequate in the prevention of noise problems, the situation would not be improved until 2020. In the latest document, the Environment Committee recommends member countries to develop comprehensive noise abatement programs and instructions about reporting information [32b].

The WHO: Based on the extensive investigation regarding the adverse effects of noise in relation to different situations, the first document was published in 1995 by the WHO and generically described environmental noise as the sounds emitted by all sources, with the exception of industrial workplace noise [33–35]. The noise limits for various activities indoors and outdoors were recommended as given in Section 7.7.2. A later report, published in 2011, declared that at least one million healthy life years were lost every year in Western Europe due to health effects arising from noise exposure to road traffic noise alone. Environmental noise was categorized as being the second worst environmental cause of ill health, coming behind air pollution [36a]. The Night Noise Guidelines published in 2009, presents a comprehensive overview about the effects of noise on sleep and health, gives guidelines and recommendations regarding assessment, thresholds and protection measures [36b]. The last document is entitled "WHO Environmental Noise Guidelines for the European Region," incorporating the new findings on the health effects of noise and with detailed guideline noise limits [36c].

8.3.1 Legislative Actions in the European Union

After the European Community (EC) and the European Parliament started to be actively involved with this issue, collaborative work among the EU countries accelerated.

The European Union's directives first established the noise-emission requirements that, together with the Environmental Noise Directive (END), intended to replace the national noise control documents. However, the flexibility of implementation at the national level was emphasized, especially concerning the choice of immission criteria for exposure to noise.

The list of the documents issued (up to 2016) are outlined below:

1) Green Paper (1996) [37].
2) Fifth Environmental Action Programme (1998) was prepared by the European Commission and the "sustainable environment concept" was adopted [38]. In this concept, the basic targets were defined in the past and the future noise policies were determined.
3) EU Directive/49/2002 [39, 40]: The European Commission declared an objective that no person should be exposed to noise which endangers health and quality of life (2003).
4) Sixth EAP Environment (2010) [41]: The European Commission accepted a knowledge-based approach to noise policy regarding the implementation of a Directive on Environmental Noise.
5) Directive EU 2015/996: On community noise assessment methods (2015) [42].

Since 1970, the EU Council has published a number of documents, focusing on establishing the harmonized maximum noise limits for various sources. The equipment/machinery noise emission limits and operating restrictions for the safety of machinery and reporting emissions under specified test conditions have been declared [43, 44]. Directives relative to the noise limits from subsonic aircraft and restrictions on flights, according to the ICAO Annex 16 [45, 46], and the limitations for noises from motor vehicles and requirements for performance of silencer systems [47, 48] have been issued. Specific documents regarding tire rolling noise [49] and other special noise sources, have been presented to provide a basis for noise abatement in the short, mid and long terms. There have been directives for the categorization of railway vehicles and in regard to traditional and high-speed trains [50, 51]. The permissible noise levels for wheeled agricultural or forest tractors and for noise from household appliances are declared in separate European regulations [52, 53].

The EU Council also published directives regarding the health and safety of labors exposed to noise-induced health risks [54, 55].

The Green Paper (1996)

As the first important document issued by the European Parliament in November 1996, the Green Paper aimed to steer the attention of member countries to future EU noise control policies, based on the following assessments [37]:

1) It was declared that 25% of the European population were complaining about noise pollution affecting their quality of life. Eighty million people (roughly 20% of all EU citizens) were living in areas with noise levels above 65 dB(A) and these communities were experiencing sleep disturbance and general annoyance, in addition to being under risk for other health problems. With regard to noise levels, about 170 million people were living in the gray zones, i.e. 55–65 L_{eq} dB(A).

2) A European environmental research project, conducted in 1995, revealed that noise ranked fifth (after traffic, air pollution, land use, and neighborhood layout and solid wastes) as a source of complaints. However, noise was declared as the only problem increasing since 1992, coinciding with the increasing desire of surrounding communities to mitigate noise.

3) Legislation in some countries was found to be inadequate and the community policies were not defined appropriately.

4) Knowledge of the present and future dimensions of noise pollution was not sufficiently disseminated.

5) Although the enforcement on the noise limits in Europe had yielded many successes in source emission levels, the situation was worsening in gray areas and populations living in those areas were increasing.

6) Noise impact increases in terms of time and space. Although the surveys revealed evidence that the major indicator of quality of life was noise, the member countries give less importance to noise problems than air and water pollution, perhaps due to the fact that the impact of noise is rather hidden but not extremely so.

7) In many cases, the source of noise pollution was not local; however, it was considered that the problem was restricted to the community.

8) In the Green Paper, there is a consensus on the acceptable highest noise levels to protect quality of life.

9) Another important point in noise policies is that any local solution demands a common responsibility. As recommended by the Green Paper, the major focus was to define the target, monitor the development, check the accuracy and standardization of the data obtained, and find a reconciliation between different actions against noise. Efficient action is associated with both local and national policies. Increasing collaboration between the member countries to provide comparability of the data and the interchangeability of knowledge were emphasized.

10) The policies implemented among the EU countries should be harmonized by describing the necessary actions. Generally, the Green Paper covers the noise limits, measurement techniques, and other technical knowledge related to the environmental noise pollution.

11) The Green Paper pointed out the importance of harmonization of the noise policies to be adopted by the EU countries, to be outlined in Chapter 9.

12) The Green Paper also declared some satisfactory indicators: the noise emissions of private cars reduced 80% from 1970 to 1995, while the reduction for trucks was 90%.The noise contours around airports decreased nine times. There was no reported reduction in the road traffic noise.

Directive 2002/49/EC (END)

After the Green Paper, the European Parliament and the Council initiated a project related to noise policies. The EU Noise Expert Network was established in 1998 and various working groups were organized.

In the fifth item of the EU Agreement, it was emphasized that fulfilling the target of high level protection of the environment and environmental health could be readily achieved by means of a collaborative action among the member states. For this purpose, the collection of data regarding environmental noise levels, organization (according to a comparable criteria), principles introduced for noise mapping, the descriptors, and the evaluation methods to be developed by the EU Committee were required of the member states.

The EU Commission declared the 5th Action program constituting the policy and action for sustainable development in 1998 [38] and issued a document in 2001 stating that the protection against noise was one of the targets to be accomplished as a part of EU policy within the context of health and the environment [39]. As suggested in the Green Paper, noise was referred to as one of the major environmental problems. The basic purpose is to develop a joint approach to be able to minimize or prevent the harmful effects of noise on human health. Such an approach should aim to spread knowledge and raise consciousness within communities regarding actual noise levels and the effects of noise on communities, in addition to developing policies to protect quiet areas The primary noise sources concerned in the Directive include motor vehicle traffic, railways, aircraft, airports, and the equipment used outdoors.

1) The main purpose of the document is to obtain reliable and comparable data for various noise sources to be able to reduce environmental noise.
2) Besides, the noise assessment methods for environmental noise are to be agreed on by the member states, along with the definition of limit values. Member states would give the numbers of a population or area which must be protected (the quiet areas) in the agglomerations for each of the limit values.
3) A descriptor for the annoyance level was selected as L_{den}, the sleep disturbance level was $L_{night,}$ as a collaborated approved descriptor. However, the states were free to use additional descriptors for specific noise situations.
4) It was made mandatory to prepare strategic noise maps (especially in concerned areas) to represent the noise data regarding perceived noise levels. Based on these factors, action plans were required to be determined by collaborating with the priorities in the planning of areas of special concern.
5) The selection of proper telecommunication methods (devices) to inform the community in an extensive and rapid manner, was emphasized.
6) The reports presenting the results of study were made mandatory to be submitted to the EU for approval in order to establish the future noise policies of EU members.
7) Evaluations regarding the implementation of Directive/49 should be made by the Commission at five years intervals
8) It was intended to develop a long-term EU strategy, which includes objectives to reduce the number of people affected by noise in the long term, and to provide a framework for developing an existing European Commission (EC) policy on noise reduction from the source.

9) When developing noise management action plans, EU member states' authorities are required to consult the public concerned.

The European Commission, working with the European Parliament and the European Council, issued Directive 49 in 2002 (also called the END), which was approved by the EU countries, then a further document was issued in 2015, regarding noise assessment methods and the health and socio-economic aspects of noise [40, 42]. The Directive, which is the final step of the environmental noise regulations in Europe, following the joint efforts of more than two decades, has targeted developing the EU strategies in the long term by focusing on the actions below:

- monitoring actions against environmental noise;
- sharing the experience on noise abatements among the member states;
- providing accurate data and standardizing data receiving;
- collaborating on different actions taken at the local and national levels;
- developing target noise levels;
- defining the obligations in implementing the measures to fulfill the targets.

To satisfy the objectives above, the tasks of the EU member states are defined as:

- monitoring the environmental noise problem;
- informing the community and consultation;
- orienting the communities to local noise issues;
- developing EU strategies in the long run.

The following studies (to be pursued by the member states) were required to reach targets starting in 2002:

- Major roads, major railways, major airports, and agglomerations:
  Strategic noise maps until 2012; noise control programs and action plans in 2014 for:
  major roads ≥ 6 million vehicle/year
  major railways $\geq 60\,000$ trains/year
  major airports $\geq 50\,000$ movements/year
  major agglomerations $\geq 250\,000$ inhabitants
- Each step to be reviewed every five years.
- Preparation of noise management action plans (based on the noise maps) within four years, opening them to the public and monitoring these plans every five years, and transferring the local data to these plans.
- Determination of the existing and future noise levels, the approved noise limits, the population assessments of those living in the areas exceeding noise limits, the number of houses in these areas, and result of the cost–benefit analyses.
- Submitting the results to the EU Commission with the target actions to reduce the number of people exposed to motor vehicles, railways, and civil aviation noises with certain emphasis on the strategies for protection of the quiet areas.
- Describing the national noise limits by using the recommended EU noise descriptors within 18 months.

In implementing the Directive, the European Commission is supported by the Noise Regulatory Committee and the Noise Expert Group, as well as the European Environment Agency (EAA).

Directive 2002/49/EC is described as a dynamic regulation because it is open to the implementations according to the needs of specific cases [56].

8.3.2 State-of-the-Art practices of Noise Regulations in the EU

- The EU countries adopted the EU Directive on Environmental Noise at the national level and it is enforced at either the central or local level. In most EU countries, additional national legislation was partly enforced centrally and partly at the regional (local) level.
- The first noise assessment reports including information from the countries requested by the EU Directive and END were submitted in 2005. The EAA issued a report in 2005 reviewing the developments on the above issues, covering the period 1999–2005 [57]. In this report, it was stated that despite some efforts by EU member countries regarding environmental policies, the developments could not proceed, especially in some key sectors (e.g. the main economic sectors) and various causes create blocking for future improvements.
- The EU's 7th EAP highlights that the majority of the EU's population living in major urban areas are exposed to high levels of noise, levels at which adverse effects frequently occur [58]. (High noise levels are defined in the document as noise levels above 55 dB(A) L_{den} and 50 dB(A) L_{night}.)
- Noise in the *Europe 2014 EEA Report* 10/2014: In order to achieve the objectives of the Directive, the Report emphasized the importance of an updated EU noise policy along with the latest scientific knowledge and the measures to reduce noise at the source, including improvements in urban design [59a]. The EEA has issued noise fact sheets (2017) on the countries' action plans based on the EU Directives. However, the EEA Report No.22/2019 presenting "an updated assessment of the population exposed to high levels of environmental noise and the associated health impacts in Europe", stressed that "Policy objectives on environmental noise have not yet been achieved" [59b].
- Evaluation of the Directive, Report WD1, 2015: By surveying the results of implementation by the member states, it was declared that in general, there was a lack of political will in local and regional laws at 52% and 49% respectively, in addition to a lack of adequate budget (58%) [60].

The Commission set up a program entitled "Regulatory Fitness and Performance Initiative" (REFIT) in 2015 [61]. The evaluation of the Directive was made with a focus on regulatory fitness with respect to the criteria, namely, effectiveness, efficiency, coherence, relevance, and the EU added value. It was reported that "the directive was highly relevant for the EU policy making" and "coherent in itself and with other relevant EU legislation." Aiming to make the EU Laws simpler and less costly, REFIT recommends reducing the unnecessary regulatory and administrative burdens and the changes in existing national regulations/laws through codification, recasting, repealing, reviewing clauses, revision, replacing EU laws with national regulations and replacing the legally binding laws with lighter alternatives [61].

8.3.3 Labeling of Noise Emissions

The declaration of noise emission information regarding products, using a uniform system is important for the benefit of the final customers. Noise labeling has been of interest in the past and present as a part of noise control policy (see Chapter 9). The EPA in the USA established a noise labeling program for producers in the 1970s. The program has been followed since 1981 except for portable air-compressors [16]. Voluntary labeling tests have been continuing by NIOSH for hand-powered tools. Labeling is mandatory in Europe through the following European Directives:

- *92/75/EEC:* The document regarding the energy labeling for household appliances also requires the sound power levels based on the measurements according to IEC 60704 series [62].
- *2006/42/EC:* The Machine Safety Directive requires documentation of noise levels at workstations for the situations of $L > 70$ dB(A) and $L_w > 85$ dB(A) [44].
- *2000/14/EC:* It is required that outdoor equipment should have a simple label with the declared sound power level [43].

Some of the EU countries apply voluntary eco-labels indicating "environmental friendliness" or environmental acceptability of products, which also cover the noise emission information. As an example, the Blue Angel noise criteria in Germany can be given (valid since 1977) for computer and construction equipment. However, this issue is still under discussion since there is a lack of consistency within product groups in countries and the information on the internet is insufficient for customers.

8.4 Development of Noise Regulations

Regulations against noise have a crucial role in the protection of human health and safety, by establishing the main rules at government or local levels. However, the deficiencies which are observed in the existing regulations throughout the world are due to the following factors:

1) They are often adapted without an expert review.
2) The main criteria and objectives are not sufficiently identified.
3) The implementation responsibilities are not clarified.
4) Weak enforcement.
5) Lack of validity.

Generally, noise regulations serve the protection of individuals (workers, the community, etc.) from health damage and discomfort due to noise exposure; however, they also protect those generating noise within the limits of the regulation against the lawsuits arising from the individuals' complaints.

8.4.1 Principles of Better Noise Regulation

The EU Inter-Institutional *Agreement on Better Law Making*, issued in 2011 by the EU Parliament, the Council of the EU and the EU Commission, addresses the questions relating to effectiveness, efficiency, coherence, and relevance. The attributes of simplicity, clarity, and consistency in the drafting of the regulations are promoted and the highest transparency of the legislative process is required [63]. The major principles to be applied in the drafting stage of a noise regulation, are outlined below:

- Being comprehensible and clear.
- Allowing stakeholders to easily understand their rights and obligations.
- Including appropriate reporting, monitoring, and evaluation requirements.
- Avoiding more stringent requirements to minimize burdens for manufacturers and administrations.
- Being practical to implement.

According to the above principles, "better noise regulation" drafting should consider the following aspects [63]:

1) Regulations should protect people with respect to health (long-term, permanent, physical damage, like high blood pressure, etc.) and safety (noise-induced fatigue, accidents, and speech interference).

2) The regulation should be open to the new and special cases possibly encountered in the future. This implies that a regulation should have a dynamic characteristic (e.g. tonal and impulsive industrial noises, and the specific noise sources, like shooting ranges, neighborhood noises, music concerts, industrial premises, etc.).

3) Regulations should measure the amount of low frequency noise (LFN), particularly high-level low frequency tonal noises, not only the broadband noises, by considering their physiological impacts and importance in health and safety.

4) A noise regulation provision should provide fair warning to local communities, but should also avoid the possibility of a misunderstanding in enforcement.

5) Regulations should provide complete protection for those exposed, but they should not be unnecessarily restrictive and should instead provide safe exposure for those participating or benefitting from an activity. Excessively restrictive regulations can also obstruct the effectiveness of possible noise abatement solutions.

6) Noise regulations should be publicized for a period to make the public aware of them, as an awareness campaign is useful at this stage. Legislators should invite the reactions of community.

7) The main rules or other authorizing procedures should be clearly identified and described, such as the body who will enforce them. The scope should be proper and limited, and the relations with existing regulations should be explained.

8) Amendments should be acknowledged, or regulations should be renewed at the request of the community, business, or the government.

9) The decision-making related to regulations should be open to all the stakeholders for their contributions.

10) The regulatory requirements should be in line with costs.

11) Regulations should be based on a noise policy and a management strategy (see Chapter 9).

The development of a national regulation on noise needs to take the following status of the country into account:

1) Institutional approach and facilities; laboratories, research potential, etc.

2) Public concern and complaints about noise.

3) The interests of various related sectors and noise control implementations.

4) Noise issues handled by the local administration (responsibilities) and government attention.

5) Regulations: those directly or indirectly related to noise, environmental and building regulations, standards, specifications on noise control in buildings.

6) Education level and training needed for environmental engineers, inspectors, architects, constructors, supervisors, etc.

7) Status of the housing industry, technological developments, and economic growth.

8.4.2 Community Noise Regulation Practices

Definition of community noise regulations is made to "establish the basic criteria for the acoustical quality of the environment and set the objective noise values that define the noise abatement purposes" [17]. The development of community noise regulations requires further consideration, since the community reactions to noise are rather complicated depending on various factors and the criteria to be transposed into the regulations as noise limits should satisfy all the requirements according to the different situations (Section 7.7.2). The prominent characteristics of community noise regulations are outlined below:

- They should enable a shared responsibility in noise abatement at the local, regional, federal, etc. levels by considering all levels of actions for each type of noise.
- They should be qualitative or quantitative in nature (i.e. define excessive noise in terms of noise levels with its characteristics).
- They should ensure a healthy and comfortable environment for all types of users.
- They may prohibit certain activities at certain locations and for certain times.
- They may require permits and licenses for noisy activities.
- They may define noise zones by setting performance standards,

The Framework of Community Noise Regulation

The following items should be found in the text in the following order:

1) Policy declaration (objectives and targets).
2) Terminology (comprehensive list of definitions of terms and technical references).
3) Enforcement authorities and responsibilities (i.e. powers and duties of noise control agency or officers) and administrative procedure. (Figure 8.1)
4) Criteria (quantitative and qualitative) for various noise sources and environments, reference time periods, and, according to the characteristics of noise such as impulsive, narrow band, etc., corrections for a rating index if necessary.
5) Assessment methods (the units and rating indexes referring to the technical standards for measurement, calculations or noise mapping).
6) Procedure to apply in comparison with the limits, containing a declaration of uncertainties.
7) Protection and abatement measures: Noise zones according to land uses and noise-sensitive activities, plans for the construction of new roads, expansion of the capacity of existing areas near noise-sensitive zones.
8) Restrictions on noisy activities (such as amplified music, public address systems, construction and machine operations) and restrictions to protect noise-sensitive buildings, recreation and leisure areas, in accordance with land-use or urban planning.
9) Limits for specific sources and licensed activities.
10) Possible violation of any provision and enforcement provisions (fines and penalty actions) to be well defined by specifying the purposes, period of time of violation, exceptions, emergencies, etc.
11) Public education, training, and information programs.
12) European regulations must comply with the European Directive/49 requirements to provide environmental noise data based on the strategic maps and the action plans in due time.

Further recommendations to ensure the quality and effectiveness of a regulation:

- It should be well written with proper wording to be comprehensible without confusion.
- It should cover all the components in a concise, clear, and precise way.
- Restrictions should be based on cost–benefit analyses, i.e. considering economic and societal prices of restrictions.

Generally, enforcement in community noise regulations is carried out by the Ministry of Environment or Environmental Departments. Further aspects to take into consideration in the development of community noise regulations are:

- Noise regulations are subject to change over the years (e.g. limit values and other requirements), therefore, they should allow a certain amount of flexibility and enable revisions in the future.
- Regulations of various types are a part of the environmental laws comprising the provisions of noise control.
- Technical memorandums or national guidelines should be prepared and published regarding the technical and administrative issues related to regulations.
- Noise regulations should be compatible with other legislations.
- The recommendations regarding outdoor noise should be compatible with urban planning regulations.
- The information about using the noise limits should be explained without causing confusion (i.e. determination of the expected values based on the measurement results with the uncertainty values, the procedure of comparison with the limits, etc.).
- It should declare the silent hours to prevent excessive noise disturbance at national and local levels.
- One of the suggestions for the new regulations is that they should provide complete protection while not being unnecessarily restrictive.
- The revisions of regulations should be made based on the experiences when implemented.

8.4.3 Validity of Noise Regulation

Noise regulation validity is defined as the degree with which a regulation succeeds in providing the benefit for which it is designed for [56, 64]. The developers of any regulation have an obligation to provide the validity of the regulation that they propose. Evaluation of existing and new regulations and comparison of the results of both, are important to ensure the proficiency in implementation which is described and quantified in the regulations. Otherwise, the following negative consequences can be observed:

- Invalid regulations can provide inadequate protection or, in some cases, reduce the effectiveness of the solutions.
- Invalid regulations provide wrong evaluations.
- Inappropriate limits may not be compatible with the primary issue and objective.
- Inappropriate limits may not be appropriate for the scientific realities.
- Setting inappropriate limits without considering the nature of sound (broadband or intense low frequencies) cannot ensure proper protection.
- An improperly designed regulation can cause harm to individuals or a community, while releasing noise producers from responsibility.
- Invalid regulations that provide inadequate protection as a result of insufficient experience in noise control, if an expert is not involved.
- Invalid regulations, without basing them on cost–benefit analyses, have less applicability.

Validation of a community noise regulation can be made by checking the effectiveness with respect to various aspects [65]: appropriateness of the criteria and limit values, cost–benefit analyses, people's satisfaction through field surveys, reduced noise levels from following noise maps, compatibility with other regulations, conflicts and problematic issues during the administrative process, applicability to new situations, etc. Based on the results of the evaluation, the invalid regulations or the provisions which are found to be ineffective, can be distinguished and removed.

8.4.4 Strengthening Enforcement of Noise Regulation

There are differences between the approaches regarding enforcement of noise regulations in reality, for example, in some countries, the regulations enacted by the local governments might not be stricter than those of federal governments. In some countries, certain actions and areas are excluded from the regulation's content, due to a national beneficiary or politicians' decisions, such as mining or transportation noises.

In the USA, the local ordinances are enforced by the local police [66]. The noise ordinances give noise control officers and police authorization to check the permits of premises, to investigate noise complaints and provide the strength of enforcement to abate the offending noise source, through shutdowns and fines after the necessary steps of warning or by immediate actions [65–68]. In the 1970s and early 1980s, the professional associations in charge of noise enforcement could be used for surveillance through site visits, such as the National Association of Noise Control Officials (NANCO) in the USA [66, 69].

The stringent measures against high and intense noise sources (like shooting ranges, amplified music, mining blasts, etc.) which influence vast areas of the surrounding neighborhoods, can be inspected by municipalities through continuous monitoring. In addition to noise officers, the municipal police are responsible for implementing the penalties stated in the noise regulation or in environmental law, however, the issue as a crime can be a judicial process that can be pursued with the other laws, such as misdemeanor law or criminal law, according to the country's legislative system.

8.4.5 Involving Complaints

The severe reactions of individuals or groups of people (as explained in Chapter 7) can emerge as an appeal to the local authorities through complaints against the noise-maker (the offender) or the offender's premises. The process of managing noise complaints should be clarified administratively and legally in the regulations. The UK document about neighborhood noise pollution, published in 2006, aiming "to suggest a structure to provide a consistent approach to the investigation of noise and nuisance complaints." In this comprehensive guideline, the officers (called *case officers* or *lead officers*) are defined as those having overall responsibility and eligibility to supervise, investigate and bring a resolution to the problem [69]. In many countries, those staff are also called "noise commissioners," "noise officers," or "noise guards"; "anti-noise" or "noise enforcement teams" often consist of a staff member making measurements and having the following responsibilities:

1) Interviewing the complainant through visits (inspections).
2) Deciding on a strategy for gathering evidence.
3) Gathering evidence also by conducting measurements.
4) Assessing and evaluating the evidence.
5) Deciding on the course of action to be taken and to implement (with provisions for abatement and orders, if necessary).
6) Keeping the complainant informed.
7) Preparation of reports.

The process for managing the noise complaints is given in Figure 8.2. The key points of this process are as follows: When the officer is satisfied with the existence of evidence, the local authority sends a notice (sometimes an abatement notice) to the person responsible for causing the nuisance. The evidence must be sufficient and substantial. The evidence is categorized as:

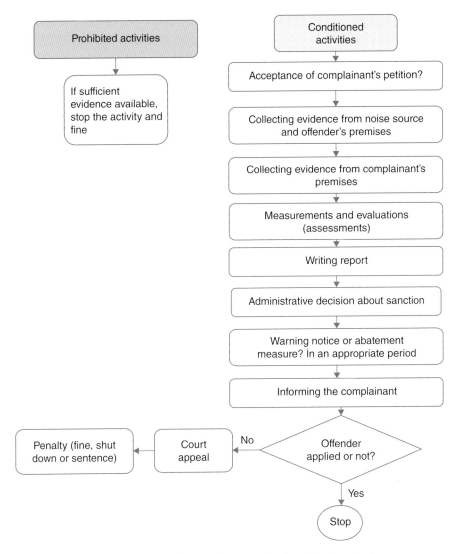

Figure 8.2 Handling complaints from environmental and neighborhood noises.

1) Definite/adequate evidence for noises continuing for long period or more than half an hour with a predictable pattern.
2) Uncertain but adequate evidence of a moderate duration (more than half an hour) or intermittent with a discernible pattern, like loud music.
3) Unlikely to be able to obtain adequate evidence, with intermittent short duration noise at random intervals (like slamming doors), or anti-social behavior, which are the most difficult and frustrating cases.

Gathering the evidence includes:

Detailed observations of the location of the complainant and the offender's premises, acoustic properties of the noise with levels, tonal components, temporal characteristics, and the effects of noise on the complainant (such as sleep disturbance and/or interference with activities), which can be obtained through noise recordings. The witness statements are then added. During this process, diary sheets about noise logging, noise time-plot, disturbances, and checklists for ensuring the correct service of notice are the documents to be prepared.

Assessing and evaluating the evidences may result in:

1) Sending a warning letter or abatement notice to the offender, acknowledging the noise limits, immediate measures, insulation requirements or recommending acoustic consultancy. A time period for compliance on a case-by-case basis is given. Notice includes either a complete stop or restrictions with hours. The notice to be given to the noise-maker should be descriptive and clear, indicating whether it is about a complete stop or limited to certain hours.
2) Appealing to a court for punishment pursuant to Environmental Protection Law against the alleged offense, if the offender continues to violate the orders. The results may be to pay compensation for the losses, penalties, or fines, which may include a prison sentence.

Different actions against different noise sources are needed, such as dealing with actual plans, licensed premises, construction site noise, poor insulated buildings, etc.

8.5 Education and Training for Proficiency in Implementation of Regulations

Successful noise management relies on good training and sufficient knowledge for those who will be involved with the preparation of the action plans and implementation at different stages. The training programs can be enforced in the noise regulations.

8.5.1 Necessity and Objectives of Training

In the past two decades, education in acoustics has been a common concern widely discussed at international levels by the academics, consultans, practicioners, etc [70, 71]. Ideally university education in acoustics and noise control is being conducted under the title of *acoustics engineering*, which is a multidisciplinary field that includes different types of engineering in which architects can also be involved after undergraduate years of education. However, it is generally admitted that it is important to increase the efficiency of educational programs, particularly in environmental noise.

Implementation of noise regulations by the local or central governments are conducted by their personnel who are trained and experienced for this purpose. A scheme of the possible tasks to be performed in an average noise regulation is given in Figure 8.3 [72]. Completion of each task has to be reported based on the evaluations of different phases of the work. Some of the reports that might be provisioned are:

- Report for the evaluation of complaints by citizens
- Acoustical Report for licensing or permitting
- Environmental Noise Evaluation Report and certification of proficiency in noise control
- Environmental Impact Assessment Report (EIA)
- Noise Mapping Evaluation report
- Noise Action Plans report.

Industrial premises requiring operating permission with regards to their noise effects (in some cases, mechanical vibrations) are normally annexed to the regulations. The licenses of the establishments and work permits are given by the local authorities for certain periods (e.g. five years, based

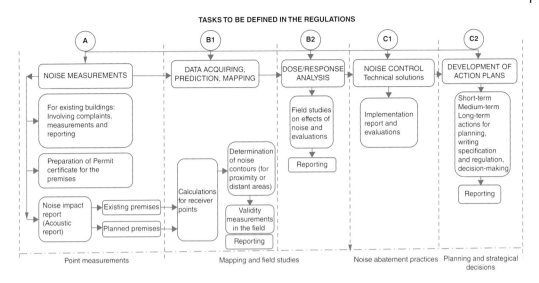

Figure 8.3 Possible tasks to be performed according to noise regulations [72].

on the Acoustical reports. The other premises for which the permissions are not required (such as entertainment locations) also need a report, such as the Environmental Noise Evaluation Report when they are located near noise-sensitive areas and when a request comes from the local municipalities/local government upon complaints of individuals or communities. The regulations clearly identify the prohibitions, prohibited actions, or circumstantial activities to be inspected (especially in the sensitive uses). The planned establishments must obtain an EIA report ensuring sufficient noise and vibration control. The formats of the above-mentioned reports are determined by the authorities. Local governments are responsible for the execution (enforcing) of the noise regulation and they must have been dealing with noise complaints and implementing sanctions according to the results of field measurements. Handling complaints related to noise first requires observations on site, problem definitions, field measurements, and the preparation of an evaluation report, which may result in restrictions or sanctions.

For assessment of noise, the field measurements are made by applying the national or international standards, as explained in Section 5.5, for example, noise level measurements according to ISO 1996-2, insulation performance of buildings in ISO 16283-1 [73, 74].

Noise mapping and action planning can be under the responsibility of the local governments, hence all the tasks can be performed within the specific departments by their officers or they can hire the services from private expert companies in order to submit these documents and reports to the upper-level authorities for approval. Their staff must be well equipped to handle these issues although they may not be directly involved in the action. (Administrative processes vary among countries.) For those who will perform some of the tasks explained above, it can be rather difficult, to comprehend the technical standards and mapping procedures which are rather sophisticated and time-consuming even for an engineer who is an expert in acoustics. It is essential to support them with the simplified guidelines for practical interpretation. On the other hand, the applications at the site supervision stage might be contradictory for noise controllers or noise inspectors due to the technical, sociological, and behavioral problems, which are rather complicated to tackle. Therefore, efficient training programs and the special courses have to be organized in an appropriate schedule.

8.5.2 Organization of Training Programs

The staff conducting the above-mentioned tasks to meet the demands of a noise regulation, must satisfy the admission criteria to be defined also in the noise regulation. Those who could be in charge of handling complaints and inspections can request technician education (a minimum of two years after high school). Those who will conduct the rest of the work (mentioned above), should have graduated from universities' technical departments and preferably with sufficient years of experience.

Those who are eligible to conduct these tasks, professionally have to receive a temporary, and then permanent license of proficiency from an upper governing body (such as the Ministry of Environment) after completing specific training courses. They might be trained as part-time experts who are working in government offices or as full-time experts whose business is this field (like an acoustic consultant). For the latter group, the advantages of such an education are clear, since the knowledge and practice they have obtained are important in pursuing their careers or in gaining a new profession, while others are able to strengthen their previous experience.

Some benefits of the training programs for the inspectors working for the local governments include approaching noise problems with more knowledge than before, and being equipped with more awareness of noise controlling possibilities. By knowing that even one decibel error in noise assessment during inspections can have a significant influence in sanctions or in victimizing noise-makers, they might be more cautious about unjustifiable decisions based on inaccurate evaluations of noise.

8.5.3 Framework of Efficient Training

With the objective of an efficient professional education and training in special topics in environmental noise management, a framework has been proposed, based on previous experiences [71, 72, 75]. The model describes the content of the training programs in relation to the tasks defined in Figures 8.3 and 8.4. In the organization of education and training programs for different kinds of participants with different backgrounds and in planning the content of training courses, the following aspects should be considered:

- The contents of the training program should serve the technical or administrative requirements of the regulation; acoustic measurements, noise mapping, and action planning, etc. Training in environmental noise in Europe should be organized according to the END requirements.
- Current and future needs should be specified.
- Implementation procedure: training programs should be conducted in universities or in a government institution or non-governmental associations e.g. acoustical societies, under the guidance of a board of directors.
- Estimation of the number of the lecturers or experts who are eligible according to certain criteria, i.e. in noise monitoring, handling of complaints, noise mapping, and action planning.

At present, it has been realized that expansion of the capacity in acoustics education and professional training is of urgent necessity in many countries for national and global implementations. The educational programs to provide competence on particular subjects, like workplace noise risk assessment or building acoustics, in addition to general diploma in acoustics and noise control, are organized also by the International Institutes like I-INCE, IOA etc.

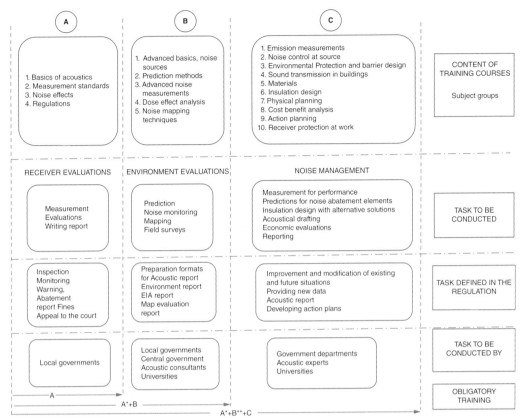

Figure 8.4 A proposal regarding the content of training programs for proficiency licenses in three categories.

8.5.4 Acoustic Consultancy

Applications of noise regulations require certain expertise and professionals in acoustics and noise problems to make assessments and to bring technical solutions. This service can be accessed from the competent companies or individuals called acoustic consultants, working privately or within a company or in government offices upon a contract. Regulations or other legal bases should be established concerning the responsibilities of acoustic consultants and their eligibilities. They are normally graduates from departments of acoustical engineering which is a branch of applied engineering dealing with the analysis, design, and control of sound and vibration in general. However, those having technical training and passing the certification programs can also be assigned as acoustic consultants. Some employers require membership of professional organizations, such as the Association of Noise Consultants in the UK. Acoustic consultancy should be a good compromise between academic and professional practice. The acoustic consultants can be categorized according to the following work in which they are engaged [75, 76]:

1) Acoustic designing based on assessments, but not having the responsibility of implementation or construction.
2) Conducting both acoustic design and construction by being aware of the risks.

At present, the acoustic consultancy companies employ different experts, since the acoustic and environmental noise problems (indoors and outdoors) cover various types of noise sources, requiring different approaches with regards to social, legal, and technical planning, etc. Generally, the required abilities of an acoustic consultant depend on the level of status in the company and the specialized field, so the eligibility of those dealing with community noise problems can be described as follows:

- Having sufficient education or training with certifications defined by the professional institutions, associations, or by the noise regulation.
- Having proficiency and experience of visiting clients, bidding, surveying, taking measurements, making design solutions and preparing the commissioning reports.
- Knowing the specifications and technical requirements, measurement standards and, if necessary, being capable of preparing those documents.
- Being trustworthy, reliable, and honest.
- Having attention, caution, and a creative and practical approach to solving acoustic problems.
- Ensuring the task is able to be performed.
- Attending meetings regularly, being eager to communicate and collaborate with other experts.
- Having knowledge of the environment and buildings in the area, as well as of the local acoustical conditions.
- Able to propose solutions by design and drafting.
- Able to give sufficient details and explanations about their design and solutions.
- Capable of calculations by using the standard prediction models.
- Able to produce reports, sharing the findings and making recommendations for action.
- Able to handle complaints and give a response in due time.
- Knowing the possible errors and tolerances and declare them in the reports.
- Being aware of the legal aspects and recent regulations.
- Having good computer usage skills.
- Having budgeting and negotiating skills.
- If necessary, equipped with project management and organizational skills.

Consultancy work should be defined clearly in the protocol between the consultant and the employee. It is generally accepted that communication with other stakeholders is key to success for an acoustic consultant during the work period. For those involved with engineering acoustics, one of the important things is that the solution must be economically as well as technically applicable.

At present, the numbers of consultancy companies working in national and international markets are increasing. They conduct various work not only described in the noise regulations, but also in other regulations demanding acoustic evaluations, e.g. building acoustics regulations, and can give complete service for all kinds of acoustic projects. Therefore, they should be specialized in different fields, which necessitates employing different specialisms, such as environmental and building acoustics, industrial noise and vibration control, mechanical services noise, occupational noise exposure and other issues relating to land planning and infrastructure practices, as well as room acoustic and audiovisual systems design. Acoustic consultants provide advice to architects, property developers, property management companies, consulting engineers, planning consultants, retail and leisure organizations, interior designers, local authorities, and other public bodies. Acoustic consultants should have adequate technical facilities, capable of field measurements according to the technical standards and should be accredited by the national or international institutions.

8.6   Regulations on Protection of Buildings Against Noise

The development of regulations against noise for the protection of people's health and well-being when they are exposed to noise in buildings, needs a different approach, though a close relationship between the environmental noise regulations (outdoors) and building noise specifications does exist, for example, the action plans against noise pay attention to façade insulation since the noise limits are difficult to be attained in many cases, particularly near transportation infrastructures or in densely populated districts (see Chapter 9). However, the requirements for indoor environments to ensure good hearing and perception conditions, peace, and tranquility for the physical and mental health of people, are generally included in the content of building regulations, by taking into account the indoor noise sources (Section 3.9), the measurement and analysis techniques of sound insulation (Section 5.9), and indoor noise criteria, including recommendations for sound insulation (Section 7.7.3). The regulations dealing with noise prevention and mitigation in buildings through enforcement and the relevant legal obligations vary significantly in the different countries, mostly focusing on acoustic performance of building elements (i.e. sound insulation).

8.6.1   An Example Regulation for the Protection of Buildings Against Noise

The new regulation entitled "Noise Protection and Sound Insulation in Buildings" issued by the Turkish Ministry of Environment and Urbanization, in 2018 based on a project conducted by Technical University of Istanbul, aims to organize the requirements and enforcement for noise control and sound insulation in all kinds of buildings against environmental noise, neighborhood noise, and noises from building services [77, 78].

The regulation comprises the following chapters:

1) Purpose, Scope, Administrative Basis, and Definitions
2) General Principles
3) Noise and Sound Insulation Descriptors to Be Used and Their Implementation
4) Required Insulation Class for Buildings and Limits for Indoor Noise Levels
5) Required Sound Insulation Values in Buildings and Principles of Implementation
6) Control Measures for Mechanical Systems Noise
7) Expertise, Evaluation, Testing, and Reporting
8) Control/Assessment/Certification (Insulation Class and Acoustic Quality Certification)
9) Evaluation of Complaints, Review, and Administration Sanctions
10) Miscellaneous and Final Provisions: Operation and Implementation

As illustrated in Figure 8.5, the regulation is intended to operate in four stages: (i) acoustic design stage; (ii) the construction stage; (iii) the post-construction stage; and (iv) assessment of complaints. It addresses all kinds of structures, buildings, facilities, and enterprises used by public institutions and organizations, private entities, and individuals. Supervision, reporting, and tests are described in the regulation. Building Acoustics Assessment Reports (BAAR) I/II/III/IV (related to the four stages of surveillance) required by the regulation, are shown in Figure 8.5. These reports have to be prepared by the experts and are submitted to the responsible authority (i.e. the municipality).

Along with the regulation, the "Guideline on Acoustical Design, Implementation, and Surveillance of Buildings" has been issued, which describes the acoustic principles applicable at

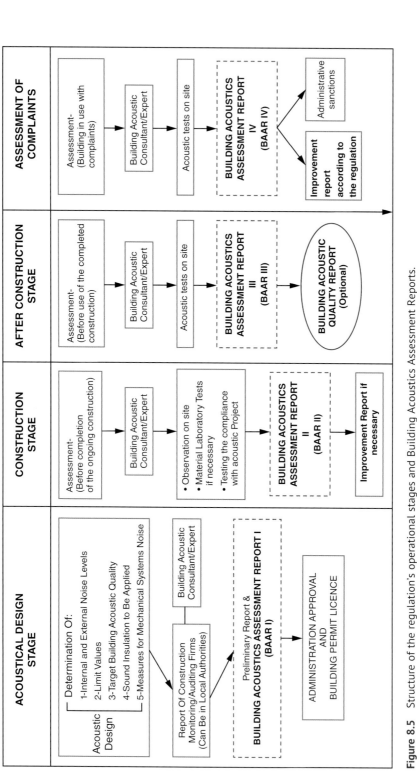

Figure 8.5 Structure of the regulation's operational stages and Building Acoustics Assessment Reports.

the design stage, the assessment procedures, the standard tests to be performed for buildings requesting an acoustic performance certificate, the checklists, and the format of the evaluation reports.

Methods to be used in the calculations and selection of the limit values are described in the guideline as follows:

- The acoustic quality class for buildings or independent units within the building will first be determined. (A minimum C class for new buildings, and D class for existing)
- The maximum permissible indoor background noise levels are determined according to the selected acoustic quality class in L_{eq}, dB(A).
- Interior background noise levels to be converted to the spectral evaluation indicator (i.e. Noise Reduction, *NR*).
- The maximum permissible indoor background noise levels for building service systems are determined according to the selected acoustic quality class in L_{eq}, dB(A).
- The reverberation time limit values for the indoor spaces are determined according to the regulation.
- Practical methodologies are introduced in the regulation and guideline for:
  - Determination of the required sound insulation values with a calculation method and a simplified method,
  - Designing the building elements conforming the criteria to be determined in (a).

The five performance parameters are introduced to evaluate the acoustic performance of buildings, conforming to the project of *COST Action TU0901: Integrating and Harmonizing Sound Insulation Aspects in Sustainable Urban Housing Constructions* (completed in 2013) and the draft technical standard ISO/AWI TS 19488 "Acoustics – Acoustic Classification Scheme for Dwellings" [79, 80]:

- indoor noise levels (not covered by the Cost project)
- airborne noise insulation (including facades)

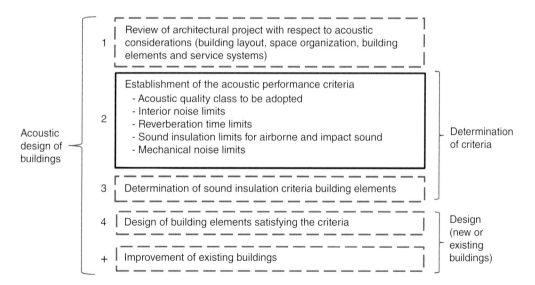

Figure 8.6 Acoustical design stages described in the regulation.
Notes: Box with straight line: subject defined in regulation; box with dotted line: subject defined in guideline.

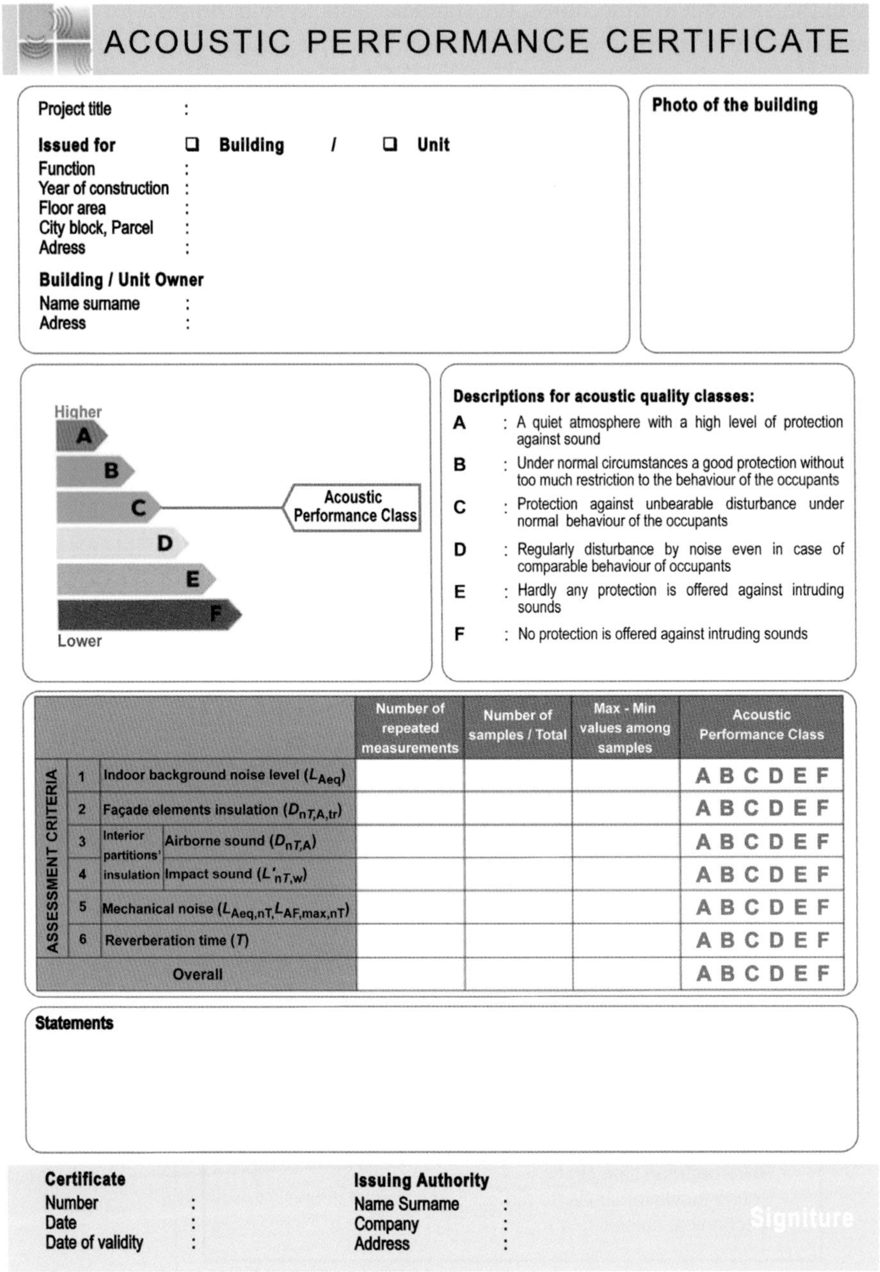

ACOUSTIC PERFORMANCE CERTIFICATE

Project title	:

Issued for ☐ **Building** / ☐ **Unit**
Function :
Year of construction :
Floor area :
City block, Parcel :
Adress :

Building / Unit Owner
Name surname :
Adress :

Photo of the building

Descriptions for acoustic quality classes:

A : A quiet atmosphere with a high level of protection against sound

B : Under normal circumstances a good protection without too much restriction to the behaviour of the occupants

C : Protection against unbearable disturbance under normal behaviour of the occupants

D : Regularly disturbance by noise even in case of comparable behaviour of occupants

E : Hardly any protection is offered against intruding sounds

F : No protection is offered against intruding sounds

Higher
A
B
C
D
E
F
Lower

Acoustic Performance Class

ASSESSMENT CRITERIA		Number of repeated measurements	Number of samples / Total	Max - Min values among samples	Acoustic Performance Class
	1 Indoor background noise level (L_{Aeq})				A B C D E F
	2 Façade elements insulation ($D_{nT,A,tr}$)				A B C D E F
	3 Interior partitions' insulation Airborne sound ($D_{nT,A}$)				A B C D E F
	4 Impact sound ($L'_{nT,w}$)				A B C D E F
	5 Mechanical noise ($L_{Aeq,nT}, L_{AF,max,nT}$)				A B C D E F
	6 Reverberation time (T)				A B C D E F
	Overall				A B C D E F

Statements

Certificate
Number :
Date :
Date of validity :

Issuing Authority
Name Surname :
Company :
Address :

Signiture

Figure 8.7 Acoustic Quality Certificate to be provided –upon demand – based on acoustic measurements in existing buildings [78].

- impact noise insulation
- building service noise levels
- reverberation times of common spaces in buildings.

The performance descriptors given in Section 5.9.1 have been accepted in determination of magnitude of performance parameters.

The principles of application of noise measures are described in the regulation and in the guideline for the indoor spaces, depending on their sensitivity to noise and as well as their capability to act as potential noise sources. Environmental noises, neighborhood noise consisting of airborne and impact sound and the noise of mechanical and electrical systems, such as heating, air conditioning, plumbing systems, etc. are taken into consideration. The guideline describes the operation and principles of the regulation. The acoustic design stages in the guideline are illustrated in Figure 8.6.

The regulation enables to provide "The Acoustic Quality Certificate," if it is requested by the building owners, and the certification system is based on the acoustic measurements in the buildings as suggested by the cost project and draft technical standard [78] (Figure 8.7).

The qualifications and tasks of the acoustic consultants specialized in building acoustics are defined in the regulation and the framework of the training programs to facilitate the application of the new regulation, and to provide certification of proficiency as expert after passing an examination, has been prepared in two parts: Basic knowledge (72 hours) and Acoustical Measurements and Standards (38 hours). The programs are open to professionals (e.g. architects and engineers) and organized by the competent institutions based upon the protocols with the Ministry of Environment and Urban Planning.

8.7 Conclusion

The development of noise regulation is an important part of noise management and action planning to protect the health and well-being of individuals and the community against environmental noises. Writing a noise regulation and compiling the proposition of its framework are rather challenging tasks at present, due to the complexity of the components involved, although some experiences were successful even in the 1970s in the USA [18].

The structure of noise regulations dealing with emission of noise sources, immissions in the environment, and hearing protection at work conditions, varies considerably in different countries and their enacting and enforcing procedures. Recently efforts have been continuing to achieve the best regulation based on a consensus, in Europe at least. Collaboration between countries has also been successful in the implementation of noise measurements, prediction models, data collection, noise mapping, etc., through the international technical standards. However, harmonization in regulations and action planning is quite difficult due to the reasons below:

- Divergence between the economic development of countries.
- Differences between the cultural levels of countries and priorities given to noise.
- Lack of available data on dose/response relationships and community reactions to noise. This issue is associated with the proposition of global noise limits.
- Unknown increase rate of noise impact in countries.
- Vanishing noise sources and emerging new ones.

- Weakness in enforcement, lack of supervision, and differences between the attention given by government institutions.
- Lack of efficient coordination between the responsible authorities.
- Varying importance given to the noise sources according to physical conditions (i.e. climate, seasonal variations, human dynamics, etc.).
- Insufficient number of trained personnel and experts in acoustics particularly in noise problems.
- Weak attention paid to noise issues by politicians.

In conclusion, legislative actions are an important part of noise policy establishment at the national and international levels (see Chapter 9). Dealing with the noise issues within the context of human rights, the noise regulations through a structural enforcement scheme, are vital in creating quiet environments and in prevention of health risks due to noise pollution. Noise regulations are more effective if the responsibilities are shared appropriately among various administrative bodies through follow-up procedures, since they address different noise sources, users, and environment scales. The success is also associated with the importance given to training people, who are in charge of the implementation of the regulations, through well-structured training programs, such as for those working in surveillance and enforcement, in administrative bodies or for acoustic consultants involved with assessment and abatement noise through engineering, planning, and design.

It can be said that noise regulations should have a dynamic characteristic and need updates and revisions even in short terms due to unforeseen circumstances in the future, changing cultural structures, economic and social levels of communities, technological advances, past experiences, international conventions, etc. in the challenging modern world.

References

1 I-INCE Technical Study Group 5 (2006). A global approach to noise control policy. I-INCE Publication 06-1. *Noise Control Engineering Journal* 54 (5).
2 I-INCE Technical Study Group 5 (2006). A global approach to noise control policy, Part 1 General. *Noise Control Engineering Journal* 54 (5): 289–346.
3 I-INCE Technical Study Group on Noise Policies and Regulations (TSG 3) (2005). Assessing the effectiveness of noise policies and regulations: phase I – policies and regulations for environmental noise, Final report 01-1.
4 I-INCE Technical Study Group on Noise Policies and Regulations (TSG 3) (2009). Survey of legislations, regulations and guidelines for control of community noise. I-INCE publication: 09-1.
5 Tachibana, H. and Lang, W.W. (2002). The I-INCE Technical Initiative No. 3, noise policies and regulations, Inter-Noise 2002, Michigan, USA (August).
6 I-INCE (2011). Guidelines for community noise impact assessment and mitigation. I-INCE publication No. 11. Final report (TSG 6).
7 Finegold, S. and Schomer, P. (2001). I-INCE Initiative No. 3, environmental noise policy assessment. Invited paper, 141st meeting of the ASA, Chicago (June).
8 Finegold, S., Gierke, H.E., McKinley, R.L. et al. (1998). Assessing the effectiveness of noise control regulations and policies, 7th International Congress on Noise as a Public Health Problem, Sydney, Australia.
9 Goldsmith, M. (2012). *Discord: The Story of Noise*. Oxford University Press.
10 Ivanov, N. (1993). Noise control problems, Proceedings of Noise-93, St Petersburg (31 May–3 June).

11 Ginjaar, L. (1981). Towards a quieter society, Inter-Noise 81, The Hague, The Netherlands.

12 Embelton, T. (1989). Noise control in the Western world over the past 2000 years. *Acta Acustica* 14 (1): 10–16.

13 Gottlob, D. (1995). Regulations for community noise. *Noise* December: 223–236.

14 Kurra, S. (2001). Key problems in implementation of Turkish noise control regulations and adaptability to EU Directives, Inter-Noise 2001, The Hague, The Netherlands (27–30 August).

15 Kurra, S. (2005). Overview of the community noise studies in Turkey and introduction to the new regulation conforming to Directive 49/EC, environmental noise control, the 2005 Congress and Exposition on Noise Control Engineering, Paper no. 1976, Rio de Janeiro, Brazil (7–10 August).

16 National Academy of Engineering (2010). Standards and regulations for product noise emissions, Chapter 6. In: *Technology for a Quieter America*. The National Academies Press https://doi.org/10.17226/12928.

17 Bento Coelho, J.L. (2008). Community noise ordinances. In: *Handbook of Noise and Vibration Control* (ed. M.J. Crocker), 1525–1537. Wiley.

18 OSHA (2000). *OSHAS's Noise Standard Defines Hazard Protection. 3M Resource Guide*: G19–G21. http://multimedia.3m.com/mws/media/918620/oshas-noise-standard-defines-hazard-protection.pdf.

19 National Environmental Policy Act (NEPA) and the Noise Pollution and Abatement Act (1969).

20 Noise Control Act of 1972 (1972).

21 Chanaud, R.C. (2014). Noise ordinances tools for enactment: modification and enforcement of a community noise ordinance, updated.

22 EPA (1975). *Model Community Noise Control Ordinance*. U.S. Environmental Protection Agency (EPA).

23 Finegold, L.S., and Brooks, B.M. (2003). Progress on developing a model noise ordinance as a national standard in the US, *Proceedings of Inter-Noise 2003*, Seoul, Korea (25–28 August).

24 Finegold, L.S., Finegold, M.S., and Maling, G.C. (2003). Overview of US national noise policy. *Noise Control Engineering* 51 (3): 162–165.

25 Tachibana, H. (2000). Recent movements of administration for environmental noise problem in Japan. *Journal of the Acoustical Society of Japan (E)* 21: 6.

26 Anon (2005). Controlling noise at work: The Control of Noise at Work, Regulations 2005, Guidance on Regulations. Health and Safety Executive.

27 Christian, F. (2012). Noise policy in Germany. *The Journal of Acoustical Society of America* 131: 3925. https://doi.org/10.1121/1.4708322.

28 (a) Technische Anleitung zum Schutz gegen Lärm (1998). TA Lärm (Technical Instructions on Noise Abatement-TA Noise), Sixth General Administrative Provision to the Federal Immission Control Act of 26;
(b) http://www.bruit.fr/tout-sur-les-bruits/transports/trafic-routier/voie-nouvelle/dispositions-reglementaires.html;
(c) Biondi, G.G. (2000). Italian legislation on noise assessment and control, Inter-Noise 2000, Nice, France;
(d) https://www.autostrade.it/la-nostra-rete/risanamento-acustico/la-normativa-italiana;
(e) https://www.der.wa.gov.au/our-work/legislative-review-regulatory-reforms/83-environmental-protection-noise-regulations-1997.

29 (a) ICAO Committee on Environmental Protection (2008). ICAO Annex 16 to the Convention on International Civil Aviation, vol. 1: Aircraft Noise;
(b) https://www.icao.int/environmental-protection/pages/Reduction-of-Noise-at-Source.aspx.

30 Alexander, A. (1975). Noise regulations in OECD countries. *Environmental Science and Technology* 9 (12): 1020–1024.

31 OECD (1985). Strengthening noise abatement policies. Final report, Environment Committee, Ad Hoc Group on Noise Abatement Policies, Environment Directorate, ENV/ N/842.

32 (a) OECD (2001). Environmental outlook;
(b) OECD (2020). Recommendation of the Council on OECD Legal Instruments Noise Abatement Policies OECD/LEGAL/0163.

33 Berglund, B., Lindwall, T., and Schwela, D.H. (1995). Community noise. Archives of the Centers for Sensory Research, Stockholm: Karolinska Institute. Document prepared for the WHO.

34 Berglund, B., Lindwall, T., and Schwela, D.H. (1999). *Guidelines for Community Noise*. World Health Organization.

35 WHO Regional Office for Europe (2013). *A European Policy Framework and Strategy for the 21st Century*. WHO Regional Office for Europe.

36 (a) WHO (2011). *Burden of Disease from Environmental Noise: Quantification of Healthy Life Years Lost in Europe*. WHO.
(b) WHO (2009). Night Noise Guidelines for Europe, WHO Regional Office for Europe
(c) WHO (2018). *Environmental Noise Guidelines for European Region*. World Health Organization Regional Office for Europe.

37 EC (1996). Green Paper on Future Noise Policy (COM(96) 540), adopted and published by the Commission.

38 EU (1998). Fifth Environmental Action Programme, "Towards sustainability." The European Community programme of policy and action in relation to the environment and sustainable development.

39 EU (2001). Common Position (EC) No. 25/2000 of 7 June 2001, adopted by the Council with a view to adopting a directive of the European Parliament and of the Council relating to the assessment and management of environmental noise.

40 EU (2002). Directive 2002/49/EC of 25 June 2002 of the European Parliament and Council relating to the assessment and management of environmental noise.

41 EU (2002). The Sixth Environment Action Programme of the European Community 2002–2012.

42 EU (2015). Commission Directive (EU) 2015/996 of 19 May 2015 establishing common noise assessment methods according to Directive 2002/49/EC of the European Parliament and of the Council.

43 EU (2000). Directive 2000/14/EC on the approximation of the laws of the member states relating to the noise emission in the environment by equipment for use outdoors (OND), 2001 and amendment 2005/88/EC.

44 EU (2006). Directive 2006/42/EC of the European Parliament and of the Council of 17 May 2006 on machinery (MD).

45 EU (2006). Directive 2006/93/EC on the regulation of the operation of aeroplanes covered by the Convention of International Civil Aviation.

46 EU (1989). Directive 89/629/EEC of the 4 December 1989 on the limitation of noise emission from civil subsonic jet aeroplanes.

47 EU (1997). Directive 97/24/EC on certain components and characteristics of two- or three-wheel motor vehicles.

48 EU (2014). Regulation EU No 540/2014 of the European Parliament and of the Council of 16 April 2014 on the sound level of motor vehicles and of replacement silencing systems and amending Directive 2007/46/EC and repealing Directive 70/157/EEC.

49 EU (2001). Directive 2001/43/EC of the European Parliament and of the Council of 27 June 2001 amending Council Directive 92/23/EEC relating to tyres for motor vehicles and their trailers and to their fitting.

50 EU (2001). Directive 2001/16/EC on the interoperability of the Trans-European conventional rail system.

51 EU (1996). Directive 96/48/EC interoperability of the Trans-European high speed rail system.

52 EU (2006). Commission Directive 2006/26/EC of 2 March 2006 amending, for the purposes of their adaptation to technical progress, Council Directives 74/151/EEC, 77/311/EEC, 78/933/EEC and 89/173/EEC relating to wheeled agricultural or forestry tractors.

53 European Commission (1986). Council Directive 86/594/EEC of 1 December 1986 on airborne noise emitted by household appliances.

54 European Commission (1986). Directive 86/188/EEC on the protection of workers from the risks related to exposure to noise at work.

55 EU (2003). Directive 2003/10/EC of the European Parliament and of the Council of 6 Feb. 2003, on the minimum health and safety requirements regarding the exposure of workers to the risks arising from physical agents (noise).

56 Maguire, D.J. (2015). Noise Regulation Validity, ICSV22, Florence, Italy.

57 European Environment Agency (2005). The European Environment: State and Outlook 2005, report no. 1/2005.

58 EU (2013). The 7th Environment Action Programme to 2020 (EAP): Annex in Decision No. 1386/2013/EU of the European Parliament and of the Council of 20 November 2013 on a General Union Environment Action Programme to 2020, "Living well, within the limits of our planet".

59 (a) European Environment Agency (2014). Noise in Europe, EEA Report, 10/2014;
(b) EEA (2019) Environmental Noise in Europe- 2020 EEA Report No.22/2019.

60 EU (2015). Evaluation Directive, 2002/49/EC relating to the assessment and management of environmental noise, Report WD1, 2015). Workshop working paper 1: The second implementation review of the END – emerging findings, Brussels (23 September).

61 EU (2016). REFIT evaluation of the Directive 2002/49/EC relating to the assessment and management of environmental noise. Commission staff working document SWD(2016) 454, final. http://ec.europa.eu/environment/noise/evaluation_en.htm.

62 European Community (1992). Council Directive 92/75/EEC of 22 September 1992 on the indication by labelling and standard product information of the consumption of energy and other resources by household appliances.

63 EU (2011). Inter-institutional Agreement on better law-making, the European Parliament, the Council of the European Union and the European Commission.

64 University of Washington (2018). Department of Global Health. http://staff.washington.edu/courses.

65 Clayton, T. and Weddington, D. (2015). Practical implementation of noise policy and standards for the permitting and regulation of industry, ICSV22, Florence, Italy.

66 Noise Pollution Clearinghouse (2019). Noise regulations & ordinances of U.S. cities, counties and towns, Montpelier, VT. http://www.nonoise.org/lawlib/cities/cities.htm.

67 Natale, R. and Luzzi, S. (2015). Noise from temporary musical events in the city of Florence: regulation, management, problems and solutions, ICSV22, Florence, Italy.

68 EU (2010). Environmental permit: decision-making process, environmental, permitting regulation 2010/75/EU Industrial Emission Directive (IED).

69 Defra (2006). Neighborhood noise policies and practice for local authorities: a management guide. Appendix 3: practice notes. Note 1: involving a complaint.

70 Kihlman, T. (1995). Education in acoustics, ICA95 15th International Congress on Acoustics, Trondheim, Norway (26–30 June).

71 Takayuki, A., Fumiaki, S., Akira, N. et al. (2006). Demonstrations for education in acoustics in Japan. *Acoustical Science and Technology* 27 (6): 344–348.

72 Kurra, S. (2008). A model for training programs on evaluation and management of environmental noise and vibration, 37th International Congress and Exposition on Noise Control Engineering, Inter-Noise 2008, Shanghai, China (26–29 October).

73 ISO (2015). ISO/DIS 1996-2: 2015 Acoustics: Description, measurement and assessment of environmental noise. Part 2: Determination of environmental noise levels.

74 ISO (2014). ISO 16283-1,2,3: 2014 Acoustics: Field measurement of sound insulation in buildings and of building elements.

75 Kurra, S. (2009). Environmental noise control and management, vol. 3, chapter 8 published by Bahcesehir University, Istanbul (in Turkish).

76 Louwers, M. (2000). The various sources of uncertainties in acoustic consultancy work. *Acoustique & Techniques* 40 http://www.bruit.fr/revues/78_11097.PDF.

77 Bayazıt, N.T., Kurra, S., Özbilen, B.Ş. et al. (2016). New regulation on noise protection for buildings and sound insulation in Turkey, 23rd International Congress on Sound and Vibration, Athens, Greece (10–14 July).

78 Bayazıt, N.T., Kurra, S., Ozbilen, B.Ş. et al. (2016). Proposed methodology for new regulation and guidelines on noise protection for buildings and sound insulation in Turkey, Inter-Noise 2016, Hamburg, Germany.

79 COST Action TU0901 (2014). Integrating and harmonizing sound insulation aspects in sustainable urban housing constructions. Building acoustics throughout Europe, vol. 1: Towards a common framework in building acoustics throughout Europe.

80 ISO ISO/AWI TS 19488: Acoustics: Acoustic classification of dwellings, (under development).

9

Environmental Noise Management

Scope

The success in community protection from the adverse effects of environmental noise is still being debated all over the world, despite the enormous steps made in the enforcement of noise control regulations, standards, and guidelines parallel to the advances in noise control technologies. The benefits of all these practices have not yet been adequately evidenced, quantitatively and qualitatively, due to various difficulties in realization, and in many countries the consequences do not satisfy expectations. The key factor in accomplishment is widely accepted as the fact that multidisciplinary actions and the substantial collaboration are needed between all the stakeholders involved in this issue, such as experts, the community, civil institutions, economists, administrators, politicians, etc. Therefore, since the second half of the twentieth century, environmental noise control has been dealt with under the concept of "noise management" to be pursued with well-defined policies on the local, national, and international stages.

A long path has been taken in the development of assessment methods, comparable techniques in evaluations, the establishment of an international network among countries (at least in Europe) to share experiences, and the work at national levels has been continuing on a schedule. In this chapter, various aspects of noise policies (which have recently appeared to be a field of expertise), will be outlined from the past to the present, and the principles of noise management will be given, based on international initiatives. The subject has been widely documented in the literature, however, this chapter will underline the key points of the issue by emphasizing the noise management practices in some countries, as examples for others developing action plans.

9.1 Concepts of Noise Policy, Strategy, and Management

The severity of the environmental noise problem and the necessity of collaborative approaches for abatements of noise were realized in the 1980s. It was well-known that the effectiveness of environmental noise control could be increased a great deal by extending the responsibilities to various sections in modern society and tackling the problem under the general context of environmental policy with strategy-led action plans [1–4]. The rationale of noise management policies is based on the evidence below, from investigations particularly in Europe [5a,b]:

1) Noise pollution is a major environmental health problem in Europe.
2) Road traffic is the most dominant source of environmental noise, with an estimated 125 million people affected by noise levels above 55 dB L_{den} (day-evening-night level). This number has

Environmental Noise and Management: Overview from Past to Present, First Edition. Selma Kurra.
© 2021 John Wiley & Sons Ltd. Published 2021 by John Wiley & Sons Ltd.

decreased to 113 million in 2019 report [5b]. Road traffic noise is the second worst form of environmental pollution in Europe, with at least 1 million healthy years of life lost each year from noise-related causes.

3) Each year, environmental noise causes at least 10 000 cases of premature death in Europe.
4) Almost 20 million adults are bothered by noise, and another 8 million suffer from sleep disturbance due to environmental noise.
5) Over 900 000 cases of hypertension are caused by environmental noise every year.
6) Noise pollution causes 43 000 hospital admissions in Europe per year.
7) Effects of noise upon the wider soundscape, including wildlife and quiet areas, need further assessment.

The report published by European Environment Agency (EEA) in 2014 declared environmental noise to be "one of the most pervasive pollutants in Europe and that drivers (such as economic growth, expanding urbanization, more extensive transport networks and increased industrial output) will present challenges to protecting the quality of the European soundscape" and acknowledged that "the health of the European ecosystems is also at risk, threatening valuable habitats and species that are particularly susceptible to noise" [5].

It has been seen that political ambitions have been increasing as shown with the initiatives of the European Union (EU) and the World Health Organization (WHO). However, the complete exposure assessment and future outlook cannot yet be accomplished by EU member states for various reasons, including the significant inconsistencies in exposure estimates and difficulty in making comparisons between countries. The European Union's *Seventh Environment Action Program* (7th EAP) forecasts a significant decrease in noise pollution by 2020, reaching levels close to those recommended by the WHO [6].

9.1.1 Noise Policy and Key Elements

Generally, the term *policy* is either defined as "a course or principle of action adopted or proposed by an organization or individual" or "a set of ideas or a plan of what to do in particular situations that has been agreed to officially by a group of people, a business organization, a government, or a political party" [7]. The frame of the policy must cover all the relevant sub-policies and be fulfilled by constituting collaboration among different parts of the community.

According to a report by the International Institute of Noise Control Engineering (I-INCE) in 2006, noise policy is described as "a high-level overall plan that includes the general goals and strategy of a national or international governmental body or agency for the control of occupational, community, and consumer product noise, as well as specific references to relevant codes" [8].

Various factors play important roles in the development of noise policy:

- the attitudes of individuals and their communities with regards to the environment;
- the expectations of a community concerning the role of government;
- the existing legislative systems and operational procedures;
- the economic and technical sources to be used for environmental noise control.

At present, the need for appropriate and adequate policies has been realized by most of the countries in the world in order to reduce the potential effects of noise impact. The WHO has emphasized that there should be a government policy as a basis for effective noise management; in addition, the WHO suggests that an efficient and successful noise management program could not be established unless a framework for adequate legislation exists [9]. A model, which is applicable in the

STEPS OF POLICY

ROLE GROUPS

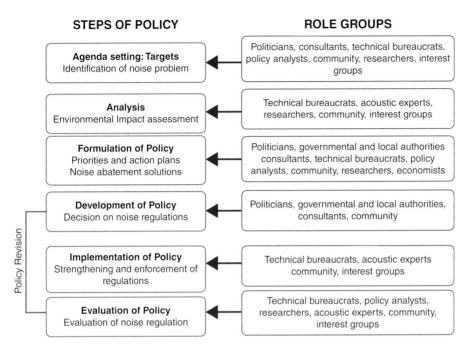

Figure 9.1 Noise policy planning and groups playing a role. (*Source:* adapted from [8]. Permission granted by INCE/USA.)

development of a legislative structure comprising the details of noise management, is given in the WHO document. The six steps in noise policy development and the implementation process within the noise management are displayed in Figure 9.1.

Definitions of the targets of noise policy are important for efficient noise management, to be followed by the establishment of an appropriate strategy and plan. The priorities should first be determined while defining the targets. The priorities vary according to the necessities and their availabilities in a community. Either the health risks should be a priority or the most extensive noise sources influencing the greater area and population should first be concentrated on during the preparation of noise policy. Figures 9.2 and 9.3 display the development of global noise policy and the elements playing a role in the process.

The following aspects should be taken into consideration in the development of noise policy:

- Noise policies should be reflected effectively in regulations and standards to minimize the health risks of noise.
- The technical standards should be stable, transparent, and publishable, ideally proposing appropriate measures against noise.
- A framework combining political, legislative and administrative aspects in noise policy is necessary.
- When developing noise policies, economic evaluation is important to facilitate decision-making.
- Policy-makers should be aware of the limitations and uncertainties since the decisions would be used in the process of noise management.
- In effective noise policy, the non-acoustic factors should be addressed.

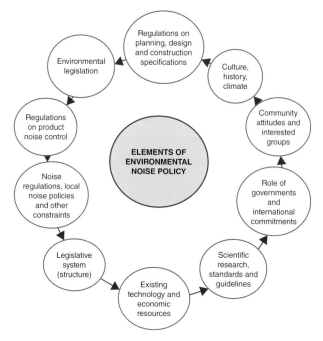

Figure 9.2 Global noise policy development. (*Source:* modified from [8]. Permission granted by INCE/USA.)

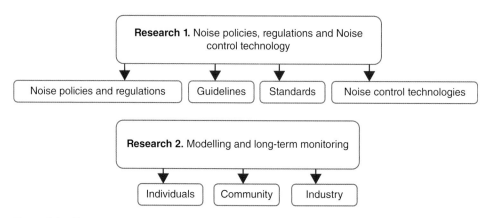

Figure 9.3 Phases of global noise policy development. (*Source:* adapted from [12]. Permission granted by ICBEN.)

- The assessment of effectiveness of a noise policy should be made in terms of the descriptors (such as a decreased population, reduction of complaints and decreased size of an area exposed to excessive noise conditions, by using socio-economic impact analyses).

Future noise policies have been examined comprehensively in various studies from the 1990s to date [10–12].

The new WHO Guidelines for the European Region, published in 2018, pay attention to the differences between the European countries in completion and regular updating of noise exposure assessments, which present difficulties in the development of noise policies, and set out the following principles [13]:

- "to reduce exposure to noise while conserving quiet areas": The recommendations are given by focusing on an overall reduction of population exposure to various environmental noises at various situations.
- "to promote interventions to reduce exposure to noise and to improve health": Assessment of the effectiveness of elements should take account not only of a single measure but also the combination of other solutions with the greatest potential in noise reduction.
- "to coordinate approaches to control noise sources and other environmental health risks": Policy development should be made through the coordinated approaches of the different sectors related to urban planning, transport, climate, and energy and the mechanisms to control other types of pollution.
- "to inform and involve communities that may be affected by a change in noise exposure." The communities should be informed about the new planning, new urban/rural developments, traffic schemes, new infrastructures, particularly in less densely populated areas and their contributions should be sought in the decision-making process.

The guidelines cover sets of recommendations as useful tools for national and local authorities, when deciding the noise measures, and allow comparisons between various options, furthermore they can be used in cost-benefit analysis (CBA) or cost-effectiveness analyses. The recommendations are categorized as "strong recommendation" that can be adopted as national policies by government, and "conditional recommendation" requiring a policy-making process after substantial debate and the involvement of various stakeholders. However, some guiding principles have been developed to incorporate all the recommendations into a policy framework.

The WHO document describes the noise mitigation measures and interventions, setting out all the actions or processes to reduce noise levels and to improve the situation so as to meet the recommended noise control criteria and limits, which are categorized as:

Type A: Source intervention
Type B: Path intervention
Type C: New/closed infrastructure
Type D: Other physical intervention
Type F: Behavior change intervention (including training)

Based on the extensive systematic overview of the health effects of noise obtained from the evidence, the results of a number of scientific investigations, and the population data, various exposure/response (or in other words; noise level/health outcomes) relationships have been presented in the guidelines in terms of:

- Typology of noise source: Road traffic noise, railway noise, aircraft noise, industry noise, wind turbine, and leisure noises for which the evidence is available.
- Typology of health outcomes:
  - adverse birth outcomes;
  - quality of life and well-being and mental health;
  - metabolic outcomes, i.e. diabetes and obesity.

The correlations between the annoyance responses and noise exposure levels yield the dose-response relationships as explained in Section 7.3. All these evidences regarding health effects of noise, should be integrated to determine the impacts of noise under the observed or expected exposure levels, in the development of action plans.

9.1.2 Noise Management

As explained in Chapter 7.7.4, the concept of noise control, mainly addressing the reduction of noise levels in terms of both source emissions and immissions in the environment, through advances in technology, optimizing procedures and by means of operational restrictions. However, the expansion of noise-polluted areas and the increasing number of parties involved, have necessitated these problems be tackled under the general concept of "environmental management," using modern methods in accordance with predefined policy and strategies.

Management is defined as "the function that coordinates the efforts of people to accomplish goals and objectives by using available resources efficiently and effectively" [14]. Generally, the system to be constituted for environmental management has been described in the international standard (ISO 14001 first issued in 2004), which gives guidance in outlining how to meet the environmental policy and objectives for all types and all sizes of organizations [15]. The standard is primarily oriented toward businesses in order to help them control the environmental impacts of their activities, products, and services. It recommends a method called "plan-do-check-act," that can be implemented in various phases of the Environmental Management System (EMS):

Plan: Establish the objectives and processes necessary to deliver results in accordance with the organization's environmental policy.

Do: Implement each process.

Check: Monitor and measure the process according to the environmental policy, objectives, targets, legal (and other) requirements and report the results.

Act: Take actions to continuously improve the performance of the environmental management scheme.

The standard, revised in 2015, aims to "increase the prominence of environmental management within the organization's planning processes" and the requirements can be constituted within the context of noise management systems. The measures against noise (suggested as "to be considered with life-cycle thinking" in the standard), are assembled under the subdivisions of engineering, planning, legislation, economic evaluation, education, etc.

The goal of noise management is declared in the WHO document as "to maintain low noise exposures, such that human health and well-being are protected. The specific objectives of noise management are to develop criteria for the maximum safe noise exposure levels, and to promote noise assessment and control as part of environmental health programs" [9]. The specific purpose of noise management is to develop the impact levels by providing the maximum reliable protection to assess noise and to take precautions as a part of the environmental health programs. The government policies should also include the noise management policies compatible with the principles of general environmental management, which are:

1) *The precautionary principle*: Noise in a place should be reduced as low as possible. This principle implies that if there is a possibility of danger to common health, it is necessary to act without waiting for scientific evidence.

2) *The "Polluter Pays" principle*: The financial responsibilities belong to those related to the noise source, including managing, monitoring, technical noise measures, and supervision.

3) *The prevention principle*: All the measures should be taken at the source when noise reduction is possible. Land use planning, impact assessment of environmental noise, and other environmental measures are considered in this context.

The important features in noise management are explained in various expert documents as follows [16–25]:

- Noise management requires the monitoring of noise in the community where people are living.
- In scanning the health and well-being of the community, it is necessary to investigate the adverse effects of noise, especially in schools, houses, and hospitals.
- It is necessary to consider the mixed environments where a number of different sources exist, the sensitive time periods (evening, night, and weekends), and the risk groups (children, the elderly, and sick people, and the hearing disabled).
- While planning land use purposes and transportation and designing a neighborhood at different scales, it is necessary to make assessments about the consequential noise conditions.
- It is essential to assess the efficiency of noise policies against the negative effects of noise.
- All types of action should be adopted when developing an acceptable acoustic environment.
- The guideline prepared by the WHO should be taken as the main target for community health and welfare. The national authorities should carry out the implementation of this guideline by incorporating the policies for community protection from noise and by developing the national methods.
- The authorities are responsible for preparing action plans whose objectives are to reduce noise levels in the short, mid, and long terms and to set the target limits in the long run.
- Also, it is recommended to conduct research on policies, perform cost-benefit analyses, make plans for environmental noise, enact specific regulations and the enforcement of the existing regulations by all countries.

Noise management, as a tool, should be able to correctly measure the performance of the optimal noise abatement solution. It gives the responsibility to the managers to supervise actions and elaborate on future assessments.

Various reports and documents have discussed the structure of noise management and the relative components to be taken into account [8, 12]. Figure 9.3 presents the phases of global noise policy development and Figure 9.4 displays the interrelations between various factors in noise management.

Figure 9.4 Interrelations between various factors in noise management.

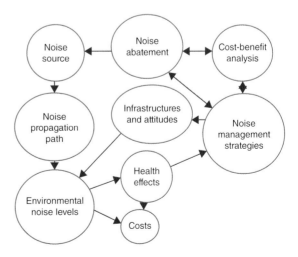

9.2 Development of Action Plans

Generally, an action plan describes the strategies needed to meet the predetermined objectives and consists of a number of action steps to be implemented. Each action step should give answers to following questions [16]:

- What actions to take?
- Who will implement them?
- When will they be implemented and for how long?
- Which resources are available? (i.e. money, staff, etc.)
- How, and to whom, will they be informed?

Giving information to the relevant parts, receiving all the needs and opinions (and even inviting them to join the action planning group), are vital when writing action plans. Action plans are efficient methods of noise control by saving time, energy, and resources in the long term. Implementation of measures in a controllable time schedule ultimately saves time, energy, and resources in the future.

9.2.1 Interaction of Noise Control Concept and Action Plans

The noise control concept, which was introduced into the literature in the 1950s, is a significant part of noise management programs. Technological solutions for individual noise sources (vehicles, industrial products, etc.) during design, manufacturing, and operating processes to reduce the sound emission values, need engineering approaches, however, noise abatement against environmental noise sources influencing communities, in large areas, requires not only engineering expertise but also planning, architectural design, with social, economic, and legal issues to be included in order to achieve the targeted reductions in immission levels.

The historical perspective of noise control engineering was reviewed in the various literature with struggles to gain recognition [26]. At present, it is widely accepted by the communities and manufacturers, the demand for engineering services is growing, resulting in a very active consulting business, international collaborations have expanded and information about new developments is readily disseminated. Courses are offered in noise control engineering by universities. Bringing a systematic approach to noise problems through the engineering, planning and design stages, including the individual protection, noise control was defined at that time, as "a system problem whose elements are sound source, propagation path (environment) and receiver (user)." Some of the basic books as pioneering publications in noise control engineering are found in [27–36].

As referred to in Section 7.7.4, noise, which is a kind of environmental pollution, can be controlled by using technology, planning, and design tools. The receiver to be protected from noise is a person using (i.e. living, working, traveling, educating, engaged in sport, etc.) in the environment (mostly in built-up areas) and in buildings of various types. The environment implies both the outdoor and the indoor environment (spaces in buildings), since the noise source and receiver can be located outside or inside. Figure 9.7 shows the earlier scheme of noise control to describe the topics involved in noise control.

Noise control has always been a complex problem which needs a well-organized management system and a well-prepared action plan establishing the priorities with commitments in a time schedule. Action plans should comprise noise control processes, by defining the responsibilities of various actors who play a role in decision-making and the implementation with good coordination. Noise control needs criteria and declared limits at the source, the environment and the user scales, i.e. the amount of noise to be decreased given in standard units, before searching for the appropriate solutions (Chapter 7). The action plans, in addition to quantified criteria, can also cover the quality descriptions of the environment to ensure a healthy and comfortable life as described by the WHO [37].

9.2.2 General Characteristics of Action Plans

The following criteria are recommended to be considered in the writing of a good action plan:

- Completeness: covering all the details of work dedicated to the objective, the order of the steps, and the relevant sections in the community.
- Clarity: transparency and description of the roles and responsibilities and time deadlines.
- Certainty: present status of the social, technological, and economic conditions of the community, availability, and the problems.

Action plans are indicators of the performance of management by emphasizing the need for good collaboration between the different disciplines. The basic principles applicable in the development of action plans are summarized below [17–26]:

- Action plans should cover all the noise control solutions to keep the environment protected against noise, which has the potential to create harmful effects on people's health (i.e. preservation of the quiet zones in urban agglomerations).
- Action plans should involve all the environmental noise aspects, problems, and effects, in line with the reduction of noise levels.
- The action plans are prepared by the responsible authorities and institutions with the collaboration of academics and those taking part in the execution of the plans. (Figure 9.5).
- Noise control in action plans should cover the areas where the limits are exceeded, decided by the responsible authority, and the priorities are identified or designated in the strategic noise maps.

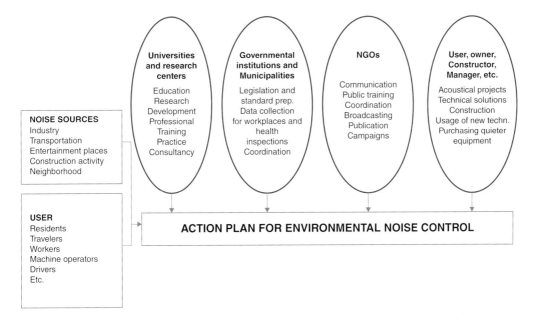

Figure 9.5 Stakeholders in the development of action plans for environmental noise control.

- Action plans should be prepared for agglomerations near major highways, railways, airports, and industrial areas.
- Action plans (if necessary) may involve new planning to avoid noise problems and to eliminate or minimize harmful effects.
- Action plans are part of the general planning and decision-making process of the local authorities.
- Noise abatement plans should be related to urban planning, which is also mandatory, and they should give motivation and support.
- The concepts in the noise action plans should be incorporated into other plans developed in a community. Figure 9.6 displays the relationships between action plans and other plans.
- The measures to be taken in the action plans should be evaluated with respect to: (i) the physical and political feasibilities; (ii) the practicability of measures and limitations; (iii) uncertainties; and (iv) possible impacts on the community (direct or indirect impacts).
- Detailed investigations of the above aspects are necessary.
- The results of the action plans should be controlled, checked, supervised, evaluated, renovated, and updated.
- It should be checked whether the applicability of solutions conflict with other environmental and social requirements related to other parameters.
- The action plan developers should have sufficient knowledge of the noise control concept and the interrelations between various topics displayed in Figure 9.7.
- Preparation of an action plan comprises a set of sub-processes briefly given below:
  a) Preparation of the draft plan.
  b) Informing the public about the policy decisions and outcomes of the action plans.
  c) Preparation of the final plan with a time schedule and list of measures, defining responsibilities in each action.
  d) Approval stage.

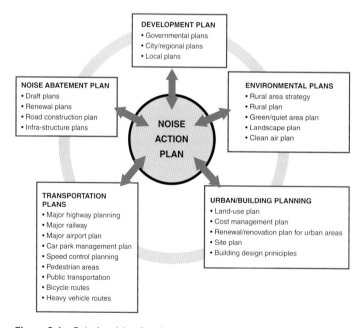

Figure 9.6 Relationship of action plans with other plans.

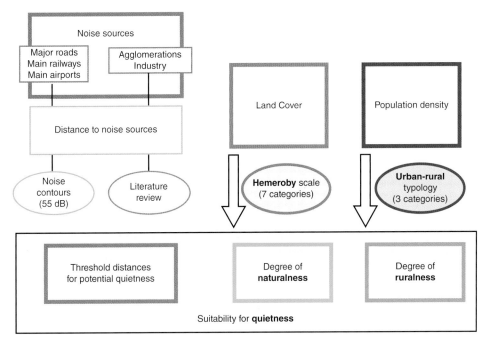

Figure 9.7 Methodological approach to obtain the Quietness Suitability Index (QSI) [81].

9.2.3 Framework of the Development of an Action Plan

The policy measures in action plans should be stated in a conceptual framework, including targeted noise reductions, and be implemented in various steps. An example of a framework applicable in the development of action plans is explained below, based on the practices in various documents [38–41]:

Phase 1. Review of existing situation (analysis and acquiring database).
Phase 2. Development of target limits and/or review of existing limits.
Phase 3. Description of administrative structures and responsibilities.
Phase 4. Technical and planning measures (target abatements at noise source, in the environment and on individuals).
Phase 5. Legal instruments.
Phase 6. Education/training/public information.
Phase 7. Economic instruments.
Phase 8. Writing an action plan, its implementation and monitoring.
Phase 9. Evaluation and revisions.

The actions displayed above can be performed in the above order or in-line processes depending on the priorities and availabilities.

Phase 1 Review of Existing Situation: Analysis and Acquiring Database

1) *Database regarding existing noise sources and operational status to be set up:* Information about industrial premises, motor vehicles, railways, aircraft, sea vessels, construction activities and equipment in use, recreational and commercial activities and transportation capacities;

numbers of vehicles, number of operations (daily, yearly, etc.) is gathered from published statistics, reports, etc. Assessments can be made through measurements, tests to be conducted on a sampling scheme, comparison with noise limits in the existing regulation, numbers exceeding the emission limits based on statistical evaluations, expectations for the future, etc. The collaborations can be established between various governmental and local departments, such as the Ministry or Departments of the Environment, the Directorate of Highways and Railways, the Department of Urban Planning, Civil Airport Managements, National Architectural and Engineering Association, National Association of Construction, and non-governmental organizations (NGOs).

2) *Database regarding environmental noise impact to be set up:* Assessment of environmental noise levels through the following actions:
   - Developing a national noise assessment model or using a standard model (see Chapter 4).
   - Preparation of noise maps (see Chapter 6).
   - Database derived from noise maps to be set up.
   - Determination of population according to noise zones (at least showing the white, gray and black zones).
   - Preparation of maps of density of construction activities.
   - Detecting and analyzing hot spots using different approaches (see Chapter 6).
   - Determination of noise sources dominant in hot spots, e.g. tire-road interaction noise, propulsion, etc., neighborhood noise, etc. Analyzing the effective factors influencing the propagation of noise.
   - Determination of quiet areas as required by END.
   - Conducting noise surveys and, based on noise data, derivation of noise control criteria and comparison with the existing limits.
   - Noise perception analysis conducting soundscape studies.

3) *Database regarding health defects to be set up:* Providing information about the health status of workers directly exposed to noise in coordination with Health Departments and Labor Departments in terms of:
   - statistical data from hearing tests.
   - psychological and physiological status.
   - estimated numbers of defected workers.
   - distribution of numbers according to the sectors, etc.

Phase 2 Development of Target Limits and/or Review of Existing Limits

1) Evaluation of existing limits and revisions on limits, establishment of new limits for specific situations/new noise sources/environments, etc.
2) Verification of the limit values (applicability evidence).

Limits should be based on the criteria explained in Section 7.7.2 and should be given in action plans, for the following:

- source emissions
- immission limits: source-specific, environment (land use-specific), and user-specific (outdoor, indoor, according to user activity, and times)
- health risk requirements.

The review of the existing limits should comprise solutions to the conflicts and discussions about the current situation (i.e. applied measures, enforcement, data availability, priorities, noise mapping, etc.). In this phase, methods to increase the stringency of the limits, particularly emission limits, should be included.

Verification of the limit values should be made by investigating the applicability in accordance with the standards and practices, by analyzing cost-effectiveness, etc.

Phase 3 Description of Administrative Structure and Responsibilities

1) Declaration of the competent authority (or leading department) coordinating the implementation of the action plans, the roles of other administrative departments, the shared responsibilities to be clearly defined.
2) Establishment of specific units/offices/staff for handling noise complaints, supervising, inspections, etc. under legal contracts.
3) Collaboration between various departments to exchange information about progress.
4) Monitoring obligations of each action groups.
5) Acoustic consultancy involvement to be clarified.
6) Revealing the body which is to be in charge of implementation of penalties.

Phase 4 Technical and Planning Measures (Target Abatement at the Noise Source, the Environment, and Individuals)

In general, technological solutions at source level can be grouped into passive noise control and active noise control. Passive noise control system implies the design and implementation using sound insulation, absorption materials, barriers, silencers, and other techniques, which are important in sustainable environment, however, these are generally ineffective at the lower frequencies. The active systems provide a significant reduction at low frequencies based on the destructive interference between the sound fields generated by the primary source and an artificial source whose acoustic output is controlled by signal processing. Fields of application of active systems are limited for noise control in mechanical systems and industry, in vehicles, however, not common in noise barriers, and sound and vibration insulation, and are not yet widely implemented in environmental acoustics.

An overview of noise measures for different types of noise sources is given in Box 9.1. Noise abatement through planning and design is outlined in Box 9.2 with the examples given in Figures 9.8a, 9.8b, 9.9–9.12. The measures for users/receivers, particularly against occupational noise in workplaces are displayed in Box 9.3. The information has been compiled from the past and present experiences, published in numbers of documents and reports.

Generally, these abatement measures, to be included in action plans, can be taken as single measures or a group of measures (combined solutions) for in-line applications. The applicability of measures should be investigated in terms of the following factors within the noise source-environment and user scheme:

1) Performance levels of measures (i.e. potential noise reductions to be assessed).
2) Technological level of communities.
3) Social and cultural status (acceptability by community).
4) Desire of implementations of all stakeholders.
5) Cost-effectiveness.
6) Other related issues concerned in further phases of the action plans.

The knowledge provided in the previous chapters are all needed upto a certain extent for decision-making in action plans and to reach a compromise between solutions: noise propagation and effects of physical factors (Chapter 2), characteristics of noise sources (Chapter 3), prediction of noise (Chapter 4), noise measurements of emission and immission (Chapter 5), noise mapping (Chapter 6), noise effects and deriving criteria for noise control (Chapter 7), and regulatory actions (Chapter 8).

Box 9.1   Source-Specific Noise Control Measures

Noise control for motorway traffic noise:

A) *Noise abatement for motor vehicles*: (Tools: noise emission tests)
- Low noise vehicles with measures in vehicle structure: in engine, radiator, fans, transmission system, brakes, and exhaust systems, etc.
- Low-noise tires to reduce tire-road interaction noise
- Interior noise control
- Maintenance and repair for existing vehicles
- Introducing age limit for vehicles in traffic
- Low-noise waste collection vehicles
- Check-on noisy vehicles in traffic.

B) *Noise abatement for roads*: (Tools: field measurements and noise maps)
- Low-noise road surface type at specific road segments (e.g. thin layer surfaces such as gross asphalt with flat top chippings, dense asphalt with high content of polymer-modified binder, double-layer porous asphalt, asphalt rubber friction source) [40]
- Maintenance of road surface: repairs, cleaning of porous surfaces, replacement of top layer (for two-layer surfaces) by taking care not to decrease the efficiency of low-noise surfaces
- Reducing roughness (unevenness) of surfaces
- Construction of humps and cushions (e.g. round top humps and narrow cushions for heavy vehicles)
- Road standards in relation to widths
- Proper road gradient
- Design of roundabouts, crossings, and traffic lights
- Depressed or elevated road infrastructures
- Design of streets to reduce speed by placing chicanes to narrow lanes, more space for pedestrians (footpaths), cyclists or parking, planting trees, narrowing lanes at junctions, strict separation between lanes.

C) *Noise abatement related to road traffic management*: (Tools: noise monitoring and noise maps)
- Providing uninterrupted flow (e.g. traffic calming by green waves: coordinated signalization at intersections)
- Reduction of traffic volume for certain road sections (note: may not be applicable for major roads)
- Reduction of heavy vehicle percentage for certain roads, bans on trucks
- Proper average speed and control (i.e. reducing and enforcing speed limits by police enforcement or automatic traffic control)
- Change in driver behavior (driving patterns) by training
- In-service controls (e.g. exhaust noise tests for motorcycles)
- Turn-off the engine scheme while waiting
- Low-noise night-time delivery using low-noise vehicles, training drivers
- Quiet delivery and garbage collection operations, other measures and charges.

Noise control for railway noise:

A) *Noise abatement for railway vehicles*: (Tool: Noise emission tests)
- Usage of low-noise trains (i.e. electric and light trains)
- Noise control with exhaust systems
- Noise control by silent brakes

- Noise control in engine and train structure
- Reduction of noise by resilient wheels, wheel dampers, and maintenance
- Low-noise trams: wheels, rails, self-ventilated traction motor, reducing fan noise by optimizing the shape of blades, improving the inlet/outlet air flow and using enclosures
- Low-noise tracks for trams
- Quietening train and tram depots (to reduce coupling noise, curve squeal, train rolling through switches, rolling noise, stationary diesel noise, air pressure noise, compressor noise, braking noise) by implementing solutions such as decreasing the number of movements, using alternatives for horn systems, using less noisy coupling, keeping delivery of equipment far from houses and operating only in the daytime).

B) *Noise abatement related to railways*: (Tools: noise monitoring and noise maps)
- Renewal of old trains by agreements with transportation providers
- Use of low-noise tracks (e.g. resilient rail fasteners, elastically mounting frogs, etc.) [77, 78]
- Modification of existing tracks by grinding, using rail friction modifiers and lubricants [79]
- Proper design of track curvature
- Use of vibration damping material and resilient rail pad
- Design of railway grade with respect to surroundings
- Vibration abatement on steel bridges
- Tunnels or semi-tunnels
- Proper design of cuttings and using absorptive treatment on the sidewalls
- Barriers and parapets along the railway line.

C) *Noise abatement related to railway traffic*: (Tools: Outdoor measurements, noise maps)
- Reducing number of diesel trains
- Adjustment of train schedules according to hours and minimizing number of night-time trains
- Management of average train speeds
- Least use of sirens, night-time ban for sirens.

Noise control for aircraft noise:

A) *Noise abatement for aircraft (air vehicles)*: (Tool: Noise emission tests)
- Low-noise engine: measures in engine structure, body, exhausts, etc.
- Measures related to take-off weights
- Management in flight profiles, tracks and routes, turns, etc. over settlements
- Management in operations during take-off and landings
- Flight procedures compliant with ICAO rules for low noise.

B) *Noise abatement related to flight traffic (flight management)*: (Tool: Noise monitoring)
- Aviation noise policies adopted by airports
- Restriction of volume of flights or certain aircraft categories (noisy flights)
- Restrictions of flight schedule (bans night-time flights)
- Airport operating rules and restrictions in use of runways, taxiways, apron areas, terminals, protection (safety) areas
- Charges/penalties for noisy aircraft operations
- Control of helicopter operations in settlements and at night.

C) *Noise abatement related to airports and ground operations*: (Tools: noise monitoring and airport noise maps)
- Airport design, number of runways, considering noise propagation in surrounding settlements
- Location of ground service areas at far distances, considering directivity of noise sources, maintenance buildings with efficient insulation
- Organization of ground operation times

- Barrier design at airport perimeter for control of noise from ground operations
- Protection of airport buildings and other sensitive areas (layout, insulations, etc.)
- Providing comfortable indoor environments by reducing noise levels and reverberation times within airport terminals also providing intelligibility of public announcement systems.

Noise control for industrial noise:

A) *Noise abatement for individual machines and equipment*: (Tool: Noise emission tests)
- Selection of low noise machine/equipment/installation
- Changing the structure and modification of machines by minimizing forces producing noise (e.g. mechanical, magnetic, friction forces)
- Reduction of surfaces radiating vibrations and noise
- Changing panel resonances (e.g. by damping)
- Changing operating techniques of machine
- Reduction of vibration forces by resilient supporting systems
- Use of damping materials on metallic surfaces
- Reduction of impacts using elastic cushions, resilient hangers/supports in contact points with building elements and other equipment
- Use of sound absorptive lining inside ducts
- Use of double-layer ducts or low noise ducts
- Prevention of turbulence noise on duct shoulders and branches
- Adjustment of fluid flow speed
- Noise tests for equipment in use
- Labeling products according to regulations.

B) *Noise control in mechanical rooms or for outdoor equipment*: (Tools: monitoring and indoor noise maps)
- Proper placement and orientation of machines and equipment
- Noise control of open-air and roof-top equipment by enclosures and screens
- Use of cabins, enclosures, or screens inside workspace to protect operators
- Selection of appropriate mounting system to eliminate noise and vibration
- Providing sufficient structural support (e.g. inertia blocks) for vibratory forces
- Proper design and configuration of industrial building and with effective sound insulation on building fabric (e.g. to eliminate airborne and structure-borne sounds through external building elements transmitted through outdoors and to nearby buildings)
- Indoor noise control within the workspace by using absorptive treatment, reducing reverberation time and interior noise
- Proper design of openings on the façades and roofs (windows, skylights, AC duct inlets/outlets)
- Prevention of sound transmission through vertical and horizontal installations by cutting direct contact with building elements, eliminating sound leakage using elastic sealants, etc.
- Increase indoor sound absorption by using absorptive treatment (e.g. suspending units, baffles, etc. from the ceiling) and to decrease reverberation times
- Use of active noise control systems for noise sources, in sound insulation and barrier design.

C) *Noise control in industrial area*: (Tools: noise monitoring, noise maps and predictions)
- Layout planning for industrial buildings and mechanical equipment in the entire industrial area and their orientation, according to the surrounding land uses
- Design of barriers with appropriate heights on the boundary of industrial premises
- Operation schedule according to times of day, restrictions at weekends
- Leaving tampon areas and green belts, considering the shape of industrial area.

Noise control for waterway noise:

A) *Noise abatement for individual waterway vehicles*: (Tool: Noise emission tests)
- Refurbishment of vessels (passenger and freight carriers)
- Modification of older vessels through renewal of diesel engine, turbines, motors, gear boxes, exhausts, ventilation and auxiliary machines and equipment
- Cutting engine room noise by efficient insulation, e.g. by construction of room-in-room principle, damping vibrations through engine mounts, resilient connections, coating installations and ducts, etc.
- Complete insulation on board, especially in relaxation spaces against airborne-and structure-borne sounds, e.g. passenger and staff cabins transmitting through HVAC equipment outlets and ducts and direct sounds from AC equipment, electrical power supply system, etc.
- Noise reduction by sound insulation between partitions, suspending ceilings, floors, eliminating flanking transmission noise from toilets, kitchens, etc.
- Noise reduction in restaurants, lounges, against kitchen noise by increasing sound absorption treatments
- Control of exhaust noise from outlets radiated to outdoor environment.

B) *Noise abatement related to waterway traffic*: (Tool: Noise monitoring and noise maps)
- Restriction of cruising speeds at certain routes at certain distances to seashore
- Management of cruise routes at certain distances to seashore
- Restriction of speedboats, heavy freight carriers at night-time
- Restriction of vessel operations and new coast lines
- Restriction of amplified music with high volume on board and while cruising at certain distances to seashore.

C) *Noise abatement related to harbors, shipyards, ports, etc.*: (Tools: monitoring and noise maps)
- Management of some noisy operations at night according to distances to settlements
- Bans on ship horns at night and other restrictions
- Layout of ports, harbors and facilities, service buildings, and heavy vehicle traffic
- Noisy activities far from the boundaries and use of obstacles from service buildings, etc.
- Shielding the noisy activities by use of service buildings, topography, and artificial barriers
- Management of smooth delivery, and traffic (railway and road traffic)
- Layout of shipyards and efficient insulation of covered shipyards.

Noise control for wind farm noise:

A) *Noise abatement for wind turbines*: (Tools: Noise emission tests and predictions)
- Selection of less noisy turbine types/size of turbine according to the requirements
- Modification of turbine structure and mountings, e.g. rotor shaft, blade design, gearbox, electric generators
- Selection of hub height considering the sound radiation to settlements
- Regular maintenance of wind turbines
- Active noise control systems.

B) *Noise abatement for wind farms*: (Tools: monitoring, noise maps)
- Layout of wind farm considering minimum distance to settlements to be determined through noise maps to reduce the negative effects of swishing sound, bursts, thumping noise and rumbling noise, e.g. preferably offshore
- Decision about number of turbines according to needs
- Distribution of wind turbines, spacing, etc. according to the noise map scenarios by taking into account the physical properties of the area, e.g. wind directions (not to be in a downwind region), with respect to the surrounding land use and buildings

- Layout planning considering the topography and shielding effect between settlements and the wind farm
- Implementation of efficient insulation on buildings against low frequency noises
- Masking sound to be determined through soundscape studies against modulated swishing noise in open air near settlements (aerodynamic noise)
- Increasing indoor sound absorption in buildings
- Increasing ground attenuation of sound by green areas with appropriately selected plantation and forestry
- Eliminating reflective surfaces in the surrounding area.

Noise control for construction noise:

A) *Noise abatement for individual construction equipment*: (Tools: emission tests for in-use equipment)
- Selection of low-noise equipment
- Noise control related to equipment structure and operations
- Use of cabins and enclosures with proper sound insulation for stationary machines
- Regular maintenance and repair of equipment
- Changing the operation technique.

B) *Noise abatement related to construction site and organization of work (site management)*: (Tools: noise monitoring and noise maps)
- Appropriate layout plan and locations of equipment (placement of noisy equipment far from the site boundaries)
- Management of operational modes for moving machines
- Restrictions of operations at certain hours and weekends
- Management of operation schedule (e.g. working with the noisy equipment when the background noises are high)
- Removing the noisy operations to outside of the construction site
- Noise control on the construction site by noise screens around the boundaries and use of mounds and material stocks as obstacles, or a portable noise screen around the equipment
- Elimination of reflective surfaces around equipment
- Continuous noise monitoring at the site perimeter
- Warning nearby residents before noisy operations.

Noise control for entertainment/leisure noise:

A) *Noise abatement for sound emitters*: (Tools: acoustic tests, or product certificates)
- Restrictions on sound pressure levels, e.g. lowering volume of loudspeakers especially bass emitters, if there are houses nearby, by using automatic monitoring and control system and warning devices
- Use of several small loudspeakers instead of a single powerful loudspeaker
- Orientation of loudspeakers toward dancing area, paying attention to bass loudspeaker locations
- Use of sound insulated partial coverage on loudspeakers not to allow sound radiation to the direction other than desired
- Appropriate placement and distribution of loudspeakers in the venue by leaving quieter sections for people to talk and dine, e.g. mounting on the tables and controllable by the clients
- Solutions for protection of hearing of staff, disc jockeys and service personnel
- Warning sign on the entrance door about the high level of music
- Orientation of loudspeakers so as the sounds are not directed toward outdoor environment or nearby buildings.

B) Noise abatement for venues: (Tool: noise monitoring on the nearest façade of noise-sensitive building)
- Time/hour restrictions for high volume music
- Restrictions on type of music
- Design of efficient noise barriers on the perimeter of the venue
- Use of retractable ceiling systems to cover the venue when the music is high
- Avoiding reflective surfaces inside and providing sound absorption especially in the relaxation areas
- Design of stage and platform layout and stage enclosure to reduce noise propagation to neighborhood (with efficient structural form and sound insulation)
- Hearing protectors for staff and clients.

Noise control for shooting ranges:

A) Noise abatement for firearms: (Tools: Noise emission tests and predictions)
- Muzzle noise reduction by using appropriate silencers
- Layout of shooting area in terms of position with heights and orientation (direction and distance considering dominant wind direction and speed.

B) Noise abatement in shooting range: (Tools: noise monitoring and noise maps)
- The minimum distance to the nearest settlements to be determined from noise maps, by taking into account the meteorological effects (wind and temperature gradients,) and other physical properties of the land
- Decision about size of range, number of shooting bouts daily, simultaneously, etc. considering the sound propagation at far distances
- Design of shooting positions according to direction and heights
- Layout planning of shooting premises by using the positive aspects of topography, forests, green areas, obstacles around, if appropriate, construction of artificial noise barriers, etc.
- Eliminating reflective surfaces around and inside the shooting range
- Providing soft ground cover in and around the premises
- Efficient insulation of semi-closed and closed buildings (e.g. walls and roofs)
- Efficient design of sheds at the shooting point (shape and form)
- Design of barriers or obstacles
- Using high sound absorption treatment within the enclosures and building
- Stringent management in shooting ranges with implementation of rules to avoid complaints from settlements nearby
- Restrictions with time schedule for open-air or semi-open ranges
- Restriction on use of some weapon types
- Restriction on number of shootings sets
- Protection of users' hearing.

Noise control for other public noise sources referred in Section 3.8
Various other noise sources in outdoors and indoors where people and the community are in contact during daily life, such as parking lots, garbage and delivery services, children's playground, playing with drones and electronic toys, sport cars, amusement parks, mining activities with blasts, sport fields, using jet skis or snowmobiles, etc. They need special attention in action plans taking into account their inclusive features, such as continuity, impulsiveness, noise levels, tonality, etc. and the measures for noise control, can be described as: (i) under land-use or urban planning phases as given in Box 9.2, and (ii) under regulatory actions (see Chapter 8)., i.e. licensing permits, restricted hours, and other stringent rules with efficient management of the spaces.

Box 9.2   Noise Control in the Environment

A) *Noise abatement through land-use planning and management*: (Tools: strategic noise maps)
   For new plans:
   - Appropriate land selection for:
   - Noise-sensitive uses: Residential, mixed areas, hotels and relaxation areas, health and welfare service areas, hotels and touristic areas, cultural and educational facilities, recreational areas, parks, cemeteries, agricultural and forestry areas, etc. compatible with noise limits.
   - Noisy areas: Commercial areas, industrial areas, harbors, etc., sporting arenas, transportation routes
   - Minimum distances to major transportation routes, airports, etc. to protect settlements from noise
   - Minimum distances to major noise sources to protect natural life
   - Selection of areas according to microclimatic factors, topography, physiological properties of land, dominant winds, acoustic shadow zones, neighboring lands, major transportation routes, airports, etc. (based on yearly statistics)
   - Preserving white zones and increasing their size.
   For revisions of existing plans:
   - Noise control by decreasing number of people living in black and gray zones, increasing size of white zones, smoothing traffic
   - Constructing infra-structures and installations or use of topography to reduce noise levels.

B) *Noise abatement through urban planning/city planning*: (Tools for existing areas: monitoring and noise maps)
   Measures at the planning phase:
   - Noise zoning to select areas for noise-sensitive buildings and for noisy activities (e.g. sporting activities are accommodated in the least-sensitive areas)
   - Noise abatement measures in environment by physical elements
   - For existing areas: hot spot determination and priority areas
   - Soundscape analysis and decision of protected areas, soundscape design in urban areas
   - Planning new routes for roads and railway lines
   - Design of transportation routes, tracks, extension of airports, industrial areas and relevant noisy facilities (terminals, airport facilities, logistic services, shipyards, etc.) by categorizing roads, design the new roads, site selections, partial tunnels, earth berms, enclosures and barriers around industrial or mechanical noise sources in the open air, etc. (Box 9.1)
   - Promoting less noisy transport systems
   - Design of parks and green spaces
   - Plant forests with evergreen trees and bushes between major noise sources and settlements
   - Tampon areas with minimum widths and distances to roads from sensitive usage areas (to be determined according to meteorological factors, topography, source type and noise levels, type of buildings), based on noise maps and calculations
   - Plantation (vegetation) in the areas between source and noise-sensitive usage areas
   - Design of pedestrian roads, bicycle paths
   - Layout of buildings considering screening and reflection effects of other buildings and surfaces (Figure 9.8a)
   - Restriction of buildings along climbing roads, at road crossings, selection of road, railway infrastructures
   - Elimination of canyon effect (multiple reflections from building walls along roads)
   - Use of the least noise-sensitive buildings as noise barriers along traffic roads and railway lines (by controlling heights to provide acoustic shadows behind).

The above actions can be implemented in coordination with the Ministries or Departments of the Environment, Urban Planning, Transportation, Industry, Forestry, Civil Airport Administrations, Railway Administration, professional associations of landscape designers and agriculture, local administrations, architects and town planners associations, universities, etc.

C) *Noise abatement in building layout design in neighborhoods*: (Tools: urban/local noise map, social surveys, and soundscape studies);
- Layout of buildings and configurations (orientation according to noise sources) (Figure 9.8a)
- Placement, orientation, and configuration of building blocks according to noise sources to provide "quiet sides" (e.g. with narrower façades exposed to noise sources)
- Accommodate the noisy activities in a neighborhood (e.g. children's playground, car parks, etc.) at certain distances
- Minimum distances between the sources and the sample façades (tampon belts)
- Use of soft ground cover, vegetation in tampon belts and around buildings
- Construction of roadside screens or use of natural barriers, walls of buildings, etc.
- Elimination of direct reflections through suitable configuration of buildings
- Controlling gap widths between building masses exposed to roads and railway tracks
- Soundscape design in nearby environment (enhancing the pleasing cultural and natural sounds, masking noise).

D) *Noise abatement by architectural design*: (Tools: building acoustic simulation models)
- Building shape and orientation of buildings
- Provision of quiet façades according to regulations (i.e. EU Directives)
- Appropriate configuration of buildings along roads, i.e. compact design, terrace houses, etc.
- Design of closed buildings with atriums or courtyards (eliminating interior reflections)
- Categorizing the functional indoor spaces according to their sensitivity level against noise
- Identifying the spaces that can generate noise and vibration in the building
- Separating noise-sensitive spaces from noisy areas both in architectural plans and sections
- Orientation of noise-sensitive rooms on the quiet side
- Location of spaces according to potential indoor noise sources
- Acoustical isolation to be provided for noise-sensitive rooms by designing tampon spaces such as corridors, lobbies, kitchens, toilets, etc.
- Outdoor noise barriers with sheds or garages next to houses of low height
- Design of massive façades to ease sound insulation
- Design of balconies and wing walls near entrances, to provide horizontal and vertical screening against outdoor noise and neighborhood noise
- Balcony protection (parapet design, sound-absorptive treatment on the balcony ceiling, etc.) (Figure 9.8b)
- Design of gardens and vegetation all around buildings, reducing hard ground surfaces, such as in parking lots and pedestrian paths
- Elimination of sound penetration from neighbors' windows with horizontal and vertical façade elements
- Noise control for outdoor equipment and installations.

The above measures can be taken by architects (building designers), constructors, supervising the architectural projects by local authorities, in coordination with professional associations, consultants, building owners, civil and mechanical engineers, etc.

E) *Noise abatement in buildings against indoor noise sources*: (Tools: building acoustic simulation models, measures against) against airborne and impact noises and vibrations from indoor noise sources (e.g. mechanical systems, neighbors, etc.)
- Design of stairwells, public entrances, corridors, lounges, etc. by reducing reverberation time and increasing indoor sound absorption using acoustic treatment

- Highly insulated service rooms (e.g. by room-in-room technique) with inner absorptions (Figure 9.9)
- Noise and vibration control for mechanical equipment (Box 9.1)
- Design of lifts, garbage chutes, with tampon spaces between dwellings
- Eliminating airborne and structure-borne noise penetration from noisy spaces (i.e. cross-talks by ducts and installations).

F) *Noise abatement in building elements (sound insulation)*: (Tools: building acoustic measurements, calculations, simulation models, indoor noise maps, residents' dissatisfaction surveys):
- Identifying the exterior and interior noise sources and the transmission paths of airborne and structure-borne sounds into the noise-sensitive areas (Figure 9.10)
- Sound insulation for building façades and roofs (smaller windows, insulated glazing, double-glazed façades with air ventilation)
- Use of closed glass façade, double-glazed wall providing air ventilation, use of acoustic shutters on the windows, sealing of perimeters of windows and AC outlets
- Design of all building construction (i.e. elements, components, selection of appropriate materials and units) to satisfy requirements of sound insulation performances based on specified indoor limits (Figures 9.11 and 9.12)
- Providing artificial ventilation in bedrooms and living rooms, when the windows are to be closed at night
- Insulation of floors against impact noises, partitions against airborne sound transmission and eliminating flanking transmission through junctions of elements
- If present, compatibility to insulation schemes according to noise zones
- Conducting socio-acoustic surveys to investigate satisfaction from sound insulation by considering non-acoustic factors
- Complaint management (see Chapter 8).

Box 9.3   Noise Control at Workplace (Individual Protection)

A) *Worker training*: (Tools: seminars, courses, published documents, internet)
Training programs about effects of noise on health and specifically on hearing risks, for workers and other staff in workplaces (outdoor or indoors). Protection techniques, and exposure times according to standards.

B) *Personal protection*: (Tools: hearing tests, noise emission tests of machines, warnings at work)
Provision of appropriate ear protectors to be selected according to acoustic characteristics of noise sources, enforcement of use at certain places, use of active noise control systems for operators directly in contact with the equipment producing excessive noise, regular hearing tests

C) *Exposure control*: (Tools: measurement of noise dose, administrative, work contracts, legislations, etc.)
Use of dosimeters to check for daily noise dose for each laborer in accordance with standards/regulations, management of exposure duration in workplace, work shifts, compensation, and other administrative measures

D) *Workplace noise control*: (Tools: monitoring and indoor noise maps)
Taking noise abatement measures to reduce noise in workplaces according to standards and regulations (i.e. EU Directives) (Box 9.1).

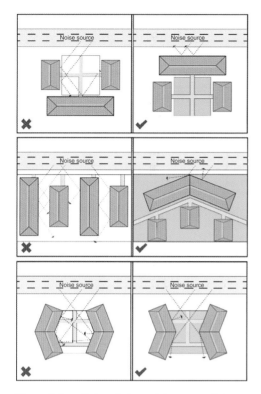

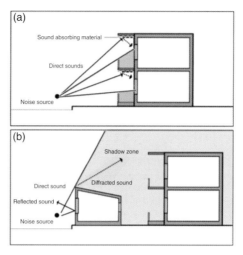

Figure 9.8b Balcony and parapet design. (a) absorptive treatment underneath of balcony floor and use of parapet as noise barrier; (b) noise abatement by non-sensitive buildings used as noise barrier (depending on their length).

Figure 9.8a Noise abatement by appropriate layout of buildings with respect to roads. (*Source:* drawn by A. Sentop in 2018.)

Phase 5 Legal Instruments

1) Review of existing regulations (governmental and municipality or local levels).
2) Search for interaction between current regulations.
3) Improvement of relevant codes of practice, relevant changes to building regulations, improvement of environmental impact regulations and noise control regulations.
4) Necessity of new regulations.
5) Organization of penalties (incentives).
6) Compatibility with the international regulations (i.e. European regulations).
7) Measurement standards.
8) Coordination and supervision.

Phase 6 Education/Training/Public Information

According to EC Directive 49 (END); the public is defined as one or more natural or legal persons and in accordance with national legislation or practice, their associates, organizations, or groups [22]. It suggests spreading the information widely through the most appropriate information channels. For this purpose, the following actions can be planned:

1) Raising public awareness by communicating noise issues with the aim of changing people's behavior: Target groups: citizens (house owners, residents, workers, tourist groups, community service dealers, shop owners), freight delivery sector (truck drivers, business people, shop owners), educational sector (children, teachers, parents), health sector (hospital staff, patients,

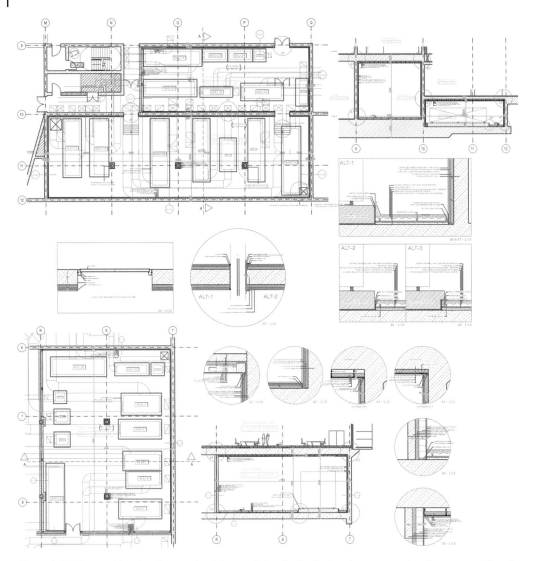

Figure 9.9 The working drawings and architectural details displaying the constructional measures taken for noise abatement in mechanical spaces (example of a public building).

public health service), media/journalists (newspapers, scientific papers, journals, etc.), NGOs, government/decision-makers, etc. The tools to be used in this regard are:

- Campaigns (special event days) on protection from noise pollution. An example is the initiative called: "INAD in Europe" organized by the EAA and the National Acoustic Societies in 2017 for International Noise Awareness Day (INAD) which started in 1995 in the USA to promote awareness of noise exposure problems and their effects on both hearing and people's health. The event comprised the presentations for the conferences and a documentary film distributed all over Europe, student competitions, etc. (Figure 9.13) [23]. Recently "International Year of Sound 2020" has been declared jointly by the International Institutes ASA, ICA, EAA, I-INCE and IIAV (See Chapter 1.1) "as a global initiative to highlight the importance of sound".

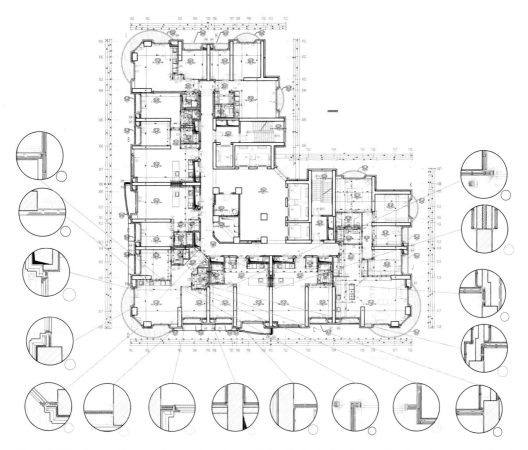

Figure 9.10 The working drawings and architectural details displaying the constructional measures taken for sound insulation in residential buildings: Example for a high rise apartment building with many dwelling units. (*Source:* S. Kurra and dBKES in 2009, simplification of drawing by B. Şan Ozbilen in 2018.)

Coordinated activities on regional, national and international levels, such as congresses, workshops, etc. are organized as can be followed through web page: www.sound2020.org

- Publications (booklets, leaflets, brochures, newspapers, etc.) targeted at raising awareness of the negative impact of noise or at specific noise problems (i.e. in schools, hospitals, etc.). Figure 9.14 shows collections of published media regarding community noise issues, even some criminal cases due to noise.
- Sound barometers (or thermometers) displaying instantaneous noise levels in streets and urban squares.
- Warning notices to be silent in public spaces or in transportation systems (Figure 9.15).
- Displaying dynamic noise maps in public squares with criterion levels on the screen.
- Organizing the information desks on hot-spot areas.
- Audial displays stressing the importance of soundscaping.
- Children's competitions regarding noise issues (essay composition, painting, etc.).
- Caricature and poster competitions and exhibitions (Figure 9.16).
- Warning notes about purchasing quieter products.
- Informing about action plans and about regulations.

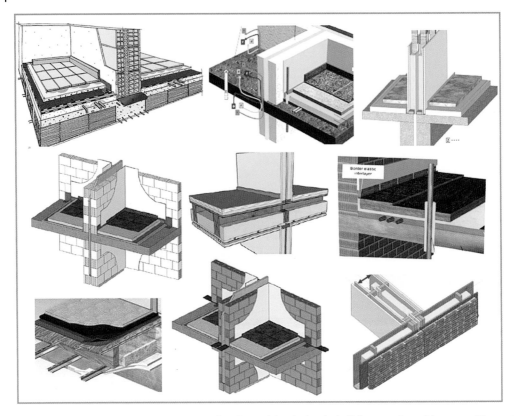

Figure 9.11 Examples of application details of sound insulation in buildings against airborne and impact noises. (*Source:* Compiled from training school documents and the company brochures)

- Communications with NGOs.
- Highlighting the importance of silence and avoiding noisy behavior.
- Informing about sound insulation and simple guidance applicable at home.
- Use of visual media (TV, cinema, community spots on TV, social media messages, etc.) to turn attention to the issue.

2) Supporting research activities on capacity assessment for implementation, new innovative technologies, social surveys, etc.

3) Organizing education and training courses/seminars to include the participation of individuals, professionals, investigators, and administrative staff (Chapter 8).

4) Consulting the public: Participation of the public in the action plans through social surveys, opinion tests through websites, targeted meetings, inviting and consulting the community groups interested in noise problems, etc.

5) Publication of relevant planning policy and policy guidelines as printed papers and distributed by electronic mail.

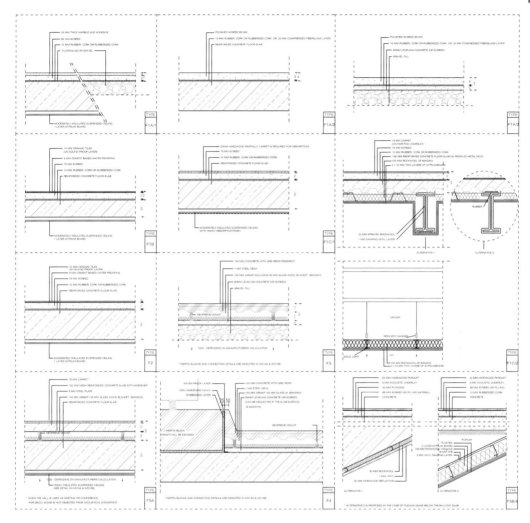

Figure 9.12 Working drawing details for sound insulated floors. (*Source:* S. Kurra and dBKES in 2006.)

Figure 9.13 The documentary film, *Sounds of My Place*, prepared by EAA and the European Societies of Acoustics for the International Noise Awareness Day, INAD 2018 [23].

Figure 9.14 Example of the media's role in increasing public awareness. Daily newspapers and journals highlighting the negative impact of noise or specific noise problems. (*Source:* Archives of S. Kurra and E. Dogukan between 1990 and 2002.)
Note: Gürültü in Turkish = noise.

(a)

(b)

(c)

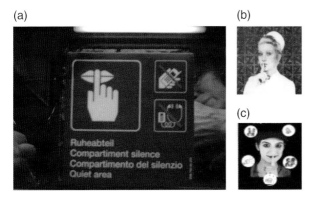

Figure 9.15   Warning about silence in public spaces, a) silent cars in Norwegian trains (*Source:* photo: S. Kurra, in the 1990s) b) all hospitals in the 1970's in Turkey b) in schools, 2019 (*Source:* M. Bulunuz).

(a)

(b)

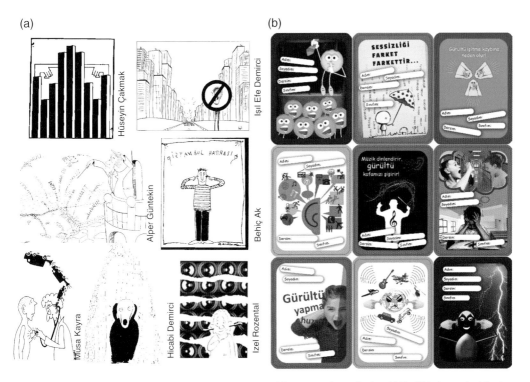

Figure 9.16   Examples from caricature and poster competitions on noise pollution (a) Turkish Acoustical Society of congress event, 2002 (*Source:* Archive of TAKDER with permission) (b) Children's event contest in 2018 (*Source:* M.Bulunuz ,Uludag University).

Phase 7 Economic Instruments

Noise control measures using economic instruments stated in the action plans are:

1) Economic incentives to encourage noise reductions.
2) Taxation revenues.
3) Sanctions: Charges/Penalties (judicial fines) according to the "polluter pays" principle.
4) Compensation (claiming damage).

The economic concerns in action plans should cover the following aspects:

- Cost assessment, including maintenance and renewal measures.
- Additional taxes according to noise emission of sources (such as landing fees, road usage fees, etc.).
- Encouraging central governments to support local governments economically.
- Financial support for research and investigations on noise control and seeking additional funds from the private sector and the community.
- Describing the economic benefits of the action plan.

Phase 8 Writing an Action Plan, Implementation, and Monitoring

This phase comprises adopting the action plans, including the approval phases, monitoring progress, and reporting the final results.

Presenting noise measures in a work plan with a time table: Proposition of abatement systems is scheduled by describing the action groups, their capacities in terms of staff, financial resources, data availability, time restrictions, etc. and ensuring that interference between other policy objectives (sustainability, air pollution, etc.) is taken into account. The measures can be ranked as short-term actions, mid-term actions, and long-term actions, which are compatible with the long-term strategies and potentials in future [41]. For example, the draft plan might serve as the basis for developing new certification processes for future products or different measures/strategies for existing situations.

Decision support systems (DSS): A transparent decision-making process is important in action planning. For this purpose, the basic document published on environmental zoning problems, based on the Geographic Information System (GIS), can be adopted to select the adequate noise abatement measures, particularly in the urban planning phase [41, 42].

Monitoring the progress: The actions can be monitored by the leading department at certain intervals, according to the schedule, through noise measurements, new noise maps, number of people affected, etc. The delays and reasons, conflicts or difficulties, potential solutions are to be investigated and presented in the progress and final reports.

Phase 9 Evaluation and Revisions of Action Plans

1) Review the achievement of the goals and sub-goals.
2) Positive and negative impact of action plan on communities and individuals.
3) Noise reduction effect (performances, number of people benefitting).
4) Public acceptance and compliance.
5) Reductions in complaints.
6) Checking measures regularly (e.g. every five years).
7) Sustainability of actions.
8) Financial constraints and need for budget allocation.
9) Cost evaluations at different phases.
10) Delays and cancelations and the reasons.

Based on the evaluations of the action plans, some revisions might be inevitable:

1) Revisions due to major/minor changes of noise situations.
2) Revision of target limits.
3) Decision about changes in the framework.
4) Revision of budget planning.
5) Final evaluations of noise policy in general.

The SILENCE Project

This is an EU-supported project conducted by groups of researchers from different countries who have produced a guidance with the aim: "Quieter surface transport in urban areas, relevant methodologies for the efficient control of noise caused by urban road and rail transport, innovative strategies for action plans on urban traffic noise abatement and practical tools for their implementation" [40]. The process of action planning is explained in steps that can be adjusted according to the needs of individual cities, with long-term strategies for noise abatement and appropriate measures. The guideline targets three main stakeholders for whom the following information is to be provided:

- *local decision-makers:* Basic information about the requirements of noise action plans, the action plan process, and noise abatement measures.
- *transport planners and urban planners:* Information about the planning process, problems, abatement measures, links to other policy fields, long-term strategies to mitigate noise.
- *transport engineers:* Detailed description about technical measures with references, experimental results, calculations, and design principles.

In the Silence Project report, the noise action planning process is accepted as part of the urban planning process and four approaches are recommended in the implementations:

1) *Participatory approach:* Involving the public during the assessment of noise, the selection of measures, evaluating the results, revisions, and adjustments.
2) *Cooperative approach:* Involving all the relevant stakeholders in the assessment, drawing up solutions, and the implementation stages.
3) *Open approach:* Providing links between noise abatement activities and other objectives like land-use planning, or air pollution control plans.
4) *Measurable approach:* Setting quantified noise reduction targets, specifying the expected reduction, implementation, and monitoring.

The Silence Project, stressing the importance of the competent authorities, makes a recommendation "to define a leader with sufficient capacities and competences to successfully set up a local noise action plan to ensure strong involvement of the stakeholders of the departments" and states "the leading department involves all relevant stakeholders to make them contribute to implementation." It may be essential to "create new organizations for steering and carrying out the work to be decided" [37]. As an outcome of the project, a handbook has been prepared covering a list of the relevant stakeholders at local and regional levels to make them aware and to enlist their participation by convincing them, drawing their attention to action planning. The document presents some examples of practices in action plans in the European countries.

Life+2008 Hush Project Results

A further research study, based on the results of the project, proposes a new methodology and a new platform for harmonization between END and national laws covering the methodological, technical, administrative, or legislative aspects, and presents an Integrated Noise Action Plan also to update the Italian legislation [43]. The proposal consists of four levels: (i) the strategic level; (ii) the project level; (iii) the executive level of interventions; and (iv) monitoring. The initial phase requires the definition of objectives for five years, the identification of the targets, the definition of noise policy, and formulation of strategic actions, etc.

1) *Strategic level:* potential synergies with other environmental policies, stakeholders' involvement, identification of effectiveness indicators to provide the consistency of the plan.
2) *Project level:* selecting intervention areas to be considered in action plans, hot spots, quiet areas, and definitions of actions, proposals of plans. Modifications after public consultations and approval phase.
3) *Executive level:* Realization of the noise abatement measures or development of the noise prevention actions.
4) *Monitoring level:* Verification of the objectives, analysis of effectiveness and coherency of the actions, benefits achieved, information to public, review of action plan.

To be used in the action plans, the methods to identify the "areas of intervention (in relation to selected type of the intervention, e.g. strategic or direct and the scales, hot spot and quiet areas" have been developed.

9.3 Overview of the Action Plans in the EU

The coordination between the EU member states on the prevention of health problems related to noise, initiated in the 1980s, has been continuing according to a schedule. Establishing working groups within the European (Economic) Community and the European Council has helped to rapidly and extensively gain the attention of the member states with regards to the environmental noise problem, by means of a number of published documents, which are mandatory in developing noise policies, noise mapping, and action planning. The Environmental Noise Directive (END) declares the aims of action plans in Article 1 of Directive 2002/49/EC [22, 44] as "... to address local noise issues, competent authorities must draw up action plans to reduce noise where necessary and maintain environmental noise quality where it is good. The directive does not set any limit values, nor does it prescribe the measures to be used in the action plans, which remain at the discretion of the competent authorities ..."

Currently, the developments and effectiveness of the noise action plans have been monitored by the EU Council to bring a common approach to avoid, prevent, and reduce the harmful effects of noise, including overall annoyance from noise. The history of the development process was summarized in Section 8.3.1.

9.3.1 State-of-the-Art Implementations of Action Plans

European Union member states have completed the first, second and third rounds of compliance with the action plans, with the principles and the schedule described in the above Directive (Chapter 8). The submitted action plans have been evaluated in the Council and by the EEA in a comparable manner and the outlook reports have been published since 2004, with the latest one published in 2014 [5]. The data required by END is accessible through a system called the

"Shared Environmental Information System" (SEIS), which constitutes a set of data management principles to access the information. In this system, the member states indicate their experiences regarding action plans, their accomplishments and failures, in various documents [45].

The recent situation has been discussed in "The Noise in Europe Conference" held in April 2017, organized by the European Commission under the "Regulatory Fitness and Performance (REFIT)" program. Some of the main statements declared in the final report, published in 2018, are as follows [46]:

Generally, Directive 49 was found to be suitable for its purpose, however, the need for better implementation was stressed, stating that "has not yet fully delivered on its potential due to delayed implementation."

Regarding the human health issues in the final report, the evidence acquired from the scientific studies, is given in Table 9.1, in terms of the number of people exposed to high noise levels compared to the noise limit given in the Directive, i.e. >55 dB(A) in L_{den}.

The report declared that the cost of the health effects of noise reached 50–100 billion Euros per year, and more action is needed, including better implementation of the Directive.

Further concerns focused on the following aspects:

- Legislation regarding noise control at source should be strengthened.
- Limit values should be established, and the existing ones should be lowered.
- Thresholds for noise mapping should be adjusted to include the other significant noise sources.
- The Directive should be extended to cover the other sources.
- Some definitions should be clarified.
- Matters regarding funding of noise reduction measures and the public share of costs should be covered in detail.
- The source abatement should be realized by balancing the other effective measures (e.g. road surface improvement, urban planning, etc.).
- Operating restrictions remain inconclusive, e.g. night-time restrictions are still being debated.
- Research and development of technical abatement should be supported, and new transportation systems should ensure sustainable transportation.
- The concrete targets at the EU level of "science-based limit values," research and innovative programs, such as smart cities, should be supported.
- More stringent noise standards should be supported also by tightening the sound limits.
- Cost-benefit analyses and public participation in the development of noise policy should be provided.

Table 9.1 The number of people exposed to environmental noise in European member states, based on the final evaluation report of the European Commission.

| Noise source | Number of people exposed to noise level above 55 dB(A) L_{den} | | |
	Numbers (in millions)	Inside agglomerations	Outside agglomerations
Road traffic noise	100	70	30
Railway noise	18	10	80
Aircraft noise	4	3	1
Industry noise	1	–	–

- The links between the strategic actions against environmental noise and other relevant urban planning actions should be established.
- Robust criteria should be developed for public procurement.
- EU members should improve implementation by strengthening enforcements.
- Public engagement in the decision-making process and in implementations should be increased.
- A noise database for Europe should be completed.

As a final remark, the report states:

"Environmental noise remains a problem in Europe. The necessary tools and solutions to address it exist. Main issues are to give more priority to their development and full implementation of legislation in place" [46].

Source-Specific Action Plans in Europe

The vast range of measures that can be implemented by the EU countries to minimize the noise impact on health might be available. The *EffNoise* Project relating to noise mitigation measures is supported by the European Commission, and reviewed the noise mitigation measures (technical, legal, and socio-economic aspects) and, in particular, those taken at EU level, and assessed the effectiveness of these measures, which is expressed in the numbers of people exposed to noise level bands of different noise mitigation packages for transportation noise [47].

Published in 2010, the report has evaluated all the action plans submitted by the member states to the EU Council, according to the three scenarios for each of the major environmental categories of noise: road, railway and aircraft noises [48]. These are:

I) Current practice
II) Policy-in-place
III) Best available technology

Based on the national reports, the measures taken, according to the above scenarios, and assessment of beneficiaries are outlined below.

a) Road traffic noise management:

Scenario I: Speed reduction (e.g. 30 km in residential areas) and the maintenance of road surface. (Results in a lower tire noise about 1.5 dB(A).)

Scenario II: Decrease in traffic intensity and percentage of heavy vehicles in agglomerations. (When the road surface is replaced by stone/mastic asphalt, the noise levels are lowered by 3 dB for tire noise and 1–2 dB for propulsion noise.)

Scenario III: New pavement on the main roads, to provide a reduction of 6 dB for cars and 4.5 dB for trucks (within five years). (The propulsion noise reduction is -5 dB and tire noise -4.5 dB. Assessment of total effectiveness of these scenarios has shown reductions of about 60% and 70%, with regards to the number of highly annoyed people and people suffering sleep disturbance.)

b) Railway noise management:

Scenario I: Track grinding, better brake blocks.

Scenario II: Extensive track grinding, low track-side barriers, wheel absorbers on some trains.

Scenario III: Extensive track grinding, medium track-side barriers, wheel absorbers, new brake blocks, bogie shrouds.

For each scenario, the calculated reduction of annoyance is about 50%.

c) Aircraft noise management:

Scenario I: Ban on all non-Chapter 3 compliant aircrafts; set up preferential runways; noise-dependent charges; add insulation.

Scenario II: In addition to Scenario I measures; night flying restrictions, narrow flight corridors, noise quotas, and relocation of residents.

Scenario III: Further restrictions within Chapter 3; incentives to implement noise abatement procedures.

The results showed that a significant reduction is obtained with Scenario II, while no further improvement is obtained with Scenario III.

Eventually, the report stated that "results demonstrate that the continuation of current practices will result in significant numbers of EU citizens remaining highly annoyed or suffering sleep disturbance due to noise." A significant reduction in the percentage of people affected by high annoyance can only be achieved if the best available technologies are implemented; however, the measures need a great amount of resources for road traffic and railway noise, and "even then, it is likely that some levels of high annoyance and sleep disturbance would remain."

For airport noise management, the 2010 ICAO report stated that aircrafts are 75% quieter than they were 50 years ago, and the number of people exposed to aircraft noise has declined [49]. However, air traffic has increased by more than 150% within the same period and will continue growing in the near future. With an expected 30% increase in passenger demand by 2017, airport noise will likely remain the major cause of annoyance in most parts of the world. The International Civil Aviation Organization (ICAO) issued a document in 2007 about "a balanced approach to aircraft noise management" to solve the aircraft noise problems at individual airports and to achieve maximum environmental benefit in a most cost-effective way [50]. The balanced approach consists of the four principal elements of: (i) reduction of noise at source; (ii) land-use planning and management; (iii) noise abatement operational procedures; and (iv) operating restrictions.

In the evaluation report for EU Directive/49, the key problems in fulfilling the concerns of END have been identified as [45]:

- delays in implementation
- non-enforcement of existing noise limit values
- poor quality of strategic noise maps
- inconsistent approaches to mapping noise
- poor quality of action plans
- confusion among responsible bodies regarding requirements.

However, in the 2014 EEA Report based on the completed data of EU countries, the above problems were not mentioned since many of those have since been overcome, but the report drew attention to the continuation of noise problems and the risks to public health. It also stressed the difficulty of making assessments to compare noise levels with WHO recommendations, especially in night-time periods. Similarly, the effectiveness of the relevant provisions regarding industrial noise cannot yet be assessed because of the wide gap in the required input data.

According to the END Article 11, the European Commission should prepare an implementation report every five years to review the acoustic environment and goals and measures to reduce environmental noise and assess the need for further community actions [46]. The document stated that the transposition of END to national regulations and the designation of the major noise sources and agglomerations had been completed, however, the delays in the required noise maps and the reporting action plans are the actual state of play. Also the failure of implementation due to the lack of priority given to noise problems, the lack of effective coordination, and the lack of comparability of noise maps, however, it is certain that the number of quiet areas has increased.

Implementation and Achievements in Some Countries

Regarding the initiatives and obligations in European countries concerning action plans against noise, some implementation of the measures is highlighted in the non-governmental papers and articles [39, 51–72].

The United Kingdom: The list of policy measures in the UK covers the stringency of permissible noise emission limits from aircraft and road vehicles, the publication of planning policy guidance and relevant codes of practice, significant changes in building regulations and implementation of noise insulation regulation, environmental impact regulations, noise control regulations, etc. [39]. Possible effects of these measures were calculated for aircraft noise, road traffic noise, and, in building regulations, construction noise. The estimated changes in noise levels as L_{eq}, 1 hour, are determined. The results showed a near 5 dB reduction for Type A roads less than 80 km/h, and came in at 2 dB for motorways.

The Noise Action Plan has been adopted as required by the Environmental Noise regulations, in 2006. The Noise Policy Statement for England (NPSE), published by Defra, that is the responsible authority making and implementing noise action plans, in March 2010, describes a policy vision to facilitate decisions regarding acceptable noise [51–53]. The NPSE is applied for three categories of noises: environmental noise, neighbor noise, and neighborhood noise (from premises used for industry or leisure purposes). The action plan describes a framework for the management of noise, to reduce noise pollution and maintain areas where noise pollution is low. A decision-making process is constituted under the approaches of "Appropriate Measures (A.M.)" and "Best Available Techniques (BAT)" to develop, agree, and disseminate good practices, aiming to draw the attention of consultants, planners, and policy-makers involving noise control.

Progress in the action plans, which have been issued for roads and urban areas separately, is monitored by Defra once every five years [52, 53]. Generally, the action plans refer to the technical measures and their expected performance in detail [54]:

Control of noise at source: Using the low noise surfaces on the new roads and during retrofitting (a thin layer of porous asphalt gives a 3–4 dB reduction in noise levels), the reduction of speed limits (a 2 dB reduction), use of silencers, tire noise limits and labelling tires, and an ultra-low emission bus scheme funded by the local governments.

Planning: Local transport and land use planning by relocating traffic from sensitive areas, use of certain roads at certain times, speed restrictions, installation of noise barriers (reduction of 10 dB (A), façade insulation (indoor noise reductions of 30 dB(A) with normal windows and 40 dB(A) with highly insulated windows).

Germany: Berlin's noise action plan involves the following actions: [55]

- To implement a strategy not only to reduce noise: Development of a strategy for noise management is integrated with the context of the urban Transport Development Plan. The main political objectives are sustainable modes of transportation and noise abatement at source.
- Cycle lanes in public streets.
- Reorganization of public street spaces and meeting zones where overlapping demands are met.

- Speed limit reductions from 50 km/h to 30 km/h (provides 3 dB(A) reduction).
- Noise reduction asphalt.
- Improve quiet areas.

Paying more attention to railway transportation, Germany has improved noise control policy for railways [56]. The German Noise Mediation program (1999) has been successfully implemented to cover noise abatement measures for noise barriers, rail grinding, sound insulation windows, replacing the existing braking systems of composite block brakes with cast-iron blocks and retrofitting wagons by providing financial support, developing programs of infrastructure to establish a model for railway noise control policy of Europe.

The German Environment Agency regularly performs investigations and carries out surveys on noise impacts. Implementation, according to the noise action plans at different levels, is checked and the achievements are delivered in terms of: the immediate results of action plans, the strategic approach through urban and transportation planning, and the approach of public participation to the action plans.

The Netherlands: By applying an integrated noise policy for railways, it was declared that noise mitigation measures were rather successful [57]. The integrated noise policy outlined economic incentives for freight wagon owners and freight operators. Regarding the ship noise action plan, the "Best Practices Guide" has been issued by the Port of Rotterdam within the Neptunes Project [58]. The list of measures with the expected values of noise level reductions are presented for low noise equipment in sea vessels in the design and construction of ships, in addition to the measures relative to port management.

Italy: The Italian Ministry of Environment reached another achievement in legislative action by issuing a regulation on market surveillance on equipment for use compatible with 2000/14/EC [59]. A pilot project on noise insulation near airports has revealed that almost a 12% reduction in the negative impact of airports was due to the improvement processes implemented on the houses and flats in the proximity [60].

Further investigations focusing on integration of the strategic action plans and noise reduction plans for cities like Florence have been conducted and some methodologies have been developed to provide an assessment instrument and an operative tool for implementation [61, 62]. In most of the cities, the areas where the noise levels are not compatible with the land use, are declared in urgent need of improvement and the noise control options are delivered.

France: Bruitparif is the center of the evaluation of environmental noise in the Ile-de-France, responsible for noise maps and implementation of action plans, through collaborations of various parts, the centralization of data and informing the public. The tools are provided to support the local authorities in promoting the implementations of noise action plans, mainly:

- A website dedicated to public authorities with a GIS tool facilitating the preparation of the noise maps and reviewing information about their territory and also covering the methodology for ranking priority areas and quiet areas.
- A workbook was prepared, including the best practices in noise management.
- A pre-written noise action plan on a template prepared for action planning. to promote implementation by the local authorities and to help them in the ranking of priority areas in noise action plans.

Training sessions are organized on the methodology in defining priority areas. Various reports have been published to improve the local authorities' knowledge concerning the reduction of noise with examples of actions taken. Technical information is included in these documents, such as relationships between the traffic parameters and the noise levels, e.g. the effect of speed reduction on noise levels [62]. Regarding noise monitoring, innovative systems have been implemented, such as the one developed in the Medusa Project, which proposes a new approach for noise management and control in an urban environment [63].

Japan: Taking noise control seriously in Japan since 1960s, the publications regarding noise abatement are issued and their enforcement is under the control of the government and local administrations, particularly focusing on the transportation systems. As an example of implementation, the noise management at Osaka International Airport (ITAMI) has yielded satisfactory achievements for the community, who has already demanded the closure of the airport and had instigated lawsuits against the government [64]. Among the measures implemented so far are strict operational restrictions, e.g. permissible aircraft type and airlines, suspension of night flights after 9 p.m., stipulating the maximum number of times of arrivals and departures, shifting the international flights to another airport, having a landing fee system based on continuous measurements of runway management, classifying the areas according to the noise level and assigning measures, including financial assistance/relocation/cooperation, creating buffer zones according to the classes, restriction of reverse thrust for jet landings on certain runways and applying a noise sharing system. Eventually it was declared that the size of the exposed affected area, with 100 000 houses, had been reduced by about 40%, however, about 40 000 houses are still affected.

The earlier and up-to-date developments in other countries can be found in various publications [65–72].

9.3.2 International Projects Contributing to Noise Management

Various European projects are being supported by the EU, oriented around the contribution of the END's concerns on noise protection.

Smart City Projects: A smart city has been defined as "a city that uses intelligent devices, makes intelligent decisions and applies intelligent solutions" and the concept is promoted by the European Commission [73]. The approach aims to reduce costs, have an environmental impact (including noise impact), reduce the consumption of resources, and improve the performance and attractiveness of the city by using all the means of information and communication technology, (i.e. how to use modern technology to improve urban space and interact with citizens to increase the quality of life) [74].

Quiet City Project: This project, funded by the European Commission under the 6th Framework program, aims [75–77]:

- to develop an integrated technology infrastructure for the control of road noise and rail noise in urban areas;
- to provide municipalities with tools to create quiet cities;
- to detect the "hot spots" in cities;
- to assist with the selection of noise measures in a specific city.

The project mainly deals with noise mapping and action plans, complaint identification, and proposals for solutions. The solutions and their efficiencies as dB reduction for each solution were given as separate work packages.

Quiet Track Project: This project, which has been conducted under the 7th Framework program of the European Commission, considers the importance of rolling noise in overall train noise. Investigations are continuing in pursuit of innovative quiet tracks for all types of trains, including metro trains with speeds of 20–200 km/h [78–80].

9.4   Quiet Areas and Quiet Façades

According to END: "It is strongly recommended that protection of quiet areas be made an integral part of the development of action plans for agglomerations rather than be treated merely as an add-on to be addressed once other issues have been resolved" [22].

Concept of Quiet Areas: A quiet area is defined in END, not "as silent, but rather one that is undisturbed by unwanted or harmful outdoor sound created by human activities" [81]. *Quietness* implies the absence of sound that interferes with activities, such as direct or indirect communication, thinking, reading, writing, learning, relaxing, resting, sleeping, etc. There are discussions about using the terms of "relaxation" or "calm," instead of "silence" or "quiet," by considering that absolute silence tends to frighten most people. The site where noise is absent, or at least not dominant, must be designated in a community for relaxation, rest, and peace of mind. Such a possibility should be provided for the people in the vicinity of their homes. A quiet area must be accessible to the general public, and thus, private gardens and private parks cannot be declared quiet areas. Briefly, a quiet area is not only related to acoustical factors but also other non-acoustic factors, such as function of area, the soundscape, public expectations, etc.

In a quiet area, the locations and time periods for different activities with low noise levels can be organized, such as playing games, listening to low levels of music, or speech. A quiet area in open country should be undisturbed by the noise of traffic. The benefits of a quiet area's designations are numerous [6, 81–85]:

- Quiet places improve the quality of a city.
- In quiet places, the inhabitants can relax and recover from their daily life and work.
- Nearby quiet zones in noisy areas reduce the annoyance, and as a result, providing access to nearby green areas seems to improve one's well-being.
- Sick people recover faster in natural surroundings, including quietness and natural sounds.
- Quiet areas improve the life of the inhabitants, i.e. also it is beneficial for biodiversity and wildlife.
- The economic value of quiet areas rises. The direct effect of the lower noise levels as seen in the increase of property values have been estimated as 0.5% per dB [86].

The importance of quietness must be taught to individuals and the quiet area to be assigned should be determined through surveys completed with the participation of the community itself.

Surveys revealed the following results, concerning the desire of the designation of quiet areas in some European countries:

Amsterdam: 75% of the population pointed out that the quiet around the house is important and 50% stated that quiet in the neighborhood is important.

The UK: 91% of the population stated that the existing quiet areas must be protected and 62% stated that these areas must be protected in London.

The Life+2008 Project explained above gives a proposal for the definition of quiet areas and the assessment methodology by using an index of quiet area (IQA) in relation to noise exposure levels, which are calculated at grid points of $10\,m \times 10\,m$ in the noise maps [43].

9.4.1   Definition of Noise Limits for Quiet Areas

The noise limits for identification of the quiet areas have not been declared by END and the WG-AEN, or the European Commission Working Group Assessment of Exposure to Noise and this subject has remained in the care of member states. However, it has been widely accepted that quiet implies very low noise levels, i.e. <45 dB L_{day} and < 40 dB L_{night}. Since such low levels cannot exist in cities, DEFRA, the Department for Environment, Food, and Rural Affairs in the UK, in 2006, recommends the identification of quiet façades based on WG-AEN's recommendation on quiet zones [86, 87].

Although the Report Task 3 declared that there was no clear definition of quiet areas in 2010, the recent EEA Report in 2014 offers a table displaying the physical and effect-oriented definitions and selection criteria for quiet areas [81, 88, 89]. In terms of size, the following is outlined: the size of quiet areas should be 100–100 000 m^2 in urban areas, and 0.1–100 km^2 outside of urban areas (or open areas); and in terms of distance, the distance should be 4–15 km from highways and 1–4 km from agglomerations. Noise range criteria were also provided in some regulations for the quiet areas in urban and open country regions.

The noise index to designate criteria for quiet areas varies according to the EU member states. The commonly accepted descriptor is L_{den}; however, supplementary indexes (such as L_d, L_e, L_n and L_{Aeq},24 hour, or its equivalent, in L_{den} or $L_{A90,\ 1hr}$) are used.

Different approaches are implemented in Europe for setting noise limits identifying the quiet areas. The statement of noise limits in countries for the designation of quiet areas varies, such as:

- 55 dB(A), as suggested by the WHO.
- L_{den} < 50 dB or L_{day} < 60 dB, based on monitoring throughout 30 days/27 nights.
- 40 dB L_{Aeq}, 24 hour as the upper noise limit.
- Most EU countries have accepted 50 dB L_{den} as an upper limit for relatively quiet areas in urban areas and stated that "If a higher 'gold standard' level is to be defined for urban areas, then it would be sensible to strive for 40 dB L_{den} as a consideration of quiet" [81].
- A 6 dB limit is accepted as the optimum difference in L_{den} between the urban center and the margin of quiet areas from its surroundings.
- Identification according to the area categorization (e.g. Class A corresponding to noise-free zone for L_{night} < 25 dB) or Class 1 specifying "Special Protected Areas" (<45 dB L_{day}, 35 dB L_{night}).
- Categorization of land areas according to the criteria of number of noise events (in Sweden) and in addition, the duration of events per day or per 15 minutes (Belgian regulations).

The subject of quiet areas had attracted great concern in the past decade and scientific studies have been conducted to search for different rating descriptors, for example.

Quietness Suitability Index (QSI): In a project on quiet areas, a methodology was proposed to identify quiet areas by using the QSI to be derived at a national level. The threshold distances to different sources and agglomerations could be determined through diagrams presented for various cases [89].

Another methodology, which is developed for the determination of suitability for quietness, is given in Figure 9.7.

Perceived Acoustic Quality/Appreciation (PAA): A model for rating the green areas by relating the noise level ($L_{Aeq,24h}$) and the perception evaluation of people about quality of environment is given in the EEA Guide [81]. Based on the surveys, the simple criteria for quiet areas have been derived as below:

$L_{Aeq,}$ or L_{day}	Subjects declaring that acoustic quality is good (%)
<45 dB(A)	100
45–55 dB(A)	50
>55 dB(A)	50–0 (sudden decrease in percentage with increasing noise)

Acoustic quality also depends on visual quality, perceived sounds (human, natural, technical), user activities and expectations. Soundscape studies provide more information about quality evaluations (Section 7.5.1).

Tranquility Rating Prediction Tool (TRAPT): This was developed under the National Planning Policy Research in the UK to quantify the impact of new developments on the quiet areas that have remained relatively undisturbed by noise and to keep planning for the conservation of such areas, or so-called "tranquil spaces," especially used for areas near wind farms [90].

9.4.2 Overview of Actions for Quiet Areas in Europe

The process of selection of quiet areas includes the following approaches:

- Review of registered areas, candidate quiet areas (QUAS) by using GIS analysis of land-use, applying a coordinated approach to quiet areas with other limitations.
- Use of results of noise mapping, such as END noise maps or global noise maps (regional maps).
- Declaration of a criterion for favorable areas or candidates, according to zones categorized as *Luxury, Comfortable, Good, Acceptable,* and *Unfavorable;* the quiet areas are identified in the first two categories (the Czech Republic).
- Use of the soundscaping approach for remodeling urban parks.
- In situ measurements, monitoring, and verification of data through digitized recordings by observations.
- Minimum size identification, e.g. area of 10 or 20 ha; locations and boundaries to be displayed in land-use plans.
- Identifying distance and accessibility criteria from road or from other noise sources, such as industrial and construction sites, or from agglomerations and recreational areas.
- Identifying sites of natural and cultural interest: recreational values, nature areas, and nature-related benefits; natural or cultural value of the landscape.
- Noting aesthetic appeal: (noting that the visual quality enhances the quietness or tranquility).
- Receiving visitors' opinions through public perception surveys, open consultation with residents and local authorities via web-tools, recording attitudes to quiet areas determined by culture and location.
- Defining the acoustic landscape, investigating the positive and negative experiences of sounds, natural sounds, perceived appropriateness of area sounds through sound walks,
- Gaining approval of city council or local authorities.

Ultimately, the process of identifying the quiet areas involves the following steps:

- Pre-selection of quiet areas from potential candidates "QUAS."
- Analyses (quantitative and qualitative data), long-term measurements and interviews.
- Selection and declaration of "QUA."
- Management plan of "QUA" (preservation plan/modification or improvement plan).

Currently, various research projects regarding quiet areas have been carried out in Europe:

- QSIDE (www.qside.eu): An acoustic model for calculating noise levels and the beneficial effect of quiet façades and areas (by implementing a soundscape approach).
- CityHush (www.cityhush.org): aims to identify the boundary conditions required to obtain Q-zones (requires simulation of traffic management).
- HUSH (http://www.hush-project.eu/en/index.html) (harmonization of urban noise reduction strategies).

- Hosannah (http://www.greener-cities.eu).
- Listen (http://tii.se/projects/Listen).
- Quadmap (www.quadmap.eu) (by implementing soundscape approach).

9.4.3 Concept of Quiet Façades

The quiet façade of buildings exposed to noise sources has been introduced to noise management through END, meeting certain criteria. The project, funded by the European Council (QSIDE), stated that "a quiet façade enables residents to sleep with their window open without being disturbed by noise. In the daytime, it allows them to leave a window open, enjoy the outdoor garden, or step out on the balcony with that façade, without undue disturbance from noise" [91].

The descriptor for the quiet façade in END is that the L_{den} value is at 4 m above ground and at 2 m in front of the façade [81]. However, the definition criteria vary according to the implementations in various countries:

- The quiet façade (with the noise coming from a special source) should be more than 20 dB lower than at the façade having a maximum noise level; for the houses with double façades, the difference between the noise levels are taken at max. 2.5 dB(A), corresponding to 10 dB(A) in L_{den}.
- The façade at which the noise level is $L_{den} < 55$ dB.
- The criteria for the definition of a quiet façade vary according to the EU member states, as some designated quiet façades are based on noise surveys. For example, in the city of Amsterdam in The Netherlands, a façade in an urban residential area can be considered quiet if:

- the noise level is not higher than other façades;
- the noise level on the façades satisfies the conditions of; $L_{den} < 55$ dB and $L_{night} > =45$ dB;
- outdoor space has sufficient physical environmental quality (i.e. gardens or parks).

The benefits of quiet façades can be outlined as:

- access to the quiet side (i.e. $L_{Aeq24h} < 45$ dB) implies a significant reduction of the exterior level;
- residents can open windows to get fresh air;
- the annoyance is considerably reduced.

Various architectural solutions are applicable for housing blocks to create quiet façades in urban areas, such as eliminating back roads, blocking the gaps between the buildings, designing quiet gardens or atriums, or sometimes, by glass-covered courtyards. For high-rise buildings, using covered balconies and loggias to provide sufficient air conditioning or simply using highly insulated façade elements can be possible to create quiet façades.

9.5 Economic Impacts of Noise Management

Noise management, which is a subject of community health policy, is closely related to the economic impact on individuals and society. In recent years, the number of investigations have increased regarding the assessment of economic impacts through the use of various tools for evaluation [92–96].

The 2010 report, published by the WHO, on the burden of disease from environmental noise outlines key points emphasized below, highlighting the general impacts [96]:

- At least one million Healthy Life Years (HLY) are lost every year in Europe due to road traffic noise alone.
- The social cost of rail and road traffic noise was estimated at €40 billion per year, 90% of which is related to passenger cars and large goods transport vehicles.

The economic impact of noise pollution is evaluated by taking into account the key elements below:

- reduction in house prices;
- reduced possibilities of land use;
- productivity losses in the workplace due to illness caused by the effects of noise pollution;
- losses due to the annoyance impact (sometimes separately valued under the quality of life aspects);
- medical expenditures needed to rectify the health impacts, including increased medicine costs.

Social costs, as a result of health impacts, emerging from noise pollution are generally estimated in terms of:

- costs related to premature deaths;
- costs related to morbidity (poor concentration, fatigue due to insufficient rest or loss of sleep, hearing problems due to stress) [88].

Using a reliable approach to assess the monetary values of health impacts due to noise, within a certain accuracy, is important. In fact, there are difficulties in measuring the economic costs of noise damage since there are other factors contributing to the clinical health effects in addition to noise.

9.5.1 Risk Valuation

Risk is simply defined as "the potential of gaining or losing something of value" [97]. The values can be physical health, social status, emotional well-being, or financial wealth. Risks include both negative and positive impacts; however, health risks often imply adverse outcomes. The OHSAS (Occupational Health & Safety Advisory Services) defines risk as the combination of the probability of a hazard, resulting in an adverse event and the severity of the event [98]. Risk valuation has vital importance for the future assessments of noise policies. Valuation of risks in health (due to environmental noise) is made by calculations of changes in community health quality by using subjective methods. In the medical sciences, the change in health risks through epidemiological studies is used to determine the effect of a factor on health through comparison tests.

Risk valuation requires communication between various stakeholders: investigators, a community, scientific teams, representatives of various sectors (including industry, building, traders, transportation, etc.), and a mutual willingness to understand the needs of every involved party. Various methods can be used in the valuation of noise on health in terms of monetary units, by considering the costs of invested or non-invested expenditures. The 2010 EEA Report gave some recommendations about the methods to be used in the valuation of noise risks [98].

The Cost-Benefit Analysis (CBA)

In order to measure the risks in terms of effects by using monetary values, systematic approaches are used. One of the commonest is cost-benefit analysis, which is an efficient tool to estimate the strengths and weaknesses of alternatives that satisfy the objectives or to determine options that provide the best approach for the adoption and practice in terms of benefits [99]. CBA is generally used to provide evidence for justification in the total expected costs for noise measures, while also enabling comparisons between the cost of each option.

In the field of health services, where it may be difficult to value the health effect, "cost-effectiveness analysis" is often used with the indicators, such as life expectancy, premature births, or by calculating the cost factors per year, per person [100]. According to the WHO, one lost year of "healthy life" is represented by one disability-adjusted life year (DALY) and the total DALY in a community is defined as the burden of disease which is a measure of difference between actual health status and ideal health status. The further metric used recently for economic evaluation of noise effects on health is the quality-adjusted life years (QALY), another descriptor of disease burden, which includes both the quality and the quantity of life lived. It assumes that a year of life lived in perfect health is worth 1 QALY (1 Year of Life $\times$ 1 Utility value $=$ 1 QALY) [101].

Evaluations of Benefits

The benefits in CBA may imply an increase in health quality, which in turn could save labor, time, and costs, in addition to the benefits of noise mitigation or the benefits of other different options. The efficiency of measures against noise are assessed with respect to their cost of investment and operations, training, maintenance, renovations, etc. The benefits of all the options proposed, according to the scenarios, are displayed as the ratio of *benefits to cost,* based on the socio-economic analyses. However, there are methodological uncertainties due to the discrepancies in the results of the objective measuring methods.

- Benefit evaluations are subjects of combined valuation approaches, requiring knowledge about economic theories and practices.
- Benefits should be carefully interpreted as absolute values. Benefit estimates value all the known possible risks, and the feedback received through the surveys from the key stakeholders.
- Benefits are compound metrics and the attitude of the public toward the monetary expression should be taken into account as uncertainty, unless contingent valuation methods are applied [88, 96].
- The sensitivity analysis should be implemented to determine the indirect effects of the measures which may be important in the assessment of both impacts and benefits; i.e. some measures will not only decrease the noise level in the city but also affect other parameters that may have an impact on the costs and benefits of this measure.
- Estimated benefits are obtained by using subjective valuation techniques (such as field surveys) to investigate change in the magnitude of annoyance from traffic noise, before and after the application of noise measures. The noise (dose) and effect (response) relationships are tools for monitoring these situations; however, expenses for these studies should also be considered in cost evaluations.

There are uncertainties in benefit evaluations:

- Difficulty in the subjective rating of the change in noise level causes problems in the valuation of the benefit of the measure. It has been revealed that the change in noise above 5 dBA is readily noticeable.
- Difficulty in the development of health valuation measures.

The estimation of benefits can be made, based on different approaches taken to attribute a value to noise reductions, expressed in decibels. In the *EffNoise* Project, the following equation is used for all noise sources [47]:

Benefit (€) = 50% of number of affected inhabitants × noise reduction in dB(A) × €25

9.5.2 Economic Impacts of Noise Management in EU Member States

In 1996, the Green Paper on future noise policy was published, and it was estimated the potential annual economic damage to the EU (due to environmental noise) was between €13 million to €30 billion, based on the cross-national studies [11].

The key actions required by the Commission to implement the recommendations, based on the noise mapping process, are explained for the EU countries. The optional measures listed for each noise source and each agglomeration type are both specified in END, such as the reduction of road vehicle noise at the source and tire noise reduction measures. The countries evaluate potential policy options by ranking these measures, evaluating a combination of measures, calculating the costs of each option, presenting the cost-benefit analysis results, and searching for the significant additional benefits in specific case studies, as required by END.

Case studies to be conducted in the European cities comprise the following data requirements:

- the population's exposure to all sources of noise (L_{den}) and to the specific noise sources;
- the number of sensitive buildings exposed to noise sources;
- the number of people exposed to a noise level higher than the limit values by noise source;
- the distribution of roads, according to noise exposure;
- the comparison of the population's exposure to noise, before and after a new transportation route (e.g. a tramway line);
- measures already implemented, or to be implemented, in the coming years;
- the comparison of a population's exposure to road noise before and after implementation of the measure;
- the population exposed to a noise level between 70 and 75 dB(A) L_{den}.

The assessments of costs associated with an exposure to noise, as indicated by some European countries, are summarized below [102]:

1) The total cost, including all the implementations according to END across 28 EU member states, was determined as between €486 million to €308 billion over the 25 years.
2) In Germany, the actions to reduce railway noise by half (i.e. 10 dB(A) according to the German national program, needed about €100 million per year. The cost of retrofitting of 180 000 wagons with quieter brakes was estimated as €300 million.
3) In Sweden, social costs due to road traffic noise over SEK16 billion,
4) In the United Kingdom, the social cost of environmental noise was £7–10 billion per year (greater than the impact of climate change). It was estimated that:
   - the most severe health effects are cardiovascular problems, costing £2–3 billion per year;
   - the effects on amenity reflecting consumer annoyance, cost £3–5 billion per year.
5) In the Czech Republic: a population of nearly 1 052 000 were found to be living in the areas of $L_{dn} > 55$ dB and the total cost of the measures was calculated as €19 988 000 per year.

Cost estimations of noise reduction measures are based on the calculations of specific materials and constructions [103]. However, there is uncertainty about the estimated costs due to the material characteristics, market conditions, contractors, and manufacturers. The updated information regarding the impacts of action plan measures has been continuously reported, based on the data collected from countries, as the results of END implementations. The 2017 overview of policy-related data for each country (Noise Fact Sheets) can be found on the webpage of the EEA: https://www.eea.europa.eu/themes/human/noise/noise-2.

9.6 Conclusion

Noise is an inescapable negative aspect in the modern world due to various reasons, e.g. growing volume of traffic, the increasing numbers of people living in densely populated cities, etc. Although health concerns in all countries have a long past, however, the awareness of the adverse effects of noise as significant enough to come under the government control, has not been at a satisfactory level. At present, prevention of all types of noise sources and environments is undertaken more consciously within organized initiatives all over the world and noise control is included in the noise management concept requiring a well-defined policy statement and action planning. The earlier action plans developed in the 1990s have been vigorously renewed with stringent implementations, based on the precise calculations of cost and health-related benefits [104–107].

The National Policy Statement for England, NPSE declares the noise policy vision is to "promote good health and good quality of life through the effective management of noise within the context of government policy on sustainable development" [52].

Noise policy aims to avoid, mitigate, or minimize the adverse effects on health and quality of life, and focus on their improvement. Possible measures against environmental noise sources, as included in the concept of "noise control," are still of value, mostly in planning and engineering; but nowadays they are involved within the new concepts of noise policy, strategies, management, and action plans. It is the decision of the policy-makers to implement these measures at a national level or promote compliance with international obligations. The policy making process requires: awareness by people and politicians, policy definition, options and scenarios, cost-benefit analysis, selecting measures, enforcement, reporting and monitoring the implementation.

The European Environment Agency, EEA Report entitled "The European Environment: State and Outlook 2015" declares that noise pollution, especially in urban areas, is the only category in which it is not possible to establish a (more than) 20-year outlook, and, therefore, an indicative assessment cannot be made [106]. Despite great efforts, it is still debated whether the expectations (suggested by the 7th EAP) by 2020 have a limited possibility of being realized in the recent findings among European member states [107].

As a final statement, it is expected that humankind will be successful in achieving the goal of living in a quiet, healthy, and pleasing environment in the future, following collaborative efforts, however, only by keeping the natural, cultural, and social features, as described in the concept of sustainable environment.

References

1 Ginjaar, L. (1981). Towards a quieter society, Inter-Noise 81, The Netherlands.

2 Anon (1985). Strengthening noise abatement policies: final report, Environment Committee Ad Hoc Group on Noise Abatement Policies, Organization for Economic Co-Operation and Development (OECD), Environment Directorate, ENV/N/842.

3 Embelton, T. (1989). Noise control in the Western world over the past 2000 years. *Acta Acustica* 14 (1): 10–16.

4 Ivanov, N. (1993). Noise control problems, Proceedings of Noise-93, St Petersburg, Russia (31 May– 3 June).

5 (a) EEA (2014). Noise in Europe, EEA Report No. 10/2014.
(b) 5(b) EEA (2019) Environmental Noise in Europe- 2020   EEA Report No.22/2019.

6 EU (2013). Decision no. 1386/2013/EU of the European Parliament and of the Council of 20 November 2013 on a General Union Environment Action Programme to 2020 'Living well, within the limits of our planet'.

7 Policy. http://dictionary.cambridge.org/dictionary/english/policy.

8 Lang, W.W. and ten Wolde, T. (2006). A global approach to noise control policy technical study group 5. *Noise Control Engineering Journal* 54 (5): 289–340.

9 Berglund, B., Lindwall, T., and Schwela, D.H. (eds.) (1999). *Guidelines for Community Noise*. World Health Organization.

10 Lambert, J. and Vallet, M. (1994). Study related to the preparation of a communication on a future EC noise policy: final report, INRETS.

11 European Commission (1996). The Green Paper on Future Noise Policy (COM(96) 540), November.

12 Finegold, L., Von Gierke, H.E., McKinley, R.L. et al. (1998). Assessing the effectiveness of noise control regulations and policies, 7th International Congress on Noise as a Public Health Problem, Sydney, Australia.

13 WHO (2018). Environmental Noise Guidelines for European Region, World Health Organization Regional Office for Europe.

14 https://en.wikipedia.org/wiki/Management.

15 ISO (2004). EN ISO 14001:2004 (revised and issued in 2015), Environmental management systems-Requirements with guidance for use.

16 Porter, N. and Knowles, A. (2016). Key issues in aviation noise management. *Acoustics Bulletin* March–April.

17 Flindell, I. and McKenzie, A.R. (2000). An inventory of current European methodologies and procedures for environmental noise management, European Environmental Agency Technical Report.

18 Finegold, S. and Schomer, P. (2001). I-INCE Initiative No. 3, Environmental noise policy assessment, invited paper, 141st Meeting of the ASA, Chicago (June).

19 Anon (2001). *Environmental Outlook*. Organisation for Economic Co-operation and Development (OECD).

20 European Community (2001). Common position (EC) No. 25/2000 of 7 June 2001 adopted by the Council with a view to adopting a directive of the European Parliament and of the Council relating to the assessment and management of environmental noise.

21 Tachibana, H. and Lang, W. (2002). The I-INCE Technical Initiative No. 3, Noise policies and regulations, Inter-Noise 2002, Michigan, USA (August).

22 EU (2002). Directive 2002/49/EC of 25 June 2002 of the European Parliament and Council relating to the assessment and management of environmental noise.

23 Luzzi, S., Natale, R., Delle Machie, S. et al. (2019). The Sound of My Place experience in the frame of the International Noise Awareness Day, Inter-Noise 2019, Madrid, Spain (16–19 June).

24 I-INCE (2005). Technical Study Group on Noise Policies and Regulations, Phase 1: Policies and regulations for environmental noise: final report 01–1.

25 European Community (2005). Fifth Environmental Action Programme, "Towards Sustainability": the European Community Programme of policy and action in relation to the environment and sustainable development.

26 Lang, W.W. (1996). Quarter century of noise control: a historical perspective. *Acoustic Bulletin* July/Aug.

27 Beranek, L.L. (ed.) (1971). *Noise and Vibration Control*. McGraw-Hill.

28 Beranek, L. and Ver, I.L. (1992). *Noise and Vibration Control Engineering:Principles and Applications*. Wiley.

29 Harris, C.M. (1957). *Handbook of Noise Control*. McGraw-Hill.

30 Harris, C.M. (1979). *Handbook of Noise Control*, 2e. McGraw-Hill.

31 Crocker, M.J. and Kessler, F.M. (1982). *Noise and Noise Control*, vol. II. CRC Press.

32 Crocker, M.J. (1984). *Noise Control*. Van Nostrand Reinhold Company Inc.

33 Crocker, M. (ed.) (2007). *Handbook of Noise and Vibration Control*. Wiley.

34 Bies, D.A. and Hansen, C.H. (2002). *Engineering Noise Control Theory and Practice*, 2e. Spon Press.

35 Harris, C.M. (1994). *Noise Control in Buildings: A Practical Guide for Architects and Engineers*. McGraw-Hill.

36 Woods, R.I. (ed.) (1972). *Noise Control in Mechanical Services*. Sound Attenuators Ltd and Sound Research Laboratories, Ltd.

37 Swela, D. (2001). World Health Guidelines on Community Noise. Presentation at the TRB Session 391, Setting an Agenda for Transportation Noise Management Policies in the United States, Washington, DC, USA (10 January).

38 Turner, S. (2016). Public health outcomes framework. *Acoustics Bulletin* 41 (2): 18–20.

39 Dryden, S. (2014). Back to the future: part 1. *Acoustics Bulletin* 39 (2): 25–28.

40 Anon (2008). Practitioner Handbook for Local Noise Action Plans, Recommendation from the SILENCE Project, Sixth Framework programme, European Commission.

41 Maffei, L. (2013). Which information for the European and local policy makers? In: *Noise Mapping in the EU: Models and Procedures* (ed. G. Licitra), 351–360. CRC Press.

42 ten Velden, H.E. and Kreuwel, G. (1990). Geographical information system-based decision support system for environmental zoning. In: *Geographical Information Systems for Urban and Regional Planning* (eds. H.J. Scholten and J.H. Stillwell), 119–128. The GeoJournal Laboratory/Springer.

43 Borchi, F., Curcuruto, S., Governi, L. et al. (2016). Life+2008 Hush project results: a new methodology and a new platform for implementing an integrated and harmonized noise action plan and proposals for updating Italian legislation and the Environment Noise Directive. *Noise Mapping* 3 (1): 71–85.

44 http://ctb.ku.edu/en/table-of-contents/structure/strategic-planning/develop-action-plans/main.

45 EU (2015). END Implementation (Milleu 2010). First round of END Implementation: Workshop working paper 1, The second implementation review of the END – emerging findings, Brussels (23 September).

46 EU (2017). Report from the Commission to theEuropean Parliament and the Council on the Implementation of the Environmental Noise Directive in accordance with Article 11 of Directive 20013/49/EC., COM (2017) 151, 30.3.2017. http://eurlex.europa.eu/legal-content/EN/TXT/?uri=CELEX:52017DC0151.

47 EffNoise (2004). Service contract relating to the effectiveness of noise mitigation measures. www.eff.org.

48 Vernon, J. (2010). Final Report on Task 3: Impact assessment and proposal of Action Plan, May 2010. prepared by RPA, Milleu and TNO Belgium.

49 ICAO (2010). Environmental Report 2010: Aviation outlook.

50 ICAO (2007). The balanced approach to aircraft noise management. https://www.icao.int/environmental-protection/Documents/Publications/Guidance_BalancedApproach_Noise.pdf.

51 Defra (2010). 2006 Noise policy statement for England, March.

52 Defra (2019). Noise Action Plan: Roads, environmental noise (England) regulations 2006.

53 Defra (2019). Noise Action Plan: Agglomerations (urban areas), environmental noise (England) regulations 2006, as amended.

54 https://www.eltis.org/resources/tools/developing-local-noise-action-plans.

55 Fabris, C. (2012). Noise policy in Germany. *The Journal of the Acoustical Society of America* 131: 3295. https://doi.org/10.1121/1.4708322.

56 Jaecker-Cueppers, M. (2016). Railway noise control in Germany: a success story and an example for Europe, Inter-Noise 2016, Berlin, Germany. http://www.internoise2016.org/satellites-in-berlin/european-noise-policy/#c751.

57 de Vos, P. (2015). An exemplary integrated approach to railway noise, ICSV22 No. 1288, Florence, Italy (12–16 July).

58 Wolfert, H. (2019). Neptunes best practices for seagoing vessels at berth, Inter-Noise 2019, Madrid, Spain (16–19 June).

59 Curcuruto, S., Atzori, D., Lanciotti, E. et al. (2015). The Italian market surveillance of outdoor machinery under Directive 2000/14/EC, ICSV22, Florence, Italy (12–16 July).

60 Baggaley, J. and Dellatore, L. (2015). The role of noise insulation schemes in airport noise management, ICSV22.

61 Bellomini, R., Borchi, F., and Luzzi, S. (2010). Integration between noise reduction plan and strategic action plan in the city of Florence, Inter-Noise 2010.

62 Echaniz, L., Mietlicki, F., and Comment, M. (2013). Designing tools to support noise action planning on a large scale: the role of a regional noise observatory, Inter-Noise 2013, Innsbruck, Austria.

63 Laetitia, N. and Fanny, M. (2019). Medusa: a new approach for noise management and control in urban environment, Inter-Noise 2019, Madrid, Spain (16–19 June).

64 Yukowa, Y. and Matsubara, K. (2019). Kansai Airports: noise control measure at ITAMI, Inter-Noise 2019, Madrid, Spain (16–19 June).

65 Kihlman, T. (1993). A Swedish action plan against noise. *Acoustic Bulletin* Nov.–Dec: 5–8.

66 Mauriz, L.E., Forssen, J., Kropp, W. et al. (2015). Traffic dynamics, road design and noise emission: a study case, Euronoise 2015, Maastricht, The Netherlands, paper no. 000282.

67 Erkelens, L.J.J. (1999). Development of noise abatement procedures in the Netherlands, NLR-TP-99386, New Aviation Technologies International Symposium, Zhukovsky, Moscow, Russia (17–22 August).

68 Finegold, L.S., Finegold, M.S., and Maling, G.C. (2002). An overview of US noise policy. *Noise News International* June: 51–63.

69 Finegold, L.S., Oliva, C., and Lambert, J. (2008). Progress on development of noise policies from 2003–2008, 9th International Congress on Noise as a Public Health Problem (ICBEN).

70 de Vos, P. (2015). Environmental noise policy: ways out of the crisis, Euronoise 2015, Mastricht, The Netherlands, paper no. 000011.

71 EC (2007). CALM, 2007. Research for a quieter Europe in 2020: an updated strategy paper of the Calm II network, September 2007, funded by DG Research of the European Commission.

72 EPA (2009). Guidance note for noise action planning for the first round of the environmental Noise Regulations 2006.

73 www.smartcitiesprojects.com.

74 Wolfret, H. (2015). Smart policies for cities, ICSV22, 2015, Florence, Italy (12–16 July).

75 Miedama, H.M.E. and Borst, H.C. (2007). Rating environmental noise on the basis of noise maps: quiet city transport deliverable, Inter-Noise 2007, Istanbul, Turkey, D.1.5.

76 Desanghere, G. (2007). Qcity: providing cities with a guide for noise action plans, Inter-Noise 2007, Istanbul, Turkey. www.qcity.org.

77 EC (2008). QTIP4-CT-2005-516420 QCITY

78 Höjer, M., Åke Nilsson, N., Sandin, Å. et al. (2008). Performance report of applied measures, Gothenburg, DELIVERABLE 5.5.

79 https://trimis.ec.europa.eu/project/quiet-tracks-sustainable-railway-infrastructures.

80 http://www.quiet-track.eu.

81 Anon (2014). Good practice guide on quiet areas, EEA Technical report No 4/2014, European Environmental Agency.

82 Bartalucci, C., Borchi, F., Carfagni, M. et al. (2015). Life+ Quadmap Project: quiet areas definition and management in action plans: results of post operam data analysis and the optimized methodology, ICSV22, Florence, Italy (12–16 July).

83 Botteldooren, D. and de Coensel, B. (2006). Quality assessment of quiet areas: a multicriteria approach, Euronoise 2006, Finland (30 May–1 June).

84 Anon (2013). Quiet places in cities, quiet façades and quiet areas in urban noise policy recommendations and examples. QSIDE coordinated by TNO, funded by EU Commission, 2013 Delft Project. http://www.qside.eu/proj/pub/QSIDE_Action5_Quiet_places_website.pdf.

85 Matsinos, Y.G., Tsaligopoulos, A., and Economou, C. Identification, prioritization and assessment of urban quiet areas: the case of Mytilene. *Global NEST Journal* 19 (1): 17–28.

86 EU (2005). WG-AEN recommendation on quiet zones by EU Working Group on the Assessment of Environmental Noise (END Annex VI 1.5).

87 Defra (2006). Neighborhood noise policies and practice for local authorities: a management guide. Chartered Institute of Environmental Health.

88 Anon (2010). Final report on Task 3: Impact assessment and proposal of Action Plan, May, prepared by RPA Ltd, Belgium.

89 http://www.eea.europa.eu/data-and-maps/figures/quietness-suitability-index-qsi/fancybox.html.

90 Watts, G.R. and Pheasant, R.J. (2015). Identifying tranquil environments and quantifying impacts. *Applied Acoustics* 89: 122–127.

91 van den Berg, F. (2013). What is a quiet place? http://www.qside.eu/definitions.

92 Volfova, J. (2015). Methodological uncertainties in the socioeconomic analysis of noise abatement, ICSV22 2015, Florence, Italy (12–16 July).

93 Millieu Ltd (2010). RPA report by Millieu Ltd for DG Environment of the European Commission under service contract No. 070307/2008/510980/SER/C3, Final Report on Task 2. Inventory measures for a better control of environmental noise: Social cost of noise.

94 Defra (2008). An economic valuation of noise pollution, developing a tool for policy appraisal. First report of Inter-departmental Group on Costs and Benefits Noise Subject Group.

95 Defra (2011). The economic value of quiet areas: final report prepared by URS/Scott Wilson for Defra.

96 WHO(2011). Burden of disease from environmental noise: Practical guidance. Report on a working group meeting, 14–15 October 2010.

97 https://en.wikipedia.org/wiki/Risk.

98 European Environment Agency (2010). EEA Good Practice Guide on noise exposure and potential health effects. EEA Technical report No. 11/2010.

99 https://en.wikipedia.org/wiki/Cost%E2%80%93benefit_analysis.

100 Weinstein, M.C., Torrance, G., and McGuire, A. (2009). QALYs: the basics. *Value in Health* 12: S5–S9. https://doi.org/10.1111/j.1524-4733.2009.00515.x.

101 Whitehead, S.J. and Ali, S. (2010). Health outcomes in economic evaluation: the QALY and utilities. *British Medical Bulletin* 96 (1): 5–21.

102 (a) EC (2015). Evaluation of Directive 2002/49/EC relating to the assessment and management of environmental noise, Workshop working paper 1 The second implementation review of the END – emerging findings, Brussels.
(b) 102(b) Decision no 1386/2013/EU of the European Parliament and of the Council of 20 november 2013 on a general union environment action programme to 2020 'living well, within the limits of our planet'.

103 Peeters, B. and van Blokland, G. (2018). Decision and cost/benefit methods for noise abatement measures in Europe. Interest Group on Traffic Noise Abatement, European Network of the Heads of Environment Protection Agencies (EPA Network).

104 Kurra, S. (1998). Noise Pollution, National Environmental Action Plan, prepared for Department of Planning of Turkey (Turkish Republic of Highway General Directorate of Highways, Ankara, Turkey (in Turkish).

105 Clayton, T. and Waddington, D. (2015). Practical implementation of noise policy and standards for the permitting and regulation of industry, ICSV22, the 22nd International Congress on Sound and Vibration, Florence, Italy, paper no. 406 (12–16 July).

106 EAA (2015). The European environment: state and outlook: synthesis report.

107 European Environment Agency https://www.eea.europa.eu/data-and-maps/indicators/exposure-to-and-annoyance-by-2/assessment-4 Created 31 Oct 2019 Published 21 Nov 2019 Last modified 21 Nov 2019.

Index

Environmental Noise and Management: Overview from Past to Present, First Edition. Selma Kurra.
© 2021 John Wiley & Sons Ltd. Published 2021 by John Wiley & Sons Ltd.